Multivariate Statistical Modeling in Engineering and Management

The book focuses on problem-solving for practitioners and model-building for academicians under multivariate situations. This book helps readers in understanding the issues, such as knowing variability, extracting patterns, building relationships, and making objective decisions. A large number of multivariate statistical models are covered in the book. The readers will learn how a practical problem can be converted to a statistical problem and how the statistical solution can be interpreted as a practical solution.

Key features:

- Provides conceptual models/approaches linking theories and practices in multivariate domain
- Provides step-by-step procedure for estimating parameters of developed models
- Provides blueprint for data-driven decision-making
- Includes practical examples and case studies relevant for intended audiences

The book will help everyone involved in data-driven problem-solving, modeling, and decision-making.

Multivariate Statistical Modeling in Engineering and Management

Jhareswar Maiti

CRC Press
Taylor & Francis Group
Boca Raton London New York

CRC Press is an imprint of the
Taylor & Francis Group, an **informa** business

First edition published 2022
by CRC Press
6000 Broken Sound Parkway NW, Suite 300, Boca Raton, FL 33487-2742

and by CRC Press
4 Park Square, Milton Park, Abingdon, Oxon, OX14 4RN

CRC Press is an imprint of Taylor & Francis Group, LLC

ISBN: 978-1-4665-6436-7 (hbk)
ISBN: 978-1-032-30018-4 (pbk)
ISBN: 978-1-003-30306-0 (ebk)

DOI: 10.1201/9781003303060

Typeset in Times
by Newgen Publishing UK

Dedication

To my father, Late Prabodh Chandra Maiti

Contents

Foreword

Multivariate model is a widely used popular statistical method that uses multiple variables for decision-making or to predict possible outcomes. Today, in the era of data-driven decision-making, where the practical phenomena are mostly multi-variate in nature, the use of multi-variate modelling and data analysis has become a natural choice to researchers and practitioners from engineering, science, and management disciplines dealing with real-word data.

In this context, the present treatise "Multivariate Statistical Modeling in Engineering and Management", authored by Prof. Jhareswar Maiti of IIT Kharagpur, is timely that provides a user-friendly and application-oriented document to help understand the subject multivariate statistical modeling in depth for intelligent use towards problem-solving and knowledge mining. This, in turn, would also enable removing the black-box character of the software, commonly used now a days. The volume has three parts, viz, prerequisites, foundations of multivariate statistics, and multivariate models. Large number of multivariate statistical models are covered. The ways of linking the theories and practices are discussed. Several practical examples and case studies are included for readers' benefit.

The author, Prof. Maiti, deserves congratulations for his long research contributions in the areas of data science, safety & quality analytics, and risk assessment & uncertainty, and bringing out the book of enormous archival value.

Sankar K. Pal
National Science Chair, and
President, Indian Statistical Institute

Calcutta
August 2022

Preface

Data-driven decision-making is the state of the art today. It spreads across all sectors of human civilization. Engineers today gather huge amounts of data and extract meaningful knowledge for interpreting behavior. Scientists do experiments under controlled conditions and analyze the results to confirm or reject hypotheses. Managers use the results of data analysis for day-to-day decision-making. Data collection and storage is an easy task today. Data-driven decision-making is a way of life now.

But there are concerns. First, most of the practical phenomena are multivariate in nature and multivariate analysis helps to capture the patterns. Analyzing individual variables separately often lead to modeling errors. For example, while measuring the performance of a system (say business performance), one needs to be aware that the performance could be measured using multiple indicators and, at the same time, it could be influenced by multiple factors synergistically. Second, statistics may fool the analyst or researcher if they are not conversant with the behavior of the phenomenon (problem) to be modeled or if they fail to use multivariate statistics when needed. The latter perhaps is more relevant in applied or application-based modeling for problem-solving. Third, there is a lot of user-friendly statistical software available for multivariate modeling and analysis. Many students, researchers, and analysts across all disciplines of engineering, science, and management often fail to avail themselves of this opportunity as they find multivariate statistical modeling a very tough subject to deal with.

This happens because there is a lack of user-friendly application-oriented textbooks on applied multivariate statistical modeling that can meet the requirements of many students, researchers, analysts, and practitioners. As today data is available in abundance and we need to extract knowledge and information from the data, we often use software as a black box for computation. This is perhaps due to a common misconception that we do not need to know how the knowledge or information are extracted from the data, rather we only need to use the findings to generate insights for population behavior. All models, including multivariate statistical models, are built on certain assumptions, on which the theoretical foundation of each of the models was established. If we do not know what are the preconditions of using a particular statistical model for solving a problem or confirming a hypothesis, results might be misleading. Similarly, insights cannot be well extracted for practical significance. This book is intended to bridge this gap.

While writing the book, I have adopted the approach of learning from practice to theory. I have assumed that the scientific developments (e.g., a theory or a model) have its origin in the real-world phenomena. So, it is better to start with a problem (phenomenon) related to a real-world process (or system) and then link statistics to define, characterize, and model it. Engineers, scientists, and managers deal with real-world problems and so it is not difficult for them to think in this line. Second, learning multivariate statistics becomes easier when we know the links between different models developed so far. For example, there are links from (i) t-distribution to Hotelling's T^2 distribution, (ii) univariate normal to multivariate normal distributions, (iii) comparing two population means (t-statistic) and mean vectors (Hoteling's T^2) to comparing several population means (ANOVA) and mean vectors (MANOVA), and (iv) multiple regression, factor model, and path model to structural equation model. In this book, these paths are explicitly spelt out for easy and comprehensive learning. The aims of this book are therefore as given below.

1. Providing guidelines to identify and describe real-life problems so that relevant data could be collected.
2. Linking data-generation process with statistical distributions, especially in the multivariate domain.

3. Linking the relationship among variables (of a process or system) and among various multivariate statistical models.
4. Providing step-by-step procedures for estimating parameters of a model.
5. Analyzing errors along with computing overall fit of a model.
6. Interpreting model results in real-life problem-solving.
7. Providing procedures for model validation.
8. Providing blueprints for data-driven decision-making.
9. Providing case studies to demonstrate the model's use and utility.

Items 1–3 fall under model conceptualization; item 4 describes parameter estimation; item 5 deals with model verification; items 6–7 convert statistical solutions to practical solutions; item 8 sums up the total process with careful decision-making; and finally, item 9 demonstrate how multivariate statistical models are built to solve real-world problems. The book will help everyone involved in data-driven problem-solving, modeling, and decision-making. The readers will learn how a practical problem can be converted to a statistical problem and how the statistical solution can be interpreted as a practical solution. As such, the reader base of the book is very broad.

The book contains 14 chapters. Chapter 1 contains the relevance of data analysis, modeling, and data-driven decision-making with illustrative examples. Useful terms and concepts are described. For example, types of variables and data, the concept of model and modeling, steps in problem-solving, caveats on overreliance on models, and links between different chapters are presented. Chapters 2 and 3 cover prerequisites (Part I). Chapter 2 highlights basic statistical concepts that are fundamentals to any statistical modeling and analysis; Chapter 3 deals with computation methods. Only those concepts that are essential to understand the computation required for the models described in this book are included in Chapter 3. These two chapters provide the relevant basics for understanding the chapters to follow, for readers not knowledgeable in basic statistics. Knowledge on Part I is essential to understand Part II (Foundations of multivariate statistics) and Part III (Multivariate models). Part II contains three chapters (Chapters 4–6). Chapter 4 deals with the multivariate descriptive statistics, namely mean vector, covariance, and correlation matrices. Zero-order, part, and partial correlations are also discussed. In addition, appropriate correlation coefficients for variables involving mixed data types are included. This is important because some of the multivariable models take correlation (or covariance) matrix as input for parameter estimation (e.g., factor model, path model, and structural equation models described in Part III). Chapter 5 describes multivariate normal distribution (MND), which is the backbone of many multivariate statistical models. The mathematical basis of MND is described, with an emphasis on the statistical and practical interpretation of these mathematics. Another important concept of multivariate modeling, called statistical distance, is also described in this chapter. Many day-to-day decisions involving multiple variables require interval estimation and hypothesis testing. For example, it includes inferences about single population mean vector, or comparing two-population mean vectors. These issues are described in Chapter 6, entitled "Multivariate inferential statistics." Part III covers the multivariate models spread over eight chapters (Chapters 7–14). Comparison of more than two population mean vectors is an extension of the comparison of more than two population means. The latter is known as analysis of variance (ANOVA) and the former is handled using multivariate analysis of variance (MANOVA). ANOVA and MANOVA are elaborately discussed in Chapter 7. In Chapter 8, the most widely used statistical model, multiple linear regression (MLR), is discussed at length. The MLR concepts—estimation, overall fit, test of parameters—if understood properly, will help in understanding the same analogously for the other models described in the subsequent chapters. The extension of MLR with multiple dependent variables (DVs) is termed as multivariate multiple linear regression (MMLR). MMLR is discussed in Chapter 9. Many times, MLR or MMLR can't be applied because data pertaining to independent variables (IVs) do not meet independent assumption and might cause multicollinearity problems. On the other hand, the explanatory variables might be correlated and the researcher wants to preserve this structure while estimating the model parameters. In this case, we use the path model described in Chapter 10. The path model gives researchers a unique opportunity to unfold the spurious relationships, if any, captured in correlation coefficients (bivariate). It also enables us to decompose the correlation matrix (for multiple variables) into causal, common, and correlated relationships. The path model is a versatile one and is used in many disciplines. For example, in social sciences literature it is better known as causal modeling of observables. Chapter 10 discusses different approaches, with more emphasis on the estimation techniques used in simultaneous equations modeling.

When the paths of the relationships (structure) is not that important, rather one is interested in prediction, then one may extract orthogonal components out of the correlated IVs using principal component analysis (PCA). Then MLR or MMLR can be used with IVs as the extracted principal components (PCs). PCA is discussed in Chapter 11. Both population and sample PCAs are considered, but emphasis is given to sample PCA. Discussions are also made regarding interval estimation of variance accounted for by each of the components extracted followed by interval estimation of a (principal) component's coefficients (loading). PCA has several interesting applications in engineering and sciences. A few of them are included in this chapter. PCA not only extracts orthogonal components, it is also used as a dimension reduction technique. Correlated data structure for multiple observed variables often is a result of few underlying hidden dimensions. PCA unfolds this. All these issues are discussed in this chapter.

It is not always possible to directly measure everything that we want to analyze. Many times, we use indirect ways of measurement (e.g., questionnaire survey). For example, most social, behavioral, administrative, management, and marketing phenomena are conceptualized in terms of latent variables (constructs or factors) that are measured using indicator (manifest) variables. Such concepts are not unusual in engineering studies. Manufacturing flexibility, supply chain coordination, continuous improvement capability, and sustainable design are a few examples that are measured using indicators. Factor models are used effectively to measure latent constructs. These are similar to PCA but differ in their focus as well as their conceptualization and uses. Factor models are discussed in Chapters 12 and 13. Both explanatory and confirmatory factor models are included. Chapter 12 deals with exploratory factor analysis (EFA), where the factors are not known *a priori*. The latent factor structure is explored from the data structure and then appropriate naming of the factors is made based on factor loadings with the manifest (indicator) variables. When the latent factors and their manifest (indicator) variables are known *a priori*, the researchers would like to confirm the factor structure. Confirmatory factor analysis (CFA) is used for this purpose, which is detailed in Chapter 13. Finally, in Chapter 14, structural equation modeling (SEM)—which is a perfect blending of factor and path models—is discussed. SEM is elaborated in light of handling the complex structure involving multiple variables that are either directly measurable or immeasurable, or both. Construction of path diagram, development of path equations, estimation of path coefficients, model adequacy test, and interpretation of results are discussed thoroughly. All the chapters in Part III are supported by a case study, which is another important treatment for understanding multivariate statistical models. It should be kept in mind that the models that are not included in this book are also valuable for data analysis, modeling, and data-driven decision-making.

The contents of this book can be taught as a postgraduate course, which can also be taken by higher level undergraduate students. It is desirable that the students should undergo a course on basic probability and statistics before attempting this subject. However, interested students without prior knowledge of basic statistics could also progress after a careful reading of Chapters 2 and 3 (Part I). The contents can be covered in a one-semester concise course on applied multivariate statistical models or in a two-semester course. For two-semester courses, the first course name could be "Introduction to multivariate statistical models" and the second course could be "Advanced statistical models for multivariate data analysis." The first course may cover up to Chapter 9 (MMLR) and in the second course, Chapter 1, Chapter 8 (in brief), and Chapters 10–14 may be covered. The practitioners can use this book upon undergoing a training program on the course.

A large number of examples covering engineering, social, and management sciences are illustrated throughout the book. All models are described with a case study that helps readers in understanding data collection procedures, analysis, and interpretation of results. The software MS Excel, R, Minitab and Matlab were used for computations. A sizable number of exercises are put at the end of every chapter to help students practicing for the concepts, models, and computations.

Jhareswar Maiti

Acknowledgments

I have been teaching applied multivariate statistical modeling as a subject to the research scholars, master's students, and higher level undergraduate students since 2004. After several years of teaching to the students from diverse disciplines, I felt that I need a book on applied multivariate statistical modeling that meets the requirements of all students across varied backgrounds. This motivated me to write this book. I deeply acknowledge all my students who have taken this course and in the process guided me in improving my knowledge and skills in this subject. Without such wonderful students over the past 17 years, this book would never have become a reality.

I am extremely thankful to NPTEL for the video lectures of the course, "Applied Multivariate Statistical Modeling," which helped in writing the book. The lectures are available at https://nptel.ac.in/courses/110/105/110105060/.

The book contains 14 chapters and each chapter contains several equations, figures, and tables. I am extremely thankful to my research scholars Ashish Garg and Souvik Das for carefully checking typographical mistakes in all the equations and formatting the tables, figures, and texts as per the publisher's requirements. Both of them spent several months with me. I am also thankful to all of my other research scholars, graduate and undergraduate students who did their thesis/projects under my supervision, and thereby directly or indirectly contributed in the preparation of the book.

Finally, my deepest gratitude to my wife Ranu Maiti for her continuous encouragement and support all through the years since 1997.

Jhareswar Maiti

Author

Jhareswar Maiti (PhD, FIE), Founder Chairman of the Centre of Excellence in Safety Engineering & Analytics (CoE-SEA) and Professor of the Department of Industrial and Systems Engineering, Indian Institute of Technology Kharagpur, has over 20 years of teaching, research, and consulting work in the fields of applied analytics and multivariate statistical modeling. He did his PhD in 1999 on "An investigation of multivariate statistical models to evaluate mine safety performance." He is a true interdisciplinary and multidisciplinary researcher, who has been working on the interfaces of engineering, management science, and statistics including analytics, where the research embodies (i) solving engineering and socio-technical problems, (ii) development of methodologies/models with innovative engineering, management science, and statistical approaches, (iii) development of novel tools and techniques using advanced statistics, data analytics, machine learning, and artificial intelligence, and (iv) application of advanced technologies. His domains of research are (i) data analytics, (ii) applied multivariate statistical modeling, (iii) safety science, engineering and analytics, (iv) quality engineering and analytics, (v) engineering ergonomics and analytics, and (vi) risk and uncertainty analysis. He is pioneer in making safety analytics a core area of research in the broad domain of safety science. He has established a unique world-class laboratory called "Safety Analytics and Virtual Reality Laboratory" at IIT Kharagpur. He authored more than 170 publications, supervised 18 PhDs, and executed several funded research and consultancy projects in the areas of data analytics, safety analytics, risk and uncertainty analysis, quality analytics, and management. He is currently serving the Editorial Board of *Safety Science*, *International Journal of Injury Control and Safety Promotion* and *Safety and Health at Work*. Professor Maiti has exceled in teaching multivariate statistical models and data analytics since 2004. His online course on "Applied Multivariate Statistical Modeling" (42-hour lecture series, available on YouTube) is a popular choice of many across the world. He has been ranked number one contributing author in the investigation of machine learning in occupational safety analysis (based on the analysis of publications from 1995 to 2019). He is the fourth most contributing author in safety research based on temporal analysis of the 30 most contributing authors of *Safety Science* from 2007 to 2011, and is overall the 17th leading contributor in safety science in the list of 50 leading authors over the past 40 years of *Safety Science* (1976–2016). A recipient of several awards including the **P. C. Mahalanobis Distinguished Educator Award** in Management Science for the year 2020, conferred by the Operational Research Society of India, Kolkata, Professor Maiti is a member of several international societies.

Part I

Prerequisites

1 Introduction

Data-driven decision-making using statistical modeling is practiced by managers, engineers, scientists, practitioners, and analysts, and its applications include scientific exploration and operational, managerial, administrative, and social decision-making. In this chapter, we demonstrate with practical applications how multivariate statistical modeling can be used in data-driven decision-making. First, the analyst needs to define the problem to be solved, set the goals or objectives of analysis, identify the variables to be measured and data to be collected. Then, they need to develop a conceptual model that mimics the relationships to be explored and find out an appropriate statistical technique that can be used to model the relationships. Upon estimation of the model parameters, the next important task is to interpret the results (findings) as the solutions to the problem being modeled. In this chapter, the readers will know different data types and their measurement scale, study design, types of modeling, and statistical approaches to model-building, examination of data, and a large number of multivariate dependence and interdependence statistical models. Several real-life examples are provided. Six case studies are highlighted and these case studies will later be described in subsequent chapters under dependence and interdependence models. Finally, the organization of the book is presented with a sequence of reading for the benefit of readers.

1.1 DATA-DRIVEN DECISION-MAKING

Every day, managers make decisions, engineers do assessments, and scientists conduct experiments and analyze the experimental data to serve the purposes intended. Objective decision-making, engineering assessment, and scientific analysis are often data-driven. For example, the twentieth century witnessed a tremendous growth in quality management owing to the development of Shewhart control charts, Taguchi methods, six-sigma, and the work of Deming, Juran, Crossby, and Feigenbaum. Data collection and analysis are an integral part of these quality management tools and approaches. Today, industries as well as service organizations keep records of almost everything. This is possible because of the development of data acquisition sensors and high-end computers. There are vast amounts of data available in today's organizations.

Acquisition and storage of data do not guarantee a path for improvement. Analyzing the data is required to extract meaningful knowledge and information for effective decision-making. Therefore, the uses of data analyses are manifold. Practitioners use it to solve day-to-day real-life problems. Academics use it to develop and validate models. Engineers seek meaningful knowledge out of it for interpreting the process behaviors. Scientists use it to confirm or reject hypotheses. Managers apply the results (findings) to make day-to-day decisions. Data provides evidence of similarity or differences, patterns, change points where the key characteristics of a phenomenon shift, relationships, etc. All these issues have made data analysis a useful and challenging task.

Let us observe the following examples.

Example 1.1

Mr. X, the manager of a courier company, wants to employ lean engineering in his organization. After effective brainstorming he found that the vital factor that governs the company's success is delivery time and one of the parameters to be monitored is variability (unevenness). He plotted delivery time data between two important cities, A and B. The plot is shown in Figure 1.1.

DOI: 10.1201/9781003303060-2

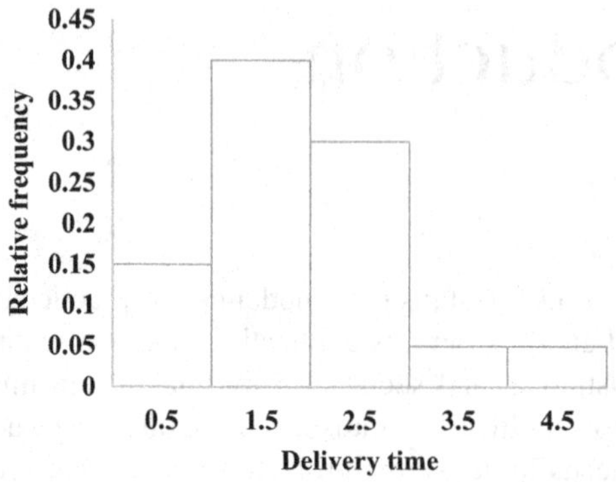

FIGURE 1.1 Histogram showing unevenness in delivery time (in days). The unevenness is measured in terms of relative frequency (probability).

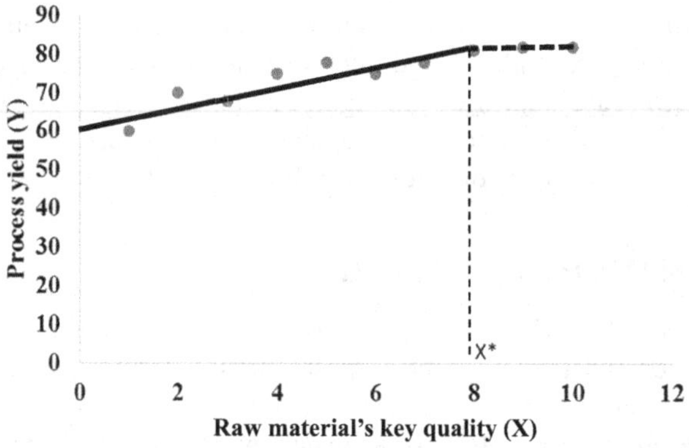

FIGURE 1.2 Relationship between process yield (Y) and raw materials quality (X). The relationship is linear for $X \leq 8$ (X^*). Beyond X^*, process yield (Y) does not depend on X.

Essentially, Figure 1.1 reveals *variability* in delivery time. It ranges from 0.50 to 4.50 days. Mr. X knows that the customers do not accept delivery time beyond three days and a large number of deliveries fall under the rejection zone. Based on this, he decides to have a meeting with his production and logistics heads.

Example 1.2

Mr. Y, an engineer entrusted for process improvement wants to reduce percentage rejection of the parts produced by a process. He knows that the process yield is affected by raw material quality and process conditions. The engineer plotted data on raw material quality and process yield for the last ten batches produced, and the pattern is shown in Figure 1.2.

Figure 1.2 shows *correlation* between process yield and raw material quality. Mr. Y knows this as it is obvious but what attracts him more is that beyond point X^* (Figure 1.2), the process yield doesn't improve. So, he can decide on fixing raw material quality in and around X^*, provided he has sufficient information on this behavior.

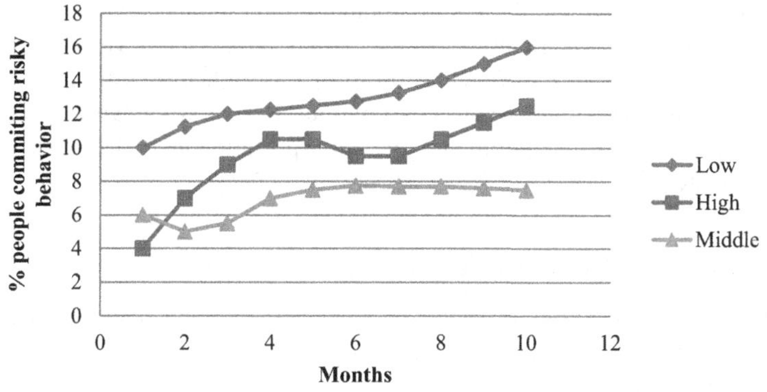

FIGURE 1.3 Risk-taking behavior of different income groups of people. The low and high income groups take higher risk in comparison to the middle income group.

Example 1.3

Mrs. Z, a public welfare officer, has been concerned about traffic accidents in a city. She conducted a survey to understand the risky behavior of city people. She plotted the observations for lower, middle, and high income groups of people for the past ten months. The differences in behavior are plotted in Figure 1.3.

Figure 1.3 depicts *differences*. Mrs. Z is surprised to see the pattern of at-risk behavior across income groups. While within a group, the percentage of risky behavior remains almost constant, between groups there is a difference. She might be interested to know whether people in the middle-income group are risk-averse.

The above three examples outline decisions to be made under the given conditions. But seldom are decisions taken with such simple analyses. For example, Mr. X will definitely be interested to know why in some cases the delivery time exceeds the due date. Similarly, Mr. Y will be interested to identify and control confounding (moderating) variables that may otherwise produce different results. Mrs. Z will not only be interested to quantify the differences in risk-taking behavior across groups but also try to identify other factors and their contributions in delineating such differences. All these cases are therefore influenced by *multiple variables*, and *multivariate analyses* are required to explore these relationships.

More examples are presented in Section 1.6. But at this point, it is important to define variables and data types as statistical analyses are very much dependent on the types of variables and data that characterize the process (system) of interest.

1.2 VARIABLE AND DATA TYPES

1.2.1 Random Variable

In statistics, variable means an entity whose value can change from one observation to another. If the measured values are known in advance with certainty, then the variable is deterministic in nature. For example, "month" in Example 1.3 above can be treated as a variable that is deterministic in nature. On the other hand, if the measured values of a variable cannot be known in advance and are governed by the laws of probability, then the variable is called a *random variable*. For example, raw material quality and process yield in Example 1.2 and delivery time in Example 1.1 are all random variables. Although the values of a random variable cannot be known in advance, its *expected value* can be obtained based on probability theory.

Let X be a random variable where $X = x$ is a particular observation and the occurrence is governed by a probability function $f(x)$. The random variable X may be a *discrete variable* such as numbers of items sold, or a *continuous variable* such as profit or loss. When X is discrete, $f(x)$ is known as *probability mass function* (*pmf*) and when X is continuous, $f(x)$ is known as *probability density function* (*pdf*). The *pdf* or *pmf* of a random variable, its expectation and other basic statistical issues are discussed in Chapter 2.

1.2.2 Measurement Scale and Data Types

Variables are measured using different scales of measurement. The scale defines the data types and accordingly variables can be named. The measurement scales are (i) nominal, (ii) ordinal, (iii) interval, (iv) integral, and (v) ratio.

Many a times, we require using some variables just to provide identity to some items or things. For example, in Example 1.3, "month" is such a variable. We identify the months in terms of January, February, etc., and so, January, February by this nomenclature does not represent anything more than identity. Further examples can be company A, company B to company M, or products A to L. When a scale is used only for identification and cannot assume any real numerical value in the sense of measurement, the scale is termed as *nominal* and the data type is also termed as *nominal*. Variables measured in nominal scale can have multiple categories (e.g., products A, B, C, etc.) but there is no order or interval among the categories.

Another class of variables which are often used in statistical modeling is *ordinal in nature*. For example, quality of a product can be judged as low, medium, or high. Performance can be judged as poor, average, good, very good, excellent, or outstanding. Items can be ordered or ranked using this type of measurement scale, known as *ordinal scale*. But the interval between any two values such as low and medium or poor and average cannot be measured. From an information content point of view, ordinal variables possess more information (i.e., order or rank) than nominal variables. However, the arithmetic operations like addition, subtraction, multiplication, and division of different values are not permissible for data measured in nominal and ordinal scales.

When the interval between two values of a variable is measurable, the values can be added or subtracted. For example, temperature can be measured either in Celsius or in Fahrenheit scale. Let, the maximum temperature of a city in a day of summer is 40°C or 104°F, and 20°C or 68°F in a day of winter. The difference in maximum temperature between the two days is 20°C and 36°F. But the difference of 20°C (= 68°F) in Celsius scale does not conform to the difference in Fahrenheit scale

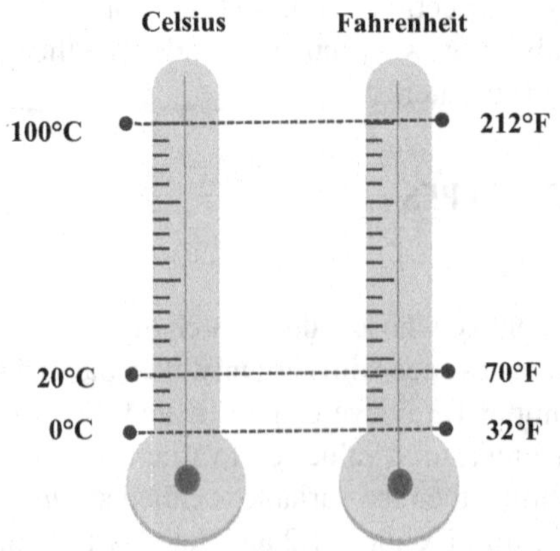

FIGURE 1.4 Temperature scales. The conversion of Celsius to Fahrenheit for three different temperature values are shown.

(36°F). Why does this happen? Because the two scales are different (see Figure 1.4 with respect to their reference point). Similarly, if we take the ratio of the maximum temperature in the two days, we will get 2 and 1.53, in Celsius and Fahrenheit scales, respectively. Therefore, *interval scale data* does not permit division.

The real life performance is often measured in counts. For example, number of units of an item sold, number of defective products, number of components failed, and number of disruptions occurred. These data are known as *count data* and the corresponding variable is known as *integral variable*. Integral variable takes values from zero to countable infinite. All arithmetic operations are possible. Count data is discrete in nature.

The highest class of data is *ratio data* that uses *ratio scale* in the sense that all the arithmetic operations are possible with this data type. For example, the delivery time in Example 1.1, process yield in Example 1.2, and percentage risky behavior in Example 1.3 are all measured in ratio scale. As both integral and ratio data are amenable to all arithmetic operations, the integral data may be grouped under ratio data.

Variables measured in nominal or ordinal scale are known as *categorical variables* and the data type is also called *non-metric data*. On the other hand, data measured in counts, interval, or ratio scales are known as *metric data*. Non-metric data only identify and describe a subject, an item, or an object, but metric data, in addition, quantify the amount or degree to which a subject, item, or object is characterized (Hair et al., 1998, 2010). Metric data is also known as *quantitative data* where as non-metric data is called as *qualitative data*. Qualitative data are also known as *attribute data*.

1.2.3 Data Sources

Data come from various sources. Data, when collected directly from the origin of source (where it is generated), is known as *primary data*. Primary data are therefore closed to the original events or phenomena. For example, Examples 1.1–1.3 dealt with primary data. Primary data are first-time information with no analysis or commentary. Sometimes, we use derived data based on the analysis of primary data collected by someone else. Such data are called *secondary data*. Secondary data might be the restatement of the primary data or the summary of primary data. *Tertiary data* are obtained from the secondary data after further refinement, such as the summary or synthesis of research using secondary data. Primary data are the best data. Secondary and tertiary data should be analyzed and interpreted carefully because these data sources are separated in time or space or both. As real-life phenomena are space- and time-dependent, the pattern extracted from the secondary and tertiary data may not represent the current behavior or may suffer from loss of information.

Another important aspect that should be kept in mind is that the secondary or tertiary data might be *aggregated information*. For example, demand for a particular product for different years is aggregate information if the demand is computed as the sum of demands collected from different market segments. Applying this information to a particular market segment may be misleading. Although this problem may also occur with primary data, the problem can be resolved as the data collection mechanism is known.

1.3 MODELS AND MODELING

Models mimic reality. They try to describe and explain the *regularity* of a phenomenon. When a model passes the test of time by predicting the regularity it intends to, it can be used as a *law* or *theory*. For example, Hooke's law describes the behavior of an elastic body under load. It says that the stress (σ) acting on an elastic body is proportional to the strain (ε) developed. Mathematically, we can write:

$$\sigma \propto \varepsilon. \quad \text{or} \quad \sigma = E\varepsilon \tag{1.1}$$

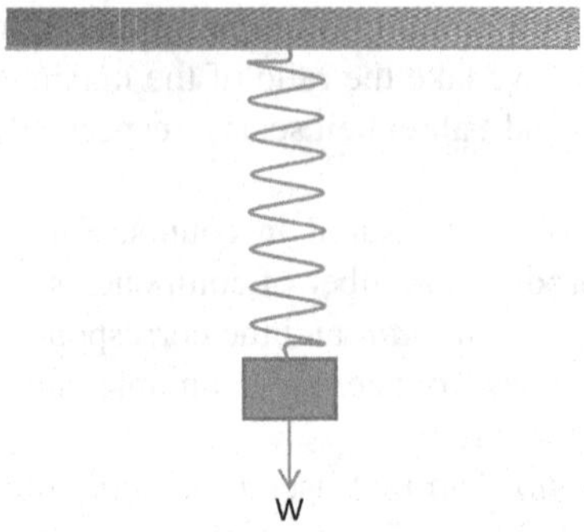

FIGURE 1.5 Physical model of Hooke's law. The behavior of the spring balance system can effectively be explained by Hooke's law.

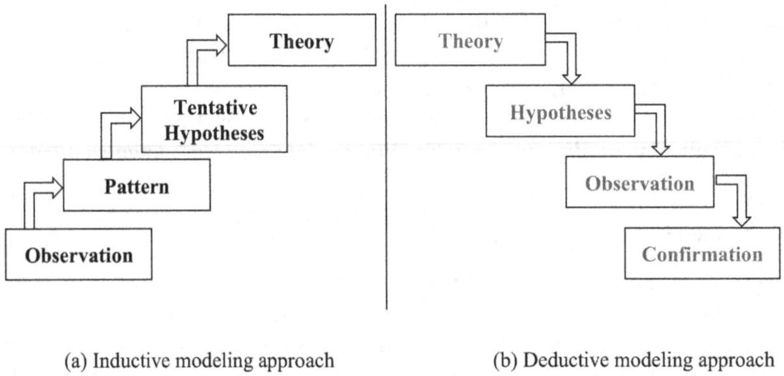

FIGURE 1.6 Types of modeling. (a) Inductive, (b) Deductive. As shown with arrows, the deductive modeling starts with a theory, then sets hypotheses, collects observations, and tries to confirm the theory. On the other hand, inductive modeling starts with data collection, then extracts patterns, sets tentative hypotheses, and validates with the available theory or proposes for a new one.

Where, E is known as the elasticity constant. Equation 1.1 is called a *mathematical model* of stress-strain behavior of an elastic body. Mathematical modeling is possible when the physics of the phenomena to be modeled is known. One may develop a *physical model* similar to Figure 1.5, using spring-balance. Now, one can do experiment in a laboratory and measure σ and ε under different load conditions. If n such trails are conducted, one gets n observations for σ and ε, which can be used to empirically estimate the relationship given in Equation 1.1. The *empirical relationships* that approximately model the reality are known as the *statistical model* (Lindsey, 2006).

Modeling is the process of building models. There are two approaches to do this. The scientific method uses deductive or inductive logic to build a model. When we follow the path from theory to hypothesis to evidence (data), we usually follow *deductive logic*. In contrast, when we extract patterns or regularity from data, propose a hypothesis, and test it, we follow *inductive logic*. We also refer the former as *analytical study* and the later as *empirical study*. Figure 1.6 describes these two approaches.

When a model is developed based on an existing theory, it is known as *theory-based model*. Data is collected and analyzed to provide evidence for this theory. Such study is known as a *confirmatory study*. On the other hand, when we observe phenomena (e.g., through observations, surveys, or experiments) and use statistical analysis to extract patterns or regularities from the data, the models so built are known as *exploratory models*. Data analysis related to the former is called ***confirmatory*** *data analysis* and the latter is called *exploratory data analysis*. In this book, both types of models are included.

Models developed through exploratory data analysis can be called *empirical statistical* and models developed through confirmatory data analysis can be called as *analytical statistical*. Interested readers may read Wacker (1998) for a lucid explanation of analytical and empirical models.

No model is perfect. They are falsifiable. Ravindran et al. (2006) mentioned ten principles to be followed during modeling. We will describe them below.

Principle 1: Do not build a complicated model when a simple one will suffice
Very often researchers fall in the trap of building complicated models with a notion that a mathematically rigorous model gives better results. It is not true. Model should be chosen based on the problem at hand. For example, the multiple regression, path model, and structural equation modeling have gradually increasing complexity. Now, if a problem can be modeled using multiple regression, it should be used instead of path model or structural equation model so long the assumptions of the chosen model are not violated. We can use higher order models if the simple model chosen cannot fit the data.

Principle 2: Beware of modeling the problem to fit the technique
"If you have a hammer, you see everything as a nail." Researchers should be aware of this syndrome. Many a times, we see a problem as if it can be solved using a particular statistical model. We collect data accordingly. This is a wrong path. Carefully defining a problem and designing the study scientifically help in choosing an appropriate model.

Principle 3: The design phase of modeling must be conducted rigorously
The nature of the problem to be modeled defines the process of literature review and type of modeling. Exploratory analysis uses inductive logic while confirmatory analysis needs deductive logic. Both logics should be perfectly spelt out and according study design should be done.

Principle 4: Model should be verified prior to validation
Verification refers to checking the adequacy of a model to fit the data used to develop the model. It also includes the test of parameters. Once a model is fit (verified), it is required to be validated where *validation* refers to the ability of a developed model to reproduce similar results under similar situations. An inadequate model cannot be used for validation.

Principle 5: A model should never be taken too literally
Very often, we describe things in terms of flow charts or other pictorial means with proposed relationships. We provide explanation of these relationships based on our literature survey or experience of modeling similar phenomena. This type of development can be termed as *conceptual model*, but doesn't qualify to be called as a *model* that mimics the reality. A conceptual model usually provides a description of the relationships but a model goes further in quantifying these relationships and many cases has the ability of predicting future behaviors.

Principle 6: A model should neither be pressed to do, nor criticized for failing to do that for which it was never intended
Every model is applicable within the boundary or scope of its development. For example, a model that is developed to explain the organization culture of a Western firm may fail to do the same for an Eastern firm. Similarly, a model may be time-dependent. Customers' demand for cosmetic products is not the same as was 30 years ago. A model developed based on data 30 years ago will not be able to forecast customer demand today.

Principle 7: Beware of overselling a model
The scope and limitations of a model (see principle 6 above) should be spelt out and the results obtained using such a model are valid for within the scope.

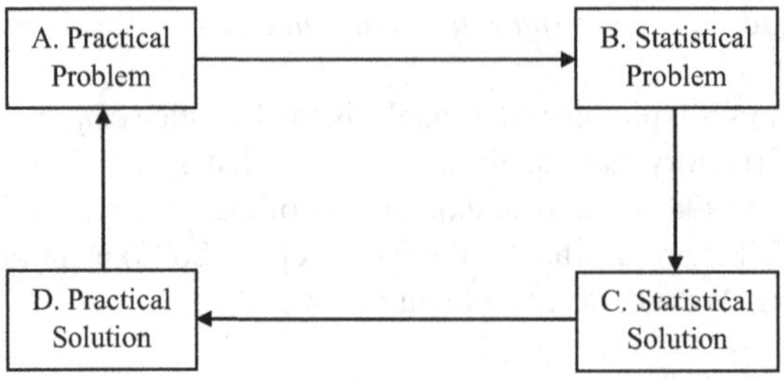

FIGURE 1.7 Statistical approaches to problem-solving. It starts with a practical problem (A) and follows the cycle ABCDA.

Principle 8: Some of primary benefits of modeling are associated with the process of developing the model
Modeling is ever-changing. A model can never be complete. Hence, while developing a model we find that it changes in its contents and quality. Researchers should not be afraid of changing a developed model.

Principle 9: A model cannot be any better than the information that goes into it
A model produces results based on its inputs. The information contents in the inputs are reshaped using a model to get meaningful interpretation. For example, in dependence models like multiple linear regression, the information in the dependent and independent variables is the variance and covariance component of the variables. The model is used to explain the variability (variance component) of the dependent variable (DV). It is never possible to explain 100% variability of the DV considered.

Principle 10: Models cannot replace decision-makers
Models are used to help decision-makers in decision-making. It is not that models define or redefine decisions. They rather approve or disprove our actions based on their capability. A decision not supported by a model is not necessarily wrong. Similarly, a decision that is favored by a model is not 100% correct. It is the decision-maker's role to consider the model findings or not. His or her judgment should be supreme.

1.4 STATISTICAL APPROACHES TO MODEL-BUILDING

Statistics is concerned with the collection, presentation, and analysis of the data and interpretation of the findings in solving real-life problems. Figure 1.7 depicts how statistics solve real-life problems.

The roots of practical problems lie in a system, where "system" can be a shop-floor, a marketing department, an engineering cell, a banking sector, or even an administration. A system is developed to serve certain functions (purposes). When the purposes are not fully served, we say the system is not functioning properly or a problem or problems exist in the system. Many times, problems cannot be identified *a priori* but can be anticipated in advance. A careful study is required to define the problem. The practical problem is then transformed to a statistical problem. In practice what is called a "system," could be termed a "population" in statistics. System characteristics are population characteristics. For example, the performance of a marketing department can be judged based on "sales volume" (i.e., number of units sold per annum). Decreasing or stagnancy in sales volumes over the years is a practical problem. A little more inquiry into the problem will lead to revealing the causes of decrease in sales. It may be due to poor marketing strategy, bad quality of product, competitors' superior performance, or something else. All these characteristics are recorded as data. In terms of a statistical problem, we say, "these characteristics" as "variables of interests." In order

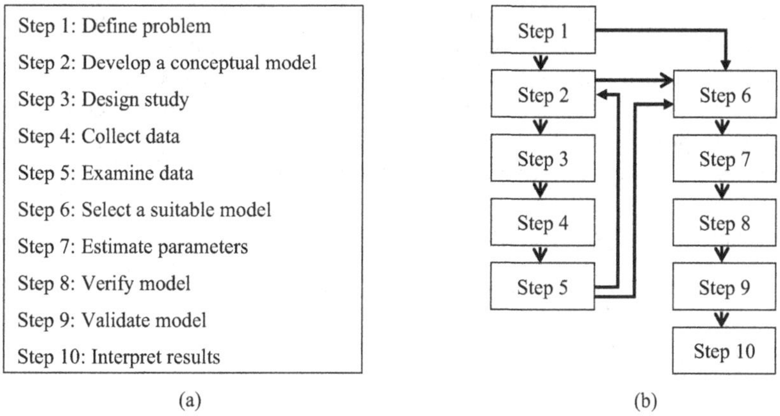

FIGURE 1.8 Steps in statistical model-building. Figure (a) represents the ten steps and Figure (b) shows the sequence as well as feedback (as a flow chart).

to achieve the statistical solution, we require building a statistical model. This is the relationships among the variables of interest. For example, we may be interested to know how sales volume can be predicted by marketing strategy, product quality, and competitors' performance, typically known as predictor variables. Quantifying the contributions of each of the predictor variables to sales volume is the solution that we want. Then, these contributions can be judged with practice. If the solution satisfies the requirements, we say that the practical solution is achieved.

A statistical solution is not obtained automatically. It requires developing an adequate statistical model. The steps to be followed for statistical model-building are given in Figure 1.8.

1.4.1 Step 1: Define Problem

As stated earlier, models mimic reality where the reality is defined by a physical or social phenomenon. It is the curiosity of humans to explain such a phenomenon usually with a purpose. For example, a process engineer may be interested to model his process to understand its mechanisms of producing good or defective parts (or products). This will help the engineer to improve process yield, which we say the purpose of modeling. By "define problem" we mean a situation when the purpose as planned is not served by the process (or system) considered. The "process" or "system" is defined in a broad sense. It can also include a super-system like the universe. Similarly, the word "problem" is also broad in the sense that it may be the unsatisfactory "yield" as stated above, or an abstract question of a scientist.

Problem definition will justify (i) the requirements of a study, (ii) the number of variables to be included, and (iii) the boundary and scope of a study.

We will illustrate the modeling steps outlined above with the help of Example 1.4.

Example 1.4

A small company, CityCan, has been doing business in the local markets and uses sales volume and profit as indicators of its business performance. The company set a medium-level technology for its production process and the process performs well when is operated and maintained properly. The measures for proper operation and maintenance, based on a previous study, are absenteeism in percent and machine breakdown in hours. Last year, the company has expanded the business beyond the city and as a precautionary measure established a marketing department in order to improve its business performance. The management of this company, being proactive, has instituted a mechanism to measure the performance of the marketing department. The management uses an indicator (M-Ratio) to measure the performance of the marketing department. M-Ratio greater than 1 indicates improved performance, whereas the M-Ratio of 1 is the desired minimum performance. The production and marketing department have a good liaison in their day-to-day work. The manager is interested to measure the variability in profit and sales volume and also wants to establish its relationships with operational and marketing performances for future actions.

The practical problem with Example 1.4 can be stated as below.

The company anticipates threat to profit and "sales volume" due to variation (possible changes), most presumably increase in absenteeism and machine breakdown (in hours), and decrease in marketing performance (M-Ratio).

1.4.2 Step 2: Develop Conceptual Model

Conceptual model describes the relationships among the variables of interest pertaining to a system under investigation, and is often represented pictorially. It reflects the researcher or analyst's knowledge and experience in building model of the problem defined in Step 1. For example, the CityCan problem can be conceptualized with the following steps:

(i) Identify the variables of interests: profit, sales volume, % absenteeism, machine breakdown in hours, and M-Ratio.
(ii) Identify response variables (*Y*): Y_1 = Profit, Y_2 = Sales volume

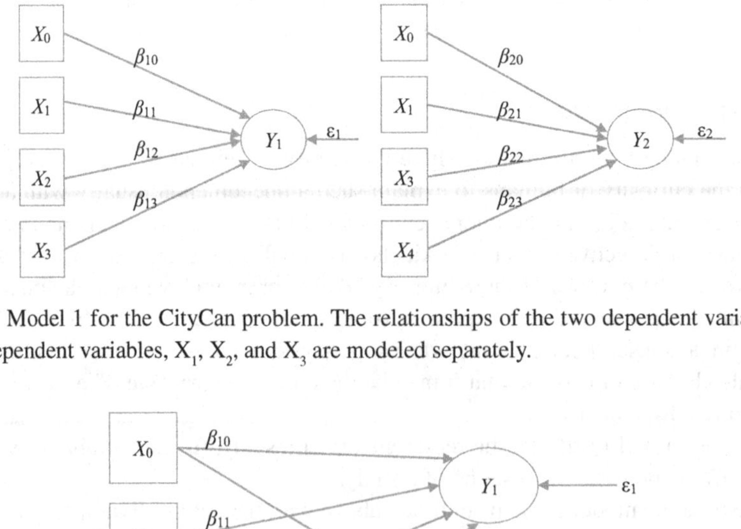

FIGURE 1.9 Model 1 for the CityCan problem. The relationships of the two dependent variables Y_1 and Y_2 with three independent variables, X_1, X_2, and X_3 are modeled separately.

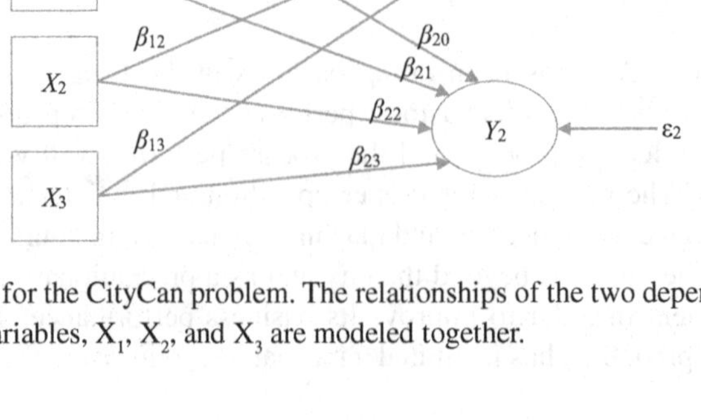

FIGURE 1.10 Model 2 for the CityCan problem. The relationships of the two dependent variables Y_1 and Y_2 with three independent variables, X_1, X_2, and X_3 are modeled together.

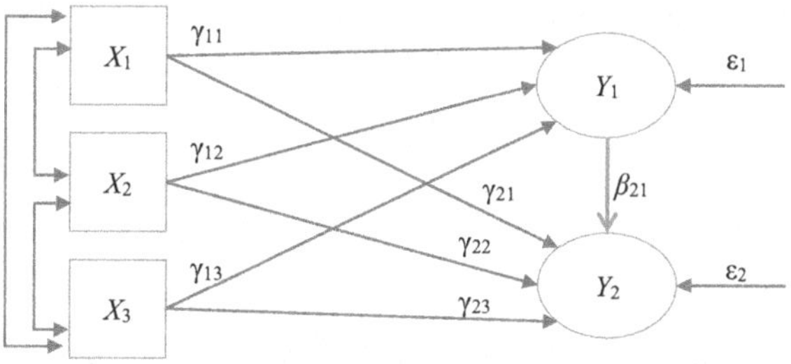

FIGURE 1.11 Model 3 for the CityCan problem. The relationships of the two dependent variables Y_1 and Y_2 with three independent variables, X_1, X_2, and X_3 are modeled together. In addition, the independent variables are allowed to covary.

(iii) Identify explanatory variables (***X***): X_1 = % absenteeism, X_2 = machine breakdown in hours and X_3 = M-Ratio.
(iv) Find out dependence relationships: $\boldsymbol{Y} = f(\boldsymbol{X})$.

So, the statistical problem for Example 1.4 is as follows:

*The changes in the response (dependent) variables, Y_1 (profit) and Y_2 (sales volume) are caused by the explanatory (independent) variables, X_1 (% absenteeism), X_2 (machine breakdown in hours) and X_3 (M-Ratio) as represented by the relationships expressed as **Y** = f(**X**). The relationship can be linear or non-linear. If we can control the variables **X** (X_1, X_2, and X_3), we can control the variations in **Y** (Y_1 and Y_2).*

The relationships are represented by Figures 1.9–1.11. In the first model (Figure 1.9), the two dependent variables, profit (Y_1) and sales volume (Y_2) are modeled separately in a sense that the explanatory variables, percentage absenteeism (X_1), machine breakdown in hours (X_2) and M-Ratio (X_3) affect Y_1 and Y_2, independently. So, the parameters of the two dependence models in Figure 1.9 can be estimated separately. If the relationships are linear, Equations 1.2 and 1.3 represent them, respectively. In the second model (Figure 1.10), the joint occurrences of profit (Y_1) and sales volume (Y_2) along with the three independent variables are captured simultaneously. So, the parameters of this model are to be estimated simultaneously. In both the models, ***X*** (X_1, X_2, and X_3) is truly independent and Y_1 doesn't affect Y_2, and vice versa. The conceptual models represented in Figures 1.9 and 1.10 are known as multiple linear regression and multivariate linear regression, respectively. For details of these types of models, please refer to Chapters 8 and 9 (Part III).

The linear relationships can be statistically modeled as:

$$Y_1 = \beta_{10} + \beta_{11}X_1 + \beta_{12}X_2 + \beta_{13}X_3 + \epsilon_1 \tag{1.2}$$

$$Y_2 = \beta_{20} + \beta_{21}X_1 + \beta_{22}X_2 + \beta_{23}X_3 + \epsilon_2 \tag{1.3}$$

Where, β_{jk}, j = 1 and 2 and k = 0, 1, 2 and 3 can be explained as average effect, and contribution of the explanatory variables X_j, j = 1, 2 and 3.

But there may be situations, when the ***X*** variables may covary and one or more of the ***Y*** may affect the others. If we require preserving this practicality, the conceptual models given in Figures 1.9 and 1.10 are insufficient. This issue is captured in a third model, given in Figure 1.11. The conceptual model represented in Figure 1.11 is known a path model (see Chapter 10). If Figure 1.11 represents the correct model, the models represented by Figures 1.9 and 1.10 are subjected to *model misspecification.*

The path equations for Figure 1.11 are shown below.

$$Y_1 = \gamma_{11}X_1 + \gamma_{12}X_2 + \gamma_{13}X_3 + \epsilon_1 \tag{1.4}$$

$$Y_2 = \beta_{21}Y_1 + \gamma_{21}X_1 + \gamma_{22}X_2 + \gamma_{23}X_3 + \epsilon_2 \tag{1.5}$$

Equations 1.2 to 1.5 describe several statistical models. For example, Equations 1.2 and 1.3 combined define the multivariate regression model (see Chapter 9). If we are interested either in Y_1 or Y_2 (Figure 1.9), then Equation 1.2 or Equation 1.3 describe the model, each one is called multiple linear regression (see Chapter 8). Equations 1.4 and 1.5 combined describe the path model (see Chapter 10). Put another way, we can say that the practical problem is formulated statistically. Now, if we can determine the values of the relationship coefficients (*parameter estimates*, see Chapters 8–10 for details), we will be in a position to plan better for the future. Once the relationship coefficients (e.g., β_{jk}) are estimated scientifically, we can say that statistical solution is obtained.

1.4.3 Step 3: Design Study

Once the conceptual model is tentatively defined, the researcher requires collecting data through appropriate study design. Study design facilitates the required data to be collected. Depending on the purpose of the study (defined in Step 1), the study design can be broadly defined as observational and experimental (Beaglehole et al., 1993). The difference between the two is that in the former, the naturally occurring relationships are explained; whereas, in the later, the relationship is manipulated in order to control the effects of one or more explanatory variables on the responses (outcomes). Further, it can be categorized as (i) naturalistic observations, (ii) case study, (ii) correlational, (iv) differential, and (v) experimental, where the first four—(i)–(iv)—fall under observational study.

At the early stages of research on a given subject, when little is known about it, it is customary to observe the behavior of objects/participants in their natural environment. This type of study design is known as *naturalistic observations*. The scope of this type of study is generally unbounded. When such a study is done in a moderately limiting environment, it is known as *case study*. The purpose of the naturalistic observations and case study research is to generate hypotheses that can be tested in future studies. In *correlational studies*, we observe the behavior of objects/participants in order to build naturally occurring relationships. This type of study design is employed at a higher stage of research, when the first two designs were employed by a large number of researchers and a number of hypotheses were built. The *differential study* is also employed at a higher stage like *correlational study* but in the former, observations are made on two or more preexisting groups for comparison. The highest level of study from causal inference point of view is experimental study. In *experimental study design*, experiments are conducted to manipulate one or more variables. This is usually done in laboratories but many times is conducted in the field also. Experimental study confirms the highest level of information and is *analytical statistical* in nature. *Correlational* and *differential* study designs are *empirical statistical* and *naturalistic observations* and *case study* designs are *descriptive* in nature.

From frequency and mode of data collection, study designs can be (i) cross-sectional, (ii) longitudinal, and (iii) cohort. In *cross-sectional design*, observations are collected at a particular point on time from a population of interest. In *longitudinal study*, data is collected in more than one point on time and it can be retrospective or prospective. A longitudinal retrospective study with differential research design is known as *case-control study*. In case-control design, two groups, one with a particular outcome (e.g., a disease) called "*case*" and the other without that outcome called "*control*" are considered. Here, the control group is also called "*reference*" group. The occurrence of a particular cause is compared between the two groups. *Cohort study* is prospective case-control study and is also called *follow-up study*. It starts at a particular time and observations are made at different points

TABLE 1.1
CityCan Data Set

Sl. No.	Months	Profit in Rs Million	Sales Volume in 1000	Absenteeism in %	Machine Breakdown in Hours	M-Ratio
1	April	10	100	9	62	1
2	May	12	110	8	58	1.3
3	June	11	105	7	64	1.2
4	July	9	94	14	60	0.8
5	August	9	95	12	63	0.8
6	September	10	99	10	57	0.9
7	October	11	104	7	55	1
8	November	12	108	4	56	1.2
9	December	11	105	6	59	1.1
10	January	10	98	5	61	1.0
11	February	11	103	7	57	1.2
12	March	12	110	6	60	1.2

on time in future. Therefore, it is prospective in nature. In cohort study, a population is selected first (e.g., all suppliers of a large manufacturing company). Let's say the manufacturer decides to help a particular group of suppliers in improving their processes. So, this group is exposed to help. Others can be considered as another group who are not exposed to help. Both groups are monitored in terms of defective lots supplied for the next year. If the help provided by the manufacturer really improves the suppliers' process, this study can test it. This type of study is known as *cohort study*.

Another important consideration in study design is *unit of analysis* and *unit of measurement*. Unit of analysis is the smallest unit for which analysis is made. A unit of analysis can be a population or group (e.g., comparing two cultures) or an individual/item/object (e.g., individual patient). On the other hand, a unit of measurement is related to measuring the variables characterizing the unit of analysis of interest.

1.4.4 Step 4: Collect Data

Once the study design is known, data collection becomes easier. Many times, data are collected in two stages as preliminary data and full data. Preliminary data upon analysis (see Step 5) gives clues for modifying the data collection process and number of observations needed. Then full data (final data) are collected. However, cost and accuracy are the two criteria that determine the stages of data collection. If full data is collected at one time, it can be split into two samples, called pilot and final for analysis. As stated earlier, data can be collected from primary, secondary, or tertiary sources (see Section 1.2.3). Data can be collected from historical records, direct observations from fields, questionnaire surveys, interviews, brainstorming, and experiments.

For example, the CityCan Company has, in appreciation of future unforeseen problems, developed a scheme for collection of data for the variables of interest. Table 1.1 describes this.

Under statistical terminologies, the activities needed to obtain Table 1.1 are known as *sampling* and Table 1.1 is the representation of a sample. A sample contains observations on the variables of interest (collectively, we say data) collected as per the study design. The number of observations, usually denoted by n, is 12 based on Table 1.1. In statistics, n is called *sample size*. It should be noted here that the data collection format used by CityCan says that CityCan records data month-wise (see column 2 of Table 1.1). If day-wise data is considered, the sample size will be 365 for the last year. But selection of period of observations and sample size depends on several practical issues such as the purpose to be served, variable measurement scheme available, manpower availability, cost of data collection, and so forth. Nevertheless, sample size (n) is an important issue and should be determined appropriately. It can also be argued that the explanatory variables are insufficient as two out of three X variables are internal that affect production rather than sales. More relevant variables are to be identified and pertinent data needs to be collected.

The key issues that should be considered during data collection are (i) what to measure and (ii) how to measure (Field, 2009). What to measure refers to three things namely, variables and data types (see Section 1.2.2), measurement error, and validity and reliability of data. How to measure refers to study design explained above in Step 3. We will describe below measurement error and validity and reliability of data collected.

Any measurement system (e.g., an instrument like a Vernier caliper, voltmeter, or a questionnaire) must possess five key characteristics. These are bias, stability, linearity, repeatability, and reproducibility (Gryna et al., 2006). *Measurement error* can be attributed to one of the five types or their combinations. *Bias* is the difference between the target measurement (reference) and the average of the measured values. *Stability* refers to measuring the same correctly at different points on time. *Linearity* is related to bias, when the scale of measurement varies. At the lower scale of measurement, the bias should be smaller and at the higher scale of measurement, the bias should be larger. *Repeatability* is the standard deviation of measured values when the same thing is measured for several times and it is also called *precision*. *Reproducibility* confirms that the instrument behaves in a similar manner when used by several people. In other words, the variability in measurements should be negligible when several people use it.

In questionnaire design, the validity and reliability of a questionnaire are very important. *Validity* refers to whether an instrument actually measures what it is intended to measure. Two important validity measures are *criterion validity* and *content validity*. Suppose that one is interested to measure supply chain coordination but uses responsiveness as one of the measures (indicators) of it. This violates criterion validity, as responsiveness is a measure of agility, not coordination. Content validity refers to the coverage of the full spectrum of a variable (construct). It relates to whether the questions administered can cover the full spectrum. Instead, if the questions cover a part of it, content validity suffers. Validity can be linked with bias, described above. The more is the bias, the less is the validity. *Reliability* is similar to precision. Reliability is the ability of an instrument to produce the same results under the same situations.

1.4.5 Step 5: Examine Data

Hair et al. (1998) outline several objectives of examining data. Among them, the objectives that must be satisfied are (i) understand relationships among the variables of interest, (ii) identify missing data, (iii) identify outliers, (iv) test the assumptions applicable to most of multivariate statistical

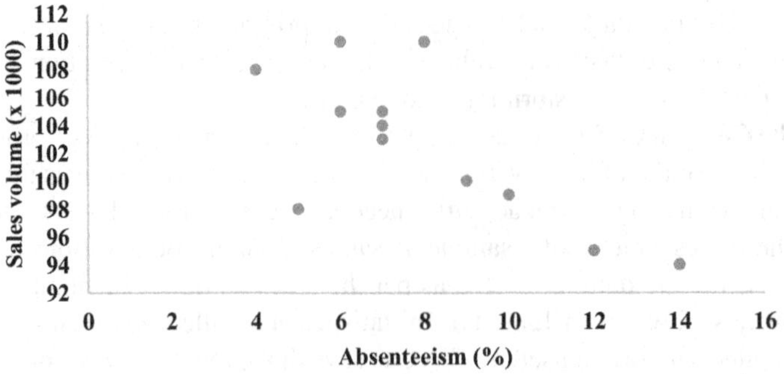

FIGURE 1.12 Scatter plot for sales volume and absenteeism in percent (Example 1.4). The scatter shows linear negative relationship between sales volume and absenteeism.

models, and (v) understand data transformation. These activities are very important and we will describe them with the statistical models considered in later chapters. However, some important concepts are explained below.

The most common relationship model is the *bivariate scatter plot*, similar to the one shown in Example 1.2 (Figure 1.2). There can be *no relation* (with observations scattered randomly), *linear relation* (either positive or negative), or *non-linear relation*. Figure 1.12 describes them. Generally, we use multiple variables of interest. For five variables, there will be ten bivariate scatter plots. Many software programs (such as MINITAB) provide matrix scatter plots to depict the bivariate relationships of the variables in a plot.

Missing data are observations not available during analysis. It is a common phenomenon that some data points are missed or unavailable due to collection or recording errors. If the number is significant (> 10%), it contributes to inexact distribution of the available data. Therefore, missing data analysis is important. Hair et al. (1998) provide several methods for missing data handling. Missing data can be *ignorable* or not. For *non-ignorable* missing data, we require to know the patterns and causes of it. Cheema (2014) reviewed missing data handling methods for education research. The authors mentioned that missing data are generated through certain mechanisms, known as "missing data mechanisms." These are *missing completely at random* (MCAR), *missed at random* (MAR), or *not missing at random* (NMAR). If possible, the missing data can be restored by using methods that are appropriate to address the missing data mechanism. For a large sample with MCAR, list-wise deletion can be done. Several imputation methods are also available. For example, individual missing values can be imputed by using the average value of the known data set for each of the variables concerned. If data is MAR, for a large sample, appropriate distribution can be employed to find the missing values. However, if there is large missing data for a variable, the variable can be deleted from the analysis, if the objectives of the study permit it. Another approach is use of

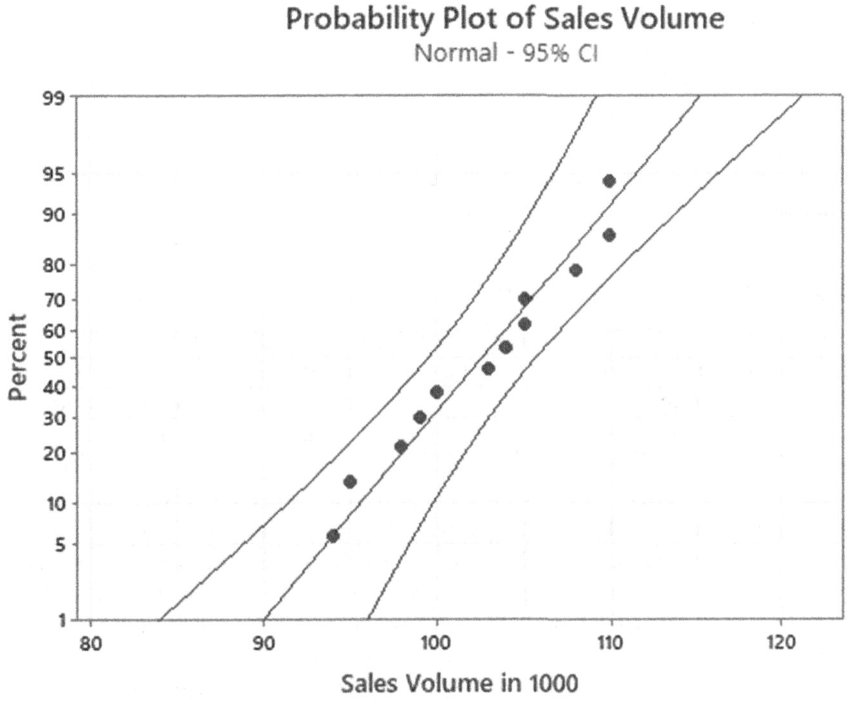

FIGURE 1.13 P-P plot for sales volume (Example 1.4) obtained using MINITAB. The plot shows that the sales volume follows normal distribution.

appropriate statistical model (e.g., regression) that can handle missing data while estimating the relationships explaining the model. For NMAR, the missing data to be handled as a part of the estimation process.

Outliers are the observations that do not belong to the general mass. It is applicable to single as well as multiple variables. For single variables, it is termed *univariate outliers*, and for multiple variables simultaneously occurring, it is called *multivariate outliers*. Univariate outliers can be detected using *dot plot*. Another approach is use of standardized values for the variable. Any standardized values more than three can be considered as outliers at 99.73% confidence level. For two variables, bivariate scatter plot can be used (e.g., see Figure 1.12). For multivariate outliers, *Mahalanobis distance* is measured for each of the observations and chi-square percentile values are used to identify outliers. Gnanadesikan and Kettenring (1972) provided a list of statistics to be used for detection of multivariate outliers where each of the statistics focuses on certain key features of outliers. For example, Mahalanobis distance detects the multivariate observations that lie far from the general mass (scatter) of data.

Most of the multivariate models assume data to be normally distributed. Although models like multivariate analysis of variance (MANOVA) (see Chapter 7) are robust from multivariate normality, normal data make model adequacy tests simpler as fit indices for multivariate normal observations are well developed. The commonly used plots of checking normality are probability-probability plots or *P-P plots* (Figure 1.13) and quantile-quantile plots or *Q-Q plots*. The commonly used indices are z-statistic, Kolmogorov-Smirnov maximum distance or K-S Dmax statistic, and chi-square statistic, Wilk and Shapiro's W (Shapiro and Wilk, 1965), and Anderson-Darling statistic (Anderson and Darling, 1952). In the multivariate domain, chi-square Q-Q plots using Mahalanobis distance are very popular. Another measure using Mahalanobis distance is *scaled residuals*. There are multivariate normality tests based on multivariate skewness and kurtosis (Mardia, 1970).

Homoscedasticity is another important assumption for all regression models (see Chapter 8). Homoscedasticity is achieved when the variance of the dependent variable (DV) doesn't change across the fixed values of independent variables (IVs). The violation of homoscedasticity is known as *heteroscedasticity*.

When data depart from normality and/or homoscedasticity, often data is transformed to satisfy these assumptions. Either the IVs or DVs or both are transformed. Usually power transformation is made. For example, for meeting homoscedasticity assumption, the DV (e.g., say Y) is transformed to square root of the DV ($\sqrt{Y}$).

All these issues will be discussed under different multivariate models in subsequent chapters, as and when needed.

1.4.6 Step 6: Select a Suitable Model

The crux of statistical modeling is explaining the variability present in a data set. For interdependence models such as principal component analysis (PCA) and factor analysis (see Section 1.5), the covariance structure of a number of variables (say, p) is examined. If there are substantial correlations among some of the p variables, at least, PCA or factor analysis is justified. The number of principle components (PCs) or factors is determined based on the percentage cumulative variance explained by the model. For details, see Chapters 11–12. For dependence models such as multiple regression, the variability of the DV (Y) is explained with the help of the IVs (p variables). The model is adequate if the percentage variance (of Y) explained by the model considered is satisfactory (e.g., 0.90 or above). This is possible only when the statistical model chosen fits the data well.

Selection of a suitable statistical model depends on the researchers' or analysts' knowledge and experience that are exercised while building the conceptual model. It is shown in Figure 1.8 with the link from Step 2 to Step 6. The discussion made in Step 2 using Figures 1.9–1.11 has explained this. The link from Step 5 to Step 6 (Figure 1.8) relates by the assumptions checked in Step 5. Many times, even though the conceptual model calls for a particular statistical model to serve the purpose, the data may not permit this. Examination of data is therefore crucial. For the CityCan case, Figure 1.11 calls for a path model as the best fit (conceptually) and data examination will certify whether such a linear model is applicable or not.

1.4.7 Step 7: Estimate Parameters

Once data are collected such as in Table 1.1, these are used to estimate the parameters of the relationship model, such as shown in Equations 1.2–1.5. For example, upon estimated, Equations 1.2 and 1.3 will take the form:

$$Y_1 = \hat{Y}_1 + \hat{\epsilon}_1 = \hat{\beta}_{10} + \hat{\beta}_{11}X_1 + \hat{\beta}_{12}X_2 + \hat{\beta}_{13}X_3 + \hat{\epsilon}_1 \tag{1.6}$$

$$Y_2 = \hat{Y}_2 + \hat{\epsilon}_2 = \hat{\beta}_{20} + \hat{\beta}_{21}X_1 + \hat{\beta}_{22}X_2 + \hat{\beta}_{23}X_3 + \hat{\epsilon}_2 \tag{1.7}$$

$\hat{Y}_1$ and $\hat{Y}_2$ represent the predicted values for the variables Y_1 and Y_2. $\hat{\beta}_{1j}$, $\hat{\beta}_{21j}$ ($j = 0,1,2$, and 3) are the estimates of parameters β_{1j} and β_{2j}. $\hat{\epsilon}_1$, and $\hat{\epsilon}_2$ are the estimated error vectors of the order $n\times 1$.

Equations 1.2 and 1.3 jointly can be termed as *multivariate regression model in theory* and Equations 1.6 and 1.7 can be termed as *multivariate regression model in practice*. The difference between the two models is that Equations 1.2 and 1.3 characterize the population (system of interest) and as a result β_{1j} and β_{2j} (j = 0, 1, 2, 3) are constant parameters of the population. On the other hand, Equations 1.6 and 1.7 represent the population based on the estimates using the sample collected from the population and should be representative of the population for decision-making. $\hat{\beta}_{1j}$ and $\hat{\beta}_{2j}$ are the estimates of parameters β_{1j} and β_{2j} and are *random variables*. If we collect a second sample and estimate $\hat{\beta}_{1j}$ and $\hat{\beta}_{2j}$, we will find that they are numerically different than the first sample. We will discuss this under sampling distribution of parameters (see Chapters 8 and 9, for example).

1.4.8 Step 8: Verify Model

The statistical solutions must be adequate enough to be used as practical solutions. A statistical model therefore must pass an adequacy test which we call *model fitting*. The model fitting issues involve verification and validation. Verification tests (i) whether the developed model is able to explain the relationships intended for based on the data set (***sample***) used in building the model or not, and (ii) which of the parameters of the model are contributing significantly. We will describe these issues for each of the multivariate models included in the book.

1.4.9 Step 9: Validate Model

Validation is the model's ability to explain the relationships incorporated into the developed model for new data sets (new samples). It refers to generalizability of the model. Three commonly used approaches for validation are (i) sliding simulation, (ii) jackknife, and (iii) bootstrapping. *Sliding simulation* is used when a large data set is collected. The sample is broken into two subsamples, one for model-building and verification and the other for validation. The second sample is called "holdout sample." *Jackknife* is a modified holdout approach used for small sample where we hold out one out of *n* observations at a time and the model is built using the remaining *n-1* observations. Then, the model is used to predict the value of the holdout observation. The iterations continue till no more observations are left to consider as a holdout observation. *Bootstrapping* is also called *resampling*, where a representative sample from a population is collected. It is assumed that as the sample (of size *n*) is representative enough, we can generate many samples of size *n* from the original sample of size *n*. The sampling method used is "sampling with replacement." We will describe these issues as and when needed in the subsequent chapters.

1.4.10 Step 10: Interpret Results

When a model is verified and validated, the solutions obtained using the model are transformed to practical solutions. By *practical solutions* we mean that the results obtained are at least able to explain the phenomenon (problem). It can describe the phenomenon fully and will also be able to predict the same in future under similar conditions. Practical solutions also examine the utility of the statistical solutions in terms of cost, quality, feasibility, and effectiveness, which were prescribed while building the conceptual models. In case deviation exists or the situation changes, the model is required to be calibrated. *Calibration* means adding or removing some variables, checking of measurement scale used, or the method of estimation employed for building the statistical model.

1.5 MULTIVARIATE MODELS

There are number of statistical models explaining *multivariate relationships.* Relationships can be linear and non-linear. However, in this book we are interested in linear models. Multivariate linear models can be of two types (i) interdependence models and (ii) dependence models. In multivariate analysis with interdependence models, a large data matrix, say $\mathbf{X}_{n\times p}$, is analyzed to identify patterns in the data set. For example, principal component analysis (PCA) is used to extract meaningful patterns in reduced dimensions (see Chapter 11). In dependence modeling, the relationship between one or more dependent variables is explored with one or more explanatory (independent) variables. Multiple linear regression is such a model. Another classification, based on the purpose of analysis, is exploratory and confirmatory models. The former relies fully on the observed data collected and selects the best-fit model for the available data set, while the latter is used to confirm hypotheses developed based on previous literatures or experience. A classic example is exploratory and confirmatory factor analysis. In this book, the following multivariate models are discussed (in Part III):

1. Multivariate analysis of variance (Chapter 7)
2. Multiple regression (Chapter 8)
3. Multivariate regression (Chapter 9)
4. Path model (Chapter 10)
5. Principal component analysis (Chapter 11)
6. Exploratory factor analysis (Chapter 12)
7. Confirmatory factor analysis (Chapter 13)
8. Structural equation modeling (Chapter 14)

In addition, analysis of variance (ANOVA, in Chapter 7) and the instrumental variable method including two- and three-stage least squares for estimating path equations (in Chapter 10) are also elaborately discussed. Part I (Chapters 1, 2 and 3) includes prerequisites for understanding the remaining chapters of the book and Part II (Chapters 4, 5 and 6) includes the foundations of multivariate statistics.

As stated earlier, a model mimics the reality. A model is used to describe, explain, and, if possible, predict the behavior of a phenomenon in a real-world situation. Each of the models described in this book does one or more of the above three things. Table 1.2 illustrates these.

1.6 ILLUSTRATIVE PROBLEMS

In addition to the examples given above, we describe below some more problems from day-to-day operations in engineering, science, medical, management, and administration. The problems are chosen to show where multivariate statistical models could be useful in modeling and problem-solving. These are only a sample of examples. Numerous such problems can be solved using multivariate statistical models.

TABLE 1.2
Utility of Multivariate Statistical Models

Models	Model utility: System Description and Data	Model utility: Explanation (Relationships)	Model utility: Prediction
Multivariate analysis of variance (MANOVA)	• Identify response variables of interest (Y) • Define factors with levels (X) • Collect data on X and Y and do preliminary analysis	• Obtain the differences in the mean vectors among the factor's levels (populations) • Quantify significant main and interaction effects	• Generally not used for prediction
Multiple linear regression (MLR)	• Identify dependent variable (say, Y) of interest • Identify independent variables (say, $X_1, X_2, \ldots, X_p$) of interest • Collect data ($\mathbf{X}_{n\times p}, \mathbf{y}_{n\times 1}$) and do preliminary analysis	• Obtain $Y = f(\boldsymbol{X})$ in the form $Y = \beta_0 + \beta_1 X_1 + \beta_2 X_2 + \cdots + \beta_p X_p + \in$ • Obtain percentage of variance of Y explained • Obtain significant explanatory (independent) variables	• Predict Y_{new} given $\boldsymbol{X}_{new}$ • Obtain forecast intervals
Multivariate linear regression (MMLR)	• Identify dependent variables (say, $Y_1, Y_2, \ldots, Y_q$) of interest • Identify independent variables (say, $X_1, X_2, \ldots, X_p$) of interest • Collect data ($\mathbf{X}_{n\times p}, \mathbf{Y}_{n\times q}$) and do preliminary analysis	• Obtain $Y_k = f(\boldsymbol{X})$ in the form $Y_k = \beta_{k0} + \beta_{k1} X_1 + \beta_{k2} X_2 + \cdots + \beta_{kp} X_p + \epsilon_k$ $k = 1, 2, \ldots, q$. • Obtain percentage of variance explained • Obtain significant explanatory (independent) variables	• Predict Y_{new} given $\boldsymbol{X}_{new}$ • Obtain forecast intervals
Path model	• Identify dependent variables of interest (say, $Y_1, Y_2, \ldots, Y_q$) • Identify independent variables of interest (say, $X_1, X_2, \ldots, X_p$) (The independent variables may be correlated) • Collect data ($\mathbf{X}_{n\times p}, \mathbf{Y}_{n\times q}$) and do preliminary analysis	• Obtain path diagram • Obtain path equations, similar to regression equations • Obtain percentage of variance explained • Obtain significant explanatory (both dependent and independent) variables • Obtain direct and indirect effects	• Usually used to explain the patterns of relationships, however can be used for prediction
Principal component analysis (PCA)	• Identify variables of interest ($X_1, X_2, \ldots, X_p$) • Collect data ($\mathbf{X}_{n\times p}$) and do preliminary analysis • Choose covariance or correlation structure to be analyzed	• Understand the covariance and/or correlation structure • Obtain principal components (PCs), usually lower than the number of variables (data reduction) • Determine number of PCs to be retained • Obtain PC scores	• Not applicable for prediction. However, the orthogonal principal components may be used as independent variables for predicting response variable(s) of interest.

(*continued*)

TABLE 1.2 (Continued)
Utility of Multivariate Statistical Models

	Model utility		
Models	**System Description and Data**	**Explanation (Relationships)**	**Prediction**
Exploratory factor model	• Identify manifest variables of interest (X_1, X_2, …, X_p) • Collect data ($\mathbf{X}_{nxp}$) and do preliminary analysis • Choose covariance or correlation structure to be analyzed	• Understand the covariance and/or correlation structure of factors • Obtain number of factors to be retained usually lower than the number of variables (data reduction) • Name the factors • Obtain factor loadings (relationships among factors and manifest variables) • Obtain factor scores	• Not applicable, but like PCA, the factors may be used in subsequent analysis including prediction.
Confirmatory factor model	• Identify factors (F_1, F_2, …, F_m) and manifest variables of interest (X_1, X_2, …, X_p) • Collect data ($\mathbf{X}_{nxp}$) and do preliminary analysis • Choose covariance or correlation structure to be analyzed	• Understand the covariance structure of factors • Obtain factor loadings (relationships among factors and manifest variables) • Obtain factor scores	• Not applicable, but the factors may be subsequently used in path model.
Structural equation model	• Identify endogenous (dependent) latent (η) and manifest variables (Y) of interest • Identify exogenous (independent) latent (ξ) and manifest variables (X) of interest • Collect data ($\mathbf{X}_{nxp}$, $\mathbf{Y}_{nxq}$) and do preliminary analysis • Choose covariance or correlation structure to be analyzed	• Understand the covariance structure of latent factors/variables (η and ξ) • Obtain path (structural) equations • Obtain percentage of variance explained • Obtain significant explanatory (both dependent and independent) latent factors/variables (η and ξ) • Obtain direct and indirect effects	• Usually used to explain the pattern of relationships, however can be used for prediction

N.B.: For notation, please refer to notations used and for further clarification, see the relevant chapters.

Example 1.5: Characterization of a Process

Company A manufactures ball bearings. Every day, more than 100,000 steel balls are required. The two important parameters that govern the quality of the balls produced are diameter (D) and compressive strength (CS). Because of large numbers of balls consumed, a very small fraction of defects owing to D and CS cause a large number of balls to be rejected. Large amounts of data on the measurement of D and CS are available. The problem is to characterize the process that produces the steel balls based on the available data on D and CS.

Example 1.6: Quality Control

A process control engineer is of the opinion that the quality variables are group-wise correlated. As a result, these can be monitored in reduced dimensions. Further, the engineer wants to prepare independent control charts for monitoring. The problem is to extract independent dimensions (lesser in number) from the original quality data and to develop control charts for each of the reduced dimensions.

Example 1.7: Robust Design

A company manufactures worm gears of two different sizes as per AGMA guidelines. One of the stages of the manufacturing process is centrifugal casting where molten metal from furnace is poured into the centrifugal casting machine and worm wheel cast is the output. One of the important quality variables of the worm wheel is hardness, which is dependent on input molten metal characteristics namely, %Cu, %Sn, %Ni and %P and casting process variables namely, preheat temperature (°C), rpm, casting time (min) including cooling. Apart from these variables, the ambient temperature and humidity play a significant role in maintaining the quality; however, they are uncontrollable and are termed as noise. The process engineer-in-charge wants to build a relationship model for hardness with molten metal, process and noise variables so that necessary changes can be made in setting the controllable variables to accommodate the changes in the noise variables over the seasons.

Example 1.8: Hotel Room Service

A city hotel is experiencing decrease in customer demand. A brainstorming session involving hotel employees at different levels reveals that customers want better room (meal) service than what the hotel provides at present. The quality variables that are of importance are delivery time (time between placing and receiving meal at a room), food quality and menu variety. In addition, daily cleaning of rooms is considered to be important. The hotel management develops a feedback form on the quality of meal service provided on a 10-point scale. The problem is to quantify the effect of the quality variables on room service provided and their effects on customers' demand.

Example 1.9: Test of Medicine

FB Pharmaceutical claims that its new medicine (D) will relieve breathing problems faster than other medicines available on the market and does not have side-effects worse than its competitors. A doctor wants to test FB's claim. He has randomly chosen 90 patients suffering from breathing problems and prescribed medicine D and other two competing medicines, A and B, to 30 patients each. After one month, he gets feedback from the patients and checks them clinically. Based on the abovementioned data, the doctor is interested to make a comparison of the new medicine, D with the existing medicines A and B.

Example 1.10: Vendor Selection

Company B assembles medical diagnostic equipment. The parts required for all assemblies are outsourced, i.e., are procured from vendors. Company B helps the vendors in maintaining the quality of the parts supplied. But the vendors differ in delivery time (the time between order placed and received) and lateness (time beyond the delivery date). Company B also wants to help the vendor in maintaining their logistics. Vendor-wise data on delivery time and lateness over the past two years are available. The problem is to compare and prioritize the vendors based on their logistic performance so that focused support can be provided to the vendors.

Example 1.11: Work-system Performance

An engineer in charge of shop-floor production is facing a problem of poor productivity and wants to optimize the performance of his workforce. He realizes that working environment and workforce expertise have strong influence on workers' performance. On further enquiry, he establishes that workers' performance solely in term of productivity measure is not exhaustive and scrap ratio (for quality) and number of injuries (for safety) may well be other indicators of performance. The shop-floor has problems with heat and humidity, vibrating equipment, and they vary from season to season. The workforce varies in education, skill, and experience. The engineer is interested to identifying influencing factors and their contributions on productivity.

Example 1.12: Safety Management

Safety management is considered to be a multidimensional problem. A meta-analysis suggests that work injuries are dependent on safety level of a workplace which in turn depends on physical hazards, safety initiatives, and management commitment. Being latent, these variables are measured using a questionnaire survey. The management is interested to test the relationships as defined in previous studies in their organization with the questionnaire data recorded over time.

Example 1.13: Work Compatibility

Work compatibility is a recent concept that defines the mismatch between demands and energizers poised on an employee while working. It varies across jobs, workplaces, and organizational hierarchies. There are 12 factors that govern demands and energizers such as physical and mental task contents, social and organizational environment, work effort, etc. These factors are usually measured through questionnaire survey. An industrial engineer administered a set of questionnaires to randomly selected employees across his organization. He wants to measure and model work compatibility across departments. He is also interested to evaluate the differences in work compatibility across job designations.

Example 1.14: Material Handling

A steel manufacturing plant utilizes a single material handling equipment for two parallel machines. Improper utilization often creates bottleneck situations in the operation of the whole plant. Designers are interested in designing a system that improves performance. Performance measure of interest is average system utilization (ASU). Factors affecting the performance are identified as machine set-up times, raw material preparation time, loading and unloading time, and maximum speed of material handling system. The problem is to identify which of these factors and which of the potential interactions between them explain the variability in the ASU.

Example 1.15: Lean Production

A large manufacturing company is successful in implementing lean engineering. The strategies that determine leanness are just-in-time (JIT) practices, worker's involvement, and automation. Thirty officials responsible for lean production at the different levels of organizational hierarchy are administered with six questions on (i) use of pull system, (ii) reduction of set-up time, (iii) on-time delivery, (iv) continuous improvement, (v) quality circles, and (vi) work standardization. The manager of the company is interested to quantify the strategies based on the responses received.

Example 1.16: Damage during Transportation

A logistics company provides long route services in delivering goods from suppliers' destination to buyers' destination. The process involves damage to goods, which is dependent on, among other things, the weight to be transported and distance traveled. Based on the preliminary analysis of the past data, the company is worried about increasing compensation for damage during transportation. The problem is to define the relationship between damage (in monetary limits) and amount of materials transported and travel distance, so that the damage can be predicted in advance and appropriate measures can be taken for improvement.

Example 1.17: Water Distribution in a City

The amount of water needed in a day (24 hours) of a city is dependent on many factors. The city is divided into three zones based on geographical differences and industrialization. The household size also varies. Another important consideration is the season of the year. The city commissioner is interested to measure the relationship between water consumed with zone-of-city, household size, and season of the year. As the city is expanding both in size and density, a model is required to predict the future demands as well as its distribution over zones and seasons.

Example 1.18: Effectiveness of Assignments

The lack of interest in reading textbooks and gradual decrease in attendance over the years are of great concern to university teachers across the world. Giving assignments is also a traditional practice. A premier institute is interested to evaluate the effect of assignments as a compulsory submission on the performance of the students. The performance parameters are grade obtained and percentage attendance in the class. As a test case, the students of a large class are divided into three sections. In section 1, submission of assignment is compulsory, in section 2 the submission is voluntary, and in section 3 no submission is needed. Based on the data obtained for the last semester, the professor in charge is required to evaluate the effect of assignments on the performance of the students.

Example 1.19: Effectiveness of Training

Training is an effective mean to upgrade the knowledge and skill of workforce in a company. A consulting firm is hired to evaluate the effect of training by company X that conducts periodic training programs on organizational excellence. Company X has five key departments. Every year, one-fifth of the employees go through specific trainings. Data is available for the past five years on amount of training imparted, number of employees trained, number of hours spent training, components of the training programs, and percentage of errors committed by the employees, where error is measured as any activity that deviates from the safe operating procedures (SOPs). The key question is whether training affects organizational excellence (percent error committed) within and across departments.

Example 1.20: Teachers' Evaluation

In any academic institution, teaching is considered to be the highest priority among all academic activities. Student feedback is considered to be the best evaluator of quality of learning imparted. All students attending a subject over a semester was requested to give their responses (score) on a 100-point scale on three aspects, namely (i) contents of the subject, (ii) teaching quality, and (iii) tutorials and assignments handled. Thirty-six students responded with a drop rate of 4%. The teacher concerned is interested to evaluate the merit of each of the three dimensions in evaluating the effect of teaching.

Example 1.21: Marketing Performance

Marketing performance measures (of a company) appear in various ways, such as sales, choice of the brands, and market shares. Different marketing instruments such as advertising, product pricing, and discounting strategies play vital roles in improving marketing performance. A cell phone manufacturing company has a database containing sales and market share data for the previous five years corresponding to different advertising and pricing strategies. The problem is to quantify the effects of marketing instruments (advertising and pricing strategy) on different marketing performance measures (sales and market shares).

Example 1.22: Purchase Intention

Purchase intention as the name suggests, is the consumers' behavior towards a new product offer. If the purchase intention is high, there are higher chances of the consumer actually buying the product. However, if the purchase intention is low, the consumer is unlikely to accept the new product offer. Among many factors, purchase intention is highly influenced by overall trust of the customers on the manufacturer of a product. The problem here is to identify the indicators of purchase intention as well as trust and to understand the relationship between them. Other aspect that is of interest to the marketing department is whether the relationship differs across income groups and geographical locations.

All these examples listed above can be modeled following the ten-step procedure given in Section 1.4. However, many of them will be discussed from modeling and analysis perspectives in subsequent chapters.

1.7 CASE DESCRIPTIONS

As stated earlier, multivariate statistical models have widespread applications from engineering to management to administration. We provide six real-life case studies in this book. The case studies are modeled with different multivariate statistical models. We describe below a brief introduction of these six cases. These cases will be used in later chapters to highlight the application of different statistical models.

1.7.1 Case 1: Job Stress Assessment among Employees in Coke Plant

The operations in coke plant are hazardous in nature and the employees are subjected to several job related stressors. The health and safety executive (HSE) provides seven job stress risk factors. They are demand (D), control (C), management support (MS), peer support (PS), relationship (RE), role (RO), and change (CH). How different groups of employees differ in experiencing job stress is important for the management to develop and implement interventions for job stress management. In Chapter 7, a case study is demonstrated using MANOVA to assess how different groups of employees differ in experiencing job stress. Under many situations, it is difficult to address the seven risk stressors provided by HSE and it may be plausible to work with lower dimensions (lower number of stressors without significant loss of information). This is a dimension reduction problem and in Chapter 11, we demonstrated the use of principal component analysis (PCA) in handling this problem.

1.7.2 Case 2: Job Demand Analysis of Underground Coal Mine Workers

Underground coal mine workers may face the difficulty of meeting the job requirements owing to many factors influencing their job performance. The job stress theory postulates that unless job demand is matched with the competency of the employees, it reduces employees' productivity. High job demand may lead to burnout and low job demand may impose less challenge, and both may lead to lower performance of employees. To assess the job demand, researchers often rely on a customized questionnaire. The customization is made to suit the prevailing local conditions governing the contexts to be analysed. There could be several underlying factors and which need to be identified and measured for design and implementation of interventions effectively. Exploratory factor analysis (EFA) is often used to solve such problems. In Chapter 12, we demonstrated a case study using EFA for this purpose.

1.7.3 Case 3: Study of the Process and Quality Characteristics and Their Relationships in Worm Gear Manufacturing

The worm wheel is a crucial part of a worm gear. It has applications in several machineries such as punching presses and rolling mills. The quality of worm gear depends on many input variables such as worm wheel quality measured in terms of hardness, Cu, Si, Ni, and P, and hobbing process variables such as speed, feed, and depth of cut. The quality variables of interest are backlash and contact percent. To produce a good quality worm gear, it is necessary to model and establish the relationships of worm wheel quality with input and process variables. In Chapters 8 and 9, the relationships are modeled using multiple liner regression (MLR) and multivariate multiple liner regression (MMLR), respectively.

1.7.4 Case 4: Study of the Process and Quality Characteristics in Cast Iron Melting Process

The quality of grey iron casting is critical from the engine performance point of view. There are three processes involved in the casting process in the foundry, namely molding, melting, and fettling. The melting section of the foundry is facing a problem of process variability and it is required to reduce the variation of the quality variables in order to provide a consistent quality to customer. During the melting process, the process characteristics are checked and recorded. There are nine process and nine quality variables. Understanding the relationships (causal and correlation) between these multiple variables may be helpful in improving the stability of the melting process and the quality of the final product. In Chapter 10, we demonstrate the use of the path model to model the relationship between several variables of a melting process in a grey iron foundry of an automobile ancillary unit.

1.7.5 Case 5: Employees Safety Practices in Mines

Mine management conduct several programs aimed at improving employees' safety performance while performing their jobs. Employees' safety practices can be a good measure of employees' safety performance. The underlying factors explaining employees' safety practices could be workers' safety practices, safety-officers' safety practices, and management safety practices. In Chapter 13, we demonstrate the use of confirmatory factor analysis (CFA) to confirm the dimensions of employees' safety practices.

1.7.6 Case 6: Modeling Causal Relationships of Job Risk Perception of EOT Crane Operators

Employees' job risk perception is important as it psychologically affects the safety behavior of workers in the workplace. An employee's job risk perception might be influenced by coworkers' attitudes to safety, supervisors' safety practices, management safety practices, and safety program effectiveness. These factors are often latent in nature and measured through soft instruments such as questionnaire surveys. Two types of quantification become necessary and these are quantification of the latent factors, and estimation of the relationships among the latent factors. Structural equation modeling is the most suited technique for this situation. In Chapter 14, we demonstrate the use of structural equation modeling (SEM) in measuring the latent factors and modeling the relationships among them.

1.8 AIMS OF THE BOOK

Most of the practical phenomena are multivariate in nature and multivariate analysis is a must to capture the patterns. Often we make the mistake of analyzing them separately. The common perception is that multivariate statistics is a difficult subject and many avoid learning this out of fear. As today data is available in abundance, we need to be able to extract knowledge and information from the data. Using software as a black box leads to another common problem that we fail to understand *how* knowledge or information are extracted from data. This may result into erroneous decision-making. All models, including multivariate statistical models, are built on certain assumptions based on which the theoretical foundations of each of the models are established. If we do not know what are the preconditions of using a particular statistical model for solving a problem or confirming a hypothesis, results must be misleading. Similarly, insight cannot be surfaced out if we do not know the theory. This book is intended to bridge this gap.

While writing the book, we assume that the students, data analysts, applied scientists, researchers and practitioners are in search of a book which blends the essential mathematical treatments of multivariate statistical models with the problem solving approaches. Most of the developments (e.g., a theory or a model) has its origin in real-world phenomena. So, it is better to start with a problem (phenomenon) of a process (or system) and then link statistics to characterize it. Most of us deal with real-world problems and so it is not difficult for us to think along these lines.

Similarly, learning multivariate statistics becomes easier when we know the link between different models developed so far. For example, there are links from (i) t-distribution to Hoteling's T^2 distribution, (ii) univariate normal distribution to multivariate normal distribution, (iii) comparing two population means (t-statistic) and mean vectors (Hoteling's T^2) to comparing several population means (ANOVA) and mean vectors (MANOVA), and (iv) multiple regression, factor model, and path model to structural equation model. In this book, these paths are explicitly spelt out for easy and comprehensive learning.

The aims of this book are therefore as given below:

1. Providing guidelines to identify and describe real life problems so that relevant data could be collected.
2. Linking data generation process with statistical distributions, especially in the multivariate domain.
3. Linking the relationship among the variables (of a process or system) with multivariate statistical models.
4. Providing step by step procedures for estimating parameters of a model developed.
5. Analyzing errors along with computing overall fit of the models,
6. Interpreting model results in real life problem-solving,
7. Providing procedures for model validation.
8. Providing blueprint for data-driven decision-making.
9. Providing case studies to demonstrate the models' use and utility.

Items 1–3 fall under model conceptualization; item 4 describes parameter estimation; item 5 deals with model verification; items 6–7 convert statistical solutions to practical solutions; item 8 sums up the total process with careful decision-making; and finally, item 9 demonstrate how multivariate statistical models are built to solve real-world problems. The book is not intended to derive or prove any statistical theory.

The book helps everyone involved in data-driven problem-solving, modeling, and decision-making. The readers learn how a practical problem can be converted to a statistical problem and how the statistical solution can be interpreted as a practical solution. As such, the reader base of the book is very broad.

1.9 ORGANIZATION OF THE BOOK

The book contains 14 chapters. In Chapter 1, we introduce the relevance of data analysis, modeling, and data-driven decision-making with illustrative examples. Useful terms and concepts are described. For example, types of variable and data, the concept of model and modeling, steps in problem-solving, caveats on over reliance on models, and links between different chapters are presented. Chapters 2 and 3 are covered under prerequisites (Part I). Chapter 2 highlights basic statistical concepts that are fundamentals to any statistical modeling and analysis. Chapter 3 deals with matrix algebra. We restrained ourselves only to those concepts that are required to understand the computations required for the models described in this book. These two chapters provide the relevant basics for understanding the chapters to follow, for readers not knowledgeable in basic statistics.

The knowledge provided in Part I is essential to understand Part II (Foundations of multivariate statistics) and Part III (Multivariate models). Part II contains three chapters (Chapters 4–6). In Chapter 4, we describe multivariate descriptive statistics namely mean vector, covariance, and correlation matrices. Zero order, part, and partial correlations are discussed. We also mention appropriate correlation coefficients for variables involving mixed data types. This is important because some of the multivariable models take correlation (or covariance) matrix as input for parameters estimation (e.g., factor model, path model, and structural equation models described in Part III). Chapter 5 describes multivariate normal distribution (MND) which is the backbone of many multivariate statistical models. We explain the mathematical basis of MND but emphasis is given to statistical interpretation of these mathematics. Another important concept of multivariate modeling called statistical distance is also described in this chapter. Many day-to-day decisions involving multiple variables require interval estimation and hypothesis testing; for example, inference about single population means vector, or comparing two population means vectors. These issues are described in Chapter 6 under multivariate inferential statistics.

Part III covers the multivariate models spread over eight chapters (Chapters 7–14). Comparison of more than two population mean vectors is an extension of the comparison of more than two population means (scalar). The latter is known as analysis of variance (ANOVA) and the former is known as multivariate analysis of variance (MANOVA). ANOVA and MANOVA are discussed in Chapter 7. In Chapter 8, the most widely used statistical model, multiple linear regression (MLR) is discussed at length. This is because the concepts, estimation, overall fit, and test of parameters, if understood properly, will help in understanding the same analogously for the other models described in the subsequent chapters. The extension of MLR with multiple dependent variables (DVs) is termed as multivariate multiple linear regression (MMLR). MMLR is discussed in Chapter 9. Many times, MLR or MMLR cannot be applied because data pertaining to independent variables do not meet independent assumption leading to multicollinearity problems. Sometimes, multicollinearity is important and we want to preserve this while estimating the parameters. In this case, we use the path model described in Chapter 10. We have discussed the concepts and estimation process deployed in social sciences and econometric literatures. The path model gives unique opportunity to researchers and analysts for unfolding spurious relationships, if any, captured in correlation coefficients (bivariate). It also enables us to decompose correlation matrix (for multiple variables) into causal, common and correlated relationships.

When the paths of the relationships (structure), as described in Chapter 10, are not important, rather one is interested in prediction, then we may extract orthogonal components (dimensions) out of the correlated variables using principal component analysis (PCA). MLR or MMLR can be used considering the extracted principal components (PCs) as independent variables. PCA is discussed in Chapter 11. Both population and sample PCAs are considered but emphasis is given to sample PCA. We also provide discussion on interval estimation of variance accounted for by each of the PCs followed by the interval estimation of (principal) component's coefficients (loading). PCA has several interesting applications in engineering and sciences. We have discussed a few of them in this chapter. PCA not only extracts orthogonal components, it is also used as a dimension reduction technique. Correlated data structure based on multiple observed variables often is a result of few underlying hidden dimensions. PCA unfolds this. We will discuss this in this book.

It is not always possible to directly measure everything that we want to analyze. Many times we use indirect way of measurement (e.g., questionnaire survey). For example, most social, behavioral, administrative, management, and marketing phenomena are conceptualized in terms of latent variables (constructs) that are measured using indicator (manifest) variables. Such concepts are not unusual in engineering studies also. Manufacturing flexibility, supply chain coordination, continuous improvement capability, and sustainable design are a few examples that we measure using indicators. Exploratory factor analysis (EFA) is used to measure latent constructs. EFA is similar to PCA but differs in its focus as well as conceptualization and uses. EFA is discussed in Chapter 12. EFA considers all the manifest (indicator) variables to identify a small number of unknown factors. On the other hand, if the latent factors and their corresponding manifest (indicator) variables are known *a priori*, the task at hand is to confirm whether the factors are truly causing the manifest (indicator) variables or not. Confirmatory factor analysis (CFA) is used for this purpose. CFA is discussed in Chapter 13. It is often in the interest of the researchers to model the relationships among several latent variables while simultaneously measuring them with the use of manifest (indicator) variables. This essentially requires two models, namely CFA and path model, to be combined in a single model. Structural equation modeling (SEM) perfectly blends the CFA and path models. In Chapter 14, we elaborate SEM in light of handling the complex structure involving multiple variables that are either directly measurable or immeasurable, or both. It should be kept in mind that there are several other multivariate statistical models such as cluster analysis, discriminant analysis, multidimensional scaling, correspondence analysis, and conjoint analysis that are also valuable for data analysis, modeling and data-driven decision-making, which are not included in this book.

A large number of solved examples are provided throughout the book. In addition, case studies are given covering all the multivariate models. As and when required, the software MS Excel, R, Minitab and Matlab are used for computations. The exercises given in each of the chapters will help students practicing for the concepts, models, and computations. The readers are also encouraged to solve the long conceptual and/or numerical questions (under exercise B), which are available in the web resources of the book.

EXERCISES

(A) Short conceptual questions

1.1 What is decision-making? How does statistical model help in decision-making?
1.2 Define variability. Can difference be a measure of variability?
1.3 Define random variable. Give example.
1.4 Define discrete and continuous random variable.
1.5 What is pdf? In what way it is different from pmf?
1.6 What is meant by scale of measurement? How does scale of measurement define information contents?
1.7 What is categorical variable? Why is categorical variable also called qualitative as well as attribute variable?
1.8 Define primary data. What ways does it differ from secondary data?
1.9 What is aggregated information? How aggregated data is generated?
1.10 Define model. Give examples.
1.11 What is physical model? How is it different from mathematical model?
1.12 Why is mathematical modeling not always possible?
1.13 When do we use statistical models?

1.14 What are inductive and deductive logic? Give example.
1.15 What are the principles of modeling?
1.16 Define statistical approach to problem-solving.
1.17 Define statistical problem. How is a practical problem converted to a conceptual model?
1.18 What is theory-based model? In what way it is different from an exploratory model?
1.19 Define exploratory and confirmatory data analysis.
1.20 What is study design? Explain different types of study design.
1.21 Define hypothesis. How do hypotheses lead to model-building?
1.22 Define "unit of analysis" and "unit of measurement."
1.23 What is a measurement error?
1.24 What is validity? Define criterion and content validity.
1.25 What is reliability? Why it is important?
1.26 Why does an analyst examine data?
1.27 What is bivariate scatter plot? What purpose does it serve?
1.28 Why missing data analysis is important? Give two popular methods of handling with missing data.
1.29 Define outlier. Explain univariate and multivariate outliers.
1.30 What is Mahalanobis distance? How it is used to detect multivariate outliers?
1.31 Why normality assumption is important in multivariate statistical modeling? How do you assess multivariate normality?
1.32 What is heteroscedasticity? How can it be removed?
1.33 What is interdependence modeling? Give examples.
1.34 What is covariance structure?

(B) **Long conceptual and/or numerical questions:** see web resources (Multivariate Statistical Modeling in Engineering and Management – 1st (routledge.com))

2 Basic Univariate Statistics

This chapter deals with the basic univariate statistics that comprises univariate descriptive as well as univariate inferential statistics. Univariate descriptive statistics measure two aspects of a variable, namely central tendency such as mean, median, and mode, and dispersion such as range and standard deviation. Inferential statistics cover estimation, confidence interval, hypothesis testing, and comparison of population means and variances. The univariate inferential statistics start with finding out the statistic(s) of interest and its distribution(s), popularly known as sampling distribution. The sampling distribution is used to compute confidence intervals, test hypothesis and to compare population means and variances. The important sampling distributions are unit normal (Z), t, χ^2, and F distributions. Z and t distributions are used to estimate the confidence interval and conduct hypothesis testing pertaining to one population mean or equality of two-population means. χ^2 and F distributions are used to estimate the confidence interval and conduct hypothesis testing pertaining to one population variance and equality of two-population variances, respectively. Knowledge about the population standard deviation and sample size (n) play important roles in estimation and hypothesis testing. Upon completion of this chapter, the reader will have the required knowledge of univariate basic statistics that are often considered the prerequisites to learn and apply multivariate statistical models.

2.1 POPULATION AND PARAMETER

The term *population* in statistics represents the totality or whole of a system of interest. Population consists of all sets of measurements representing the system (universe). For example, if 10,000 units are produced in a year by a manufacturing process for which the analyst is interested, then the population size is 10,000. If the analyst is interested to see the production pattern of the same manufacturing process over several years, say five, the population size would be 50,000.[1] Similarly, the analyst may be interested in considering the quality of items produced by several similar manufacturing processes (e.g., identical machines) in a year. If there are five identical machines each producing 10,000 units, then the population size could be 50,000. For uncountable numbers of identical machines or indefinite periods, the population size could be infinite. The above discussion considers two things to define a population: the system boundary and quantum of measurements. When the analyst is interested in all the items produced in a year by a machine, the system boundary is defined by one machine as space boundary and one year as time boundary. If the analyst considers five identical machines for one year, the space boundary changes from one to five. The quantum of measurements considers all possible measurements of interest within the system boundary. Based on the quantum of measurements, a population is called a finite population with finite size, say N, or an infinite population.

Statistics is concerned with unfolding the behavior of a population with certain characteristics of interest. Accordingly, a population is said to be univariate if its behavior can be measured with one variable or multivariate when several variables simultaneously govern the behavior of the population. Suppose the customers are interested in the quality of the steel washers produced by a manufacturing process. The quality of the steel washers can be represented by its dimensional properties, such as outer diameter, thickness, and physio-mechanical properties like strength. Depending on the requirements, if customers specify the strength value only, the population (steel washer manufacturing process for a specified time) behavior can be judged based on the strength values of the

DOI: 10.1201/9781003303060-3

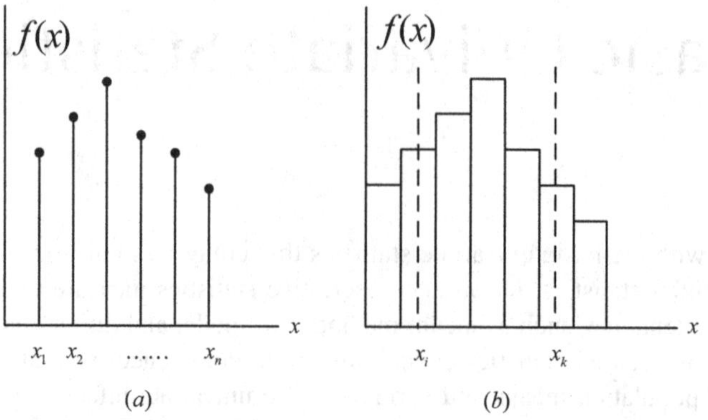

FIGURE 2.1 (a) pmf of X and (b) pdf of X. pmf is used when X is a discrete variable and f(x) for every x represents its probability. pdf is used when X is a continuous variable and f(x) represents its density, where $p(x_i \le x \le x_k) = f(x_k) - f(x_i)$; p stands for probability.

steel washers produced within the system boundary and the analyst can treat the population under the stated condition as univariate. However, if the customers specify quality of the steel washers in terms of both the dimensional and physio-mechanical properties simultaneously, the population under the stated conditions is multivariate. In reality, all populations are multivariate in nature. Summarily, we can assume the following:

- A population has a system boundary.
- A population consists of all the measurements of interest within the system boundary and can be finite if the quantum of measurements is finite, or infinite otherwise.
- Each of the measurements can be univariate or multivariate in nature. Accordingly, a population is said to be univariate or multivariate population.

Let X be a random variable whose measured values are governed by a probability function $f(x)$. X can be a discrete variable like number of defective steel washers produced by the manufacturing process or a continuous variable like outer diameter, or thickness, or strength. If X is discrete, $f(x)$ is termed as *probability mass function* (pmf) and looks like Figure 2.1(a). On the other hand, for continuous X, $f(x)$ is termed as *probability density function* (pdf) and looks like Figure 2.1(b).

Figure 2.1 is a description of a population defined by the variable X. If we look at Figure 2.1 carefully, we find two important aspects to be considered to describe $f(x)$, namely central tendency and dispersion of X. Mean, denoted by μ, is a measure of the central tendency and standard deviation, denoted as σ, is a measure of the spread or dispersion. Measures, such as μ and σ, used to define a population, are treated as parameters of the population. As X is random, unless known we cannot accurately say the exact value of μ or σ, rather we can expect the exact values. That's why μ and σ are represented in terms of expected values of X as shown below.

When X is a discrete random variable,

$$\mu = E(x) = \sum_{all\ x} x f(x) \tag{2.1}$$

$$\sigma^2 = \mathrm{E}(x-\mu)^2 = \sum_{\text{all } x} (x-\mu)^2 f(x) \tag{2.2}$$

When X is a continuous random variable,

$$\mu = \mathrm{E}(x) = \int_{-\infty}^{\infty} x f(x)\,\mathrm{d}x \tag{2.3}$$

and
$$\sigma^2 = \mathrm{E}(x-\mu)^2 = \int_{-\infty}^{\infty} (x-\mu)^2 f(x)\mathrm{d}x \tag{2.4}$$

Some other measures of central tendency are median and mode and for dispersion are skewness and kurtosis.

$E(x)$ and $E(x-\mu)^2$ as defined in Equations 2.1 to 2.4, are also known as first moment about the origin and second moment about the mean, respectively. Although these two quantities are usually sufficient to characterize a random variable X, there are situations when one needs to go beyond mean and variance to higher moments up to k-th order. For a random variable X, the k-th order moment is defined below.

Let m_k be the k-th moment. Then,

$$m_k = \begin{cases} \sum_{all\,x} x^k f(x), & \text{when } X \text{ is discrete} \\ \int_{-\infty}^{+\infty} x^k f(x)dx, & \text{when } X \text{ is continuous} \end{cases} \tag{2.5}$$

When $k = 0$, $m_0 = 1$. For $k = 1$, $m_1 = \mu$ (see Equations 2.1 and 2.3). If we put $k = 2$ in Equation 2.5, the moment m_2 is called the second moment of X. But m_2 in Equation 2.5 is not equal to the quantity (i.e., σ^2) given in Equations 2.2 and 2.4. The necessary modifications define another moment, called central moment or moment around mean (μ). If m_k is the k-th moment around μ (or k-th central moment), then

$$m_k = \begin{cases} \sum_{all\,x} (x-\mu)^k f(x), & \text{when } X \text{ is discrete} \\ \int_{-\infty}^{+\infty} (x-\mu)^k f(x)dx, & \text{when } X \text{ is continuous} \end{cases} \tag{2.6}$$

Here, $m_0 = 1, m_1 = 0$ and $m_2 = \sigma^2$ (see Equations 2.2 and 2.4). In similar manner, we can define m_3, m_4,...., and so on. If we divide m_k by σ^k (where σ is the standard deviation of X as defined in Equations 2.2 and 2.4), the ratio is called as the k-th normalized central moment, say m_k^*. Then,

$$m_k^* = {m_k}/{\sigma^k} \tag{2.7}$$

In connection with moments, another important concept of interest is moment generating function (MGF). As the name suggests, MGF is used to generate moment of a distribution, although it has several other uses in statistics.

Let $M(t)$ be the MGF of X. Then by definition,

$$M(t) = E(e^{tx}) \tag{2.8}$$

Equation 2.8 holds if the expectation exists for t in the vicinity of the origin. Equation 2.8 can also be written as follows:

$$M(t) = \begin{cases} \sum_{all\,x} e^{tx} f(x), & \text{when } X \text{ is discrete} \\ \int_{-\infty}^{+\infty} e^{tx} f(x)dx, & \text{when } X \text{ is continuous} \end{cases} \tag{2.9}$$

Using series expansion of e^{tx}, Equation 2.8 can be expanded as below:

$$M(t) = E(e^{tx}) = E\left[1 + tx + \frac{t^2x^2}{2!} + \frac{t^3x^3}{3!} + \cdots + \frac{t^kx^k}{k!} + \cdots\right]$$
$$= 1 + tE(x) + \frac{t^2}{2!}E(x^2) + \frac{t^3}{3!}E(x^3) + \cdots + \frac{t^k}{k!}E(x^k) + \cdots$$
$$= 1 + tm_1 + \frac{t^2}{2!}m_2 + \frac{t^3}{3!}m_3 + \cdots + \frac{t^k}{k!}m_k + \cdots$$

Where m_k = k-th moment of X.

2.2 DEFINING POPULATION: THE PROBABILITY DISTRIBUTIONS

As stated earlier, the parameters μ and σ are used to define a population. By definition, we mean obtaining an appropriate probability distribution using the population parameters that can mimic the stated behavior of the process (system) under consideration. Some of the commonly used probability distributions are mentioned below.

a) Discrete distributions
   - Poison distribution
   - Binomial distribution
   - Negative binomial distribution
   - Geometric distribution
   - Hypergeometric distribution

b) Continuous distribution
   - Exponential distribution
   - Normal distribution
   - Lognormal distribution
   - Weibul distribution

However, we will describe in this section only the normal distribution as a large number of random variables occurring in practice can be approximated to normal distribution. That is why enormous use of normal distribution is found in statistics. For other distributions, interested readers may consult Hines et al. (2003) and Johnson (2002).

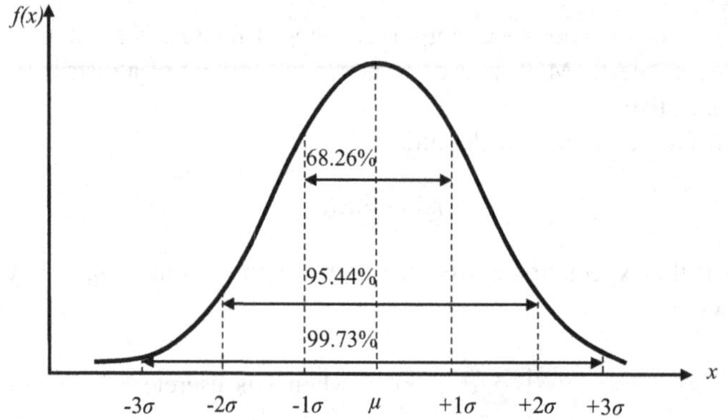

FIGURE 2.2 Univariate normal pdf with mean (μ) and standard deviation (σ). The areas under the curve (f(x)) within intervals μ ± 1σ, μ ± 2σ, and μ ± 3σ are 68.28%, 95.44%, and 99.73%, respectively. Each area represents the percentage of observations falling under the corresponding interval.

2.2.1 Univariate Normal Distribution

Univariate normal distribution of a random variable X is a two parameter continuous distribution whose pdf is

$$f(x)=\frac{1}{\sqrt{2\pi\sigma^2}}e^{-\frac{1}{2}\left(\frac{x-\mu}{\sigma}\right)^2}, -\infty < x < \infty \quad (2.10)$$

Where μ is the mean and σ^2 is the variance of X.

The pdf of a normal random variable X is shown in Figure 2.2. From Figure 2.2, it is seen that 68.26% of the total observations falls within $\mu \pm 1\sigma$ limits, 95.44% of the total observation falls within $\mu \pm 1.96\sigma$ limits and 99.73% of the total observations falls within $\mu \pm 3\sigma$ limits.

The notation used to describe univariate normal distribution with mean μ and variance σ^2 is N (μ, σ^2). For example, if X is normally distributed with mean 50 and variance 16, then $X \sim$ N (50, 4^2).

Example 2.1

The quality of service provided, measured on a 100 point scale, at three service centers A, B and C is normally distributed as N (80, 3^2), N (80, 4^2) and N (90, 3^2), respectively. Comment on the performance of the three service centers.

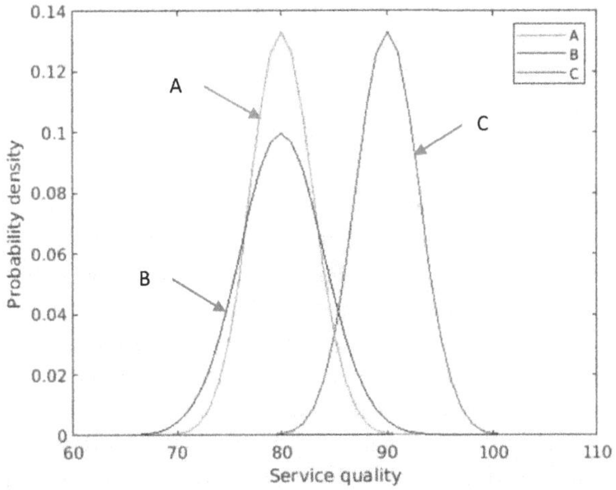

FIGURE 2.3 Performance of three service centers (Example 2.1). The performance of all the three processes (A, B, and C) follows normal distribution. As service quality is "higher the better" type variable, process C's performance is superior in comparison to other two processes.

TABLE 2.1
Summary of Quality of Service Provided at the Three Service Centers A, B, and C

Service Centers	Mean (μ)	Variability (σ^2)	Comment
A	80	9	C's performance is better than A and B.
B	80	16	
C	90	9	

Solution:

Figure 2.3 shows the normal density functions for the quality of service provided by the service centers A, B, and C, respectively. We see that the mean service score is equal for A and B but more for C. On the other hand, from a variability point of view, A and C perform equally but B's variability is more than A and C.

Quality of service is a "larger the better" type characteristic. So, the higher the mean value, the better the performance. Further, variability should be as low as possible. Table 2.1 summarizes it for the three service centers A, B, and C.

Service center C performs better than A from mean service point of view and better than B from both mean service and its variability point of view. Service center A performs better than B from variability point of view.

2.3 SAMPLE AND STATISTICS

Equations 2.1–2.4 define the parameters of a population characterized by a single variable X. In reality, we estimate these parameters using sample statistics where a sample is a set of data for the variable X collected through appropriate sampling strategy.[2] Let us consider that based on certain sampling strategy n observations will be collected on X. In vector notation, we can write the n observations (data) as:

$$\mathbf{x}_{\text{nx1}} = \begin{bmatrix} x_1 \\ x_2 \\ \cdot \\ \cdot \\ \cdot \\ x_n \end{bmatrix}$$

It is to be noted here that $\mathbf{x}_{\text{nx1}}$ is a random vector of observations so long it is not collected. Each of the observations, $x_1, x_2, \ldots, x_n$, is not known in advance and their occurrence is governed by the laws of probability. So, these observations are random[3] in nature. Once the observations are collected, these represent fixed values, and are also known as sample of observations or sample of size n.

As stated earlier, a sample is used to compute statistics where a statistic is a numerical measure of a sample that can be used to estimate the corresponding parameter of the population from which the sample is collected. The procedure is known as estimation and is covered under statistical inference. The other use of such samples is to summarize the data to visualize or describe the population of the interest, which is known as descriptive statistics. Descriptive statistics cover percentiles, measures of central tendency, measure of dispersion (variability), and grouping and displaying of data using histogram and charts.

2.3.1 Measures of Central Tendency

The sample of observations $x_1, x_2, \ldots., x_n$ can be ordered according to their magnitudes, which enable us to locate the relative position of a given data point or a value of X. For example, one may be interested to know, in an ordered sample, where the average value lies or what the value of X is at the middle point of the ordered data set. Similarly, what is the value of X at which the occurrence (frequency) is the maximum? In statistical terms, the answers to these three questions are mean, median, and mode, the common measures of central tendency.

Let the ordered sample of observations is $x_{(1)}, x_{(2)}, \ldots, x_{(n)}$ in such a way that $x_{(1)} \leq x_{(2)} \leq \ldots \leq x_{(n)}$. So, the minimum value is $x_{(1)}$, the maximum value is $x_{(n)}$ and the range is $x_{(n)} - x_{(1)}$. We can divide the range into several equal segments, called quantities, where each quantile represents a range of values of X. The first quantile represents a subset with the smallest values and the last quantile contains a subset with the largest values. It should be noted that the number of data points in each quantile is equal in size.

If we divide the data set into m equal segments, m quantiles are formed. When $m = 100$, the quantiles are known as percentiles. The $m-th$ percentile of a data set is that value below which lie m percentage of observations of the data set. The position of the $m-th$ percentile is computed by $(n + 1) \times m/100$, where n is the total number of observations in the data set.

When m = 4, the data set is divided into four equal segments and each segment is known as quartile. The first quartile is the 25th percentile, below which lie one-fourth of the data set. The second and third quartiles are the 50th and 75th percentile, respectively. When m = 2, the data set is divided into two equal segments and the location which divides the data set into two equal parts is known as the median of the data set. The median is the 50th percentile or the second quartile.

Example 2.2

The following sample (n=12) is collected from a population. Obtain its first quartile, second quartile (median) and the third quartile values.

10, 12, 11, 9, 9, 10, 11, 12, 11, 10, 11, 12

Solution:

Step 1: Obtain the Ordered Data Set

The ordered data set is 9, 9, 10, 10, 10, 11, 11, 11, 11, 12, 12, 12

The data set varies from 9 to 12 and there are n =12 data points.

Step 2: Find out the Desired Locations

The locations of interest are the first, second and the third quartiles. With respect to the 12 data points, the corresponding locations are given below.

Location of the 1st quartile = 25th percentile = $(n + 1) \times m/100$ (where $n = 12$ and $m = 25$)
= (12 + 1) × 25/100 = 13/4 = 3.25th position of ordered data.

Similarly,

2nd quartile = (12 + 1) × 50/100 = 13/2 = 6.50th position, and

3rd quartile = (12 + 1) × 75/100 = 13/4 × 3 = 9.75th position

Step 3: Obtain the Desired Values.

Counting from the smallest to largest values of the ordered data set, we find that the third value is 10 and the fourth value is also 10, so, the 3.25-th position (25th percentile) data value is also 10. Accordingly, the first quartile data value is 10. The second quartile value is at 6.50th

TABLE 2.2
Frequency Table

Value	Frequency
9	2
10	3
11 ← Mode	4 ← Max freq
12	3

position. The data values at the 6th and 7th positions are 11 only. So, the value at the 6.50th position is also 11. Hence, the second quartile or median of the data set is 11. Finally, the data values at the 9th and 10th positions are 11 and 12, respectively. So, the data value at the 9.75th position (75th percentile) is 11.75 as the data value lies 0.75 units from 11.

The data set in Example 2.2 can be arranged in a different way. The arrangement is shown in Table 2.2. Table 2.2 represents the frequencies at which each of the observed values occurred. For example, the observations 9 occurred twice and 10 occurred thrice. In addition, we observe from Table 2.2 that the value 11 occurred for the maximum times, i.e., with the maximum frequency of 4. It is known as the mode of the data set. Therefore, the mode of a data set is the value that occurs most frequently.

Now, we will compute the arithmetic average or mean of a data set. Here, we consider the original data set, $x_1, x_2, \ldots, x_n$. The sample mean is denoted by $\bar{x}$ (x bar) where

$$\bar{x} = \frac{1}{n}\sum_{i=1}^{n} x_i = \frac{x_1 + x_2 + .. + x_n}{n} \tag{2.11}$$

What does the mean of a data set represent? If we see Equation 2.11, we find that mean is nothing but a point where each of the observations is concentrated with equal weight (by $\frac{1}{n}$). It is the center of the mass of n data points. Often it is required to compute average of a set of n data points where the observations are weighted differently. If the desired weights are $w_1, w_2, \ldots, w_n$ for the observations $x_1, x_2, \ldots, x_n$, respectively, the arithmetic mean is computed as below.

$$\bar{x} = \frac{\sum_{i=1}^{n} w_i x_i}{\sum_{i=1}^{n} w_i} = \frac{w_1 x_1 + w_2 x_2 + \cdots + w_n x_n}{w_1 + w_2 + \cdots + w_n} \tag{2.12}$$

Equation 2.12 gives weighted mean of a given data set. When the data collected are grouped, the occurrences for each group are stored as frequency weighted data as shown in Table 2.2. Then, Equation 2.12 is modified to compute mean as

$$\bar{x} = \sum_{i=1}^{n} \frac{f_i x_i}{\sum fi} = \frac{f_1 x_1 + f_2 x_2 + \cdots + f_n x_n}{f_1 + f_2 + \cdots + f_n} \tag{2.13}$$

Example 2.3

Compute mean for the data set given in Example 2.2.

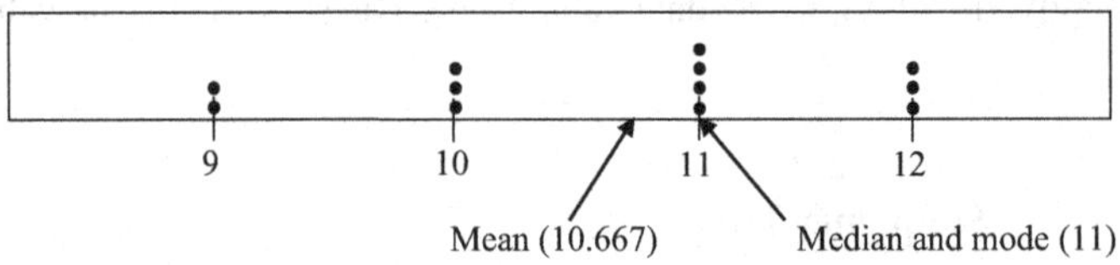

FIGURE 2.4 Mean, median and mode for Example 2.2. While mean value is 10.667, the median and mode coincide at 11.

Solution:

$$\begin{aligned}\text{Mean} = \bar{x} &= \frac{1}{12}\sum_{i=1}^{12} x_i \\ &= \frac{10+12+11+9+9+10+11+12+11+10+11+12}{12} \\ &= 10.667\end{aligned}$$

Figure 2.4 depicts the mean, median and mode for the data set shown in Table 2.2.

2.3.2 Measures of Dispersion

Dispersion is the measure of variability of a given data set. Range, inter-quartile range, and standard deviation are some of the measures of variability. Range is defined as the difference between the maximum and minimum value of a data set. That is,

$$\text{Range} = x_{\max} - x_{\min}$$

If we denote the first, second and third quartiles as Q_1, Q_2, and Q_3, then the inter-quartile range (IQR) is Q_3–Q_1. But the measure that is widely used to define variability is standard deviation. Standard deviation is the square root of variance of a given data set where variance is the average squared deviation of the data points from their mean. Consider a sample of size n with mean $\bar{x}$. The variance of the sample, denoted by s^2, is computed using the following formula:

$$s^2 = \frac{1}{n-1}\sum_{i=1}^{n}\left(x_i - \bar{x}\right)^2 \tag{2.14}$$

Subsequently, the standard deviation s is

$$s = \sqrt{\frac{1}{n-1}\sum_{i=1}^{n}\left(x_i - \bar{x}\right)^2} \tag{2.15}$$

s^2 is known as sample variance and s as sample standard deviation. In computing s^2, we add n squared deviation but while making average, the sum of squared deviations is divided by $n - 1$. Why we do so? Two explanations can be given. The first one is based on degrees of freedom (DOF) and the second one is based on unbiased estimation. We will describe the first explanation here. In order to understand DOF, consider the following equation.

$$x_1 + x_2 + x_3 = 20$$

Now, you are asked to choose three data values x_1, x_2, and x_3. How many DOF do you have? If you choose $x_1 = 5$ and $x_2 = 11$, x_3 is automatically computed as 4 (=20–11–5). So, essentially, you have two DOF as you can choose freely only two observations out of three. Similar restriction is put in Equations 2.14 and 2.15. It is known that the sum of mean subtracted observations is zero. As we require using mean subtracted observations in Equations 2.14 and 2.15, actually we have n – 1 independent choices for the n mean subtracted observations. In other words, we have $n - 1$ DOF. Hence, while computing the average of sum squared deviations, the sum is divided by $n - 1$.

Example 2.4

Compute variance and standard deviation of the data set given in Example 2.2.

TABLE 2.3
Sum of Squared Mean Standard Deviations (MSD) for Data Given in Example 2.2

Observation No.	1	2	3	4	5	6	7	8	9	10	11	12
Observed value	10	12	11	9	9	10	11	12	11	10	11	12
Mean subtracted deviation (MSD)	–0.67	1.33	0.33	–1.67	–1.67	–0.67	0.33	1.33	0.33	–0.67	0.33	1.33
Squared (MSD)	0.44	1.78	0.11	2.78	2.78	0.44	0.11	1.78	0.11	0.44	0.11	1.78
Sum of squared MSD	12.67											

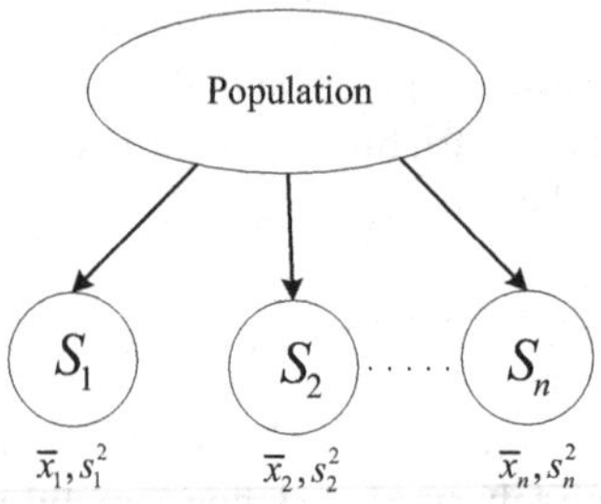

FIGURE 2.5 Repeated sampling from a population. The number of samples taken is n and the size of each sample is also n. S_1, S_2, …, S_n represent the samples.

Solution:

The mean of the data set is 10.667 (Example 2.3) and the sum of the squared mean subtracted deviation (MSD) is 12.67 (Table 2.3).

$$\therefore \text{Variance} = \left(\frac{1}{12-1}\right) \times \text{sum of squared MSD}$$

$$= (1/11) \times 12.67 = 1.15$$

$$\therefore \text{Standard deviation} = \sqrt{1.15} = 1.07$$

2.4 SAMPLING DISTRIBUTION

We have seen earlier that a sample is used to compute statistics (e.g., mean, standard deviation). We also know that a sample has a fixed size, say n and the sample is collected through appropriate sampling strategy. Now let us have a situation as depicted in Figure 2.5, where n random samples of size n are collected from the same population. The n random samples S_1, S_2, …, S_n may be collected by a person or a group over time or collected independently by n persons or groups. So, we have n samples of equal size (= n). Now, if we compute sample mean ($\bar{x}$) or sample variance (s^2) for each sample, we will have n values for the two statistics, $\bar{x}$ and s^2. If we denote the n sample means as $\bar{x}_1, \bar{x}_2, \ldots \bar{x}_n$ and n sample variances as $s_1^2, s_2^2, \ldots, s_n^2$, it is most likely that these values vary randomly. So, $\bar{x}$ and s^2 are random variables and each follow certain probability distribution.

Now, we will define sampling distribution. Probability distribution of a sample statistic (e.g., $\bar{x}$ or s^2) is known as sampling distribution. The distributions are used in inferential statistics for finding out confidence intervals as well as facilitating hypothesis testing. The common probability distributions with which the sample statistics generally fit in are as follows:

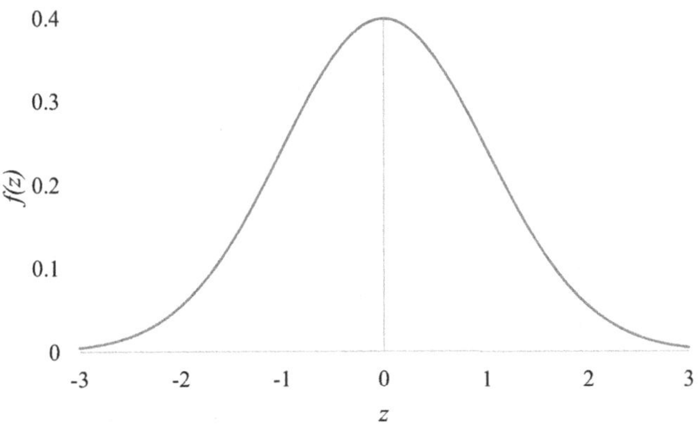

FIGURE 2.6 Unit normal distribution, Z = N(0, 1).

- Standard normal distribution (Z)
- Chi-square distribution (χ^2)
- *t*-distribution
- F-distribution

2.4.1 Standard Normal Distribution

We have already discussed univariate normal distribution. A special class of normal distribution is called standard normal distribution and is defined below.

Let $X \sim N(\mu, \sigma^2)$. If we transform x to z in such a way that $z = \frac{x-\mu}{\sigma}$, then $Z \sim N(0,1)$ is a unit normal distribution, where Z is unit normal random variable.

The pdf of unit normal distribution is

$$f(z) = \frac{1}{\sqrt{2\pi}} e^{-z^2/2}, -\infty < z < \infty \tag{2.16}$$

Figure 2.6 shows the unit normal distribution. The conversion of $X \sim N(\mu, \sigma^2)$ to $Z \sim N(0,1)$ helps in using standard normal table for all practical purposes. The z-value represents the number of standard deviations that a value x is above or below its mean (μ). For example, for $z = -2$, the value of x is 2 standard deviations below the mean, μ and for $z = +3$, it is 3 standard deviations above μ.

Example 2.5

Assume that the delivery time for a courier service is normally distributed with mean of seven days and standard deviation of three days. What is the probability that a service ordered will be delivered within ten days?

Solution:

Let the delivery time be denoted by a normal random variable X, where $X \sim N(7, 3^2)$. The problem at hand is to find out $p(x \leq 10)$. Now, we compute z for $x = 10$. That is,

$$z = \frac{x-\mu}{\sigma} = \frac{10-7}{3} = 1$$

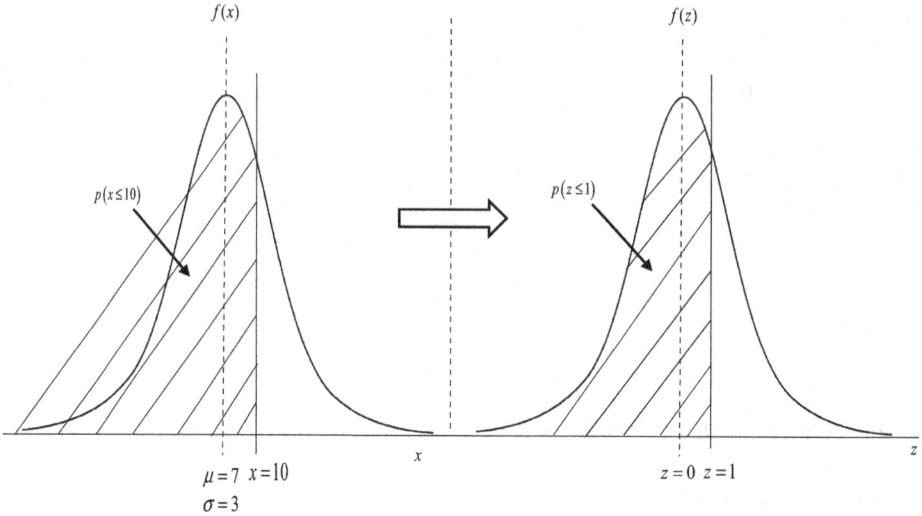

FIGURE 2.7 Conversion of normal X (N(7, 3^2)) to unit normal Z (N(0, 1)). This conversion is made to compute p(x ≤ 10) using unit normal pdf or from normal distribution table. The area under the hatched portion of the figure represents the probability.

So,

$$p(x \le 10) = p\left(\frac{x-\mu}{\sigma} \le \frac{10-\mu}{\sigma}\right)$$
$$= p\left(z \le \frac{10-7}{3}\right) = p(z \le 1)$$

Figure 2.7 shows the equivalence of $p(z \le 1)$.

Now, looking at the standard normal table (Appendix A1), $p(z \le 1)$ is 0.8413. That is, there is 84.13% probability that the service will be delivered within ten days.

In the above discussion, we have shown the utility of Z-distribution for a normal random variable. From sampling distribution point of view, Z-distribution is also used to demonstrate the utility of the sampling distribution of $\bar{x}$. It can be proved that $\bar{x}$ is normally distributed with mean μ and standard deviation σ^2/n when sampled from normal population. Let $z = (\bar{x} - \mu)/(\sigma^2/n)$, which is unit normal (Z). For non-normal population and further discussion, see Section 2.5 (Central Limit theorem).

2.4.2 Chi-Square Distribution

When we calculate sample variance, we sum up the squares of the mean subtracted observations and divide the summation over n (number of observations) by n – 1, if population mean (μ) is unknown and estimated from the sample collected. That is

$$\text{Variance} = s^2 = \frac{1}{n-1}\left[(x_1-\mu)^2 + (x_2-\mu)^2 + \cdots + (x_n-\mu)^2\right] \tag{2.17}$$

Now, let the observations $x_1, x_2, \ldots, x_n$ are random (before data collection) and follows normal distribution, $N(\mu, \sigma^2)$. If we divide equation 2.17 by σ^2 and rearrange, we get

$$\frac{(n-1)s^2}{\sigma^2} = \left[\left(\frac{x_1-\mu}{\sigma}\right)^2 + \left(\frac{x_2-\mu}{\sigma}\right)^2 + \cdots + \left(\frac{x_n-\mu}{\sigma}\right)^2\right]$$
$$= z_1^2 + z_2^2 + \cdots + z_n^2 \tag{2.18}$$

Now, what will be the distribution of $\frac{(n-1)s^2}{\sigma^2}$? It follows χ^2 distribution with $(n-1)$ degrees of freedom (DOF). It was proven in statistical theory. The general rule is, if $Z \sim N(0,1)$ and $z_1, z_2, \ldots\ldots, z_k$ are the k independent observations from $Z \sim N(0,1)$, then $\sum_{i=1}^{n} z_i^2$ follows χ_k^2. In Equation 2.18, although there are n number of z^2 summed up, but it contains only $n-1$ independent observations (See Section 2.3.2 for DOF). Hence, for Equation 2.18, $k = n - 1$. It should be noted here that if μ is known, $(n-1)$ will be replaced by n in equation 2.17 and 2.18, and the statistic of interest will be ns^2/σ^2 instead of $(n-1)s^2/\sigma^2$.

The probability density function (pdf) of a Chi-square variable (y) with k degrees of freedom is

$$f(y) = \frac{1}{2^{k/2}\,\Gamma\left(\frac{k}{2}\right)} y^{(k/2)-1} e^{-y/2}, \text{ for } y > 0 \tag{2.19}$$

with mean equals to k (the degrees of freedom) and variance equals to $2k$.

Example 2.6

Let $X \sim N(\mu, \sigma^2)$. Further, assume that two samples of size 10 and 20 observations are collected from $X \sim N(\mu, \sigma^2)$. Develop χ^2 distribution for the variability (s^2) observed in the sample data sets.

Solution:

Here two samples of size $n_1 = 10$ and $n_2 = 20$, respectively are collected. Let the corresponding variances are s_1^2 and s_2^2, separately. From Equation 2.18, we know that $\frac{(n-1)s^2}{\sigma^2}$ is χ_{n-1}^2, where n is the sample size. Further, from equation (2.19), we know that the mean and variance of a χ^2 distribution with $k(=n-1)$ degrees of freedom is k and $2k$. So, with reference to the given problem, the required χ^2 variables are $(n_1-1)s_1^2/\sigma_1^2$ and $(n_2-1)s_2^2/\sigma_2^2$, respectively. Now, $\frac{(n_1-1)s_1^2}{\sigma_1^2} \sim \chi_{n_1-1}^2$, and $\frac{(n_2-1)s_2^2}{\sigma_2^2} \sim \chi_{n_2-1}^2$. Putting the values of n_1 and n_2, the required χ^2 distributions are χ_9^2 and χ_{19}^2.

2.4.3 t-DISTRIBUTION

Now, let us do another transformation.

Let, $t = \frac{\bar{x}-\mu}{s/\sqrt{n}} = \frac{\bar{x}-\mu}{\sqrt{s^2/n}} = \frac{\sqrt{n}(\bar{x}-\mu)}{\sqrt{s^2}}$

Dividing both numerator and denominator by σ, we get

$$t = \frac{\sqrt{n}(\bar{x}-\mu)/\sigma}{\sqrt{\frac{s^2}{\sigma^2}}}$$

$$= \frac{\sqrt{n}(\bar{x}-\mu)/\sigma}{\sqrt{\frac{(n-1)s^2}{\sigma^2}\cdot\frac{1}{(n-1)}}}$$

So, using $\bar{x} \sim N\left(\mu, \frac{\sigma^2}{n}\right)$ and Equation 2.18, we can write

$$t = \frac{Z}{\sqrt{\chi^2_{n-1}/n-1}} \tag{2.20}$$

which is a ratio of Z ~ N (0, 1) to square root of $\chi^2_{n-1} / n-1$.

The quantity t follows t distribution with $n - 1$ degrees of freedom. In general, t distribution with k degrees of freedom has the following pdf.

$$f(\mathrm{t}) = \frac{\Gamma\left(\frac{k+1}{2}\right)}{\sqrt{k\pi}\ \Gamma\left(\frac{k}{2}\right)} \frac{1}{\left[(\mathrm{t}^2/k)+1\right]^{\frac{k+1}{2}}}, \quad -\infty < t < \infty \tag{2.21}$$

with mean equals to zero and standard deviation equals to $\frac{k}{k-2}$ for $k > 2$, respectively.

The general rule for t-distribution can be written as follows: if a random quantity is the ratio of a unit normal variable in the numerator by the square root of a weighted χ^2_k variable (with weight $= \frac{1}{k}$) in the denominator, the random quantity follows t-distribution with k degrees of freedom. Note that in equation 2.20, $k = n - 1$.

Example 2.7

Consider Example 2.5. Suppose the standard of deviation the population is not known. A sample of size n (= 10) is collected and the sample standard deviation is 2.5 days. Develop a *t*-distribution for the average delivery time. What is the probability that the average delivery time for the service will be more than eight days?

Solution:

Considering Equations 2.20 and 2.21, it can be said that the required *t*-distribution is t_{n-1}, i.e., $t_{10-1} = t_9$ with mean of 0 (zero) and standard deviation of 1.286 ($= \frac{k}{k-2} = \frac{n-1}{n-1-2} = \frac{9}{7}$). Figure 2.8 shows the distribution.

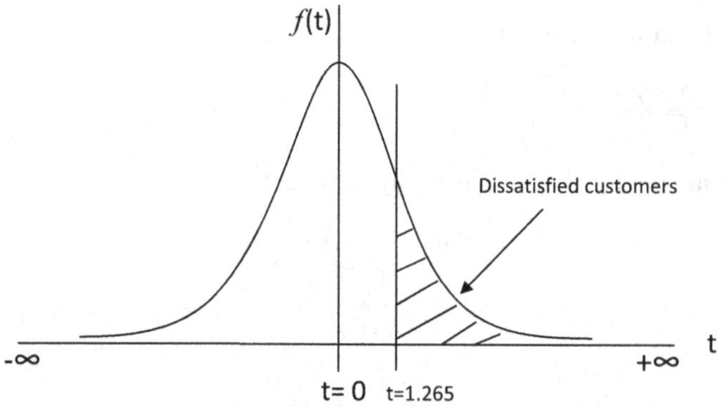

FIGURE 2.8 t-distribution for the sample mean. t=0 shows the population mean. The computed t (=1.265) is away from the population mean. The area under the curve right to t (=1.265) shows p(t > 1.265).

For sample mean of 8 and standard deviation of 2.5, the corresponding t value is

$$t_9 = \frac{\bar{x}-\mu}{s/\sqrt{n}} = \frac{8-7}{2.5/\sqrt{10}} = 1.265. \text{ [note: n} - 1 = 9\text{]}$$

Now, looking at the t-table (Appendix A2), $p(t_9 > 1.265)$ is 0.119. So, the probability that the average delivery time will be more than 8 days is 0.119.

2.4.4 F-Distribution

Let us consider two univariate populations defined by $N(\mu_1, \sigma_1^2)$ and $N(\mu_2, \sigma_2^2)$, respectively. Samples of size n_1 and n_2 are collected from populations 1 and 2 and the computed variances are s_1^2 and s_2^2, respectively. We want to know the distribution of $\dfrac{s_1^2/\sigma_1^2}{s_2^2/\sigma_2^2}$.

Now,

$$\frac{s_1^2}{s_2^2} = \frac{(n_1-1)s_1^2/\sigma_1^2}{(n_2-1)s_2^2/\sigma_2^2} \cdot \frac{(n_2-1)\sigma_1^2}{(n_1-1)\sigma_2^2}$$

$$= \frac{\chi^2_{n_1-1}/n_1-1}{\chi^2_{n_2-1}/n_2-1} \cdot \frac{\sigma_1^2}{\sigma_2^2} \text{(from Section 2.4.2)}$$

$$\text{or } \frac{s_1^2/\sigma_1^2}{s_2^2/\sigma_2^2} = \frac{\chi^2_{n_1-1}/n_1-1}{\chi^2_{n_2-1}/n_2-1} \tag{2.22}$$

From statistical theory, it is known that in general, $f(k_1, k_2) = \dfrac{\chi^2_{k_1}/k_1}{\chi^2_{k_2}/k_2}$ follows F distribution with k_1 numerator degrees of freedom and k_2 denominator degrees of freedom. The pdf of F distribution is

$$f(w) = \frac{\Gamma\left(\frac{k_1+k_2}{2}\right)\left(\frac{k_1}{k_2}\right)^{k_1/2} w^{\left(\frac{k_1}{2}\right)-1}}{\Gamma\left(\frac{k_1}{2}\right)\Gamma\left(\frac{k_2}{2}\right)\left[\left(\frac{k_1}{k_2}\right)w+1\right]^{\frac{k_1+k_2}{2}}}, \quad 0 < w < \infty \tag{2.23}$$

So, the statistic $\dfrac{s_1^2/\sigma_1^2}{s_2^2/\sigma_2^2}$ follows F-distribution with $k_1 = n_1 - 1$ numerator and $k_2 = n_2 - 1$ denominator degrees of freedom.

Example 2.8

Consider Example 2.6. Develop F-distribution for the case.

Solution:

Considering Equation 2.22, the quantity $\dfrac{s_1^2/\sigma_1^2}{s_2^2/\sigma_2^2}$ follows F-distribution with 9 (=10–1) and 19 (=20–1) DOF. The F-distribution is $F_{9,19}$.

All the examples shown under sampling distribution (Section 2.4) show the practical uses of these distributions. The usefulness will further be explored under estimation (Section 2.6) and hypothesis testing (Section 2.7).

2.5 CENTRAL LIMIT THEOREM

A population can be normal or non-normal. The distribution of a sample statistic drawn from a normal population follows certain known probability distribution. For example, the mean of a sample drawn from a normal population with known variance follows normal distribution irrespective of the size of the sample. But the same is not true if the population is non-normal or the population variance is not known. The central limit theorem gives a solution to the latter case. *It states that for a sufficiently large sample drawn from any population (normal or non-normal) with mean μ and standard distribution σ, the sample mean, $\bar{x}$ follows normal distribution with mean μ and standard deviation $\frac{\sigma}{\sqrt{n}}$*. That is for large *n*,

$$\bar{x} \sim \mathrm{N}(\mu_{\bar{x}}, \sigma_{\bar{x}}^2) = N\left(\mu, \frac{\sigma^2}{n}\right) \tag{2.24}$$

Now, we will derive the expected value and variance of $\bar{x}$ in Equation 2.24. Here, we treat $\bar{x}$ as a random variable. So, the expected value of $\bar{x}$, i.e., $E(\bar{x})$ is

$$\begin{aligned} E(\bar{x}) &= E\left(\sum_{i=1}^{n}\frac{x_i}{n}\right) = \frac{1}{n}E(x_1 + x_2 + \cdots + x_n) \\ &= \frac{1}{n}[E(x_1) + E(x_2) + \cdots + E(x_n)] = \frac{1}{n}[\mu + \mu + \cdots + \mu] \\ &= \frac{n\mu}{n} = \mu \end{aligned} \tag{2.25}$$

Let the variance of $\bar{x}$ be $\sigma^2_{\bar{x}}$. So,

$$\begin{aligned} \sigma^2_{\bar{x}} &= V(\bar{x}) = V\left[\frac{1}{n}\sum_{i=1}^{n} x_i\right] \\ &= V\left[\frac{1}{n}\{x_1 + x_2 + \cdots + x_n\}\right] \\ &= \frac{1}{n^2}\left[V(x_1) + V(x_2) + \cdots + V(x_n)\right] \\ &= \frac{1}{n^2}.n\sigma^2 \\ &= \frac{\sigma^2}{n} \end{aligned} \tag{2.26}$$

Central limit theorem can also be described as below.

If $\bar{x}$ be the sample mean of a random sample of size n drawn from any population, then the random variable z, where $z = \frac{\bar{x} - E(\bar{x})}{\sigma_{\bar{x}}} = \frac{\bar{x} - \mu}{\frac{\sigma}{\sqrt{n}}}$ follows unit normal distribution, N(0, 1) for large n.

For univariate case, $n \geq 30$ is considered to be a large sample.

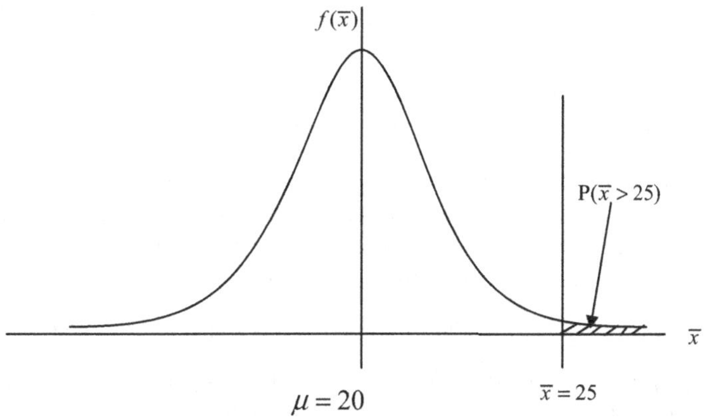

FIGURE 2.9 Probability distribution of $\bar{x}$ with mean (=20) and standard deviation (=5/4). The area under the hatched portion of the figure represents the probability that sample average (number of absentees) is greater than 25.

Example 2.9

In a production shop, the average numbers of absentees in a day is known to be 20, with a standard deviation of 5. A work sampling (random) study is conducted for 36 different days. What is the probability that the sample average (number of absentees) is greater than 25?

Solution:

Here, the population is non-normal with mean $\mu = 20$ and standard deviation $\sigma = 5$. The problem is to determine $p(\bar{x} > 25)$. Figure 2.9 represents the situation.

Using central limit theorem, we can write

$$\begin{aligned} p(\bar{x} > 25) &= p\left\{\frac{\bar{x}-\mu}{\sigma/\sqrt{n}} > \frac{25-20}{5/\sqrt{36}}\right\} \\ &= p\left\{z > \frac{5\times 6}{5}\right\} \\ &= p\{z > 6\} \\ &= 1 - p\{z \le 6\} \\ &= 0.000000001 \end{aligned}$$

So, the probability that sample average absentees are greater than 25 is 1 in 10^9.

2.6 ESTIMATION

One of the important uses of statistics is making inferences about population parameters using sample statistics. In order to do so, we require two things. First, computing a point estimate of the parameter of interest and then determining its appropriate sampling distribution. It is to be noted here that the sample statistics $\bar{x}$ and s^2 are the point estimates of population parameters μ and σ^2, respectively. Both $\bar{x}$ and s^2 are random variable. Next, we determine the desired confidence interval, and the process is known as interval estimation. In the following discussions, first we provide the confidence intervals for mean (μ) and variance (σ^2) of a single population. Then, we describe the confidence intervals for the differences in two-population means and the ratio of two-population variances.

2.6.1 Confidence Interval for Single Population Mean (μ)

We will define first the term "confidence interval." We know that the statistic $\bar{x}$ is used to estimate the parameter μ. We have also seen that $\bar{x}$ is a random variable, i.e., it can assume many values at random. But μ is constant. The question that comes to everybody's mind is that how a random variable $\bar{x}$ can be used as an estimate for the parameter μ, which is a constant?

Let us consider Figure 2.10. It is the distribution of $\bar{x}$. As stated in Equation 2.24, $\bar{x} \sim N\left(\mu, \frac{\sigma^2}{n}\right)$. Let a sample of size n has the sample average of $\bar{x}$. In Figure 2.10 see the position of $\bar{x}$ and the population mean μ along the X-axis. The difference between the two points is $\bar{x} - \mu$. Is the difference significant, i.e., sufficiently large? If so, $\bar{x}$ is not an estimate of μ.

But in reality we don't know the value of μ. So, we cannot accurately measure the difference $\bar{x} - \mu$. However, there is a way out if we know the distribution of $\bar{x} - \mu$ and we sacrifice a certain amount of accuracy in estimating the difference. It has been shown earlier that $\frac{\bar{x} - \mu}{\sigma_{\bar{x}}}$ follows standard normal (Z) distribution.

Now, assume that we don't want all values of $\frac{\bar{x} - \mu}{\sigma_{\bar{x}}}$, rather we are interested in an interval which contains 100 (1–α) % of all $\frac{\bar{x} - \mu}{\sigma_{\bar{x}}}$ values, where α is a predetermined probability value that determines the percentage of observation not included in the interval. The α value is usually considered to be of very small value; typically 0.05 or 0.01. α is also known as level of significance. If the lower and upper values of the interval are ℓ and u, then,

$p\left\{l < \frac{\bar{x} - \mu}{\sigma_{\bar{x}}} < u\right\} = 1 - \alpha$, which is depicted in Figure 2.11.

As $\frac{\bar{x} - \mu}{\sigma_{\bar{x}}}$ follows Z distribution, which is symmetric over mean 0, l and u are equidistance from 0. We can define $\ell = -Z_{\alpha/2}$ and $u = +Z_{\alpha/2}$. Now, the interval l to u can be defined as

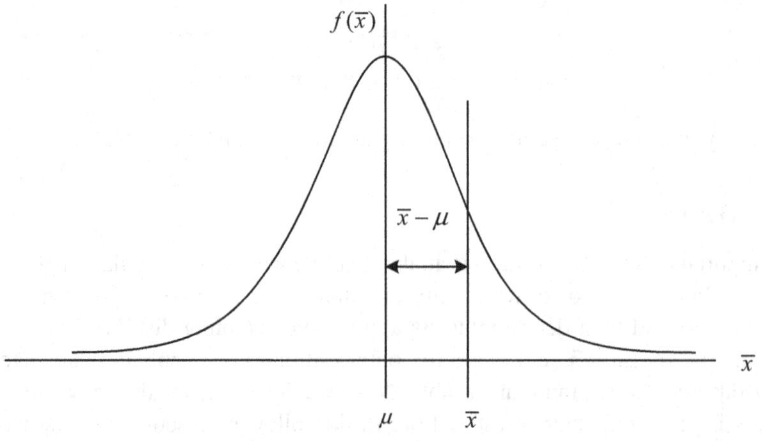

FIGURE 2.10 Difference between sample and population means.

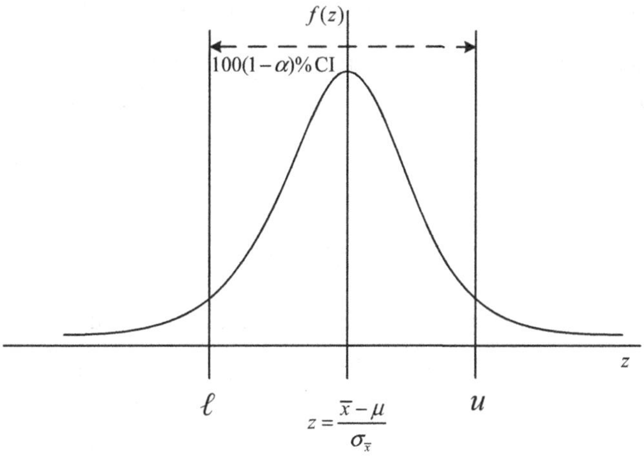

FIGURE 2.11 100 (1–α) % confidence interval of μ.

$$-Z_{\alpha/2} \le \frac{\bar{x}-\mu}{\sigma_{\bar{x}}} \le Z_{\alpha/2}$$

$$\Rightarrow \quad -Z_{\alpha/2}\,\sigma_{\bar{x}} \le \bar{x}-\mu \le Z_{\alpha/2}\cdot\sigma_{\bar{x}}$$

$$\Rightarrow \quad -\bar{x}-Z_{\alpha/2}\,\sigma_{\bar{x}} \le -\mu \le -\bar{x}+Z_{\alpha/2}\cdot\sigma_{\bar{x}}$$

$$\Rightarrow \quad \bar{x}-Z_{\alpha/2}\,\sigma_{\bar{x}} \le \mu \le \bar{x}+Z_{\alpha/2}\cdot\sigma_{\bar{x}} \tag{2.27}$$

What does Equation 2.27 mean? It tells us that the interval $\ell = \bar{x}-Z_{\alpha/2}\,\sigma_{\bar{x}}$ and $u = \bar{x}+Z_{\alpha/2}\,\sigma_{\bar{x}}$ will contain the population mean μ with a probability value of 1–α. The quantity α is known as *confidence coefficient* or *significance level* or *level of significance* and $\bar{x} \pm Z_{\alpha/2}\,\sigma_{\bar{x}}$ is known as 100 (1–α)% confidence interval. For α = 0.05, the confidence interval is $\bar{x} \pm 1.96\sigma_{\bar{x}}$ and confidence level is 1 – 0.05 = 0.95. So, with 95% confidence (probability) the interval $\bar{x} \pm 1.96\sigma_{\bar{x}}$ contains the population mean μ. The quantity $Z_{\alpha/2}\,\sigma_{\bar{x}}$ (used in Equation 2.27) is called *reliability coefficient* or margin of error (on the estimate, $\bar{x}$).

What is the value of $\sigma_{\bar{x}}$? Can we use $Z_{\alpha/2}$ in all cases? The central limit theorem says that $\frac{\bar{x}-\mu}{\sigma_{\bar{x}}} = \frac{\bar{x}-\mu}{\sigma/\sqrt{n}}$ is unit normal when n is large irrespective of the population distribution. Further, we have seen that in order to compute the confidence interval of μ, the value of σ is to be known. Accordingly, the following situations will arise:

(i) Sampling from normal population with large sample and σ is known.
(ii) Sampling from normal population with small sample and σ is known.
(iii) Sampling from normal population with large sample and σ is unknown.
(iv) Sampling from normal population with small sample and σ is unknown.
(v) Sampling from non-normal population with large sample and σ is known.
(vi) Sampling from non-normal population with small sample and σ is known.
(vii) Sampling from non-normal population with large sample and σ is unknown.
(viii) Sampling from non-normal population with small sample and σ is unknown.

We will now discuss on confidence interval of the population mean for all the eight situations.

Situations (i) and (ii): Sampling from normal population with known σ.

$\frac{\bar{x}-\mu}{\sigma/\sqrt{n}}$ follows Z distribution and the $100(1-\alpha)\%$ confidence interval (CI) for μ is $\bar{x} \pm Z_{\alpha/2}\frac{\sigma}{\sqrt{n}}$.

So,

$$\bar{x} - Z_{\alpha/2}\frac{\sigma}{\sqrt{n}} \le \mu \le \bar{x} + Z_{\alpha/2}\frac{\sigma}{\sqrt{n}} \quad (2.28)$$

Situation (iii): Sampling from normal population with large sample and σ is unknown.

The sample standard deviation s is considered in place of σ and $\frac{\bar{x}-\mu}{s/\sqrt{n}}$ follows Z distribution. The $100(1-\alpha)\%$ CI for μ is

$$\bar{x} - Z_{\alpha/2}\frac{s}{\sqrt{n}} \le \mu \le \bar{x} + Z_{\alpha/2}\frac{s}{\sqrt{n}} \quad (2.29)$$

Situation (iv): Sampling from normal population with small sample and σ is unknown.

s replaces σ and the quantity $\frac{\bar{x}-\mu}{s/\sqrt{n}}$ follows t distribution with n–1 degrees of freedom. The $100(1-\alpha)\%$ CI for μ is

$$\bar{x} - t_{n-1}^{(\alpha/2)}\frac{s}{\sqrt{n}} \le \mu \le \bar{x} + t_{n-1}^{(\alpha/2)}\frac{s}{\sqrt{n}} \quad (2.30)$$

Situation (v): Sampling from non-normal population with large sample and σ is known.

The quantity $\frac{\bar{x}-\mu}{\sigma/\sqrt{n}}$ follows Z distribution and the $100(1-\alpha)\%$ confidence interval for μ is $\bar{x} \pm Z_{\alpha/2}\frac{\sigma}{\sqrt{n}}$, which is the same as Equation 2.28.

Situation (vii): Sampling from non-normal population with large sample and σ is unknown.

The quantity $\frac{\bar{x}-\mu}{s/\sqrt{n}}$ follows Z distribution and the $100(1-\alpha)\%$ CI for μ is $\bar{x} \pm Z_{\alpha/2}\frac{s}{\sqrt{n}}$, which is the same as Equation 2.29.

Situations (vi) and (viii): Sampling from non-normal population with small sample. No parametric distribution is available and we use non-parametric methods.

Example 2.10

Musculoskeletal disorder (MSD) is a serious problem of crane operators in heavy industries. In a survey to assess crane operators' MSD, they were asked approximately how long in a month they took unscheduled rest due to body pain. A random sample of 76 responses yielded a mean of seven hours. Let the population standard deviation be three hours. Construct a 95% confidence internal for the mean of the unscheduled rest time of an operator in a month the due to body pain. Assume normal population.

Solution:

Here $\bar{x} = 7$, $\sigma = 3$ and $n = 76$.

So, $\sigma_{\bar{x}} = \frac{\sigma}{\sqrt{n}} = \frac{3}{\sqrt{76}}$

Further, $1-\alpha = 0.95$ as 95% confidence interval (CI) is chosen.

$$\therefore \alpha = 1 - 0.95 = 0.05$$

From normal table $Z_{\alpha/2} = Z_{0.025} = 1.96$

Now, using Equation (2.28) and putting the values of $\bar{x}$, $\sigma_{\bar{x}}$ and $Z_{\alpha/2}$, we get the desired 95% CI as

$$\bar{x} - Z_{\alpha/2} \cdot \frac{\sigma}{\sqrt{n}} \le \mu \le \bar{x} + Z_{\alpha/2} \cdot \frac{\sigma}{\sqrt{n}}$$

$$\text{or} \quad 7 - 1.96 \times \frac{3}{\sqrt{76}} \le \mu \le 7 + 1.96 \times \frac{3}{\sqrt{76}}$$

$$\text{or} \quad 6.33 \le \mu \le 7.67$$

Example 2.11

Consider Example 2.10. Suppose the population standard deviation is not known. Rather, the sample standard deviation is 4. Construct 95% CI for the mean of unscheduled rest time per month.

Solution:

The procedure remains same as in Example 2.10 but only the population standard deviation σ will be replaced by its sample estimate, i.e., $\hat{\sigma} = s = 4$. Accordingly, the $100(1-\alpha)\%$ CI for μ is (using Equation 2.29)

$$7 - 1.96 \times \frac{4}{\sqrt{76}} \le \mu \le 7 + 1.96 \times \frac{4}{\sqrt{76}}$$

$$or \quad 6.10 \le \mu \le 7.90$$

Example 2.12

Consider Example 2.10 again. Let us assume that the population standard deviation (σ) is unknown and the sample size (n) is 20 (<30). If the sample mean and standard deviation are 7.5 and 4.5, respectively, obtain 95% CI for the mean. Compare the CI obtained here with the CI obtained in Examples 2.10 and 2.11.

Solution:

As the population is normal, its standard deviation is unknown and sample size (n) is small (<30), we use equation 2.30 (situation iv).

Now, $\bar{x} = 7.5$, $s = 4.5$ and $n = 20$.

Further, $1-\alpha = 0.95$, so, $\alpha = 0.05$

Form t-distribution table, $t_{n-1}^{(\alpha/2)} = t_{19}^{(0.025)} = 2.093$.

Putting all these values in Equation 2.30, we get 95% CI as

$$7.5 - 2.093 \times \frac{4.5}{\sqrt{20}} \le \mu \le 7.5 + 2.093 \times \frac{4.5}{\sqrt{20}}$$

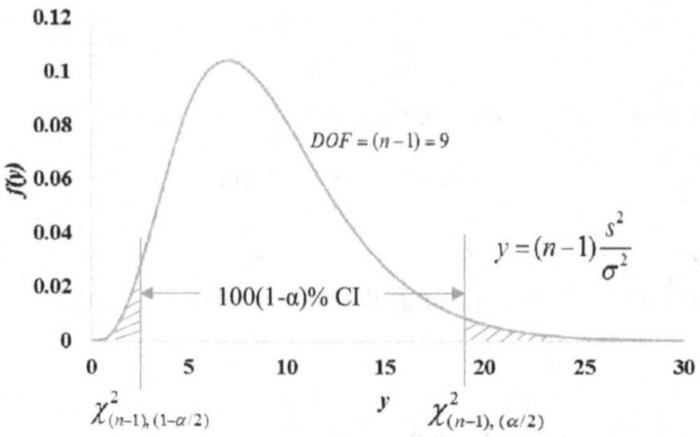

FIGURE 2.12 Sampling distribution of (n-1)s²/σ². The 100(1-α)% confidence interval for (n-1)s²/σ² ranges between $\chi^2_{n-1,1-\alpha/2}$ and $\chi^2_{n-1,\alpha/2}$. The sample size, n = 10 and DOF=9.

$$\text{or} \quad 5.39 \le \mu \le 9.61$$

The 95% CI for μ increases in this case as compared to that in Examples 2.10 and 2.11.

2.6.2 Confidence Interval for Single Population Variance (σ^2)

Here we will develop the $100(1-\alpha)\%$ confidence interval for single population variance (σ^2). We have seen in Section 2.4.2 that the quantity $\frac{(n-1)s^2}{\sigma^2}$ follows χ^2_{n-1} distribution (Equation 2.18). Following Figure 2.12, we can say that

$$P\left\{\chi^2_{n-1,1-\alpha/2} \le \frac{(n-1)s^2}{\sigma^2} \le \chi^2_{n-1,\alpha/2}\right\} = 1-\alpha \tag{2.31}$$

So, the $100(1-\alpha)\%$ CI for σ^2 is

$$\chi^2_{n-1,1-\alpha/2} \le \frac{(n-1)s^2}{\sigma^2} \le \chi^2_{n-1,\alpha/2}$$

$$\Rightarrow \frac{\chi^2_{n-1,1-\alpha/2}}{(n-1)s^2} \le \frac{1}{\sigma^2} \le \frac{\chi^2_{n-1,\alpha/2}}{(n-1)s^2}$$

$$\Rightarrow \frac{(n-1)s^2}{\chi^2_{n-1,\alpha/2}} \le \sigma^2 \le \frac{(n-1)s^2}{\chi^2_{n-1,1-\alpha/2}} \tag{2.32}$$

Example 2.13

A company manufactures worm wheels for worm gears. One of the critical to quality (CTQ) variables is hardness, which is normally distributed. The quality control engineer wants to control its variability. A random sample of 30 worm wheels are tested that yielded mean hardness of 100 (measured using Brinnel hardness number) with standard deviation of 5. Develop 90% confidence interval for the population variance (σ^2).

Solution:

Here, $n = 30$, $\bar{x} = 100$ and $s = 5$.

Further, $1-\alpha = 0.90$.
or $\alpha = 0.10$, i.e., $\alpha/2 = 0.05$.
Using Equation 2.32, we get $\dfrac{(30-1)\cdot 5^2}{\chi^2_{29,0.05}} \le \sigma^2 \le \dfrac{(30-1)\cdot 5^2}{\chi^2_{29,0.95}}$

From χ^2-table we get $\chi^2_{29,0.05} = 42.56$ and $\chi^2_{29,0.95} = 17.71$.

So, the 90% CI for σ^2 is $\dfrac{29\times 25}{42.56} \le \sigma^2 \le \dfrac{29\times 25}{17.71}$

or $17.03 \le \sigma^2 \le 40.94$

2.6.3 Confidence Interval for the Difference Between Two-population Means

Many times we need to compare the differences between two-population means or to test the equality of them. For example, there may be two machines producing a product of interest (e.g., steel washers). Suppose the inner diameter of the washers is of quality concern. Because one machine is comparatively older, the management is interested to know the interval for the mean difference (of inner diameter) of the washers produced by the two machines. If the interval is small, then the manufacturer will mix the washers produced by the two machines for onward supply to the customers.

The above example calls for finding out the distribution of $\bar{x}_1 - \bar{x}_2$, where $\bar{x}_1$ and $\bar{x}_2$ are the means of the samples of size n_1 and n_2, respectively collected from the machines 1 and 2. Essentially, we need to find out the mean of the sample mean differences $\bar{x}_1 - \bar{x}_2$ and its variance, denoted by $\mu_{\bar{x}_1-\bar{x}_2}$ and $\sigma^2_{\bar{x}_1-\bar{x}_2}$. Let the population (machine) means be μ_1 and μ_2 and variances are σ_1^2 and σ_2^2 for machines 1 and 2, respectively.

Now,

$$\begin{aligned}\mu_{\bar{x}_1-\bar{x}_2} &= E(\bar{x}_1-\bar{x}_2) = E(\bar{x}_1) - E(\bar{x}_2)\\ &= \mu_1 - \mu_2\end{aligned}$$

and

$$\begin{aligned}\sigma^2_{\bar{x}_1-\bar{x}_2} &= V(\bar{x}_1-\bar{x}_2)\\ &= V(\bar{x}_1)+V(\bar{x}_2),\ \text{as } \bar{x}_1 \text{ and } \bar{x}_2 \text{ are independent}\\ &= \frac{\sigma_1^2}{n_1}+\frac{\sigma_2^2}{n_2}\ (\text{using equation 2.26})\end{aligned} \tag{2.33}$$

Again, using central limit theorem,

$$\frac{(\bar{x}_1-\bar{x}_2)-\mu_{\bar{x}_1-\bar{x}_2}}{\sigma_{\bar{x}_1-\bar{x}_2}} \sim N(0,1)$$

Therefore, $100(1\text{-}\alpha)\%$ CI for $\mu_1-\mu_2$ is

$$-Z_{\alpha/2} \le \frac{(\bar{x}_1-\bar{x}_2)-\mu_{\bar{x}_1-\bar{x}_2}}{\sigma_{\bar{x}_1-\bar{x}_2}} \le Z_{\alpha/2}$$

$$\therefore -Z_{\alpha/2} \le \frac{(\bar{x}_1-\bar{x}_2)-(\mu_1-\mu_2)}{\sqrt{\dfrac{\sigma_1^2}{n_1}+\dfrac{\sigma_2^2}{n_2}}} \le Z_{\alpha/2}$$

$$\text{or} \quad (\bar{x}_1 - \bar{x}_2) - Z_{\alpha/2}\sqrt{\frac{\sigma_1^2}{n_1} + \frac{\sigma_2^2}{n_2}} \le \mu_1 - \mu_2 \le (\bar{x}_1 - \bar{x}_2) + Z_{\alpha/2}\sqrt{\frac{\sigma_1^2}{n_1} + \frac{\sigma_2^2}{n_2}} \tag{2.34}$$

Equation 2.34 can be used under the following conditions:

- Both n_1 and n_2 are greater than or equal to 30 and σ_1 and σ_2 are known, irrespective of population distribution.
- Both populations are normally distributed and σ_1 and σ_2 are known.

Some special cases:

(i) Normal populations with unknown but equal variance

Let, $\sigma_1^2 = \sigma_2^2 = \sigma^2$, but σ^2 is unknown. We calculate pooled variances (s_p^2) as an estimate of σ^2. The pooled variance (s_p^2) is

$$s_p^2 = \frac{(n_1 - 1)s_1^2 + (n_2 - 1)s_2^2}{n_1 + n_2 - 2} \tag{2.35}$$

Under this special condition, t-distribution is widely used.

Now, the statistic t is

$$t = \frac{(\bar{x}_1 - \bar{x}_2) - \mu_{(\bar{x}_1 - \bar{x}_2)}}{\sigma_{(\bar{x}_1 - \bar{x}_2)}} = \frac{(\bar{x}_1 - \bar{x}_2) - (\mu_1 - \mu_2)}{s_p\sqrt{\frac{1}{n_1} + \frac{1}{n_2}}} \tag{2.36}$$

t follows t-distribution with $n_1 + n_2 - 2$ degrees of freedom.

The $100(1-\alpha)\%$ confidence interval for $\mu_1 - \mu_2$ is

$$(\bar{x}_1 - \bar{x}_2) - t^{(\alpha/2)}_{n_1+n_2-2} \cdot s_p\sqrt{\frac{1}{n_1} + \frac{1}{n_2}} \le \mu_1 - \mu_2 \le (\bar{x}_1 - \bar{x}_2) + t^{(\alpha/2)}_{n_1+n_2-2} \cdot s_p\sqrt{\frac{1}{n_1} + \frac{1}{n_2}} \tag{2.37}$$

(ii) Normal populations with unknown and unequal variances

Here, $\sigma_1^2 \neq \sigma_2^2$.

The $100(1-\alpha)\%$ confidence interval for $\mu_1 - \mu_2$ is (Daniel, 2000)

$$(\bar{x}_1 - \bar{x}_2) - \frac{\frac{s_1^2}{n_1} t_1 + \frac{s_2^2}{n_2} \cdot t_2}{\frac{s_1^2}{n_1} + \frac{s_2^2}{n_2}}\sqrt{\frac{s_1^2}{n_1} + \frac{s_2^2}{n_2}} \le \mu_1 - \mu_2 \le (\bar{x}_1 - \bar{x}_2) + \frac{\frac{s_1^2}{n_1} t_1 + \frac{s_2^2}{n_2} \cdot t_2}{\frac{s_1^2}{n_1} + \frac{s_2^2}{n_2}}\sqrt{\frac{s_1^2}{n_1} + \frac{s_2^2}{n_2}} \tag{2.38}$$

where, $t_1 = t_{n_1-1}(\alpha/2)$ and $t_2 = t_{n_2-1}(\alpha/2)$.

TABLE 2.4
Average Marks Obtained by Students of the Two Groups

Section	Method of Teaching	No of Students	Average Marks Obtained	Population Standard Deviation
Section 1	Method A: "chalk and talk"	30	80	5
Section 2	Method B: "PPT and talk"	30	70	10

Example 2.14

Teaching is considered to be a difficult and evolving job. Professor X, being innovative, adopts different teaching methods. Recently, he adopted two methods, A and B. The method A is the traditional one where he used "chalk and talk" policy and method B is another extreme where he adopted only PowerPoint presentations (PPT). The two methods were administered to two sections of students for the same subject. In order to eliminate the bias from the students' perspective, Professor M divided students equally to the two sections based on the students' background and prior performance. Both sections contain 30 students. The average marks (on a 100 point scale) obtained by the students of the two groups along with other details are given in Table 2.4.

Develop 95% confidence interval for the difference of the means of the two teaching methods.

Solution:

Given

$$\bar{x}_1 = 80 \;\; \bar{x}_2 = 70, \;\; n_1 = 30, \;\; n_2 = 30$$
$$\sigma_1 = 5 \;\; and \;\; \sigma_2 = 10$$

We use Equation 2.34, considering $\alpha = 0.05$ (as CI=95%). The 95% CI for the difference between means of the two teaching methods is

$$(80-70) - Z(0.025)\sqrt{\frac{5^2}{30}+\frac{10^2}{30}} \le \mu_1 - \mu_2 \le (80-70) + Z(0.025)\sqrt{\frac{5^2}{30}+\frac{10^2}{30}}$$

$$\text{or} \quad 10 - 1.96\sqrt{\frac{125}{30}} \le \mu_1 - \mu_2 \le 10 + 1.96\sqrt{\frac{125}{30}}$$

$$\text{or} \quad 6 \le \mu_1 - \mu_2 \le 14$$

Example 2.15

Asthmatic patients often require medicine to get relief from breathing problems. A company launches two competing medicines, A and B, which the manufacturer claims perform equally well. The critical to quality variable here is the time to get relief after the medicine is taken. A doctor prescribed medicine A to 20 patients and medicine B to 25 different patients and recorded the time taken to get the relief after taking the prescribed dose. The average relief time (and its standard deviation) for A and B computed from the samples are 2(2) and 3(2) hours, respectively. Construct 99% confidence interval for the difference between the performance of the two medicines A and B.

Solution:

Here, for medicine A, $\bar{x}_1 = 2,\ s_1 = 2,\ n_1 = 20$
and for medicine B, $\bar{x}_2 = 3,\ s_2 = 2,\ n_2 = 25$
Further, $1-\alpha = .99$.
So, $\alpha = 0.01$

As the manufacturer claims that the two medicines perform equally, we assume that the population standard deviation for A and B are same, i.e., $\sigma_1^2 = \sigma_2^2 = \sigma^2$. In this case, we need to compute pooled variance s_p^2. Using Equation 2.35, we get

$$s_p^2 = \frac{(n_1-1)s_1^2 + (n_2-1)s_2^2}{n_1+n_2-2} = \frac{(20-1)\times 2^2 + (25-1)\times 2^2}{20+25-2}$$

$$= \frac{19\times 4 + 24\times 4}{43} = 4$$

Let the population means for A and B be μ_1 and μ_2, respectively.
Now, using Equation 2.37, we get

$$(2-3) - t_{43}^{(0.005)} \times s_p\sqrt{\frac{1}{20}+\frac{1}{25}} \le \mu_1 - \mu_2 \le (2-3) + t_{43}^{(0.005)} \times s_p\sqrt{\frac{1}{20}+\frac{1}{25}}$$

Form t-table, $t_{43}^{(0.005)} = 2.696$
So, the 99% confidence interval for $\mu_1 - \mu_2$ is

$$-1 - 2.696\times 2\sqrt{\frac{1}{20}+\frac{1}{25}} \le \mu_1 - \mu_2 \le -1 + 2.696\times 2\sqrt{\frac{1}{20}+\frac{1}{25}}$$

$$\text{or} \quad -2.61 \le \mu_1 - \mu_2 \le 0.62$$

Example 2.16

A company with more than 1,000 shop-floor employees conducts training for the workers to upgrade their knowledge and skills. As such training is usually 10–20% effective, retraining is recommended. In order to facilitate the process faster, the company hired two consultants to train their employees on principles and practices of standard operating procedure (SOP). The bottom result for this training is to reduce the "percentage of errors committed by the employees" while following SOP. Consultant A has trained 25 employees and consultant B has trained 21 employees in the last month. As a follow up of the effectiveness of the training imparted, the percentage of error committed by each of the employees trained during this month was recorded. The average error percentage (with standard deviation) for the two groups of employees are 7% (5%) and 10% (3%), respectively. Compute 95% confidence interval for the mean difference between the effectiveness of the training imparted by the two consultants. Assume unequal population variance as the consultant may differ in their competency and ability in imparting training.

Solution:

For consultant A, $n_1 = 25,\ \bar{x}_1 = 7,\ s_1 = 5$ and for consultant B, $n_2 = 21,\ \bar{x}_2 = 10,\ s_2 = 3$. Let population means for consultants A and B are μ_1 and μ_2, respectively.

We require to compute 95% CI for $\mu_1 - \mu_2$. So, $\alpha = 0.05$. Here, sample sizes are small (<30), and population variances are unequal (given). So, to compute the desired CI, we need to use Equation 2.38.

Now,

$$\bar{x}_1 - \bar{x}_2 = 7 - 10 = -3, \text{and}$$

$$\frac{\frac{s_1^2}{n_1}t_1 + \frac{s_2^2}{n_2}t_2}{\frac{s_1^2}{n_1} + \frac{s_2^2}{n_2}}$$

$$= \frac{\frac{5^2}{25}t_{24}^{(0.025)} + \frac{3^2}{21}t_{20}^{(0.025)}}{\frac{5^2}{25} + \frac{3^2}{21}}$$

$$= \frac{\frac{25}{25}\times 2.064 + \frac{9}{21}\times 2.086}{\frac{25}{25} + \frac{9}{21}}$$

$$= \frac{2.96}{1.43}$$

$$= 2.07$$

Further, $\sqrt{\frac{s_1^2}{n_1} + \frac{s_2^2}{n_2}} = \sqrt{\frac{25}{25} + \frac{9}{21}} = 1.20$

So, the 95% confidence interval for $\mu_1 - \mu_2$ is (using Equation 2.38).

$$-3 - 2.07 \times 1.20 \le \mu_1 - \mu_2 \le -3 + 2.07 \times 1.20$$

$$\text{or } -5.484 \le \mu_1 - \mu_2 \le -0.516$$

2.6.4 Confidence Interval for the Ratio Of Two-population Variances

In this section, we derive the confidence interval for the ratio of two-population variances. We have seen earlier that $(n-1)s^2/\sigma^2$ follows χ^2 distribution with n–1 degrees of freedom. We also know that the ratio of two χ^2 variables divided by their respective degrees of freedom yields a random variable which follows F-distribution.

So, the quantity $\frac{(n_1-1)s_1^2}{\sigma_1^2}$ and $\frac{(n_2-1)s_2^2}{\sigma_2^2}$ follow $\chi^2_{n_1-1}$ and $\chi^2_{n_2-1}$ distributions, respectively. The ratio $\frac{s_1^2/\sigma_1^2}{s_2^2/\sigma_2^2} = \frac{(n_1-1)s_1^2/\sigma_1^2(n_1-1)}{(n_2-1)s_2^2/\sigma_2^2(n_2-1)}$ follows F distribution with n_1–1 numerator degrees of freedom and n_2 – 1 denominator degrees of freedom.

Following Figure 2.13, we can write

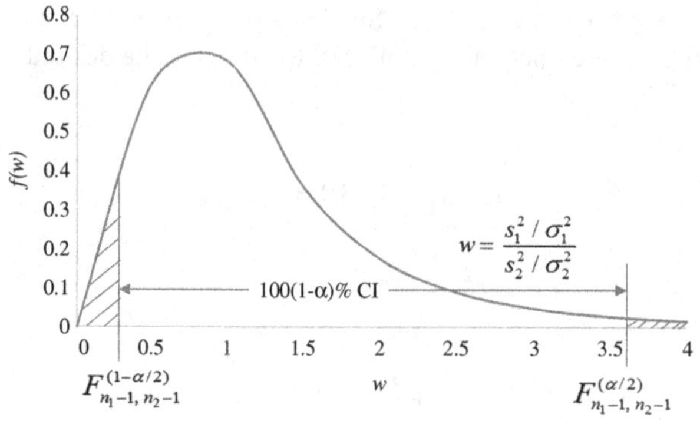

FIGURE 2.13 Sampling distribution of $\frac{s_1^2/\sigma_1^2}{s_2^2/\sigma_2^2}$.

$$F^{(1-\alpha/2)}_{n_1-1,n_2-1} \le \frac{s_1^2/\sigma_1^2}{s_2^2/\sigma_2^2} \le F^{(\alpha/2)}_{n_1-1,n_2-1}$$

$$\text{or} \quad \frac{F^{(1-\alpha/2)}_{n_1-1,n_2-1}}{s_1^2/s_2^2} \le \frac{\sigma_2^2}{\sigma_1^2} \le \frac{F^{(\alpha/2)}_{n_1-1,n_2-1}}{s_1^2/s_2^2}$$

$$\text{or} \quad \frac{s_1^2/s_2^2}{F^{(\alpha/2)}_{n_1-1,n_2-1}} \le \frac{\sigma_1^2}{\sigma_2^2} \le \frac{s_1^2/s_2^2}{F^{(1-\alpha/2)}_{n_1-1,n_2-1}} \tag{2.39}$$

It should be noted here that Equation 2.39 has two critical points, as also shown in Figure 2.13. The right-hand critical point $F^{(\alpha/2)}_{n_1-1,n_2-1}$ can be obtained from F(α/2) table. The left-hand critical point can be obtained from F(α/2) table with the following manipulation:

$$F^{(1-\alpha/2)}_{n_1-1,n_2-1} = \frac{1}{F^{(\alpha/2)}_{n_1-1,n_2-1}}.$$

Example 2.17

Consider Example 2.15 and construct 90% CI for the ratio of variances between the performance of medicines A and B.

Solution:

We need to use Equation 2.39. The given data are $s_1^2 = 4,\ s_2^2 = 4,\ \alpha = 0.10,\ n_1 = 20$ and $n_2 = 25$. From F Table, we get $F^{(0.95)}_{19,24} = 0.48$ and $F^{(0.05)}_{19,24} = 2.08$.

$$\text{So,} \quad \frac{s_1^2/s_2^2}{F^{(\alpha/2)}_{n_1-1,n_2-1}} \le \frac{\sigma_1^2}{\sigma_2^2} \le \frac{s_1^2/s_2^2}{F^{(1-\alpha/2)}_{n_1-1,n_2-1}} \quad \text{is}$$

$$\frac{4/4}{2.08} \le \frac{\sigma_1^2}{\sigma_2^2} \le \frac{4/4}{0.48}$$

$$\text{or} \quad 0.48 \le \frac{\sigma_1^2}{\sigma_2^2} \le 2.08.$$

2.7 HYPOTHESIS TESTING

2.7.1 Hypothesis Testing Concerning Single Population Mean

Hypothesis testing refers to refuting or accepting the statement about population parameter(s) based on the estimation of sample statistic(s). We will describe the concept with the help of an example. The manufacturer of a mobile handset claims that the mean recharge period of its newly launched mobile set is seven days, beyond which it has to be recharged. As a busy person traveling frequently, Mr. R found it interesting but he wanted to be assured that the statement was true. He collected data from 25 users of the newly launched mobile and the data so obtained is given in Table 2.5.

As Mr. R has to make a decision whether to buy the new mobile set or not, this is a decision-making problem. There will be four decision-making scenarios under this situation. The scenarios are depicted in Table 2.6.

Scenarios I and IV reveal that Mr. R will be right in his decision but for scenarios II and III, Mr. R will be a loser. In scenario II, he will not purchase the mobile though his purpose would be served. As he does not purchase, the risk is borne by the manufacturer (loss of a customer although the product is of right quality). In decision-making, this is a type I risk and the error committed is a type I or α error. In scenario III, Mr. R is a loser because he purchases the mobile but it will not serve the purpose it is intended for. Here, the risk is borne by Mr. R (the customer). In decision-making, this is a type II risk and the error committed is a type II or β error.

Can Mr. R come to a decision based on the data collected (Table 2.5)? What will be the risk of his decision? Hypothesis testing is used to answer these two questions.

Let the mean of the sample data collected (Table 2.5) is $\bar{x}$ and variance is s^2. Here the sample size is n (= 25). The manufacture claims that the mean of the population (μ) is μ_0(= 7 days). In order to arrive at a decision, Mr. R puts forward two alternate suppositions namely null hypothesis (H_0) and alternate hypothesis (H_1).

TABLE 2.5
Recharge Period for Mobile Handset in Days

Observation No.	1	2	3	4	5	6	7	8	9	10
Recharge period in days	5.00	5.50	7.00	3.00	4.50	6.00	5.00	4.00	4.50	7.50
Observation no.	**11**	**12**	**13**	**14**	**15**	**16**	**17**	**18**	**19**	**20**
Recharge period in days	3.50	6.00	5.00	4.50	7.00	7.50	6.50	5.50	6.00	6.50
Observation no.	**21**	**22**	**23**	**24**	**25**					
Recharge period in days	4.50	4.00	3.50	7.00	5.00					

TABLE 2.6
Decision-making Scenarios

	Decision	
Status	Accept Statement	Reject Statement
Statement is true	Right decision Scenario I	Type-I error (α) Scenario II
Statement is false	Type-II error (β) Scenario III	Right decision Scenario IV

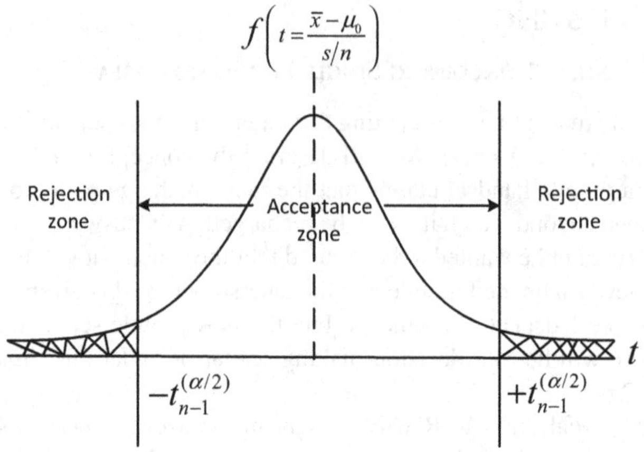

FIGURE 2.14 Rejection zone for H_0; $\mu = \mu_0$. The shaded area (on both sides) collectively represent α, the type-I error.

$$H_0 \;:\; \mu = \mu_0$$
$$H_1 \;:\; \mu \neq \mu_0$$

Now Mr. R requires to know the distribution of $\bar{x}$ when H_0 is true, i.e., $\mu = \mu_0$.

When $\mu = \mu_0$, the quantity $\frac{\bar{x} - \mu_0}{s/\sqrt{n}}$ follows t distribution with (n–1) degrees of freedom with an assumption that population is normal and sample size is small. Following Figure 2.14, if the value of the quantity $\frac{\bar{x} - \mu_0}{s/\sqrt{n}}$ falls beyond $\left|t_{n-1}^{(\alpha/2)}\right|$, then H_0 is not true.

The decision in this case will be reject H_0 (i.e., $\mu = \mu_0$). In order to determine the zone of rejection, a probability value α (usually α = 0.05) is chosen which is known as type I error. If $\frac{\bar{x} - \mu_0}{s/\sqrt{n}}$ follows t distribution with (n – 1) degrees of freedom, there is 100 (1 – α)% chance that it will be within $-t_{n-1}^{(\alpha/2)}$ to $t_{n-1}^{(\alpha/2)}$. $\left|t_{n-1}^{(\alpha/2)}\right|$ will be obtained from t-distribution table. In order to operationalize the hypothesis testing of single population mean, the following steps are performed.

Step 1: Set hypothesis as

$$H_0 \;:\; \mu = \mu_0$$
$$H_1 \;:\; \mu \neq \mu_0$$

Step 2: Obtain the value of the quantity $\frac{\bar{x} - \mu_0}{s/\sqrt{n}}$. Let it be t_c.

Step 3: Find threshold value of $t_{n-1}^{(\alpha/2)}$ from t-distribution table. Let it be t.

Step 4: If $\left|t_c\right| > t$, reject H_0 (i.e., $\mu = \mu_0$). Otherwise we say that we fail to reject H_0.

Example 2.18

For the mobile handset case (Table 2.5), test the hypothesis $H_0: \mu = 7$ days against $H_1: \mu \neq 7$ days for $\alpha = 0.05$. What will be Mr. R's decision?

Solution:

From Table 2.5, we get $\bar{x} = 5.36$ and $s = 1.30$. Now, the computed t is

$$t_c = \frac{\bar{x} - \mu_0}{s/\sqrt{n}} = \frac{5.36 - 7}{1.30/5} = -6.31.$$

The threshold t value is $t_{\text{tabulated}} = t_{n-1}^{(\alpha/2)} = t_{24}^{0.025} = 2.064$. As $|t_c| > t_{tabulated}$ i.e., 6.31>2.064, we reject H_0.

So, Mr. R will not purchase the mobile handset.

Based on the above, the decision-making scenarios (Table 2.6) in hypothesis testing can be reformulated as shown in Table 2.7.

In the above discussions, we have tested the hypothesis on the basis of two-tailed test statistics (the rejection zones are in the two tails of the t-distribution, Figure 2.14). Two-tailed test statistics are useful for a variable which is nominal type. In general, we encounter three types of variables. They are:

1. Nominal is the best, such as diameter of steel washers.
2. The smaller the better, such as absenteeism.
3. The larger the better, such as battery recharge period (Mr. R's case).

Note that we have used two-tailed test for Mr. R's case, whereas an appropriate one-tailed test is better as the random variable is "the larger the better" type. Readers may conduct the one-tailed test for the mean recharge period and compare the results.

The appropriate set of hypotheses for the three categories of variables is given in Table 2.8. Here, we consider those cases where *t*-test is applicable (see Section 2.4.3).

In the above discussion, we have considered only one sampling case, i.e., sampling from normal population with unknown population variance and the sample size is small. This is the most general situation. The other situations that we may come across are:

- Sampling from normal population with large sample and σ is known.
- Sampling from normal population with small sample and σ is known.
- Sampling from normal population with large sample and σ is unknown.
- Sampling from non-normal population with large sample and σ is known.
- Sampling from non-normal population with large sample and σ is unknown.

In all the above situations, use *Z* instead of t-distribution.

TABLE 2.7
Hypothesis Testing and Decision-making

	Decision	
Status	**Accept H_0**	**Reject H_0**
H_0 is true	Right decision	Type-I error (α)
H_0 is false	Type-II error (β)	Right decision

TABLE 2.8
Variable Type versus Hypothesis Testing Using t-statistics

Variable Type	Appropriate Hypothesis	Test Statistic	Rejection Region for H_0*
Nominal is the best	$H_0: \mu = \mu_0$ $H_1: \mu \neq \mu_0$ [two-tailed test]	$t_c = \dfrac{\bar{x} - \mu_0}{s/\sqrt{n}}$	$\lvert t_c \rvert \geq t_{n-1}^{(\alpha/2)}$ 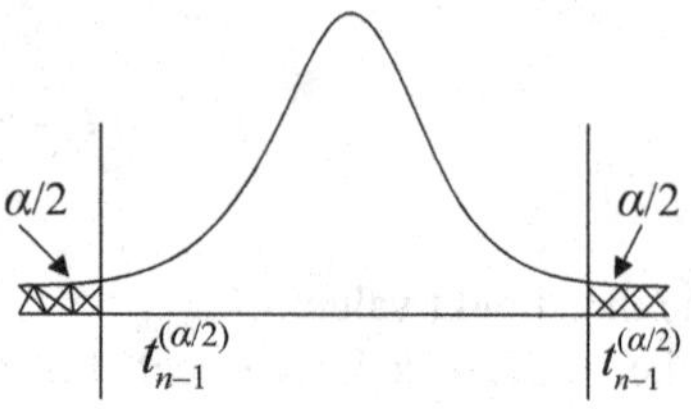
The smaller the better	$H_0: \mu = \mu_0$ $H_1: \mu > \mu_0$ [upper one-tailed test]	$t_c = \dfrac{\bar{x} - \mu_0}{s/\sqrt{n}}$	$t_c \geq t_{n-1}^{(\alpha)}$ 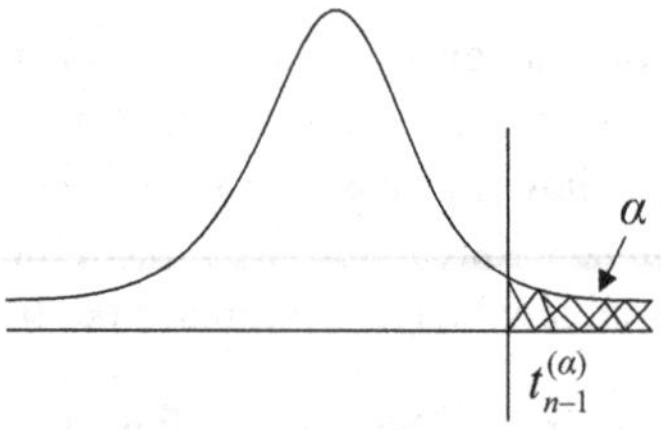
The larger the better	$H_0: \mu = \mu_0$ $H_1: \mu < \mu_0$ [lower one-tailed test]	$t_c = \dfrac{\bar{x} - \mu_0}{s/\sqrt{n}}$	$t_c \leq -t_{n-1}^{(\alpha)}$ 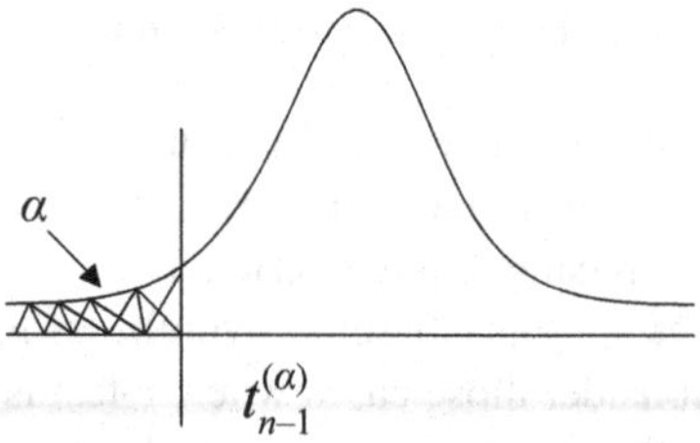

* Note: For "the smaller the better" type random variable, the upper one-tailed test is preferred, whereas, for "the larger the better" type random variable, the lower one-tailed test is preferred.

2.7.2 Hypothesis Testing Concerning Single Population Variance

Let's say the random variable X has mean μ and variance σ^2. We would like to test whether σ^2 is equal to a predefined value, say σ_0^2 or not.

The steps to be followed are given below.

Step 1: Set hypothesis as

$$H_0 \; : \; \sigma^2 = \sigma_0^2$$

$$H_1 \; : \; \begin{array}{l} (i)\; \sigma^2 \neq \sigma_0^2 \text{ for two-tailed test} \\ (ii)\; \sigma^2 > \sigma_0^2 \text{ for upper one-tailed test} \\ (iii)\; \sigma^2 < \sigma_0^2 \text{ for lower one-tailed test} \end{array}$$

Step 2: Obtain the value of the quantity $\frac{(n-1)s^2}{\sigma_0^2}$. Let it be χ_c^2.

Step 3: Find threshold value of χ_{n-1}^2 from χ^2 distribution table.

Step 4: Reject H_0 (i.e., $\sigma^2 = \sigma_0^2$), if

(i) $\chi_c^2 < \chi_{n-1,1-\alpha/2}^2$ or $\chi_c^2 > \chi_{n-1,\alpha/2}^2$ for a two-tailed test
(ii) $\chi_c^2 > \chi_{n-1,\alpha}^2$ for an upper one-tailed test
(iii) $\chi_c^2 < \chi_{n-1,1-\alpha}^2$ for a lower one-tailed test

Figure 2.15 depicts the step-4 pictorially.

Example 2.19

Consider the mobile handset case (Table 2.5). Test the hypothesis that $H_0: \sigma^2 = 1.5^2$ against $H_1: \sigma^2 > 1.5^2$.

Solution:

From Table 2.5, we get $s^2 = 1.30^2$. Now, the computed χ^2 is

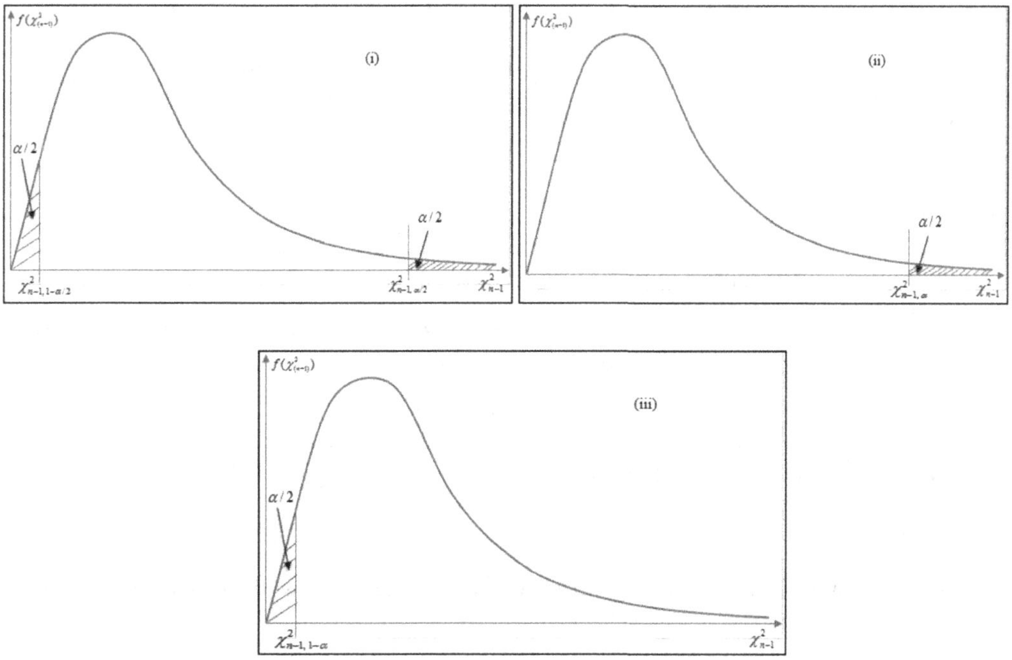

FIGURE 2.15 One and two-tailed χ^2 tests for single population variance.

$$\chi_c^2 = \frac{(n-1)s^2}{\sigma_0^2} = \frac{(25-1)\times 1.3^2}{1.5^2} = 18.03.$$

The situation requires the upper one-tailed test for decision-making.

The threshold chi-square value is $\chi^2_{n\text{-}1,\alpha/2} = \chi^2_{24,0.05} = 36.42$. As $\chi_c^2 < \chi^2_{24,0.05}$, we fail to reject H_0.

2.7.3 Hypothesis Testing Concerning Equality Of Two-population Means

We will limit our discussion here to the following situations:

- Sampling from normal population.
- Both large and small samples.
- Both known and unknown population variance.

Let's say populations 1 and 2 are characterized by $N(\mu_1, \sigma_1^2)$ and $N(\mu_2, \sigma_2^2)$, respectively. Two samples, one of size n_1 from population 1 and another of size n_2 from population 2 are collected. The sample means and standard deviations are $\bar{x}_1, s_1$ and $\bar{x}_2, s_2$, respectively from populations 1 and 2.

Scenario I

Sampling from two normal populations and population variances σ_1^2 and σ_2^2 are known.

Step 1: Set the hypothesis: H_0: $\mu_1 - \mu_2 = 0$

H_1: (i) $\mu_1 - \mu_2 \neq 0$, for two-tailed test

(ii) $\mu_1 - \mu_2 > 0$, for upper one-tailed test

(iii) $\mu_1 - \mu_2 < 0$, for lower one-tailed test

Step 2: Compute test statistic: $Z_c = \dfrac{\bar{x}_1 - \bar{x}_2}{\sqrt{\sigma_1^2/n_1 + \sigma_2^2/n_2}}$

Note Z_c indicates computed Z.

Step 3: Set the rejection regions for H_0

Alternative hypothesis	Rejection zone for H_0
(i) H_1: $\mu_1 - \mu_2 \neq 0$	$Z_c < -Z_{\alpha/2}$ *or* $Z_c > Z_{\alpha/2}$
(ii) H_1: $\mu_2 - \mu_2 > 0$	$Z_c > Z_\alpha$
(iii) H_1: $\mu_1 - \mu_2 < 0$	$Z_c < -Z_\alpha$

Step 4: Reject H_0, if H_0 falls in the rejection zone as stated above.

Example 2.20

Consider Example 2.14. Test the hypothesis that the two teaching methods have same mean performance. Take α=0.05.

Solution: From data we have

$$\bar{x}_1 = 80,\ \bar{x}_2 = 70,\ n_1 = 30,\ n_2 = 30,\ \sigma_1 = 5 \ and\ \sigma_2 = 10$$

Here, the H_0 and H_1 are

$$H_0 : \mu_1 = \mu_2 \quad \text{or} \quad \mu_1 - \mu_2 = 0$$

$$H_1 : \mu_1 \neq \mu_2 \quad \text{or} \quad \mu_1 - \mu_2 \neq 0$$

Computed Z is

$$Z_c = \frac{\bar{x}_1 - \bar{x}_2}{\sqrt{\sigma_1^2/n_1 + \sigma_2^2/n_2}} = \frac{80-70}{\sqrt{\frac{25}{30}+\frac{100}{30}}} = 4.90.$$

Threshold Z (0.025) = 1.96.

As computed Z (4.90) > threshold Z (1.96), we reject H_0. So, there is mean difference in performance of the two teaching methods adopted. The method 1 (chalk and talk) is better than the method 2 (only PPT), as the mean difference (=10) is positive.

Scenario II

Sampling from two normal populations with unknown variances but sample sizes are large. Follow Steps 1–3 in Scenario I. The test statistics in Step 2 will be

$$Z_c = \frac{\bar{x}_1 - \bar{x}_2}{\sqrt{\frac{s_1^2}{n_1}+\frac{s_2^2}{n_2}}}$$

Note that $\sigma_j^2, j = 1, 2$ is replaced by $s_j^2, j = 1, 2$.

Example 2.21

The quality of service of maintenance can be measured on a 100-point scale. There are two competing maintenance shops in a town. A researcher wanted to know whether they are truly competitive or not. She gathered data on maintenance service quality for the past year from both the shops, with sample sizes of 100 and 90, respectively from shop 1 and shop 2. The sample mean service time of shop 1 is 24 hours with standard deviation of eight hours and that of shop 2 is 30 hours with standard deviation of seven hours. What can the researcher conclude about their competitiveness?

Solution:

From data we have

$$\bar{x}_1 = 24, \ \bar{x}_2 = 30, \ n_1 = 100, \ n_2 = 90, \ s_1 = 8 \ \textit{and} \ s_2 = 7$$

As the population standard deviations are unknown and sample size is large, we use Z statistics with sample variances as estimates of their respective population variances.

$$\text{Computed } Z_c = \frac{\bar{x}_1 - \bar{x}_2}{\sqrt{s_1^2/n_1 + s_2^2/n_2}} = \frac{24-30}{\sqrt{\frac{64}{100}+\frac{49}{90}}} = -5.51.$$

The appropriate H_0 and H_1 are

$$H_0 : \mu_1 - \mu_2 = 0$$
$$H_1 : \mu_1 - \mu_2 \neq 0$$

Assuming α = 0.05, the threshold $Z(0.025)$ is 1.96.

As computed Z (=–5.51) in absolute terms is greater than the threshold Z (=1.96), we reject H_0. So, there is mean difference in performance of the two maintenance shops as far as service time is concerned. The researcher can conclude that statistically the shop 1, with mean service time less than shop 2 outperforms shop 2. But the key question remains unanswered, owing to the consideration of appropriate variable to be investigated. It is not clear whether lower service time is a better measure of service quality or not.

Scenario III

Sampling from two normal populations of unknown but equal variances and sample sizes are small.

Step 1: Set the hypothesis: $H_0: \mu_1 - \mu_2 = 0$

H_1: (i) $\mu_1 - \mu_2 \neq 0$, for two-tailed test
(ii) $\mu_1 - \mu_2 > 0$, for upper one-tailed test
(iii) $\mu_1 - \mu_2 < 0$, for lower one-tailed test

Step 2: Compute test statistic: $t_c = \dfrac{\bar{x}_1 - \bar{x}_2}{s_p\sqrt{\dfrac{1}{n_1} + \dfrac{1}{n_2}}}$

where $s_p^{\,2} = \dfrac{(n_1 - 1)s_1^2 + (n_2 - 1)s_2^2}{n_1 + n_2 - 2}$ $\left[Note : \sigma_1^2 = \sigma_2^2 = \sigma^2\right]$

Step 3: Set the rejection region for H_0

Alternative hypothesis	**Rejection zone for H_0**
(i) $\mu_1 - \mu_2 \neq 0$	$t_c < -t^{(\alpha/2)}_{n_1+n_2-2}$ or $t_c > +t^{(\alpha/2)}_{n_1+n_2-2}$
(ii) $\mu_1 - \mu_2 > 0$	$t_c > +t^{(\alpha)}_{n_1+n_2-2}$
(iii) $\mu_1 - \mu_2 < 0$	$t_c < -t^{(\alpha)}_{n_1+n_2-2}$

Step 4: Reject H_0, if H_0 falls in the rejection zone as stated above.

Example 2.22

Consider Example 2.15. Test the hypothesis that the two competing medicines (A and B) are performing equally. Take α = 0.05.

Solution: From data we have

$$\bar{x}_1 = 2,\ \bar{x}_2 = 3,\ n_1 = 20,\ n_2 = 25,\ s_1 = 2 \ and \ s_2 = 2$$

The appropriate H_0 and H_1 are

$$H_0: \mu_1 - \mu_2 = 0 \text{ and } H_1: \mu_1 - \mu_2 \neq 0.$$

Assume equal population variance. So, the test statistics is $t = \dfrac{\bar{x}_1 - \bar{x}_2}{s_p\sqrt{\dfrac{1}{n_1} + \dfrac{1}{n_2}}}$

Here,

$$s_p^2 = \frac{(n_1-1)s_1^2 + (n_2-1)s_2^2}{n_1+n_2-2} = \frac{(20-1)\times 2^2 + (25-1)\times 2^2}{20+25-2}$$
$$= \frac{19\times 4 + 24\times 4}{43} = 4$$

$$\text{Computed } t = \frac{\overline{x}_1 - \overline{x}_1}{s_p\sqrt{1/n_1 + 1/n_2}} = \frac{2-3}{2\sqrt{\frac{1}{20}+\frac{1}{25}}} = -1.67.$$

Threshold $t_{43}(0.025) = 2.02$.

As the computed t (=–1.67) in absolute term < threshold t (=2.02), we fail to reject H_0. So, there is no significant difference in mean performance of the two competing medicines. We can conclude that they might be performing equally.

2.7.4 Hypothesis Testing Concerning Equality of Two-population Variances

We have seen earlier that the ratio $\frac{s_1^2/\sigma_1^2}{s_2^2/\sigma_2^2}$ follows F distribution with $n_1 - 1$ numerator degrees of freedom and $n_2 - 1$ denominator degrees of freedom. Under null hypothesis $\left(\text{i.e.}, H_0 : \sigma_1^2 = \sigma_2^2\right)$, the ratio becomes $\frac{s_1^2}{s_2^2}$. The concerned hypothesis can be tested following the steps given below.

Step 1: Set hypothesis

H_0 : $\sigma_1^2 = \sigma_2^2$

H_1: (i) $\sigma_1^2 \neq \sigma_2^2$ for two-tailed test
(ii) $\sigma_1^2 > \sigma_2^2$ for upper one-tailed test
(iii) $\sigma_1^2 < \sigma_2^2$ for lower one-tailed test

Step 2: Compute the test statistics

Alternative hypothesis	Test statistics
(i) H_1: $\sigma_1^2 \neq \sigma_2^2$	$F = \begin{cases} \frac{s_1^2}{s_2^2} & \text{if } s_1^2 > s_2^2 \text{ or} \\ \frac{s_2^2}{s_1^2} & \text{if } s_2^2 > s_1^2 \end{cases}$
(ii) H_1: $\sigma_1^2 > \sigma_2^2$	$F = \frac{s_1^2}{s_2^2}$
(iii) H_1: $\sigma_1^2 < \sigma_2^2$	$F = \frac{s_2^2}{s_1^2}$

Step 3: Set the rejection region for H_0

(i) For $H_1 : \sigma_1^2 \neq \sigma_2^2$

If $F=\frac{s_1^2}{s_2^2}$, then rejection region is $F > F^{\alpha/2}_{n_1-1,n_2-1}$

or for $F=\frac{s_2^2}{s_1^2}$, the rejection region is $F > F^{\alpha/2}_{n_2-1,n_1-1}$

(ii) For $H_1 : \sigma_1^2 > \sigma_2^2$, the rejection region is $F > F^{\alpha}_{n_1-1,n_2-1}$
(iii) For $H_1 : \sigma_1^2 < \sigma_2^2$, the rejection region is $F > F^{\alpha}_{n_2-1,n_1-1}$

Example 2.23

Consider Example 2.22. We have concluded that the two medicines perform equally. However, in order to arrive at this conclusion, we have assumed that the two populations have same variance. Therefore, it is customary to test first the equality of population variances. Let, test the hypothesis that the two populations (producing medicines A and B) have equal variance. Take $\alpha = 0.05$.

Solution:

From data we have $\bar{x}_1 = 2$, $\bar{x}_2 = 3$, $n_1 = 20$, $n_2 = 25$, $s_1 = 2$ *and* $s_2 = 2$. Note that $s_1 = s_2 = 2$. The appropriate H_0 and H_1 are

$$H_0 : \sigma_1^2 = \sigma_2^2 \text{ and } H_1 : \sigma_1^2 \neq \sigma_2^2.$$

The computed $F=\frac{s_1^2}{s_2^2}=\frac{4}{4}=1$

The threshold $F^{\alpha/2}_{n_1-1,n_2-1} = F^{0.025}_{19,24} = 2.35$.

As the computed F (=1) < threshold F (2.35), we fail to reject H_0. So, the two population variances might be the same at 0.05 probability level of significance.

2.8 LEARNING SUMMARY

Understanding univariate statistics is a must in order to proceed to learning multivariate statistics. In this chapter, the basic concepts of sampling distribution, central limit theorem, both point and interval estimation and hypothesis testing are discussed.

In summary:

- The term population represents the totality or whole of a system of interest. Population is characterized by one or more variables of interest and termed as univariate when one variable is of interest, bivariate when two variables are of interest, and multivariate when two or more variables are of interest.
- The behavior of a population is measured in terms of probability distribution of the random variable of interest, and its central tendency and dispersion measures (usually called the parameters of the population).
- If the random variable (say *X*) is continuous in nature, the corresponding probability distribution is called continuous probability distribution (e.g., normal, exponential, and lognormal distributions). On the other hand, if the random variable (say *X*) is discrete in nature, the corresponding probability distribution is called discrete probability distribution (e.g., binomial, Poisson, and negative binomial distributions).

- The parameter of a population is usually unknown but constant and is measured through sample data collected from the said population.
- A sample is characterized by its statistics, for example, sample average or sample standard deviation (in univariate population). A statistic is usually used to estimate the corresponding population parameter. The statistic is known but is a random quantity.
- The probability distribution of a sample statistic, such as sample average, is also called sampling distribution. The important sampling distributions are unit normal, Chi-square, t-, and F-distributions.
- Often, we encounter non-normal populations and/or populations with unknown standard deviation. Under such situations, the central limit theorem provides reasonable approximation of the distribution of large sample statistics.
- The estimation of population parameters comprises computation of point estimate followed by interval estimation with the help of appropriate sampling distribution. Depending on the availability of information about population distribution, population standard deviation and sample size, different scenarios for estimation will arise (see Section 2.6 for details).
- Equality of two population means or two population variances are also of interest under different situations. Depending on the availability of information about population distribution, population standard deviation, and sample size, different scenarios will arise and using appropriate sampling distribution, interval estimation is made (see Section 2.6 for details).
- A hypothesis is a statement that is yet to be proven. In inferential statistics, the statement about population parameters is investigated through hypothesis test. The procedure involves (i) setting up null (H_0) and alternate (H_1) hypotheses, (ii) finding out the appropriate test statistic and its sampling distribution, (iii) computation of test statistic value and comparison with threshold (theoretical) value for certain significance level (say $\alpha = 0.05$), and (iv) decision-making about the null hypothesis (H_0).
- In hypothesis testing, there could be three decision-making scenarios as (i) right decision: acceptance of H_0 when it is true or rejection of H_0 when it is false, (ii) type-I error (α): rejection of H_0 when it is true, and (iii) type-II error (β): acceptance of H_0 when it is false.

EXERCISES

(A) Short conceptual questions

2.1 Define population. How do you create system boundary for population?

2.2 Define parameter of a population. If *X* is a random variable, define population mean, μ and variance, σ^2 when *X* is continuous. What will happen to μ and σ^2 when *X* is discrete?

2.3 Name four discrete and four continuous probability distributions. Define univariate normal pdf. Convert it to standard normal *Z*. What benefit do you get by this conversion?

2.4 What is sampling distribution? Name four commonly used sampling distributions.

2.5 If *X* is normally distributed with mean μ and standard deviation σ, show that $\frac{x-\mu}{\sigma}$ follows standard normal distribution.

2.6 A sample of size *n* is to be collected from a normal population with mean μ and standard deviation σ. Let the sample standard deviation is *s*. Show that $(n-1)s/\sigma^2$ follows χ^2 distribution with $n-1$ degrees of freedom.

2.7 Show that the ratio of a standard normal variable to the square root of a χ_v^2 variable weighted by $1/v$ follows *t*-distribution. What will be the DOF for the *t*-distribution?

2.8 Show that the relationship between *t* and *F* distributions is $t^2 = F$.

2.9 What is central limit theorem? Why it is important? A sample of size n is collected on a random variable X. Show that the sample variance $\sigma_{\bar{x}}^2$ is equal to σ^2 / n where σ^2 is the population variance.

2.10 What is estimation? Why interval estimation is needed? Construct $(1-\alpha)100\%$ confidence interval for population mean (μ).

2.11 Suppose you have sampled from non-normal population with small sample size. Can you develop CI through parametric methods? If not, what will be the solution?

2.12 What is hypothesis testing? What are type-I and type-II errors? What is their practical significance?

2.13 Explain types of random variables from decision-making point of view. How do you set rejection zone for each variable type?

(B) Long conceptual and/or numerical questions: see web resources (Multivariate Statistical Modeling in Engineering and Management – 1st (routledge.com))

NOTES

1 Assuming 10,000 units produced every year. Assuming equal production rate per year with a hope that the population characteristics do not change.

2 Throughout the book, we consider random sampling strategy. We will describe sampling strategy later.

3 If random sampling scheme is adopted. In statistical modeling, this is the mostly used strategy.

N.B.: R and MS Excel software are used for computation, as and when required.

3 Basic Computations

Multivariate statistical modeling involves large amount of computations. For example, multivariate descriptive statistics require computation of mean vector and covariance matrix involving n observations. Hypothesis testing for multivariate mean vector requires Hotelling's T^2 to be computed. Parameter estimation in multiple regression is done using the ordinary least squares (OLS) method. Principal component analysis (PCA) involves eigenvalue-eigenvector decomposition of covariance matrix. In factor analysis and structural equation modeling, parameters may be estimated using maximum likelihood estimation (MLE). Fortunately, linear algebra provides necessary foundations for the basic computations in statistics. We describe in Section 3.1, the most commonly used concepts and computations of matrix algebra that are useful for understanding multivariate computations. Least square methods are described in Section 3.2 followed by maximum likelihood estimation in Section 3.3. Many times, the statistical analyst requires data to be generated or aggregated. The generation of univariate data is described in Section 3.4. We also discuss two resampling techniques, namely jackknife and bootstrap in Section 3.5.

3.1 MATRIX ALGEBRA

3.1.1 Data as a Matrix

We will use CityCan data set (Table 1.1 of Chapter 1) to describe the concepts. For clarity in presentation, we reproduce the data here in Table 3.1. We consider the variables profit, sales volume, % absenteeism, breakdown (hr) and M-ratio as X_1, X_2, X_3, X_4 and X_5, respectively.

Table 3.1 has 12 rows and 5 columns (excluding the first column). It can be said a rectangle array of numbers with 12 × 5 dimensions. Such a rectangular array is called a matrix. So, the CityCan data in matrix form looks as below.

$$\mathbf{X}_{12\times 5} = \begin{bmatrix} 10 & 100 & 9 & 62 & 1 \\ 12 & 110 & 8 & 58 & 1.3 \\ 11 & 105 & 7 & 64 & 1.2 \\ 9 & 94 & 14 & 60 & 0.8 \\ 9 & 95 & 12 & 63 & 0.8 \\ 10 & 99 & 10 & 57 & 0.9 \\ 11 & 104 & 7 & 55 & 1 \\ 12 & 108 & 4 & 56 & 1.2 \\ 11 & 105 & 6 & 59 & 1.1 \\ 10 & 98 & 5 & 61 & 1 \\ 11 & 103 & 7 & 57 & 1.2 \\ 12 & 110 & 6 & 60 & 1.2 \end{bmatrix}$$

In a similar manner, if we collect n observations on p-variables, we get a data matrix of the order $n \times p$. The general form of an $n \times p$ matrix is shown below. Hence forth, we denote data matrix as $X_{n\times p}$.

DOI: 10.1201/9781003303060-4

TABLE 3.1
CityCan Data Set

	Variables				
Observation No.	X_1	X_2	X_3	X_4	X_5
1	10	100	9	62	1
2	12	110	8	58	1.3
3	11	105	7	64	1.2
4	9	94	14	60	0.8
5	9	95	12	63	0.8
6	10	99	10	57	0.9
7	11	104	7	55	1
8	12	108	4	56	1.2
9	11	105	6	59	1.1
10	10	98	5	61	1.0
11	11	103	7	57	1.2
12	12	110	6	60	1.2

$$\mathbf{X}_{n\times p} = \begin{bmatrix} x_{11} & x_{12} & \cdots & x_{1p} \\ x_{21} & x_{22} & \cdots & x_{2p} \\ x_{31} & x_{32} & \cdots & x_{3p} \\ \vdots & \vdots & \vdots & \vdots \\ x_{i1} & x_{i2} & \cdots & x_{ip} \\ \vdots & \vdots & \vdots & \vdots \\ x_{n1} & x_{n2} & \cdots & x_{np} \end{bmatrix}$$

3.1.2 Row and Column Vectors

Let us rewrite $\mathbf{X}_{n\times p}$ as

$$\mathbf{X}_{n\times p} = \begin{bmatrix} \mathbf{x}_1 \\ \mathbf{x}_2 \\ \mathbf{x}_3 \\ \vdots \\ \mathbf{x}_i \\ \vdots \\ \mathbf{x}_n \end{bmatrix} = \begin{bmatrix} x_{11} & x_{12} & \cdots & x_{1p} \\ x_{21} & x_{22} & \cdots & x_{2p} \\ x_{31} & x_{32} & \cdots & x_{3p} \\ \vdots & \vdots & \vdots & \vdots \\ x_{i1} & x_{i2} & \cdots & x_{ip} \\ \vdots & \vdots & \vdots & \vdots \\ x_{n1} & x_{n2} & \cdots & x_{np} \end{bmatrix}$$

Where,

$\mathbf{x}_1, \mathbf{x}_2, \mathbf{x}_3, \ldots, \mathbf{x}_i, \ldots, \mathbf{x}_n$ are row vectors of size $1\times p$. The $i-th$ row vector is represented by $\mathbf{x}_i$.

$\mathbf{x}_i^T$ represents transpose of the vector $\mathbf{x}_i$, where row and column elements are interchanged.

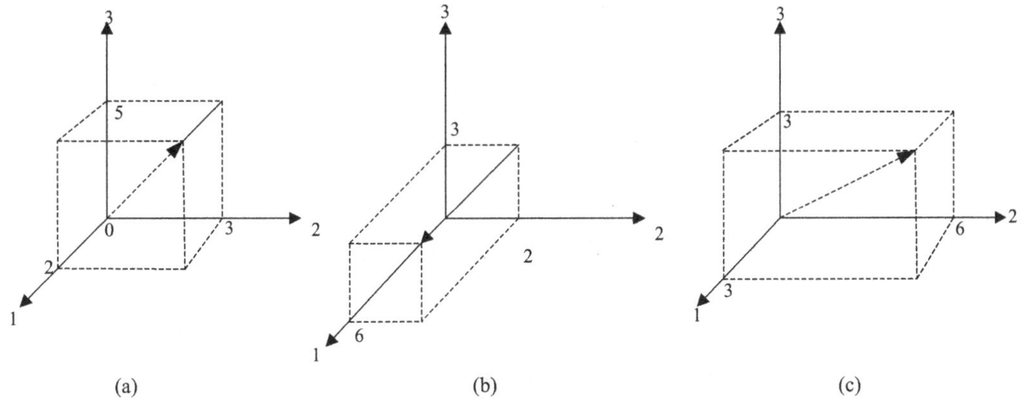

FIGURE 3.1 Geometric representation of a row vector. Figures (a), (b), and (c) represent the row vectors, $\mathbf{x}_1 = [2 \quad 3 \quad 5]$, $\mathbf{x}_2 = [6 \quad 2 \quad 3]$, and $\mathbf{x}_3 = [3 \quad 6 \quad 3]$, respectively.

Example 3.1

Obtain row vectors for the following data matrix

$$X_{3\times 3} = \begin{bmatrix} 2 & 3 & 5 \\ 6 & 2 & 3 \\ 3 & 6 & 3 \end{bmatrix}$$

Solution:

There are three row vectors:

$$\mathbf{x}_1 = [2 \quad 3 \quad 5]$$
$$\mathbf{x}_2 = [6 \quad 2 \quad 3]$$
$$\mathbf{x}_3 = [3 \quad 6 \quad 3]$$

Geometrically, the row vectors $\mathbf{x}_1, \mathbf{x}_2$ and $\mathbf{x}_3$ can be represented as directed lines in $p(= 3)$ dimensions as shown in Figures 3.1(a) to 3.1(c).

Key question: what do the row vectors signify in multivariate data representation?

Answer: Each row vector depicts a point (position) in p-dimensional coordinate system where p represents the number of variables simultaneously observed. In Example 3.1, $p = 3$. Therefore, geometrically $\mathbf{X}_{3\times 3}$ (Example 3.1) can be plotted as Figure 3.2.

Figure 3.2 can be conceptually extended to n simultaneous observations on p-variables, where the p-variables represent p-dimensions and n observations represent n row vectors. The total space created by the n vectors is known as multivariate space.

Like row vectors, $\mathbf{X}_{n\times p}$ can also be represented by column vectors. Let us rewrite $\mathbf{X}_{n\times p}$ as

$$\mathbf{X}_{n\times p} = [\mathbf{x}_1 \quad \mathbf{x}_2 \quad \ldots \quad \mathbf{x}_j \quad \ldots \quad \mathbf{x}_p]$$

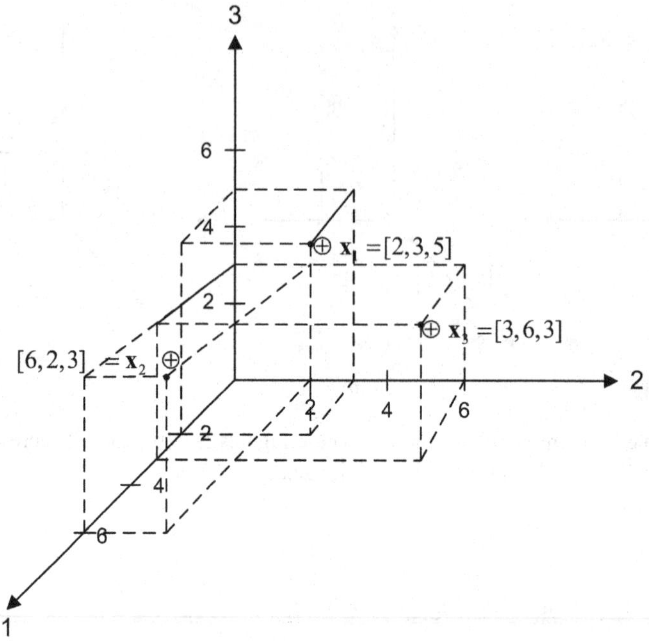

FIGURE 3.2 Vector representation of multivariate observations on p variables (here, p = 3). Each vector depicts a point (position) in the three-dimensional coordinate system. Each axis represents a variable.

$$= \begin{bmatrix} x_{11} & x_{12} & \cdots & x_{1j} & \cdots & x_{1p} \\ x_{21} & x_{22} & \cdots & x_{2j} & \cdots & x_{2p} \\ \vdots & \vdots & \vdots & \vdots & \vdots & \vdots \\ x_{i1} & x_{i2} & \cdots & x_{ij} & \cdots & x_{ip} \\ \vdots & \vdots & \vdots & \vdots & \vdots & \vdots \\ x_{n1} & x_{n2} & \cdots & x_{nj} & \cdots & x_{np} \end{bmatrix}$$

$$\mathbf{x}_j = \begin{matrix} x_{1j} \\ x_{2j} \\ \vdots \\ x_{ij} \\ \vdots \\ x_{nj} \end{matrix}$$

Where $\mathbf{x}_1, \mathbf{x}_2, \ldots, \mathbf{x}_j, \ldots, \mathbf{x}_p$ are called column vectors of size $n \times 1$. The $j-th$ column vector is represented by $\mathbf{x}_j$, where

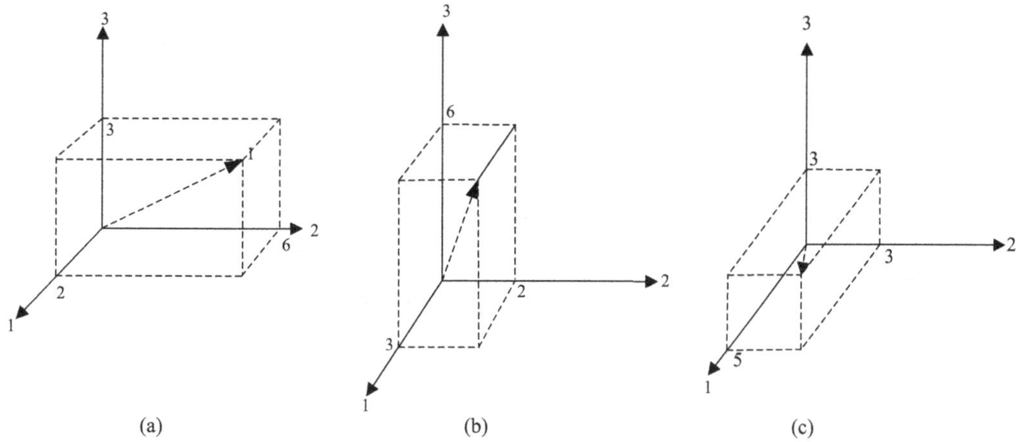

FIGURE 3.3 Geometric representation of a column vector. Figures (a), (b), and (c) represent the column vectors, $\mathbf{x}_1 = [2,\ 6,\ 3]^T$, $\mathbf{x}_2 = [3,\ 2,\ 6]^T$, and $\mathbf{x}_3 = [5,\ 3,\ 3]^T$, respectively.

$$\mathbf{x}_j = \begin{bmatrix} x_{1j} \\ x_{2j} \\ \vdots \\ x_{ij} \\ \vdots \\ x_{nj} \end{bmatrix} \text{ and j } = 1, 2, \ldots, \text{p}.$$

Example 3.2

Obtain column vectors for the data matrix given in Example 2.1.

Solution:

The data matrix is

$$\mathbf{X}_{3\times 3} = \begin{bmatrix} 2 & 3 & 5 \\ 6 & 2 & 3 \\ 3 & 6 & 3 \end{bmatrix}, \text{ therefore } \begin{matrix} \mathbf{x}_1 = [2 \quad 6 \quad 3]^T \\ \mathbf{x}_2 = [3 \quad 2 \quad 6]^T \\ \mathbf{x}_3 = [5 \quad 3 \quad 3]^T \end{matrix}$$

Geometrically, the column vectors $\mathbf{x}_1, \mathbf{x}_2$ and $\mathbf{x}_3$ can be represented as directed lines in $n(=3)$ dimensions as shown in Figures 3.3(a) to 3.3(c).

Key question: What do the column vectors signify in multivariate data representation?

Answer: Each column vector depicts a point (position) in n dimensional coordinate system where n represents the number of observations taken on p-variables. In Example 3.2, $n = 3$. Therefore, geometrically, $\mathbf{X}_{3\times 3}$ (for Example 3.2) can be plotted as shown in Figure 3.4.

Similar to Figure 3.2, Figure 3.4 can be extended to p-variable vectors of size $p \times 1$ where n represents number of observations on each variable.

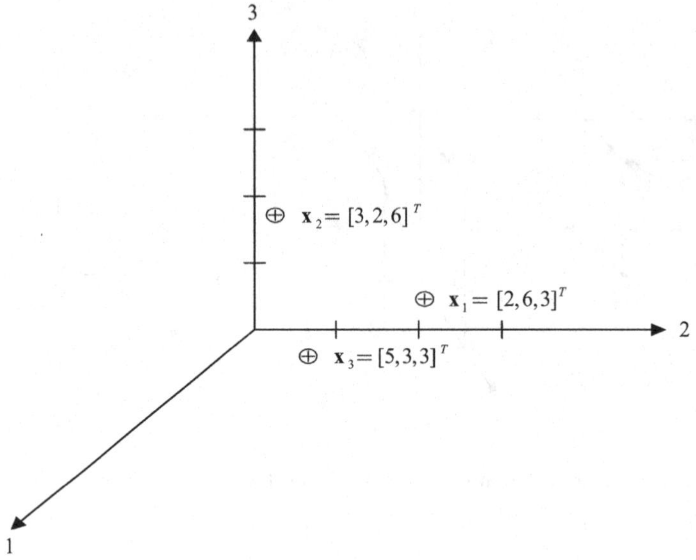

FIGURE 3.4 Vector representation of multiple observations (n) on a single variable (here, n = 3). Each vector depicts a point (position) in the three-dimensional coordinate system. Each axis represents a multivariate observation.

TABLE 3.2
Data Matrix for Example 3.3

Observation Number	X_1	X_2
1	10	7
2	15	9
3	5	6

Key question: What is the difference between Figure 3.2 and Figure 3.4?

Answer: Both represent the same multivariate data set. But while Figure 3.2 depicts the respective positions of n multivariate observations, Figure 3.4 depicts the positions the p-variable vectors (both n and p are 3). If we measure distance between any two points in Figure 3.2, it is the distance between two observations while the same in Figure 3.4 is the distance between two variables. The concepts explained in Figure 3.2 and Figure 3.4 are explored in cluster analysis and factor analysis, respectively.

Example 3.3

Obtain row and column vector for the data matrix in Table 3.2.

Solution:

In this example, there are three observations and two variables. So, there are three row vectors measured in two dimensions ($p = 2$ for two variables X_1 and X_2) and two column vectors measured in three dimensions ($n = 3$). Figures 3.5 and 3.6 show the row and column vectors, respectively.

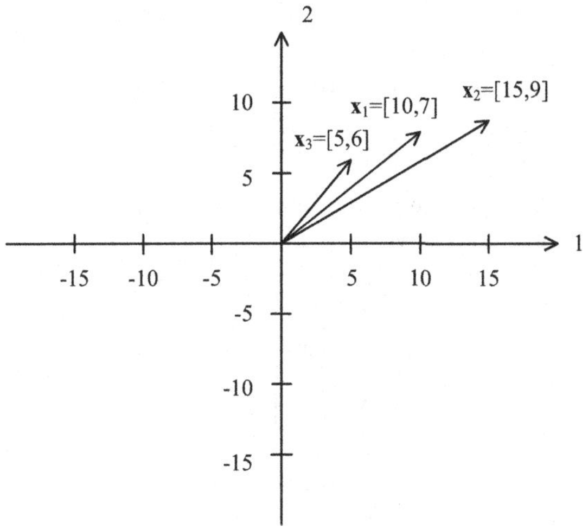

FIGURE 3.5 Row vectors $\mathbf{x}_1, \mathbf{x}_2$ and $\mathbf{x}_3$. Each axis represents a variable and each vector represents a multivariate observation.

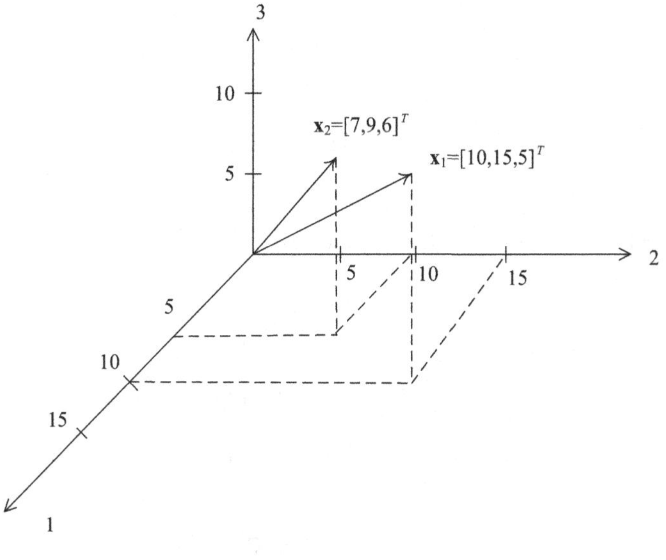

FIGURE 3.6 Column vectors $\mathbf{x}_1$ and $\mathbf{x}_2$. Each axis represents a multivariate observation and each vector represents a variable.

Row vectors:

$$\mathbf{x}_1 = \begin{bmatrix} 10 & 7 \end{bmatrix}$$
$$\mathbf{x}_2 = \begin{bmatrix} 15 & 9 \end{bmatrix}$$
$$\mathbf{x}_3 = \begin{bmatrix} 5 & 6 \end{bmatrix}$$

Column vectors:

$$\mathbf{x}_1 = \begin{bmatrix} 10 & 15 & 5 \end{bmatrix}^T$$
$$\mathbf{x}_2 = \begin{bmatrix} 7 & 9 & 6 \end{bmatrix}^T$$

Many times, we need to aggregate (e.g., sum total of observations) a vector of observations. For example, let $\mathbf{x}_{n\times 1}$ is a n-element column vector.

$$\mathbf{x} = \begin{bmatrix} x_1 \\ x_2 \\ \vdots \\ x_i \\ \vdots \\ x_n \end{bmatrix}_{n\times 1}$$

We want to compute sum of the elements of $\mathbf{x}_{n\times 1}$ and average of the elements of $\mathbf{x}_{n\times 1}$. We can do so as follows.

Let $\mathbf{1}_{n\times 1}$ is a unit vector whose all elements are 1 only.

$$\mathbf{1} = \begin{bmatrix} 1 \\ 1 \\ \vdots \\ 1 \\ \vdots \\ 1 \end{bmatrix}_{n\times 1}$$

The sum of $(\mathbf{x}) = \mathbf{x} \bullet \mathbf{1} = \mathbf{x}^T \mathbf{1}$ (see dot product is used),

$$= \begin{bmatrix} x_1 & x_2 & \cdots & x_n \end{bmatrix}_{1\times n} \begin{bmatrix} 1 \\ 1 \\ \vdots \\ 1 \\ \vdots \\ 1 \end{bmatrix}_{n\times 1}$$

$$= x_1 + x_2 + \cdots + x_n = \sum_{i=1}^{n} x_i \tag{3.1}$$

$$\begin{aligned} \text{Average of } (\mathbf{x}) &= \frac{1}{n} \text{ sum of } (\mathbf{x}) \\ &= \frac{1}{n} \sum_{i=1}^{n} x_i \\ &= \frac{1}{n} \mathbf{x}^T \mathbf{1} \end{aligned} \tag{3.2}$$

The operation ' • ' is known as dot and the product of $\mathbf{x} \bullet \mathbf{1}$ is known as dot product of the vectors **x** and **1**, denoted by $\mathbf{x}^T\mathbf{1}$.

Example 3.4

Compute average of the following observations.

$$\mathbf{x} = \begin{bmatrix} 10 & 5 & 20 & 15 & 10 \end{bmatrix}^T$$

Solution:

There are five observations $n = 5$. Let the average of (**x**) be $\bar{x}$. Then,

$$\bar{x} = \frac{1}{n}\mathbf{x}.\mathbf{1} = \frac{1}{n}\mathbf{x}^T\mathbf{1}$$

$$= \frac{1}{5}\begin{bmatrix} 10 & 5 & 20 & 15 & 10 \end{bmatrix}\begin{bmatrix} 1 \\ 1 \\ 1 \\ 1 \\ 1 \end{bmatrix}$$

$$= \frac{1}{5}[10+5+20+15+10]$$

$$= \frac{1}{5}\times 60$$

$$= 12$$

3.1.3 Orthogonal Vectors

The foregoing discussion (Section 3.1.1) reveals that a variable with n observations can be treated as a column vector of size $n \times 1$. In multivariate modeling, we use several variables simultaneously. Each of them can be treated as an individual column vector. We have seen earlier in Section 3.1.2 that a multivariate data matrix $\mathbf{X}_{n\times p}$ can be decomposed into p column vectors. So, $\mathbf{X}_{n\times p} = [\mathbf{x}_1 \ \ \mathbf{x}_2 \ \cdots \ \mathbf{x}_j \cdots \ \ \mathbf{x}_p]$ where $\mathbf{x}_j$, $j = 1, 2, \ldots, p$ is vector of size $n \times 1$.

In dependence multivariate models such as multiple linear regression, one of the assumptions is that the independent variables should be mutually independent. If two variable vectors are orthogonal, then they are independent.

Two vectors $\mathbf{x}_j$ and $\mathbf{x}_k$ of the order $n \times 1$ is said to be orthogonal, if their dot product equals to zero. So,

$$\mathbf{x}_j \bullet \mathbf{x}_k = \mathbf{x}_j^T\mathbf{x}_k = 0 \tag{3.3}$$

Example 3.5

Consider the following two vectors and assess their independency.

$$\mathbf{x}_j = \begin{bmatrix} 2 \\ 4 \\ -9 \end{bmatrix} \quad \text{and} \quad \mathbf{x}_k = \begin{bmatrix} 11 \\ -1 \\ 2 \end{bmatrix}$$

Solution:

$$\mathbf{x}_j^T \mathbf{x}_k = \begin{bmatrix} 2 & 4 & -9 \end{bmatrix} \begin{bmatrix} 11 \\ -1 \\ 2 \end{bmatrix} = 22 - 4 - 18 = 0$$

So, $\mathbf{x}_j$ and $\mathbf{x}_k$ are independent (orthogonal).

Two vectors are said to be orthogonal when they form an angle $\theta (= 90^\circ)$ between them. In order to compute the angle between two vectors, we require understanding first the length of a vector. Geometrically, the length of a vector of size 2×1 (2-diamensions) can be viewed as the hypotenuse of a right-angled triangle. Figure 3.7 depicts this.

The length of a vector is the norm of the vector.

So, length of $\mathbf{x} = L_\mathbf{x} = |\mathbf{x}| = \sqrt{x_1^2 + x_2^2}$

For a vector of size $n \times 1$, the length (norm) is

$$L_\mathbf{x} = |\mathbf{x}| = \sqrt{x_1^2 + x_2^2 + \cdots + x_n^2} = \sqrt{\sum_{i=1}^{n} x_i^2} \tag{3.4}$$

$$\text{In matrix notation, } L_\mathbf{x} = \sqrt{\mathbf{x}^T \mathbf{x}} \tag{3.5}$$

In Equations 3.4 and 3.5, the length is computed from the origin. In statistics, we often need to compute the length (distance) from the average (mean). The length of vector from its mean (average) position is

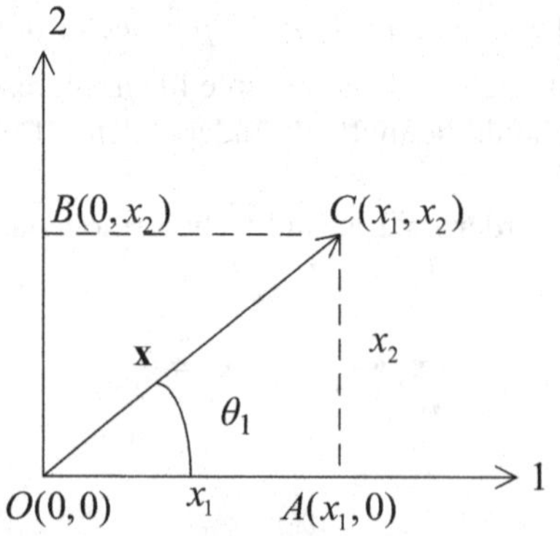

FIGURE 3.7 Geometric representation of the length of a vector ($\mathbf{x}_{2\times 1}$). The hypotenuse (OC) of the triangle OAC represents the length of $\mathbf{x}$.

$$L_{\mathbf{x}m} = |\mathbf{x}_m| = \sqrt{\sum_{i=1}^{n}(x_i - \bar{x})^2} \tag{3.6}$$

$$\text{In matrix notation, } L_{\mathbf{x}m} = \sqrt{(\mathbf{x}-\bar{\mathbf{x}})^T(\mathbf{x}-\bar{\mathbf{x}})} \tag{3.7}$$

Now, if we divide the vector **x** by its length, we get a normalized vector with unit magnitude. This vector is called unit vector. So, unit vector $= \dfrac{\mathbf{x}}{L_{\mathbf{x}}} = L_{\mathbf{x}}^{-1}\mathbf{x}$.

Example 3.6

Compute the length of the vector $\mathbf{x} = [5 \quad 10 \quad 15 \quad 20]^T$ from origin and obtain its unit vector.

Solution:

$\mathbf{x} = [5 \quad 10 \quad 15 \quad 20]^T$

$$\begin{aligned}\text{Norm of } \mathbf{x} = |\mathbf{x}| &= \sqrt{5^2+10^2+15^2+20^2}\\ &= \sqrt{5^2+10^2+15^2+20^2}\\ &= \sqrt{750}\\ &= 5\sqrt{30}\end{aligned}$$

So, the length of $\mathbf{x} = L_{\mathbf{x}} = 5\sqrt{30}$

Now the unit vector of $\mathbf{x} = L_{\mathbf{x}}^{-1}\mathbf{x}$

$$= \frac{1}{5\sqrt{30}}\begin{bmatrix}5\\10\\15\\20\end{bmatrix} = \begin{bmatrix}\frac{1}{\sqrt{30}}\\ \frac{2}{\sqrt{30}}\\ \frac{3}{\sqrt{30}}\\ \frac{4}{\sqrt{30}}\end{bmatrix}$$

Again, the length of the unit vector of $\mathbf{x} = 1$.

If the vector **x** makes an angle θ_1 with dimension 1 (Figure 3.7), then

$$\cos\theta_1 = \frac{x_1}{L_{\mathbf{x}}} = \frac{x_1}{\sqrt{\mathbf{x}^T\mathbf{x}}} \text{ and } \sin\theta_1 = \frac{x_2}{L_{\mathbf{x}}} = \frac{x_2}{\sqrt{\mathbf{x}^T\mathbf{x}}} \tag{3.8}$$

Now, consider two vectors **x** and **y** of size $n \times 1$. Let they make an angle θ between them as shown in Figure 3.8.

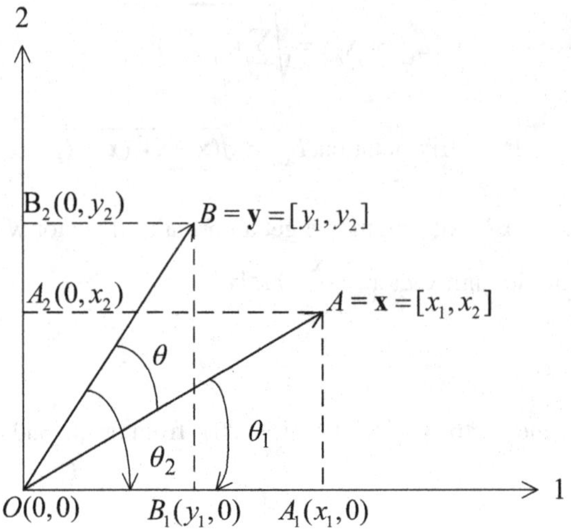

FIGURE 3.8 The angle θ between vectors $\mathbf{x}$ and $\mathbf{y}$. $\theta = \angle BOA = \angle BOB_1 - \angle AOA_1 = \theta_2 - \theta_1$.

Also consider the angles the vectors $\mathbf{x}$ and $\mathbf{y}$ make with the dimension 1 is θ_1 and θ_2 respectively. So,

$$\begin{aligned}
\cos\theta_1 &= \frac{x_1}{L_{\mathbf{x}}} = \frac{x_1}{\sqrt{\mathbf{x}^T\mathbf{x}}} \\
\sin\theta_1 &= \frac{x_2}{L_{\mathbf{x}}} = \frac{x_2}{\sqrt{\mathbf{x}^T\mathbf{x}}} \\
\cos\theta_2 &= \frac{y_1}{L_{\mathbf{y}}} = \frac{y_1}{\sqrt{\mathbf{y}^T\mathbf{y}}} \\
\sin\theta_2 &= \frac{y_2}{L_{\mathbf{y}}} = \frac{y_2}{\sqrt{\mathbf{y}^T\mathbf{y}}},
\end{aligned} \tag{3.9}$$

Again, from Figure 3.8, we get $\theta = \theta_2 - \theta_1$

$$\begin{aligned}
\therefore \cos\theta &= \cos(\theta_2 - \theta_1) = \cos\theta_2 \cos\theta_1 + \sin\theta_2 \sin\theta_1 \\
&= \frac{x_1}{L_{\mathbf{x}}} \cdot \frac{y_1}{L_{\mathbf{y}}} + \frac{x_2}{L_{\mathbf{x}}} \cdot \frac{y_2}{L_{\mathbf{y}}} \\
&= \frac{x_1 y_1 + x_2 y_2}{L_{\mathbf{x}} L_{\mathbf{y}}}
\end{aligned} \tag{3.10}$$

As **x** and **y** are vectors of two dimensions,

$$\mathbf{x}^T\mathbf{y} = \begin{bmatrix} x_1 & x_2 \end{bmatrix}\begin{bmatrix} y_1 \\ y_2 \end{bmatrix}$$
$$= x_1 y_1 + x_2 y_2$$

$$\text{So, } \cos\theta = \frac{\mathbf{x}^T\mathbf{y}}{\sqrt{\mathbf{x}^T\mathbf{x}}\sqrt{\mathbf{y}^T\mathbf{y}}} \quad (3.11)$$

When $\theta = 90^\circ$ or 270°, **x** and **y** becomes perpendicular (orthogonal). In this case, $\cos\theta$ is $\cos 90^\circ$ or $\cos 270^\circ$ which takes a value of zero (0).

Then, $\mathbf{x}^T\mathbf{y} = 0$ or $\mathbf{x} \bullet \mathbf{y} = 0$.

So, when the dot product of two vectors is 0, the vectors are called orthogonal or perpendicular or linearly independent.

Example 3.7

Consider the following two vectors and find out the angle between them. Comment on the result.

$$\mathbf{x} = \begin{bmatrix} 5 \\ 7 \\ 10 \end{bmatrix}$$
$$\mathbf{y} = \begin{bmatrix} 1 \\ 10 \\ 3 \end{bmatrix}$$

Solution:

Let the angle between **x** and **y** be θ. So,

$$\cos\theta = \frac{\mathbf{x}^T\mathbf{y}}{\sqrt{\mathbf{x}^T\mathbf{x}}\sqrt{\mathbf{y}^T\mathbf{y}}}$$

Now,

$$\mathbf{x}^T\mathbf{x} = 5^2 + 7^2 + 10^2 = 25 + 49 + 100 = 174$$
$$\mathbf{y}^T\mathbf{y} = 1^2 + 10^2 + 3^2 = 1 + 100 + 9 = 110$$
$$\mathbf{x}^T\mathbf{y} = 5\times 1 + 7\times 10 + 10\times 3 = 5 + 70 + 30 = 105.$$

$$\therefore \cos\theta = \frac{105}{\sqrt{174}\sqrt{110}} = \frac{105}{\sqrt{19{,}140}} = 0.759$$
$$\therefore \theta = \cos^{-1}(0.759) = 40.62^0$$

The two vectors are non-orthogonal (oblique) in nature.

3.1.4 Linear Dependency of a Set of Vectors

Now, we generalize the linear dependency of vectors. A set of vectors $\mathbf{x}_1, \mathbf{x}_2, ..., \mathbf{x}_p$ having same dimension is said to be linearly dependent if there exist constants $a_1, a_1, ..., a_p$, not all zero such that

$$a_1\mathbf{x}_1 + a_2\mathbf{x}_2 + \cdots\cdots + a_p\mathbf{x}_p = \mathbf{0} \tag{3.12}$$

For linearly independent vectors of same dimension, the vectors do not satisfy Equation 3.12.

Consider a pair of vectors $\mathbf{x}_1$ and $\mathbf{x}_2$. Let $\mathbf{z}$ be their linear combination. So, $\mathbf{z} = a_1\mathbf{x}_1 + a_2\mathbf{x}_2$, where a_1 and a_2 are unique and not equal to zero. Following Kraus (2002), we compute a_1 and a_2.

Taking dot product of $\mathbf{z}$ with $\mathbf{x}_1$ and $\mathbf{x}_2$ we get (See Section 3.1.9 for details),

$$\mathbf{z}\cdot\mathbf{x}_1 = a_1\mathbf{x}_1\cdot\mathbf{x}_1 + a_2\mathbf{x}_2\cdot\mathbf{x}_1$$
$$\mathbf{z}\cdot\mathbf{x}_2 = a_1\mathbf{x}_1\cdot\mathbf{x}_2 + a_2\mathbf{x}_2\cdot\mathbf{x}_2$$

Which in matrix form

$$\begin{bmatrix} \mathbf{x}_1\cdot\mathbf{x}_1 & \mathbf{x}_2\cdot\mathbf{x}_1 \\ \mathbf{x}_1\cdot\mathbf{x}_2 & \mathbf{x}_2\cdot\mathbf{x}_2 \end{bmatrix}\begin{bmatrix} a_1 \\ a_2 \end{bmatrix} = \begin{bmatrix} \mathbf{z}\cdot x_1 \\ \mathbf{z}\cdot x_2 \end{bmatrix}$$

To have unique solution of a_1 and a_2, the matrix

$\mathbf{G} = \begin{bmatrix} \mathbf{x}_1\cdot\mathbf{x}_1 & \mathbf{x}_2\cdot\mathbf{x}_1 \\ \mathbf{x}_1\cdot\mathbf{x}_2 & \mathbf{x}_2\cdot\mathbf{x}_2 \end{bmatrix} = \begin{bmatrix} \mathbf{x}_1{}^T\mathbf{x}_1 & \mathbf{x}_2{}^T\mathbf{x}_1 \\ \mathbf{x}_1{}^T\mathbf{x}_2 & \mathbf{x}_2{}^T\mathbf{x}_2 \end{bmatrix}$ must be non-singular. So, $|\mathbf{G}| \neq 0$, where $|\mathbf{G}|$ is known as Gram determinant or Grammian.

Now,

$$\begin{aligned} |\mathbf{G}| &= \det\begin{bmatrix} \mathbf{x}_1\cdot\mathbf{x}_1 & \mathbf{x}_2\cdot\mathbf{x}_1 \\ \mathbf{x}_1\cdot\mathbf{x}_2 & \mathbf{x}_2\cdot\mathbf{x}_2 \end{bmatrix} \\ &= |\mathbf{x}_1|^2|\mathbf{x}_2|^2 - |\mathbf{x}_1\cdot\mathbf{x}_2|^2 \\ &= |\mathbf{x}_1|^2|\mathbf{x}_2|^2 - (|\mathbf{x}_1||\mathbf{x}_2|\cos\theta)^2 \\ &= |\mathbf{x}_1|^2|\mathbf{x}_2|^2(1-\cos^2\theta) \\ &= |\mathbf{x}_1|^2|\mathbf{x}_2|^2\sin^2\theta \end{aligned}$$

For collinear $\mathbf{x}_1$ and $\mathbf{x}_2$, θ will be zero. So, $|\mathbf{G}| = 0$ for linearly dependent $\mathbf{x}_1$ and $\mathbf{x}_2$.

The above discussion can be extended to p vectors. For linear dependence of p-vector, the Grammian will become zero (Kraus, 2002).

Example 3.8

Consider Example 3.7 and show that $\mathbf{x}$ and $\mathbf{y}$ are linearly independent.

$$\mathbf{x} = \begin{bmatrix} 5 \\ 7 \\ 10 \end{bmatrix}, \quad \mathbf{y} = \begin{bmatrix} 1 \\ 10 \\ 3 \end{bmatrix}$$

Solution:

$$\mathbf{x}^T\mathbf{x} = 5^2 + 7^2 + 10^2 = 174$$
$$\mathbf{y}^T\mathbf{x} = \mathbf{x}^T\mathbf{y} = 5\times1+7\times10+10\times3 = 105$$
$$\mathbf{y}^T\mathbf{y} = 1^2 + 10^2 + 3^2 = 110$$

$$\text{So}, \mathbf{G} = \begin{bmatrix} \mathbf{x}^T\mathbf{x} & \mathbf{y}^T\mathbf{x} \\ \mathbf{x}^T\mathbf{y} & \mathbf{y}^T\mathbf{y} \end{bmatrix} = \begin{bmatrix} 174 & 105 \\ 105 & 110 \end{bmatrix}$$

Now,

$$\begin{aligned} |\mathbf{G}| &= \det\begin{bmatrix} 174 & 105 \\ 105 & 110 \end{bmatrix} \\ &= 174\times110 - 105^2 \\ &= 19140 - 11025 \\ &= 8115 \end{aligned}$$

Because $|\mathbf{G}| \neq 0$, the vectors $\mathbf{x}$ and $\mathbf{y}$ are linearly independent.

3.1.5 The Gram-Schmidt Orthogonalization Process

Consider a set of p linearly independent vectors $\mathbf{x}_1, \mathbf{x}_2, ..., \mathbf{x}_p$. We want to make a set of p orthogonal and p orthonormal vectors out of the original p linearly independent vectors. Orthonormal vector is an orthogonal vector with unit length. The Gram-Schmidt orthogonalization process is explained below.

Let $\mathbf{z}_1, \mathbf{z}_2, ..., \mathbf{z}_p$ be the orthogonal vectors that can be derived. Consider $\mathbf{z}_1 = \mathbf{x}_1$. Similarly, let $\mathbf{u}_1, \mathbf{u}_2, ..., \mathbf{u}_p$ be the orthonormal vectors for, $\mathbf{z}_1, \mathbf{z}_2, ..., \mathbf{z}_p$, respectively.

For example, in Figure 3.9, two vectors $\mathbf{x}_1$ and $\mathbf{x}_2$ of size 2 × 1 are shown. OA and OB represent $\mathbf{x}_1$ and $\mathbf{x}_2$, respectively. The orthogonal vector $\mathbf{z}_1$ is considered to be equal to $\mathbf{x}_1$ and corresponding unit vector is $\mathbf{u}_1$. OC represents α times $\mathbf{u}_1$, where α is a scalar whose value is required to be determined. We are interested to find out the vector CB that is perpendicular to OA ($\mathbf{z}_1$). CB is denoted by $\mathbf{z}_2$.

So, the first orthogonal and orthonormal vectors are:

$$\left.\begin{aligned} \mathbf{z}_1 &= \mathbf{x}_1 \\ \mathbf{u}_1 &= \frac{\mathbf{z}_1}{\sqrt{\mathbf{z}_1^T\mathbf{z}_1}} \end{aligned}\right\} \quad (3.13)$$

From vector geometry, we know (from Figure 3.9)

$$\text{CB} = \text{OB} - \text{OC}$$
$$\therefore \mathbf{z}_2 = \mathbf{x}_2 - \alpha\mathbf{u}_1$$

Now, as $\mathbf{z}_2$ is orthogonal to $\mathbf{u}_1$, their dot product must be zero.

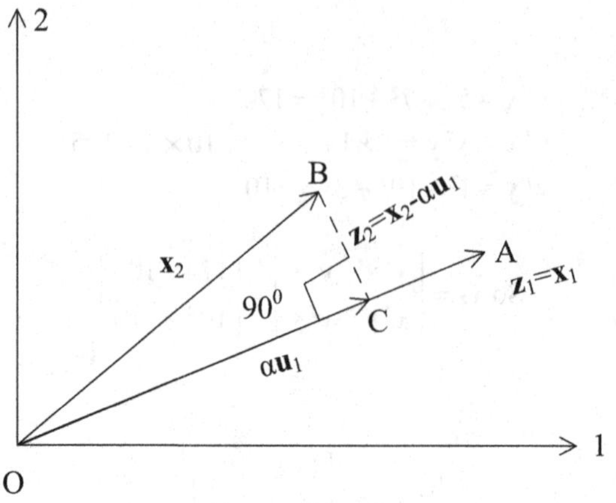

FIGURE 3.9 Geometric representation of orthogonalization of two vectors $\mathbf{x}_1$ and $\mathbf{x}_2$ (each of size 2×1).

So,

$$\begin{aligned} &\mathbf{u}_1 \cdot \mathbf{z}_2 = 0 \\ &\Rightarrow \mathbf{u}_1 \cdot (\mathbf{x}_2 - \alpha\mathbf{u}_1) = 0 \\ &\Rightarrow \mathbf{u}_1 \cdot \mathbf{x}_2 - \alpha\mathbf{u}_1 \cdot \mathbf{u}_1 = 0 \\ &\Rightarrow \mathbf{u}_1{}^T\mathbf{x}_2 - \alpha\mathbf{u}_1{}^T\mathbf{u}_1 = 0 \end{aligned}$$

Further $\mathbf{u}_1$ is unit vector and so $\mathbf{u}_1^T\mathbf{u}_1 = 1$.
So,

$$\begin{aligned} &\mathbf{u}_1^T\mathbf{x}_2 - \alpha \cdot 1 = 0 \\ &\Rightarrow \alpha = \mathbf{u}_1^T\mathbf{x}_2 \end{aligned}$$

And putting the value of α, we get $\mathbf{z}_2 = \mathbf{x}_2 - (\mathbf{u}_1^T\mathbf{x}_2)\mathbf{u}_1$
So, the second orthogonal and orthonormal vectors are

$$\left.\begin{aligned} \mathbf{z}_2 &= \mathbf{x}_2 - (\mathbf{u}_1^T\mathbf{x}_2)\mathbf{u}_1 \\ \mathbf{u}_2 &= \frac{\mathbf{z}_2}{\sqrt{\mathbf{z}_2{}^T\mathbf{z}_2}} \end{aligned}\right\} \tag{3.14}$$

The next step is to consider $\mathbf{x}_3$ and find a vector $\mathbf{z}_3$ that is orthogonal to both $\mathbf{u}_1$ and $\mathbf{u}_2$.
Let $\mathbf{z}_3 = \mathbf{x}_3 - \beta_1\mathbf{u}_1 - \beta_2\mathbf{u}_2$.
We require to find out the values of β_1 and β_2. As $\mathbf{z}_3$ is orthogonal to both $\mathbf{u}_1$ and $\mathbf{u}_2$, their dot product must be equal to zero. So,

$$\mathbf{u}_1 \cdot \mathbf{z}_3 = \mathbf{u}_1 \cdot \left(\mathbf{x}_3 - \beta_1\mathbf{u}_1 - \beta_2\mathbf{u}_2\right) = 0, \text{ and } \mathbf{u}_2 \cdot \mathbf{z}_3 = \mathbf{u}_2 \cdot \left(\mathbf{x}_3 - \beta_1\mathbf{u}_1 - \beta_2\mathbf{u}_2\right) = 0$$

By simplification,

$$\begin{aligned}
&\mathbf{u}_1^T\mathbf{z}_3 = \mathbf{u}_1^T\mathbf{x}_3 - \beta_1\mathbf{u}_1^T\mathbf{u}_1 - \beta_2\mathbf{u}_1^T\mathbf{u}_2 = 0\\
&\Rightarrow \mathbf{u}_1^T\mathbf{x}_3 - \beta_1 1 - 0 = 0 \text{ [as } \mathbf{u}_1^T\mathbf{u}_1 = 1 \text{ and } \mathbf{u}_1^T\mathbf{u}_2 = 0]\\
&\Rightarrow \beta_1 = \mathbf{u}_1^T\mathbf{x}_3
\end{aligned}$$

Similarly, $\beta_2 = \mathbf{u}_2{}^T\mathbf{x}_3$

So, the third orthogonal and orthonormal vectors are

$$\left.\begin{aligned}
\mathbf{z}_3 &= \mathbf{x}_3 - (\mathbf{u}_1^T\mathbf{x}_3)\mathbf{u}_1 - (\mathbf{u}_2^T\mathbf{x}_3)\mathbf{u}_2\\
\mathbf{u}_3 &= \frac{\mathbf{z}_3}{\sqrt{\mathbf{z}_3^T\mathbf{z}_3}}
\end{aligned}\right\}$$

The general formula for Gram-Schmidt orthogonalization process is

$$\begin{aligned}
&\mathbf{z}_1 = \mathbf{x}_1\\
&\mathbf{z}_k = \mathbf{x}_k - \sum_{i=1}^{k-1}\left(\mathbf{u}_i^T\mathbf{x}_k\right)\mathbf{u}_i,\ k = 2,3,\ldots.,p\\
&\textit{and}\quad \mathbf{u}_k = \frac{\mathbf{z}_k}{\sqrt{\mathbf{z}_k{}^T\mathbf{z}_k}}
\end{aligned} \tag{3.16}$$

In Equation 3.16, the term $(\mathbf{u}_i^T\mathbf{x}_k)\mathbf{u}_i$ is the projection of $\mathbf{x}_k$ on $\mathbf{u}_i$ and the term $\sum_{i=1}^{k-1}\left(\mathbf{u}_i^T\mathbf{x}_k\right)\mathbf{u}_i,\ k = 2,3,\ldots.,p$ is the projection of $\mathbf{x}_k$ on the linear span of $\mathbf{x}_1, \mathbf{x}_2, \ldots, \mathbf{x}_{k-1}$ (Johnson and Wichern, 2002).

Example 3.9

Consider Example 3.7 and construct a set of orthogonal and orthonormal vectors using the Gram-Schmidt process.

$$\mathbf{x}_1 = \begin{bmatrix}5\\7\\10\end{bmatrix} \text{ and } \mathbf{x}_2 = \begin{bmatrix}1\\10\\3\end{bmatrix}$$

Solution:

Let the orthogonal vectors are $\mathbf{z}_1$ and $\mathbf{z}_2$ and the corresponding orthonormal vectors are $\mathbf{u}_1$ and $\mathbf{u}_2$. So, $\mathbf{z}_1 = \mathbf{x}_1 = [5 \quad 7 \quad 10]^T$ and

$$\begin{aligned}
\mathbf{u}_1 &= \frac{\mathbf{z}_1}{\sqrt{\mathbf{z}_1{}^T\mathbf{z}_1}} = \frac{1}{\sqrt{25+49+100}}[5 \quad 7 \quad 10]^T\\
&= \frac{1}{\sqrt{174}}\begin{bmatrix}5\\7\\10\end{bmatrix} = \begin{bmatrix}0.38\\0.53\\0.76\end{bmatrix}
\end{aligned}$$

And, $\mathbf{z}_2 = \mathbf{x}_2 - \left(\mathbf{u}_1^T\mathbf{x}_2\right)\mathbf{u}_1$

Now,

$$\mathbf{u}_1^T\mathbf{x}_2 = \frac{1}{\sqrt{174}}\begin{bmatrix}5 & 7 & 10\end{bmatrix}_{1\times 3}\begin{bmatrix}1\\10\\3\end{bmatrix}$$

$$= \frac{1}{\sqrt{174}}\cdot 105 = \frac{105}{\sqrt{174}}$$

$$\mathbf{z}_2 = \begin{bmatrix}1\\10\\3\end{bmatrix} - \frac{105}{\sqrt{174}}\cdot\begin{bmatrix}0.38\\0.53\\0.76\end{bmatrix} = \begin{bmatrix}-2\\5.8\\-3\end{bmatrix}$$

Again,

$$\mathbf{z}_2{}^T\mathbf{z}_2 = 4 + 33.64 + 9 = 46.64$$

$$\mathbf{u}_2 = \frac{\mathbf{z}_2}{\sqrt{\mathbf{z}_2{}^T\mathbf{z}_2}} = \frac{1}{\sqrt{46.64}}\begin{bmatrix}-2\\5.8\\-3\end{bmatrix} = \begin{bmatrix}-0.29\\0.85\\-0.44\end{bmatrix}$$

3.1.6 Projection of One Vector on Another

Consider Figure 3.10. Draw a perpendicular line (dashed line AC in Figure 3.10) from $\mathbf{x}$ on $\mathbf{y}$. AC generates a vector $\mathbf{p}(= \text{OC})$ which is a scalar multiple of $\mathbf{y}$ such that $\mathbf{p} = \alpha\mathbf{y}$, where α is a scalar. The vector $\mathbf{p}$ is known as projection of $\mathbf{x}$ on $\mathbf{y}$. The scalar α can be determined as follows (Kraus, 2002):

From Figure 3.10, it is seen that the vector represented by the dashed line is $\mathbf{x} - \mathbf{p}$, which is perpendicular to $\alpha\mathbf{y}$.

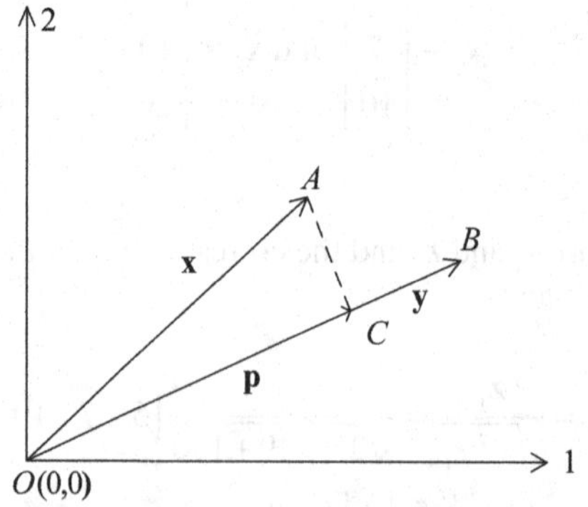

FIGURE 3.10 The projection of vector $\mathbf{x}$ (= OA) on vector $\mathbf{y}$ (= OB). The projection vector is $\mathbf{p}$ (= OC). AC is perpendicular to OB.

Now, $\mathbf{x}-\mathbf{p}=\mathbf{x}-\alpha\mathbf{y}$ and $(\mathbf{x}-\alpha\mathbf{y}).\mathbf{y}=0$

$$\Rightarrow \mathbf{x}\cdot\mathbf{y}-\alpha\mathbf{y}\cdot\mathbf{y}=0$$
$$\Rightarrow \mathbf{x}^T y-\alpha\mathbf{y}^T\mathbf{y}=0$$
$$\Rightarrow \alpha=\frac{\mathbf{x}^T\mathbf{y}}{\mathbf{y}^T\mathbf{y}}$$

$$\text{Therefore, } \mathbf{p}=\alpha\mathbf{y}=\frac{\mathbf{x}^T\mathbf{y}}{\mathbf{y}^T\mathbf{y}}y \tag{3.17}$$

Example 3.10

Consider Example 3.7 and obtain projection of $\mathbf{x}$ on $\mathbf{y}$.

Solution:

Let the projection vector be $\mathbf{p}$. So,

$$\mathbf{p}=\frac{\mathbf{x}^T\mathbf{y}}{\mathbf{y}^T\mathbf{y}}\mathbf{y}, \mathbf{x}^T\mathbf{y}=105 \text{ and } \mathbf{y}^T\mathbf{y}=110 \text{ (from Example 3.7)}$$

$$\text{So, } \mathbf{p}=\frac{105}{110}\begin{bmatrix}1\\10\\3\end{bmatrix}=0.95\begin{bmatrix}1\\10\\3\end{bmatrix}=\begin{bmatrix}0.95\\9.50\\2.85\end{bmatrix}$$

The projection vector $\mathbf{p}$ has important applications in multivariate modeling. Its utility can be seen in multiple linear regression (Chapter 8) and principle component analysis (Chapter 11).

3.1.7 Basic Matrices

A matrix is a rectangular array of numbers with a size; for example $n\times p$, as shown in the data matrix $\mathbf{X}_{n\times p}$. When $n=1$, $\mathbf{X}_{n\times p}$ becomes a row vector $\mathbf{x}_{1\times p}$ of size p, which we have defined earlier as a vector of simultaneous observations on p variables. Similarly, when $p=1$, $\mathbf{X}_{n\times p}$ becomes a column vector $\mathbf{x}_{n\times 1}$ of size n, which we have defined earlier as a vector of n observations on a particular variable, say X. So, vectors are also the special cases of a matrix. In this section, we describe a matrix in its most general form and highlight its different properties.

Consider a system of n algebraic equations with p unknown, $x_1, x_2, \ldots, x_p$ in the following form:

$$\begin{aligned}
&a_{11}x_1+a_{12}x_2+\cdots+a_{1p}x_p=y_1\\
&a_{21}x_1+a_{22}x_2+\cdots+a_{2p}x_p=y_2\\
&\vdots\\
&a_{i1}x_1+a_{i2}x_2+\cdots+a_{ip}x_p=y_i\\
&\vdots\\
&a_{n1}x_1+a_{n2}x_2+\cdots+a_{np}x_p=y_n
\end{aligned}$$

Which can be rewritten in matrix form as

$$\begin{bmatrix} a_{11} & a_{12} & \cdots & a_{1p} \\ a_{21} & a_{22} & \cdots & a_{2p} \\ \vdots & \vdots & \vdots & \vdots \\ a_{i1} & a_{i2} & \cdots & a_{ip} \\ \vdots & \vdots & \vdots & \vdots \\ a_{n1} & a_{n2} & \cdots & a_{np} \end{bmatrix}_{n\times p} \begin{bmatrix} x_1 \\ x_2 \\ \vdots \\ x_i \\ \vdots \\ x_p \end{bmatrix}_{p\times 1} = \begin{bmatrix} y_1 \\ y_2 \\ \vdots \\ y_i \\ \vdots \\ y_n \end{bmatrix}_{n\times 1}$$

$$\text{or} \quad \mathbf{Ax} = \mathbf{y} \tag{3.18}$$

Note that in Equation 3.18, $\mathbf{x} = [x_1, x_2, ..., x_p]$ is a vector of p unknowns, whereas **A** and y are known. The data matrix $\mathbf{X}_{n\times p}$, discussed earlier, has values similar to **A** in Equation 3.18. We are using a in **A** in place of x in $\mathbf{X}_{n\times p}$ as once the sample $\mathbf{X}_{n\times p}$ is collected from a population, it represents fixed values (like a) only. The general form of a matrix of order $n \times p$ is

$$\mathbf{A} = \begin{bmatrix} a_{11} & a_{12} & \cdots & a_{1p} \\ a_{21} & a_{22} & \cdots & a_{2p} \\ \vdots & \vdots & \vdots & \vdots \\ a_{i1} & a_{i2} & \cdots & a_{ip} \\ \vdots & \vdots & \vdots & \vdots \\ a_{n1} & a_{n2} & \cdots & a_{np} \end{bmatrix}_{n\times p}$$

a_{ij} is the $ij-th$ element of **A**. By changing i and j, $i = 1,2,...,n$ and $j = 1,2,..., p$, we will get all elements of **A**.

Square matrix: The matrix **A** becomes a square matrix when $n = p$. A square matrix has equal number of rows and columns. For example, the 3 × 3 matrix $\begin{bmatrix} 2 & 3 & 5 \\ -1 & 4 & 3 \\ 6 & 5 & 1 \end{bmatrix}$ is square.

Null matrix: A matrix is said to be null when all its elements equal to zero. For example, the 2 × 3 matrix $\begin{bmatrix} 0 & 0 & 0 \\ 0 & 0 & 0 \end{bmatrix}$ is null.

Diagonal matrix: When all the off-diagonal elements of a square matrix becomes zero, i.e., $a_{ij} = 0$ for $i \neq j$, the matrix is said to be diagonal. For example, the 4 × 4 matrix

$$\begin{bmatrix} a_{11} & 0 & 0 & 0 \\ 0 & a_{22} & 0 & 0 \\ 0 & 0 & a_{33} & 0 \\ 0 & 0 & 0 & a_{44} \end{bmatrix} \text{ is diagonal.}$$

Identity matrix: When the diagonal elements of a diagonal matrix equal to 1, i.e., $a_{ij} = 1$ for $i = j$ and $a_{ij} = 0$ for $i \neq j$, the matrix is said to be identity matrix. For example, the 4 × 4 matrix

$$\begin{bmatrix} 1 & 0 & 0 & 0 \\ 0 & 1 & 0 & 0 \\ 0 & 0 & 1 & 0 \\ 0 & 0 & 0 & 1 \end{bmatrix} \text{ is an identity matrix.}$$

Symmetric matrix: A square matrix is symmetric if for all $i \neq j$, $a_{ij} = a_{ji}$. For example, the square matrix

$$\mathbf{A} = \begin{bmatrix} a_{11} & a_{12} & a_{13} \\ a_{12} & a_{22} & a_{23} \\ a_{13} & a_{23} & a_{33} \end{bmatrix} \text{ is symmetric.}$$

One of the properties of the symmetric matrix $\mathbf{A}$ is that its transpose $\mathbf{A}^T$ is equal to $\mathbf{A}$.

$\therefore \mathbf{A}^T = \mathbf{A}$ if $\mathbf{A}$ is symmetric.

Trace of a matrix: The sum of the diagonal elements of a square matrix $\mathbf{A}$ is called the trace of $\mathbf{A}$.

So, $trace(\mathbf{A}) = \sum_{k=1}^{n} a_{kk}$ where $\mathbf{A}$ is a square matrix of the order $n \times n$.

A square matrix $\mathbf{A}$ is said to be orthogonal if its transpose ($\mathbf{A}^T$) is equal to its inverse ($\mathbf{A}^{-1}$). So, for orthogonal matrix $\mathbf{A}^T = \mathbf{A}^{-1}$. See Section 3.1.8 for matrix transposition and Section 3.1.10 for matrix inversion.

3.1.8 Basic Matrix Operations

Golub and Loan (2007) included transposition, addition, scalar matrix multiplication, and matrix-matrix multiplication as the basic matrix operations and the building blocks of matrix computations. In addition, we discuss matrix equality and matrix subtraction here.

Matrix equality: Two matrices $\mathbf{A}_{n\times p}$ and $\mathbf{B}_{n\times p}$ of the same order $n \times p$ is said to be equal if and only if every element of $\mathbf{A}(a_{ij})$ is equal to the corresponding element of $\mathbf{B}(b_{ij})$. That is

$$a_{ij} = b_{ij}, i = 1, 2, ..., n; j = 1, 2, ..., p.$$

For example, the two matrices $\mathbf{A}$ and $\mathbf{B}$ shown below are not equal as $a_{23} \neq b_{23}$.

$$\mathbf{A} = \begin{bmatrix} 5 & 3 & 2 \\ 4 & 6 & -5 \\ 1 & -2 & 4 \end{bmatrix} \text{ and } \mathbf{B} = \begin{bmatrix} 5 & 3 & 2 \\ 4 & 6 & 6 \\ 1 & -2 & 4 \end{bmatrix}$$

Transpose of a matrix: When the row and column elements of a matrix are interchanged, the resultant matrix is called the transpose of the original matrix. For example, the transpose of the 2 × 3 matrix

$$\begin{bmatrix} 3 & 2 & 9 \\ 6 & 5 & -2 \end{bmatrix} \text{ is } \begin{bmatrix} 3 & 6 \\ 2 & 5 \\ 9 & -2 \end{bmatrix}.$$

The element a_{ij} of the original matrix becomes a_{ji} in the transposed matrix. The transpose of a matrix **A** is denoted by $\mathbf{A}^T$, when T represents transpose. When we take the transpose of a transposed matrix, we get back the original matrix, i.e., $(\mathbf{A}^T)^T = \mathbf{A}$.

Matrix addition: Two matrices $\mathbf{A}_{n \times p}$ and $\mathbf{B}_{n \times p}$ of the same order $n \times p$ can be added to form a matrix $\mathbf{C}_{n \times p}$ of the same order where the elements of **C** will be sum of the corresponding elements of **A** and **B**. That is,

$$c_{ij} = a_{ij} + b_{ij},\ i = 1,2,...,n \text{ and } j = 1,2,..., p.$$

Matrix subtraction: Two matrices $\mathbf{A}_{n \times p}$ and $\mathbf{B}_{n \times p}$ of the same order $n \times p$ can be subtracted to form a matrix $\mathbf{C}_{n \times p}$ of the same order where the elements of **C** will be the difference between the corresponding elements of **A** and **B**. That is

$$c_{ij} = a_{ij} - b_{ij},\ i = 1,2,...,n \text{ and } j = 1,2,..., p.$$

Scalar multiplication: When a scalar (α) is multiplied to a matrix $\mathbf{A}_{n \times p}$, each element of **A**, a_{ij}, is multiplied by the scalar (α). For example,

$$\alpha \begin{bmatrix} a_{11} & a_{12} \\ a_{21} & a_{22} \end{bmatrix} = \begin{bmatrix} \alpha a_{11} & \alpha a_{12} \\ \alpha a_{21} & \alpha a_{22} \end{bmatrix}$$

Matrix multiplication: For multiplication of two matrices, it is required that two matrices should be dimensionally compatible for multiplication. By dimensional compatibility we mean, if the matrices **A** and **B** are to be multiplied to form **AB**, the column dimension of **A** must be equal to the row dimension of **B**. For example, $\mathbf{A}_{n \times p}$ and $\mathbf{B}_{p \times m}$ are dimensionally compatible as column dimension of **A** (i.e., p) is equal to the row dimension of **B** (i.e., p). But $\mathbf{A}_{n \times p}$ and $\mathbf{B}_{m \times p}$ cannot be multiplied as the column and row dimension of **A** and **B**, respectively are not compatible.

For dimensionally compatible matrices $\mathbf{A}_{n \times p}$ and $\mathbf{B}_{p \times m}$, when multiplied from a matrix $\mathbf{C}_{n \times m}$, where

$$c_{ij} = \sum_{k=1}^{p} a_{ik} b_{kj},\ i = 1,2,...,n \text{ and } j = 1,2,...,m \tag{3.19}$$

Example 3.11

Multiply the following two matrices.

$$\mathbf{A} = \begin{bmatrix} 5 & 10 \\ 2 & 6 \end{bmatrix}_{2 \times 2} \quad \mathbf{B} = \begin{bmatrix} 1 & 6 & 9 \\ 4 & 3 & 10 \end{bmatrix}_{2 \times 3}$$

Solution:

Here, $\mathbf{A}_{n \times p} = \mathbf{A}_{2 \times 2}$ and $\mathbf{B}_{p \times m} = \mathbf{B}_{2 \times 3}$

So, $\mathbf{C}_{n \times m}$ will be of the order $\mathbf{C}_{2 \times 3}$.

Now, from **A**, $a_{11} = 5, a_{12} = 10, a_{21} = 2$ and $a_{22} = 6$

From **B**, $b_{11} = 1, b_{12} = 6, b_{13} = 9, b_{21} = 4, b_{22} = 3$ and $b_{23} = 10$

we know $c_{ij} = \sum_{k=1}^{p} a_{ik} b_{kj}$, $i = 1,2$, $j = 1,2,3$ and $p = 2$.

$$\begin{aligned} c_{11} &= \sum_{k=1}^{2} a_{1k} b_{k1} \\ &= a_{11}b_{11} + a_{12}b_{21} \\ &= 5\times 1 + 10\times 4 \\ &= 5 + 40 \\ &= 45 \end{aligned}$$

Similarly,

$$c_{12} = 60; c_{13} = 145; c_{21} = 26; c_{22} = 30; c_{23} = 78.$$

So,

$$\mathbf{C} = \begin{bmatrix} 45 & 60 & 145 \\ 26 & 30 & 78 \end{bmatrix}$$

The following rules are useful in matrix multiplication.

- $\alpha(\mathbf{AB}) = (\alpha\mathbf{A})\mathbf{B}$, where α is scalar
- $\mathbf{A}(\mathbf{BC}) = (\mathbf{AB})\mathbf{C}$
- $\mathbf{A}(\mathbf{B}+\mathbf{C}) = \mathbf{AB}+\mathbf{AC}$
- $(\mathbf{B}+\mathbf{C})\mathbf{A} = \mathbf{BA}+\mathbf{CA}$
- $(\mathbf{AB})^T = \mathbf{B}^T\mathbf{A}^T$

3.1.9 Determinants

Any square matrix (e.g., **A**) can be converted into a uniquely defined scalar called determinant and is expressed as

$$\det(\mathbf{A}) = |\mathbf{A}|$$

Only square matrices possess determinants. For example, of the following two matrices **A** and **B**

$$\mathbf{A} = \begin{bmatrix} 4 & 5 \\ 6 & 4 \end{bmatrix}, \mathbf{B} = \begin{bmatrix} 1 & 2 & 3 \\ 6 & 3 & 4 \end{bmatrix}$$

A has a determinant, but **B** does not, as **B** is not a square matrix.

Consider a $p \times p$ matrix, **A**

$$\mathbf{A} = \begin{bmatrix} a_{11} & a_{12} & \cdots & a_{1p} \\ a_{21} & a_{22} & \cdots & a_{2p} \\ \vdots & \vdots & \vdots & \vdots \\ a_{i1} & a_{i2} & \cdots & a_{ip} \\ \vdots & \vdots & \vdots & \vdots \\ a_{p1} & a_{p2} & \cdots & a_{pp} \end{bmatrix}_{p\times p}$$

The following formula can be used to obtain $\det(\mathbf{A}) = |\mathbf{A}|$ (Kraus, 2002):

$$\begin{aligned} &|\mathbf{A}| = a_{11} \text{ if } p = 1 \\ &\text{or } |\mathbf{A}| = \sum_{j=1}^{p} a_{ij} |\mathbf{A}_{ij}| (-1)^{1+j} \text{ if } p > 1 \end{aligned} \tag{3.20}$$

Where $\mathbf{A}_{ij}$ is the $(p-1)\times(p-1)$ matrix obtained by deleting the $i-th$ row and $j-th$ column of $\mathbf{A}$.

Example 3.12

Compute the determinant of the following matrix

$$\mathbf{A} = \begin{bmatrix} a_{11} & a_{12} & a_{13} \\ a_{21} & a_{22} & a_{23} \\ a_{31} & a_{32} & a_{33} \end{bmatrix}$$

Solution:

Using Equation 3.20, we get

$$\det(\mathbf{A}) = |\mathbf{A}| = \begin{vmatrix} a_{11} & a_{12} & a_{13} \\ a_{21} & a_{22} & a_{23} \\ a_{31} & a_{32} & a_{33} \end{vmatrix} = a_{11}|\mathbf{A}_{11}|(-1)^{1+1} + a_{12}|\mathbf{A}_{12}|(-1)^{1+2} + a_{113}|\mathbf{A}_{13}|(-1)^{1+3}$$

$$= a_{11}\begin{vmatrix} a_{22} & a_{23} \\ a_{32} & a_{33} \end{vmatrix}(-1)^{1+1} + a_{12}\begin{vmatrix} a_{21} & a_{23} \\ a_{31} & a_{33} \end{vmatrix}(-1)^{1+2} + a_{13}\begin{vmatrix} a_{21} & a_{22} \\ a_{31} & a_{32} \end{vmatrix}(-1)^{1+3}$$

Now,

$$|\mathbf{A}_{11}| = \begin{vmatrix} a_{22} & a_{23} \\ a_{32} & a_{33} \end{vmatrix} = a_{22} \cdot a_{33} - a_{23} \cdot a_{32}$$

Similarly,

$$|\mathbf{A}_{12}| = \begin{vmatrix} a_{21} & a_{23} \\ a_{31} & a_{33} \end{vmatrix} = a_{21} \cdot a_{33} - a_{23} \cdot a_{31}$$

And

$$|\mathbf{A}_{13}| = \begin{vmatrix} a_{21} & a_{22} \\ a_{31} & a_{32} \end{vmatrix} = a_{21} \cdot a_{32} - a_{22} \cdot a_{31}$$

So,

$$\begin{aligned} |\mathbf{A}| &= a_{11}(a_{22}a_{33} - a_{23}a_{32}) - a_{12}(a_{21}a_{33} - a_{23}a_{31}) + a_{13}(a_{21}a_{32} - a_{22}a_{31}) \\ &= a_{11}a_{22}a_{33} - a_{11}a_{23}a_{32} - a_{12}a_{21}a_{33} + a_{12}a_{23}a_{31} + a_{13}a_{21}a_{32} - a_{13}a_{22}a_{31} \\ &= a_{11}a_{22}a_{33} + a_{12}a_{23}a_{31} + a_{13}a_{21}a_{32} - a_{11}a_{23}a_{32} - a_{12}a_{21}a_{33} - a_{13}a_{22}a_{31} \end{aligned}$$

Example 3.13

Compute determinant of $\mathbf{A}$, where

$$\mathbf{A} = \begin{bmatrix} 5 & 2 & 1 \\ 4 & 10 & 3 \\ 10 & 5 & 20 \end{bmatrix}$$

Solution:

$$\begin{aligned} |\mathbf{A}| &= 5\begin{vmatrix} 10 & 3 \\ 5 & 20 \end{vmatrix}(-1)^{1+1} + 2\begin{vmatrix} 4 & 3 \\ 10 & 20 \end{vmatrix}(-1)^{1+2} + 1\begin{vmatrix} 4 & 10 \\ 10 & 5 \end{vmatrix}(-1)^{1+3} \\ &= 5(200-15) - 2(80-30) + 1(20-100) \\ &= 5 \times 185 - 2 \times 50 - 80 \\ &= 925 - 100 - 80 \\ &= 745 \end{aligned}$$

The determinant of a matrix of size $p \times p$ equals the volume of a box in the p dimensional space (Strang, 2006). For example, in Example 3.13, the determinant is 745, which measures the volume of a box in three dimensions.

If the determinant of a matrix $\mathbf{A}$ is zero, $\mathbf{A}$ is said to be *singular*. Johnson and Wichern (2002) mentioned singularity as follows: for a non-zero vector $\mathbf{x}$, a matrix $\mathbf{A}$ is said to be singular if the relationship $\mathbf{Ax} = 0$ holds.

Interestingly, determinants possess several important characteristics. A few of them that are relevant to multivariate computations are stated below. For a complete picture, the reader may consult Kraus (2002).

- The value of the determinant of a diagonal matrix is the product of the diagonal elements.
- The determinants of a matrix $(\mathbf{A})$ and its transpose $(\mathbf{A}^T)$ are equal; i.e., $\det(\mathbf{A}) = \det(\mathbf{A}^T)$.

- The determinant of the product of two matrices **A** and **B** is equal to the product of their determinants; i.e., $\det(\mathbf{AB}) = \det(\mathbf{A}) \cdot \det(\mathbf{B})$.
- If a matrix $\mathbf{A}_{m\times n}$ is pre-multiplied a matrix $\mathbf{A}_{n\times m}$, where $m > n$, then in general, the product **AB** will be singular.

3.1.10 Rank of a Matrix

As we have seen earlier, a matrix $\mathbf{X}_{n\times p}$ can be represented by n row vectors of the order $p \times 1$ or p column vectors of the order $n \times 1$. Further, we also know the conditions for linearly independent vectors. Using the same procedure, we can find out the number of independent row vectors, known as row rank or the number of independent column vectors, known as column rank. Further, as the rows of matrix are the columns of its transpose, it can be proved that the row and column ranks of a matrix are equal. So, the rank of a matrix is the row rank (= the column rank) of the matrix.

There are many methods to compute the rank of a matrix. For example, Gaussian elimination and singular value decomposition (SVD) are used. MATLAB uses SVD to compute the rank of a matrix. We will describe SVD in Section 3.1.14.

3.1.11 Inverse of a Matrix

For square matrix $\mathbf{A}_{p\times p}$, if there exists another square matrix $\mathbf{B}_{p\times p}$ of the same order such that $\mathbf{BA} = \mathbf{AB} = \mathbf{I}$, where $\mathbf{I}$ is the identity matrix of order $p \times p$, the matrix **B** is called the inverse of **A**, and is denoted by $\mathbf{A}^{-1}$.

For example, let $\mathbf{A} = \begin{bmatrix} 5 & 5 \\ 7 & 9 \end{bmatrix}$ and $\mathbf{B} = \begin{bmatrix} 0.9 & -0.5 \\ -0.7 & 0.5 \end{bmatrix}$

Now,

$$\mathbf{AB} = \begin{bmatrix} 5 & 5 \\ 7 & 9 \end{bmatrix} \begin{bmatrix} 0.9 & -0.5 \\ -0.7 & 0.5 \end{bmatrix}$$
$$= \begin{bmatrix} 1 & 0 \\ 0 & 1 \end{bmatrix} = \mathbf{I}$$

Similarly,

$$\mathbf{BA} = \begin{bmatrix} 0.9 & -0.5 \\ -0.7 & 0.5 \end{bmatrix} \begin{bmatrix} 5 & 5 \\ 7 & 9 \end{bmatrix} = \begin{bmatrix} 1 & 0 \\ 0 & 1 \end{bmatrix} = \mathbf{I}$$

Therefore, **B** is inverse of **A** and is denoted as $\mathbf{A}^{-1}$.

The precondition for a square matrix **A** to have its inverse ($\mathbf{A}^{-1}$) is that $\det(\mathbf{A}) \neq 0$. When the determinant of a matrix is non-zero, the matrix is said to be non-singular. A matrix whose determinant is zero is called a singular matrix. For non-singular matrices the following properties hold:

(i) $(\mathbf{AB})^{-1} = \mathbf{B}^{-1}\mathbf{A}^{-1}$
(ii) If $\mathbf{AB} = \mathbf{AC}$, then $\mathbf{B} = \mathbf{C}$
(iii) $(\mathbf{A}^{-1})^{-1} = \mathbf{A}$

(iv) $(\mathbf{A}^T)^{-1} = (\mathbf{A}^{-1})^T$
(v) $(\mathbf{A}^{-1})^n = (\mathbf{A}^n)^{-1} = \mathbf{A}^{-n}$
(vi) $\mathbf{A}^m \mathbf{A}^n = \mathbf{A}^{mn}$
(vii) $(\mathbf{A}^m)^n = \mathbf{A}^{mn}$

For (v) n and for (vi) and (vii) both m and n are integers.

There are many methods for obtaining inverse of a matrix such as (i) method of adjoint matrix, (ii) method of elementary transformation, (iii) sweep-out method, (iv) method of partitioning a matrix, and (v) Cayley-Hamilton theorem (Kraus, 2002). We will describe here the method of adjoint matrix. In order to employ the method of adjoint matrix, we require defining the following terms:

Cofactor: The cofactor of a square matrix **A** of the order $n \times n$ whose elements are $\{a_{ij}\}, i = 1,2,...,n$ and $j = 1,2,...,n$, termed as A_{ij} is obtained by the following formula:

$$A_{ij} = (-1)^{i+j} M_{ij}. \tag{3.21}$$

Where M_{ij} is known as the $(n-1)th$ order minor that is obtained by taking the determinant of a matrix obtained by deleting the $i-th$ row and $j-th$ column of the original matrix **A**.

For example, let **A** is a 3 × 3 matrix shown below.

$$\mathbf{A} = \begin{bmatrix} a_{11} & a_{12} & a_{13} \\ a_{21} & a_{22} & a_{23} \\ a_{31} & a_{32} & a_{33} \end{bmatrix}$$

The $(n-1)th$ [3-1=2] order cofactors are

$$\begin{aligned}
A_{11} &= (-1)^{1+1} M_{11} \\
A_{12} &= (-1)^{1+2} M_{12} \\
A_{13} &= (-1)^{1+3} M_{13} \\
A_{21} &= (-1)^{2+1} M_{21} \\
A_{22} &= (-1)^{2+2} M_{22} \\
A_{23} &= (-1)^{2+3} M_{23} \\
A_{31} &= (-1)^{3+1} M_{31} \\
A_{32} &= (-1)^{3+2} M_{32} \\
A_{33} &= (-1)^{3+3} M_{33}
\end{aligned}$$

Now, M_{11} is obtained by deleting the first row and first column of **A**. So,

$$A_{11} = (-1)^{1+1} \begin{vmatrix} a_{22} & a_{23} \\ a_{32} & a_{33} \end{vmatrix} = a_{22}a_{33} - a_{23}a_{32}$$

Similarly,

$$A_{12} = (-1)^{1+2}\begin{vmatrix} a_{21} & a_{23} \\ a_{31} & a_{33} \end{vmatrix} = -a_{21}a_{33} + a_{23}a_{31}$$

$$A_{13} = (-1)^{1+3}\begin{vmatrix} a_{21} & a_{22} \\ a_{31} & a_{32} \end{vmatrix} = a_{21}a_{32} - a_{22}a_{31}$$

$$A_{21} = (-1)^{2+1}\begin{vmatrix} a_{12} & a_{13} \\ a_{32} & a_{33} \end{vmatrix} = -a_{12}a_{33} + a_{13}a_{32}$$

$$A_{22} = (-1)^{2+2}\begin{vmatrix} a_{11} & a_{13} \\ a_{31} & a_{33} \end{vmatrix} = a_{11}a_{33} - a_{13}a_{31}$$

$$A_{23} = (-1)^{2+3}\begin{vmatrix} a_{11} & a_{12} \\ a_{31} & a_{32} \end{vmatrix} = -a_{11}a_{32} + a_{12}a_{31}$$

$$A_{31} = (-1)^{3+1}\begin{vmatrix} a_{12} & a_{13} \\ a_{22} & a_{23} \end{vmatrix} = a_{12}a_{23} - a_{13}a_{22}$$

$$A_{32} = (-1)^{3+2}\begin{vmatrix} a_{11} & a_{13} \\ a_{31} & a_{33} \end{vmatrix} = -a_{11}a_{33} + a_{13}a_{31}$$

$$A_{33} = (-1)^{3+3}\begin{vmatrix} a_{11} & a_{12} \\ a_{21} & a_{22} \end{vmatrix} = a_{11}a_{22} - a_{12}a_{21}$$

The foregoing discussion shows that for every element of a square matrix, there is one cofactor. So, for a $n \times n$ matrix, there will be $n \times n$ cofactors and we can form a square cofactor matrix ($\mathbf{A}^C$) of the order $n \times n$ as follows:

$$\mathbf{A}^C = \begin{bmatrix} A_{11} & A_{12} & \cdots & A_{1j} & \cdots & A_{1n} \\ A_{21} & A_{22} & \cdots & A_{2j} & \cdots & A_{2n} \\ \vdots & \vdots & & \vdots & & \vdots \\ A_{i1} & A_{i2} & \cdots & A_{ij} & \cdots & A_{in} \\ \vdots & \vdots & & \vdots & & \vdots \\ A_{n1} & A_{n2} & \cdots & A_{nj} & \cdots & A_{nn} \end{bmatrix}_{n \times n}$$

Where A_{ij} is defined in Equation 3.21.

Example 3.14: Obtain cofactor matrix of **A** given below

$$\mathbf{A} = \begin{bmatrix} 5 & 3 & 2 \\ 4 & 6 & 0 \\ 2 & 0 & 3 \end{bmatrix}$$

Solution:

A is 3 × 3 matrix. So, there are nine cofactors.

$$A_{11} = (-1)^{1+1}\begin{vmatrix} 6 & 0 \\ 0 & 3 \end{vmatrix} = 18$$

$$A_{12} = (-1)^{1+2}\begin{vmatrix} 4 & 0 \\ 2 & 3 \end{vmatrix} = -12$$

Similarly,

$$\begin{aligned} A_{13} &= -12 \\ A_{21} &= -9 \\ A_{22} &= 11 \\ A_{23} &= 6 \\ A_{31} &= -12 \\ A_{32} &= 8 \\ A_{33} &= 18 \end{aligned}$$

So, the cofactor matrix is $\mathbf{A}^C = \begin{bmatrix} 18 & -12 & -12 \\ -9 & 11 & 6 \\ -12 & 8 & 18 \end{bmatrix}$

Adjoint of a matrix: The adjoint of a matrix **A**, termed as adj(**A**), is defined as the transpose of the cofactor matrix $\mathbf{A}^C$.

$$\text{adj}(\mathbf{A}) = (\mathbf{A}^C)^T$$

The inverse of a square matrix **A**, termed as $\mathbf{A}^{-1}$, is obtained as

$$\mathbf{A}^{-1} = \frac{1}{|\mathbf{A}|}\text{adj}\ (\mathbf{A}) \tag{3.22}$$

Example 3.15

Compute $\mathbf{A}^{-1}$ for the matrix **A** shown in Example 3.14.

Solution:

From Example 3.14, $\mathbf{A}^C = \begin{bmatrix} 18 & -12 & -12 \\ -9 & 11 & 6 \\ -12 & 8 & 18 \end{bmatrix}$

Now, determinant of **A** is

$$\begin{aligned}\det(\mathbf{A}) &= 5\times A_{11} + 3\times A_{12} + 2\times A_{13}\\ &= 5\times 18 - 3\times 12 - 2\times 12\\ &= 90 - 36 - 24\\ &= 90 - 60\\ &= 30\end{aligned}$$

So, $\mathbf{A}^{-1} = \frac{1}{|\mathbf{A}|}(\mathbf{A}^C)^T$

$$= \frac{1}{30}\begin{bmatrix} 18 & -9 & -12 \\ -12 & 11 & 8 \\ -12 & 6 & 18 \end{bmatrix} = \begin{bmatrix} 0.60 & -0.30 & -0.40 \\ -0.40 & 0.37 & 0.27 \\ -0.40 & 0.20 & 0.60 \end{bmatrix}$$

3.1.12 Eigenvalues and Eigenvectors

Consider a square matrix **A** of the order $p \times p$. Let there be a scalar λ and non-zero vector **e** of the order $p \times 1$ such that the following holds

$$\mathbf{Ae} = \lambda\mathbf{e} \tag{3.23}$$

Then λ is called the eigenvalue of **A** and **e** is called the eigenvector of **A**. With little manipulation, Equation 3.23 can be written as

$$(\mathbf{A} - \lambda\mathbf{I})\mathbf{e} = \mathbf{0} \tag{3.24}$$

Where **I** is an identity matrix of the order $p \times p$. As **e** is a non-zero vector, to satisfy equation (3.24), $\mathbf{A} - \lambda\mathbf{I}$ must be singular; i.e.,

$$|\mathbf{A} - \lambda\mathbf{I}| = 0 \tag{3.25}$$

This is the characteristic equation. The left hand side of Equation 3.25 is termed as characteristic polynomial of $p-th$ order. The roots of Equation 3.25 are the eigenvalues. Essentially, for a $p \times p$ square matrix of rank p, there will be p number of eigenvalues.

In order to find out the eigenvalues and eigenvectors of a square matrix $\mathbf{A}_{p\times p}$, the following steps are followed (Strang, 2006):

Step 1: Compute the determinant of $\mathbf{A} - \lambda\mathbf{I}$. As **A** is a $p \times p$ matrix, the determinant of $\mathbf{A} - \lambda\mathbf{I}$ is a p degree polynomial.

Step 2: Obtain the roots of the polynomial using Equation 3.25. These roots are the eigenvalues of **A**.

Step 3: In order to get the eigenvector **e**, the following equation is solved for each of the eigenvalues obtained in Step 2.

$$(\mathbf{A} - \lambda\mathbf{I})\mathbf{e} = \mathbf{0}$$

Please note that **e** is a non-zero vector of the order $p \times 1$.

Example 3.16

Obtain eigenvalues and eigenvectors of the matrix given below.

$$\mathbf{A} = \begin{bmatrix} 10 & 5 \\ 4 & 6 \end{bmatrix}$$

Solution:

Step 1: Compute the determinant of $\mathbf{A} - \lambda\mathbf{I}$.

$$\begin{aligned} \mathbf{A} - \lambda\mathbf{I} &= \begin{bmatrix} 10 & 5 \\ 4 & 6 \end{bmatrix} - \lambda \begin{bmatrix} 1 & 0 \\ 0 & 1 \end{bmatrix} \\ &= \begin{bmatrix} 10 & 5 \\ 4 & 6 \end{bmatrix} - \begin{bmatrix} \lambda & 0 \\ 0 & \lambda \end{bmatrix} \\ &= \begin{bmatrix} 10-\lambda & 5 \\ 4 & 6-\lambda \end{bmatrix} \end{aligned}$$

So,

$$\begin{aligned} |\mathbf{A} - \lambda\mathbf{I}| &= \begin{vmatrix} 10-\lambda & 5 \\ 4 & 6-\lambda \end{vmatrix} \\ &= (10-\lambda)(6-\lambda) - 20 \\ &= 60 - 10\lambda - 6\lambda + \lambda^2 - 20 \\ &= \lambda^2 - 16\lambda + 40 \end{aligned}$$

Step 2: Obtain the roots of $\lambda^2 - 16\lambda + 40 = 0$.

$\lambda^2 - 16\lambda + 40 = 0$ is a 2nd degree polynomial of λ and hence has two roots; say, λ_1 and λ_2. The roots of the polynomial are

$$\begin{aligned} \lambda &= \frac{-(-16) \pm \sqrt{(-16)^2 - 4 \cdot 1 \cdot 40}}{2 \cdot 1} \\ &= \frac{16 \pm \sqrt{256 - 160}}{2} \\ &= \frac{16 \pm 4\sqrt{6}}{2} \\ &= 8 \pm 2\sqrt{6} \end{aligned}$$

So,

$$\begin{aligned} \lambda_1 &= 8 + 2\sqrt{6} = 12.90 \\ \lambda_2 &= 8 - 2\sqrt{6} = 3.10 \end{aligned}$$

Step 3: Obtain eigenvectors

For every eigenvalue, there is one eigenvector. In this case, there are two eigenvectors; one for $\lambda_1 = 12.90$ and another for $\lambda_2 = 3.10$, denoted by $\mathbf{e}_1$ and $\mathbf{e}_2$, respectively. Both $\mathbf{e}_1$ and $\mathbf{e}_2$ are of the order of 2 × 1.

$$\text{So, } \mathbf{e}_1 = \begin{bmatrix} e_{11} \\ e_{12} \end{bmatrix} \text{ and } \mathbf{e}_2 = \begin{bmatrix} e_{21} \\ e_{22} \end{bmatrix}$$

Now, for the first eigenvector $\mathbf{e}_1$, we put the value of λ_1 in equation 3.24. So,

$$(\mathbf{A} - \lambda_1 \mathbf{I})\mathbf{e}_1 = \mathbf{0}$$

$$\begin{bmatrix} 10 - 12.90 & 5 \\ 4 & 6 - 12.90 \end{bmatrix} \begin{bmatrix} e_{11} \\ e_{12} \end{bmatrix} = \begin{bmatrix} 0 \\ 0 \end{bmatrix}$$

$$\Rightarrow \begin{bmatrix} -2.90 & 5 \\ 4 & -6.90 \end{bmatrix} \begin{bmatrix} e_{11} \\ e_{12} \end{bmatrix} = \begin{bmatrix} 0 \\ 0 \end{bmatrix}$$

which yields

$$-2.90e_{11} + 5e_{12} = 0 \ldots\ldots\ldots\ldots\ldots\ldots\ldots\ldots (i)$$
$$4e_{11} - 6.90e_{12} = 0 \ldots\ldots\ldots\ldots\ldots\ldots\ldots\ldots (ii)$$

We are now in a situation where there is no unique solution to $\mathbf{e}_1$. It can take infinite number of values. For example, if we arbitrarily choose $e_{11} = 1$, then

$$e_{12} = \frac{2.90}{5.0} \times 1 = 0.58 \text{ (from equation i) and}$$
$$e_{12} = \frac{4}{6.90} \times 1 = 0.58 \text{ (from equation ii)}$$

Essentially, both the equations produce same set of eigenvectors. So, for any eigenvalue, there is an eigenspace which contains an infinite number of eigenvectors. Which eigenvector is to be considered is dependent on the purpose of the analysis. For example, in principal component analysis orthonormal (unit orthogonal) eigenvectors are obtained. Table 3.3 shows some of the eigenvectors for $\lambda_1 = 12.90$ and $\lambda_2 = 3.10$.

Similarly, for the second eigenvector, $\mathbf{e}_2$, we put $\lambda_2 = 3.10$ in Equation 3.24. So,

$$(\mathbf{A} - \lambda_2 \mathbf{I})\mathbf{e}_2 = \mathbf{0}$$

$$\begin{bmatrix} 10 - 3.10 & 5 \\ 4 & 6 - 3.10 \end{bmatrix} \begin{bmatrix} e_{21} \\ e_{22} \end{bmatrix} = \begin{bmatrix} 0 \\ 0 \end{bmatrix}$$

$$\Rightarrow \begin{bmatrix} 6.90 & 5 \\ 4 & 2.90 \end{bmatrix} \begin{bmatrix} e_{21} \\ e_{22} \end{bmatrix} = \begin{bmatrix} 0 \\ 0 \end{bmatrix}$$

TABLE 3.3
Eigenvectors for Example 3.16

Eigenvector No.	$\lambda_1 = 12.90\ 3$ Eigenvector Values e_{11}	e_{12}	$\lambda_2 = 3.10$ Eigenvector Values e_{21}	e_{22}
1	1	0.58	1	–1.38
2	0.5	0.29	0.5	–0.69
3	2	1.16	2	–2.76
4	1.5	0.87	1.5	–2.07
5	2.5	1.45	2.5	–3.45

Which yields,

$$6.90\ e_{21} + 5.00\ e_{22} = 0$$
$$4\ e_{21} + 2.90\ e_{22} = 0$$

$$\text{or}\quad e_{22} = \frac{-6.90}{5.00} e_{21} = -1.38 e_{21}$$

Assuming $e_{21} = 1$, e_{22} is -1.38.

Similarly for $\lambda_2 = 3.10$, we will get a eigenspace which contains an infinite number of eigenvectors. Five eigenvectors are shown in Table 3.3.

Determining the characteristic polynomial of the order p ($p > 2$) as described above is a difficult task. We will describe Bocher's method for this purpose. Bocher's method uses $j-th$ power of $A, j = 1,2,\ldots,p$ and corresponding traces to compute the characteristic polynomial $|\mathbf{A} - \lambda\mathbf{I}|$. The characteristic polynomial $|\mathbf{A} - \lambda\mathbf{I}|$ can be written as

$$f(\lambda) = \lambda^p + \beta_1\lambda^{p-1} + \beta_2\lambda^{p-2} + \cdots + \beta_{p-1}\lambda + \beta_p \tag{3.26}$$

We require to obtain the values of $\beta_1, \beta_2, \ldots, \beta_p$. The following procedure can be used (Kraus 2002):

Step 1: Compute $j-th$ power of **A**, $\mathbf{A}^j, j = 1,2,\ldots,p$ where p is the order of matrix **A**.

Step 2: Compute $T_j = trace(\mathbf{A}^j), j = 1,2,\ldots,p$.

Step 3: Follow the sequences of operations given below to obtain $\beta_1, \beta_2, \ldots, \beta_p$.

$$\beta_1 = -T_1$$
$$\beta_2 = -\frac{T_1\beta_1 + T_2}{2}$$
$$\beta_3 = -\left(\frac{T_1\beta_2 + T_2\beta_1 + T_3}{3}\right)$$
$$\vdots$$
$$\beta_p = -\left(\frac{T_1\beta_{p-1} + T_2\beta_{p-2} + \ldots + T_{p-1}\beta_1 + T_p}{p}\right)$$

Example 3.17

Compute the characteristic polynomial of the matrix **B** given below.

$$\mathbf{B} = \begin{bmatrix} 1 & 0 & 5 \\ 0 & 5 & 4 \\ 3 & 0 & 1 \end{bmatrix}$$

Solution:

B is a 3 × 3 matrix. So, the characteristic polynomial is of the order 3.

Steps 1 & 2: Find out $\mathbf{B}^j \; and \, T_j, j = 1,2,3$

$$\mathbf{B}^1 = \begin{bmatrix} 1 & 0 & 5 \\ 0 & 5 & 4 \\ 3 & 0 & 1 \end{bmatrix}, \; T_1 = trace(\mathbf{B}^1) = 1+5+1 = 7$$

$$\mathbf{B}^2 = \mathbf{B}^1 \cdot \mathbf{B}^1 = \begin{bmatrix} 1 & 0 & 5 \\ 0 & 5 & 4 \\ 3 & 0 & 1 \end{bmatrix}\begin{bmatrix} 1 & 0 & 5 \\ 0 & 5 & 4 \\ 3 & 0 & 1 \end{bmatrix} = \begin{bmatrix} 16 & 0 & 10 \\ 12 & 25 & 24 \\ 6 & 0 & 16 \end{bmatrix}$$

$$T_2 = trace(\mathbf{B}^2) = 16+25+16 = 57$$

$$\mathbf{B}^3 = \mathbf{B}^1 \cdot \mathbf{B}^2 = \begin{bmatrix} 1 & 0 & 5 \\ 0 & 5 & 4 \\ 3 & 0 & 1 \end{bmatrix}\begin{bmatrix} 16 & 0 & 10 \\ 12 & 25 & 24 \\ 6 & 0 & 16 \end{bmatrix} = \begin{bmatrix} 46 & 0 & 90 \\ 84 & 125 & 184 \\ 54 & 0 & 46 \end{bmatrix}$$

$$T_3 = trace(\mathbf{B}^3) = 46+125+46 = 217$$

Step 3: Compute $\beta_1, \beta_2, \text{and}\, \beta_3$

$$\beta_1 = -T_1 = -7$$

$$\beta_2 = -\left(\frac{T_1\beta_1 + T_2}{2}\right) = -\left(\frac{-7\times 7+57}{2}\right) = -4$$

$$\beta_3 = -\frac{1}{3}\left[T_1\beta_2 + T_2\beta_1 + T_3\right] = -\frac{1}{3}\left[7\times(-4)+57\times(-7)+217\right] = 70$$

So, the characteristic polynomial is $f(\lambda) = \lambda^3 - 7\lambda^2 - 4\lambda + 70$.

It may so happen that some of the eigenvalues of a square matrix become equal or they repeat. This implies that some of the resultant eigenvectors may be linearly dependent. The foregoing discussion reveals that a square matrix **A** of the order $p \times p$ may contain p distinct eigenvalues and p distinct eigenvectors of the order $p \times 1$. If we combine all the eigenvectors, we get a $p \times p$ matrix of eigenvectors where the first column denotes the eigenvector corresponding to λ_1. Similarly, second to the $p-th$ columns represent the eigenvectors for $\lambda_1, \lambda_2, ..., \lambda_p$, respectively.

For example, let us consider the eigenvalues and eigenvectors obtained in Example 3.16. The original matrix **A**, its eigenvalues and eigenvectors (first row of Table 3.3) are shown below.

$$\mathbf{A} = \begin{bmatrix} 10 & 5 \\ 4 & 6 \end{bmatrix}$$

$$\mathbf{\Lambda} = \begin{bmatrix} 12.90 & 0 \\ 0 & 3.10 \end{bmatrix}$$

$$\mathbf{E} = \begin{bmatrix} \mathbf{e}_1 & \mathbf{e}_2 \end{bmatrix} = \begin{bmatrix} e_{11} & e_{21} \\ e_{12} & e_{22} \end{bmatrix} = \begin{bmatrix} 1 & 1 \\ 0.58 & -1.38 \end{bmatrix}$$

Note that **E** is a matrix of 2 × 2 where the first column contains the eigenvector related to the first eigenvalue λ_1 and the second column represents the eigenvector for λ_2. $\mathbf{\Lambda}$ is a diagonal matrix containing eigenvalues as the diagonal elements. For a $p \times p$ square matrix $\mathbf{A}, \mathbf{\Lambda}$ and $\mathbf{E}$ can be written as below.

$$\mathbf{A} = \begin{bmatrix} a_{11} & a_{12} & \cdots & a_{1p} \\ a_{21} & a_{22} & \cdots & a_{2p} \\ \vdots & \vdots & \vdots & \vdots \\ a_{p1} & a_{p2} & \cdots & a_{pp} \end{bmatrix}_{p \times p}$$

$$\mathbf{\Lambda} = \begin{bmatrix} \lambda_1 & 0 & 0 & 0 \\ 0 & \lambda_2 & 0 & 0 \\ \vdots & \vdots & \vdots & \vdots \\ 0 & 0 & \ldots & \lambda_p \end{bmatrix}_{p \times p} \text{ and } \mathbf{E} = \begin{bmatrix} e_{11} & e_{21} & \cdots & e_{p1} \\ e_{12} & e_{22} & \cdots & e_{p2} \\ \vdots & \vdots & \vdots & \vdots \\ e_{1p} & e_{2p} & \cdots & e_{pp} \end{bmatrix}_{p \times p}$$

These three matrices $\mathbf{A}, \mathbf{\Lambda}$ and $\mathbf{E}$ hold an important relationship.

$$\mathbf{E}^{-1}\mathbf{A}\mathbf{E} = \mathbf{\Lambda} = \begin{bmatrix} \lambda_1 & 0 & \ldots & 0 \\ 0 & \lambda_2 & \ldots & 0 \\ 0 & 0 & \ldots & \lambda_p \end{bmatrix} \tag{3.27}$$

This is an important result from eigenvalues and eigenvectors decomposition. The relationship states that eigenvectors diagonalize a matrix. The matrices **E** and $\mathbf{\Lambda}$ are also known as model matrix and spectral matrix, respectively.

Example 3.18

Consider Example 3.16 and show that eigenvectors diagonalize the matrix **A**.

Solution:

We obtained

$$\mathbf{A}=\begin{bmatrix}10 & 5\\ 4 & 6\end{bmatrix}, \mathbf{\Lambda}=\begin{bmatrix}12.90 & 0\\ 0 & 3.10\end{bmatrix}, \mathbf{E}=\begin{bmatrix}1 & 1\\ 0.58 & -1.38\end{bmatrix}$$

Now,

$$|\mathbf{E}| = 1\times(-1.38) - 1\times(0.58) = -1.38 - 0.58 = -1.96$$

$$\mathbf{E}^{-1} = \frac{1}{|\mathbf{E}|}\text{adj }(\mathbf{E}) = \frac{1}{-1.96}\begin{bmatrix}-1.38 & -1\\ -0.58 & 1\end{bmatrix} = \frac{1}{1.96}\begin{bmatrix}1.38 & 1\\ 0.58 & -1\end{bmatrix}$$

Now,

$$\mathbf{AE} = \begin{bmatrix}10 & 5\\ 4 & 6\end{bmatrix}\begin{bmatrix}1 & 1\\ 0.58 & -1.38\end{bmatrix} = \begin{bmatrix}12.9 & 3.10\\ 7.48 & -4.28\end{bmatrix}$$

So,

$$\mathbf{E}^{-1}\mathbf{AE} = \frac{1}{1.96}\begin{bmatrix}1.38 & 1\\ 0.58 & -1\end{bmatrix}\begin{bmatrix}12.9 & 3.10\\ 7.48 & -4.28\end{bmatrix} = \frac{1}{1.96}\begin{bmatrix}25.28 & 0\\ 0 & 6.08\end{bmatrix} = \begin{bmatrix}12.9 & 0\\ 0 & 3.10\end{bmatrix} = \mathbf{\Lambda}$$

The eigenvalue and eigenvector decompositions of a matrix $\mathbf{A}_{p\times p}$ gives some useful results. Some of them are mentioned below.

(i) $trace(\mathbf{A}) = \sum_{j=1}^{p} \lambda_j$.

(ii) The eigenvalues of $\mathbf{A}$ equal the eigenvalues of its transpose, $\mathbf{A}^T$.

(iii) $\mathbf{A}_{p\times p}$ matrix with p distinct eigenvalues has p-linearly independent rows and columns.

(iv) Any matrix with distinct eigenvalues can be diagonalized.

(v) If $(\lambda_j, \mathbf{e}_j), j = 1,2,...,p$ be the j-th pair of eigenvalue and eigenvector of $\mathbf{A}_{p\times p}$, the eigenvalue and eigenvector pair of $\mathbf{A}^k$ ($k-th$ power of $\mathbf{A}_{p\times p}$) are $(\lambda_j^k, \mathbf{e}_j), j = 1,2,...,p$.

(vi) If $\mathbf{A}_{p\times p}$ is invertible and has eigenvalues $\lambda_j, j = 1,2,...,p$, then $\mathbf{A}^{-1}$ has eigenvalues $1/\lambda_j, j = 1,2,...,p$.

Eigenvalues and eigenvectors have several uses in multivariate statistical models. For a clear application, readers may refer to principal component analysis (Chapter 11).

3.1.13 Spectral Decomposition

If $\mathbf{A}_{p\times p}$ is square symmetric, i.e., $\mathbf{A} = \mathbf{A}^T$ with $(\lambda_j, \mathbf{e}_j), j = 1,2,...,p$ pairs of eigenvalues and eigenvectors, $\mathbf{A}$ can be expressed as

$$\mathbf{A} = \mathbf{E}\Lambda\mathbf{E}^T = \sum_{j=1}^{p} \lambda_j \mathbf{e}_j \mathbf{e}_j^T \tag{3.28}$$

where the eigenvectors $\mathbf{e}_j$ are scaled to unit length.

Example 3.19

Consider the following matrix and explain spectral decomposition

$$\mathbf{A} = \begin{bmatrix} 4 & 5 \\ 5 & 9 \end{bmatrix}$$

Solution:

To obtain $(\lambda_j, \mathbf{e}_j), j = 1,2,$ we require doing the following:

$$|\mathbf{A} - \lambda\mathbf{I}| = 0$$

$$\begin{aligned}
&\Rightarrow \begin{vmatrix} 4-\lambda & 5 \\ 5 & 9-\lambda \end{vmatrix} = 0 \\
&\Rightarrow (4-\lambda)(9-\lambda) - 25 = 0 \\
&\Rightarrow 36 - 4\lambda - 9\lambda + \lambda^2 - 25 = 0 \\
&\Rightarrow \lambda^2 - 13\lambda + 11 = 0 \\
&\therefore \lambda = \frac{13 \pm \sqrt{169 - (4\times 11)}}{2\cdot 1} \\
&\quad = \frac{13 \pm \sqrt{125}}{2} = \frac{13 \pm 5\sqrt{5}}{2} \\
&\quad = \frac{13 \pm 11.18}{2} \\
&\therefore \lambda_1 = 12.09 \\
&\quad \lambda_2 = 0.91
\end{aligned}$$

To find out $\mathbf{e}_1$, we use λ_1 as below.

$$(\mathbf{A} - \lambda_1\mathbf{I})\mathbf{e}_1 = \mathbf{0}$$

$$\therefore \begin{bmatrix} 4-12.09 & 5 \\ 5 & 9-12.09 \end{bmatrix} \begin{bmatrix} e_{11} \\ e_{12} \end{bmatrix} = \begin{bmatrix} 0 \\ 0 \end{bmatrix}$$

Which yields,

$$-8.09e_{11} + 5e_{12} = 0 \text{ and}$$
$$5e_{11} - 3.09e_{12} = 0$$

So, $e_{12} = \frac{8.09}{5}e_{11} = 1.618e_{11}$

Let $e_{11} = 1$, arbitrarily. Then $e_{12} = 1.618$ *or* $\mathbf{e}_1 = \begin{bmatrix} e_{11} \\ e_{12} \end{bmatrix} = \begin{bmatrix} 1 \\ 1.618 \end{bmatrix}$

Now,

$$\mathbf{e}_1{}^T\mathbf{e}_1 = \begin{bmatrix} 1 & 1.618 \end{bmatrix}\begin{bmatrix} 1 \\ 1.618 \end{bmatrix} = 1 + 2.617 = 3.617$$

So, the unit length eigenvector $\mathbf{e}_1$ is $\mathbf{e}_1^1 = \frac{1}{\sqrt{3.617}}\begin{bmatrix} 1 \\ 1.618 \end{bmatrix}$

We find similarly, the second eigenvector $\mathbf{e}_2$.

$$\begin{bmatrix} 4-0.91 & 5 \\ 5 & 9-0.91 \end{bmatrix}\begin{bmatrix} e_{21} \\ e_{22} \end{bmatrix} = \begin{bmatrix} 0 \\ 0 \end{bmatrix}$$

which yields

$$3.09e_{21} + 5e_{22} = 0 \text{ and}$$
$$5e_{21} + 8.09e_{22} = 0$$
$$or \quad e_{21} = -\frac{8.09}{5}e_{22} = -1.618e_{22}.$$

Again taking arbitrarily $e_{21} = 1$, we get $e_{22} = -0.618$

So, $\mathbf{e}_2 = \begin{bmatrix} e_{21} \\ e_{22} \end{bmatrix} = \begin{bmatrix} 1 \\ -0.618 \end{bmatrix}$

The unit length eigenvector of $\mathbf{e}_2$ is $= \mathbf{e}_2^1 = \frac{1}{\sqrt{1.382}}\begin{bmatrix} 1 \\ -0.618 \end{bmatrix}$

Now, we reconstruct the matrix **A** using $(\lambda_1, \mathbf{e}_1^1)$ and $(\lambda_2, \mathbf{e}_2^1)$ in Equation 3.28.

$$
\begin{aligned}
\mathbf{A}_{2\times2} &= \sum_{j=1}^{2}\lambda_j(\mathbf{e}_j^1)(\mathbf{e}_j^1)^T \\
&= \lambda_1(\mathbf{e}_1^1)(\mathbf{e}_1^1)^T + \lambda_2(\mathbf{e}_2^1)(\mathbf{e}_2^1)^T \\
&= \frac{12.09}{3.617}\begin{bmatrix}1\\1.618\end{bmatrix}\begin{bmatrix}1 & 1.618\end{bmatrix} + \frac{0.91}{1.382}\begin{bmatrix}1\\-0.618\end{bmatrix}\begin{bmatrix}1 & -0.618\end{bmatrix} \\
&= 3.34\begin{bmatrix}1 & 1.618\\1.618 & 2.617\end{bmatrix} + 0.659\begin{bmatrix}1 & -0.618\\-0.618 & 0.382\end{bmatrix} \\
&= \begin{bmatrix}3.34 & 5.40\\5.40 & 8.74\end{bmatrix} + \begin{bmatrix}0.659 & -0.40\\-0.40 & 0.251\end{bmatrix} \\
&= \begin{bmatrix}4.00 & 5.00\\5.00 & 9.00\end{bmatrix} = \mathbf{A}
\end{aligned}
$$

3.1.14 Singular Value Decomposition (SVD)

The prerequisite to eigenvalue and eigenvector factorization of a matrix **A** is that **A** must be square. But we often deal with rectangular matrices for which eigenvalues are undefined. Singular value decompositions (SVD) can factorize a rectangular matrix of the order $n \times p$.

Let **A** be a $n \times p$ rectangular matrix. SVD decompose **A** as

$$\mathbf{A}_{n\times p} = \mathbf{U}_{n\times n}\mathbf{S}_{n\times p}\mathbf{V}^T_{p\times p} \tag{3.29}$$

Where

$\mathbf{U} =$ a $n \times n$ orthogonal matrix whose entries are the eigenvectors of $\mathbf{AA}^T$
$\mathbf{V} =$ a $p \times p$ orthogonal matrix whose entries are the eigenvectors of $\mathbf{A}^T\mathbf{A}$
$\mathbf{S} =$ a $n \times p$ diagonal matrix with r singular values $s_1, s_2, \ldots, s_r$ on the diagonal of **S** if r is the rank of the matrix **A**. Note, $r \le \min(n, p)$. The remaining diagonal elements are zero.

The following steps are to be followed to obtain singular values:

Step 1: Compute $\mathbf{AA}^T$ and obtain its eigenvalues, $\mathbf{\Lambda}$ and eigenvectors **U**.
Step 2: Compute $\mathbf{A}^T\mathbf{A}$ and obtain its eigenvalues $\mathbf{\Lambda}$ and eigenvectors **V**.

[Note that both $\mathbf{AA}^T$ and $\mathbf{A}^T\mathbf{A}$ have same eigenvalues, $\mathbf{\Lambda}$ as the eigenvalues of a square matrix are equal to the eigenvalues of its transpose.]

Step 3: The singular values of A is the square roots of $\mathbf{\Lambda}$. That is $\mathbf{S} = \mathbf{\Lambda}^{1/2}$ (see the proof below).

From Equation 3.29,

$$
\begin{aligned}
&\mathbf{A} = \mathbf{USV}^T \text{ so, } \mathbf{AA}^T = \left(\mathbf{USV}^T\right)\left(\mathbf{VS}^T\mathbf{U}^T\right) = \mathbf{USV}^T\mathbf{VS}^T\mathbf{U}^T = \mathbf{USS}^T\mathbf{U}^T \\
&as \;\; \mathbf{V}^T\mathbf{V} = \mathbf{VV}^T = I, \text{(note, } \mathbf{V} \text{ is orthogonal).}
\end{aligned}
$$

Similarly, we can prove $\mathbf{A}^T\mathbf{A} = \mathbf{VS}^T\mathbf{SV}^T$.

As Λ is the eigenvalue matrix (diagonal) of both $\mathbf{AA}^T$ and $\mathbf{A}^T\mathbf{A}$, and from spectral decomposition (Equation 3.28) we can write $\mathbf{AA}^T = \mathbf{U\Lambda U}^T$ and $\mathbf{A}^T\mathbf{A} = \mathbf{V\Lambda V}^T$, which essentially reveals that $\Lambda = \mathbf{SS}^T = \mathbf{S}^T\mathbf{S}$ where $\mathbf{S}, \mathbf{S}^T$ and Λ are diagonal matrices. If $s_1, s_2, ..., s_r$, are the diagonal elements of $\mathbf{S}$, then $s_1^2, s_2^2, ..., s_r^2$ will be the diagonal elements of $\mathbf{SS}^T$ as well as $\mathbf{S}^T\mathbf{S}$. There will be $(n-r)$ and $(p-r)$ zeros in the diagonal of the $\mathbf{SS}^T$ and $\mathbf{S}^T\mathbf{S}$, respectively. Further, Λ contains r non-zero eigenvalues. So, we can write $s_j^2 = \lambda_j, j = 1,2,...,r$.

Example 3.20

Obtain singular values for the matrix given below.

$$\mathbf{A} = \begin{bmatrix} 1 & 0 & 2 \\ 3 & 1 & 0 \end{bmatrix}$$

Solution:

Step 1: Compute $\mathbf{AA}^T$ and find eigenvalues and eigenvectors

$$\mathbf{AA}^T = \begin{bmatrix} 1 & 0 & 2 \\ 3 & 1 & 0 \end{bmatrix} \begin{bmatrix} 1 & 3 \\ 0 & 1 \\ 2 & 0 \end{bmatrix} = \begin{bmatrix} 5 & 3 \\ 3 & 10 \end{bmatrix}$$

$$|\mathbf{AA}^T| = 5 \times 10 - 3^2 \neq 0$$

To find eigenvalues, we put $|\mathbf{AA}^T - \lambda\mathbf{I}| = 0$. So,

$$\begin{vmatrix} 5-\lambda & 3 \\ 3 & 10-\lambda \end{vmatrix} = 0$$
$$\Rightarrow (5-\lambda)(10-\lambda) - 9 = 0$$
$$\Rightarrow 50 - 15\lambda + \lambda^2 - 9 = 0$$
$$\Rightarrow \lambda^2 - 15\lambda + 41 = 0$$

$$\text{So, } \lambda = \frac{+15 \pm \sqrt{225 - 4\cdot 1\cdot 41}}{2\cdot 1} = \frac{15 \pm \sqrt{61}}{2}$$

$$or,\ \lambda_1 = \frac{15+\sqrt{61}}{2} = 11.40, \text{ and}$$
$$\lambda_2 = \frac{15-\sqrt{61}}{2} = 3.60$$

Step 2: Compute $\mathbf{A}^T\mathbf{A}$ and its eigenvalues

$$\mathbf{A}^T\mathbf{A} = \begin{bmatrix} 1 & 3 \\ 0 & 1 \\ 2 & 0 \end{bmatrix} \begin{bmatrix} 1 & 0 & 2 \\ 3 & 1 & 0 \end{bmatrix} = \begin{bmatrix} 10 & 3 & 2 \\ 3 & 1 & 0 \\ 2 & 0 & 4 \end{bmatrix}$$

Putting $|\mathbf{A}^T\mathbf{A}-\lambda\mathbf{I}|=0$, we get

$$\begin{bmatrix} 10-\lambda & 3 & 2 \\ 3 & 1-\lambda & 0 \\ 2 & 0 & 4-\lambda \end{bmatrix} = 0$$

$$\Rightarrow (10-\lambda)\begin{vmatrix} 1-\lambda & 0 \\ 0 & 4-\lambda \end{vmatrix} - 3\begin{vmatrix} 3 & 0 \\ 2 & 4-\lambda \end{vmatrix} + 2\begin{vmatrix} 3 & 1-\lambda \\ 2 & 0 \end{vmatrix} = 0$$
$$\Rightarrow (10-\lambda)(1-\lambda)(4-\lambda) - 3\times 3(4-\lambda) - 2\times 2(1-\lambda) = 0$$
$$\Rightarrow (10-\lambda)(4-\lambda-4\lambda+\lambda^2) - 9(4-\lambda) + 4(\lambda-1) = 0$$
$$\Rightarrow 40-10\lambda-40\lambda+10\lambda^2-4\lambda+\lambda^2+4\lambda^2-\lambda^3-36+9\lambda-4+4\lambda = 0$$
$$\Rightarrow -\lambda^3+10\lambda^2+5\lambda^2-50\lambda-4\lambda+9\lambda+4\lambda = 0$$
$$\Rightarrow -\lambda^3+15\lambda^2-41\lambda = 0$$
$$\Rightarrow -\lambda(\lambda^2-15\lambda+41) = 0$$

Therefore, either $\lambda = 0$ or $\lambda^2-15\lambda+41=0$. Considering $\lambda^2-15\lambda+41=0$, we get

$$\lambda_1 = \frac{15+\sqrt{61}}{2} = 11.40$$
$$\lambda_2 = \frac{15-\sqrt{61}}{2} = 3.60$$

Therefore, the three roots of the characteristic equation obtained above are $\lambda_1 = 11.40, \lambda_2 = 3.60, \lambda_3 = 0$. So, the singular values are $\sqrt{11.40}, \sqrt{3.60}$, and 0, i.e., 3.38, 1.90 and 0. The resultant singular value matrix is $\mathbf{S}_{2\times 3} = \begin{bmatrix} 3.38 & 0 & 0 \\ 0 & 1.90 & 0 \end{bmatrix}$.

3.1.15 Positive Definite Matrices

In order to understand positive definite matrices, we need to know the quadratic form of a matrix. Let's say there is a 2 × 2 square matrix **A** and 2 × 1 column vector **x** as given below.

$$\mathbf{A} = \begin{bmatrix} a_{11} & a_{12} \\ a_{21} & a_{22} \end{bmatrix} \text{ and } \mathbf{x} = \begin{bmatrix} x_1 \\ x_2 \end{bmatrix}$$

Let us compute $\mathbf{x}^T\mathbf{A}\mathbf{x}$.

$$\mathbf{x}^T\mathbf{A}\mathbf{x} = \begin{bmatrix} x_1 & x_2 \end{bmatrix}\begin{bmatrix} a_{11} & a_{12} \\ a_{21} & a_{22} \end{bmatrix}\begin{bmatrix} x_1 \\ x_2 \end{bmatrix}$$
$$= \begin{bmatrix} x_1 & x_2 \end{bmatrix}\begin{bmatrix} a_{11}x_1 + a_{12}x_2 \\ a_{21}x_1 + a_{22}x_2 \end{bmatrix}$$
$$= a_{11}x_1^2 + a_{12}x_1x_2 + a_{21}x_1x_2 + a_{22}x_2^2$$
$$= a_{11}x_1^2 + (a_{12}+a_{21})x_1x_2 + a_{22}x_2^2$$

The right hand side of the above equation is the quadratic form and can be expressed as the well-known quadratic form in two dimensions,

$$ax^2 + 2bxy + cy^2, \text{ where } a = a_{11},\ b = \frac{a_{12} + a_{21}}{2},\ c = a_{22},\ x_1 = x \text{ and } x_2 = y.$$

So, we can say $\mathbf{x}^T\mathbf{Ax}$ represents the quadratic form of **A**.

Now, if we take a generalized square matrix **A** of the order $p \times p$ and a generalized vector **x** of the order $p \times 1$, the quadratic form of **A** is

$$\mathbf{x}^T\mathbf{Ax} = \begin{bmatrix} x_1 & x_2 & \dots & x_p \end{bmatrix} \begin{bmatrix} a_{11} & a_{12} & \dots & a_{1p} \\ a_{21} & a_{22} & \dots & a_{2p} \\ \vdots & \vdots & \ddots & \vdots \\ a_{p1} & a_{p2} & \dots & a_{pp} \end{bmatrix} \begin{bmatrix} x_1 \\ x_2 \\ \vdots \\ x_p \end{bmatrix} \tag{3.30}$$

$$= \sum_{j=1}^{p}\sum_{k=1}^{p} a_{jk}x_j x_k$$

For symmetric square matrices $a_{jk} = a_{kj}$, Equation 3.30 takes the pure quadratic form as

$$\mathbf{x}^T\mathbf{Ax} = a_{11}x_1^2 + 2a_{12}x_1x_2 + 2a_{13}x_1x_3 + \dots + a_{pp}x_p^2 \tag{3.31}$$

If $p = 2$,

$$\mathbf{x}^T\mathbf{Ax} = a_{11}x_1^2 + 2a_{12}x_1x_2 + a_{22}x_2^2$$

$\mathbf{x}^T\mathbf{Ax}$ possesses certain properties. Its value can be positive, zero, or negative. Accordingly, the matrix **A** is defined (see Table 3.4).

It should be noted here that the condition $\mathbf{x}^T\mathbf{Ax} > 0$ is one of the criteria required for a symmetric matrix to become positive definite. The necessary and sufficient conditions for a matrix to become positive definite are (Strang, 2003):

(i) $\mathbf{x}^T\mathbf{Ax} > 0$ for $\mathbf{x} \neq \mathbf{0}$.
(ii) All of the eigenvalues ($\lambda_j, j = 1,2,..,p$) of **A** must be greater than zero.
(iii) All the upper left sub-matrices of **A** must have positive determinants.
(iv) All pivots must be greater than zero.

For clarification of the conditions (iii) and (iv), interested readers may follow Chapter 6 of Strang (2003).

TABLE 3.4
The Properties of the Symmetric Square Matrix A Based on the Value of $\mathbf{x}^T\mathbf{Ax}$

Value of $\mathbf{x}^T\mathbf{Ax}$	The Symmetric Square Matrix A is
$\mathbf{x}^T\mathbf{Ax}>0$	Positive definite
$\mathbf{x}^T\mathbf{Ax} \geq 0$	Positive semi-definite
$\mathbf{x}^T\mathbf{Ax}<0$	Negative definite
$\mathbf{x}^T\mathbf{Ax} \leq 0$	Negative semi-definite

The quadratic forms of a matrix as well as the property of positive definiteness play a central role in multivariate statistics. In Chapter 5, we will introduce the concept of statistical distance and the well-known multivariate normal distribution. We will demonstrate the use of $\mathbf{x}^T\mathbf{A}\mathbf{x}$ there.

3.2 METHODS OF LEAST SQUARES

The primary computational issue in multivariate statistical modeling is parameter estimation. For example, in multiple or multivariate regression, we need to estimate the regression coefficients; in the path model, we need to estimate path coefficients; and so on for the other models described in the book. One of the most widely used methods for parameter estimation is least squares. In this section, we describe some of the most widely used methods of least squares.

Let us consider a situation when we observe the behavior of two variables X and Y, where Y depends on X. The behavior can be best represented by a two-dimensional scatter diagram. Further, assume that we have only two observations (x_1, y_1) and (x_2, y_2). In Figure 3.11, these two observations are denoted by P and Q in $X-Y$ coordinates. A straight line PQ can be drawn whose equation can be derived as below.

$$\frac{y-y_1}{y_1-y_2}=\frac{x-x_1}{x_1-x_2}$$

$$\Rightarrow y-y_1=\frac{y_1-y_2}{x_1-x_2}\cdot(x-x_1)$$

$$\Rightarrow y=y_1-x_1\cdot\frac{y_1-y_2}{x_1-x_2}+\frac{y_1-y_2}{x_1-x_2}\cdot x=\frac{y_1(x_1-x_2)-x_1(y_1-y_2)}{x_1-x_2}+\frac{y_1-y_2}{x_1-x_2}\cdot x$$

As x_1, x_2, y_1 *and* y_2 are constants, we write

$$y=\beta_0+\beta_1 x \tag{3.32}$$

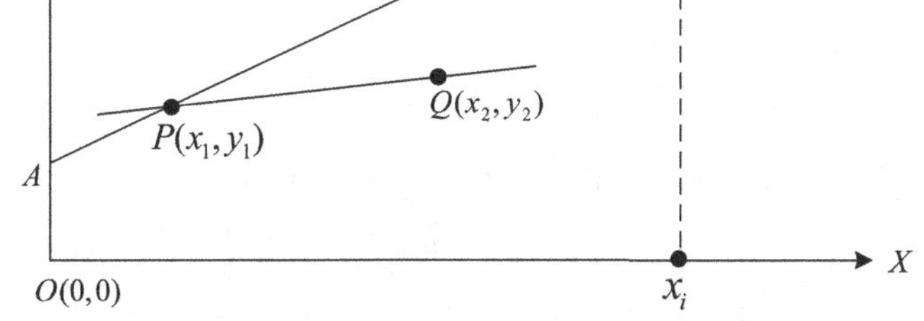

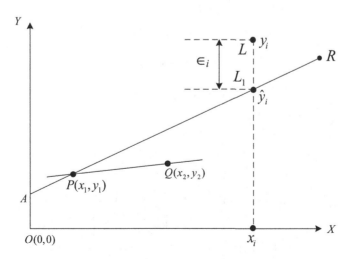

FIGURE 3.11 Linear fit between Y and X. The line AR represents the linear fit equation. OA is the intercept. x_i and y_i are the i-th observation on X and Y, respectively. $\hat{y}_i$ is the fitted value of Y when X = x_i.

Where,

$$\beta_0 = \frac{y_1(x_1 - x_2) - x_1(y_1 - y_2)}{x_1 - x_2} \text{ and } \beta_1 = \frac{y_1 - y_2}{x_1 - x_2}$$

Equation 3.32 is the equation of a straight line where β_0 is known as the intercept and β_1 is the slope of the straight line. As is seen, a perfect straight line can be fitted with two points. Now, if more observations, say n, are collected, what will happen to the line fitted? The answer is simple. We will not get a perfect line. For example, if we collect $n(=15)$ observations as shown in Figure 3.11, the line PQ doesn't represent the relationship perfectly. Rather, a better fit could be the line AR. How do we get the line AR ? Is AR the best linear fit? How much error is encountered in obtaining the relationship between Y and X? The method of least square estimation (LSE) answers these questions.

3.2.1 Ordinary Least Squares (OLS)

The algebraic procedure of the method of least squares was first published by Legendre in 1805 and Gauss justified it as a statistical procedure in 1809 (Bjorck, 1996). It is a method to obtain linear fit for an overdetermined linear system of equations where the number of unknowns are less than the number of equations. For example, for a linear fit between Y and X (Figure 3.11) there are two unknowns β_0 and β_1 (see Equation 3.32). If we collect $n(> 2)$ observations, we get more than two equations for two unknowns and the resultant linear system of equations is overdetermined. For an overdetermined linear system of equations, there cannot be perfect fit in reality. An error is inevitable. So, based on Gauss-Markov theorem, Equation 3.32 can be written as

$$y_i = \beta_0 + \beta_1 x_i + e_i,\ i = 1,2,\ldots.,n \tag{3.33}$$

Where, ϵ_i is the error for the i-th observation. For example, in Figure 3.11, LL_1 represents the i-th error, ϵ_i. ϵ_i is a random variable with mean zero and standard deviation σ_i^2.

For n observations, we get n equations:

$$\begin{aligned} y_1 &= \beta_0 + \beta_1 x_1 + \epsilon_1 \\ y_2 &= \beta_0 + \beta_1 x_2 + \epsilon_2 \\ &\vdots \\ y_i &= \beta_0 + \beta_1 x_i + \epsilon_i \\ &\vdots \\ y_n &= \beta_0 + \beta_1 x_n + \epsilon_n \end{aligned}$$

In matrix form:

$$\begin{bmatrix} y_1 \\ y_2 \\ \vdots \\ y_i \\ \vdots \\ y_n \end{bmatrix}_{n\times 1} = \begin{bmatrix} 1 & x_1 \\ 1 & x_2 \\ \vdots & \vdots \\ 1 & x_i \\ \vdots & \vdots \\ 1 & x_n \end{bmatrix}_{n\times 2} \begin{bmatrix} \beta_0 \\ \beta_1 \end{bmatrix}_{2\times 1} + \begin{bmatrix} \epsilon_1 \\ \epsilon_2 \\ \vdots \\ \epsilon_i \\ \vdots \\ \epsilon_n \end{bmatrix}_{n\times 1}$$

$$\text{or} \quad \mathbf{y} = \mathbf{X}\boldsymbol{\beta} + \boldsymbol{\epsilon} \tag{3.34}$$

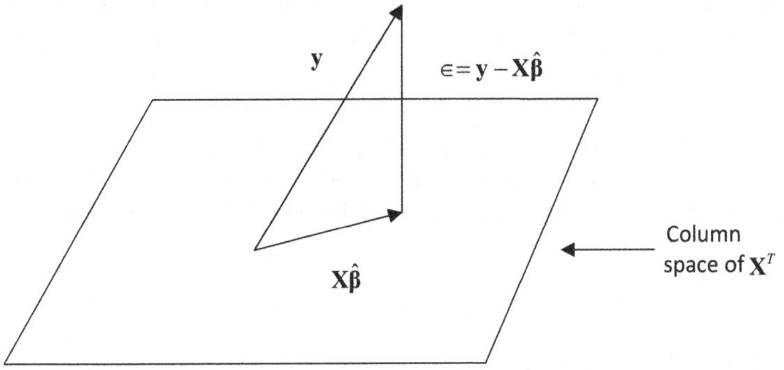

FIGURE 3.12 Geometric interpretation of least square solution $\hat{\boldsymbol{\beta}}$. The rhombus represents the column space for $\mathbf{X}^{\mathrm{T}}$. $\mathbf{X}\hat{\boldsymbol{\beta}}$ is the projection of $\mathbf{y}$ in the space created by $\mathbf{X}^{\mathrm{T}}$. The error vector ($\in = \mathbf{y} - \mathbf{X}\hat{\boldsymbol{\beta}}$) is perpendicular to the column space of $\mathbf{X}^{\mathrm{T}}$.

$\mathbf{y} = n \times 1$ vector of dependent observations (known)
$X = n \times 2$ matrix of independent observations including constant (known)
$\boldsymbol{\beta} = 2 \times 1$ vector of unknown parameters
$\in =$ Random error vector of the order $n \times 1$ with mean zero and variance σ^2, i.e., Var ($\in$) = $\sigma^2 \mathrm{I}$

Method of least squares minimize the squared error term, i.e., $\in^T \in$.

$$\text{Now,} \in^T \in = (\mathbf{y} - \mathbf{X}\boldsymbol{\beta})^T (\mathbf{y} - \mathbf{X}\boldsymbol{\beta}) \tag{3.35}$$

$\in^T \in$ is the squared distance form $\mathbf{y}$ to the point $\mathbf{X}\boldsymbol{\beta}$. As least square solution $\hat{\boldsymbol{\beta}}$ minimizes $\in^T \in$, $\mathbf{X}\hat{\boldsymbol{\beta}}$ has the least distance from $\mathbf{y}$ where $\mathbf{X}\hat{\boldsymbol{\beta}}$ is the projection of $\mathbf{y}$ in the space created by $\mathbf{X}^T$. Geometrically, least squares solution can be explained by Figure 3.12.

For the two-variable situation (Y and X), $\mathbf{X}$ has two columns

$\mathbf{x}_0 = [1, 1, 1, \cdots, 1]^T$ and $\mathbf{x}_1 = [x_1, x_2, x_3, \cdots, x_n]^T$. These two columns must be perpendicular to the error vector $\in$. Hence,

$$
\begin{aligned}
&\mathbf{x}_0{}^T (\mathbf{y} - \mathbf{X}\hat{\boldsymbol{\beta}}) = 0 \\
&\mathbf{x}_1{}^T (\mathbf{y} - \mathbf{X}\hat{\boldsymbol{\beta}}) = 0 \\
&or \quad \begin{bmatrix} \mathbf{x}_0{}^T \\ \mathbf{x}_1{}^T \end{bmatrix} \begin{bmatrix} \mathbf{y} - \mathbf{X}\hat{\boldsymbol{\beta}} \end{bmatrix} = 0 \\
&or \quad \mathbf{X}^T \left(\mathbf{y} - \mathbf{X}\hat{\boldsymbol{\beta}}\right) = 0. \qquad (3.36) \\
&\Rightarrow \mathbf{X}^T\mathbf{y} = \mathbf{X}^T\mathbf{X}\hat{\boldsymbol{\beta}} \\
&\Rightarrow (\mathbf{X}^T\mathbf{X})^{-1}\mathbf{X}^T\mathbf{y} = (\mathbf{X}^T\mathbf{X})^{-1}(\mathbf{X}^T\mathbf{X})\hat{\boldsymbol{\beta}} \\
&\Rightarrow \hat{\boldsymbol{\beta}} = (\mathbf{X}^T\mathbf{X})^{-1}\mathbf{X}^T\mathbf{y}
\end{aligned}
$$

Equation 3.36 can also be obtained by taking first-order derivative of Equation 3.35 with respect to $\boldsymbol{\beta}$, i.e., $\partial(\in^T \in)/\partial\boldsymbol{\beta}$ and equating $\partial(\in^T \in)/\partial\boldsymbol{\beta} = 0$.

Equation 3.36 is a general equation where $\mathbf{X}$ can be of any order matrix, say $n \times (p+1)$, $\mathbf{y}$ is a $n \times 1$ vector and $\hat{\boldsymbol{\beta}}$ will then be $(p+1) \times 1$ vector. When there is only one variable, X, $\mathbf{X}\hat{\boldsymbol{\beta}}$ represents a line but for higher dimensions (2 or more X variables), $\mathbf{X}\hat{\boldsymbol{\beta}}$ represents a surface (plane). In summary, OLS provides orthogonal projection of dependent data vector ($\mathbf{y}$) onto the space defined by the independent variable vectors.

Some important information emerges from Equation 3.36. The matrix $\mathbf{X}^T\mathbf{X}$ is symmetric. $\hat{\boldsymbol{\beta}}$ is the projection multiplier and $\mathbf{X}\hat{\boldsymbol{\beta}} = \mathbf{X}(\mathbf{X}^T\mathbf{X})^{-1}\mathbf{X}^T\mathbf{y}$ is the projection of $\mathbf{y}$ on X. The equation $\mathbf{X}^T\mathbf{X}\hat{\boldsymbol{\beta}} = \mathbf{X}^T\mathbf{y}$ is called normal equations. The estimate $\hat{\boldsymbol{\beta}}$ is known as ordinary least squares (OLS) estimate.

The foregoing discussions answer the question, how do we obtain the line AR shown in Figure 3.11? The other two questions remain unanswered are: (i) Is AR the best linear fit? (ii) How much error is encountered in obtaining the linear fit? These will be discussed in Chapter 8.

Example 3.21

Compute OLS estimate for the following set of observations

$$\mathbf{y}^T = \begin{bmatrix} 5 & 10 & 15 \end{bmatrix}$$

$$\mathbf{X}^T = \begin{bmatrix} 1 & 1 & 1 \\ 1 & 3 & 5 \end{bmatrix}$$

Solution:

The OLS estimate in $\hat{\boldsymbol{\beta}} = (\mathbf{X}^T\mathbf{X})^{-1}\mathbf{X}^T\mathbf{y}$

Step 1: Compute $\mathbf{X}^T\mathbf{X}$

$$\mathbf{X}^T\mathbf{X} = \begin{bmatrix} 1 & 1 & 1 \\ 1 & 3 & 5 \end{bmatrix}\begin{bmatrix} 1 & 1 \\ 1 & 3 \\ 1 & 5 \end{bmatrix}$$

$$= \begin{bmatrix} 3 & 9 \\ 9 & 35 \end{bmatrix}$$

Step 2: Compute $(\mathbf{X}^T\mathbf{X})^{-1}$

$$(\mathbf{X}^T\mathbf{X})^{-1} = \frac{1}{|\mathbf{X}^T\mathbf{X}|} \, adj\ (\mathbf{X}^T\mathbf{X})$$

Now,

$$|\mathbf{X}^T\mathbf{X}| = \begin{vmatrix} 3 & 9 \\ 9 & 35 \end{vmatrix} = 105 - 81 = 24 \text{ and } adj\ (\mathbf{X}^T\mathbf{X}) = \begin{bmatrix} 35 & -9 \\ -9 & 3 \end{bmatrix}$$

$$(\mathbf{X}^T\mathbf{X})^{-1} = \frac{1}{24}\begin{bmatrix} 35 & -9 \\ -9 & 3 \end{bmatrix}$$

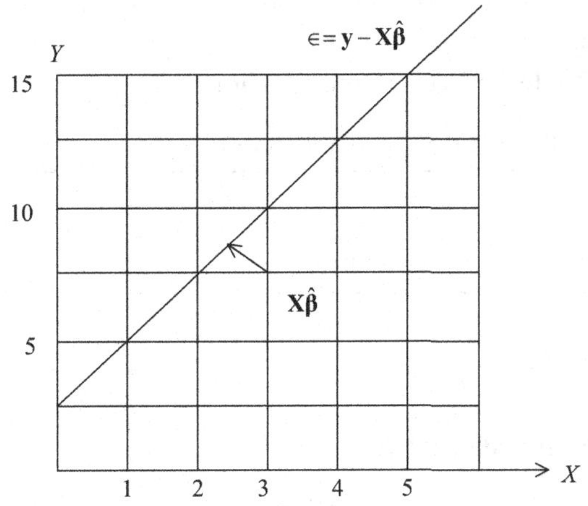

FIGURE 3.13 OLS linear fit for Example 3.21.

Step 3: Compute $\mathbf{X}^T\mathbf{y}$

$$\mathbf{X}^T\mathbf{y} = \begin{bmatrix} 1 & 1 & 1 \\ 1 & 3 & 5 \end{bmatrix}_{2\times 3} \begin{bmatrix} 5 \\ 10 \\ 15 \end{bmatrix}_{2\times 1} = \begin{bmatrix} 30 \\ 110 \end{bmatrix}$$

Step 4: Compute $\hat{\boldsymbol{\beta}} = (\mathbf{X}^T\mathbf{X})^{-1}\mathbf{X}^T\mathbf{y}$

$$\hat{\boldsymbol{\beta}} = \frac{1}{24}\begin{bmatrix} 35 & -9 \\ -9 & 3 \end{bmatrix}\begin{bmatrix} 30 \\ 110 \end{bmatrix} = \frac{1}{24}\begin{bmatrix} 60 \\ 60 \end{bmatrix} = \begin{bmatrix} 2.5 \\ 2.5 \end{bmatrix}$$

So, $Y = 2.5 + 2.5X$

The projection of $\mathbf{y}$ on $\mathbf{X}$ is $\mathbf{X}\hat{\boldsymbol{\beta}}$ which is

$$\mathbf{X}\hat{\boldsymbol{\beta}} = \begin{bmatrix} 1 & 1 \\ 1 & 3 \\ 1 & 5 \end{bmatrix}_{3\times 2} \begin{bmatrix} 2.5 \\ 2.5 \end{bmatrix}_{2\times 1} = \begin{bmatrix} 5 \\ 10 \\ 15 \end{bmatrix}_{x_{1\times p}}$$

The error in the estimate $= \hat{\in}$

$$\hat{\in} = \mathbf{y} - \mathbf{X}\hat{\boldsymbol{\beta}} = \begin{bmatrix} 5 \\ 10 \\ 15 \end{bmatrix} - \begin{bmatrix} 5 \\ 10 \\ 15 \end{bmatrix} = \begin{bmatrix} 0 \\ 0 \\ 0 \end{bmatrix}$$

Now, if we plot the OLS linear fit, we get Figure 3.13.

TABLE 3.5
Data Table for WLS Demonstration

Observation no. (i)	x_0(constant)	x	y
1	1	x_1	y_1
2	1	x_2	y_2
3	1	x_3	y_3

TABLE 3.6
Weighted Data Matrix

Observation No. (i)	x_0 (Constant)	x	y
1	$\sqrt{w_1}$	$\sqrt{w_1 x_1}$	$\sqrt{w_1 y_1}$
2	$\sqrt{w_2}$	$\sqrt{w_2 x_2}$	$\sqrt{w_2 y_2}$
3	$\sqrt{w_3}$	$\sqrt{w_3 x_3}$	$\sqrt{w_3 y_3}$

3.2.2 Weighted Least Squares (WLS)

In OLS, each of the data points (observation vectors) carries equal importance in determining the OLS estimates. This is done with the assumption that each data point equally provides precise information on the relationship explained. This may not be true for all phenomena we explain. For example, it may happen that the variability in Y may differ across the value of X, which is known as heteroscedasticity in linear regression. Under such conditions, it is desirable to weight each data point based on the information they provide. Reliable or precise data points may be given more weight than unreliable or less precise information. The OLS so adjusted with weights is known as weighted least squares (WLS).

How does WLS work? We show this with an example. Let's say we have collected three observations on two variables, X and Y, to get a linear fit $X\hat{\beta} = Y$. The data is shown in Table 3.5.

There are three data points for $i = 1, 2,$ and 3, respectively, called row vectors. We assume that each row (data point) contains different information, so they need to be weighted. Let the weight vector is $\sqrt{w_i}, i = 1,2,3$. Then the weighted data matrix will be as given in Table 3.6.

Using Equation 3.34, we can write $\mathbf{y} = \mathbf{X}\boldsymbol{\beta} + \epsilon$

or

$$\begin{bmatrix} \sqrt{w_1 y_1} \\ \sqrt{w_2 y_2} \\ \sqrt{w_3 y_3} \end{bmatrix} = \begin{bmatrix} \sqrt{w_1} & \sqrt{w_1 x_1} \\ \sqrt{w_2} & \sqrt{w_2 x_2} \\ \sqrt{w_3} & \sqrt{w_3 x_3} \end{bmatrix} \begin{bmatrix} \beta_0 \\ \beta_1 \end{bmatrix} + \begin{bmatrix} \sqrt{w_1}\ \epsilon_1 \\ \sqrt{w_2}\ \epsilon_2 \\ \sqrt{w_3}\ \epsilon_3 \end{bmatrix}$$

With little matrix manipulation, the above equation can be written as

$$\begin{bmatrix} \sqrt{w_1} & 0 & 0 \\ 0 & \sqrt{w_2} & 0 \\ 0 & 0 & \sqrt{w_3} \end{bmatrix} \begin{bmatrix} y_1 \\ y_2 \\ y_3 \end{bmatrix} = \begin{bmatrix} \sqrt{w_1} & 0 & 0 \\ 0 & \sqrt{w_2} & 0 \\ 0 & 0 & \sqrt{w_3} \end{bmatrix} \begin{bmatrix} 1 & x_1 \\ 1 & x_2 \\ 1 & x_3 \end{bmatrix} \begin{bmatrix} \beta_0 \\ \beta_1 \end{bmatrix} + \begin{bmatrix} \sqrt{w_1} & 0 & 0 \\ 0 & \sqrt{w_2} & 0 \\ 0 & 0 & \sqrt{w_3} \end{bmatrix} \begin{bmatrix} \epsilon_1 \\ \epsilon_2 \\ \epsilon_3 \end{bmatrix}$$

Or

$$\mathbf{W}^{1/2}\mathbf{y} = \mathbf{W}^{1/2}\mathbf{X}\boldsymbol{\beta} + \mathbf{W}^{1/2}\epsilon \tag{3.37}$$

Where $\mathbf{W}^{1/2}$ is a $n \times n(n = 3$, in the example given above) diagonal matrix with weight $\sqrt{w_i}$ as the $i-th$ diagonal element. Equation 3.37 is called weighted linear relations. Here, $\text{Var}(\epsilon) = \sigma^2\,\mathbf{W}^{-1}$.

Equation 3.37 can be written as $\mathbf{y}_\mathrm{w} = \mathbf{X}_\mathrm{w}\boldsymbol{\beta} + \epsilon_\mathrm{w}$ where $\mathbf{y}_\mathrm{w} = \mathbf{W}^{1/2}\mathbf{y}$, $\mathbf{X}_\mathrm{w} = \mathbf{W}^{1/2}\mathbf{X}$ and $\epsilon_\mathrm{w} = \mathbf{W}^{1/2}\epsilon$. Here, $\text{Var}(\epsilon_\mathrm{w}) = \sigma^2\mathbf{I}$. So, for this equation, we can use OLS (Equation 3.36) to estimate $\boldsymbol{\beta}$.

Now, using Equation 3.36, the estimate of $\boldsymbol{\beta}$, i.e., $\hat{\boldsymbol{\beta}}$ is $\hat{\boldsymbol{\beta}} = (\mathbf{X}_\mathrm{w}^T\mathbf{X}_\mathrm{w})^{-1}\mathbf{X}_\mathrm{w}^T\mathbf{y}_\mathrm{w}$

Again, $\mathbf{X}_\mathrm{w}^T = (\mathbf{W}^{1/2}\mathbf{X})^T = \mathbf{X}^T\mathbf{W}^{1/2}$, as $\mathbf{W}$ is a diagonal matrix.

$$\text{So, } \hat{\boldsymbol{\beta}} = (\mathbf{X}^T\mathbf{W}^{1/2}\mathbf{W}^{1/2}\mathbf{X})^{-1}\mathbf{X}^T\mathbf{W}^{1/2}\mathbf{W}^{1/2}\mathbf{y} = (\mathbf{X}^T\;\mathbf{W}\;\mathbf{X})^{-1}\;\mathbf{X}^T\;\mathbf{W}\mathbf{y} \tag{3.38}$$

The next question to be answered is how can we determine the weights $\sqrt{w_i}, i = 1,2,..,n$. It is a difficult task. If we have *a priori* knowledge on the pattern of weights for each data point, we should use the information. For example, if ϵ_i is proportional to x_i, then w_i could be x_i^{-1}. Similarly, if y_i is the average of n_i observations such as done in designed experiments, then w_i could be n_i. If $\epsilon_i, i = 1,2,..,n$ are independent with various σ_i^2, then $\sqrt{w_i}$ could be $1/\sigma_i$.

Example 3.22

Consider the data given in Example 3.21. If weights (w_i) are inversely proportional to x_i observations, compute WLS estimates.

Solution:

The following data are given

$$\mathbf{X} = \begin{bmatrix} 1 & 1 \\ 1 & 3 \\ 1 & 5 \end{bmatrix}, \mathbf{y} = \begin{bmatrix} 5 \\ 10 \\ 15 \end{bmatrix}, \mathbf{W} = \begin{bmatrix} 1 & 0 & 0 \\ 0 & 1/3 & 0 \\ 0 & 0 & 1/5 \end{bmatrix}$$

Please note that $w_i \propto \dfrac{1}{x_i}$ (given). So, we have considered $w_i = x_i^{-1}$

Step 1: Compute $\mathbf{X}_\mathrm{w}$ and $\mathbf{X}_\mathrm{w}^T$

$$\mathbf{X}_\mathrm{w} = \mathbf{W}^{1/2}\mathbf{X} = \begin{bmatrix} 1 & 0 & 0 \\ 0 & 1/\sqrt{3} & 0 \\ 0 & 0 & 1/\sqrt{5} \end{bmatrix}\begin{bmatrix} 1 & 1 \\ 1 & 3 \\ 1 & 5 \end{bmatrix} = \begin{bmatrix} 1 & 1 \\ 1/\sqrt{3} & \sqrt{3} \\ 1/\sqrt{5} & \sqrt{5} \end{bmatrix}$$

So,

$$\mathbf{X}_\mathrm{w}^T = \begin{bmatrix} 1 & 1/\sqrt{3} & 1/\sqrt{5} \\ 1 & \sqrt{3} & \sqrt{5} \end{bmatrix}$$

Step 2: Compute $(\mathbf{X}_W^T\mathbf{X}_W)^{-1}$

$$(\mathbf{X}_W^T\mathbf{X}_W) = \begin{bmatrix} 1 & 1/\sqrt{3} & 1/\sqrt{5} \\ 1 & \sqrt{3} & \sqrt{5} \end{bmatrix} \begin{bmatrix} 1 & 1 \\ 1/\sqrt{3} & \sqrt{3} \\ 1/\sqrt{5} & \sqrt{5} \end{bmatrix} = \begin{bmatrix} 1.533333 & 3 \\ 3 & 13 \end{bmatrix}$$

$$\left|\mathbf{X}_W^T\mathbf{X}_W\right| = \begin{vmatrix} 1.533333 & 3 \\ 3 & 9 \end{vmatrix} = 4.8$$

$$\left(\mathbf{X}_W^T\mathbf{X}_W\right)^{-1} = \frac{1}{\left|\mathbf{X}_W^T\mathbf{X}_W\right|} adj\left(\mathbf{X}_W^T\mathbf{X}_W\right)$$

$$= \begin{bmatrix} 1.875 & -0.625 \\ -0.625 & 0.319444 \end{bmatrix}$$

Step 3: Compute $\mathbf{X}_W^T\mathbf{y}_W$

$$\mathbf{X}_W^T Y_W = \begin{bmatrix} 34/3 \\ 30 \end{bmatrix}$$

Step 4: $\hat{\beta} = (\mathbf{X}_W^T\mathbf{X}_W)^{-1}\mathbf{X}_W^T\mathbf{y}_W$

$$\hat{\beta} = \begin{bmatrix} 2.5 \\ 2.5 \end{bmatrix}$$

If we use Equation 3.38, we get the same values for $\hat{\beta}$ which is exactly the same as obtained using OLS (see Example 3.21). It should be noted here that the number of observations (sample size) is only three, a very low value. So, there is no scope for checking the difference in variability of y across the observations of x. So, comparison cannot be made. Nevertheless, for large sample, WLS and OLS give different estimates unless $w_i = 1, i = 1,2,..,n$.

3.2.3 Iteratively Reweighted Least Squares (IRLS)

The main disadvantage of WLS is that we need to know the weights, w_i, $i = 1,2,..,n$, *a priori*. For example, it may so happen that the error variance σ_i, $i = 1,2,..,n$ varies across x_i, $i = 1,2,..,n$ but the values of σ_i^2 are not completely known. So, we cannot use the weight $\sqrt{w_i} = 1/\sigma_i$. Under such situation, it is reasonable to assume σ_i^2 as a linear function of x_i and the weights are adjusted iteratively using OLS estimates of β and γ where β relates y to x and γ captures the linear relationships between σ_i^2 and x_i. So, essentially, we have two linear equations. For example, in two dimensions,

$$y_i = \beta_0 + \beta_1 x_i + \varepsilon_i,$$
$$\sigma_i^2 = \gamma_0 + \gamma_1 x_i, \ and$$
$$\sqrt{w_i} = \frac{1}{\sigma_i}$$

Now, σ_i^2 is not known completely. As ϵ_i is assumed to follow $N(0, \sigma_i^2)$, σ_i^2 can be approximated to ϵ_i^2 for computation of γ and w_i. So, the second equation above becomes $\epsilon_i^2 = \gamma_0 + \gamma_1 x_i$.

The steps to be followed for IRLS are:

(i) Select $w_i = 1$
(ii) Compute $\hat{\beta}$ using OLS
(iii) Compute $\epsilon_i^2, i = 1,2,...,n$, the square of the residuals based on step (i)
(iv) Regress ϵ_i^2 on x to compute $\hat{\gamma}$
(v) Re-compute w_i and go to step (ii)
(vi) Repeat the process until convergence is achieved

3.2.4 Generalized Least Squares (GLS)

The method of least squares minimizes the squared error function as shown below. For example,

(i) OLS minimizes $(\mathbf{y} - \mathbf{X\beta})^T(\mathbf{y} - \mathbf{X\beta})$
(ii) WLS minimizes $(\mathbf{y} - \mathbf{X\beta})^T \mathbf{W}(\mathbf{y} - \mathbf{X\beta})$
(iii) IRLS minimizes $(\mathbf{y} - \mathbf{X\beta})^T \mathbf{W}^*(\mathbf{y} - \mathbf{X\beta})$

where $\mathbf{W}^*$ is a diagonal matrix of weights w_i, $i = 1,2,..,n$ and w_i is adjusted after every iteration.

One of the assumptions of OLS is that the error variance (σ_i^2), $\text{var}(\epsilon) = \sigma^2\mathbf{I}$ is equal across x_i, $i = 1,2,..,n$. So, each of the data points are weighted equally. The OLS can be written as WLS where $\mathbf{W} = \mathbf{I}$ (Identity matrix). WLS relates the assumption by proposing that the data point provides different level of information and hence, they should be weighted. Another important assumption of OLS is that the error terms are uncorrelated, i.e., have zero correlation (or covariance). Generalized least squares (GLS) relaxes the above assumption and minimizes

$$(\mathbf{y} - \mathbf{X\beta})^T \mathbf{\Sigma}_E^{-1}(\mathbf{y} - \mathbf{X\beta})$$

where $\mathbf{\Sigma}_E$ is the relative variance-covariance matrix between errors. By the word *relative*, we mean that their absolute values are not known. The GLS estimate $\hat{\boldsymbol{\beta}}$ is

$$\hat{\boldsymbol{\beta}} = \left(\mathbf{X}^T\mathbf{\Sigma}_E^{-1}\mathbf{X}\right)^{-1}\mathbf{X}\mathbf{\Sigma}_E^{-1}\mathbf{y} \tag{3.39}$$

The OLS, GLS, and IRLS are the variants of WLS. Recall Equation 3.38.

$$\hat{\boldsymbol{\beta}} = (\mathbf{X}^T\mathbf{W}\mathbf{X})^{-1}\mathbf{X}^T\mathbf{W}\mathbf{y}$$

(i) If $\mathbf{W} = \mathbf{I}, \hat{\boldsymbol{\beta}} = (\mathbf{X}^T\mathbf{X})^{-1}\mathbf{X}^T\mathbf{y}$, which is the traditional OLS estimate.
(ii) If $\mathbf{W} = \mathbf{\Sigma}_E^{-1}$, $\hat{\boldsymbol{\beta}} = \left(\mathbf{X}^T\mathbf{\Sigma}_E^{-1}\mathbf{X}\right)^{-1}\mathbf{X}\mathbf{\Sigma}_E^{-1}\mathbf{y}$, which is the GLS estimate.
(iii) IRLS differ from WLS in the computational aspects where the weight matrix **W** is updated iteratively.

There are more variants of least squares methods applied in different applications and different situations. Some of the most widely used methods, apart from those described here, are non-linear least squares (NLS), partial least squares (PLS), and two-stage least square (2SLS). For detailed discussions on the methods of least squares, readers may consult Gentle (2009), Bjorck (1996), and Carrol and Ruppert (1988). For 2SLS, see Chapter 10, Section 10.4.2.

3.3 MAXIMUM LIKELIHOOD METHOD

There are two general methods of parameter estimation, namely least squares estimation (LSE) and maximum likelihood estimation (MLE) (Myung, 2003). LSE is described is Section 3.2. In this section, we will describe MLE.

MLE was originally developed by R.A. Fisher in 1920s to fit probability models to observed data. However, it is appropriate for higher level models also. For example, in statistical modeling we collect sample data and assume a suitable model to explain the behavior of the population. In simple linear regression, we assume that the dependent variable Y is linearly dependent on independent variable X and the relationship model contains two parameters, intercept (β_0) and slope (β_1). The population model is $Y = \beta_0 + \beta_1 X + \in$. Now, if we collect data on X and Y, our aim is to find the best values of β_0 and β_1, that make the observations on X and Y the most likely. Further, in regression it is assumed that the error $\in$ follows normal distribution with mean zero and standard deviation σ; i.e., $\in\sim N(0,\sigma)$. MLE finds out the best possible normal distribution that makes the observed error data most likely. In summary, given a model function and the observations made, MLE provides the range of possible values of the parameters of the model (population). Myung (2003) and Lindsey (2006) have given a lucid explanation of MLE.

3.3.1 Probability Function

MLE assumes that the observations are random and follow a certain probability function. If the observations are continuous, the probability function is the probability density function (pdf) and for discrete observations, the probability function is the probability mass function (pmf). The MLE procedure is same for pmf and pdf. Let's say n random observations, $y_1, y_2, \ldots, y_n$ were obtained from a process having a probability function $f(\mathbf{y}/\boldsymbol{\theta})$, where $\boldsymbol{\theta}$ is the parameter of the distribution. $\boldsymbol{\theta}$ can be scalar or a vector. The joint probability function of $y_1, y_2, \ldots, y_n$ can be written as $f(y_1, y_2, \ldots, y_n/\boldsymbol{\theta})$. Now, if $y_1, y_2, \ldots, y_n$ are independent and identically distributed $(iid) f(y_1, y_2, \ldots, y_n/\boldsymbol{\theta})$ can be the multiple of the n individual probability functions, $f(y_i/\boldsymbol{\theta})$, $i = 1, 2, \ldots, n$. That is

$$\begin{aligned} f(y_1, y_2, \ldots y_n/\boldsymbol{\theta}) &= f(y_1/\boldsymbol{\theta}) \times f(y_2/\boldsymbol{\theta}) \times \ldots \times f(y_n/\boldsymbol{\theta}) \\ &= \prod_{i=1}^{n} f(y_i/\boldsymbol{\theta}) \end{aligned} \tag{3.40}$$

To illustrate the above concept, we follow the approaches given by Lindsey (2006) and Myung (2003). Let us assume that $f(y/\boldsymbol{\theta})$ follows binomial distribution such as,

$$f(y/n,p) = {}^{n}C_{y}\, p^{y}(1-p)^{n-y}, y \geq 0 \tag{3.41}$$

Here, $\boldsymbol{\theta} = [n \quad p]^T$, a two-parameter vector and the binomial process can be completely expressed by Equation 3.41 with parameters n and p. It should be noted here that binomial distribution with parameters n and p represents n Bernoulli trials (e.g., tossing a coin once is one trial), i.e., y can take discrete values ranging from 0 to n such as 0, 1, 2, …, n, and p represent the probability of success (e.g., the proportion of heads in the n trials) that lies within 0 and 1, i.e., $0 \leq p \leq 1$. Let us further assume that we have conducted $n = 5$ trials and the probability of success (p) is 0.50. Then

$$f(y/5, 0.50) = {}^{5}C_{y}(0.5)^{y}(0.5)^{5-y} = {}^{5}C_{y}(0.5)^{5} \tag{3.42}$$

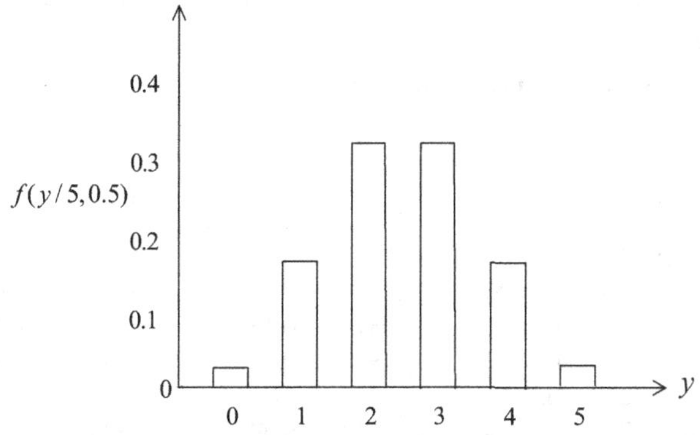

FIGURE 3.14 Binomial pmf for $n = 5$ and $p = 0.50$.

Now, putting the various values of $y(= 0, 1, \ldots, 5)$, we get the following probability mass function.

y	$f(y/5, 0.50) = {}^5C_y (0.5)^5$
0	0.0312
1	0.1562
2	0.3152
3	0.3152
4	0.1562
5	0.0312

The shape of the probability mass function is shown in Figure 3.14.

Now, if we vary p values within 0–1, we get different pmf (e.g., Figure 3.14) for different values of p. Table 3.7 shows binomial pmfs under different p values. Each row represents a pmf for the corresponding parameter value of p ($n = 5$).

Example 3.23

Draw binomial pmf for ($n = 5$) and $p = 0.9$

Solution:

From Table 3.7, the binomial pmf values for $n = 5$ and $p = 0.9$ are shown in Table 3.8.

The corresponding pmf is shown in Figure 3.15.

3.3.2 Likelihood Function

Consider Table 3.7 again. The row values represent the pmf for specific parameters (p and n) of the binomial distribution given the data observed (**y** values). What do the column values represent? The column values determine the probability of observing a particular y value (e.g., $y = 0$ for column 2) for given parameter values, p and n. In this case n is constant and equal to 5 whereas p varies from 0.10 to 0.90 with increment of 0.1. If we move across column-wise, we find a cell with the maximum pmf value. For example, for $y = 0$, the maximum pmf value is 0.5905 which corresponds to $p = 0.10$. Similarly, for the values of $y = 1, 2, 3, 4, \text{and } 5$, the corresponding maximum pmf

TABLE 3.7
Binomial pmf under Different p Values ($n = 5$)

y / *p*	0	1	2	3	4	5
0.10	**0.5905**	0.3280	0.0729	0.0081	0.0004	.0000
0.20	0.3277	**0.4096**	0.2048	0.0512	0.0064	.0003
0.30	0.1681	0.3601	0.3087	0.1323	0.0283	.0024
0.40	0.0778	0.2592	**0.3456**	0.2304	0.0768	0.0102
0.50	0.0312	0.1562	0.3152	0.3152	0.1562	0.0312
0.60	0.0102	0.0768	0.2304	**0.3456**	0.2592	0.0778
0.70	0.0024	0.0283	0.1323	0.3087	0.3601	0.1681
0.80	0.0003	0.0064	0.0512	0.2048	**0.4096**	0.3277
0.90	0.0000	0.0004	0.0081	0.0729	0.3280	**0.5905**

TABLE 3.8
Binomial pmf Values for $n = 5$ and $p = 0.9$

y	0	1	2	3	4	5
$f(y/5,0.90)$	0.0000	0.0004	0.0081	0.0729	0.3280	0.5905

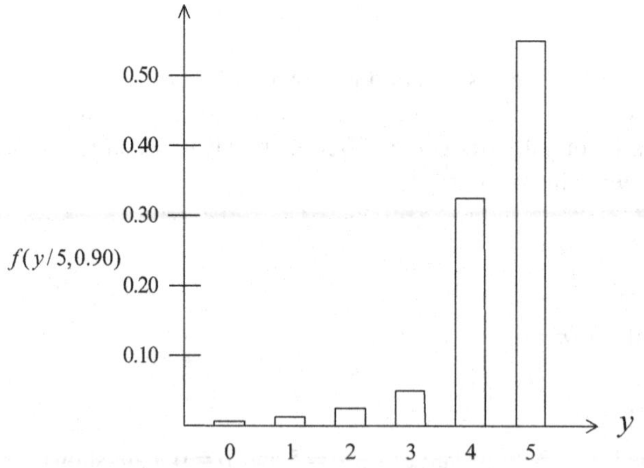

FIGURE 3.15 Binomial pmf for $n = 5$ and $p = 0.90$.

values are 0.4096, 0.3456, 0.3456, 0.4096, and 0.5905, respectively. These pmf values correspond to $p = 0.20, 0.40, 0.60, 0.80$ and 0.90, respectively. So, we can argue that there is a specific parameter (p) value for which the data observed is the most likely. Now, if we assume that the plausible p values for Table 3.7 are 0.10 to 0.90 with increment 0.10, then given $n = 5$ the parameter (p) values that makes the **y** observations most likely are 0.10 for $y = 0$, 0.20 for $y = 1$, 0.40 for $y = 2$, 0.60 for $y = 3$, 0.80 for $y = 4$ and 0.90 for $y = 5$. The parameter value for which an observation is most likely is known as MLE estimate. The pmf values for a specific y across different parameter (p) values are likelihood values. In summary, the values in a row of Table 3.7 represent pmf for given data and the values in a column represent the likelihood function of an observation for given parameters.

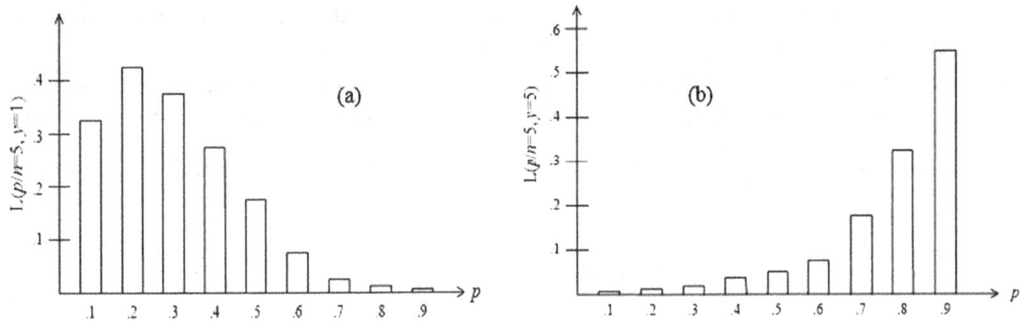

FIGURE 3.16 (a) Likelihood function for $y = 1$, and (b) Likelihood function for $y = 5$.

Now, we will mathematically define likelihood function. The likelihood function of n random observations $y_1, y_2, ..., y_n$ is the joint probability function (pdf or pmf) like Equation 3.40 but has differences in nomenclature and interpretation. The role of data vector $\mathbf{y}$ and parameter vector $\boldsymbol{\theta}$ in likelihood functions is the reverse of the probability function. If we define the likelihood function as $L(\boldsymbol{\theta} / \mathbf{y})$, then the relationship is

$$L(\boldsymbol{\theta} / \mathbf{y}) = f(\mathbf{y} / \boldsymbol{\theta}) \tag{3.43}$$

For n observations, $L(\boldsymbol{\theta} / \mathbf{y})$ is $L(\boldsymbol{\theta} / y_1, y_2, .., y_n)$ and for *iid* data,

$$L(\boldsymbol{\theta} / y_1, y_2, .., y_n) = \prod_{i=1}^{n} L(\boldsymbol{\theta} / y_i) = \prod_{i=1}^{n} f(y_i / \boldsymbol{\theta}) \tag{3.44}$$

So, the difference of probability function and likelihood function in practice is that, given a particular set of parameters, the probability function computes the pdf or pmf of different known observations and is a function of the data, defined on the data scale as shown in Figures 3.14 and 3.15, whereas the likelihood function is a function of the parameters, defined on the parametric scale and computes the likelihood of a particular parameter value for a given data set (Myung, 2003; Lindsey, 2006). The likelihood functions for $y = 1$ and $y = 5$ shown in Table 3.7 are plotted in Figures 3.16 (a) and (b).

3.3.3 Maximum Likelihood Estimation

The maximum likelihood estimation (MLE) determines the parameter vector that makes the observations most likely. So, it is a maximization problem. The steps to be followed are given below.

Step 1: Obtain the likelihood function $L(\boldsymbol{\theta} / \mathbf{y})$ and find out its log likelihood, i.e., $\ell nL(\boldsymbol{\theta} / \mathbf{y})$. As $L(\boldsymbol{\theta} / \mathbf{y})$ and $\ell nL(\boldsymbol{\theta} / \mathbf{y})$ have their maxima at the same value of $\boldsymbol{\theta}$ (Mood, 1974), for convenience to differentiation $\ell nL(\boldsymbol{\theta} / \mathbf{y})$ is considered.

Step 2: Obtain first derivative of $\ell nL(\boldsymbol{\theta} / \mathbf{y})$ with respect to θ_j which is equal to zero at θ_{MLE} if it exists. So,

$$\frac{\partial \ell nL(\boldsymbol{\theta} / \mathbf{y})}{\partial \theta_j} = 0, j = 1, 2, ..., p \tag{3.45}$$

Where p is the number of parameters to be estimated. $j = 1$ if $L(\boldsymbol{\theta} / \mathbf{y})$ contain one parameter. The solution to Equation 3.45 gives the MLE estimates of the parameter vector.

Step 3: Step 2 doesn't completely satisfy conditions that $\ell nL(\boldsymbol{\theta} / \mathbf{y})$ is the maximum at the θ_{MLE}. One additional condition needs to be satisfied. We require to test the second derivative of $\ell nL(\boldsymbol{\theta} / \mathbf{y})$, i.e., $\frac{\partial^2 \ell nL(\boldsymbol{\theta}/\mathbf{y})}{\partial^2 \theta_j}, j = 1,2,..,p$. For maximization of $\ell nL(\boldsymbol{\theta} / \mathbf{y})$, $\frac{\partial^2 \ell nL(\boldsymbol{\theta}/\mathbf{y})}{\partial^2 \theta_j}$ should be negative. So,

$$\frac{\partial^2 \ell nL(\boldsymbol{\theta}/\mathbf{y})}{\partial^2 \theta_j} < 0, j = 1,2,....,p \tag{3.46}$$

It should be noted here that the steps described above may fall under the trap of local maxima. When the closed-form solution of Equation 3.45 is not available, we require iterative numerical or heuristic methods to obtain the MLE estimates. An initial starting feasible solution is therefore crucial and careful analysis is required.

Example 3.24

Let's say n random observations $y_1, y_2, ..., y_n$ were collected from a normal population with mean μ and standard deviation σ. Obtain the MLE estimate of μ and σ^2.

Solution:

Let the pdf of a normal population with mean μ and variance σ^2 be denoted by $f(y/\mu, \sigma^2)$.

So, for the observation y_i,

$$f(y_i / \mu, \sigma^2) = \frac{1}{\sqrt{2\pi\sigma^2}} e^{-\frac{1}{2}\left(\frac{y_i - \mu}{\sigma}\right)^2}, -\infty < y_i < \infty$$

Step 1: Let's say $y_1, y_2, ..., y_n$ are iid. So,

$$\begin{aligned} L(\mu, \sigma^2 \mid y_1, y_2, ..., y_n) &= \prod_{i=1}^{n} L(\mu, \sigma \mid y_i) \\ &= \prod_{i=1}^{n} f(y_i \mid \mu, \sigma) \\ &= \left(\frac{1}{\sqrt{2\pi\sigma^2}}\right)^n .e^{-\frac{1}{2}\sum_{i=1}^{n}\left(\frac{y_i - \mu}{\sigma}\right)^2} \\ &= \left(\frac{1}{2\pi\sigma^2}\right)^{n/2} .e^{-\frac{1}{2\sigma^2}\sum_{i=1}^{n}\left(y_i - \mu\right)^2} \end{aligned}$$

So, the $\ell nL(\mu, \sigma^2 \mid \mathbf{y})$ is

$$\ell nL(\mu,\sigma^2 \mid \mathbf{y}) = \frac{-n}{2}\ell n2\pi - \frac{n}{2}\ell n\sigma^2 - \frac{1}{2\sigma^2}\sum_{i=1}^{n}(y_i-\mu)^2$$

Step 2: There are two parameters, μ and σ^2. So, we require differentiating $\ell nL(\mu,\sigma^2 \mid \mathbf{y})$ with respect to μ and σ^2, respectively. Now,

$$\frac{\partial \ell nL(\mu,\sigma^2 \mid \mathbf{y})}{\partial \mu} = -\frac{1}{\sigma^2}\sum_{i=1}^{n}(y_i-\mu)$$

and
$$\frac{\partial \ell nL(\mu,\sigma^2 \mid \mathbf{y})}{\partial \sigma^2} = -\frac{n}{2}\cdot\frac{1}{\sigma^2} + \frac{1}{2\sigma^4}\sum_{i=1}^{n}(y_i-\mu)^2$$

Putting the first derivative equal to zero, we get

$$\frac{-1}{\sigma^2}\sum_{i=1}^{n}(y_i-\mu) = 0.$$

$$or \quad \hat{\mu} = \frac{1}{n}\sum_{i=1}^{n}y_i = \bar{y}$$

For σ^2

$$-\frac{n}{2}\cdot\frac{1}{\sigma^2} + \frac{1}{2\sigma^4}\sum_{i=1}^{n}(y_i-\mu)^2 = 0$$

$$\Rightarrow n = \frac{1}{\sigma^2}\sum_{i=1}^{n}(y_i-\mu)^2$$

$$\Rightarrow \hat{\sigma}^2 = \frac{1}{n}\sum_{i=1}^{n}(y_i-\mu)^2$$

Step 3: Compute second derivatives and show that they are negative at $\mu = \hat{\mu}$ and $\sigma^2 = \hat{\sigma}^2$. It is easy to prove and is left for the readers for practice.

The conditions for maximization given in Equation 3.46 are a little conservative as they are used for estimating one parameter at a time. Instead, we should use a Hessian matrix that takes the following form:

$$\mathbf{H}(\boldsymbol{\theta}) = \begin{bmatrix} \frac{\partial^2 L(\boldsymbol{\theta}\mid y)}{\partial^2 \theta_1} & \frac{\partial^2 L(\boldsymbol{\theta}\mid y)}{\partial \theta_1 \partial \theta_2} & \cdots & \frac{\partial^2 L(\boldsymbol{\theta}\mid y)}{\partial \theta_1 \partial \theta_p} \\ \frac{\partial^2 L(\boldsymbol{\theta}\mid y)}{\partial \theta_2 \partial \theta_1} & \frac{\partial^2 L(\boldsymbol{\theta}\mid y)}{\partial^2 \theta_2} & \cdots & \frac{\partial^2 L(\boldsymbol{\theta}\mid y)}{\partial \theta_2 \partial \theta_p} \\ \vdots & \vdots & & \vdots \\ \frac{\partial^2 L(\boldsymbol{\theta}\mid y)}{\partial \theta_p \partial \theta_1} & \frac{\partial^2 L(\boldsymbol{\theta}\mid y)}{\partial \theta_p \partial \theta_2} & \cdots & \frac{\partial^2 L(\boldsymbol{\theta}\mid y)}{\partial^2 \theta_p} \end{bmatrix} \tag{3.47}$$

For maximization, for a non-zero vector $\boldsymbol{\theta}$ the quantity $\boldsymbol{\theta}^T\mathbf{H}(\boldsymbol{\theta})\boldsymbol{\theta} < \mathbf{0}$, i.e., $\mathbf{H}(\boldsymbol{\theta})$ must be negative definite.

3.4 GENERATION OF RANDOM VARIABLE

Random variables play a key role in statistical data analysis. Many times, we need to generate random variables. For example, it may so happen that the population distribution is known in advance and we want to explain the relationships of the population characteristics or want to predict some of them based on the relationships. In this case, we can simulate random observations of the identified population characteristic and subsequently the data can be used for further modeling such as finding dependence relationships. In general, there are three situations when we require data to be generated. First, the population distribution and the parameters of the distribution are known. Second, population distribution is known but the parameter values are not known in advance. Finally, the population distribution is unknown. In this section, we describe generation of data from univariate normal populations. The reader should note that a population can be other than normal but for all practical purposes normality can be established through appropriate transformation of data.

3.4.1 Generation of Univariate Normal Observations

The basic steps involved are (i) generation of pseudorandom numbers form a uniform distribution and (ii) generation of non-uniform random numbers (Gentle, 2009). For generating pseudorandom numbers, the most widely used method is the linear congruential method proposed by Lehmer (1951). The linear congruential method works with the following relationship.

$$x_{i+1} = (ax_i + c) \bmod m, i = 0,1,2,\ldots,\mathrm{n} \tag{3.48}$$

Where $x_1, x_2, \ldots, x_n$ are the sequence of integers between 0 and $(n-1)$, x_0, a, c and m are respectively called the seed, the multiplier, the increment, and the modulus.

For generating pseudorandom numbers between 0 and 1 (i.e., uniform random numbers), the following equation is used:

$$R_i = \frac{x_i}{m}, i = 1,2,\ldots,n \tag{3.49}$$

The R_i so developed has the following properties (Banks et al., 2007):

$$f_R(x) = \begin{cases} 1 & 0 \le x \le 1 \\ 0 & \text{otherwise} \end{cases} \tag{3.50}$$

and the cdf

$$F_R(x) = \begin{cases} 0 & x < 0 \\ x & 0 \le x \le 1 \\ 1 & x > 1 \end{cases} \tag{3.51}$$

Now, we will discuss generating normal random numbers. It is known that if a random variable X has the cdf F_X then the variable $u = F_X(x)$ has a uniform distribution. So,

$$X = F_X^{-1}(\mathrm{u}) \tag{3.52}$$

Equation 3.52 defines inverse cdf method. This is applicable when cdf is easy to compute. But for normal distribution, the cdf cannot be written in closed form. So, inverse cdf is not recommended.

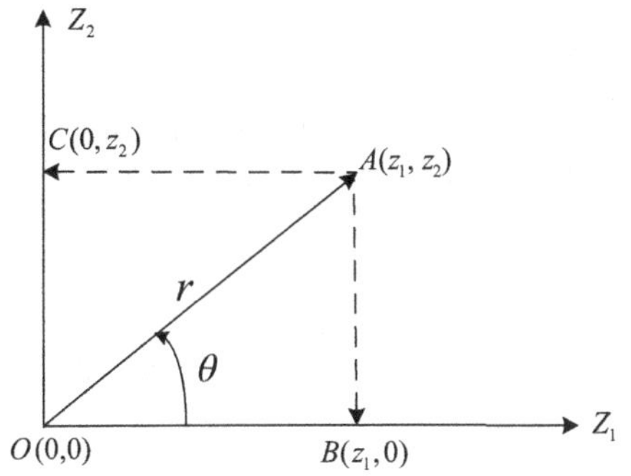

FIGURE 3.17 Polar (r, θ) transformation of Cartesian coordinates (z_1, z_2).

The mostly used method is the Box and Muller (1958) transformation method. We will discuss it below.

Let us consider two standard normal variables Z_1 and Z_2, Figure 3.17 represents the polar transformation of the two standard normal variables Z_1 and Z_2.

From Figure 3.17,

$$\begin{aligned} z_1 &= r\cos\theta \\ z_2 &= r\sin\theta \end{aligned} \tag{3.53}$$

$$\text{and} \quad r^2 = z_1^2 + z_2^2$$

In order to know the values of z_1 and z_2, we require knowing r and θ. Now, θ can vary from 0 to 2π.

r can be found using the following formula:

$$r = (-2\ell nR)^{1/2} \tag{3.54}$$

where R is uniform random numbers.

Now, the formula used to compute two independent standard normal variates z_1 and z_2 are

$$\left.\begin{aligned} z_1 &= (-2\ell nR_1)^{1/2}\cos(2\pi R_2) \\ z_2 &= (-2\ell nR_1)^{1/2}\sin(2\pi R_2) \end{aligned}\right\} \tag{3.55}$$

where R_1 and R_2 are two random numbers. We can convert the standard normal variates z_i, $i = 1,2,...n$ to general normal variates x_i, $i = 1,2,...$ n using a simple transformation given below.

$$x_i = \mu + \sigma z_i \tag{3.56}$$

where x_i follows normal distribution with parameters μ and σ.

Example 3.25

Let X be normally distributed with mean 20 and standard deviation 5. Generate four normal variates.

Solution:

Let us create first three random numbers R_1, R_2 and R_3 using linear congruential method. Further, let us assume the following data (see Equation 3.48):

$x_0 = 21$
$a = 34$
$c = 7$
$m = 65$

$$x_1 = (ax_0 + c) \bmod m$$

$$= (34 \times 21 + 7) \bmod 65$$

$$= 721 \bmod 65$$

$$= 6$$

So, $R_1 = \frac{x_1}{m} = \frac{6}{65} = 0.09$

Similarly, $x_2 = (ax_1 + c) \bmod m$

$$= (34 \times 6 + 7) \bmod 65$$

$$= 211 \bmod 65$$

$$= 16$$

So, $R_2 = \frac{x_2}{m} = \frac{16}{65} = 0.25$

Similarly, $x_3 = (ax_2 + c) \bmod m$

$$= (34 \times 16 + 7) \bmod 65$$

$$= 551 \bmod 65$$

$$= 31$$

So, $R_3 = \frac{31}{65} = 0.48.$

Now, consider R_1 and R_2 and find out z_1 and z_2 using Equation 3.55.

$$z_1 = (-2\ell n 0.09)^{1/2} \cos(2\pi \times 0.25) = 0$$
$$z_2 = (-2\ell n 0.09)^{1/2} \sin(2\pi \times .25) = 2.19$$

Similarly using R_2 and R_3, we may get z_3 and z_4, where

$$z_3 = (-2\ell n 0.25)^{1/2} \cos(2\pi \times 0.48) = -1.65$$
$$z_4 = (-2\ell n 0.25)^{1/2} \sin(2\pi \times 0.48) = 0.21$$

Now, let x_1, x_2, x_3 and x_4 are the four normal variates. Then using Equation 3.56, we get

$$x_1 = \mu + \sigma z_1 = 20 + 5 \times 0 = 20$$
$$x_2 = \mu + \sigma z_2 = 20 + 5 \times 2.19 = 30.95$$
$$x_3 = \mu + \sigma z_3 = 20 - 1.65 \times 5 = 11.75$$
$$x_4 = \mu + \sigma z_4 = 20 + 5 \times 0.21 = 21.05$$

3.4.2 Generating Multivariate Normal Observations

There are three commonly used methods for generating multivariate normal samples. They are the rotation method, conditional method, and triangular factorization method (Barr and Slezak, 1972). As the discussion requires knowledge on multivariate normal distribution in addition to matrix decompositions such as Cholesky factorization, we restrict our discussion to the generation of univariate normal variates.

3.5 RESAMPLING METHODS

Many times we could select a small amount of data on which reliable statistical tests cannot be performed. The situation gets tougher when we do not know the underlying distribution of the populations. For example, let's say we have collected n, a small number of data points $x_1, x_2, \ldots x_n$ from an unknown distribution F_X, where X is the random variable. Further, let's say $\theta = \mu_X$, the population mean, and $\hat{\theta} = \hat{\mu}_X = \frac{1}{n}\sum_{i=1}^{n} x_i$. Now, we may be interested to know the followings (Zoubir and Iskandler, 2007):

(i) What is the distribution of $\hat{\theta}$?
(ii) How variable is the parameter estimator $\hat{\theta}$?
(iii) Is θ is different from θ_0 (hypothesis test)?

As n is small, F_X is unknown and a repeat experiment is impossible, we cannot use central limit theorem (Zoubir and Iskandler, 2007). Under such situations, resampling methods can be useful to answer the questions given above. There are many methods of resampling, but the two best-known methods are jackknife and bootstrap.

3.5.1 Jackknife

In this method, given a sample of size n, n new samples of size $n-1$ are created. The procedures involve eliminating one observation in turn from the sample (of size n) collected. As we have n new samples, we can compute n values of $\hat{\theta}$, denoted by $\hat{\theta}_{(i)}$ where $\hat{\theta}_{(i)}$ is the estimated value of θ when the $i-th$ observation is removed to obtain the sample. Now, with $n\hat{\theta}_{(i)}$, $i = 1, 2, \ldots n$ we can develop the histogram of $\hat{\theta}$ to answer question (i) given above. For reliable estimate of $\hat{\theta}$, we can compute accuracy (bias) and precision using the following two formulas, respectively (Wehrens et al., 2000; Efron and Gong, 1983).

$$\hat{\theta}_{bias} = (n-1)\left(\hat{\theta}_{(i)} - \hat{\theta}\right)$$
$$\hat{\theta}_{SE} = \sqrt{\frac{n-1}{n}\sum_{i=1}^{n}\left(\hat{\theta}_{(i)} - \hat{\theta}\right)^2} \tag{3.57}$$

where $\hat{\theta}$, is the mean of $\hat{\theta}_{(i)}$, $i = 1,2,\ldots n$.

The jackknife method works well when the statistic, e.g., $\hat{\theta}$, is not sensitive to small changes in the data (Wehrens et al., 2000). Other drawback of jackknife is that it is not consistent for the median values obtained from the jackknife samples.

3.5.2 Bootstrap

Bootstrap method was developed by Efron (1992). It is a resampling method and is applicable in the situations given in Section 3.5. We will describe below the bootstrap method based on Efron and Gong (1983). Suppose we have a sample of n observations $x_1, x_2, \ldots x_n$, whose distribution is known. Also assume that we are interested to know the sampling distribution of a statistic $\hat{\theta}$ that can be obtained through the sample collected. As shown in Section 3.5.1, we cannot use theoretical distribution because it is unknown, we require relying on a set of new samples (say B) of size n and then compute the statistic of interest $\hat{\theta}_j, j = 1,2,\ldots B$, for each of the samples thus collected. The bootstrap method works as follows (Efron and Gong, 1983):

Step 1: Using $x_1, x_2, \ldots x_n$, construct the empirical probability distribution $\hat{F}$, where $\hat{F}$ is the "mass $\frac{1}{n}$ on each observed data points x_i, $i = 1,2,\ldots n$."

Step 2: Draw a bootstrap sample of n observations from $\hat{F}$ through n random draws with replacement from $\{x_1, x_2, \ldots x_n\}$ (e.g., from $\hat{F}$).

Step 3: Compute the statistic of interest, $\hat{\theta}_1$.

Step 4: Repeat Steps 2 and 3 for a large number of times, say B.

So, we have B values of $\hat{\theta}$, i.e., $\hat{\theta}_1, \hat{\theta}_2, \ldots \hat{\theta}_B$. Using these B number of computed values, we can estimate the statistic of interest, for example, mean and standard deviation $\hat{\theta}$ as follows:

$$\hat{\theta}_{Bmean} = \frac{1}{B}\sum_{j=1}^{B}\hat{\theta}_j \tag{3.58}$$

$$\hat{\theta}_{Bsd} = \sqrt{\frac{1}{B-1}\sum_{j=1}^{B}(\hat{\theta}_j - \hat{\theta}_{Bmean})^2}$$

where B stands for bootstrapped.

3.6 LEARNING SUMMARY

This chapter contains the necessary details of matrix algebra and other computational issues needed to understand multivariate statistical modeling of data. The important concepts are highlighted below.

- The simple vector and matrix operations help in computing basic descriptive statistics like mean, mean vectors, variance, covariance, and covariance matrix.

- As orthogonalization of variables is important for many statistical models, vector-based orthogonalization processes are described.
- Many models like PCA and factor analysis use decomposition of covariance or correlation matrix. Three important matrix decomposition techniques namely eigenvalue-eigenvector, spectral, and singular value decompositions are included in this chapter.
- For parameter estimation, methods of least squares and maximum likelihood method are described.
- For generating random numbers from univariate normal distributions, the useful procedures are presented.
- Finally, two resampling methods, namely jackknife and bootstrap, are described.

EXERCISES

(A) Short conceptual questions

3.1 Define data matrix. Obtain row and column vectors for the data matrix, $\mathbf{A} = \begin{bmatrix} 2 & 2 & 3 \\ -2 & 5 & 4 \end{bmatrix}$.

3.2 Compute average of the data vector $\begin{bmatrix} 10 & 20 & 30 & 50 & 70 & 90 & 110 & 130 & 150 & 200 \end{bmatrix}^T$.

3.3 What is orthogonal vector? Consider the following two vectors **a** and **b** and show that they are orthogonal.

$$\mathbf{a} = \begin{bmatrix} 5 \\ 3 \\ -8 \end{bmatrix} \text{ and } \mathbf{b} = \begin{bmatrix} 8 \\ 8 \\ 8 \end{bmatrix}$$

3.4 Show that the length of a vector **x** from origin is $\sqrt{\mathbf{x}^T\mathbf{x}}$.

3.5 Consider the following two vectors and find out the angle between them.

$$\mathbf{x} = \begin{bmatrix} 5 & 3 & 9 & 2 & 7 \end{bmatrix}^T \text{ and } \mathbf{y} = \begin{bmatrix} -10 & 3 & 6 & 3 & 8 \end{bmatrix}^T$$

3.6 What is Grammian? If the Gram determinant or Grammian of two vectors is zero, what can be concluded about the relation between the two vectors?

3.7 Explain the Gram-Schmidt orthogonalization process.

3.8 Consider the two vectors **x** and **y** given in question 3.5. Construct a set of orthogonal vectors using Gram-Schmidt process.

3.9 Define the square matrix, null matrix, diagonal matrix, identity matrix, and symmetric matrix.

3.10 Define the following with respect to matrix operations.

(i) Trace of a matrix
(ii) Equality of matrices
(iii) Transpose of a matrix
(iv) Addition of k number of matrices
(v) Subtraction of k number of matrices
(vi) Multiplication of two matrices

3.11 What is determinant? For a matrix **A** of size $m \times n$, can you compute det(**A**)?

3.12 What is rank of a matrix? For a matrix $\mathbf{X}_{n\times p}$, use Gaussion elimination to compute rank of **X**.

3.13 Define inverse of a square matrix. Consider the matrix **A** given in Example 3.1. Obtain $\mathbf{A}^{-1}$.

3.14 Define cofactor. Obtain all the cofactors for $\mathbf{A} = \begin{bmatrix} a & b & c \\ b & c & a \\ c & a & b \end{bmatrix}$.

3.15 What are eigenvalues and eigenvectors of a square matrix? How do you compute them?

3.16 Show that the characteristics equation of a square matrix **A** is $|\mathbf{A} - \lambda\mathbf{I}| = 0$, where λ denotes eigenvalue and **I** denotes an identity matrix.

3.17 What is spectral decomposition? Explain spectral decomposition for the matrix $\mathbf{\Sigma} = \begin{bmatrix} 4 & 2 \\ 2 & 9 \end{bmatrix}$.

3.18 What is singular value decomposition? Why it is important? What is the relationship between singular value and eigenvalue of a matrix?

3.19 What is ordinary least squares (OLS) method? Why it is needed?

3.20 Explain the geometry of least squares solution.

3.21 What is weighted least squares (WLS)? In what way it is different from OLS?

3.22 What is iteratively reweighted least squares (IRLS)? Explain the steps followed in IRLS.

3.23 What is generalized least squares (GLS)? Obtain its relationships with OLS, WLS, and IRLS.

3.24 What is probability function? In what way it is different from likelihood function?

3.25 What is maximum likelihood estimation (MLE)? Explain the steps. Using MLE obtain the mean and standard deviation of y that is normally distributed.

3.26 What is Hessian? Why it is important?

3.27 How do you compute normal random numbers?

3.28 Explain one method of generating univariate normal observations.

3.29 What is bootstrap? Why it is important? What is jackknife? Explain the steps.

3.30 Name three methods for generating multivariate normal observations.

(B) Long conceptual and/or numerical questions: see web resources (Multivariate Statistical Modeling in Engineering and Management – 1st (routledge.com))

N.B.: R and MS Excel software are used for computation, as and when required.

Part II

Foundations of Multivariate Statistics

4 Multivariate Descriptive Statistics

In this chapter, we discuss multivariate descriptive statistics. Univariate descriptive statistics is primarily concerned with measures of central tendency and measures of dispersion. The popular measures of central tendency are mean, median, and mode and dispersion is measured using range and variance. The analogous measures in the multivariate domain are mean vectors and covariance matrices. The covariance matrix not only considers the variance of each of the random variables (X) considered but also includes a measure called covariance that describes how two random variables covary with each other. In addition, standardized covariance matrix is also used, which is known as correlation matrix. The term correlation defines one variable's association with another. There are different variants of correlation coefficient based on data types; for example, Pearson correlation between two numeric variables, and biserial correlation between a continuous and a dichotomous variable; and polyserial correlation between a continuous and an ordinal variable (with more than two categories). With dependence structure, part and partial correlations are used. All these concepts with examples are described in this chapter.

4.1 MULTIVARIATE OBSERVATIONS

Let assume a population is characterized by p number of variables ($\boldsymbol{X}$) and the variable vector can be denoted as

$$\boldsymbol{X} = \begin{bmatrix} X_1 \\ X_2 \\ . \\ X_j \\ . \\ X_{\text{p}} \end{bmatrix}_{p\times 1}$$

The j-th variable from $\boldsymbol{X}$ is termed as X_{j}, $j = 1, 2 \ldots p$. Suppose we are interested in collecting n observations for the p-variable vector (Table 4.1). These n observations on the p variables constitute n multivariate observations.

In matrix form, we can reproduce the data set of Table 4.1 as

$$\mathbf{X}_{n\times p} = \begin{bmatrix} x_{11} & x_{12} & \cdots & x_{1j} & \cdots & x_{1p} \\ x_{21} & x_{22} & \cdots & x_{2j} & \cdots & x_{2p} \\ . & . & . & . & . & . \\ . & . & . & . & . & . \\ . & . & . & . & . & . \\ x_{i1} & x_{i2} & \cdots & x_{ij} & \cdots & x_{ip} \\ . & . & . & . & . & . \\ . & . & . & . & . & . \\ . & . & . & . & . & . \\ x_{n1} & x_{n2} & \cdots & x_{nj} & \cdots & x_{np} \end{bmatrix}$$

DOI: 10.1201/9781003303060-6

TABLE 4.1
Data Table for Multivariate Observations

Observation Number	Variables X_1	X_2	....	X_j	....	X_p
1	x_{11}	x_{12}	....	x_{1j}	....	x_{1p}
2	x_{21}	x_{22}	....	x_{2j}	....	x_{2p}
....	....	....	....	....	....	....
....	....	....	....	....	....	....
i	x_{i1}	x_{i2}	....	x_{ij}	....	x_{ip}
....	....	....	....	....	....	....
....	....	....	....	....	....	....
n	x_{n1}	x_{n2}	....	x_{nj}	....	x_{np}

and $\mathbf{x}_i = \begin{bmatrix} x_{i1} \\ x_{i2} \\ . \\ . \\ x_{ij} \\ . \\ . \\ x_{ip} \end{bmatrix}_{p\times 1}$ [Note that the i-th observation vector is represented as a column vector of $p \times 1$].

$\mathbf{x}_i$ represents the i-th multivariate observations on p variables. Before the n multivariate observations are collected, each of the observation vector, $\mathbf{x}_i$, $i = 1, 2, \ldots, n$ is random. After data collection, the sample comprises n multivariate observations that are of fixed values (as already collected). Similarly, the data matrix $\mathbf{X}_{n\times p}$, planned to be collected is a random matrix and when these data are collected, $\mathbf{X}_{n\times p}$ represents a matrix with observed values (fixed values).

Example 4.1

A quality control engineer is assigned to control the dimensional characteristics of ball bearings produced in their manufacturing shop. The dimensional characteristics are inner and outer diameters, and surface finish, denoted by ID, OD, and SF, respectively. Develop a random as well as fixed matrix for n multivariate observations.

Solution:

Let's say ID = X_1, OD = X_2 and SF = X_3. If we plan to collect n observations on X_1, X_2, and X_3, we get the following random data matrix

$$\mathbf{X}_{n\times 3} = \begin{bmatrix} x_{11} & x_{12} & x_{13} \\ x_{21} & x_{22} & x_{23} \\ \vdots & \vdots & \vdots \\ x_{i1} & x_{i2} & x_{i3} \\ \vdots & \vdots & \vdots \\ x_{n1} & x_{n2} & x_{n3} \end{bmatrix}$$

The matrix $\mathbf{X}_{n\times3}$ is random as it is planned to be sampled. So, all values of this matrix are random and can take any value that is realistic. However, when a sample of size n is collected, all the elements of $\mathbf{X}_{n\times3}$ will be fixed. Suppose the following five observations are collected.

$$\mathbf{X}_{n\times3} = \begin{bmatrix} 20 & 35 & 3 \\ 25 & 40 & 2 \\ 23 & 38 & 4 \\ 22 & 36 & 2 \\ 20 & 40 & 3 \end{bmatrix}$$

The above matrix represents a matrix with observed values (fixed values) with sample size n=5.

4.2 MEAN VECTORS

Consider a multivariate population characterized by p random variables. Further, consider a particular variable, the $j-th$ variable from the variable vector $\boldsymbol{X}$, termed as X_j. The n observations to be collected on X_j are[1]

$$\mathbf{x}_j = \begin{bmatrix} x_{1j} \\ x_{2j} \\ . \\ . \\ x_{ij} \\ . \\ . \\ x_{nj} \end{bmatrix}_{n\times1}$$

Now, we express the mean value of X_j, termed as μ_j. X_j can be discrete with pmf, $f(x_j)$ or continuous with pdf, $f(x_j)$. Then the mean (μ_j) for the variable X_j is

$$\mu_j = E\left(X_j\right) = \begin{cases} \sum_{all\, x_j} x_j\, f\left(x_j\right) & \text{for discrete } X_j \\ \int x_j\, f\left(x_j\right) dx_j & \text{for continuous } X_j \end{cases} \tag{4.1}$$

As there are p variables (j = 1, 2, …, p), we will get p mean values that form a mean vector as shown below.

$$\boldsymbol{\mu} = \begin{bmatrix} \mu_1 \\ \mu_2 \\ \cdot \\ \cdot \\ \mu_j \\ \cdot \\ \cdot \\ \mu_p \end{bmatrix} = \begin{bmatrix} \mathrm{E}(X_1) \\ \mathrm{E}(X_2) \\ \cdot \\ \cdot \\ \mathrm{E}(X_j) \\ \cdot \\ \cdot \\ \mathrm{E}(X_p) \end{bmatrix} \tag{4.2}$$

The population mean vector μ is unknown and is estimated using sample data. If a sample of n multivariate observations are collected on p variables, the estimate of the population mean vector ($\hat{\mu}$) is the sample mean vector, denoted by $\bar{\mathbf{x}}$. The computation procedure for $\bar{\mathbf{x}}$ is given below.

As we have seen in univariate case, the sample mean for a variable X_j is $\bar{x}_j = \frac{1}{n}\sum_{i=1}^{n} x_{ij}$.

If we compute $\bar{x}_j$ for $j = 1, 2, \ldots, p$, we will get the sample mean vector as.

$$\bar{\mathbf{x}} = \begin{bmatrix} \bar{x}_1 \\ \bar{x}_2 \\ \cdot \\ \cdot \\ \bar{x}_j \\ \cdot \\ \cdot \\ \bar{x}_p \end{bmatrix}_{p\times 1} = \begin{bmatrix} \frac{1}{n}\sum_{i=1}^{n} x_{i1} \\ \frac{1}{n}\sum_{i=1}^{n} x_{i2} \\ \cdot \\ \cdot \\ \frac{1}{n}\sum_{i=1}^{n} x_{ij} \\ \cdot \\ \cdot \\ \frac{1}{n}\sum_{i=1}^{n} x_{ip} \end{bmatrix}_{p\times 1} \tag{4.3}$$

In matrix operation

$$\bar{\mathbf{x}} = \frac{1}{n}\mathbf{X}^T\mathbf{1} \tag{4.4}$$

where $\boldsymbol{X}_{n\times p}$ is the multivariate data matrix of size $n \times p$ and $\mathbf{1}$ is column vector of 1 with size $n \times 1$.

Example 4.2

Three multivariate observations on two variables, X_1 and X_2 are collected as below.

$$\mathbf{X}_{3\times 2} = \begin{bmatrix} 5 & 30 \\ 9 & 20 \\ 15 & 10 \end{bmatrix}$$

Compute the sample mean vector.

Solution:

Let $\bar{\mathbf{x}} = \begin{bmatrix} \bar{x}_1 \\ \bar{x}_2 \end{bmatrix}$.

Using Equation 4.4, we get

$$\begin{aligned}
\bar{\mathbf{x}} &= \frac{1}{n}\mathbf{X}^T\mathbf{1} \\
&= \frac{1}{3}\begin{bmatrix} 5 & 9 & 15 \\ 30 & 20 & 10 \end{bmatrix}_{2\times 3} \begin{bmatrix} 1 \\ 1 \\ 1 \end{bmatrix}_{3\times 1} \\
&= \frac{1}{3}\begin{bmatrix} 29 \\ 60 \end{bmatrix} = \begin{bmatrix} 9.67 \\ 20.00 \end{bmatrix} = \begin{bmatrix} \bar{x}_1 \\ \bar{x}_2 \end{bmatrix}
\end{aligned}$$

Example 4.3

Compute mean vector for the data given in Example 4.1.

Solution:

The data matrix is

$$\mathbf{X}_{5\times 3} = \begin{bmatrix} 20 & 35 & 3 \\ 25 & 40 & 2 \\ 23 & 38 & 4 \\ 22 & 36 & 2 \\ 20 & 40 & 3 \end{bmatrix}$$

$$\text{So,} \quad \mathbf{1} = \begin{bmatrix} 1 \\ 1 \\ 1 \\ 1 \\ 1 \end{bmatrix}$$

$$\begin{aligned}
\bar{\mathbf{x}} &= \frac{1}{n}\mathbf{X}^T\mathbf{1} \\
&= \frac{1}{5}\begin{bmatrix} 20 & 25 & 23 & 22 & 20 \\ 35 & 40 & 38 & 36 & 40 \\ 3 & 2 & 4 & 2 & 3 \end{bmatrix}\begin{bmatrix} 1 \\ 1 \\ 1 \\ 1 \\ 1 \end{bmatrix} \\
&= \begin{bmatrix} 22.0 \\ 37.8 \\ 2.8 \end{bmatrix}
\end{aligned}$$

4.3 COVARIANCE MATRIX

Covariance is the measure of association between two variables. For example, someone may be interested to know how his expenditure varies with the variation of income as the two are closely connected. Similarly, the quality control engineer (see Example 4.1) may be interested to know how inner diameter (ID) and outer diameter (OD) of the ball bearings covary. To understand covariance, we need to first understand the variance of a variable. The variance of a variable X_j is

$$\sigma_j^2 = \sigma_{jj} = E\left[\left(X_j - \mu_j\right)^2\right] = \begin{cases} \sum_{all\ x_j} \left(x_j - \mu_j\right)^2 f\left(x_j\right) & \text{for discrete } X_j \\ \int_{-\infty}^{\infty} \left(x_j - \mu_j\right)^2 f\left(x_j\right) & \text{for continuous } X_j \end{cases} \tag{4.5}$$

Similarly, we can write the p variances for p variables, in a column vector of $p \times 1$.

$$V(\boldsymbol{X}) = \begin{bmatrix} \sigma_{11} \\ \sigma_{22} \\ \vdots \\ \sigma_{33} \\ \vdots \\ \sigma_{pp} \end{bmatrix}_{p\times 1} = \begin{bmatrix} \sigma_1^2 \\ \sigma_2^2 \\ \vdots \\ \sigma_3^2 \\ \vdots \\ \sigma_p^2 \end{bmatrix}_{p\times 1} \tag{4.6}$$

Now, suppose two discrete random variables X_j and X_k have joint *pmf* $f_{jk}(x_j, x_k)$. The covariance between X_j and X_k is defined as

$$\begin{aligned} Cov\left(X_j, X_k\right) &= \sigma_{jk} = E\left[\left(X_j - \mu_j\right)\left(X_k - \mu_k\right)\right] \\ &= \sum_{all\ x_j} \sum_{all\ x_k} \left(x_j - \mu_j\right)\left(x_k - \mu_k\right) f_{jk}\left(x_j, x_k\right) \end{aligned} \tag{4.7}$$

Similarly, if X_j and X_k are continuous random variables with joint *pdf* $f_{jk}(x_j, x_k)$, then

$$Cov\left(X_j, X_k\right) = \int_{-\infty}^{\infty} \int_{-\infty}^{\infty} \left(x_j - \mu_j\right)\left(x_k - \mu_k\right) f_{jk}\left(x_j, x_k\right) dx_j\, dx_k \tag{4.8}$$

If $j = k$, then

$$Cov\left(X_j, X_k\right) = Cov\left(X_j, X_j\right) = Var\left(X_j\right) = \sigma_{jj} = \sigma_j^2 \tag{4.9}$$

From the above-mentioned definitions, for a p variable vector, there can be p unique variances and $p(p-1)/2$ unique covariances which collectively define the variance-covariance matrix, popularly known as covariance matrix, $\boldsymbol{\Sigma}$, where

$$\Sigma = \begin{bmatrix} \sigma_{11} & \sigma_{12} & \cdots & \sigma_{1j} & \cdots & \sigma_{1p} \\ \sigma_{12} & \sigma_{22} & \cdots & \sigma_{2j} & \cdots & \sigma_{2p} \\ . & & \cdots & & \cdots & \\ . & & \cdots & & \cdots & \\ \sigma_{1j} & \sigma_{2j} & \cdots & \sigma_{ij} & \cdots & \sigma_{jp} \\ . & . & \cdots & & \cdots & \\ . & . & \cdots & & \cdots & \\ \sigma_{1p} & \sigma_{2p} & \cdots & \sigma_{jp} & \cdots & \sigma_{pp} \end{bmatrix}_{p\times p}$$

It is to be noted here that the diagonal elements of $\mathbf{\Sigma}$ represent the variances of the p variables and off-diagonal elements represent the covariances between the respective pair of variables. As we are dealing with real variables (i.e., the variables assume real values), the covariance coefficient σ_{jk} is equal to σ_{kj} (i.e., the order of variables does not change the covariance coefficient).

It should be noted here that the mean vector and covariance matrix of a population are constant in nature and are usually unknown. We estimate them using sample statistics namely sample mean vector and sample covariance matrix, denoted by $\overline{\mathbf{x}}$ and $\mathbf{S}$, respectively.

The sample covariance matrix $\mathbf{S}$ is shown below.

$$\mathbf{S} = \begin{bmatrix} s_{11} & s_{12} & \cdots & s_{1j} & \cdots & s_{1p} \\ s_{12} & s_{22} & \cdots & s_{2j} & \cdots & s_{2p} \\ . & & \cdots & & \cdots & \\ . & & \cdots & & \cdots & \\ s_{1j} & s_{2j} & \cdots & s_{jj} & \cdots & s_{jp} \\ . & . & \cdots & & \cdots & \\ . & . & \cdots & & \cdots & \\ s_{1p} & s_{2p} & \cdots & s_{jp} & \cdots & s_{pp} \end{bmatrix}_{p \times p}$$

where $s_{jj} = s_j^2$ = sample variance of the $j-th$ variable, X_j and s_{jk} = sample covariance between the variables X_j and X_k.

Given a sample of n observations on p-variables, $\mathbf{X}_{n \times p}$, s_{jj} and s_{jk} can be computed as follows:

$$s_{jj} = s_j^2 = \frac{1}{n-1}\sum_{i=1}^{n}\left(x_{ij} - \overline{x}_j\right)^2 \tag{4.10}$$

and in matrix notation

$$s_{jj} = s_j^2 = \frac{1}{n-1}\left(\mathbf{x}_j - \mathbf{1}\overline{x}_j\right)^{\mathrm{T}}\left(\mathbf{x}_j - \mathbf{1}\overline{x}_j\right) \tag{4.11}$$

Where $\mathbf{x}_j = \begin{bmatrix} x_{1j} \\ x_{2j} \\ . \\ . \\ x_{ij} \\ . \\ . \\ x_{nj} \end{bmatrix}_{n \times 1}$ and $\mathbf{1} = \begin{bmatrix} 1 \\ 1 \\ 1 \\ . \\ 1 \\ . \\ . \\ 1 \end{bmatrix}$

The covariance coefficient between X_j and X_k is

$$s_{jk} = \frac{1}{n-1}\sum_{i=1}^{n}\left(x_{ij} - \overline{x}_j\right)\left(x_{ik} - \overline{x}_k\right) \tag{4.12}$$

and in matrix notation

$$s_{jk} = \frac{1}{n-1}\left(\mathbf{x}_j - \mathbf{1}\bar{x}_j\right)^{\mathrm{T}}\left(\mathbf{x}_k - \mathbf{1}\bar{x}_k\right) \tag{4.13}$$

Example 4.4

Compute covariance coefficient s_{12} between the variables X_1 and X_2 with the data set given in Example 4.2.

Solution:

$$s_{jk} = \frac{1}{n-1}\left(\mathbf{x}_j - \mathbf{1}\bar{x}_j\right)^{\mathrm{T}}\left(\mathbf{x}_k - \mathbf{1}\bar{x}_k\right)$$

$$\therefore \quad s_{12} = \frac{1}{3-1}\left(\mathbf{x}_1 - \mathbf{1}\bar{x}_1\right)^{\mathrm{T}}\left(\mathbf{x}_2 - \mathbf{1}\bar{x}_2\right)$$

$$\text{Now, } \mathbf{x}_1 = \begin{bmatrix} 5 \\ 9 \\ 15 \end{bmatrix}, \mathbf{x}_2 = \begin{bmatrix} 30 \\ 20 \\ 10 \end{bmatrix}$$

$$\bar{\mathbf{x}} = \begin{bmatrix} \bar{x}_1 \\ \bar{x}_2 \end{bmatrix} = \begin{bmatrix} 9.67 \\ 20 \end{bmatrix} \qquad \text{(from example 4.2)}$$

So,

$$\mathbf{1}\bar{x}_1 = \begin{bmatrix} 9.67 \\ 9.67 \\ 9.67 \end{bmatrix} \text{ and } \mathbf{1}\,\bar{x}_2 = \begin{bmatrix} 20 \\ 20 \\ 20 \end{bmatrix}$$

Now,

$$\mathbf{x}_1 - \mathbf{1}\bar{x}_1 = \begin{bmatrix} 5-9.67 \\ 9-9.67 \\ 15-9.67 \end{bmatrix} = \begin{bmatrix} -4.67 \\ -0.67 \\ 5.33 \end{bmatrix}$$

$$\mathbf{x}_2 - \mathbf{1}\bar{x}_2 = \begin{bmatrix} 30-20 \\ 20-20 \\ 10-20 \end{bmatrix} = \begin{bmatrix} 10 \\ 0 \\ -10 \end{bmatrix}$$

$$\therefore \quad s_{12} = \frac{1}{2}\left[\left(\mathbf{x}_1 - \mathbf{1}\bar{x}_1\right)^{T}\left(\mathbf{x}_2 - \mathbf{1}\bar{x}_2\right)\right]$$

$$= \frac{1}{2}\begin{bmatrix} -4.67 & -0.67 & 5.33 \end{bmatrix}\begin{bmatrix} 10 \\ 0 \\ -10 \end{bmatrix}$$

$$= \frac{1}{2}\times(-46.7-53.3) = \frac{1}{2}\times(-100.00) = -50$$

Further,

$$s_{11} = \frac{1}{2}\begin{bmatrix}-4.67 & -0.67 & 5.33\end{bmatrix}\begin{bmatrix}-4.67\\-0.67\\5.33\end{bmatrix}$$

$$= \frac{1}{2}(21.8089 + 0.4489 + 28.4089)$$

$$= \frac{1}{2}(50.67) = 25.34$$

and $$s_{22} = \frac{1}{2}\begin{bmatrix}10 & 0 & -10\end{bmatrix}\begin{bmatrix}10\\0\\-10\end{bmatrix}$$

$$= \frac{1}{2}(100 + 100) = 100$$

So, variance-covariance matrix $\mathbf{S}$ is

$$\mathbf{S} = \begin{bmatrix} s_{11} & s_{12} \\ s_{12} & s_{22} \end{bmatrix} = \begin{bmatrix} 25.34 & -50 \\ -50 & 100 \end{bmatrix}$$

Now, we see how through matrix manipulation, one can directly compute $\mathbf{S}_{p\times p}$ in one go.

We know

$$\bar{\mathbf{x}} = \frac{1}{n}\mathbf{X}^T\mathbf{1} = \begin{bmatrix}\bar{x}_1\\ \bar{x}_2\\ \cdot\\ \cdot\\ \cdot\\ \bar{x}_p\end{bmatrix}_{p\times 1} \quad \text{and} \quad \mathbf{X}_{n\times p} = \begin{bmatrix} x_{11} & x_{12} & \cdots & x_{1p}\\ x_{21} & x_{22} & \cdots & x_{2p}\\ \cdot & \cdot & \cdots & \cdot\\ \cdot & \cdot & \cdots & \cdot\\ \cdot & \cdot & \cdots & \cdot\\ x_{n1} & x_{n2} & \cdots & x_{np}\end{bmatrix}_{n\times p}$$

Now, we want to compute $\mathbf{X} - \bar{\mathbf{X}}$ as a $n \times p$ matrix.

Let $$\mathbf{1} = \begin{bmatrix}1\\1\\ \cdot\\ \cdot\\ \cdot\\ 1\end{bmatrix}_{n\times 1}$$

Then $$\mathbf{1}\,\bar{\mathbf{x}}^T = \begin{bmatrix}1\\1\\ \cdot\\ \cdot\\ \cdot\\ 1\end{bmatrix}_{n\times 1} \begin{bmatrix}\bar{x}_1 & \bar{x}_2 & \cdots & \bar{x}_p\end{bmatrix}_{1\times p} = \begin{bmatrix} \bar{x}_1 & \bar{x}_2 & \cdots & \bar{x}_p\\ \bar{x}_1 & \bar{x}_2 & \cdots & \bar{x}_p\\ \cdot & \cdot & \cdots & \cdot\\ \cdot & \cdot & \cdots & \cdot\\ \cdot & \cdot & \cdots & \cdot\\ \bar{x}_1 & \bar{x}_2 & \cdots & \bar{x}_p\end{bmatrix}_{n\times p}$$

$\mathbf{X}-\bar{\mathbf{X}}$ is actually the simplified form of $\mathbf{X}-\mathbf{1}\bar{\mathbf{x}}^T$. Now,

$$\mathbf{X}-\mathbf{1}\bar{\mathbf{x}}^T=\begin{bmatrix} x_{11}-\bar{x}_1 & x_{12}-\bar{x}_2 & \cdots & x_{1p}-\bar{x}_p \\ x_{21}-\bar{x}_1 & x_{22}-\bar{x}_2 & \cdots & x_{2p}-\bar{x}_p \\ \cdot & \cdot & \cdots & \cdot \\ \cdot & \cdot & \cdots & \cdot \\ \cdot & \cdot & \cdots & \cdot \\ x_{n1}-\bar{x}_1 & x_{n2}-\bar{x}_2 & \cdots & x_{np}-\bar{x}_p \end{bmatrix}_{n\times p}$$

Let,

$$\mathbf{S}=\frac{1}{n-1}\left[\left(\mathbf{X}-\mathbf{1}\bar{\mathbf{x}}^T\right)^T_{p\times n}\left(\mathbf{X}-\mathbf{1}\bar{\mathbf{x}}^T\right)_{n\times p}\right]_{p\times p}$$

Further, for simplicity, let there are two variables X_1 and X_2. Then,

$$\left(\mathbf{X}-\mathbf{1}\bar{\mathbf{x}}^T\right)^T\left(\mathbf{X}-\mathbf{1}\bar{\mathbf{x}}^T\right)$$

$$=\begin{bmatrix} x_{11}-\bar{x}_1 & x_{21}-\bar{x}_1 & \cdots & x_{n1}-\bar{x}_1 \\ x_{12}-\bar{x}_2 & x_{22}-\bar{x}_2 & \cdots & x_{n2}-\bar{x}_2 \end{bmatrix}_{2\times n}\begin{bmatrix} x_{11}-\bar{x}_1 & x_{12}-\bar{x}_2 \\ x_{21}-\bar{x}_1 & x_{22}-\bar{x}_2 \\ \cdot & \cdot \\ \cdot & \cdot \\ \cdot & \cdot \\ x_{n1}-\bar{x}_1 & x_{n2}-\bar{x}_2 \end{bmatrix}_{n\times 2}$$

$$=\begin{bmatrix} \sum_{i=1}^{n}\left(x_{i1}-\bar{x}_1\right)^2 & \sum_{i=1}^{n}\left(x_{i1}-\bar{x}_1\right)\left(x_{i2}-\bar{x}_2\right) \\ \sum_{i=1}^{n}\left(x_{i1}-\bar{x}_1\right)\left(x_{i2}-\bar{x}_2\right) & \sum_{i=1}^{n}\left(x_{i2}-\bar{x}_2\right)^2 \end{bmatrix}$$

$$=\begin{bmatrix} (n-1)s_{11} & (n-1)s_{12} \\ (n-1)s_{12} & (n-1)s_{22} \end{bmatrix}=(n-1)\begin{bmatrix} s_{11} & s_{12} \\ s_{12} & s_{22} \end{bmatrix}$$

$$=(n-1)\mathbf{S}$$

So, $$\mathbf{S}=\frac{1}{n-1}\left[\left(\mathbf{X}-\mathbf{1}\bar{\mathbf{x}}^T\right)^T\left(\mathbf{X}-\mathbf{1}\bar{\mathbf{x}}^T\right)\right]=\frac{1}{n-1}\left[\left(\mathbf{X}-\bar{\mathbf{X}}\right)^T\left(\mathbf{X}-\bar{\mathbf{X}}\right)\right] \quad (4.14)$$

Example 4.5

Compute $\mathbf{S}$ for $\mathbf{X}_{3x2}$ given in Example 4.2 using Equation 4.14.

Solution:

Given,

$$\mathbf{X}_{3\times2}=\begin{bmatrix}5 & 30\\ 9 & 20\\ 15 & 10\end{bmatrix} \text{ and } \bar{\mathbf{x}}=\begin{bmatrix}\bar{x}_1\\ \bar{x}_2\end{bmatrix}=\begin{bmatrix}9.67\\ 20\end{bmatrix} \quad \text{(from example 4.2)}$$

$$\text{Let, } \mathbf{1}=\begin{bmatrix}1\\1\\1\end{bmatrix}$$

$$\text{So}, \mathbf{1}\bar{\mathbf{x}}^T=\begin{bmatrix}1\\1\\1\end{bmatrix}_{3\times1}\begin{bmatrix}9.67 & 20\end{bmatrix}_{1\times2}=\begin{bmatrix}9.67 & 20\\ 9.67 & 20\\ 9.67 & 20\end{bmatrix}$$

$$\mathbf{X}-\mathbf{1}\bar{\mathbf{x}}^T=\begin{bmatrix}5 & 30\\ 9 & 20\\ 15 & 10\end{bmatrix}-\begin{bmatrix}9.67 & 20\\ 9.67 & 20\\ 9.67 & 20\end{bmatrix}=\begin{bmatrix}-4.67 & 10\\ -0.67 & 0\\ 5.33 & -10\end{bmatrix}$$

Now,

$$\left(\mathbf{X}-\mathbf{1}\bar{\mathbf{x}}^T\right)^T\left(\mathbf{X}-\mathbf{1}\bar{\mathbf{x}}^T\right)=\begin{bmatrix}-4.67 & -0.67 & 5.33\\ 10 & 0 & -10\end{bmatrix}\begin{bmatrix}-4.67 & 10\\ -0.67 & 0\\ 5.33 & -10\end{bmatrix}=\begin{bmatrix}50.67 & -100\\ -100 & 200\end{bmatrix}.$$

$$So, (n-1)\,\mathbf{S}=\begin{bmatrix}50.67 & -100\\ -100 & 200\end{bmatrix}$$

$$or,\ 2\mathbf{S}=\begin{bmatrix}50.67 & -100\\ -100 & 200\end{bmatrix}$$

$$or,\ \mathbf{S}=\begin{bmatrix}25.34 & -50\\ -50 & 100\end{bmatrix}$$

4.4 CORRELATION MATRIX

So far we have discussed multivariate mean vector and variance-covariance matrix for both population and sample. The other widely used multivariate descriptive statistic is correlation matrix. The population correlation matrix is denoted by $\boldsymbol{\rho}$ and the sample correlation matrix is denoted by $\mathbf{R}$. $\boldsymbol{\rho}$ and $\mathbf{R}$ take the following forms:

$$\boldsymbol{\rho}=\begin{bmatrix}\rho_{11} & \rho_{12} & \cdots & \rho_{1j} & \cdots & \rho_{1p}\\ \rho_{12} & \rho_{22} & \cdots & \rho_{2j} & \cdots & \rho_{2p}\\ \vdots & \vdots & \vdots & \vdots & \vdots & \vdots\\ \rho_{1j} & \rho_{2j} & \cdots & \rho_{jj} & \cdots & \rho_{jp}\\ \vdots & \vdots & \vdots & \vdots & \vdots & \vdots\\ \rho_{1p} & \rho_{2p} & \cdots & \rho_{jp} & \cdots & \rho_{pp}\end{bmatrix},\ \rho_{jj}=1 \text{ for } j=1,2,\ldots..,p.$$

and

$$\mathbf{R} = \begin{bmatrix} r_{11} & r_{12} & \cdots & r_{1j} & \cdots & r_{1p} \\ r_{12} & r_{22} & \cdots & r_{2j} & \cdots & r_{2p} \\ \vdots & \vdots & \vdots & \vdots & \vdots & \vdots \\ r_{1j} & r_{2j} & \cdots & r_{jj} & \cdots & r_{jp} \\ \vdots & \vdots & \vdots & \vdots & \vdots & \vdots \\ r_{1p} & r_{2p} & \cdots & r_{jp} & \cdots & r_{pp} \end{bmatrix}, \quad r_{jj} = 1 \text{ for } j = 1,2,\ldots,p.$$

Where

$$\rho_{jk} = \frac{\sigma_{jk}}{\sqrt{\sigma_{jj} \cdot \sigma_{kk}}} = \frac{\sigma_{jk}}{\sigma_j \cdot \sigma_k} \quad \text{or} \quad \sigma_{jk} = \rho_{jk}\sigma_j\sigma_k \tag{4.15}$$

and

$$r_{jk} = \frac{s_{jk}}{\sqrt{s_{jj} \cdot s_{kk}}} = \frac{s_{jk}}{s_j \cdot s_k} \quad \text{or} \quad s_{jk} = r_{jk}s_js_k \tag{4.16}$$

Where $j = k$

$$\rho_{jk} = \rho_{jj} = \frac{\sigma_{jj}}{\sigma_j \cdot \sigma_j} = \frac{\sigma_j^2}{\sigma_j^2} = 1,$$

and

$$r_{jk} = r_{jj} = \frac{s_{jj}}{s_j \cdot s_j} = \frac{s_j^2}{s_j^2} = 1$$

That is why the diagonal elements of $\boldsymbol{\rho}$ and $\mathbf{R}$ are 1.

Now let us see how we can compute the sample correlation matrix from sample data $\mathbf{X}_{n\times p}$. We have seen that in order to compute $\mathbf{S}_{p\times p}$, we transformed $\mathbf{X}_{n\times p}$ into its mean subtracted matrix, $\mathbf{X}_{n\times p} - \mathbf{1}\bar{\mathbf{x}}^T$. If we denote $\mathbf{X}^* = \mathbf{X}_{n\times p} - \mathbf{1}\bar{\mathbf{x}}^T$, then from Equation 4.14 we can write that

$$\mathbf{X}^{*\mathrm{T}}\mathbf{X}^* = (n-1)\mathbf{S} \tag{4.17}$$

Let $\tilde{\mathbf{X}}_{n\times p}$ be the transformed matrix of $\mathbf{X}_{n\times p}$ in such a way that each element x_{ij} of $\mathbf{X}_{n\times p}$ is subtracted from its corresponding variable mean and divided by its corresponding standard deviation s_j. If $\tilde{x}_{ij}$ denote an element of $\tilde{\mathbf{X}}_{n\times p}$, then

$$\tilde{x}_{ij} = \frac{x_{ij} - \bar{x}_j}{s_j} = \frac{x_{ij} - \bar{x}_j}{\sqrt{s_{jj}}}$$

We get $\tilde{\mathbf{X}}$ as below.

$$\tilde{\mathbf{X}} = \begin{bmatrix} \tilde{x}_{11} & \tilde{x}_{12} & \cdots & \tilde{x}_{1p} \\ \tilde{x}_{21} & \tilde{x}_{22} & \cdots & \tilde{x}_{2p} \\ \cdot & \cdot & \cdots & \cdot \\ \cdot & \cdot & \cdots & \cdot \\ \cdot & \cdot & \cdots & \cdot \\ \tilde{x}_{n1} & \tilde{x}_{n2} & \cdots & \tilde{x}_{np} \end{bmatrix}_{n\times p}$$

Then, $$(n-1)\mathbf{R} = \tilde{\mathbf{X}}^T\tilde{\mathbf{X}} \tag{4.18}$$

Example 4.6

Compute **R** for $\mathbf{X}_{3x2}$ given in Example 4.2

Solution:

$$\mathbf{X}_{3\times2} = \begin{bmatrix} 5 & 30 \\ 9 & 20 \\ 15 & 10 \end{bmatrix} \text{ and } \overline{\mathbf{x}} = \begin{bmatrix} 9.67 \\ 20 \end{bmatrix} \quad (\text{from Example 4.2})$$

$$\text{and} \quad \mathbf{S} = \begin{bmatrix} 25.34 & -50 \\ -50 & 100 \end{bmatrix} \quad (\text{from Example 4.4})$$

$$\text{So,} \quad \tilde{\mathbf{X}} = \begin{bmatrix} \frac{5-9.67}{\sqrt{25.34}} & \frac{30-20}{\sqrt{100}} \\ \frac{9-9.67}{\sqrt{25.34}} & \frac{20-20}{\sqrt{100}} \\ \frac{15-9.67}{\sqrt{25.34}} & \frac{10-20}{\sqrt{100}} \end{bmatrix} = \begin{bmatrix} -0.93 & 1 \\ -0.13 & 0 \\ 1.06 & -1 \end{bmatrix}$$

Now,

$$\tilde{\mathbf{X}}^T\tilde{\mathbf{X}} = \begin{bmatrix} -0.93 & -0.13 & 1.06 \\ 1.0 & 0.00 & -1.0 \end{bmatrix} \begin{bmatrix} -0.93 & 1.00 \\ -0.13 & 0.00 \\ 1.06 & -1.00 \end{bmatrix} = \begin{bmatrix} 2.00 & -1.99 \\ -1.99 & 2.00 \end{bmatrix}$$

$$\therefore \quad (3-1)\mathbf{R} = \begin{bmatrix} 2.00 & -1.99 \\ -1.99 & 2.00 \end{bmatrix}$$

$$\therefore \quad \mathbf{R} = \begin{bmatrix} 1.00 & -0.995 \\ -0.995 & 1.00 \end{bmatrix}$$

Comparing Equations 4.14 and 4.18, it can be seen that the correlation matrix **R** is the standardized covariance matrix **S**, where each element of **S**, i.e., s_{jk}, $j = 1,2,\ldots,p$ and $k = 1,2,\ldots,p$, is divided by the respective standard deviations, i.e., s_j and s_k (see Equation 4.16). The same can be done using matrix manipulation. The steps are given below.

Step 1: Compute $\mathbf{S}_{p\times p}$ using Equation 4.14.
Step 2: Obtain $\mathbf{D}_S$ where $\mathbf{D}_S$ is $p\times p$ diagonal matrix with diagonal elements from **S**. So,

$$\mathbf{D}_S = \begin{bmatrix} s_{11} & 0 & 0\cdots0 \\ 0 & s_{22} & 0\cdots0 \\ \vdots & & \vdots \\ 0 & 0 & 0\cdots s_{pp} \end{bmatrix}_{p\times p}$$

Step 3: Obtain inverse of $\mathbf{D}_S$, i.e., $\mathbf{D}_S^{-1}$

Step 4: Obtain square root of $\mathbf{D}_S^{-1}$, i.e., $\mathbf{D}_S^{-1/2}$
Step 5: Compute **R** using the following equation

$$\mathbf{R} = \mathbf{D}_S^{-1/2}\ \mathbf{S}\ \mathbf{D}_S^{-1/2} \tag{4.19}$$

Example 4.7

Consider the covariance matrix **S** from Example 4.5. Find out the correlation matrix **R** using Equation 4.19.

Solution:

Step 1: Given $\mathbf{S} = \begin{bmatrix} 25.34 & -50 \\ -50 & 100 \end{bmatrix}$ from Example 4.5

Step 2: $\mathbf{D}_S = \begin{bmatrix} 25.34 & 0 \\ 0 & 100 \end{bmatrix}$

Step 3: Obtain $\mathbf{D}_S^{-1}$

The inverse of a diagonal matrix is obtained by making reciprocal of the diagonal elements. So,

$$\mathbf{D}_S^{-1} = \begin{bmatrix} \frac{1}{25.34} & 0 \\ 0 & \frac{1}{100} \end{bmatrix} = \begin{bmatrix} 0.039 & 0 \\ 0 & 0.01 \end{bmatrix}$$

Step 4: Obtain square root of $\mathbf{D}_S^{-1}$, i.e., $\mathbf{D}_S^{-1/2}$
$\mathbf{D}_S^{-1/2}$ can be obtained by taking square root of each of the elements. So,

$$\mathbf{D}_S^{-1/2} = \begin{bmatrix} \sqrt{0.039} & 0 \\ 0 & \sqrt{0.01} \end{bmatrix} = \begin{bmatrix} 0.20 & 0 \\ 0 & 0.10 \end{bmatrix}$$

Step 5: Compute **R**, where

$$\begin{aligned}
\mathbf{R} &= \mathbf{D}_S^{-1/2}\ \mathbf{S}\ \mathbf{D}_S^{-1/2} \\
&= \begin{bmatrix} 0.20 & 0 \\ 0 & 0.10 \end{bmatrix} \begin{bmatrix} 25.34 & -50 \\ -50 & 100 \end{bmatrix} \begin{bmatrix} 0.20 & 0 \\ 0 & 0.10 \end{bmatrix} \\
&= \begin{bmatrix} 0.20 \times 25.34 & -50 \times 0.20 \\ -50 \times 0.10 & 100 \times 0.10 \end{bmatrix} \begin{bmatrix} 0.20 & 0 \\ 0 & 0.10 \end{bmatrix} \\
&= \begin{bmatrix} 0.20 \times 25.34 \times 0.20 & -50 \times 0.20 \times 0.10 \\ -50 \times 0.20 \times 0.10 & 100 \times 0.10 \times 0.10 \end{bmatrix} \\
&= \begin{bmatrix} 1.01 & -1.0 \\ -1.0 & 1.0 \end{bmatrix} \\
&= \begin{bmatrix} 1.0 & -1.0 \\ -1.0 & 1.0 \end{bmatrix}
\end{aligned}$$

See the difference in correlation coefficient value (–1.0) with that in Example 4.6 (–0.995). It is because of rounding off errors.

In a similar manner, if we know **R** and $\mathbf{D}_S$, we can calculate **S** by using the following formula.

$$\mathbf{S} = \mathbf{D}_S^{1/2}\ \mathbf{R}\ \mathbf{D}_S^{1/2} \tag{4.20}$$

The symmetric matrices $\mathbf{X}^{*T}\ \mathbf{X}^{*}$ (Equation 4.17) and $\tilde{\mathbf{X}}^{T}\ \tilde{\mathbf{X}}$ (Equation 4.18) are known as sum squares and cross product (SSCP) matrices. Another SSCP matrix which is most frequently used in multivariate statistics is $\mathbf{X}^{T}\ \mathbf{X}$ as shown below.

$$\mathbf{X}^{T}\mathbf{X} = \begin{bmatrix} \sum_{i=1}^{n} x_{i1}^2 & \sum_{i=1}^{n} x_{i1}x_{i2} & \cdots \sum_{i=1}^{n} x_{i1}x_{ip} \\ \sum_{i=1}^{n} x_{i1}x_{i2} & \sum_{i=1}^{n} x_{i2}^2 & \cdots \sum_{i=1}^{n} x_{i2}x_{ip} \\ \cdot & \cdot & \cdots \quad \cdot \\ \cdot & \cdot & \cdots \quad \cdot \\ \cdot & \cdot & \cdots \quad \cdot \\ \sum_{i=1}^{n} x_{i1}x_{ip} & \sum_{i=1}^{n} x_{i2}x_{ip} & \cdots \sum_{i=1}^{n} x_{ip}^2 \end{bmatrix}$$

Example 4.8

Using the CityCan data set, perform the following tasks:

(i) Compute mean vector $\bar{\mathbf{x}}$, and SSCP matrices namely $\mathbf{X}^{T}\ \mathbf{X}$, $\mathbf{X}^{*T}\ \mathbf{X}^{*}$ and $\tilde{\mathbf{X}}^{T}\ \tilde{\mathbf{X}}$.
(ii) Compute covariance matrix **S** and correlation matrix **R**.
(iii) Convert **S** to **R** and vice versa.

Solution:

(i) The CityCan data matrix, $\mathbf{X}_{12\times5}$ is given below.

$$\mathbf{X} = \begin{bmatrix} 10 & 100 & 9 & 62 & 1 \\ 12 & 110 & 8 & 58 & 1.3 \\ 11 & 105 & 7 & 64 & 1.2 \\ 9 & 94 & 14 & 60 & 0.8 \\ 9 & 95 & 12 & 63 & 0.8 \\ 10 & 99 & 10 & 57 & 0.9 \\ 11 & 104 & 7 & 55 & 1 \\ 12 & 108 & 4 & 56 & 1.2 \\ 11 & 105 & 6 & 59 & 1.1 \\ 10 & 98 & 5 & 61 & 1 \\ 11 & 103 & 7 & 57 & 1.2 \\ 12 & 110 & 6 & 60 & 1.2 \end{bmatrix}$$

Suppose, $\mathbf{1} = \begin{bmatrix} 1 \\ 1 \\ 1 \\ 1 \\ 1 \\ 1 \\ 1 \\ 1 \\ 1 \\ 1 \\ 1 \\ 1 \end{bmatrix}$

So, the mean vector $\bar{\mathbf{x}}$ is

$$\bar{\mathbf{x}} = \frac{1}{12}\mathbf{X}^T\mathbf{1}$$

$$= \frac{1}{12}\begin{bmatrix} 10 & 12 & 11 & 9 & 9 & 10 & 11 & 12 & 11 & 10 & 11 & 12 \\ 100 & 110 & 105 & 94 & 95 & 99 & 104 & 108 & 105 & 98 & 103 & 110 \\ 9 & 8 & 7 & 14 & 12 & 10 & 7 & 4 & 6 & 5 & 7 & 6 \\ 62 & 58 & 64 & 60 & 63 & 57 & 55 & 56 & 59 & 61 & 57 & 60 \\ 1 & 1.3 & 1.2 & 0.8 & 0.8 & 0.9 & 1 & 1.2 & 1.1 & 1 & 1.2 & 1.2 \end{bmatrix}\begin{bmatrix} 1 \\ 1 \\ 1 \\ 1 \\ 1 \\ 1 \\ 1 \\ 1 \\ 1 \\ 1 \\ 1 \\ 1 \end{bmatrix}$$

$$= \begin{bmatrix} 10.67 \\ 102.58 \\ 7.92 \\ 59.33 \\ 1.06 \end{bmatrix}$$

The transpose of $\mathbf{X}$, i.e., $\mathbf{X}^{\mathrm{T}}$ is

$$\mathbf{X}^T = \begin{bmatrix} 10 & 12 & 11 & 9 & 9 & 10 & 11 & 12 & 11 & 10 & 11 & 12 \\ 100 & 110 & 105 & 94 & 95 & 99 & 104 & 108 & 105 & 98 & 103 & 110 \\ 9 & 8 & 7 & 14 & 12 & 10 & 7 & 4 & 6 & 5 & 7 & 6 \\ 62 & 58 & 64 & 60 & 63 & 57 & 55 & 56 & 59 & 61 & 57 & 60 \\ 1 & 1.3 & 1.2 & 0.8 & 0.8 & 0.9 & 1 & 1.2 & 1.1 & 1 & 1.2 & 1.2 \end{bmatrix}$$

So, $\mathbf{X}^T\mathbf{X}$ is

$$\mathbf{X}^T\mathbf{X} = \begin{bmatrix} 1378 & 13194 & 987 & 7580 & 137.3 \\ 13194 & 126605 & 9622 & 72980 & 1312 \\ 987 & 9622 & 845 & 5663 & 96.6 \\ 7580 & 72980 & 5663 & 42334 & 752.4 \\ 137.3 & 1312 & 96.6 & 752.4 & 13.75 \end{bmatrix}$$

Now, $\mathbf{X}^*$ denotes the mean subtracted $\mathbf{X}$, i.e., $\mathbf{X} - \overline{\mathbf{X}}$ where $\overline{\mathbf{X}} = \mathbf{1}\overline{\mathbf{x}}^T$.

$$\mathbf{X}^* = \begin{bmatrix} -0.67 & -2.58 & 1.08 & 2.67 & -0.06 \\ 1.33 & 7.42 & 0.08 & -1.33 & 0.24 \\ 0.33 & 2.42 & -0.92 & 4.67 & 0.14 \\ -1.67 & -8.58 & 6.08 & 0.67 & -0.26 \\ -1.67 & -7.58 & 4.08 & 3.67 & -0.26 \\ -0.67 & -3.58 & 2.08 & -2.33 & -0.16 \\ 0.33 & 1.42 & -0.92 & -4.33 & -0.06 \\ 1.33 & 5.42 & -3.92 & -3.33 & 0.14 \\ 0.33 & 2.42 & -1.92 & -0.33 & 0.04 \\ -0.67 & -4.58 & -2.92 & 1.67 & -0.06 \\ 0.33 & 0.42 & -0.92 & -2.33 & 0.14 \\ 1.33 & 7.42 & -1.92 & 0.67 & 0.14 \end{bmatrix}$$

So,

$$\mathbf{X}^{*T} = \begin{bmatrix} -0.67 & 1.33 & 0.33 & -1.67 & -1.67 & -0.67 & 0.33 & 1.33 & 0.33 & -0.67 & 0.33 & 1.33 \\ -2.58 & 7.42 & 2.42 & -8.58 & -7.58 & -3.58 & 1.42 & 5.42 & 2.42 & -4.58 & 0.42 & 7.42 \\ 1.08 & 0.08 & -0.92 & 6.08 & 4.08 & 2.08 & -0.92 & -3.92 & -1.92 & -2.92 & -0.92 & -1.92 \\ 2.67 & -1.33 & 4.67 & 0.67 & 3.67 & -2.33 & -4.33 & -3.33 & -0.33 & 1.67 & -2.33 & 0.67 \\ -0.06 & 0.24 & 0.14 & -0.26 & -0.26 & -0.16 & -0.06 & 0.14 & 0.04 & -0.06 & 0.14 & 0.14 \end{bmatrix}$$

Therefore,

$$\mathbf{X}^{*T}\mathbf{X}^* = \begin{bmatrix} 12.67 & 63.33 & -26.33 & -14.67 & 1.83 \\ 63.33 & 324.92 & -123.42 & -59.33 & 9.19 \\ -26.33 & -123.42 & 92.92 & 26.33 & -3.94 \\ -14.67 & -59.33 & 26.33 & 88.67 & -1.13 \\ 1.83 & 9.19 & -3.94 & -1.13 & 0.31 \end{bmatrix}$$

Again, let $\tilde{\mathbf{X}}$ denotes the standardized $\mathbf{X}$ and is

$$\tilde{\mathbf{X}} = \begin{bmatrix} -0.62 & -0.48 & 0.37 & 0.94 & -0.35 \\ 1.24 & 1.36 & 0.03 & -0.47 & 1.44 \\ 0.31 & 0.44 & -0.32 & 1.64 & 0.85 \\ -1.55 & -1.58 & 2.09 & 0.23 & -1.54 \\ -1.55 & -1.40 & 1.40 & 1.29 & -1.54 \\ -0.62 & -0.66 & 0.72 & -0.82 & -0.94 \\ 0.31 & 0.26 & -0.32 & -1.53 & -0.35 \\ 1.24 & 1.00 & -1.35 & -1.17 & 0.85 \\ 0.31 & 0.44 & -0.66 & -0.12 & 0.25 \\ -0.62 & -0.84 & -1.00 & 0.59 & -0.35 \\ 0.31 & 0.08 & -0.32 & -0.82 & 0.85 \\ 1.24 & 1.36 & -0.66 & 0.23 & 0.85 \end{bmatrix}$$

Then,

$$\tilde{\mathbf{X}}^T = \begin{bmatrix} -0.62 & 1.24 & 0.31 & -1.55 & -1.55 & -0.62 & 0.31 & 1.24 & 0.31 & -0.62 & 0.31 & 1.24 \\ -0.48 & 1.36 & 0.44 & -1.58 & -1.40 & -0.66 & 0.26 & 1.00 & 0.44 & -0.84 & 0.08 & 1.36 \\ 0.37 & 0.03 & -0.32 & 2.09 & 1.40 & 0.72 & -0.32 & -1.35 & -0.66 & -1.00 & -0.32 & -0.66 \\ 0.94 & -0.47 & 1.64 & 0.23 & 1.29 & -0.82 & -1.53 & -1.17 & -0.12 & 0.59 & -0.82 & 0.23 \\ -0.35 & 1.44 & 0.85 & -1.54 & -1.54 & -0.94 & -0.35 & 0.85 & 0.25 & -0.35 & 0.85 & 0.85 \end{bmatrix}$$

So,

$$\tilde{\mathbf{X}}^T\tilde{\mathbf{X}} = \begin{bmatrix} 11.00 & 10.86 & -8.44 & -4.81 & 10.19 \\ 10.86 & 11.00 & -7.81 & -3.85 & 10.09 \\ -8.44 & -7.81 & 11.00 & 3.19 & -8.09 \\ -4.81 & -3.85 & 3.19 & 11.00 & -2.38 \\ 10.19 & 10.09 & -8.09 & -2.38 & 11.00 \end{bmatrix}$$

(ii) From Equation 4.17, we know that $(n-1)\mathbf{S} = \mathbf{X}^{*\mathrm{T}}\mathbf{X}^{*}$. So,

$$\mathbf{S} = \frac{1}{12-1}\mathbf{X}^{*\mathrm{T}}\mathbf{X}^{*} = \frac{1}{11}\begin{bmatrix} 12.67 & 63.33 & -26.33 & -14.67 & 1.83 \\ 63.33 & 324.92 & -123.42 & -59.33 & 9.19 \\ -26.33 & -123.42 & 92.92 & 26.33 & -3.94 \\ -14.67 & -59.33 & 26.33 & 88.67 & -1.13 \\ 1.83 & 9.19 & -3.94 & -1.13 & 0.31 \end{bmatrix}$$

$$= \begin{bmatrix} 1.15 & 5.76 & -2.39 & -1.33 & 0.17 \\ 5.76 & 29.54 & -11.22 & -5.39 & 0.84 \\ -2.39 & -11.22 & 8.45 & 2.39 & -0.36 \\ -1.33 & -5.39 & 2.39 & 8.06 & -0.10 \\ 0.17 & 0.84 & -0.36 & -0.10 & 0.03 \end{bmatrix}$$

Similarly, from Equation 4.18, we know that $(n-1)\mathbf{R} = \tilde{\mathbf{X}}^T\tilde{\mathbf{X}}$. So,

$$\mathbf{R} = \frac{1}{12-1}\tilde{\mathbf{X}}^T\tilde{\mathbf{X}} = \frac{1}{11}\begin{bmatrix} 11.00 & 10.86 & -8.44 & -4.81 & 10.19 \\ 10.86 & 11.00 & -7.81 & -3.85 & 10.09 \\ -8.44 & -7.81 & 11.00 & 3.19 & -8.09 \\ -4.81 & -3.85 & 3.19 & 11.00 & -2.38 \\ 10.19 & 10.09 & -8.09 & -2.38 & 11.00 \end{bmatrix}$$

$$= \begin{bmatrix} 1.00 & 0.99 & -0.77 & -0.44 & 0.93 \\ 0.99 & 1.00 & -0.71 & -0.35 & 0.92 \\ -0.77 & -0.71 & 1.00 & 0.29 & -0.74 \\ -0.44 & -0.35 & 0.29 & 1.00 & -0.22 \\ 0.93 & 0.92 & -0.74 & -0.22 & 1.00 \end{bmatrix}$$

(iii) $\mathbf{R} = \mathbf{D}_S^{-1/2}\ \mathbf{S}\ \mathbf{D}_S^{-1/2}$

$$\mathbf{D}_S = \begin{bmatrix} 1.15 & 0 & 0 & 0 & 0 \\ 0 & 29.54 & 0 & 0 & 0 \\ 0 & 0 & 8.45 & 0 & 0 \\ 0 & 0 & 0 & 8.06 & 0 \\ 0 & 0 & 0 & 0 & 0.03 \end{bmatrix}$$

So,

$$\mathbf{D}_S^{-1/2} = \begin{bmatrix} 0.93 & 0 & 0 & 0 & 0 \\ 0 & 0.18 & 0 & 0 & 0 \\ 0 & 0 & 0.34 & 0 & 0 \\ 0 & 0 & 0 & 0.35 & 0 \\ 0 & 0 & 0 & 0 & 5.96 \end{bmatrix}$$

So, $$\mathbf{D}_S^{-1/2}\ \mathbf{S}\ \mathbf{D}_S^{-1/2} = \begin{bmatrix} 0.93 & 0 & 0 & 0 & 0 \\ 0 & 0.18 & 0 & 0 & 0 \\ 0 & 0 & 0.34 & 0 & 0 \\ 0 & 0 & 0 & 0.35 & 0 \\ 0 & 0 & 0 & 0 & 5.96 \end{bmatrix}$$

$$* \begin{bmatrix} 1.15 & 5.76 & -2.39 & -1.33 & 0.17 \\ 5.76 & 29.54 & -11.22 & -5.39 & 0.84 \\ -2.39 & -11.22 & 8.45 & 2.39 & -0.36 \\ -1.33 & -5.39 & 2.39 & 8.06 & -0.10 \\ 0.17 & 0.84 & -0.36 & -0.10 & 0.03 \end{bmatrix}$$

$$* \begin{bmatrix} 0.93 & 0 & 0 & 0 & 0 \\ 0 & 0.18 & 0 & 0 & 0 \\ 0 & 0 & 0.34 & 0 & 0 \\ 0 & 0 & 0 & 0.35 & 0 \\ 0 & 0 & 0 & 0 & 5.96 \end{bmatrix}$$

$$
= \begin{bmatrix} 1.00 & 0.99 & -0.77 & -0.44 & 0.93 \\ 0.99 & 1.00 & -0.71 & -0.35 & 0.92 \\ -0.77 & -0.71 & 1.00 & 0.29 & -0.74 \\ -0.44 & -0.35 & 0.29 & 1.00 & -0.22 \\ 0.93 & 0.92 & -0.74 & -0.22 & 1.00 \end{bmatrix}
$$

The reader may exercise $\mathbf{S} = \mathbf{D}_S^{1/2}\ \mathbf{R}\ \mathbf{D}_S^{1/2}$.

4.5 TYPES OF CORRELATION

The correlation coefficient obtained through Equation 4.15 or 4.16 is known as Pearson product moment correlation. For Pearson product moment correlation between two variables, both the variables must be measured using ratio or interval scale. We have seen earlier (in Chapter 1) that in real life, many variables are measured using nominal or ordinal scale. For example, let us consider all variables used by CityCan, i.e., month, profit, sales volume, absenteeism, breakdown-hour, and M-ratio (marketing performance). The variable month is nominal in nature. In addition, suppose marketing performance is measured using ordinal scale such as low (M ratio < 1), medium (1 ≤ M-ratio ≤ 1.1) high (M-ratio >1.1). Similarly, let's say absenteeism is also measured in ordinal scale as low, medium, and high. Further, assume that there are three vendors V_1, V_2, and V_3 supplying raw materials to CityCan.

So, the situation involves mixed data types as given below.

- Ratio data: profit, sales-volume, breakdown-hour
- Ordinal data: M-ratio, absenteeism
- Nominal data: month, vendor

The question is how to measure correlation coefficient when such mixed data is to be analyzed. The following rules can be adopted:

- For correlation between two variables, both measured in ratio or interval scale (continuous scale), use Pearson product moment correlation.
- For correlation between two variables, both measured in ordinal scale, use Spearman rank order correlation (Spearman rho) or Kendal Tau.
- For correlation between two variables, both measured in nominal scale, use χ^2-based association such as phi (ϕ).
- For correlation between two variables, one measured in continuous scale and the other in ordinal scale, use biserial and polyserial correlations, depending on whether the ordinal variable is dichotomous or polytomous in nature.
- For correlation between two variables, one measured in continuous scale and other in nominal scale, use point polyserial or point biserial correlation.
- For correlation between two variables, one measured in ordinal scale and the other in nominal scale, use gamma.

It should be noted here that there are also other correlation measures available in the literature for the situations described above. Interested readers may consult Pearson (1909), Tate (1955), Olkin and Tate (1961), Olsson et al. (1982), and Kraemer (2006).

4.5.1 Correlation between Two Ordinal Variables

In this section we will explain Spearman rank order correlation (Spearman rho) and gamma. Spearman rho is used for continuous ordinal data. Continuous ordinal data represents a situation when each of the observations can have unique rank or there is no or a very small number of tied rankings. Tied rankings arise when two or more observations have equal value. The second type of ordinal data is called collapsed ordinal data, extensively used in survey research, where the variable is measured on a Likert type scale. For example, employee satisfaction can be measured with a five-point Likert scale with the order of measurement as very low, low, medium, high, and very high. A large number of tied observations are obtained in such situations. Therefore, Spearman rho is not applicable. One can use gamma coefficient.

Spearman's rho

The steps to compute Spearman's rho for two ordinal variables X and Y are given below.

Step 1: Arrange observations on X (i.e., x_i, $i = 1, 2, \ldots, n$) in ascending order ($x_{(i)}$, $i = 1, 2, \ldots, n$) and rank them. If there are n observations with no tied observations, there will be n ranks. For tied observations, use equal rank as the average of the positions they hold in the spectrum from smallest to highest. For example, if the second and third observations are tied, their rank will be $\frac{2+3}{2} = 2.5$.

Step 2: Develop a table of ranked observations for X, where $R(x_{(i)})$ is the rank of the ordered i-th observation ($x_{(i)}$) of X, for $i = 1, 2, \ldots, n$.

Step 3: Rearrange the data in the original order of observations on X and assign the corresponding ranks obtained in Step 2. Denote the rank of x_i as $R(x_i)$.

Step 4: Repeat Steps 1 to 3 for variable Y.

Step 5: Obtain differences in the ranks of x_i and y_i, i.e., d_i, where $d_i = R(x_i) - R(y_i)$

Step 6: Obtain Spearman's rank correlation or Spearman rho (ρ) as

$$\rho = 1 - \frac{6\sum_{i=1}^{n} d_i^2}{n(n^2-1)} \tag{4.21}$$

The interpretation of Spearman's ρ is similar to Pearson's correlation coefficient.

- $-1 \le \rho \le 1$
- $\rho = 0$ signifies no correlation
- $|\rho| = 1$ signifies perfect correlation
- $|\rho| \ge 0.90$ is desirable and $|\rho| \ge 0.70$ is acceptable for most of the practical situations.

Example 4.9

A training program was attended by ten participants. In order to evaluate the improvement due to the training imparted, two-stage evaluations were done, i.e., pre- and post-training evaluation. The performances of the ten participants are given in Table 4.2. Obtain Spearman ρ and comment on the result.

Solution:

Let the pre-training performance be termed as X and that for the post training be termed as Y.

Following Steps 1 to 3, discussed above, the $R(x_{(i)})$ and $R(x_i)$ are obtained and given in Tables 4.3 and 4.4.

As there is tie in the second and third ordered observations, i.e., $x_{(2)}$ and $x_{(3)}$, their rank is $\frac{2+3}{2} = 2.5$. So, x_i, $i = 1,2,...,10$ with rank $R(x_i)$ is given in Table 4.4.

Similarly, $R(y_{(i)})$ and $R(y_i)$ are obtained and shown in Tables 4.5 and 4.6, respectively.

As there is no tied observation, $R(y_{(i)})$ is equal to its rank position in the data spectrum. The y_i, $i = 1,2,...10$ with rank $R(y_i)$ is given in Table 4.6.

Following Steps 5 and 6, d_i and ρ are computed as shown in Table 4.7.

The Spearman ρ is 0.91 which indicates high positive correlation between pre-training performance and post-training performance of the participants. As the post-training performance

TABLE 4.2
Data for Example 4.9

Participant	1	2	3	4	5	6	7	8	9	10
Pre-training Performance	35	49	69	35	60	54	29	42	56	45
Post-training Performance	54	66	80	65	86	68	55	63	71	59

TABLE 4.3
Rank of the Ordered i-th Observation, $R(x_{(i)})$

i	1	2	3	4	5	6	7	8	9	10
x_i	35	49	69	35	60	54	29	42	56	45
$x_{(i)}$	29	35	35	42	45	49	54	56	60	69
Rank Position	1	2	3	4	5	6	7	8	9	10
$R(x_{(i)})$	1	2.5	2.5	4	5	6	7	8	9	10

TABLE 4.4
Rank of Original Order of Observations, $R(x_i)$

i	1	2	3	4	5	6	7	8	9	10
x_i	35	49	69	35	60	54	29	42	56	45
$R(x_i)$	2.5	6	10	2.5	9	7	1	4	8	5

TABLE 4.5
Rank of the Ordered i-th Observation, $R(y_{(i)})$

	1	2	3	4	5	6	7	8	9	10
y_i	54	66	80	65	86	68	55	63	71	59
$y_{(i)}$	54	55	59	63	65	66	68	71	80	86
Rank Position	1	2	3	4	5	6	7	8	9	10
$R(y_{(i)})$	1	2	3	4	5	6	7	8	9	10

TABLE 4.6
Rank of Original order of Observations, $R(y_i)$

i	1	2	3	4	5	6	7	8	9	10
y_i	54	66	80	65	86	68	55	63	71	59
$R(y_i)$	1	6	9	5	10	7	2	4	8	3

TABLE 4.7
Rank Difference and Spearman Rank Correlation (ρ)

i	$R(x_i)$	$R(y_i)$	d_i	d_i^2	Spearman ρ
1	2.5	1	1.5	2.25	$\rho = 1 - \frac{6\sum_{i=1}^{n} d_i^2}{n(n^2-1)}$
2	6	6	0	0	
3	10	9	1	1	$= 1 - \frac{6 \times 15.5}{10(100-1)}$
4	2.5	5	−2.5	6.25	
5	9	10	−1	1	$= 0.91$
6	7	7	0	0	
7	1	2	−1	1	
8	4	4	0	0	
9	8	8	0	0	
10	5	3	2	4	
Total				15.5	

of each participant (y_i) is better than pre-training performance (x_i) and ρ is high, we conclude that the training is effective.

Gamma Coefficient

For ordinal variables, a monotone trend association is a common phenomenon. That is for ordinal X and Y, as the level of X increases, the response on Y will increase or decrease towards higher or lower levels of Y. For example, consider operators' expertise (X) versus errors committed (Y) in performing an operation. Let's say X is measured as low, medium, and high and Y is measured as less, average, and high. It is quite likely that an operator with low expertise commits higher number of errors and vice versa. Under such situations, gamma coefficient is used as a measure of correlation that basically measures the degree to which the relationship (between X and Y) is monotone.

To measure gamma, the following steps are to be followed (Agresti, 2002):

Step 1: Develop contingency table between X and Y.

Step 2: Compute the number of concordant pairs (N_c). A pair is said concordant when a subject (respondent) ranked higher on X also ranks higher on Y.

Step 3: Compute the number of discordant pairs (N_d). A pair is said to be discordant if a subject having higher ranking on X ranks lower on Y. If a subject has same ranking both on X and Y, the pair is said to be "tied."

Step 4: Compute gamma (γ) using the following formula (Goodman & Kruskal, 1954):

$$\gamma = \frac{N_c - N_d}{N_c + N_d} \tag{4.22}$$

Consider the following general contingency table (Table 4.8)

TABLE 4.8
Contingency Table of Two Ordinal Variables

x_i \ y_i	y_1	y_2	...	y_b
x_1	n_{11}	n_{12}	...	n_{1b}
x_2	n_{21}	n_{22}	...	n_{2b}
$\vdots$	$\vdots$	$\vdots$		$\vdots$
x_a	n_{a1}	n_{a2}	...	n_{ab}

TABLE 4.9
Contingency Table of *X* (Expertise) and *Y* (Error Committed)

	Errors (*Y*)			
Expertise (*X*)	**Less**	**Average**	**High**	**Total**
Low	10	15	10	**35**
Medium	15	12	8	**35**
High	20	8	2	**30**
Total	**45**	**35**	**20**	**100**

X has a ordinal categories, say, $x_1 < x_2 < \cdots < x_a$ and Y has b ordinal categories, say, $y_1 < y_2 < \cdots < y_b$. The number of concordant (N_c) and discordant (N_d) pairs are

$$N_c = \sum_i \sum_j n_{ij} \left(\sum_{h>i} \sum_{k>j} n_{hk} \right) \tag{4.23}$$

$$N_d = \sum_i \sum_j n_{ij} \left(\sum_{h>i} \sum_{k<j} n_{hk} \right) \tag{4.24}$$

Example 4.10

Obtain γ using the following contingency table of X (expertise) and Y (error committed) (Table 4.9).

Solution:

As the contingency table between X and Y is provided, we start with Step 2.

Step 2: Number of concordant pairs, N_c

Both X and Y have three levels. So, there are $3 \times 3 = 9$ cells. Now, consider the cell LL (low, less), the top-left corner cell which has ten respondents (subjects). Similarly, the other cells row-wise can be denoted as LA, LH, ML, MA, MH, HL, HA, and HH. If we consider one respondent at LL, then any pair formed with any respondents in cells MA, MH, HA, and HH is concordant. This is because the subjects in MA, MH, HA, or HH have higher ordinal response than the subjects on LL on both X and Y. So, for one subject in LL, there are $12 + 8 + 8 + 2 = 30$

TABLE 4.10
Concordant Pairs for X (Expertise) and Y (Error Committed)

(i, j) cell	Contribution to N_c $n_{ij}\left(\sum_{h>i}\sum_{k>j} n_{hk}\right)$		Contribution to N_d $n_{ij}\left(\sum_{h>i}\sum_{k<j} n_{hk}\right)$	
LL (1,1)	$10 \times (12+8+8+2) =$	300	$10 \times 0 =$	0
LA (1,2)	$15 \times (8+2) =$	150	$15 \times (15+20) =$	525
LH (1,3)	$10 \times 0 =$	0	$10 \times (15+12+20+8) =$	550
ML (2,1)	$15 \times (8+2) =$	150	$15 \times 0 =$	0
MA (2,2)	$12 \times 2 =$	24	$12 \times (20) =$	240
MH (2,3)	$8 \times 0 =$	0	$8 \times (20+8) =$	224
HL (3,1)	$20 \times 0 =$	0	$20 \times 0 =$	0
HA (3,2)	$8 \times 0 =$	0	$8 \times 0 =$	0
HH (3,3)	$2 \times 0 =$	0	$2 \times 0 =$	0
Total	$N_c = 624$		$N_d = 1539$	

(total cell counts of MA, MH, HA, and HH) concordant pairs. As there are ten respondents in cell LL, the total number of concordant pairs is equal to $10 \times 30 = 300$. Similarly, the concordant pairs for other cells can be formulated. See Table 4.10.

Step 3: Number of discordant pairs

Following Equation 4.24, we compute N_d. The contributions of each cell to N_d and the resultant value of N_d is shown in Table 4.10.

Step 4: Compute γ.

Using Equation 4.22, we get

$$\gamma = \frac{N_c - N_d}{N_c + N_d} = \frac{624 - 1539}{624 + 1539}$$
$$= \frac{-915}{2163}$$
$$= -0.42.$$

So, as expertise increases, the number of errors committed decreases.

4.5.2 Correlation between Two Nominal Variables

Let's say X and Y are two nominal variables with a and b categories, respectively. The general contingency table is given in Table 4.11.

Let n_{ij} be the frequency count in the cell (i, j) of Table 4.11, $i = 1,2,..,a$ and $j = 1,2,..,b$. Let n_{i+} and n_{+j} represent the $i-th$ row total (marginal row) and $j-th$ column total (marginal column) of the contingency table and n represent the grand total. To measure the association between X and Y, one can use χ^2-based statistic. We will describe here Pearson χ^2-statistic. The Pearson χ^2, say Q_P, can be computed using the following equation:

TABLE 4.11
Contingency Table of Two Nominal Variables

x_i \ y_i	y_1	y_2	...	y_b	Row Total
x_1	n_{11}	n_{12}	...	n_{1b}	n_{1+}
x_2	n_{21}	n_{22}	...	n_{2b}	n_{2+}
⋮	⋮	⋮		⋮	⋮
x_a	n_{a1}	n_{a2}	...	n_{ab}	n_{a+}
Column Total	n_{+1}	n_{+2}	...	n_{+b}	n

$$Q_P = \sum_{i=1}^{a} \sum_{j=1}^{b} \left(n_{ij} - m_{ij}\right)^2 / m_{ij} \tag{4.25}$$

Where m_{ij} is the expected value of n_{ij} when the null hypothesis (H_0) of no association between X and Y is true.

$$\text{So, } m_{ij} = E\left(n_{ij} / H_0\right) = \frac{n_{i+}\, n_{+j}}{n} \tag{4.26}$$

For sufficiently large sample (i.e., $m_{ij} \geq 5$), Q_P approximately follows χ^2-distribution with $(a-1)(b-1)$ degrees of freedom (Stokes et al., 2012). If there is ideally no association between X and Y, Q_P will become zero. A higher value of χ^2 indicates association. Statistically, if the computed $Q_P > \chi^2_{(a-1)(b-1)}(\alpha)$, then the association between X and Y is significant at the significance level, α. α is usually considered as 0.05.

Example 4.11

Consider the following 3 × 2 contingency table (Table 4.12). Obtain Q_P and comment on the relationship between X and Y.

Solution:

Using Equation 4.26, the expected cell counts are shown in Table 4.13.

$$m_{ij} = \frac{n_{i+}\, n_{+j}}{n}$$

The Pearson χ^2, i.e., Q_P is (from Equation 4.25)

$$\begin{aligned} Q_P &= \frac{(20-13.88)^2}{13.88} + \frac{(7-13.12)^2}{13.12} + \frac{(150-128.56)^2}{128.56} \\ &+ \frac{(100-121.44)^2}{121.44} + \frac{(10000-10027.56)^2}{10027.56} + \frac{(9500-9472.44)^2}{9472.44} \\ &= 2.70 + 2.85 + 3.58 + 3.79 + 0.08 + 0.08 \\ &= 13.08. \end{aligned}$$

TABLE 4.12
Contingency Table for Example 4.11

x_i \ y_i	1	2	Row Total
1	20	7	**27**
2	150	100	**250**
3	10000	9500	**19500**
Column Total	**10170**	**9607**	**19777**

TABLE 4.13
Expected Cell Counts

x_i \ y_i	1	2	Row Total (Original)
1	13.88	13.12	**27**
2	128.56	121.44	**250**
3	10027.56	9472.44	**19500**
Column Total (Original)	**10170**	**9607**	**19777**

Now,

$$\chi^2_{(a-1)(b-1)}(\alpha) \text{ with } \alpha = 0.05 = \chi^2_{(3-1)(2-1)}(0.05) = \chi^2_2(0.05) = 5.99.$$

As $Q_P > \chi^2_2(0.05)$, the no association between X and Y is ruled out.

The χ^2-statistic tests association between two nominal variables but it is difficult to say something objectively on the strength of the association, unlike the Pearson correlation coefficient $(-1 \le \rho \le +1)$. A similar measure is needed. For 2×2 contingency table, phi (ϕ) is used where

$$\phi^2 = \frac{\chi^2}{n} \tag{4.27}$$

where n is the total number of observations.

ϕ ranges from -1 to $+1$. However the extreme values of $|\phi| = 1$ can only be attained when both the row totals as well as the column totals are equal. In other words, when every marginal proportion is equal to 0.50 with empty diagonal cells, $|\phi| = 1$. When the contingency table has higher dimensions such as $p \times q$, $p > 2$ and $q > 2$, ϕ is not used. For a square contingency table $(p \times p,\ p \ge 3)$, contingency coefficient, C is used and for a rectangular contingency table $(p \times q,\ p > 2,\ q > 2 \text{ and } p \ne q)$ Cramer's V is used.

The formula for contingency coefficient C is

$$C = \sqrt{\frac{\chi^2}{n + \chi^2}} \tag{4.28}$$

The Cramer's coefficient V is (Cramer, 1946)

$$V = \sqrt{\frac{\chi^2}{n(k-1)}} \tag{4.29}$$

Where $k = \min(p, q)$.

Example 4.12

Consider Example 4.11. Compute Cramer's V and comment on the result.

Solution:

The formula for Cramer's V is

$$V = \sqrt{\frac{\chi^2}{n(k-1)}}$$

In this case $n = 19777$, $k = \min(2,3) = 2$, and $\chi^2 = 13.08$ (from Example 4.11).
So,

$$V = \sqrt{\frac{13.08}{19777 \times 1}} = 0.026$$

As $V = 0.026$, a very low value, the correlation between the two variables can be neglected. The result is contradicting with χ^2-test.

4.5.3 Correlation between One Continuous and One Ordinal Variable

To measure the correlation between a continuous and an ordinal variable, we can use biserial or polyserial correlation. Biserial correlation is used when the ordinal variable is dichotomous, i.e., having two levels such as low and high, and polyserial correlation is used when the ordinal variable assumes more than two levels, such as a five-point Likert scale. In both the cases, the other variable is continuous, i.e., measured either in ratio or interval scale. The biserial correlation between a dichotomous variable Y and a continuous variable X is (Cohen et al., 2003)

$$r_b = \frac{(\bar{x}_1 - \bar{x}_o)P_oP_1}{h\, s_x} \tag{4.30}$$

where $\bar{x}_1$ = mean of X when $Y = 1$
$\bar{x}_0$ = mean of X when $Y = 0$
P_1 = Proportion of data points when $Y = 1$
$P_o = 1 - P_1$ = Proportion of data points when $Y = 0$.
s_x = The sample standard deviation for the variable X.

$$s_x = \sqrt{\frac{1}{n}\sum_{i=1}^{n}(x_i - \bar{x})^2} = \sum_{i=1}^{n} x_i^2 - \frac{\left(\sum_{i=1}^{n} x_i\right)^2}{n}$$

Where, $\bar{x} = \dfrac{n_0\bar{x}_0 + n_1\bar{x}_1}{n_0 + n_1}$, $n = n_0 + n_1$, n_0 and n_1 are the number of observations when $Y = 0$ and $Y = 1$, respectively.

h = the height of the standardized normal distribution (z) at w such that $p(z < w) = P_o$ and $p(z > w) = P_1$. So, h is the ordinate of the standard normal curve at the point at which the area is divided into P_0 and P_1 proportions.

r_b may exceed 1 in magnitude but such an event rarely occurs (Tate, 1955). So, in most of the practical cases $-1 \le r_b \le 1$. In *polyserial correlation*, Y has more than two ordered categories such as $y_1 < y_2 < \cdots y_m$. For details, please see Olsson et al. (1982).

Example 4.13

It is known that motivation helps to improve performance. An experiment was conducted where five people were subjected to low levels of motivation and another five people to high levels of motivation. So, motivation is an ordinal dichotomous variable such as $Y = 0$ and $Y = 1$ representing a person subjected to low or high motivation, respectively. Let X represent the performance. The data obtained are shown in Table 4.14. Compute biserial correlation between X and Y.

Solution:

The biserial correlation between X and Y is

$$r_b = \frac{(\bar{x}_1 - \bar{x}_o)P_oP_1}{s_x \cdot h}$$

$$\bar{x}_o = \frac{70+60+50+65+75}{5} = 64$$

$$\bar{x}_1 = \frac{90+80+85+75+90}{5} = 84$$

$$\bar{x} = \frac{64\times5+84\times5}{10} = 74$$

$$\sum_{i=1}^{10} x_i = 70+60+50+65+75+90+80+85+75+90 = 740$$

$$s_x^2 = \sum_{i=1}^{10} x_i^2 - \frac{\left(\sum_{i=1}^{10} x_i\right)^2}{10}$$

$$= 70^2 + 60^2 + 50^2 + 65^2 + 75^2 + 90^2 + 80^2 + 85^2 + 75^2 + 90^2 - \frac{740^2}{10}$$

$$= 56300 - 54760 = 1540$$

So,

$$s_x = \sqrt{1540} = 39.24$$

Now, $P_o = \dfrac{5}{10} = 0.5$ and $P_1 = \dfrac{5}{10} = 0.50$

Again, h is the ordinate of the standard normal curve at the point at which the area is divided into P_o and P_1 proportions.

h = height at $z = 0$ as $P_o = P_1 = 0.50$
$= 0.399$

TABLE 4.14
Data Table for Example 4.13

Subject	Y	X	
1	0	70	
2	0	60	
3	0	50	$\overline{x}_0$
4	0	65	
5	0	75	
6	1	90	
7	1	80	
8	1	85	$\overline{x}_1$
9	1	75	
10	1	90	

So, $r_b = ((84-64)(0.50 \times 0.50))/(39.24 \times 0.399) = 0.32.$

4.5.4 Correlation between One Continuous and One Nominal Variable

When a continuous variable is correlated with a nominal variable, point biserial or point polyserial correlation can be used, with the former being used when the nominal variables has two categories, i.e. dichotomous.

The point biserial correlation coefficient r_{pb} is (Lev, 1949)

$$r_{pb} = \frac{\overline{x}_1 - \overline{x}_o}{s_x}\sqrt{P_o P_1} \tag{4.31}$$

where $\overline{x}_1$, $\overline{x}_o$, s_x, P_o, and P_1 are the same as defined with Equation 4.30.

The relation between biserial and point biserial correlation is $r_{pb} = \dfrac{h\, r_b}{\sqrt{P_o P_1}}$.

Example 4.14

The standard hours per day worked by two groups of workers, group 1 (who enjoy incentive for better performance) and group 2 (who does not have any incentive scheme), are shown in Table 4.15.

Is there any relation of incentive scheme on worker performance?

Solution:

Let $Y = 0$ represents a worker without incentive scheme and $Y = 1$ represents a worker with incentive scheme. Similarly, let X represents standard hours worked per day by a worker. So, Y is a dichotomous nominal variable and X is a continuous variable. We need to calculate point biserial correlation r_{pb} to find the stated relationship.

Now, $r_{pb} = \dfrac{\overline{x}_1 - \overline{x}_o}{s_x}\sqrt{P_o P_1}$

TABLE 4.15
Data Table for Example 4.14

Worker No.	Incentives	Standard Hrs. Worked
1	Yes	9
2	Yes	10
3	Yes	11
4	Yes	10
5	No	8
6	No	7
7	No	6

Again, P_1 = Proportion of workers with incentive scheme

$$= \frac{4}{7} = 0.571$$

$$P_0 = 1 - P_1 = 1 - 0.571 = 0.429$$

$$\bar{x}_1 = \text{Mean of } X \text{ when } Y = 1$$

$$= \frac{9+10+11+10}{4} = \frac{40}{4} = 10$$

$$\bar{x}_o = \text{Mean of } X \text{ when } Y = 0$$

$$= \frac{21}{3} = 7$$

$$\bar{x} = \frac{1}{7}(40+21) = \frac{61}{7} = 8.71$$

$$s_x = \sqrt{\frac{1}{n}(x_i - \bar{x})^2}$$

$$= \sqrt{\frac{1}{7}\left[(9-8.71)^2 + (10-8.71)^2 + (11-8.71)^2 + (10-8.71)^2 + (8-8.71)^2 + (7-8.71)^2 + (6-8.71)^2\right]}$$

$$= \sqrt{\frac{1}{7}(0.0841+1.6641+5.2441+1.6641+0.5041+2.9241+7.3441)}$$

$$= \sqrt{\frac{1}{7} \times 19.4287}$$

$$= \sqrt{2.776}$$

So,

$$r_{pb} = \frac{(10-7)\sqrt{0.429 \times 0.571}}{\sqrt{2.776}}$$

$$= 3 \times \sqrt{0.088}$$

$$= 0.89$$

4.5.5   CORRELATION BETWEEN ONE ORDINAL AND ANOTHER NOMINAL VARIABLE

For correlation between two variables, one measured in ordinal scale (X) and another in binary nominal (Y), rank biserial correlation coefficient is used. However, when the nominal variable is polytomous, rank polyserial correlation can be used. The rank biserial correlation r_{rb} is

$$r_{rb} = \frac{2\left(\bar{x}_1 - \bar{x}_o\right)}{n} \tag{4.32}$$

where, $\bar{x}_1$ = mean of X when $Y = 1$
$\quad\quad \bar{x}_0$ = mean of X when $Y = 0$
$\quad\quad n = n_o + n_1$
$\quad\quad n_o$ = Number of observation when $Y = 0$
$\quad\quad n_1$ = Number of observation when $Y = 1$

4.6   CORRELATION WITH DEPENDENCE STRUCTURE

Let's say X_1, X_2, and Y are three variables characterizing a population in such a manner that Y is dependent on X_1 and X_2, and X_1 and X_2 may or may not covary. Such a relationship is called dependence structure. The variable Y is called the dependent variable (DV) and X_1 and X_2 are called independent variables (IVs). If Y is dependent on X_1 and X_2, Y must be correlated with X_1 and X_2. Pictorially, the situation can be represented as shown in Figure 4.1. In order to understand Figure 4.1, let us assume that X_1, X_2, and Y are converted to unit normal (standardized) variables with variability 1. The corresponding circle in Figure 4.1 represents the variability. The shaded portions in Figures 4.1(a)–4.1(c) represent the shared variability between the two variables considered, respectively.

In Figures 4.1(a)–4.1(c), r_{y1}^2, r_{y2}^2, and r_{12}^2 represent the common (shared) variability between Y and X_1, Y and X_2, and X_1 and X_2, respectively. When one variable is dependent on the other, for example

(a) Shared variance   (b) Shared variance   (c) Shared variance
of Y and X_1, r_{y1}^2 of Y and X_2, r_{y2}^2 of X_1 and X_2, r_{12}^2

$$r_{y1}^2 = a + c$$
$$r_{y2}^2 = b + c$$
$$r_{12}^2 = c + f$$
$$R_{y.12}^2 = a + b + c$$

(d) The ballantine for X_1 and X_2 with Y (Cohen et al., 2003)

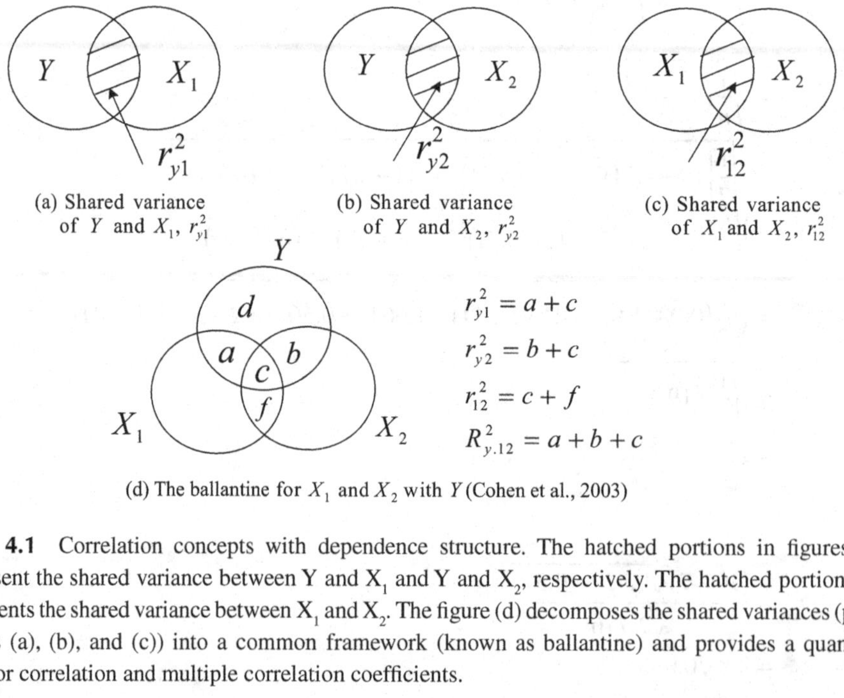

FIGURE 4.1   Correlation concepts with dependence structure. The hatched portions in figures (a) and (b) represent the shared variance between Y and X_1 and Y and X_2, respectively. The hatched portion in figure (c) represents the shared variance between X_1 and X_2. The figure (d) decomposes the shared variances (presented in figures (a), (b), and (c)) into a common framework (known as ballantine) and provides a quantification scheme for correlation and multiple correlation coefficients.

Y on X_1 or X_2, the shared portion represents the variability of the DV explained by the IV. So, r^2_{y1}, r^2_{y2} can be treated as explained variance of Y by X_1 and X_2, respectively. On the other hand, r^2_{12}, the shared variability of X_1 and X_2, represents the amount of (standardized) covariance between X_1 and X_2. Readers may refer to Cohen et al., (2003).

The square root of the common variance, i.e., r_{y1}, r_{y2}, and r_{12} represent the zero-order correlation between Y and X_1, Y and X_2, and X_1 and X_2, respectively. By zero-order we mean that r_{y1}, r_{y2}, and r_{12} represent Pearson product moment correlation (Equation 4.15 or 4.16) between the respective two variables without controlling the presence of other variables in the system.

Now, using Equation 4.16, we can write

$$r_{y1} = \frac{s_{y1}}{s_y \cdot s_1} = s_{y1} \text{ as } s_y = 1 \text{ and } s_1 = 1 (\text{standardized variable})$$

Similarly, $r_{y2} = s_{y2}$, and $r_{12} = s_{12}$.

So, for standardized variables, the correlation coefficient is equal to their covariance.

Figure 4.1(d) depicts a different picture. Here Y's variability is shared by both X_1 and X_2. Let,

a = variability of Y uniquely explained by X_1

b = variability of Y uniquely explained by X_2

c = variability of Y shared by both X_1 and X_2 together

d = variability of Y explained by none of the two IVs.

f = shared variability of X_1 and X_2 that is not a part of Y's variability.

Then, we can define the following:

$$r^2_{y1} = a + c$$

$$r^2_{y2} = b + c$$

$$r^2_{12} = c + f$$

$$R^2_{y \cdot 12} = a + b + c$$

$R^2_{y \cdot 12}$ is the proportion of the DV's variance (s^2_y) shared with the optimally weighted IV$_s$, X_1 and X_2 (Cohen et al., 2003). The squared root of $R^2_{y.12}$, i.e., $R_{y \cdot 12}$ is called "multiple correlation coefficient" and is computed by the following equation:

$$R_{y \cdot 12} = \sqrt{\frac{r^2_{y1} + r^2_{y2} - 2r_{y1} r_{y2} r_{12}}{1 - r^2_{12}}} \tag{4.33}$$

$R_{y \cdot 12}$ indicates how a group of IVs together explain Y. $R_{y \cdot 12}$ takes values between 0 and 1, indicating no or perfect relationship, respectively. It is to be noted here that $R_{y.12}$ will never become negative. When there are p IVs explaining the variability of one DV, the multiple correlation coefficient is denoted as $R_{y.123...p}$.

Example 4.15

In Example 4.8, a correlation matrix using CityCan data is computed. A part of the matrix involving correlation coefficients of sales volume (Y), absenteeism (X_1) and breakdown hours (X_2) is shown below.

$$\mathbf{R} = \begin{bmatrix} 1.00 & -0.72 & -0.37 \\ -0.72 & 1.00 & 0.29 \\ -0.37 & 0.29 & 1.00 \end{bmatrix}$$

Compute multiple correlation coefficient for y.

Solution:

Here

$$r_{y1} = -0.72$$

$$r_{y2} = -0.37$$

$$r_{12} = 0.29$$

Using Equation 4.33, $R_{y\cdot12}$ is

$$R_{y\cdot12} = \sqrt{\frac{r_{y1}^2 + r_{y2}^2 - 2r_{y1}\cdot r_{y2}\cdot r_{12}}{1 - r_{12}^2}}$$

$$= \sqrt{\frac{(-0.72)^2 + (-0.37)^2 - 2\times(-0.72)(-0.37)\times 0.29}{1-(0.29)^2}}$$

$$= 0.739.$$

It indicates moderate relationship. The value of $R_{y\cdot12}^2$ is $(0.739)^2 = 0.547$, which indicates that 54.70% of Y's variability is explained by X_1 and X_2 together.

4.6.1 Part Correlation

The coefficient of multiple correlation ($R_{y\cdot12.......p}$) for Y with p IVs explains how the group of p IVs collectively explains Y. One may be interested to know what is the contribution of X_j to $R_{y\cdot12.......p}$. Part correlation answers this. Another name for part correlation is semi-partial correlation (sr).

Consider Figure 4.1(d). The area labeled a and b represents the portion of Y's variability uniquely explained by X_1 and X_2, respectively. Therefore, a and b are called squared part or semi-partial correlation between Y and X_1 and Y and X_2, respectively.

Let,

$$s\,r_1^2 = \text{squared semi-partial correlation between } Y \text{ and } X_1$$

$$s\,r_2^2 = \text{squared semi-partial correlation between } Y \text{ and } X_2$$

Then,

$$s\,r_1^2 = a = (a+b+c)-(b+c) = R_{y\cdot12}^2 - r_{y2}^2$$

$$s\,r_2^2 = b = (a+b+c)-(a+c) = R_{y\cdot12}^2 - r_{y1}^2$$

$$\text{Now, } sr_1^2 = R^2_{y\cdot 12} - r^2_{y2}$$

$$= \frac{r^2_{y1} + r^2_{y2} - 2r_{y1}\cdot r_{y2}\cdot r_{12}}{1 - r^2_{12}} - r^2_{y2}$$

$$= \frac{r^2_{y1} + r^2_{y2} - 2r_{y1}\, r_{y2}\, r_{12} - r^2_{y2} + r^2_{y2}\cdot r^2_{12}}{1 - r^2_{12}}$$

$$= \frac{r^2_{y1} - 2r_{y1}\, r_{y2}\, r_{12} + \left(r_{y2}\, r_{12}\right)^2}{1 - r^2_{12}}$$

$$= \frac{\left(r_{y1} - r_{y2}\, r_{12}\right)^2}{1 - r^2_{12}}$$

$$\text{or, } sr_1 = \frac{r_{y1} - r_{y2}\, r_{12}}{\sqrt{1 - r^2_{12}}} \tag{4.34}$$

Similarly,

$$sr_2 = \frac{r_{y2} - r_{y1}\, r_{12}}{\sqrt{1 - r^2_{12}}} \tag{4.35}$$

The *sr* between a pair of DV and IV is computed after the effect of all other IVs is removed from the IV considered.

Example 4.16

Consider Example 4.15. Compute sr_1 and sr_2.

Solution:

$$r_{y1} = -0.72$$
$$r_{y2} = -0.37$$
$$r_{12} = +0.29$$

$$\text{So, } sr_1 = \frac{r_{y1} - r_{y2}\cdot r_{12}}{\sqrt{1 - r^2_{12}}}$$

$$= \frac{-0.72 - (-0.37)\times 0.29}{\sqrt{1 - 0.29^2}} = -0.64$$

Similarly,

$$sr_2 = \frac{r_{y2} - r_{y1}\cdot r_{12}}{\sqrt{1 - r^2_{12}}}$$

$$= \frac{-0.37 - (-0.72)\times 0.29}{\sqrt{1 - 0.29^2}} = -0.17$$

Cohen et al. (2003) pointed out that sr_1^2, i.e., the area a in Figure 4.1(d), and sr_2^2, i.e., the area b in Figure 4.1(d), represent the proportion of variance of Y uniquely explained by X_1 and X_2. They cautioned that such interpretation cannot be made for c, the variance of Y commonly shared by X_1 and X_2, as there is no mathematical way of preventing c becoming negative in some situations. This type of spurious relationship has to be dealt with using more advanced models.

4.6.2 Partial Correlation

Consider Figure 4.1(d) again. Suppose the variability of Y explained by X_2, i.e., r_{y2}^2 is removed. The remaining variability of Y is $a+d$ $(= \mathrm{a+b+c+d-b-c})$. Now, of the Y's remaining variability, how much is explained by X_1? The answer is a. The above fact is used to compute partial correlation (pr). Let's say the partial correlation between Y and X_1 is pr_1.

Then,

$$pr_1^2 = \frac{\text{Contribution of } X_1 \text{ in uniquely explaining } Y\text{'s variability}}{\text{Remaining variance of y,when the contribution of other IVs (except } X_1\text{) is removed}}$$

$$= \frac{a}{a+d}$$

$$= \frac{(\mathrm{a+b+c})-(\mathrm{b+c})}{(\mathrm{a+b+c+d})-(\mathrm{b+c})}$$

$$= \frac{R_{y.12}^2 - r_{y2}^2}{1-r_{y2}^2} \quad \text{as } a+b+c+d=1 \text{ (total variability of } Y\text{)}$$

$$= \frac{\dfrac{r_{y1}^2 + r_{y2}^2 - 2r_{y1}r_{y2}r_{12}}{1-r_{12}^2} - r_{y2}^2}{1-r_{y2}^2}$$

$$= \frac{(\mathrm{r}_{y1} - \mathrm{r}_{y2}\mathrm{r}_{12})^2}{(1-r_{y2}^2)(1-\mathrm{r}_{12}^2)}$$

So,

$$p\mathrm{r}_1 = \frac{r_{y1} - r_{y2}r_{12}}{\sqrt{1-r_{y2}^2}\sqrt{1-r_{12}^2}} \tag{4.36}$$

$$= sr_1 \Big/ \sqrt{1-r_{y2}^2}$$

Similarly,

$$p\mathrm{r}_2^2 = \frac{b}{b+d}$$

$$= \frac{(\mathrm{a+b+c})-(\mathrm{a+c})}{(\mathrm{a+b+c+d})-(\mathrm{a+c})}$$

$$= \frac{R_{y.12}^2 - r_{y1}^2}{1-r_{y1}^2}$$

$$= \frac{(\mathrm{r}_{y2} - \mathrm{r}_{y1}\mathrm{r}_{12})^2}{(1-\mathrm{r}_{y1}^2)(1-\mathrm{r}_{12}^2)}$$

$$\therefore pr_2 = \frac{r_{y2} - r_{y1}r_{12}}{\sqrt{1-r_{y1}^2}\sqrt{1-r_{12}^2}} \tag{4.37}$$

$$= sr_2 \Big/ \sqrt{1-r_{y1}^2}$$

As $\sqrt{1-r_{y2}^2}$ or $\sqrt{1-r_{y1}^2}$ cannot exceed 1, $|pr_1| \geq |sr_1|$ and $|pr_2| \geq |sr_2|$.

Example 4.17

Consider Example 4.16. Obtain pr_1 and pr_2 and compare the results with sr_1 and sr_2.

Solution:

$$r_{y1} = -0.72$$
$$r_{y2} = -0.37$$
$$r_{12} = +0.29$$

Now,

$$pr_1 = \frac{r_{y1} - r_{y2}\, r_{12}}{\sqrt{1-r_{y2}^2}\sqrt{1-r_{12}^2}}$$
$$= \frac{-0.72 - (-0.37 \times 0.29)}{\sqrt{1-(-0.37)^2}\sqrt{1-(0.29)^2}}$$
$$= \frac{-0.6127}{0.93 \times 0.96}$$
$$= -0.69$$

Similarly,

$$pr_2 = \frac{r_{y2} - r_{y1}\, r_{12}}{\sqrt{1-r_{y1}^2}\sqrt{1-r_{12}^2}}$$
$$= \frac{-0.37 - (-0.72 \times 0.29)}{\sqrt{1-(-0.72)^2}\sqrt{1-(0.29)^2}}$$
$$= \frac{-0.1612}{0.69 \times 0.96}$$
$$= -0.24$$

It is seen that $|pr_1| = 0.69$ which is more than $|sr_1| = 0.64$. Similarly, $|pr_2| = 0.24$ is more than $|sr_2| = 0.17$.

Although the coefficient of multiple correlation ($R_{y.12}$), part (sr) and partial correlation (pr) computed above uses one DV and two IVs, the same concept can be extrapolated to multiple IVs, say $X_1, X_2, \ldots, X_p$. If the coefficient of multiple correlation involving p IVs is $R^2_{y.123\ldots p}$ and that without the IV, X_j is $R^2_{y.123\ldots(j)..p}$, the sr_j^2 and pr_j^2 can be computed as follows:

$$\left.\begin{aligned} sr_j^2 &= R^2_{y.123\ldots p} - R^2_{y.123\ldots(j)\ldots p}, \\ pr_j^2 &= \frac{sr_j^2}{1-R^2_{y.123\ldots(j)\ldots p}}, j = 1,2,3,..p. \end{aligned}\right\} \tag{4.38}$$

4.7 LEARNING SUMMARY

In this chapter multivariate descriptive statistics from concepts to computation are discussed. Several numerical examples are solved. In summary, the highlights of the chapter are as follows:

- Multivariate data matrices include n number of multivariate observations on p variables, denoted by $\mathbf{X}_{n\times p}$.
- Multivariate descriptive statistics include sample mean vectors, sample covariance matrices, and sample correlation matrices.
- An analogy between univariate and multivariate descriptive statistics could be established with a difference in the computation. While univariate statistics are scalar quantities, multivariate descriptive statistics are vector quantities. The computation in the multivariate domain therefore involves matrix algebra.
- If $X^T = [X_1, X_2, \ldots, X_p]_{p\times 1}$ represents p number of variables with population mean vector μ and population covariance matrix $\boldsymbol{\Sigma}$, then $\mu^T = [\mu_1, \mu_2, \ldots, \mu_p]_{p\times 1}$ and $\boldsymbol{\Sigma}$ is a $p \times p$ matrix given in Section 4.3.
- The sample covariance matrix, $\mathbf{S} = \frac{1}{n-1}\left[\left(\mathbf{X}-\mathbf{1}\overline{\mathbf{X}}^T\right)^T_{p\times n}\left(\mathbf{X}-\mathbf{1}\overline{\mathbf{X}}^T\right)_{n\times p}\right]_{p\times p}$.
- The correlation matrix among p variables is the standardized covariance matrix. The sample correlation matrix ($\mathbf{R}$) is computed as $(n-1)\mathbf{R} = \tilde{\mathbf{X}}^T\tilde{\mathbf{X}}$, where $\tilde{\mathbf{X}}$ is standardized data matrix of size $n\times p$ and n is sample size.
- The relation between covariance matrix ($\mathbf{S}$) and correlation matrix ($\mathbf{R}$) is $\mathbf{S} = \mathbf{D}_S^{1/2}\ \mathbf{R}\ \mathbf{D}_S^{1/2}$, $\mathbf{D}_s$ is a diagonal matrix with diagonal element of $\mathbf{S}$.
- Multivariate statistics often use sum squares and cross product (SSCP) matrices such as $\mathbf{X}^T\mathbf{X}, \mathbf{X}^{*T}\mathbf{X}^*$ and $\tilde{\mathbf{X}}^T\tilde{\mathbf{X}}$, where $\mathbf{X}^*$and $\tilde{\mathbf{X}}$ are the mean subtracted and standardized data matrices of size $n \times p$, respectively (n = sample size).
- Depending on the data types (based on scale of measurement), there could be different correlation coefficient between two variables such as Pearson product moment correlation, Spearman rho correlation, biserial, and polyserial correlations (see Section 4.5 for details).
- When the variables exhibit dependent structure, part and partial correlations are used (see Section 4.6 for details).

EXERCISES

(A) Short conceptual questions

4.1 Define multivariate observations. In what way it is different from univariate observations? If you collect n univariate observations on p variables, each at different times, does the data set represent multivariate observations?

4.2 When we collect n univariate observations and compute its mean, it is a scalar quantity. But in multivariate case we call it mean vector—why?

4.3 If $\mathbf{S}$ is the sample covariance matrix, show that $\mathbf{S} = \frac{1}{n-1}\left[\left(\mathbf{X}-\mathbf{1}\bar{\mathbf{x}}^T\right)^T\left(\mathbf{X}-\mathbf{1}\bar{\mathbf{x}}^T\right)\right]$ where n is sample size, $\mathbf{X}$ is $n\times p$ data matrix, $\bar{\mathbf{x}}$ is $p\times 1$ mean vector and $\mathbf{1}$ is $n\times 1$ vector of one.

4.4 Define correlation. In what way it is different from covariance?

4.5 Show that the sample correlation matrix $\mathbf{R}$ is related to sample covariance matrix $\mathbf{S}$ as $\mathbf{R} = \mathbf{D}_S^{-1/2}\ \mathbf{S}\ \mathbf{D}_S^{-1/2}$ where $\mathbf{D}_s$ is a diagonal matrix with diagonal elements of $\mathbf{S}$.

4.6 Define SSCP. If $\mathbf{X}$ is $n\times p$ data matrix, define $\mathbf{X}^T\mathbf{X}, \mathbf{X}^{*T}\mathbf{X}^*$ and $\tilde{\mathbf{X}}^T\tilde{\mathbf{X}}$ (follow Section 4.4 for general notations). Also compute covariance matrix $\mathbf{S}$ and correlation matrix $\mathbf{R}$ using *SSCP* matrices.

4.7 Explain different types of correlation coefficients with examples.
4.8 What is Spearman rank correlation? Explain with an example.
4.9 How do you compute correlation between two ordinal variables?
4.10 What is gamma coefficient (γ)? Explain $-1 \leq \gamma \leq +1$.
4.11 What is phi? When it is used?
4.12 How do you compute correlation between ordinal and continuous variables? What is point biserial correlation?
4.13 What is part correlation? Give examples.
4.14 Define partial correlation with an example.

(B) Long conceptual and/or numerical questions: see web resources (Multivariate Statistical Modeling in Engineering and Management – 1st (routledge.com))

N.B.: R and MS Excel software are used for computation, as and when required.

NOTE

1 Both the i-th multivariate observation on p-variables (i.e., $\mathbf{x}_i$) and n observations on the variable X_j (i.e., $\mathbf{x}_j$) are represented as column vectors of size $p\times1$ and $n\times1$, respectively. So, in the data matrix $\mathbf{X}_{n\times p}$, the i-th row vector represents the i-th multivariate observation ($\mathbf{x}_i^T$) and the j-th column represents the n observations ($\mathbf{x}_j$) on the j-th variable X_j. Throughout the book, unless otherwise mentioned, $\mathbf{x}_{k\times1}$ represents a column vector of size $k\times1$ and $\mathbf{x}^T{}_{1\times k}$ represents a row vector of size $1\times k$.

5 Multivariate Normal Distribution

In Chapter 2 (Section 2.2.1), we described univariate normal distribution. In this chapter, we define its multivariate counterpart, the multivariate normal distribution (MND). In order to grasp the physical significance of the MND, the concept of statistical distance is discussed first. Then, we demonstrate how univariate normal distribution can be extended to bivariate normal distribution (BND) and then analogously, we generalize it to MND. Most of the physical interpretations of MND are explained in terms of BND as graphical displays are possible in two dimensions. We also discuss the properties of MND in this chapter. As multivariate normality is one of the important assumptions for many multivariate statistical models, how to assess the multivariate normality of a given dataset is also discussed.

5.1 STATISTICAL DISTANCE

Statistical distance plays a significant role in multivariate data analysis. In order to understand it fully, we start with Euclidian distance and then gradually get into the deeper meaning of statistical distance. The Euclidean distance is a measure of the distance between two points in a Cartesian coordinate system. For example, the distance between two cities can be measured with two-dimensional Cartesian coordinates. Denoted by X_1 and X_2 axes, the Cartesian coordinate system can be represented as Figure 5.1. The distance between the origin O (0,0) and a point P (x_1, x_2) can be written as:

$$d(OP) = \sqrt{x_1^2 + x_2^2} \tag{5.1}$$

Equation 5.1 is an equation of a circle and *d(OP)* is the radius of the circle. All points on the circle are equidistance from the origin O.

For a p-dimensional coordinate system, Equation 5.1 becomes

$$d(OP) = \sqrt{x_1^2 + x_2^2 + \cdots\cdots + x_p^2} \tag{5.2}$$

If one is interested to measure the distance from a fixed point Q with coordinate $(q_1, q_2, ..., q_p)$ to P with coordinates (x_1, x_2, …, x_p), the distance *d(PQ)* is

$$d(PQ) = \sqrt{(x_1 - q_1)^2 + (x_2 - q_2)^2 + \cdots\cdots\cdots\cdots + (x_p - q_p)^2} \tag{5.3}$$

Equations 5.2 and 5.3 are equations of a sphere centered at origin O and some other arbitrary point Q, respectively. All points on the surface of a sphere are equidistance from its center.

Let us now consider four processes A, B, C, and D, each characterized with two variables X_1 and X_2. Let one observed n data points for each of the four processes and the scatter plots of the n points are shown in Figure 5.2. What can one interpret about the four processes based on the scatter plots? Every point on the scatter is a bivariate observation and is measured in the two-dimensional statistical system (X_1, X_2). The data for process A (Figure 5.2, top left) exhibit a circular pattern that is indicative of equal spread (variance) for the two variables X_1 and X_2. There is no correlation between

DOI: 10.1201/9781003303060-7

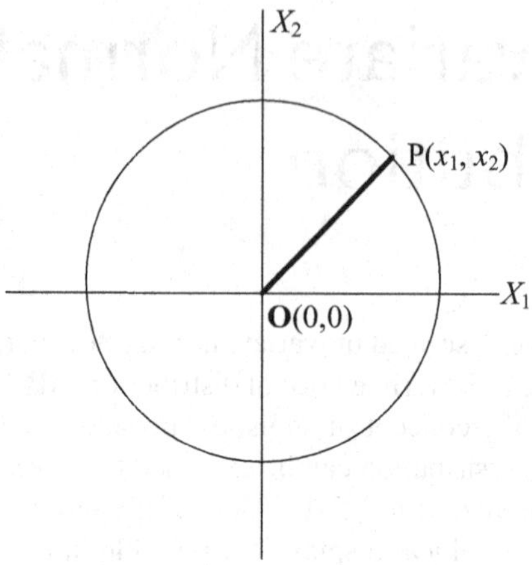

FIGURE 5.1 The physical meaning of Euclidean distance. Any point on the circumference of a circle is equidistant from the origin.

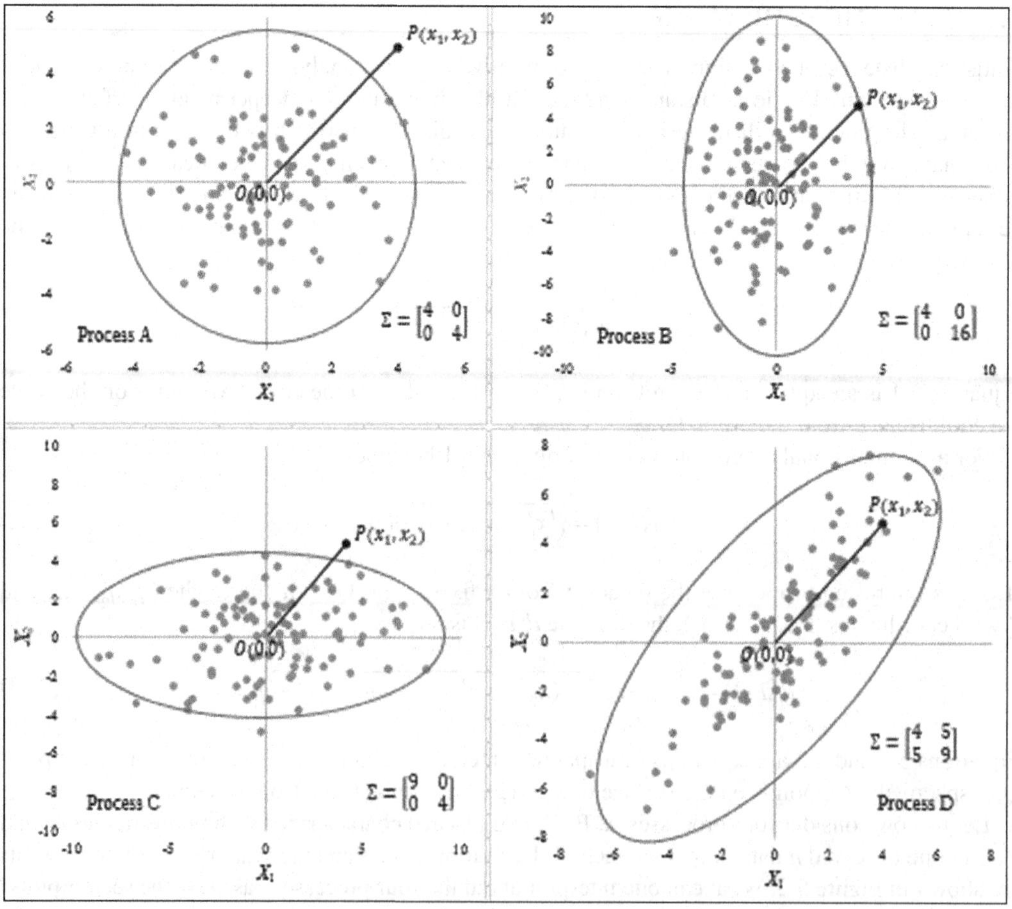

FIGURE 5.2 Bivariate scatter plots between X_1 and X_2 for processes A, B, C, and D. The scatter plots and the shapes inscribing the observation vectors on X_1 and X_2 demonstrate the joint behavior of the four processes, A, B, C, and D in terms of variance and covariance (relationship) of X_1 and X_2.

the two variables. The data for process B (Figure 5.2, top right) exhibit an elliptical pattern and also shows no correlation between X_1, X_2 but the variability across X_2 is more than the variability across X_1. Process C (Figure 5.2, bottom left) also exhibits an elliptical pattern and no correlation similar to the process B but the variability in X_1 is more than the variability in X_2. Process D (Figure 5.2, bottom right) exhibits unequal variability across X_1 and X_2 and correlation between X_1 and X_2. It exhibits the most general pattern of bivariate data, which is oblique elliptical.

The most interesting point that can be observed from Figure 5.2 is the position of the observation $\mathbf{x} = [x_1, x_2]^T$, denoted as $P(x_1, x_2)$. In all the four processes A, B, C, and D, the Euclidean distance $d(OP)$ is same, which can be obtained using Equation 5.1. But from process behavior point of view, does the observation $P(x_1, x_2)$ represent the same thing for all the four processes? If we believe that the behavior of a process can be traced from the pattern shown by the data (e.g., circular or elliptical) where the pattern accounts for the general mass of the data, the observation $P(x_1, x_2)$ doesn't represent the same thing for all the four processes. For the processes A and C, the observation $P(x_1, x_2)$ doesn't represent the general mass and is therefore termed an outlier whereas for the processes B and D, the observation $P(x_1, x_2)$ belongs to the general mass. If we further assume that distance is a good measure of association among observations, it can be argued that the Euclidean distance is incapable of representing the general mass (hence, the process behavior) and we require some other measure of distance that can capture the variance-covariance structure of multiple variables characterizing a process or system. Statistical distance comes to the rescue.

Based on the above discussion, for computing statistical distance the following assumptions can be made:

- Every process is characterized by one or more variables.
- The variables, if more than one, may or may not have equal variance.
- The variables, if more than one, may or may not be correlated.
- The variance-covariance relationships define the pattern in the data.
- The pattern can be captured through appropriate distance measure.

Now let us consider Figure 5.2 (bottom left for Process C). The pattern that defines the data is an ellipse. Here the major and minor axes of the ellipse are parallel to the $X-Y$ ordinate system. If we consider the center is at the origin (0,0), we can develop the following equation for the ellipse based on the approach given below.

Let's say the variance-covariance matrix of $X = (X_1, X_2)^T$ is Σ.

So, for Process C, $\Sigma = \begin{bmatrix} \sigma_{11} & 0 \\ 0 & \sigma_{22} \end{bmatrix}$ as both X_1 and X_2 are uncorrelated, and assuming X_1 and X_2 are independent.

From Figure 5.2 (bottom left for Process C), it is seen that $\sigma_{11} > \sigma_{22}$. In order to obtain the distance of $P(x_1, x_2)$ from O (0,0), we first require the coordinates to be on equal footing. This is possible by the following transformation:

Let $$x_1^* = \frac{x_1}{\sqrt{\sigma_{11}}} \quad \text{and} \quad x_2^* = \frac{x_2}{\sqrt{\sigma_{22}}}$$

So, using Euclidean distance formula (Equation 5.1), we can write

$$d(OP) = \sqrt{x_1^{*2} + x_2^{*2}}$$

Assuming $d(OP) = d$, we get

$$x_1^{*2} + x_2^{*2} = d^2$$

$$\text{or} \quad \left(\frac{x_1}{\sqrt{\sigma_{11}}}\right)^2 + \left(\frac{x_2}{\sqrt{\sigma_{22}}}\right)^2 = d^2$$

$$\text{or} \quad \frac{x_1^2}{\sigma_{11}} + \frac{x_2^2}{\sigma_{22}} = d^2 \tag{5.4}$$

Equation 5.4 is the equation of an ellipse, centered at the origin (0, 0).

Now, let us do some elementary matrix manipulation with **x** and Σ as given below.

$$\Sigma^{-1} = \begin{bmatrix} 1/\sigma_{11} & 0 \\ 0 & 1/\sigma_{22} \end{bmatrix}$$

$$\mathbf{x} = \begin{bmatrix} x_1 \\ x_2 \end{bmatrix}, \ \mathbf{x}^T = \begin{bmatrix} x_1, & x_2 \end{bmatrix}$$

$$\text{So, } \mathbf{x}^T \ \boldsymbol{\Sigma}^{-1}\mathbf{x} = \begin{bmatrix} x_1 & x_2 \end{bmatrix} \begin{bmatrix} 1/\sigma_{11} & 0 \\ 0 & 1/\sigma_{22} \end{bmatrix} \begin{bmatrix} x_1 \\ x_2 \end{bmatrix}$$

$$= \frac{x_1^2}{\sigma_{11}} + \frac{x_2^2}{\sigma_{22}}$$

which is equal to the left hand side (LHS) of Equation 5.4.

So, we can write

$$\mathbf{x}^T \ \Sigma^{-1}\mathbf{x} = d^2 \tag{5.5}$$

What will happen to Equations 5.4 and 5.5 if the center of the ellipse is not in the origin (0, 0)?

Let the center be at $\mu = [\mu_1, \mu_2]^T$. Then, the transformed variables are

$$x_1^* = (x_1 - \mu_1) / \sqrt{\sigma_{11}}$$

$$x_2^* = (x_2 - \mu_2) / \sqrt{\sigma_{22}}$$

So, $x_1^{*2} + x_2^{*2} = d^2$ becomes

$$\left(\frac{x_1 - \mu_1}{\sqrt{\sigma_{11}}}\right)^2 + \left(\frac{x_2 - \mu_2}{\sqrt{\sigma_{22}}}\right)^2 = d^2 \tag{5.6}$$

In matrix notation,

$$(\mathbf{x} - \mu)^T \ \boldsymbol{\Sigma}^{-1} (\mathbf{x} - \mu) = d^2 \tag{5.7}$$

Now consider Figure 5.2 (bottom right for Process D). The major and minor axes of the ellipse are not parallel to the $X_1 - X_2$ plane, rather makes an angle θ as shown in Figure 5.3.

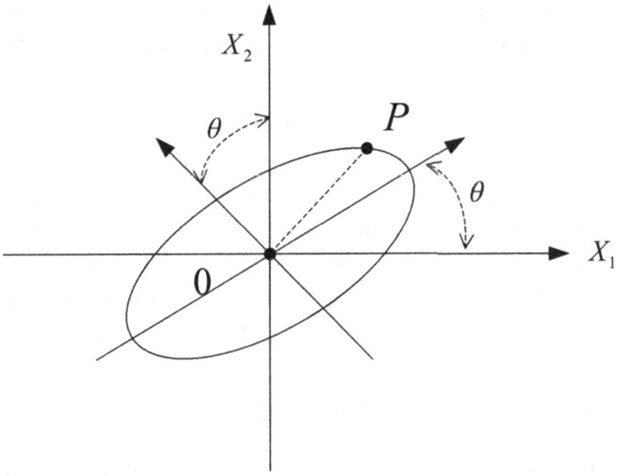

FIGURE 5.3 Bivariate ellipse representing the data behavior for process D. The two variables X_1 and X_2 are correlated and their degree of correlation can be measured using θ, where θ is the angle between the major axis of the ellipse and the horizontal axis, represented by variable X_1.

$$\text{Here } \mathbf{x} = \begin{bmatrix} x_1 \\ x_2 \end{bmatrix} \text{ and } \mathbf{\Sigma} = \begin{bmatrix} \sigma_{11} & \sigma_{12} \\ \sigma_{12} & \sigma_{22} \end{bmatrix}$$

$$\text{So, } \mathbf{\Sigma}^{-1} = \frac{1}{\sigma_{11}\sigma_{22} - \sigma_{12}^2} \begin{bmatrix} \sigma_{22} & -\sigma_{12} \\ -\sigma_{12} & \sigma_{11} \end{bmatrix}$$

So, using Equation 5.5, the distance $d(OP)$ is

$$\mathbf{x}^T\ \Sigma^{-1}\ \mathbf{x} = d^2 \ \text{ or } [x_1\ x_2] \begin{bmatrix} \frac{\sigma_{22}}{\sigma_{11}\sigma_{22} - \sigma_{12}^2} & \frac{-\sigma_{12}}{\sigma_{11}\sigma_{22} - \sigma_{12}^2} \\ \frac{-\sigma_{12}}{\sigma_{11}\sigma_{22} - \sigma_{12}^2} & \frac{\sigma_{11}}{\sigma_{11}\sigma_{22} - \sigma_{12}^2} \end{bmatrix} \begin{bmatrix} x_1 \\ x_2 \end{bmatrix} = d^2 \tag{5.8}$$

If the ellipse is centered at $\mu = [\mu_1,\ \mu_2]^T$, then equation 5.8 becomes

$$(\mathbf{x} - \mu)^T\ \Sigma^{-1}(\mathbf{x} - \mu) = d^2 \ \text{ or } [x_1 - \mu_1\ x_2 - \mu_2] \begin{bmatrix} \frac{\sigma_{22}}{\sigma_{11}\sigma_{22} - \sigma_{12}^2} & \frac{-\sigma_{12}}{\sigma_{11}\sigma_{22} - \sigma_{12}^2} \\ \frac{-\sigma_{12}}{\sigma_{11}\sigma_{22} - \sigma_{12}^2} & \frac{\sigma_{11}}{\sigma_{11}\sigma_{22} - \sigma_{12}^2} \end{bmatrix} \begin{bmatrix} x_1 - \mu_1 \\ x_2 - \mu_2 \end{bmatrix} = d^2 \tag{5.9}$$

Equations 5.4–5.9 are the equation of an ellipse in different forms under different situations. The d^2, measured in those equations, is known as squared statistical distance. Statistical distance takes care of the variance-covariance relationship of a given process, and hence is very different from the Euclidean distance. Multivariate statistics largely owe to statistical distance.

Equation 5.9, if extended to p number of variables, is the general form of computing statistical distance. If there are n data points, we can calculate $d_i^2, i = 1, 2, \ldots n$ using Equation 5.9 as given below.

$$(\mathbf{x}_i - \mu)^T\ \mathbf{\Sigma}^{-1}(\mathbf{x}_i - \mu) = d_i^2 \tag{5.10}$$

However, in Equation 5.10, the population mean vector μ and covariance matrix Σ are seldom known. So, we use their sample estimates $\bar{\mathbf{x}}$ and **S**. Then, putting these estimates in Equation 5.10, we get

$$d_i^2 = \left(\mathbf{x}_i - \bar{\mathbf{x}}\right)^T \mathbf{S}^{-1} \left(\mathbf{x}_i - \bar{\mathbf{x}}\right) \tag{5.11}$$

The d_i^2, computed using equation 5.11, is known as squared Mahalanobis distance (MD^2) which was developed by Prasanta Chandra Mahalanobis in the 1930s. Mahalanobis distance plays a very important role in multivariate statistics. It has multiple uses such as (i) detection of multivariate outliers, (ii) multivariate quality control, (iii) selection of calibration samples for a large set of measurements, (iv) clustering based pattern recognition, (v) fault detection, etc. (De Maesschalck et al., 2000).

5.2 BIVARIATE NORMAL DENSITY FUNCTION

Let us assume that X is a bivariate variable vector with random variables X_1 and X_2. Both X_1 and X_2 are individually normally distributed with mean μ_1 and μ_2, and standard deviation σ_1 and σ_2, respectively. Their probability density functions, $f(x_1)$ and $f(x_2)$ are shown below.[1]

$$f\left(x_1\right) = \frac{1}{\sqrt{2\pi\sigma_1^2}} \mathrm{e}^{-\frac{1}{2}\left(\frac{x_1-\mu_1}{\sigma_1}\right)^2} ; -\infty < x_1 < +\infty \tag{5.12}$$

$$f\left(x_2\right) = \frac{1}{\sqrt{2\pi\sigma_2^2}} \mathrm{e}^{-\frac{1}{2}\left(\frac{x_2-\mu_2}{\sigma_2}\right)^2} ; -\infty < x_2 < +\infty \tag{5.13}$$

Both Equations 5.12 and 5.13 have two components namely, the *constant* terms, $\frac{1}{\sqrt{2\pi\sigma_1^2}}$ and $\frac{1}{\sqrt{2\pi\sigma_2^2}}$, and the *exponent* terms, $\left[-\frac{1}{2}\left(\frac{x_1-\mu_1}{\sigma_1}\right)^2\right]$ and $\left[-\frac{1}{2}\left(\frac{x_2-\mu_2}{\sigma_2}\right)^2\right]$.

Let us further assume that X_1 and X_2 are independent. Then their joint pdf, $f(x_1, x_2)$ can be written as:

$$f\left(x_1, x_2\right) = f\left(x_1\right) \times f\left(x_2\right)$$

Putting $f(x_1)$ and $f(x_2)$ from Equations 5.12 and 5.13, we get

$$\begin{aligned} f\left(x_1, x_2\right) &= \frac{1}{\sqrt{2\pi\sigma_1^2}} \mathrm{e}^{-\frac{1}{2}\left(\frac{x_1-\mu_1}{\sigma_1}\right)^2} \times \frac{1}{\sqrt{2\pi\sigma_2^2}} \mathrm{e}^{-\frac{1}{2}\left(\frac{x_2-\mu_2}{\sigma_2}\right)^2} \\ &= \frac{1}{(2\pi)\left(\sigma_1^2\sigma_2^2\right)^{1/2}} \cdot \mathrm{e}^{-\frac{1}{2}\left[\left(\frac{x_1-\mu_1}{\sigma_1}\right)^2 + \left(\frac{x_2-\mu_2}{\sigma_2}\right)^2\right]} \end{aligned} \tag{5.14}$$

where, $-\infty < x_1 < \infty$ and $-\infty < x_2 < \infty$.

See the exponent of Equation 5.14 which is an equation of an ellipse. ***Is it not similar to statistical distance?*** Now we will do a certain amount of matrix manipulation for Equation 5.14. Let's say the bivariate observation vector ($\mathbf{x}$), its mean vector ($\boldsymbol{\mu}$), and covariance matrix ($\boldsymbol{\Sigma}$) are as follows:

$$\mathbf{x} = \begin{bmatrix} x_1 \\ x_2 \end{bmatrix}$$

$$\boldsymbol{\mu}_{2\times1} = \begin{bmatrix} \mu_1 \\ \mu_2 \end{bmatrix}, \text{ and}$$

$$\boldsymbol{\Sigma}_{2\times2} = \begin{bmatrix} \sigma_1^2 & 0 \\ 0 & \sigma_2^2 \end{bmatrix}$$

Note that the off-diagonal elements of Σ are zero (0) as X_1 and X_2 are independent.

The determinant and inverse of Σ are

$$|\boldsymbol{\Sigma}| = \begin{vmatrix} \sigma_1^2 & 0 \\ 0 & \sigma_2^2 \end{vmatrix} = \sigma_1^2\,\sigma_2^2$$

$$\boldsymbol{\Sigma}^{-1} = \frac{1}{|\boldsymbol{\Sigma}|} \times (\text{transpose of cofactor of } \Sigma)$$

$$= \frac{1}{\sigma_1^2\,\sigma_2^2}\begin{bmatrix} \sigma_2^2 & 0 \\ 0 & \sigma_1^2 \end{bmatrix}$$

$$= \begin{bmatrix} \dfrac{1}{\sigma_1^2} & 0 \\ 0 & \dfrac{1}{\sigma_2^2} \end{bmatrix}$$

Reconsider Equation 5.14. The *constant term* of Equation 5.14 is $\dfrac{1}{(2\pi)^{2/2}\left(\sigma_1^2\sigma_2^2\right)^{1/2}}$, which is equal to $\dfrac{1}{(2\pi)^{2/2}\left(|\Sigma|\right)^{1/2}}$.

Similarly, the exponent of $f(x_1, x_2)$ of Equation 5.14 can be written in terms of matrix multiplication. The general exponent of a normal pdf f (x) is $-\dfrac{1}{2}\left(\dfrac{x-\mu}{\sigma}\right)^2$, which can be rewritten as

$$-\frac{1}{2}\left(\frac{x-\mu}{\sigma}\right)^2 = -\frac{1}{2}(x-\mu)(\sigma^2)^{-1}(x-\mu) \tag{5.15}$$

In order to get bivariate equivalence of the univariate exponent in Equation 5.15, we need to understand the equivalence of each of the terms of Equation 5.15, which are shown below.

Univariate X	Bivariate X (X_1 and X_2 independent)
pdf : $X \sim N(\mu, \sigma^2)$	$X \sim N_2(\mu, \Sigma)$
Mean subtracted value: $[x-\mu]$	$\mathbf{x}-\mu = \begin{bmatrix} x_1-\mu_1 \\ x_2-\mu_2 \end{bmatrix}$
Inverse of variance: $(\sigma^2)^{-1}$	$\Sigma^{-1} = \begin{bmatrix} \frac{1}{\sigma_1^2} & 0 \\ 0 & \frac{1}{\sigma_2^2} \end{bmatrix}$

So, for bivariate case, the scalar multiplication form of Equation 5.15 takes a matrix multiplication form and the bivariate exponent for X_1 and X_2 is

$$\begin{aligned} & -\frac{1}{2}(\mathbf{x}-\mu)^T \Sigma^{-1}(\mathbf{x}-\mu) \\ & = -\frac{1}{2}\begin{bmatrix} x_1-\mu_1 \\ x_2-\mu_2 \end{bmatrix}^T \begin{bmatrix} \frac{1}{\sigma_1^2} & 0 \\ 0 & \frac{1}{\sigma_2^2} \end{bmatrix} \begin{bmatrix} x_1-\mu_1 \\ x_2-\mu_2 \end{bmatrix} \\ & = -\frac{1}{2}\begin{bmatrix} x_1-\mu_1 & x_2-\mu_2 \end{bmatrix} \begin{bmatrix} \frac{1}{\sigma_1^2} & 0 \\ 0 & \frac{1}{\sigma_2^2} \end{bmatrix} \begin{bmatrix} x_1-\mu_1 \\ x_2-\mu_2 \end{bmatrix} \\ & = -\frac{1}{2}\begin{bmatrix} \frac{x_1-\mu_1}{\sigma_1^2} & \frac{x_2-\mu_2}{\sigma_2^2} \end{bmatrix} \begin{bmatrix} x_1-\mu_1 \\ x_2-\mu_2 \end{bmatrix} \\ & = -\frac{1}{2}\left[\left(\frac{x_1-\mu_1}{\sigma_1}\right)^2 + \left(\frac{x_2-\mu_2}{\sigma_2}\right)^2\right] \end{aligned} \tag{5.16}$$

which is the equation of an ellipse whose major and minor axes are parallel to X_1 and X_2 axes, respectively. Readers may relate it with the scatters for the processes B and C (Figure 5.2).

Equation 5.16 represents the exponent of $f(x_1, x_2)$ given in Equation 5.14.

$$\text{So, } f(x_1, x_2) = \frac{1}{(2\pi)^{2/2}|\Sigma|^{1/2}} e^{-\frac{1}{2}(\mathbf{x}-\mu)^T \Sigma^{-1}(\mathbf{x}-\mu)} \tag{5.17}$$

Now, let consider a general bivariate case, when X_1 and X_2 are not independent, i.e., $\sigma_{12} \neq 0$ and $\sigma_1 \neq \sigma_2$. Then,

$$\Sigma = \begin{bmatrix} \sigma_1^2 & \sigma_{12} \\ \sigma_{12} & \sigma_2^2 \end{bmatrix}$$

$$\therefore |\Sigma| = \sigma_1^2\sigma_2^2 - \sigma_{12}^2 = \sigma_1^2\sigma_2^2 - \sigma_1^2\sigma_2^2\rho_{12}^2 = \sigma_1^2\sigma_2^2(1-\rho_{12}^2)$$

$$[\text{as } \sigma_{jk} = \sigma_j\sigma_k\rho_{jk}]$$

$$\text{Again, } \mathbf{\Sigma}^{-1} = \frac{1}{|\mathbf{\Sigma}|}[\text{Transpose of cofactor of } \mathbf{\Sigma}]$$

$$= \frac{1}{\sigma_1^2\sigma_2^2 - \sigma_{12}^2}\begin{bmatrix} \sigma_2^2 & -\sigma_{12} \\ -\sigma_{12} & \sigma_1^2 \end{bmatrix}$$

$$\text{Now, } -\frac{1}{2}(\mathbf{x}-\boldsymbol{\mu})^T \Sigma^{-1}(\mathbf{x}-\boldsymbol{\mu})$$

$$= -\frac{1}{2}\begin{bmatrix} x_1-\mu_1 & x_2-\mu_2 \end{bmatrix}\frac{1}{\sigma_1^2\sigma_2^2-\sigma_{12}^2}\begin{bmatrix} \sigma_2^2 & -\sigma_{12} \\ -\sigma_{12} & \sigma_1^2 \end{bmatrix}\begin{bmatrix} x_1-\mu_1 \\ x_2-\mu_2 \end{bmatrix}$$

$$= \left(-\frac{1}{2}\right)\cdot\frac{1}{\sigma_1^2\sigma_2^2-\sigma_{12}^2}\left[\sigma_2^2(x_1-\mu_1)-\sigma_{12}(x_2-\mu_2) \quad -\sigma_{12}(x_1-\mu_1)+\sigma_1^2(x_2-\mu_2)\right]\begin{bmatrix} x_1-\mu_1 \\ x_2-\mu_2 \end{bmatrix} \tag{5.18}$$

$$= \left(-\frac{1}{2}\right)\frac{1}{\sigma_1^2\sigma_2^2-\sigma_{12}^2}\left[\sigma_2^2(x_1-\mu_1)^2-2\sigma_{12}(x_1-\mu_1)(x_2-\mu_2)+\sigma_1^2(x_2-\mu_2)^2\right]$$

$$= \left(-\frac{1}{2}\right)\cdot\frac{\sigma_1^2\sigma_2^2}{\sigma_1^2\sigma_2^2-\sigma_{12}^2}\left[\left(\frac{x_1-\mu_1}{\sigma_1}\right)^2-2\sigma_{12}\left(\frac{x_1-\mu_1}{\sigma_1^2}\right)\left(\frac{x_2-\mu_2}{\sigma_2^2}\right)+\left(\frac{x_2-\mu_2}{\sigma_2}\right)^2\right]$$

Again, we know that $\sigma_{12} = \sigma_1\sigma_2\rho_{12}$. So Equation 5.18 becomes

$$-\frac{1}{2}(\mathbf{x}-\boldsymbol{\mu})^T \Sigma^{-1}(\mathbf{x}-\boldsymbol{\mu})$$

$$= \left(-\frac{1}{2}\right)\frac{\sigma_1^2\sigma_2^2}{\sigma_1^2\sigma_2^2-\sigma_1^2\sigma_2^2\rho_{12}^2}\left[\left(\frac{x_1-\mu_1}{\sigma_1}\right)^2-\frac{2\sigma_1\sigma_2\rho_{12}}{\sigma_1^2\sigma_2^2}\times\left(\frac{x_1-\mu_1}{\sigma_1}\right)\left(\frac{x_2-\mu_2}{\sigma_2}\right)+\left(\frac{x_2-\mu_2}{\sigma_2}\right)^2\right] \tag{5.19}$$

$$= \left(-\frac{1}{2}\right)\frac{1}{1-\rho_{12}^2}\left[\left(\frac{x_1-\mu_1}{\sigma_1}\right)^2-2\rho_{12}\left(\frac{x_1-\mu_1}{\sigma_1}\right)\left(\frac{x_2-\mu_2}{\sigma_2}\right)+\left(\frac{x_2-\mu_2}{\sigma_2}\right)^2\right]$$

which is the equation of an ellipse whose major and minor axes are not parallel to X_1 and X_2 axes because $\rho_{12} \neq 0$. Readers may relate it with the scatter of Process D in Figure 5.2.

Therefore, the general bivariate normal pdf is

$$f(x_1,x_2) = \frac{1}{2\pi\sqrt{\sigma_1^2\sigma_2^2(1-\rho_{12}^2)}}e^{\left(-\frac{1}{2}\right)\frac{1}{1-\rho_{12}^2}\left[\left(\frac{x_1-\mu_1}{\sigma_1}\right)^2-2\rho_{12}\left(\frac{x_1-\mu_1}{\sigma_1}\right)\left(\frac{x_2-\mu_2}{\sigma_2}\right)+\left(\frac{x_2-\mu_2}{\sigma_2}\right)^2\right]} \tag{5.20}$$

$$-\infty < x_1 < \infty \text{ and } -\infty < x_2 < \infty$$

A bivariate normal density is given in Figure 5.4.

Now, let $z_1 = \dfrac{x_1-\mu_1}{\sigma_1}$ and $z_2 = \dfrac{x_2-\mu_2}{\sigma_2}$, when z_1 and z_2 are standardized x_1 and x_2. Then Equation 5.20 becomes

$$f(z_1,z_2) = \frac{1}{2\pi\sqrt{(1-\rho_{12}^2)}}e^{\frac{-1}{2(1-\rho_{12}^2)}\left[z_1^2-2\rho_{12}z_1z_2+z_2^2\right]} \tag{5.21}$$

ρ_{12} is the correlation coefficient between X_1 and X_2.

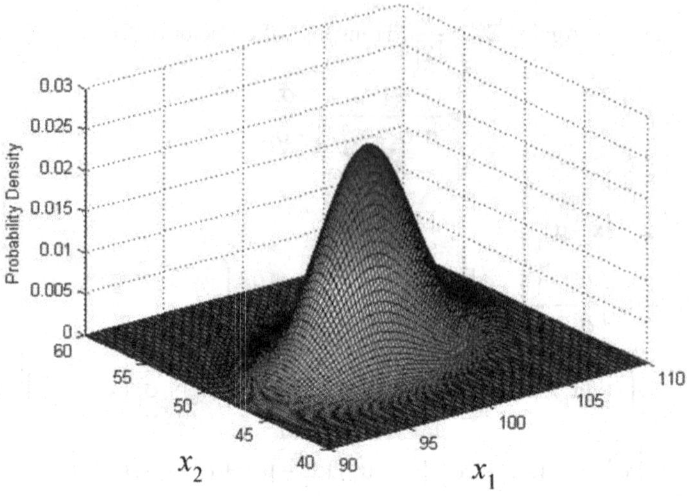

FIGURE 5.4 A bivariate normal density plot. The probability density represents a hill shape. The hill shape is formed to accommodate the joint densities for X_1 and X_2.

Example 5.1

A process $X \sim N_2(\boldsymbol{\mu}, \Sigma)$ is designed to produce laminar aluminum sheet of length X_1 and breadth X_2 with the following population parameters:

$$\boldsymbol{\mu} = \begin{bmatrix} \mu_1 \\ \mu_2 \end{bmatrix} = \begin{bmatrix} 100 \\ 50 \end{bmatrix} \quad \text{and } \Sigma = \begin{bmatrix} \sigma_1^2 & \sigma_{12} \\ \sigma_{12} & \sigma_2^2 \end{bmatrix} = \begin{bmatrix} 10 & 0 \\ 0 & 5 \end{bmatrix}$$

Obtain its bivariate normal distribution.

Solution:

We will use Equation 5.14, as $\sigma_{12} = 0$.

$$f(x_1, x_2) = \frac{1}{2\pi\left(\sigma_1^2\,\sigma_2^2\right)^{\frac{1}{2}}} e^{-\frac{1}{2}\left[\left(\frac{x_1-\mu_1}{\sigma_1}\right)^2 + \left(\frac{x_2-\mu_2}{\sigma_2}\right)^2\right]}$$

Here, $\mu_1 = 100$, $\mu_2 = 50$, $\sigma_1^2 = 10$ and $\sigma_2^2 = 5$

Putting these values in Equation 5.14, we get,

$$f(x_1, x_2) = \frac{1}{2\pi(10 \times 5)^{\frac{1}{2}}} e^{-\frac{1}{2}\left[\left(\frac{x_1-100}{\sqrt{10}}\right)^2 + \left(\frac{x_2-50}{\sqrt{5}}\right)^2\right]}$$

$$= 0.023\,e^{-\frac{1}{20}\left[(x_1-100)^2 + 2(x_2-50)^2\right]}$$

The $f(x_1, x_2)$ obtained in Example 5.1 when plotted, resembles a bell-shaped three-dimensional plot shown in Figure 5.5.

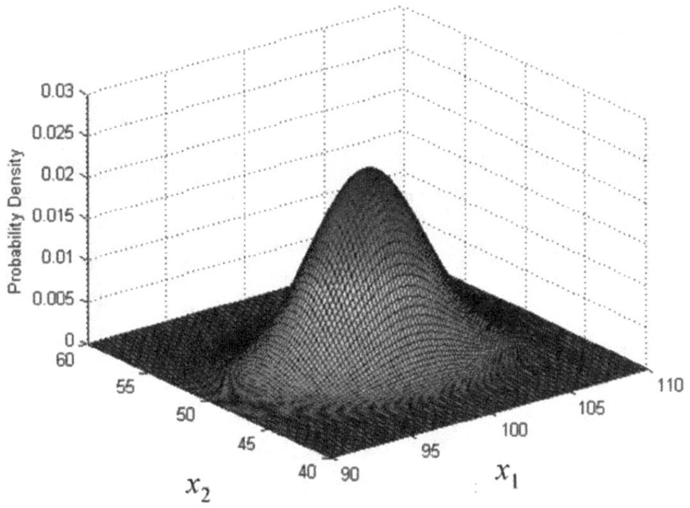

FIGURE 5.5 Bivariate normal density for Example 5.1.

5.3 MULTIVARIATE NORMAL DENSITY FUNCTION

Multivariate normal density function[1] is the extension of bivariate pdf shown in Equation 5.20. Let, $X = [X_1, X_2, ..., X_p]^T$, a p-variable vector follows multivariate normal distribution with parameters $\mu_{p\times 1}$ and $\Sigma_{p\times p}$ and denoted by $N_p(\mu, \Sigma)$. Then, the multivariate normal pdf, $f(X)$ is

$$f(X) = f(x_1, x_2, ..., x_p) = \frac{1}{(2\pi)^{\frac{p}{2}}(|\Sigma|)^{1/2}} e^{-\frac{1}{2}\left[(\mathbf{x}-\mu)^T \Sigma^{-1}(\mathbf{x}-\mu)\right]}, \quad (5.22)$$

$$-\infty < x_j < \infty, \ j = 1, 2, ..., p.$$

where,

$$\mathbf{x} = \begin{bmatrix} x_1 \\ x_2 \\ . \\ . \\ . \\ x_p \end{bmatrix}, \ \mu = \begin{bmatrix} \mu_1 \\ \mu_2 \\ . \\ . \\ . \\ \mu_p \end{bmatrix} \text{ and } \Sigma = \begin{bmatrix} \sigma_{11} & \sigma_{12} & .... & \sigma_{1p} \\ \sigma_{12} & \sigma_{22} & .... & \sigma_{2p} \\ . & & .... & \\ . & & .... & \\ . & & .... & \\ \sigma_{1p} & \sigma_{2p} & .... & \sigma_{pp} \end{bmatrix}$$

The exponent $\left(\text{without} -\frac{1}{2}\right)$ of Equation 5.22 is very important and therefore requires further elaboration. When p = 2 (bivariate case), $(\mathbf{x}-\mu)^T \Sigma^{-1}(\mathbf{x}-\mu)$ becomes

$$d^2 = \frac{1}{1-\rho_{12}^2}\left[\left(\frac{x_1-\mu_1}{\sigma_1}\right)^2 - 2\rho_{12}\left(\frac{x_1-\mu_1}{\sigma_1}\right)\left(\frac{x_2-\mu_2}{\sigma_2}\right) + \left(\frac{x_2-\mu_2}{\sigma_2}\right)^2\right] \quad (5.23)$$

We will have three situations for d^2 shown in Figure 5.6.

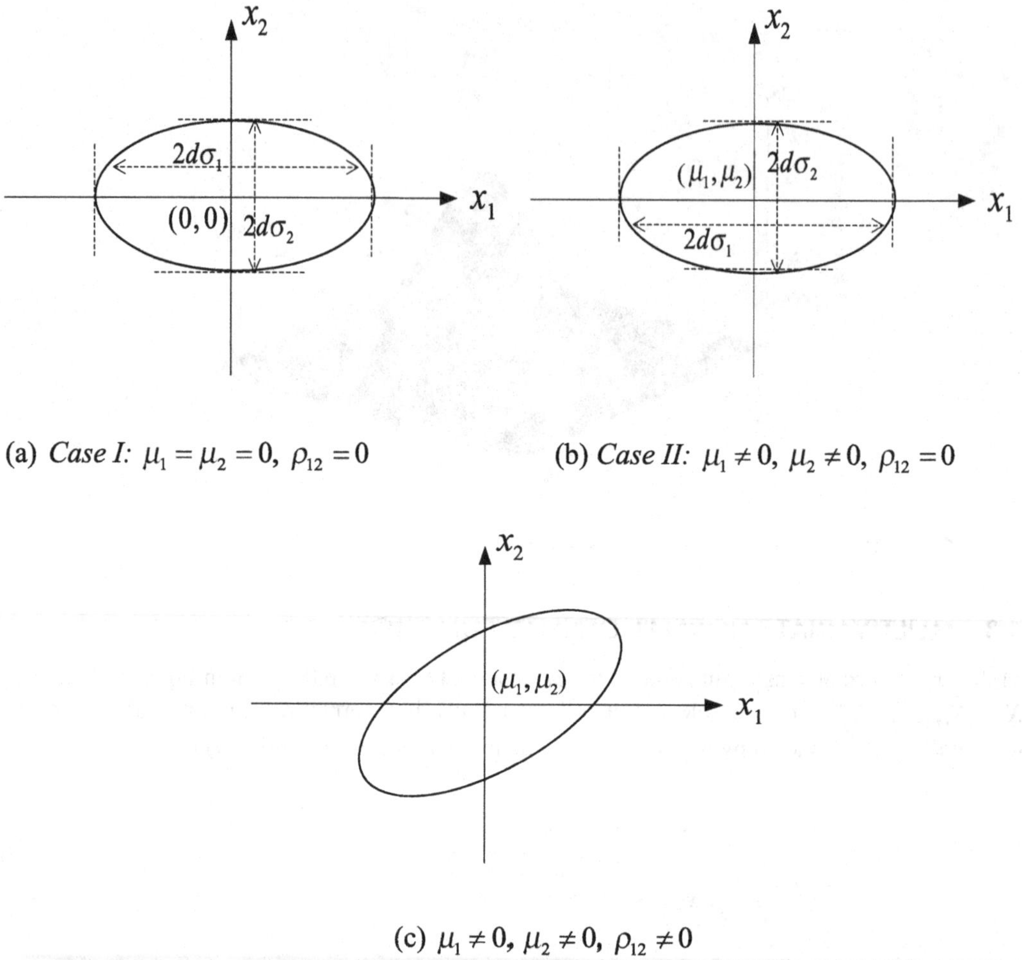

FIGURE 5.6 Illustration of the exponent of bivariate normal distribution: (a) The figure represents an ellipse with center at (0, 0), variance of X_1 (σ_1^2) is greater than the variance of X_2 (σ_2^2) and the two variables X_1 and X_2 are independent ($\rho_{12} = 0$); (b) Similar to (a) but the center is shifted from (0, 0) to (μ_1, μ_2); (c) the center of the ellipse is at (μ_1, μ_2) and the two variables X_1 and X_2 are not independent ($\rho_{12} \neq 0$).

Case I: $\mu_1 = \mu_2 = 0$ and $\rho_{12} = 0$
Case II: $\rho_{12} = 0$
Case III: None of the three is zero.

Case I: $\mu_1 = \mu_2 = 0,\ \rho_{12} = 0$

Putting the values $\mu_1 = \mu_2 = 0,\ \rho_{12} = 0$ in Equation 5.23, we get

$$d^2 = \frac{x_1^2}{\sigma_1^2} + \frac{x_2^2}{\sigma_2^2}, \text{ or, } \frac{x_1^2}{(d\sigma_1)^2} + \frac{x_2^2}{(d\sigma_2)^2} = 1 \tag{5.24}$$

which is the equation of an ellipse, centered at (0, 0) with major and minor axes of length $2d\sigma_1$ and $2d\sigma_2$, respectively or vice versa. Figure 5.6(a) represents the ellipse.

Case II: $\rho_{12} = 0$

Putting $\rho_{12} = 0$ in Equation 5.23, we get

$$\left(\frac{x_1 - \mu_1}{\sigma_1}\right)^2 + \left(\frac{x_2 - \mu_2}{\sigma_2}\right)^2 = d^2 \tag{5.25}$$

which is also an equation of an ellipse as shown in Figure 5.6(b) where the origin of the ellipse shifted from (0, 0) to (μ_1, μ_2).

Case III: None of the parameters are zero

When none of μ_1, μ_2 and ρ_{12} is zero, Equation 5.23 prevails its present form, which is the general equation of an ellipse. One of the possible shapes of the ellipse is given in Figure 5.6(c).

Note the similarity of these ellipses with the scatters presented in Figure 5.2.

The next question to be answered is what will be the shape of $(\mathbf{x}-\boldsymbol{\mu})^T\Sigma^{-1}(\mathbf{x}-\boldsymbol{\mu})$ when X is p-variable vector. For p-dimension, the shape will be an ellipsoid centered at $\boldsymbol{\mu} = [\mu_1\, \mu_2 \ldots. \mu_p]^T$. So for the most general case, we can rewrite equation 5.23 as

$$(\mathbf{x}-\boldsymbol{\mu})^T \Sigma^{-1} (\mathbf{x}-\boldsymbol{\mu}) = d^2 \tag{5.26}$$

which is an ellipsoid of p-axes.

Now we will see what d^2 is. d^2 is known as squared statistical distance of $\mathbf{x}$ from $\boldsymbol{\mu}$. That $(\mathbf{x}-\boldsymbol{\mu})^T\Sigma^{-1}(\mathbf{x}-\boldsymbol{\mu})$ follows chi-square (χ^2) distribution with p degrees of freedom (Mardia et al., 1979). The *pdf* for χ^2 distribution is shown in Figure 5.7. Figure 5.7 shows the *pdf* of χ^2 for 2, 4, 6, and 10 degrees of freedom. Figure 5.8 shows the *pdf* of d^2 for 6 degrees of freedom with cut off d^2 for $\alpha = 0.05$.

Let us consider the upper $100(1 - \alpha)^{th}$ percentile of the d^2 distribution. The value of d^2 at $100(1 - \alpha)^{th}$ percentile value represents all multivariate-observations of $\boldsymbol{X}$ such that

$$(\mathbf{x}-\boldsymbol{\mu})^T \Sigma^{-1} (\mathbf{x}-\boldsymbol{\mu}) \le \chi^2_p(\alpha) \text{ with a probability } 1-\alpha. \tag{5.27}$$

When $\alpha = 0.05$, we get a d^2 value at 95th percentile point (see Figure 5.8) which is $\chi^2_6(0.05)$.

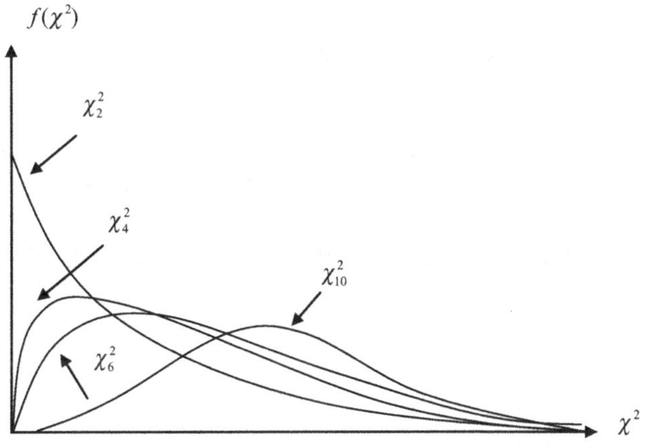

FIGURE 5.7 χ^2 distribution for 2, 4, 6, and 10 degrees of freedom. It shows that the shape and size of χ^2 distribution is dependent on its degrees of freedom.

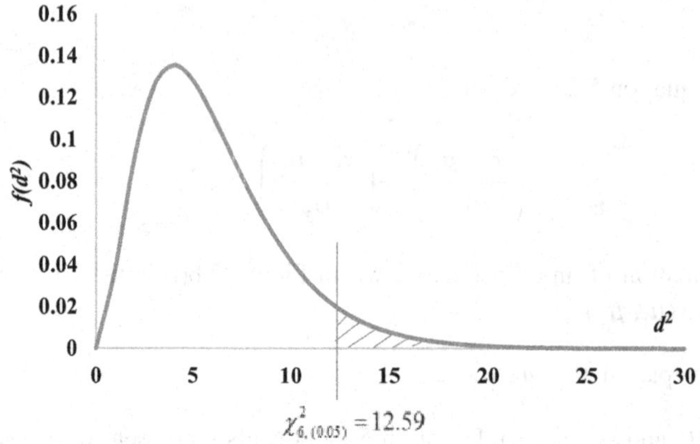

FIGURE 5.8 The pdf of d^2 for p = 6 with cut-off point. The χ^2 value for six degrees of freedom and 0.05 probability level of significance is 12.59. The hatched area, right to this χ^2 value, is 0.05 (the probability level of significance).

Once the value of d^2 is determined, Equation 5.26 can be used to construct the ellipsoid. When $p = 2$, the ellipsoid becomes an ellipse, like Figure 5.6. What does the ellipse mean? It says that all observations lying on the surface of the ellipse are equidistance from the origin.

Example 5.2

Consider the data given in Example 5.1. Obtain the value of d^2 for α = 0.05 and α = 0.01.

Solution:

$$\begin{aligned} d^2 &= (\mathbf{x}-\boldsymbol{\mu})^T \Sigma^{-1} (\mathbf{x}-\boldsymbol{\mu}) \\ &= \left(\frac{x_1 - 100}{\sqrt{10}}\right)^2 + \left(\frac{x_2 - 50}{\sqrt{5}}\right)^2 \\ &\sim \chi_2^2 \end{aligned}$$

In this example, d^2 follows $\chi 2$ with two degrees of freedom as p = 2.
Now, for α = 0.05, $\chi_2^2\,(0.05) = 5.99$ and for α = 0.01, $\chi_2^2\,(0.01) = 9.21$.

So, the value of d^2 and corresponding elliptical equations for α = 0.05 and α = 0.10 are

$$\left(\frac{x_1 - 100}{\sqrt{(10)}}\right)^2 + \left(\frac{x_2 - 50}{\sqrt{(5)}}\right)^2 = 5.99 \quad \text{and} \quad \left(\frac{x_1 - 100}{\sqrt{(10)}}\right)^2 + \left(\frac{x_2 - 50}{\sqrt{(5)}}\right)^2 = 9.21$$

Figure 5.9 shows the two ellipses represented by the above two equations.

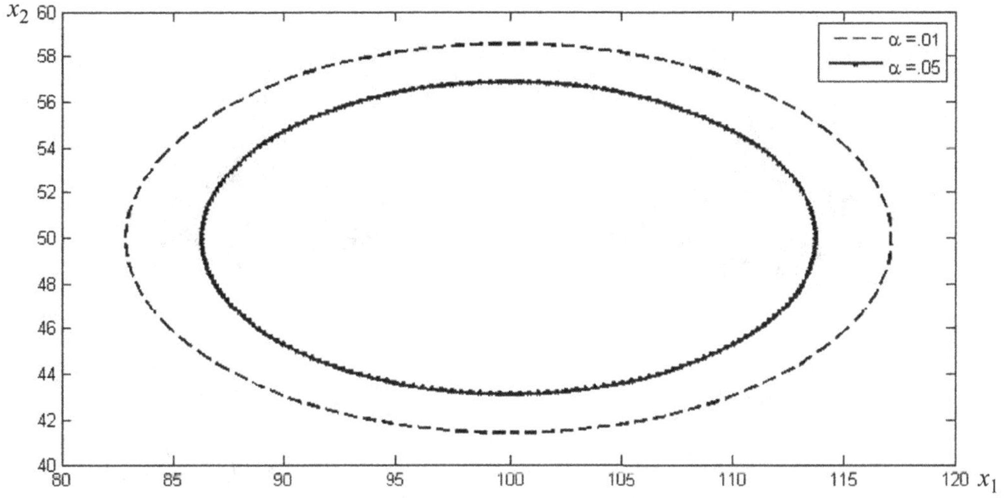

FIGURE 5.9 Ellipses for Example 5.2. The ellipse with a solid line represents 95% ellipse and the one with a dotted line represents 99% ellipse.

5.4 CONSTANT DENSITY CONTOURS

Contour is a path joining points equidistance from a reference base. For example, consider the outer surface of a hill. Now, if you draw a curve joining all points of equal altitude on the periphery of the hill, you get a contour of that altitude. There may be temperature contour, pressure contour, or contours based on some other characteristics. The key property of a contour is that all points on the contour path have equal property based on which the contour is plotted. Analogously constant density contour is a path such that all points on this curve have equal probably density, represented by $f(\boldsymbol{X})$, where $\boldsymbol{X}$ is vector of p variables.

Consider Figure 5.10. The top two figures (5.10(a) and 5.10(b)) represent the bivariate pdf for two different situations. Figure 5.10(a) represents a bivariate pdf of two independent random variables X_1 and X_2 while Figure 5.10(b) shows the pdf when X_1 and X_2 are correlated. In both situations, the vertical axis (y axis) represents the probability density.

Now, if we take a cross-section of the pdf at a particular density, we get curves like Figures 5.10(c) and 5.10(d). For example, in Figure 5.10(c) the outermost ellipse contains 99% observations while the innermost one contains 50% of the total observations. See Equation 5.27 for further insights. Correspondingly, cross-sections are taken at a defined density level. From an interpretation point of view, all points on the surface of the ellipse have equal probably density value. The same is true for Figure 5.10(d) with a difference that the ellipses are oblique. Note that each of the ellipses in Figures 5.10 (c) and 5.10 (d) is centered at its respective $\mu = [\mu_1, \mu_2]^T$.

5.5 PROPERTIES OF MULTIVARIATE NORMAL DENSITY FUNCTION

Like univariate normal distribution, multivariate normal distribution is widely used as it possesses many interesting and useful properties. Some of them are listed below (Mardia et al., 1979; Johnson and Wichern, 2013).

Property (i): $\boldsymbol{X} \sim N_p(\boldsymbol{\mu}, \boldsymbol{\Sigma})$ is completely defined by its first and second moments only.

Property (ii): If $\boldsymbol{X} \sim N_p(\boldsymbol{\mu}, \boldsymbol{\Sigma})$, the Mahalanobis transformation $\boldsymbol{\Sigma}^{-1/2}(\boldsymbol{X} - \boldsymbol{\mu})$ is $N_p(\mathbf{0}, \mathbf{I})$, where $\boldsymbol{\Sigma}^{-1/2}$ is the square root of $\boldsymbol{\Sigma}^{-1}$. Further, each element of $\boldsymbol{\Sigma}^{-1/2}(\boldsymbol{X} - \boldsymbol{\mu})$, say $\mathbf{y}_j$, $j = 1, 2, \ldots, p$ is $N(0,1)$.

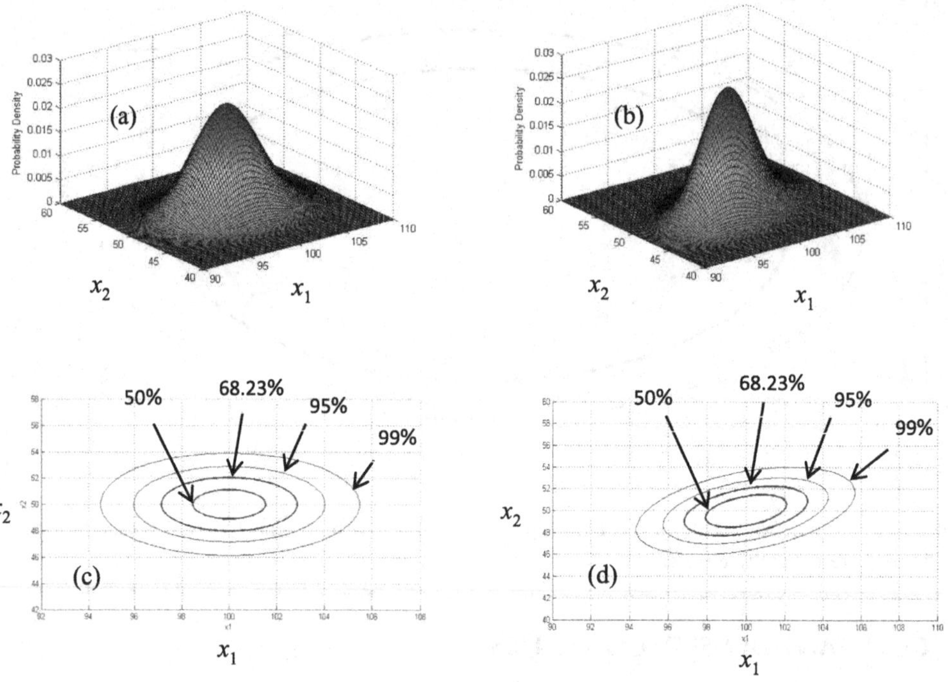

FIGURE 5.10 Illustration of constant density contours. It represents paths joining points with equal densities. The 50% ellipse includes 50% of the observations. Similarly, 68.23%, 95%, and 99% ellipses include 68.23%, 95%, and 99% of the observations, respectively.

Property (iii): If $X \sim N_p(\mu, \Sigma)$, $(\mathbf{x}-\mu)^T \Sigma^{-1}(\mathbf{x}-\mu) \sim \chi^2_p$. We have discussed this in Section 5.3.

Property (iv): Let $X = \begin{bmatrix} X_1 \\ X_2 \end{bmatrix} \sim N_p(\mu, \Sigma)$ and $\Sigma = \begin{bmatrix} \Sigma_{11} & \Sigma_{12} \\ \Sigma_{21} & \Sigma_{22} \end{bmatrix}$, where X_1 and X_2 are subsets of X. If X_1 and X_2 are independent, then $\Sigma_{12} = \mathbf{0}$.

Property (v): If $X_{p\times 1} \sim N_p(\mu, \Sigma)$, then X_j is $N(\mu_j, \sigma_j^2)$ for all X_j, j = 1, 2, … p.

Property (vi): If $X_{p\times 1} \sim N_p(\mu, \Sigma)$, then the subset of $X_{p\times 1}$, i.e., $X_{q\times 1} (q < p)$ is $N_q(\mu_q, \Sigma_{qq})$, where

$$\mu = \begin{bmatrix} \mu_1 \\ \mu_2 \\ \vdots \\ \mu_q \\ \vdots \\ \mu_p \end{bmatrix} = \begin{bmatrix} \mu_q \\ \mu_{p-q} \end{bmatrix}$$

and

$$\boldsymbol{\Sigma} = \begin{bmatrix} \sigma_{11} & \sigma_{12} & \cdots & \sigma_{1q} & \cdots & \sigma_{1p} \\ \sigma_{12} & \sigma_{22} & \cdots & \sigma_{2q} & \cdots & \sigma_{2p} \\ . & & \cdots & & \cdots & \\ . & & \cdots & & \cdots & \\ \sigma_{1q} & \sigma_{2q} & \cdots & \sigma_{qq} & \cdots & \sigma_{qp} \\ . & . & \cdots & & \cdots & \\ . & . & \cdots & & \cdots & \\ \sigma_{1p} & \sigma_{2p} & \cdots & \cdots & \cdots & \sigma_{pp} \end{bmatrix}$$

$$= \begin{bmatrix} \boldsymbol{\Sigma}_{qq} & \vdots & \boldsymbol{\Sigma}_{q\times(p-q)} \\ \cdots & \vdots & \cdots \\ \boldsymbol{\Sigma}_{(p-q)\times q} & \vdots & \boldsymbol{\Sigma}_{(p-q)\times(p-q)} \end{bmatrix}$$

Example 5.3

Let $\boldsymbol{X}_{3\times1} \sim N_3(\boldsymbol{\mu}, \boldsymbol{\Sigma})$ with $\boldsymbol{\mu}$ and $\boldsymbol{\Sigma}$ as given below.

$\boldsymbol{X}_{3\times1} = \begin{bmatrix} X_1 \\ X_2 \\ X_3 \end{bmatrix}$, $\boldsymbol{\mu} = \begin{bmatrix} 10 \\ 15 \\ 9 \end{bmatrix}$ and $\boldsymbol{\Sigma} = \begin{bmatrix} 9 & -8 & 6 \\ -8 & 16 & -7 \\ 6 & -7 & 4 \end{bmatrix}$. What is the distribution of

$\boldsymbol{X}_1 = \begin{bmatrix} X_1 & X_2 \end{bmatrix}^T$?

Solution:

Let us partition the information given as follows:

$$\boldsymbol{X} = \begin{bmatrix} \boldsymbol{X}_1 \\ \cdots \\ \boldsymbol{X}_2 \end{bmatrix} = \begin{bmatrix} X_1 \\ X_2 \\ \cdots \\ X_3 \end{bmatrix}, \quad \boldsymbol{\mu} = \begin{bmatrix} \mu_1 \\ \mu_2 \\ \cdots \\ \mu_3 \end{bmatrix} = \begin{bmatrix} 10 \\ 15 \\ \cdots \\ 9 \end{bmatrix} = \begin{bmatrix} \boldsymbol{\mu}_1 \\ \cdots \\ \boldsymbol{\mu}_2 \end{bmatrix}$$

$$\text{and} \quad \boldsymbol{\Sigma} = \begin{bmatrix} \sigma_{11} & \sigma_{12} & \vdots & \sigma_{13} \\ \sigma_{21} & \sigma_{22} & \vdots & \sigma_{23} \\ \cdots & \cdots & \vdots & \cdots \\ \sigma_{31} & \sigma_{32} & \vdots & \sigma_{33} \end{bmatrix} = \begin{bmatrix} 9 & -8 & \vdots & 6 \\ -8 & 16 & \vdots & -7 \\ \cdots & \cdots & \vdots & \cdots \\ 6 & -7 & \vdots & 4 \end{bmatrix} = \begin{bmatrix} \boldsymbol{\Sigma}_{11} & \vdots & \boldsymbol{\Sigma}_{12} \\ \cdots & \vdots & \cdots \\ \boldsymbol{\Sigma}_{21} & \vdots & \boldsymbol{\Sigma}_{22} \end{bmatrix}$$

The required random variable vector is $\boldsymbol{X}_1 = [X_1 \quad X_2]^T$ and the corresponding mean vector and covariance matrix are

$$\boldsymbol{\mu}_1 = \begin{bmatrix} \mu_1 \\ \mu_2 \end{bmatrix} = \begin{bmatrix} 10 \\ 15 \end{bmatrix} \text{ and } \boldsymbol{\Sigma}_{11} = \begin{bmatrix} \sigma_{11} & \sigma_{12} \\ \sigma_{21} & \sigma_{22} \end{bmatrix} = \begin{bmatrix} 9 & -8 \\ -8 & 16 \end{bmatrix}$$

As per property (ii), $\boldsymbol{X}_1$ follows $N_2(\boldsymbol{\mu}_1, \boldsymbol{\Sigma}_{11})$ with $\boldsymbol{\mu}_1$ and $\boldsymbol{\Sigma}_{11}$ given above.

Property (vii): If $\boldsymbol{X}_{p\times 1} \sim N_p(\boldsymbol{\mu}, \boldsymbol{\Sigma})$, then the linear combination of X_j, $(j = 1, 2, \ldots, p)$ is univariate normal.

Let $\mathbf{c} = [\mathrm{c}_1\ \mathrm{c}_2 \ldots \mathrm{c}_p]^T$, a vector where $\mathrm{c}_1, \mathrm{c}_2, \ldots, \mathrm{c}_p$ are constant values. Then the linear combination of p-variable is

$$\mathbf{c}^{\mathrm{T}} \boldsymbol{X} = \mathrm{c}_1 X_1 + \mathrm{c}_2 X_2 + \cdots + \mathrm{c}_j X_j + \cdots + \mathrm{c}_p X_p$$

Now, mean of $\mathbf{c}^{\mathrm{T}} \boldsymbol{X}$ is

$$\begin{aligned} E(\mathbf{c}^{\mathrm{T}} \boldsymbol{X}) &= E\left(\mathrm{c}_1 X_1 + \mathrm{c}_2 X_2 + \cdots + \mathrm{c}_p X_p\right) \\ &= \mathrm{c}_1 E(X_1) + \mathrm{c}_2 E(X_2) + \cdots + \mathrm{c}_p \mathrm{E}(X_p) \\ &= \mathrm{c}_1 \mu_1 + \mathrm{c}_2 \mu_2 + \cdots + \mathrm{c}_p \mu_p \\ &= \mathbf{c}^{\mathrm{T}} \boldsymbol{\mu} \end{aligned} \tag{5.28}$$

Now, variance of $\mathbf{c}^{\mathrm{T}} \boldsymbol{X}$ is

$$\mathrm{Var}\,(\mathbf{c}^{\mathrm{T}} \boldsymbol{X}) = \mathbf{c}^{\mathrm{T}} \mathrm{Cov}(\boldsymbol{X}) \mathbf{c} = \mathbf{c}^{\mathrm{T}} \boldsymbol{\Sigma} \mathbf{c} \tag{5.29}$$

So, $\mathbf{c}^{\mathrm{T}} \boldsymbol{X}$ follows $\mathrm{N}\,(\mathbf{c}^{\mathrm{T}} \boldsymbol{\mu}, \mathbf{c}^{\mathrm{T}} \boldsymbol{\Sigma} \mathbf{c})$.

Example 5.4

Consider $\boldsymbol{X}_{3\times 1} \sim N_3(\boldsymbol{\mu}, \boldsymbol{\Sigma})$ with the data given in Example 5.3. Let an arbitrary vector be $\mathbf{c}^T = [\mathrm{c}_1\ \ \mathrm{c}_2\ \ \mathrm{c}_3]$. What is the distribution of $\mathbf{c}^T \boldsymbol{X}$? If $\mathbf{c}^T = [-2\ \ 1\ \ 1]$, what will be the distribution of $\mathbf{c}^T \boldsymbol{X}$?

Solution:

From property (iii), we know if $\boldsymbol{X}_{p\times 1} \sim N_p(\boldsymbol{\mu}, \boldsymbol{\Sigma})$, $\mathbf{c}^T \boldsymbol{X}$ is univariate with mean $\mathbf{c}^T \boldsymbol{\mu}$ and variance $\mathbf{c}^T \boldsymbol{\Sigma} \mathbf{c}$. For the given problem, the mean and variance of $\mathbf{c}^T \boldsymbol{X}$ are as follows:

$$\mathrm{E}(\mathbf{c}^T \boldsymbol{X}) = \mathbf{c}^T \boldsymbol{\mu} = \begin{bmatrix} \mathrm{c}_1 & \mathrm{c}_2 & \mathrm{c}_3 \end{bmatrix} \begin{bmatrix} \mu_1 \\ \mu_2 \\ \mu_3 \end{bmatrix} = [\mathrm{c}_1 \mu_1 + \mathrm{c}_2 \mu_2 + \mathrm{c}_3 \mu_3] = [10\mathrm{c}_1 + 15\mathrm{c}_2 + 9\mathrm{c}_3]$$

$$\text{and } \mathrm{V}(\mathbf{c}^T \boldsymbol{X}) = \mathbf{c}^T \boldsymbol{\Sigma} \mathbf{c} = \begin{bmatrix} \mathrm{c}_1 & \mathrm{c}_2 & \mathrm{c}_3 \end{bmatrix} \begin{bmatrix} \sigma_{11} & \sigma_{12} & \sigma_{13} \\ \sigma_{12} & \sigma_{22} & \sigma_{23} \\ \sigma_{13} & \sigma_{23} & \sigma_{33} \end{bmatrix} \begin{bmatrix} \mathrm{c}_1 \\ \mathrm{c}_2 \\ \mathrm{c}_3 \end{bmatrix}$$

$$= \begin{bmatrix} \mathrm{c}_1 & \mathrm{c}_2 & \mathrm{c}_3 \end{bmatrix} \begin{bmatrix} 9 & -8 & 6 \\ -8 & 16 & -7 \\ 6 & -7 & 4 \end{bmatrix} \begin{bmatrix} \mathrm{c}_1 \\ \mathrm{c}_2 \\ \mathrm{c}_3 \end{bmatrix} = [9\mathrm{c}_1^2 + 16\mathrm{c}_2^2 + 4\mathrm{c}_3^2 - 16\mathrm{c}_1\mathrm{c}_2 + 12\mathrm{c}_1\mathrm{c}_3 - 14\mathrm{c}_2\mathrm{c}_3]$$

So, the distribution of $\mathbf{c}^T X$ is

$$N(10c_1 + 15c_2 + 9c_3,\ 9c_1^2 + 16c_2^2 + 4c_3^2 - 16c_1c_2 + 12c_1c_3 - 14c_2c_3).$$

When $\mathbf{c}^T = [-2\ 1\ 1]$, $\mathbf{c}^T \boldsymbol{\mu} = 10\times(-2) + 15\times 1 + 9\times 1 = 4$
and $\mathbf{c}^T \boldsymbol{\Sigma} \mathbf{c} = 9(-2)^2 + 16(1)^2 + 4(1)^2 - 16(-2\times 1) + 12(-2\times 1) - 14(1\times 1) = 50$
So, $\mathbf{c}^T X \sim N(4, 50)$, i.e., univariate normal distribution with mean = 4 and variance = 50.

Property (vii): If $X_{p\times 1} \sim N_p(\boldsymbol{\mu}, \boldsymbol{\Sigma})$, then q linear combinations of X_j, (j = 1, 2, …, p) is multivariate (q-dimensions) normal.

$$\text{Let, } \mathbf{C} = \begin{bmatrix} c_{11} & c_{12} & \cdots & c_{1p} \\ c_{21} & c_{22} & \cdots & c_{2p} \\ . & . & \cdots & . \\ . & . & \cdots & . \\ . & . & \cdots & . \\ c_{q1} & c_{q2} & \cdots & c_{qp} \end{bmatrix}_{q\times p}$$

Then, the q-linear combination of X is

$$\mathbf{C}X = \begin{bmatrix} c_{11} & c_{12} & \cdots & c_{1p} \\ c_{21} & c_{22} & \cdots & c_{2p} \\ . & . & \cdots & . \\ . & . & \cdots & . \\ . & . & \cdots & . \\ c_{q1} & c_{q2} & \cdots & c_{qp} \end{bmatrix} \begin{bmatrix} X_1 \\ X_2 \\ . \\ . \\ . \\ X_p \end{bmatrix}$$

$$= \begin{bmatrix} c_{11}X_1 + & c_{12}X_2 & + \cdots + & c_{1p}X_p \\ c_{21}X_1 + & c_{22}X_2 & + \cdots + & c_{2p}X_p \\ \cdots & \cdots & \cdots & \cdots \\ c_{q1}X_1 + & c_{q2}X_2 & + \cdots + & c_{qp}X_p \end{bmatrix}$$

So, mean of $\mathbf{C}X$ *is*

$$E(\mathbf{C}X) = \mathbf{C}\,E(X) = \mathbf{C}\boldsymbol{\mu} \tag{5.30}$$

Similarly, the covariance of $\mathbf{C}X$ is

$$\text{Cov}(\mathbf{C}X) = \mathbf{C}\,\text{Cov}(X)\,\mathbf{C}^T = \mathbf{C}\,\boldsymbol{\Sigma}\,\mathbf{C}^T \tag{5.31}$$

So, $\mathbf{C}X$ follows $N_q(\mathbf{C}\boldsymbol{\mu}, \mathbf{C}\,\boldsymbol{\Sigma}\,\mathbf{C}^T)$.

Example 5.5

Consider Example 5.3 again, and the matrix **C** given below. What is the distribution of the ***CX*** ?

$$\mathbf{C} = \begin{bmatrix} c_{11} & c_{12} & c_{13} \\ c_{21} & c_{22} & c_{33} \end{bmatrix} = \begin{bmatrix} -2 & 1 & 1 \\ 1 & 2 & -2 \end{bmatrix}$$

Solution:

From Property (vi), we can write

$$\boldsymbol{\mu}_c = E(\mathbf{CX}) = \mathbf{C}\boldsymbol{\mu} = \begin{bmatrix} -2 & 1 & 1 \\ 1 & 2 & -2 \end{bmatrix}\begin{bmatrix} \mu_1 \\ \mu_2 \\ \mu_3 \end{bmatrix} = \begin{bmatrix} -2 & 1 & 1 \\ 1 & 2 & -2 \end{bmatrix}\begin{bmatrix} 10 \\ 15 \\ 9 \end{bmatrix} = \begin{bmatrix} 4 \\ 22 \end{bmatrix}$$

$$\boldsymbol{\Sigma}_c = Cov(\mathbf{CX}) = \mathbf{C\Sigma\, C}^T = \begin{bmatrix} -2 & 1 & 1 \\ 1 & 2 & -2 \end{bmatrix}\begin{bmatrix} 9 & -8 & 6 \\ -8 & 16 & -7 \\ 6 & -7 & 4 \end{bmatrix}\begin{bmatrix} -2 & 1 \\ 1 & 2 \\ 1 & -2 \end{bmatrix}$$

$$= \begin{bmatrix} -20 & 25 & -15 \\ -19 & 38 & -16 \end{bmatrix}\begin{bmatrix} -2 & 1 \\ 1 & 2 \\ 1 & -2 \end{bmatrix} = \begin{bmatrix} 50 & 60 \\ 60 & 89 \end{bmatrix}$$

So, $\boldsymbol{CX} \sim N_2(\boldsymbol{\mu}_c, \boldsymbol{\Sigma}_c)$, where $\boldsymbol{\mu}_c$ and $\boldsymbol{\Sigma}_c$ are as computed above.

5.6 ASSESSING MULTIVARIATE NORMALITY

Multivariate normality is one of the key assumptions of many multivariate statistical models. Multivariate normality assumption states that the population generating multivariate data of the form $\mathbf{X}_{n\times p}$ is multivariate normal. For developing statistical models using $\mathbf{X}_{n\times p}$, it is mandatory to check whether the data are being generated from a multivariate normal population or not. In order to grasp it fully, we first describe the tests of univariate normality followed by the tests of multivariate normality.

5.6.1 Tests of Univariate Normality

There are many tests of univariate normality such as probability-probability (P-P) plot, quantile-quantile (Q-Q) plot, z-test, sample coefficients of kurtosis and skewness, Wilk and Shapiro's W, Anderson-Darling statistic, etc. We will describe here the P-P plot. For other tests, interested readers may consult D'Agostino (1986), Wilk and Gnanadesikan (1968), and Bowman and Shenton (1975).

Suppose X is univariate normal, i.e., $X = N(\mu, \sigma^2)$, with mean μ and variance σ^2. Let $x_1, x_2, \ldots x_n$ be n observations collected randomly from the population. We want to test whether the data is normal or not. So, the problem boils down to hypothesis testing as shown below.

H_0: X is normal, i.e., $X = N(\mu, \sigma^2)$
H_1: X is not normal, i.e., $X \neq N(\mu, \sigma^2)$

The P-P plot is done using the following steps:

1. Arrange the data in ascending order. Let the ordered data be $x_{(1)}, x_{(2)}, \ldots x_{(i)}, \ldots x_{(n)}$.
2. Compute cumulative probability for each of the ordered data. The cumulative probability corresponding to the i-th ordered observation is

$$F\left(x_{(i)}\right) = 100 \times \left(\frac{i - 0.50}{n}\right) \tag{5.32}$$

where $i = 1, 2, \ldots, n$.

3. Plot $F(x_{(i)})$ versus $x_{(i)}$. If the plot resembles a straight line, the data is univariate normal.

Example 5.6

Ten random observations on a random variable X were collected from a population (see Table 5.1). Is the population univariate normal?

Solution:

Using Steps 1 and 2 described above, we get Table 5.2.

Now following Step 3, we get the P-P plot between $F(x_{(i)})$ and $x_{(i)}$ as shown in Figure 5.11.

As the plot resembles a straight line relationship between $F(x_{(i)})$ and $x_{(i)}$, we can conclude that the data was generated from a univariate normal population.

5.6.2 Tests of Multivariate Normality

Let $\boldsymbol{X} \sim N_p(\boldsymbol{\mu}, \boldsymbol{\Sigma})$ be multivariate normal. n number of multivariate observations, $\mathbf{x}_i = [x_{i1}, x_{i2}, \ldots x_{ip}]^T$ on p variables, $i = 1, 2, \ldots, n$ were collected. To test multivariate normality, we have the following hypothesis.

$$H_0 : \boldsymbol{X} \sim N_p(\boldsymbol{\mu}, \boldsymbol{\Sigma})$$
$$H_1 : \boldsymbol{X} \text{ does not follow } N_p(\boldsymbol{\mu}, \boldsymbol{\Sigma})$$

There are informal plots such as Q-Q plot and formal hypothesis tests such as χ^2-test for assessing multivariate normality. We discuss first χ^2 Q-Q plot. For χ^2 Q-Q plot, the steps to be followed are:

TABLE 5.1
Data for Example 5.6

i	1	2	3	4	5	6	7	8	9	10
x_i	110	15	60	45	90	120	145	125	134	75

TABLE 5.2
Ordered data

$x_{(i)}$	15	45	60	75	90	110	120	125	134	145
$F(x(i))$	5	15	25	35	45	55	65	75	85	95

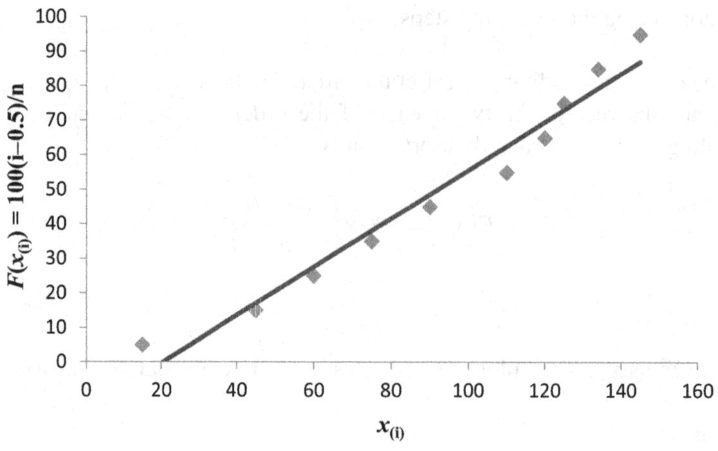

FIGURE 5.11 P-P plot for Example 5.3. As the observations closely follow a straight line, these are obtained from a normal population.

1. Compute squared Mahalanobis distance (MD[2]) for each and every observation vector $\mathbf{x}_i$, $i = 1,2,...,n$ Reproducing Equation 5.11, MD[2] for the i-th observation is $d_i^2 = (\mathbf{x}_i - \bar{\mathbf{x}})^T S^{-1}(\mathbf{x}_i - \bar{\mathbf{x}})$, $i = 1,2,...,n$.
2. Arrange d_i^2, $i = 1,2,...,n$ in ascending order. Let it be $d_{(i)}^2$, $i = 1,2,...,n$ such that $d_{(1)}^2 \le d_{(2)}^2 \le ..... \le d_{(n)}^2$.
3. d_i^2 follows χ^2 distribution with p degrees of freedom. So, obtain $100\left(i - \frac{1}{2}\right)/n$ quantile of χ_p^2, which is $\chi_p^2\left[\left(n - i + \frac{1}{2}\right)/n\right]$.
4. Plot $\chi_p^2\left[\left(n - i + \frac{1}{2}\right)/n\right]$ versus $d_{(i)}^2$.

The plot should exhibit a straight line with slope 1 and pass through the origin (0,0). If one or more observations fall far away from the straight line, H_0 can be rejected, warranting further investigation.

Example 5.7

The following ten bivariate random observations were collected from a population (see Table 5.3). Is the population bivariate normal?

Solution:

Using the steps described above, the solution obtained is given below (see Table 5.4 and Figure 5.12).

TABLE 5.3
Data for Example 5.7

i	1	2	3	4	5	6	7	8	9	10
$\mathbf{x}_1$	93	94	95	96	97	98	99	100	101	102
$\mathbf{x}_2$	52.33	53.46	54.18	54.69	55.04	55.29	56.28	59.48	61.81	63.71

TABLE 5.4
Squared Mahalanobis Distance (MD²)

$\chi_2^2\left((n-i+1/2)/n\right)$	0.1	0.33	0.58	0.86	1.2	1.6	2.1	2.77	3.79	5.99
$d_{(i)}^2$	0.26	0.57	0.68	0.81	1.80	1.99	2.04	2.70	3.05	4.12

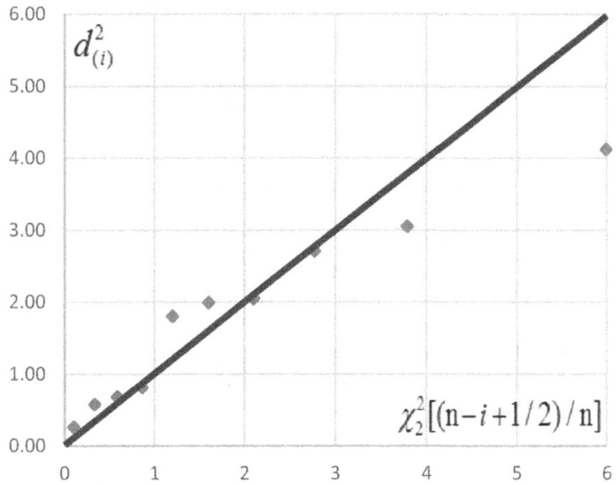

FIGURE 5.12 Q-Q plot for bivariate normal data. The squared Mahalanobis distances $d_{(i)}^2$ follow a straight line with slope 1 passing through the origin, exhibiting that the observed data follow bivariate normal distribution. Note that one observation is reasonably away from the straight line.

To demonstrate the use of Q-Q plot in testing multivariate normality, a process characterized by three random variables is considered. The mean vector and covariance matrix of the population are given below. Using "rmvnorm" function of R software, 100 multivariate observations were generated and, using the procedure described above, the Chi-square Q-Q plot (Figure 5.13) was constructed. As expected, the test exhibits perfect multivariate normality. Note that in the Q-Q plot, the straight line with slope 1 passes through the origin.

$$\boldsymbol{\mu}=\begin{bmatrix}50.85\\29.57\\10.11\end{bmatrix} \quad \text{and} \quad \boldsymbol{\Sigma}=\begin{bmatrix}14.35 & -3.05 & 2.7\\-3.05 & 16.68 & -9.08\\2.70 & -9.08 & 6.14\end{bmatrix}$$

For formal tests, we describe multivariate skewness and multivariate kurtosis tests. Multivariate skewness (MS) and kurtosis (MK) can be defined as follows (Mardia 1970; Mardia et al. 1979; Krzanowski and Marriott, 2014a)

$$\text{MS} = \text{E}\left[\left\{(\mathbf{x}-\boldsymbol{\mu})^T\,\boldsymbol{\Sigma}^{-1}(\mathbf{y}-\boldsymbol{\mu})\right\}^3\right] \tag{5.33}$$

$$\text{MK} = \text{E}\left[\left\{(\mathbf{x}-\boldsymbol{\mu})^T\,\boldsymbol{\Sigma}^{-1}(\mathbf{x}-\boldsymbol{\mu})\right\}^2\right] \tag{5.34}$$

where **x** and **y** are iid.

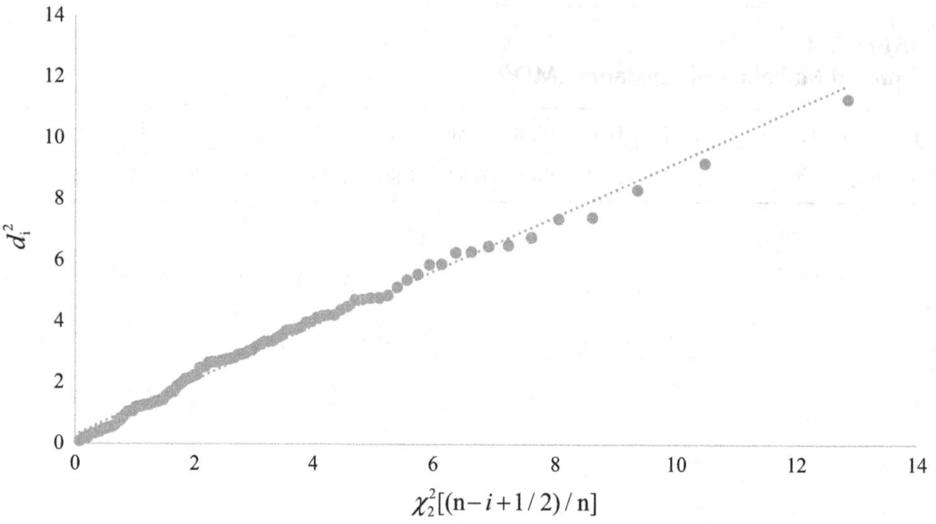

FIGURE 5.13 Chi-squared Q-Q plot for multivariate normal data. The Q-Q plot is a straight line passing through the origin with slope 1, exhibiting that the observed data are from a multivariate normal population.

The sample estimates are

$$\hat{\text{MS}} = \frac{1}{n^2}\sum_{i=1}^{n}\sum_{j=1}^{n}\left[\{(\mathbf{x}_i - \bar{\mathbf{x}})^T \mathbf{S}^{-1}(\mathbf{x}_j - \bar{\mathbf{x}})\}^3\right] \tag{5.35}$$

$$\hat{\text{MK}} = \frac{1}{n}\sum_{i=1}^{n}\left[\{(\mathbf{x}_i - \bar{\mathbf{x}})^T \mathbf{S}^{-1}(\mathbf{x}_i - \bar{\mathbf{x}})\}^2\right] \tag{5.36}$$

If $\boldsymbol{X} \sim N_p(\mu, \Sigma)$, then MS = 0 and MK = $p(p+2)$. Further, $\frac{\hat{\text{MS}}.n}{6}$ and $\hat{\text{MK}}$ for large sample follow $\chi^2_{p(p+1)(p+2)/6}$ and $N\left(p(p+2), \frac{8p(p+2)}{n}\right)$, respectively.

Example 5.8

Consider Example 5.7 and conduct multivariate normality test using multivariate skewness and kurtosis.

Solution:

$$\bar{\mathbf{x}}_1 = 97.5, \quad \bar{\mathbf{x}}_2 = 56.63, \quad n = 10$$

$$\mathbf{S} = \begin{bmatrix} 9.17 & 10.69 \\ 10.69 & 14.20 \end{bmatrix}$$

$$\mathbf{S}^{-1} = \begin{bmatrix} 0.890 & -0.670 \\ -0.670 & 0.574 \end{bmatrix}$$

Following Equations 5.35 and 5.36, we get $\hat{\text{MS}} = 1.79$ and $\hat{\text{MK}} = 4.59$. $\frac{\hat{\text{MS}}.n}{6}$ will follow χ_4^2. The $\chi_4^2(0.05) = 9.49$. As the computed statistic is less than the tabulated threshold, the multivariate skewness test supports multivariate normality of the data.

The statistic $\hat{\text{MK}}$ will follow N(8, 6.4). Accordingly, the computed Z-value is –4.26 and for $\alpha = 0.05$, the tabulated threshold Z is 1.96. This does not support multivariate normality. Note that the sample size (ten) is very low and the above two statistics are recommended for large samples. So, conclusion cannot be made for the above case. This example only demonstrates how to use multivariate skewness and kurtosis tests.

5.6.3 Remedy to Violation of Multivariate Normality

Many times, if data do not exhibit multivariate normality, then check for presence of outliers. Squared Mahalanobis distance (MD^2) can be used for this purpose. As we have seen earlier, for MND data MD^2 follows χ^2 distribution and the χ^2 Q-Q plot exhibits a straight line passing through the origin. The data points that do not lie on or close to the straight line can be treated as outliers. It is customary to remove these outliers and redraw the χ^2 Q-Q plot with the reduced data set. Krzanowski and Marriott (2014a) reviewed several approaches, both informal such as plots and formal tests for outlier detection. One approach is ordering of multivariate data, analogous to natural ordering of data used in univariate case. Barnett (1976) stated that finding out a single value for each of the multivariate observations (i.e., the vector of observations on p-variable) as a scalar quantity is the most helpful.

Gnanadesikan and Kettenring (1972) tested several statistics where each multivariate observation is reduced to a single value. The Squared Mahalanobis distance (MD^2), discussed earlier, is one such statistic. Another approach, as discussed in Krzanowski and Marriott (2014a), is tests based on how outliers are generated (outlier mechanism). Hawkins (1980) discussed two mechanism as data from long-tailed distributions or from a mixture of distributions. Hypothesis tests are available for such outliers' detection.

It may so happen that the outlier removal process continues without significant improvement. It indicates that data might be generated from non-normal processes. To make the data normal, data can be transformed. Three popular approaches are as follows (Krzanowski and Marriott, 2014a):

(i) Treat each variable separately and transform the non-normal data to normal using a proper univariate data transformation method such as Box-Cox transformation. This approach satisfies marginal normality as all the variables individually are univariate normal. However, this does not guarantee multivariate normality.

(ii) Consider all the variables simultaneously (together) and transform the data matrix $\mathbf{X}_{n\times p}$, suitably such that each row vector satisfies $N_p(\boldsymbol{\mu}, \boldsymbol{\Sigma})$. This approach satisfies joint normality, which is the most desired.

(iii) It may so happen that $\mathbf{X}_{n\times p}$ could exhibit multivariate normality in some other directions different from the coordinate system represented by the original p-variables. Then, the data could be projected to the directions that exhibit multivariate normality. This approach satisfies directional normality.

Readers may refer to Chapter 8, Section 8.11.2 for the procedure of Box-Cox transformation.

5.7 LEARNING SUMMARY

In this chapter, the multivariate normal distribution (MND) and its properties are discussed. The other important concepts include statistical distance, Mahalanobis distance, constant density contours, and the chi-square quantile-quantile plot for assessing multivariate normality.

In summary,

- The multivariate normal distribution (MND) is the building block of most of the multivariate models. An important aspect to understand the utility of the MND is statistical distance. In fact the exponent of multivariate normal pdf is a distance measure. When such distance is measured for every observation using sample statistics, namely sample mean vector ($\overline{\mathbf{x}}$) and sample covariance matrix (**S**), the distance measure is known as squared Mahalanobis distance, $d_i^2 = (\mathbf{x}_i - \overline{\mathbf{x}})^T \mathbf{S}^{-1}(\mathbf{x}_i - \overline{\mathbf{x}})$.
- Geometrically, the exponent of a multivariate normal pdf defines an ellipsoid, which becomes an ellipse for two-variable situation (bivariate case).
- Another important concept that requires attention is constant density contours. It represents a paths joining points with equal densities.
- The MND possesses certain useful properties that make it widely applicable.
- As the parameters of the MND ($\boldsymbol{\mu}$, $\boldsymbol{\Sigma}$) are seldom known, it is customary to test multivariate data using different statistical tools to confirm whether the underlying population is multivariate normal or not. χ^2 quantile-quantile plot is frequently used for this purpose.
- If data is not multivariate normal, several options can be adopted to convert it to near multivariate normal. First, remove outliers using Mahalanobis distance. If it doesn't satisfy the requirement, power transformation of original data is recommended. Box-Cox method of transformation can also be performed in this regard.

EXERCISES

(A) Short conceptual questions

5.1 Define statistical distance. How does it differ from Euclidian distance?

5.2 Statistical distance is weighted Euclidian distance. Do you agree? Explain.

5.3 The scatter plot of a two-dimensional data set exhibits elliptical pattern where the major and minor axes of the ellipse are parallel to the $X_1 - X_2$ axes, respectively. Comment on the behavior of the data.

5.4 Show that for p-variable case, the statistical distance is $(\mathbf{x}-\boldsymbol{\mu})^{T}\Sigma^{-1}(\mathbf{x}-\boldsymbol{\mu})$ where $\boldsymbol{\mu}$ denotes $p \times 1$ mean vector and $\boldsymbol{\Sigma}$ denotes $p \times p$ covariance matrix.

5.5 Define bivariate normal distribution (BND). Show that the exponent of the BND represents the equation of an ellipse.

5.6 Derive multivariate normal *pdf*.

5.7 The exponent of the multivariate normal *pdf* follows χ_p^2 distribution where p is the number of variables characterizing the population. Explain.

5.8 What is constant density contour? What is its physical significance?

5.9 State the important properties of multivariate normal distribution.

5.10 If $\boldsymbol{X}_{p\times 1} \sim N_p(\boldsymbol{\mu}, \boldsymbol{\Sigma})$, then show that the q-linear combinations of $X_j\ (j = 1,2,...p)$ is also multivariate normal.

5.11 A data set $\mathbf{X}_{n\times p}$ is thought to be generated from a multivariate normal process. How do you test it?

5.12 What is χ^2 Q-Q plot? Explain its steps with an example.

5.13 Why does multivariate data differ from normality? Is there any remedy to make data normal?

5.14 Define outliers. Mention a few methods to detect outliers.

5.15 What is Box-Cox transformation? How do you determine the transformation coefficient?

(B) **Long conceptual and/or numerical questions:** see web resources (Multivariate Statistical Modeling in Engineering and Management – 1st (routledge.com))

N.B.: R, MATLAB and MS Excel software are used for computation, as and when required.

NOTE

1. Attempt is made to make readers understand the nature and use of bivariate and multivariate normal distributions; not to derive or prove it using statistical theory.

6 Multivariate Inferential Statistics

In univariate inferential statistics, we have discussed point and interval estimation and hypothesis testing concerning single population mean and variance, differences between two population means, and ratio of two population variances. The multivariate counterparts of the above are (i) inference about single population mean vector, (ii) inference about the difference between two population mean vectors, and (iii) inference about covariance and correlation matrices. We will describe them in this chapter. We start with Hotelling's T^2, which is the foundation of multivariate inferential statistics. It is the multivariate counterpart of t-distribution. As we deal with two or more variables simultaneously, the concept of confidence interval (CI) (in univariate statistics) is no longer valid, whose analogue in multivariate domain is confidence region (CR). As it becomes difficult to make decisions based on confidence regions involving more than three variables, simultaneous confidence interval (SCI) for each of the variables needs to be developed. For computation of SCI, both the linear combination and Bonferroni approaches are described in this chapter. Finally, hypothesis testing for mean vectors for both single and two populations, and single population covariance matrix is discussed.

6.1 ESTIMATION OF PARAMETERS OF MULTIVARIATE NORMAL DISTRIBUTION

In Chapter 5, we described the multivariate normal distribution (MND) and its various properties. As discussed earlier, the MND of a random vector of the form $\mathbf{x} \sim N_p(\boldsymbol{\mu},\boldsymbol{\Sigma})$ is sufficiently defined by its parameters $\boldsymbol{\mu} = [\mu_1, \mu_2, \ldots, \mu_p]^T$ and $\boldsymbol{\Sigma} = \{\sigma_{jk}\}$, $j = 1,2,\ldots, p$ and $k = 1,2,\ldots, p$, where $\{\}$ represents the set of all elements of $\boldsymbol{\Sigma}$. Both $\boldsymbol{\mu}$ and $\boldsymbol{\Sigma}$ are seldom known and are usually estimated from sample data. The most commonly used method of estimation is maximum likelihood estimation (MLE). For fundamentals of MLE, see Chapter 3 (Section 3.3.3). In the section below we demonstrate how MLE can be used to estimate $\boldsymbol{\mu}$ and $\boldsymbol{\Sigma}$.

Let $\mathbf{x}_1, \mathbf{x}_2, \ldots, \mathbf{x}_n$ be the n multivariate observations collected from a multivariate population defined by $\mathbf{x} \sim N_p(\boldsymbol{\mu},\boldsymbol{\Sigma})$. Let all $\mathbf{x}_i$, $i = 1, 2, \ldots, n$ are independent and identically distributed as $N_p(\boldsymbol{\mu},\boldsymbol{\Sigma})$. The likelihood function for the sample of size n is given by

$$L(\boldsymbol{\mu},\boldsymbol{\Sigma}) = \prod_{i=1}^{n} \frac{1}{(2\pi)^{p/2}\,|\boldsymbol{\Sigma}|^{1/2}} e^{-\frac{1}{2}(\mathbf{x}_i-\boldsymbol{\mu})^T \boldsymbol{\Sigma}^{-1}(\mathbf{x}_i-\boldsymbol{\mu})}$$

$$= (2\pi)^{\frac{-np}{2}}\,|\boldsymbol{\Sigma}|^{\frac{-n}{2}}\, e^{-\frac{1}{2}\sum_{i=1}^{n}\left[(\mathbf{x}_i-\boldsymbol{\mu})^T \boldsymbol{\Sigma}^{-1}(\mathbf{x}_i-\boldsymbol{\mu})\right]} \tag{6.1}$$

Taking log of the above, we get

$$\ln L(\boldsymbol{\mu},\boldsymbol{\Sigma}) = -\frac{np}{2}\ln(2\pi) - \frac{n}{2}\ln|\boldsymbol{\Sigma}| - \frac{1}{2}\sum_{i=1}^{n}\left[(\mathbf{x}_i-\boldsymbol{\mu})^T \boldsymbol{\Sigma}^{-1}(\mathbf{x}_i-\boldsymbol{\mu})\right] \tag{6.2}$$

The RHS of Equation 6.2 contains three terms. The first term is constant, the second term involves $\boldsymbol{\Sigma}$, and the third term comprises both $\boldsymbol{\mu}$ and $\boldsymbol{\Sigma}$. The third term can be simplified as follows:

DOI: 10.1201/9781003303060-8

From matrix algebra,

$$\sum_{i=1}^{n}\left[(\mathbf{x}_i-\boldsymbol{\mu})^T\boldsymbol{\Sigma}^{-1}(\mathbf{x}_i-\boldsymbol{\mu})\right]=trace\left[\boldsymbol{\Sigma}^{-1}\sum_{i=1}^{n}\left[(\mathbf{x}_i-\boldsymbol{\mu})(\mathbf{x}_i-\boldsymbol{\mu})^T\right]\right]$$

By subtracting and adding $\bar{\mathbf{x}}$ in the right hand side (RHS) of the above, we get

$$\sum_{i=1}^{n}\left[(\mathbf{x}_i-\boldsymbol{\mu})^T\boldsymbol{\Sigma}^{-1}(\mathbf{x}_i-\boldsymbol{\mu})\right]=trace\left[\boldsymbol{\Sigma}^{-1}\sum_{i=1}^{n}(\mathbf{x}_i-\bar{\mathbf{x}}+\bar{\mathbf{x}}-\boldsymbol{\mu})(\mathbf{x}_i-\bar{\mathbf{x}}+\bar{\mathbf{x}}-\boldsymbol{\mu})^T\right]$$

$$=trace\left[\boldsymbol{\Sigma}^{-1}\left\{\sum_{i=1}^{n}(\mathbf{x}_i-\bar{\mathbf{x}})(\mathbf{x}_i-\bar{\mathbf{x}})^T+n(\bar{\mathbf{x}}-\boldsymbol{\mu})(\bar{\mathbf{x}}-\boldsymbol{\mu})^T+(\bar{\mathbf{x}}-\boldsymbol{\mu})\sum_{i=1}^{n}(\mathbf{x}_i-\bar{\mathbf{x}})^T+(\bar{\mathbf{x}}-\boldsymbol{\mu})^T\sum_{i=1}^{n}(\mathbf{x}_i-\bar{\mathbf{x}})\right\}\right]$$

Now, $\sum_{i=1}^{n}(\mathbf{x}_i-\bar{\mathbf{x}})=\sum_{i=1}^{n}\mathbf{x}_i-n\bar{\mathbf{x}}=n\bar{\mathbf{x}}-n\bar{\mathbf{x}}=\mathbf{0}$ and $\sum_{i=1}^{n}(\mathbf{x}_i-\bar{\mathbf{x}})(\mathbf{x}_i-\bar{\mathbf{x}})^T \sim n\mathbf{S}$ (as n is large).

$$\text{So, }\sum_{i=1}^{n}\left[(\mathbf{x}_i-\boldsymbol{\mu})^T\boldsymbol{\Sigma}^{-1}(\mathbf{x}_i-\boldsymbol{\mu})\right]=trace\left[\boldsymbol{\Sigma}^{-1}\left\{n\mathbf{S}+n(\bar{\mathbf{x}}-\boldsymbol{\mu})(\bar{\mathbf{x}}-\boldsymbol{\mu})^T\right\}\right]$$

Therefore, Equation 6.2 can be written as

$$\ln L(\boldsymbol{\mu},\boldsymbol{\Sigma})=-\frac{np}{2}\ln(2\pi)-\frac{n}{2}\ln|\boldsymbol{\Sigma}|-\frac{n}{2}trace\left[\boldsymbol{\Sigma}^{-1}\mathbf{S}\right]+\frac{n}{2}trace\left[\boldsymbol{\Sigma}^{-1}(\bar{\mathbf{x}}-\boldsymbol{\mu})(\bar{\mathbf{x}}-\boldsymbol{\mu})^T\right] \quad (6.3)$$

Equation 6.3 is a function of $\boldsymbol{\mu}$ and $\boldsymbol{\Sigma}$. If the MLE estimate of $\boldsymbol{\mu}$ is $\hat{\boldsymbol{\mu}}$ and that of $\boldsymbol{\Sigma}$ is $\hat{\boldsymbol{\Sigma}}$, then $\hat{\boldsymbol{\mu}}$ and $\hat{\boldsymbol{\Sigma}}$ are those values of the mean vector, $\boldsymbol{\mu}$ and covariance matrix, $\boldsymbol{\Sigma}$ that make the observed data, $\mathbf{X}_{n\times p}$, the most likely. The $\hat{\boldsymbol{\mu}}$ and $\hat{\boldsymbol{\Sigma}}$ can be obtained by maximizing Equation 6.3 with respect to $\boldsymbol{\mu}$ and $\boldsymbol{\Sigma}$, respectively. However, the process is not as easy as seen in the univariate case. For mathematical details, see Mardia et al. (1979) and Krzanowski and Marriott (2014a). The $\hat{\boldsymbol{\mu}}$ and $\hat{\boldsymbol{\Sigma}}$ are computed as below.

Taking derivative of $\ln L(\boldsymbol{\mu},\boldsymbol{\Sigma})$ with respect to $\boldsymbol{\mu}$ in Equation 6.3 yields the following equation.

$$\frac{\partial \ln(\boldsymbol{\mu},\boldsymbol{\Sigma})}{\partial\boldsymbol{\mu}}=n\boldsymbol{\Sigma}^{-1}(\bar{\mathbf{x}}-\boldsymbol{\mu})=\mathbf{0}$$

This gives $\hat{\boldsymbol{\mu}}=\bar{\mathbf{x}}$.

Alternatively, Krzanowski and Marriott (2014a) stated that the quantity within the third bracket of the fourth term of Equation 6.3 should be non-negative. So, to maximize $\ln L(\boldsymbol{\mu},\boldsymbol{\Sigma})$ in Equation 6.3, $\hat{\boldsymbol{\mu}}=\bar{\mathbf{x}}$ and this makes $\ln L(\hat{\boldsymbol{\mu}},\boldsymbol{\Sigma})\geq\ln L(\boldsymbol{\mu},\boldsymbol{\Sigma})$.

Now, putting $\hat{\boldsymbol{\mu}}=\bar{\mathbf{x}}$ in Equation 6.3, we get

$$\ln L(\hat{\boldsymbol{\mu}},\boldsymbol{\Sigma})=-\frac{np}{2}\ln(2\pi)-\frac{n}{2}\left[\ln|\boldsymbol{\Sigma}|+trace\left(\boldsymbol{\Sigma}^{-1}\mathbf{S}\right)\right] \quad (6.4)$$

Using the derivative of a random matrix, and assuming a new parameter matrix $\mathbf{V}=\boldsymbol{\Sigma}^{-1}$, Mardia et al. (1979) showed that

$$\frac{\partial lnL(\boldsymbol{\mu},\boldsymbol{\Sigma})}{\partial\mathbf{V}}=\frac{n}{2}(2\mathbf{M}-Diag\,\mathbf{M}) \quad (6.5)$$

where $\mathbf{M}=\boldsymbol{\Sigma}-\mathbf{S}-(\bar{\mathbf{x}}-\boldsymbol{\mu})(\bar{\mathbf{x}}-\boldsymbol{\mu})^T$

Putting $\frac{\partial lnL(\boldsymbol{\mu},\mathbf{V})}{\partial \mathbf{V}} = \mathbf{0}$, implies $\mathbf{M} = \mathbf{0}$.

$$\text{So, } \mathbf{M} = \boldsymbol{\Sigma} - \mathbf{S} - (\overline{\mathbf{x}} - \boldsymbol{\mu})(\overline{\mathbf{x}} - \boldsymbol{\mu})^T = \mathbf{0} \tag{6.6}$$

Putting $\hat{\boldsymbol{\mu}} = \overline{\mathbf{x}}$ in Equation 6.6, we get

$$\hat{\boldsymbol{\Sigma}} = \mathbf{S}$$

Alternatively, the second term within third bracket of Equation 6.4 is a function of $\boldsymbol{\Sigma}$ and can be written as (Krzanowski and Marriott, 2014a):

$$f(\boldsymbol{\Sigma}) = \ln|\boldsymbol{\Sigma}| + trace[\boldsymbol{\Sigma}^{-1}\mathbf{S}] \text{and}$$

$$f(\boldsymbol{\Sigma}) - f(\mathbf{S}) = -\ln|\boldsymbol{\Sigma}^{-1}\mathbf{S}| + trace(\boldsymbol{\Sigma}^{-1}\mathbf{S}) - p$$
$$= \sum_{j=1}^{p}\left(-\ln\lambda_j + \lambda_j - 1\right) \geq 0$$

where λ_j is the *j-th* eigenvalue of $\Sigma^{-1}\mathbf{S}$ and all λ_j, $j = 1, 2, \ldots$, p are positive

Note: Krzanowski and Marriott (2014a) used $\mathbf{A}$ instead of $\mathbf{S}$, where $\mathbf{A} = (n-1)\mathbf{S}$ and showed that $\hat{\boldsymbol{\Sigma}} = \frac{1}{n}\mathbf{A} = \frac{n-1}{n}\mathbf{S} \sim \mathbf{S}$ for large n.

Using this result, Krzanowski and Marriott (2014a) further stated that the maximum of Equation 6.4 is achieved when $\hat{\boldsymbol{\Sigma}} = \mathbf{S}$, and $\ln L(\hat{\boldsymbol{\mu}},\hat{\boldsymbol{\Sigma}}) \geq \ln L(\hat{\boldsymbol{\mu}},\boldsymbol{\Sigma}) \geq \ln L(\boldsymbol{\mu},\boldsymbol{\Sigma})$ for all $\boldsymbol{\mu}$ and $\boldsymbol{\Sigma}$. The statistics $\overline{\mathbf{x}}$ and $\mathbf{S}$ are independent. These two are the sufficient statistics for a multivariate normal distribution as these two statistics contain all the information about $\boldsymbol{\mu}$ and $\boldsymbol{\Sigma}$ in $\mathbf{X}_{n\times p}$ (Johnson and Wichern, 2013).

These two statistics are also used to summarize data coming from multivariate non-normal distribution. However, these statistics are sensitive to outliers. It is better to have outliers removed before finalizing these estimates. Alternatively, robust estimation methods may be used but these are accurate for small deviations only.

6.2 SAMPLING DISTRIBUTION OF $\overline{\mathbf{x}}$ AND S

In Section 2.4, we discussed the sampling distribution of $\overline{x}$ and s^2, i.e., sample mean and sample variance of the univariate random variable X. The same approach can be used for deriving the sampling distribution of $\overline{\mathbf{x}}$, the sample mean vector, and $(n\text{-}1)\mathbf{S}$, where $\mathbf{S}$ is the sample covariance matrix.

If $\mathbf{x}_i$ comes from multivariate normal, $N_p(\boldsymbol{\mu},\boldsymbol{\Sigma})$, and all $\mathbf{x}_i$, $i = 1, 2, \ldots, n$ are iid, then

$$E(\overline{\mathbf{x}}) = \frac{1}{n}\sum_{i=1}^{n} E(\mathbf{x}_i) = \frac{1}{n}(\boldsymbol{\mu} + \boldsymbol{\mu} + \ldots\ldots + \boldsymbol{\mu}) = \frac{n\boldsymbol{\mu}}{n} = \boldsymbol{\mu}$$

$$V(\overline{\mathbf{x}}) = \frac{1}{n^2}\left[\sum_{i=1}^{n} V(\mathbf{x}_i) + \sum_{i\neq k} Cov(\mathbf{x}_i,\mathbf{x}_k)\right]$$
$$= \frac{1}{n^2}[n\boldsymbol{\Sigma} + \mathbf{0}] = \frac{1}{n}\boldsymbol{\Sigma}$$

and the sampling distribution of $\overline{\mathbf{x}}$ is $N_p\left(\boldsymbol{\mu}, \frac{1}{n}\boldsymbol{\Sigma}\right)$.

The expected value of **S**, i.e., $E(\mathbf{S})$ can be written as

$$E(\mathbf{S}) = \frac{1}{n}E\left[\sum_{i=1}^{n}(\mathbf{x}_i - \bar{\mathbf{x}})(\mathbf{x}_i - \bar{\mathbf{x}})^T\right] = \frac{1}{n}(n-1)\boldsymbol{\Sigma} = \frac{n-1}{n}\boldsymbol{\Sigma} \tag{6.7}$$

Here,

$$\begin{aligned}
\sum_{i=1}^{n}(\mathbf{x}_i - \bar{\mathbf{x}})(\mathbf{x}_i - \bar{\mathbf{x}})^T &= \sum_{i=1}^{n}(\mathbf{x}_i - \bar{\mathbf{x}})\mathbf{x}_i^T + \sum_{i=1}^{n}(\mathbf{x}_i - \bar{\mathbf{x}})(-\bar{\mathbf{x}})^T \\
&= \sum_{i=1}^{n}\mathbf{x}_i\mathbf{x}_i^T - \sum_{i=1}^{n}\bar{\mathbf{x}}\mathbf{x}_i^T + \sum_{i=1}^{n}\mathbf{x}_i(-\bar{\mathbf{x}})^T - \sum_{i=1}^{n}\bar{\mathbf{x}}(-\bar{\mathbf{x}})^T \\
&= \sum_{i=1}^{n}\mathbf{x}_i\mathbf{x}_i^T - n\bar{\mathbf{x}}\bar{\mathbf{x}}^T
\end{aligned}$$

$$\begin{aligned}
\text{or, } E\left[\sum_{i=1}^{n}(\mathbf{x}_i - \bar{\mathbf{x}})(\mathbf{x}_i - \bar{\mathbf{x}})^T\right] &= \sum_{i=1}^{n}E(\mathbf{x}_i\mathbf{x}_i^T) - nE(\bar{\mathbf{x}}\bar{\mathbf{x}}^T) = n\Sigma + n\boldsymbol{\mu}\boldsymbol{\mu}^T - \frac{n}{n}\Sigma - n\boldsymbol{\mu}\boldsymbol{\mu}^T \\
&= (n-1)\Sigma
\end{aligned}$$

The sampling distribution of $(n-1)\mathbf{S}$ is a Wishart distribution with n-1 degrees of freedom. Wishart *pdf* doesn't exist for $n \le p$. For $p = 1$, Wishart distribution coincides with χ^2 distribution. So, $p \ge 2$, Wishart distribution represents multivariate generalization of χ^2 distribution.

Recall from Chapter 4, the *SSCP* matrix, $\mathbf{X}^{*T}\mathbf{X}^*$, where $\mathbf{X}^*_{n\times p}$ is the mean subtracted data matrix and we have seen that $\mathbf{X}^{*T}\mathbf{X}^* = (n-1)\mathbf{S}$. Alternatively, $\mathbf{X}^*_{n\times p}$ can be thought of a data matrix, sampled from a multivariate normal population whose mean is zero (i.e., $\boldsymbol{\mu} = \mathbf{0}$) and covariance matrix is $\boldsymbol{\Sigma}$. What will be the distribution of $\mathbf{X}^{*T}\mathbf{X}^*$? Wishart (1928) developed a distribution for random matrix of this nature, which was named after him.

Let $\mathbf{X}_{n\times p}$, is a data matrix from $N_p(\mathbf{0}, \boldsymbol{\Sigma})$. Then, $\mathbf{M} = \mathbf{X}^T\mathbf{X}$ follows Wishart distribution, $W_p(\boldsymbol{\Sigma}, n)$ with parameters $\boldsymbol{\Sigma}$ matrix as scale matrix and degrees of freedom n (Mardia et al., 1979). **M** is a square and symmetric matrix of the order $p \times p$. $W_p(\boldsymbol{\Sigma}, n)$ has several important properties and we list a few of them below. For proof and further reading, interested readers may see Section 3.4 of Mardia et al (1979).

Property I: If $p = 1$, $\mathbf{M} \sim W_p(\boldsymbol{\Sigma}, n)$ is $W_1(\sigma^2, n) \sim \sigma^2\chi^2_n$ distribution. That is for univariate case ($p = 1$), Wishart distribution coincides with χ^2 distribution.

Property II: When $\boldsymbol{\Sigma}_{p\times p} = \mathbf{I}_{p\times p}$, $\mathbf{M} \sim W_p(\mathbf{I}, n)$ is called Wishart distribution in standard form.

Property III: If $\mathbf{M} \sim W_p(\boldsymbol{\Sigma}, n)$ and **A** is a $p \times m$ matrix, then $\mathbf{A}^T\mathbf{M}\mathbf{A} \sim W_m(\mathbf{A}^T\boldsymbol{\Sigma}\mathbf{A}, n)$. From this, if $\mathbf{A} = \boldsymbol{\Sigma}^{-1/2}$, then $\boldsymbol{\Sigma}^{-1/2}\mathbf{M}\boldsymbol{\Sigma}^{-1/2} \sim W_p(\mathbf{I}, n)$. Note that $\left(\boldsymbol{\Sigma}^{-1/2}\right)^T = \boldsymbol{\Sigma}^{-1/2}$ as $\boldsymbol{\Sigma}^{-1}$ is a square symmetric matrix.

Property IV: Let $\mathbf{M} \sim W_p(\boldsymbol{\Sigma}, n)$, $\mathbf{c}^T = [c_1, c_2, \ldots\ldots, c_p]$, a fixed p-vector, and $\mathbf{c}^T\boldsymbol{\Sigma}\mathbf{c} \ne \mathbf{0}$, then $\dfrac{\mathbf{c}^T\mathbf{M}\mathbf{c}}{\mathbf{c}^T\boldsymbol{\Sigma}\mathbf{c}} \sim \chi_n^2$.

Property V: Let $\mathbf{M}_1 \sim W_p(\boldsymbol{\Sigma}, n_1)$, $\mathbf{M}_2 \sim W_p(\boldsymbol{\Sigma}, n_2)$ and $\mathbf{M}_1$ and $\mathbf{M}_2$ are independent. Then, $\mathbf{M}_1 + \mathbf{M}_2 \sim W_p(\boldsymbol{\Sigma}, n_1 + n_2)$.

Property VI: Let $\mathbf{X}_{n\times p}$ comes from $N_p(\mathbf{0}, \boldsymbol{\Sigma})$ and $\mathbf{C}_{m\times m}$ is a square symmetric matrix. Then, $\mathbf{X}^T\mathbf{C}\mathbf{X} \sim \sum_{i=1}^{n}\lambda_i W_p(\boldsymbol{\Sigma}, 1)$, where $\lambda_i, i = 1, 2, \ldots\ldots, n$, is the *i-th* eigenvalue of **C**. If **C** is idempotent, then, $\mathbf{X}^T\mathbf{C}\mathbf{X} \sim W_p(\boldsymbol{\Sigma}, \mathrm{r})$, where $\mathrm{r} = trace(\mathbf{C}) = rank\ \mathbf{C}$.

TABLE 6.1
Multivariate Central Limit Theorem

Univariate Case	Multivariate Case
(i) Random variable: X	(i) Random vector: $\boldsymbol{X}_{p\times 1}$
(ii) Distribution of X: Any	(ii) Distribution of $\boldsymbol{X}$: Any
(iii) Parameters of interest: μ and σ^2	(iii) Parameters of interest: $\boldsymbol{\mu}_{p\times 1}$ and $\boldsymbol{\Sigma}_{p\times p}$
(iv) Sample size: n (large)	(iv) Sample size: n (large)
(v) Sample statistics: $\bar{x}$	(v) Sample statistics: $\bar{\mathbf{x}}_{p\times 1}$
(vi) CLT: $\frac{\bar{x}-\mu}{\sigma/\sqrt{n}} \rightarrow Z(0,1) \quad as\, n \rightarrow \infty$	(vi) Multivariate CLT: $\sqrt{n}(\bar{\mathbf{x}}-\boldsymbol{\mu}) \rightarrow N_p(\mathbf{0},\boldsymbol{\Sigma}) \quad as\, n \rightarrow \infty$
(vii) Distribution of $\bar{x}$: $\bar{x} \sim N\left(\mu, \frac{\sigma^2}{n}\right)$	(vii) Distribution of $\bar{\mathbf{x}}$: $\bar{\mathbf{x}} \sim N_p\left(\boldsymbol{\mu}, \frac{1}{n}\boldsymbol{\Sigma}\right)$

From Property VI and $n\mathbf{S} = \mathbf{X}^T\mathbf{H}\mathbf{X}$, where $\mathbf{H} = \mathbf{I} - \frac{1}{n}\mathbf{1}\mathbf{1}^T$, an idempotent matrix and also known as centering matrix, it can be shown that $n\mathbf{S} \sim W_p(\boldsymbol{\Sigma}, n-1)$.

6.3 MULTIVARIATE CENTRAL LIMIT THEOREM

In Section 2.5, we described univariate central limit theorem, where we stated that for large n, $\frac{\bar{x}-\mu}{\sigma/\sqrt{n}} \sim Z(0,1)$. The generalization of this in the multivariate domain is given in Table 6.1.

6.4 HOTELLING'S T² DISTRIBUTION

Let, $X \sim N_p(\boldsymbol{\mu}, \boldsymbol{\Sigma})$ and the sample statistics are $\bar{\mathbf{x}}$ and $\mathbf{S}$. Then, $\bar{\mathbf{x}} \sim N_p(\boldsymbol{\mu}, \boldsymbol{\Sigma}/n)$, and $n\mathbf{S}$ is distributed as Wishart random matrix with $n-1$ degrees of freedom when sample mean $\bar{\mathbf{x}}$ and covariance $\mathbf{S}$ are independent (Johnson and Wichern, 2002). Further, the central limit theorem says that for large sample, $\sqrt{n}(\bar{\mathbf{x}}-\boldsymbol{\mu}) \sim N_p(0,\boldsymbol{\Sigma})$. Based on the work of Hotelling (1931), the general distribution of the quadratic form $n\,\mathbf{y}^T_{1\times p}\mathbf{M}^{-1}_{p\times p}\mathbf{y}_{p\times 1}$ where $\mathbf{y} \sim N_p(\mathbf{0},\mathbf{I})$ and $\mathbf{M} \sim W_p(\mathbf{I}, n)$, is Hotelling's T^2 with parameters p and n, denoted by $T^2(p,n)$ (Mardia et al., 1979). Based on the above definition, it can be shown that

$$n(\bar{\mathbf{x}}-\boldsymbol{\mu})^T\,\mathbf{S}^{-1}(\bar{\mathbf{x}}-\boldsymbol{\mu}) \sim T^2(p, n-1) \tag{6.8}$$

There is a relationship between univariate t and Hotelling's T^2. We know that

$$t = \frac{\bar{x}-\mu}{\frac{s}{\sqrt{n}}} = \frac{\sqrt{n}(\bar{x}-\mu)}{s}$$
$$\therefore t^2 = n(\bar{x}-\mu)(s^2)^{-1}(\bar{x}-\mu) \tag{6.9}$$

Therefore, in multivariate case, the right hand side of Equation 6.9 can be written as

$$n(\bar{\mathbf{x}}-\boldsymbol{\mu})^T\,\mathbf{S}^{-1}(\bar{\mathbf{x}}-\boldsymbol{\mu}).$$

There are certain useful results of Hotelling's $T^2(p,m)$ distribution (Mardia et al.,1979).

- For $p=1$, $T^2(1,m)=t_m^2$. Again, $t_m^2 = F_{1,m}$. So, $T^2(1,m)=F_{1,m}$. Putting $m=n-1$, we get $T^2(1,n-1)=F_{1,n-1}$.
- $T^2(p,m)=\dfrac{mp}{m-p+1}F_{p,m-p+1}$. Putting $m=n-1$, we get $T^2(p,n-1)=\dfrac{(n-1)p}{n-p}F_{p,n-p}$. So,

$$n(\bar{\mathbf{x}}-\boldsymbol{\mu})^T\mathbf{S}^{-1}(\bar{\mathbf{x}}-\boldsymbol{\mu}) \sim T^2(p,n-1)=\frac{(n-1)p}{n-p}F_{p,n-p} \tag{6.10}$$

Now let us assume that $p = 2$. Then,

$$T^2 = n(\bar{\mathbf{x}}-\boldsymbol{\mu})^T\mathbf{S}^{-1}(\bar{\mathbf{x}}-\boldsymbol{\mu})$$

Where $\bar{\mathbf{x}}=\begin{bmatrix}\bar{x}_1\\ \bar{x}_2\end{bmatrix}$ and $\mathbf{S}=\begin{bmatrix}s_1^2 & s_{12}\\ s_{12} & s_2^2\end{bmatrix}$.

So, $\mathbf{S}^{-1}=\dfrac{1}{s_1^2s_2^2-s_{12}^2}\begin{bmatrix}s_2^2 & -s_{12}\\ -s_{12} & s_1^2\end{bmatrix}$

and $\bar{\mathbf{x}}-\boldsymbol{\mu}=\begin{bmatrix}\bar{x}_1-\mu_1\\ \bar{x}_2-\mu_2\end{bmatrix}$

Therefore,

$$\begin{aligned}
T^2 &= n\begin{bmatrix}\bar{x}_1-\mu_1 & \bar{x}_2-\mu_2\end{bmatrix}_{1\times2}\frac{1}{s_1^2s_2^2-s_{12}^2}\begin{bmatrix}s_2^2 & -s_{12}\\ -s_{12} & s_1^2\end{bmatrix}_{2\times2}\begin{bmatrix}\bar{x}_1-\mu_1\\ \bar{x}_2-\mu_2\end{bmatrix}_{2\times1}\\
&= n\begin{bmatrix}\bar{x}_1-\mu_1 & \bar{x}_2-\mu_2\end{bmatrix}_{1\times2}\frac{1}{s_1^2s_2^2-s_{12}^2}\begin{bmatrix}s_2^2(\bar{x}_1-\mu_1)-s_{12}(\bar{x}_2-\mu_2)\\ -s_{12}(\bar{x}_1-\mu_1)+s_1^2(\bar{x}_2-\mu_2)\end{bmatrix}_{2\times1}\\
&= \frac{n}{s_1^2s_2^2-s_{12}^2}\begin{bmatrix}s_2^2(\bar{x}_1-\mu_1)^2-s_{12}(\bar{x}_1-\mu_1)(\bar{x}_2-\mu_2)\\ -s_{12}(\bar{x}_1-\mu_1)(\bar{x}_2-\mu_2)+s_1^2(\bar{x}_2-\mu_2)^2\end{bmatrix}\\
&= \frac{ns_1^2s_2^2}{s_1^2s_2^2-s_{12}^2}\left[\left(\frac{\bar{x}_1-\mu_1}{s_1}\right)^2-2\frac{s_{12}}{s_1s_2}\left(\frac{\bar{x}_1-\mu_1}{s_1}\right)\left(\frac{\bar{x}_2-\mu_2}{s_2}\right)+\left(\frac{\bar{x}_2-\mu_2}{s_2}\right)^2\right]
\end{aligned} \tag{6.11}$$

Putting $s_{12}=s_1\ s_2\ r_{12}$ into Equation 6.11, we get

$$T^2=\frac{n}{1-r_{12}^2}\left[\left(\frac{\bar{x}_1-\mu_1}{s_1}\right)^2-2r_{12}\left(\frac{\bar{x}_1-\mu_1}{s_1}\right)\left(\frac{\bar{x}_2-\mu_2}{s_2}\right)+\left(\frac{\bar{x}_2-\mu_2}{s_2}\right)^2\right] \tag{6.12}$$

Example 6.1

A random sample with $n = 20$ was collected from a bivariate normal process $\boldsymbol{X} \sim N_p(\boldsymbol{\mu},\boldsymbol{\Sigma})$. The sample mean vector and covariance matrix are

$$\bar{\mathbf{x}}=\begin{bmatrix}10\\20\end{bmatrix} \text{ and } \mathbf{S}=\begin{bmatrix}40 & -50\\ -50 & 100\end{bmatrix}$$

(a) Obtain Hotelling's T^2.
(b) What will be the distribution of T^2?

Solution:

Here $\bar{\mathbf{x}} = \begin{bmatrix} \bar{x}_1 \\ \bar{x}_2 \end{bmatrix} = \begin{bmatrix} 10 \\ 20 \end{bmatrix}$ and $\mathbf{S} = \begin{bmatrix} s_1^2 & s_{12} \\ s_{12} & s_2^2 \end{bmatrix} = \begin{bmatrix} 40 & -50 \\ -50 & 100 \end{bmatrix}$

From Equation 6.11, we get

$$T^2 = \frac{ns_1^2 s_2^2}{s_1^2 s_2^2 - s_{12}^2}\left[\left(\frac{\bar{x}_1 - \mu_1}{s_1}\right)^2 - 2\frac{s_{12}}{s_1 s_2}\left(\frac{\bar{x}_1 - \mu_1}{s_1}\right)\left(\frac{\bar{x}_2 - \mu_2}{s_2}\right) + \left(\frac{\bar{x}_2 - \mu_2}{s_2}\right)^2\right]$$

Putting the values of $n, \bar{x}_1, \bar{x}_2, s_1, s_2,$ we get

$$T^2 = \frac{20 \times 40 \times 100}{40 \times 100 - 50^2}\left[\left(\frac{10-\mu_1}{\sqrt{40}}\right)^2 + 2\frac{50}{\sqrt{40 \times 100}}\left(\frac{10-\mu_1}{\sqrt{40}}\right)\left(\frac{20-\mu_2}{\sqrt{100}}\right) + \left(\frac{20-\mu_2}{\sqrt{100}}\right)^2\right]$$
$$= 1.33(10-\mu_1)^2 + 1.33(10-\mu_1)(20-\mu_2) + 0.53(20-\mu_2)^2$$

We know that T^2 follows $T^2(p, n-1)$, which is again $\frac{(n-1)p}{n-p}F_{p,n-p}$ (see Equation 6.10). So, in this case T^2 follows $\frac{(20-1)\times 2}{18}F_{2,18} = \frac{38}{18}F_{2,18}$.

6.5 INFERENCE ABOUT SINGLE POPULATION MEAN VECTOR

In order to draw inferences about population mean vector, we need to know the sampling distribution of the point estimate $\bar{\mathbf{x}}$, the sample mean vector, which is $N_p(\boldsymbol{\mu}, \boldsymbol{\Sigma}/n)$. In Chapter 2, under univariate inferential statistics, we have estimated the confidence interval of the population mean μ (a scalar quantity) with the help of sampling distribution of sample mean $\bar{x}$ (a scalar quantity) which is $N(\mu, \sigma^2/n)$. In multivariate case, $\bar{\mathbf{x}}$ is not a scalar quantity, it is a vector of the order $p \times 1$. Further, all the variables are assumed to occur simultaneously (joint distribution). For every mean ($\mu_j, j = 1,2,.., p$), we can get a confidence interval and all the means (mean vector) collectively define a confidence region (CR). For example, consider a bivariate $\boldsymbol{X} \sim N_2(\boldsymbol{\mu}, \boldsymbol{\Sigma})$. Figure 6.1 shows the hypothetical confidence interval for each mean as well as the confidence region for the mean vector ($\boldsymbol{\mu}$).

Let AB and CD in Figure 6.1 show confidence intervals for μ_1 and μ_2, the means of X_1 and X_2, respectively at a certain significance level. These two confidence intervals form a rectangular region EFGH, can be termed as rectangular confidence region. The ellipses I and II, represent two hypothetical confidence regions for μ_1 and μ_2, for two different situations, and can be termed as confidence ellipses. Confidence ellipse I has major and minor axes parallel to the axes of the original variables X_1 and X_2, respectively. This situation represent a case when X_1 and X_2 are independent. In confidence ellipse II, X_1 and X_2 are correlated and the correlation is reflected by a shift (rotation) in major and minor axes with respect to the axes of the variables X_1 and X_2. The two-variable confidence region, popularly known as confidence ellipse, can be extended to the general p-variable confidence region, which is termed as confidence ellipsoid.

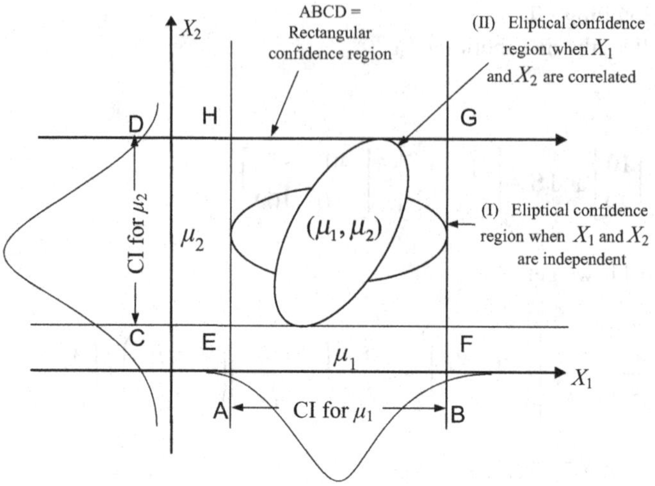

FIGURE 6.1 Confidence region for $\mathbf{\mu} = [\mu_1\ \mu_2]^T$. AB and CD represent the confidence intervals for μ_1 and μ_2, respectively at a certain significance level. Ellipses (I) and (II) represent the confidence regions for μ when X_1 and X_2 are independent and X_1 and X_2 are correlated, respectively. If individual CI for μ_1 and μ_2 are considered, the resulting CR is the rectangle EFGH. If EFGH is considered (wrongly) as the CR instead of the confidence ellipse, it gives false impression of the CR.

6.5.1 Confidence Region

What does the confidence region (CR) mean? The interpretation is similar to interpretation of confidence interval in univariate case. For example, for a multivariate sample, $\mathbf{X} = [\mathbf{x}_1, \mathbf{x}_2, \cdots \mathbf{x}_n]^T$, the sampling distribution of $\bar{\mathbf{x}}$ will form a $100(1-\alpha)\%$ confidence region, $R(\bar{\mathbf{x}})$ for $\mathbf{\mu}$ vector in such a way that

$$P\left[R(\bar{\mathbf{x}}) \text{ will contain } \mathbf{\mu}\right] = 1-\alpha. \tag{6.13}$$

We will consider the following three scenarios for obtaining confidence region (CR) for single population mean vector:

Scenario 1: Sampling from multivariate normal population, $X \sim N_p(\mathbf{\mu}, \mathbf{\Sigma})$ when $\mathbf{\Sigma}$ is known.
Scenario 2: Sampling from $X \sim N_p(\mathbf{\mu}, \mathbf{\Sigma})$ when $\mathbf{\Sigma}$ is unknown but sample size is large.
Scenario 3: Sampling from $X \sim N_p(\mathbf{\mu}, \mathbf{\Sigma})$ when $\mathbf{\Sigma}$ is unknown and sample size is small to medium.

For all the above three situations, the steps to be followed to identify CR are given below. For mathematical details, readers may refer Johnson and Wichern (2013) and Mardia et al. (1979).

Step 1: Define population of interest. It should be multivariate normal. Otherwise, appropriate transformation should be used to convert data to multivariate normal.
Step 2: Obtain sample statistic.
Step 3: Identify appropriate sampling distribution.
Step 4: Develop CR

Scenario 1: Sampling from $X \sim N_p(\mathbf{\mu}, \mathbf{\Sigma})$ and Σ is known

- Population : $X \sim N_p(\mathbf{\mu}, \mathbf{\Sigma})$, $\mathbf{\Sigma}$ known
- Sample statistic: $T^2 = n(\bar{\mathbf{x}} - \mathbf{\mu})^T \mathbf{\Sigma}^{-1} (\bar{\mathbf{x}} - \mathbf{\mu})$

- Sampling distribution: χ_p^2 [from multivariate central limit theorem, the Property (III) of multivariate normal distribution (Section 5.5) and property of χ^2 distribution]
- $100(1-\alpha)\%$ confidence region for μ is characterized by

$$n(\bar{\mathbf{x}}-\boldsymbol{\mu})^T \boldsymbol{\Sigma}^{-1}(\bar{\mathbf{x}}-\boldsymbol{\mu}) \le \chi_p^2(\alpha) \tag{6.14}$$

The confidence region is an ellipsoid centered at $\bar{\mathbf{x}}$ and axes are defined by $\sqrt{\chi_p^2(\alpha)}$, where α is the probability level of significance. For 95 % confidence region $\alpha = 0.05$.

Example 6.2

A bivariate process is characterized by the following parameters:

$$\boldsymbol{\mu} = \begin{bmatrix} \mu_1 \\ \mu_2 \end{bmatrix} \text{ and } \boldsymbol{\Sigma} = \begin{bmatrix} \sigma_1^2 & \sigma_{12} \\ \sigma_{12} & \sigma_2^2 \end{bmatrix}$$

The covariance matrix $\boldsymbol{\Sigma}$ is known as $\boldsymbol{\Sigma} = \begin{bmatrix} 10 & 3 \\ 3 & 2 \end{bmatrix}$

A sample of size $n = 30$ was collected and the sample mean vector is $\bar{\mathbf{x}} = [10 \quad 15]^T$. Obtain 99% CR for the process mean vector $[\mu_1\ \mu_2]^T$.

Solution:

Population: $X \sim N_2(\boldsymbol{\mu}, \boldsymbol{\Sigma})$ and $\boldsymbol{\Sigma}$ is known

Sample static: $T^2 = n(\bar{\mathbf{x}}-\boldsymbol{\mu})^T \boldsymbol{\Sigma}^{-1}(\bar{\mathbf{x}}-\boldsymbol{\mu})$

From Equation 6.11, by replacing the elements of $\mathbf{S}$ by that of $\boldsymbol{\Sigma}$ we get

$$\begin{aligned} T^2 &= n(\bar{\mathbf{x}}-\boldsymbol{\mu})^T \boldsymbol{\Sigma}^{-1}(\bar{\mathbf{x}}-\boldsymbol{\mu}) \\ &= \frac{n\sigma_1^2\sigma_2^2}{\sigma_1^2\sigma_2^2-\sigma_{12}^2}\left[\left(\frac{\bar{x}_1-\mu_1}{\sigma_1}\right)^2 - 2\frac{\sigma_{12}}{\sigma_1\sigma_2}\left(\frac{\bar{x}_1-\mu_1}{\sigma_1}\right)\left(\frac{\bar{x}_2-\mu_2}{\sigma_2}\right)+\left(\frac{\bar{x}_2-\mu_2}{\sigma_2}\right)^2\right] \\ &= \frac{30\times10\times2}{10\times2-9}\left[\left(\frac{10-\mu_1}{\sqrt{10}}\right)^2 - \frac{2\times3}{\sqrt{20}}\left(\frac{10-\mu_1}{\sqrt{10}}\right)\left(\frac{15-\mu_2}{\sqrt{2}}\right)+\left(\frac{15-\mu_2}{\sqrt{2}}\right)^2\right] \\ &= 5.45(10-\mu_1)^2 - 16.36(10-\mu_1)(15-\mu_2)+27.27(15-\mu_2)^2 \end{aligned}$$

Sampling distribution: χ_2^2 as population $\boldsymbol{\Sigma}$ is known and $p = 2$

Confidence region (CR): $100(1-\alpha)\%$ CR for $\boldsymbol{\mu}$, $\alpha = 0.01$ is

$$5.45(10-\mu_1)^2 - 16.36(10-\mu_1)(15-\mu_2)+27.27(15-\mu_2)^2 \le \chi_2^2(0.01) = 9.21.$$

Scenario 2: Sampling from $X \sim N_p(\boldsymbol{\mu}, \boldsymbol{\Sigma})$ with unknown $\boldsymbol{\Sigma}$ and large sample

- Population : $X \sim N_p(\boldsymbol{\mu}, \boldsymbol{\Sigma})$, $\boldsymbol{\Sigma}$ unknown but n is large
- Sample statistic: $\mathrm{n}(\bar{\mathrm{x}}-\mu)^{\mathrm{T}} \mathrm{S}^{-1}(\bar{\mathrm{x}}-\mu)$

  [Note: $\boldsymbol{\Sigma}$ is replaced by $\mathbf{S}$]

- Sampling distribution: χ_p^2 [From multivariate CLT, property of χ^2, and law of large sample]
- $100(1-\alpha)\%$ confidence region for $\boldsymbol{\mu}$ is characterized by

$$n(\bar{\mathbf{x}}-\boldsymbol{\mu})^T \mathbf{S}^{-1}(\bar{\mathbf{x}}-\boldsymbol{\mu}) \le \chi_p^2(\alpha) \tag{6.15}$$

The CR is an ellipsoid centered at $\bar{\mathbf{x}}$ and axes are defined by $\sqrt{\chi_p^2 \cdot (\alpha)}$.

Now what value of n is large? Yang and Trewn (2003) state that $n-p>40$ is considered a large sample.

Example 6.3

A chemical process is characterized by two variables and is bivariate normal. Being a production process, its population parameters $\boldsymbol{\mu}$ and $\boldsymbol{\Sigma}$ are unknown. A sample of size 100 was collected from the process and the summary statistics are given below.

$$\bar{\mathbf{x}} = \begin{bmatrix} \bar{x}_1 \\ \bar{x}_2 \end{bmatrix} = \begin{bmatrix} 20 \\ 30 \end{bmatrix}, \mathbf{S} = \begin{bmatrix} s_1^2 & s_{12} \\ s_{12} & s_2^2 \end{bmatrix} = \begin{bmatrix} 9 & 0 \\ 0 & 4 \end{bmatrix}$$

Obtain 95% CR for the process mean vector $\boldsymbol{\mu} = [\mu_1 \ \mu_2]^T$.

Solution:

Population: $\mathbf{X} \sim N_2(\boldsymbol{\mu}, \boldsymbol{\Sigma}), \boldsymbol{\Sigma}$ unknown but $n-p>40$

Sample statistic: $T^2 = n(\bar{\mathbf{x}}-\boldsymbol{\mu})^T \mathbf{S}^{-1}(\bar{\mathbf{x}}-\boldsymbol{\mu})$

From Equation 6.11, we can write

$$T^2 = \frac{n s_1^2 s_2^2}{s_1^2 s_2^2 - s_{12}^2}\left[\left(\frac{\bar{x}_1-\mu_1}{s_1}\right)^2 - 2\frac{s_{12}}{s_1 s_2}\left(\frac{\bar{x}_1-\mu_1}{s_1}\right)\left(\frac{\bar{x}_2-\mu_2}{s_2}\right) + \left(\frac{\bar{x}_2-\mu_2}{s_2}\right)^2\right]$$

Now, putting the values of n, $\bar{x}_1$, $\bar{x}_2$, s_1, s_2 and s_{12}, we get

$$T^2 = \frac{100\times 9\times 4}{9\times 4-0}\left[\left(\frac{20-\mu_1}{3}\right)^2 - \frac{2\times 0}{3\times 2}\left(\frac{20-\mu_1}{3}\right)\left(\frac{30-\mu_2}{2}\right) + \left(\frac{30-\mu_2}{2}\right)^2\right]$$
$$= 100\left[\left(\frac{20-\mu_1}{3}\right)^2 + \left(\frac{30-\mu_2}{2}\right)^2\right]$$

Sampling distribution: As $n = 100$ is large, T^2 is distributed as χ_2^2.

Confidence region (CR): $100(1-\alpha)\%$ CR for μ, $\alpha = 0.05$, is

$$100\left[\left(\frac{20-\mu_1}{3}\right)^2 + \left(\frac{30-\mu_2}{2}\right)^2\right] \le \chi_2^2(0.05) = 5.99.$$

Scenario 3: Sampling from $\boldsymbol{X} \sim N_p(\boldsymbol{\mu}, \boldsymbol{\Sigma})$ with unknown $\boldsymbol{\Sigma}$ and small to medium size sample

- Population : $\boldsymbol{X} \sim N_p(\boldsymbol{\mu}, \boldsymbol{\Sigma}), \boldsymbol{\Sigma}$ unknown and sample size is small to medium
- Sample statistic: $n(\bar{\mathbf{x}} - \boldsymbol{\mu})^T \mathbf{S}^{-1}(\bar{\mathbf{x}} - \boldsymbol{\mu})$
- Sampling distribution: $\frac{(n-1)p}{n-p} F_{p,n-p}$ [from one sample Hotelling's T²—see Section 6.4]
- $100(1-\alpha)\%$ CR for $\boldsymbol{\mu}$ is characterized by

$$n(\bar{\mathbf{x}} - \boldsymbol{\mu})^T \mathbf{S}^{-1}(\bar{\mathbf{x}} - \boldsymbol{\mu}) \le \frac{(n-1)p}{n-p} F_{p,n-p}^{(\alpha)} \tag{6.16}$$

The CR is an ellipsoid centered at $\bar{\mathbf{x}}$ and axes are defined by $\sqrt{\frac{(n-1)p}{n-p} F_{p,n-p}^{(\alpha)}}$.

Example 6.4

Consider Example 6.1, where a random sample with n = 20 was collected from a bivariate normal process $\boldsymbol{X} \sim N_2(\boldsymbol{\mu}, \boldsymbol{\Sigma})$. The sample mean vector and covariance matrix are:

$$\bar{\mathbf{x}} = \begin{bmatrix} 10 \\ 20 \end{bmatrix} \text{ and } \mathbf{S} = \begin{bmatrix} 40 & -50 \\ -50 & 100 \end{bmatrix}.$$

Obtain 95% CR for the process means.

Solution:

Population: $\boldsymbol{X} \sim N_2(\boldsymbol{\mu}, \boldsymbol{\Sigma})$, $\boldsymbol{\Sigma}$ unknown and $n - p = 20 - 2 = 18 < 40$

Sample statistic: $T^2 = n(\bar{\mathbf{x}} - \boldsymbol{\mu})^T \mathbf{S}^{-1}(\bar{\mathbf{x}} - \boldsymbol{\mu})$

From Equation 6.11, we can write

$$\begin{aligned} T^2 &= \frac{n s_1^2 s_2^2}{s_1^2 s_2^2 - s_{12}^2}\left[\left(\frac{\bar{x}_1 - \mu_1}{s_1}\right)^2 - 2\frac{s_{12}}{s_1 s_2}\left(\frac{\bar{x}_1 - \mu_1}{s_1}\right)\left(\frac{\bar{x}_2 - \mu_2}{s_2}\right) + \left(\frac{\bar{x}_2 - \mu_2}{s_2}\right)^2\right] \\ &= 1.33(10-\mu_1)^2 + 1.33(10-\mu_1)(20-\mu_2) + 0.53(20-\mu_2)^2 \end{aligned}$$

Sampling distribution:

$$\frac{(n-1)p}{n-p} F_{p,n-p} = \frac{38}{18} F_{2,18}$$

Confidence region (CR): $100(1-\alpha)$ CR = 95% CR is

$$1.33(10-\mu_1)^2 + 1.33(10-\mu_1)(20-\mu_2) + 0.53(20-\mu_2)^2 \le \frac{38}{18} F_{2,18}^{(0.05)} = 7.51$$

TABLE 6.2
Changes in d^2 with Scenarios

Scenario	d^2
Sampling from $X \sim N_p(\mu, \Sigma)$, Σ Known	$\chi_p^2(\alpha)$
Sampling from $X \sim N_p(\mu, \Sigma)$, Σ Unknown but Large Sample	$\chi_p^2(\alpha)$
Sampling from $X \sim N_p(\mu, \Sigma)$, Σ Unknown and Small to Medium Sample	$\frac{(n-1)p}{n-p} F_{p,n-p}^{(\alpha)}$

For all three scenarios, the questions to be answered now are:

- What are the directions of the axes of the confidence ellipsoid?
- What are the relative lengths of the axes of the confidence ellipsoid?

For the most general case (Scenario 3), let us write the confidence region as

$$n(\bar{\mathbf{x}} - \mu)^{\mathbf{T}} \mathbf{S}^{-1} (\bar{\mathbf{x}} - \mu) \leq d^2 \tag{6.17}$$

where, d^2 varies as per the scenarios considered and is shown in Table 6.2.

We can rewrite Equation 6.17 as

$$(\bar{\mathbf{x}} - \mu)^T \mathbf{S}^{-1} (\bar{\mathbf{x}} - \mu) \leq \left(\frac{d}{\sqrt{n}}\right)^2 \tag{6.18}$$

Equations 6.17 and 6.18 are equations of ellipsoids having p-axes. The direction of the p-axes are according to the directions of the p eigenvectors of $\mathbf{S}_{p \times p}$ and the weighted lengths correspond to the p eigenvalues of $\mathbf{S}$(weighted by $\left|\frac{d}{\sqrt{n}}\right|$). For scenario 1, $\mathbf{S}_{p \times p}$ will be replaced by $\Sigma_{p \times p}$.

Let, $\lambda_1 \geq \lambda_2 \geq \ldots \geq \lambda_p$ be the p eigenvalues and $e_1, e_2, \cdots e_p$ are the corresponding eigenvectors of $\mathbf{S}(\text{or } \Sigma)$. Then,

- The lengths of the p-axes are $\sqrt{\lambda_1}\left|\frac{d}{\sqrt{n}}\right|, \sqrt{\lambda_2}\left|\frac{d}{\sqrt{n}}\right| \cdots \sqrt{\lambda_p}\left|\frac{d}{\sqrt{n}}\right|$, respectively, and
- The directions of the p-axes are the directions of $e_1, e_2, \cdots e_p$, respectively.

Example 6.5

Consider Example 6.4. Obtain the length and direction of the axes of the confidence ellipse. Consider $\alpha = 0.05$.

Solution:

The problem falls under Scenario 3, where the population Σ is unknown and sample is small. We require using Equation 6.18. Following the solution given in Example 6.4, the left hand side (LHS) of Equation 6.17 for this case can be obtained by dividing the LHS by n, i.e., by 20.

So, $(\bar{\mathbf{x}}-\boldsymbol{\mu})^T \mathbf{S}^{-1}(\bar{\mathbf{x}}-\boldsymbol{\mu})$ for this case is

$$0.0665(10-\mu_1)^2+0.0665(10-\mu_1)(20-\mu_2)+0.0265(20-\mu_2)^2$$

Now, the value of d^2 is $\frac{(n-1)p}{n-p}F_{p,n-p}^{(\alpha)} = \frac{38}{18}F_{2,18}^{(0.05)} = 7.51.$

Again, in order to get length and direction of the axes of the of the confidence ellipse, we need to do eigenvalue and eigenvector decomposition of **S**, the sample covariance matrix. We will follow the procedures developed in Section 3.1.12 (Chapter 3).

$$|\mathbf{S}-\lambda\mathbf{I}|=0.$$

So,

$$\begin{vmatrix} 40-\lambda & -50 \\ -50 & 100-\lambda \end{vmatrix}=0$$

or

$$(40-\lambda)(100-\lambda)-2500=0$$
$$\Rightarrow 4000-140\lambda+\lambda^2-2500=0$$
$$\Rightarrow \lambda^2-140\lambda+1500=0$$

$$\therefore \lambda=\frac{140\pm\sqrt{140^2-4.1500.1}}{2.1}$$
$$=\frac{140\pm116.62}{2}$$

∴ The eigenvalues are

$$\lambda_1=\frac{140+116.62}{2}=128.31$$
$$\lambda_2=\frac{140-116.62}{2}=11.69$$

As $\lambda_1>\lambda_2$, λ_1 defines the magnitude of the major axis of the ellipse and λ_2 defines the minor axis.

For every eigenvalue there is one eigenvector of the order $p\times1$ (here p = 2). Let the eigenvectors are e_1 for λ_1 and e_2 for λ_2. Then , $e_1=\begin{bmatrix} e_{11} \\ e_{12} \end{bmatrix}$ and $e_2=\begin{bmatrix} e_{21} \\ e_{22} \end{bmatrix}$.

Now, considering (λ_1, e_1), we can write (see Equation 3.24 for details):

$$(\mathbf{S}-\lambda_1\mathbf{I})e_1=0$$
$$\Rightarrow \begin{pmatrix} 40-128.31 & -50 \\ -50 & 100-128.31 \end{pmatrix}\begin{pmatrix} e_{11} \\ e_{12} \end{pmatrix}=\begin{pmatrix} 0 \\ 0 \end{pmatrix}$$
$$\Rightarrow \begin{pmatrix} -88.31e_{11} & -50e_{12} \\ -50e_{11} & -28.31e_{12} \end{pmatrix}=\begin{pmatrix} 0 \\ 0 \end{pmatrix}$$

Which yields two equations as given below

$$-88.31e_{11}-50e_{12}=0$$
$$-50e_{11}-28.31e_{12}=0$$

With little manipulation, we get from both the above equations

$$e_{11} = -0.566e_{12}$$

Now, in order to get unique solution, we normalize the vector e_1, such that $e_1^T e_1 = 1$.

$$\text{So, } \begin{bmatrix} e_{11} & e_{12} \end{bmatrix} \begin{bmatrix} e_{11} \\ e_{12} \end{bmatrix} = 1$$

or $e_{11}^2 + e_{12}^2 = 1$

putting $e_{11} = -0.566e_{12}$, we get

$$\begin{aligned} &(-0.556e_{12})^2 + e_{12}^2 = 1 \\ &\Rightarrow 0.32e_{12}^2 + e_{12}^2 = 1 \\ &\Rightarrow e_{12} = \pm 0.87 \end{aligned}$$

So,

$$\begin{aligned} e_{11} &= -0.566\ e_{12} \\ &= -0.566 \times (\pm 0.87) \\ &= \mp 0.49 \end{aligned}$$

$$\therefore e_1 = \begin{bmatrix} e_{11} \\ e_{12} \end{bmatrix} = \begin{bmatrix} \pm 0.87 \\ \mp 0.49 \end{bmatrix}$$

$$\text{or} \quad e_1 = \begin{bmatrix} 0.87 \\ -0.49 \end{bmatrix} \text{or} \begin{bmatrix} -0.87 \\ 0.49 \end{bmatrix}$$

As seen earlier, λ_1 defines the magnitude of the major axis of the confidence ellipse, the corresponding eigenvector, e_1 defines the direction of the major axis.

So, the magnitude of the major axis is $\sqrt{\lambda_1}\left|\frac{d}{\sqrt{n}}\right| = \sqrt{128.31} \times \left|\frac{\sqrt{7.51}}{\sqrt{20}}\right| = 6.94$ and the direction is along e_1 as given above. The readers may do self-exercise for (λ_2, e_2).

6.5.2 Simultaneous Confidence Intervals

The confidence region (CR) for $\boldsymbol{\mu}$ for $p = 2$ is an ellipse and for $p \geq 3$, it is an ellipsoid. For $p > 3$, the CR cannot be visualized pictorially. The other important aspect to look into is what the CR infers on the individual mean's confidence interval. In other words, we need to find out the confidence interval for μ_j, $j = 1, 2, \ldots, p$, which simultaneously satisfies the CR, and is therefore termed as simultaneous confidence interval (SCI). The SCI provides a readily visualized interval for each μ_j with a specified probability in such a way that all these separate intervals simultaneously hold the desired 100(1-α)% confidence for the CR of the $\boldsymbol{\mu}$ vector. There are two popularly known approaches to compute simultaneous confidence intervals:

1. Linear combination approach.
2. Bonferroni approach.

Johnson and Wichren (2013) explained the fundamentals behind these approaches. We will primarily focus on the application of these two approaches for the three scenarios described earlier in Section 6.5.1. For mathematical details, readers may refer to Johnson and Wichern (2013).

Scenario 1: Sampling from $X \sim N_p(\mu, \Sigma)$ When Σ is Known

We know that if $X \sim N_p(\mu, \Sigma)$, then $\bar{\mathbf{x}} \sim N_p(\mu, \Sigma/n)$. Let **a** be an arbitrary vector of $p \times 1$, then the linear combination, $\mathbf{a}^T \bar{\mathbf{x}}$ assumes univariate normal distribution as $N(\mathbf{a}^T \mu,\ \mathbf{a}^T \Sigma \mathbf{a}/n)$. See Property (vii) under Section 5.5. Let Z be the standardized variable as defined below.

$$Z = \frac{\mathbf{a}^T \bar{\mathbf{x}} - E(\mathbf{a}^T \bar{\mathbf{x}})}{\sqrt{V(\mathbf{a}^T \bar{\mathbf{x}})}} = \frac{\mathbf{a}^T \bar{\mathbf{x}} - \mathbf{a}^T \mu}{\sqrt{\mathbf{a}^T \Sigma \mathbf{a} / n}} \sim Z(0,1) \tag{6.19}$$

As Z is N(0, 1), from Equation 6.19, we can make the confidence interval (like in univariate case) as shown below.

$$\mathbf{a}^T \bar{\mathbf{x}} - Z_{\alpha/2} \sqrt{\frac{\mathbf{a}^T \Sigma \mathbf{a}}{n}} \le \mathbf{a}^T \mu \le \mathbf{a}^T \bar{\mathbf{x}} + Z_{\alpha/2} \sqrt{\frac{\mathbf{a}^T \Sigma \mathbf{a}}{n}} \tag{6.20}$$

which for an arbitrary $\mathbf{a}^T = [0,0, \cdots,1,\cdots,0,0]$, where j th reading is 1, else 0, becomes

$$\bar{x}_j - Z_{\alpha/2} \sqrt{\frac{\sigma_{jj}}{n}} \le \mu_j \le \bar{x}_j + Z_{\alpha/2} \sqrt{\frac{\sigma_{jj}}{n}}, \quad j = 1,2,\ldots,p \tag{6.21}$$

Equation 6.21 satisfies 100(1-α)% CI for the mean of the j- th variable. We get p such equations for p-variable case, where each equation individually satisfy $100(1-\alpha)\%$ CI. But collectively the confidence interval is $100(1-\alpha)^p\,\%$ which is much less than the required $100(1-\alpha)\%$ CR, even for a relatively less number of variables. So, we need to have a vector **a** (in Equation 6.20) such that all the individual SCIs, when superimposed over the $100(1-\alpha)\%$ CR, will lie within the CR. This is essentially a search for the vector $\mathbf{a} \neq \mathbf{0}$ that maximizes Equation 6.19 and can be obtained by squaring Equation 6.19 and using maximization lemma from matrix algebra (Johnson and Wichern, 2013). Squaring Equation 6.19, we get

$$Z^2 = \frac{n\{\mathbf{a}^T (\bar{\mathbf{x}} - \mu)\}^2}{\mathbf{a}^T \Sigma \mathbf{a}} \tag{6.22}$$

Now, the vector $\mathbf{a} \neq \mathbf{0}$ that maximizes Z^2 in Equation 6.22 also maximizes Z in Equation 6.19. For definite matrix $\Sigma_{p\times p}$ and a given vector $\Upsilon_{p\times 1}$, if we consider an arbitrary non-zero vector $\mathbf{a}_{p\times 1}$, then from maximization lemma

$$\max_{\mathbf{a}\neq \mathbf{0}} \frac{\{\mathbf{a}^T \Upsilon\}^2}{\mathbf{a}^T \Sigma \mathbf{a}} = \Upsilon^T \Sigma^{-1} \Upsilon \tag{6.23}$$

Considering $\Upsilon = \bar{\mathbf{x}} - \mu$, Equation 6.23 becomes

$$\max_{\mathbf{a}\neq \mathbf{0}} \frac{\{\mathbf{a}^T (\bar{\mathbf{x}} - \mu)\}^2}{\mathbf{a}^T \Sigma \mathbf{a}} = (\bar{\mathbf{x}} - \mu)^T \Sigma^{-1} (\bar{\mathbf{x}} - \mu)$$

So, Equation 6.22 can be written as

$$Z^2 = n(\bar{\mathbf{x}} - \mu)^T \Sigma^{-1} (\bar{\mathbf{x}} - \mu) \tag{6.24}$$

We have seen earlier from the properties of multivariate normal distribution (see Property (iii) in Section 5.5) the quantity $n(\bar{\mathbf{x}} - \mu)^T \Sigma^{-1} (\bar{\mathbf{x}} - \mu)$ follows χ_p^2 distribution.

Further, from Equation 6.14, the CR $n(\bar{\mathbf{x}}-\boldsymbol{\mu})^T \boldsymbol{\Sigma}^{-1}(\bar{\mathbf{x}}-\boldsymbol{\mu}) \le \chi_p^2(\alpha)$ implies,

$$\frac{n\{\mathbf{a}^T(\bar{\mathbf{x}}-\boldsymbol{\mu})\}^2}{\mathbf{a}^T \boldsymbol{\Sigma} \mathbf{a}} \le \chi_p^2(\alpha) \text{ for every } \mathbf{a}, \text{ or}$$

$$\mathbf{a}^T \bar{\mathbf{x}} - \sqrt{\chi_p^2(\alpha)}\sqrt{\frac{\mathbf{a}^T \boldsymbol{\Sigma} \mathbf{a}}{n}} \le \mathbf{a}^T \boldsymbol{\mu} \le \mathbf{a}^T \bar{\mathbf{x}} + \sqrt{\chi_p^2(\alpha)}\sqrt{\frac{\mathbf{a}^T \boldsymbol{\Sigma} \mathbf{a}}{n}} \tag{6.25}$$

which for $\mathbf{a}^T = [0,0,\cdots,1,\cdots,0,0]$, where only the j-th element attains the value 1 and others are zero, becomes

$$\bar{x}_j - \sqrt{\chi_p^2(\alpha)}\sqrt{\frac{\sigma_{jj}}{n}} \le \mu_j \le \bar{x}_j + \sqrt{\chi_p^2(\alpha)}\sqrt{\frac{\sigma_{jj}}{n}};\, j = 1,2,\ldots,p \tag{6.26}$$

Note that the critical value in equation (6.26) is $\sqrt{\chi_p^2(\alpha)}$, not $Z_{\alpha/2}$.

Example 6.6

Consider Example 6.2. Obtain 99% simultaneous confidence intervals for the means of the variables X_1 and X_2.

Solution:

Here, the population is multivariate (precisely bivariate) normal and its covariance matrix $\boldsymbol{\Sigma}$ is known. Using Equation 6.26, we get

$$\bar{x}_1 - \sqrt{\chi_p^2(\alpha)}\sqrt{\frac{\sigma_{11}}{n}} \le \mu_1 \le \bar{x}_1 + \sqrt{\chi_p^2(\alpha)}\sqrt{\frac{\sigma_{11}}{n}}$$

$$\bar{x}_2 - \sqrt{\chi_p^2(\alpha)}\sqrt{\frac{\sigma_{22}}{n}} \le \mu_2 \le \bar{x}_2 + \sqrt{\chi_p^2(\alpha)}\sqrt{\frac{\sigma_{22}}{n}}$$

as the simultaneous confidence intervals for the means of the variables X_1 and X_2, respectively. From Example 6.2, we know that

$$n = 30,\ \bar{x}_1 = 10,\ \bar{x}_2 = 15,\ \sigma_{11} = 10,\ \sigma_{22} = 2, \text{ and } \chi_2^2(0.01) = 9.21.$$

Putting appropriate values for the SCI of μ_1 we get

$$10 - \sqrt{\chi_2^2(0.01)}\sqrt{\frac{10}{30}} \le \mu_1 \le 10 + \sqrt{\chi_2^2(0.01)}\sqrt{\frac{10}{30}}$$

$$\text{or} \quad 10 - \sqrt{9.21}\sqrt{\frac{10}{30}} \le \mu_1 \le 10 + \sqrt{9.21}\sqrt{\frac{10}{30}}$$

$$\text{or} \quad 8.25 \le \mu_1 \le 11.75$$

Similarly, putting appropriate values for the SCI for μ_2, we get

$$15 - \sqrt{9.21}\sqrt{\tfrac{2}{30}} \le \mu_2 \le 15 + \sqrt{9.21}\sqrt{\tfrac{2}{30}}$$

$$\text{or} \quad 14.22 \le \mu_2 \le 15.78$$

Using the Bonferroni approach

In the Bonferroni approach, instead of modifying the statistic to be tested (Z^2 in Equation 6.24, not Z in Equation 6.19), each of the variables' CI is tightened in such a manner that the resultant intervals will jointly satisfy the overall significance level α. The overall significance level α is divided into p parts as given below.

$$\sum_{j=1}^{p} \alpha_j = \alpha \tag{6.27}$$

Now, one can use Equation 6.20 with the following modification:

$$\mathbf{a}^T \bar{\mathbf{x}} - Z_{(\alpha_j/2)}\sqrt{\frac{\mathbf{a}^T \boldsymbol{\Sigma} \mathbf{a}}{n}} \le \mathbf{a}^T \boldsymbol{\mu} \le \mathbf{a}^T \bar{\mathbf{x}} + Z_{(\alpha_j/2)}\sqrt{\frac{\mathbf{a}^T \boldsymbol{\Sigma} \mathbf{a}}{n}} \tag{6.28}$$

Now, choosing appropriate $\mathbf{a}^T$, we get the simultaneous confidence intervals for the p means.

$$\begin{aligned}
&\bar{x}_1 - Z_{(\alpha_1/2)}\sqrt{\frac{\sigma_{11}}{n}} \le \mu_1 \le \bar{x}_1 + Z_{(\alpha_1/2)}\sqrt{\frac{\sigma_{11}}{n}} \\
&\bar{x}_2 - Z_{(\alpha_2/2)}\sqrt{\frac{\sigma_{22}}{n}} \le \mu_2 \le \bar{x}_2 + Z_{(\alpha_2/2)}\sqrt{\frac{\sigma_{22}}{n}} \\
&\cdot \\
&\cdot \\
&\cdot \\
&\bar{x}_p\ Z_{(\alpha_p/2)}\sqrt{\frac{\sigma_{pp}}{n}} \le \mu_p \le \bar{x}_p + Z_{(\alpha_p/2)}\sqrt{\frac{\sigma_{pp}}{n}}
\end{aligned} \tag{6.29}$$

Note the difference in Equations 6.26 and 6.29.

Example 6.7

Construct 99% SCI for the Example 6.6 problem using the Bonferroni approach.

Solution:

Using Equation 6.29, we can write the SCI for μ_1 and μ_2, respectively as

$$\begin{aligned}
&\bar{x}_1 - Z_{(\alpha_1/2)}\sqrt{\frac{\sigma_{11}}{n}} \le \mu_1 \le \bar{x}_1 + Z_{(\alpha_1/2)}\sqrt{\frac{\sigma_{11}}{n}} \\
&\bar{x}_2 - Z_{(\alpha_2/2)}\sqrt{\frac{\sigma_{22}}{n}} \le \mu_2 \le \bar{x}_2 + Z_{(\alpha_2/2)}\sqrt{\frac{\sigma_{22}}{n}}
\end{aligned}$$

Now, $\alpha = 0.01$. Giving equal weight to both α_1 and α_2, we can write $\alpha_1 = \alpha_2 = \frac{\alpha}{2} = \frac{0.01}{2} = 0.005$.

From Z Table, $Z_{(0.005/2)} = Z_{(0.0025)} = 2.81$.

Now, putting appropriate values for the SCI for μ_1, we get

$$10 - 2.81\sqrt{\frac{10}{30}} \le \mu_1 \le 10 + 2.81\sqrt{\frac{10}{30}}$$

$$\text{Or} \quad 8.38 \le \mu_1 \le 11.62.$$

Similarly, for μ_2, the 99% SCI is

$$15-2.81\sqrt{\frac{2}{30}} \le \mu_2 \le 15+2.81\sqrt{\frac{2}{30}}$$

$$\text{Or} \quad 14.27 \le \mu_2 \le 15.73.$$

Scenario 2: Σ is unknown and sample size is large

Equations 6.26 and 6.29 give simultaneous confidence intervals for single population mean vector when population covariance matrix Σ is known. This is for Scenario 1 explained earlier. For Scenario 2, where Σ is unknown and sample is large, the sampling distribution is χ_p^2. Accordingly, the simultaneous confidence intervals are as follows.

The 100(1 – α)% SCI for μ_j using linear combination approach is

$$\bar{x}_j - \sqrt{\chi_p^2(\alpha)}\,\sqrt{s_{jj}/n} \le \mu_j \le \bar{x}_j + \sqrt{\chi_p^2(\alpha)}\,\sqrt{s_{jj}/n};\quad j=1,2,\ldots,p \tag{6.30}$$

and the Bonferroni 100(1 – α)% SCI is

$$\bar{x}_j - Z(\alpha_j/2)\sqrt{s_{jj}/n} \le \mu_j \le \bar{x}_j + Z(\alpha_j/2)\sqrt{s_{jj}/n},;\quad j=1,2,\ldots,p \tag{6.31}$$

where $\sum_{j=1}^{p} \alpha_j = \alpha.$

Example 6.8

Consider Example 6.3. Obtain 95% SCI for each of process means μ_1 and μ_2 using (a) linear combination approach and (b) the Bonferroni approach.

Solution:

Here, the population is bivariate normal and the population covariance matrix Σ is unknown. But the sample size is large. So, we can use Equation 6.30 for linear combination approach and 6.31 for the Bonferroni approach.

Using Linear Combination Approach

The required intervals are

$$\bar{x}_1 - \sqrt{\chi_2^2(.05)}\sqrt{\frac{s_{11}}{n}} \le \mu_1 \le \bar{x}_1 + \sqrt{\chi_2^2(.05)}\sqrt{\frac{s_{11}}{n}}$$

$$\bar{x}_2 - \sqrt{\chi_2^2(.05)}\sqrt{\frac{s_{22}}{n}} \le \mu_2 \le \bar{x}_2 + \sqrt{\chi_2^2(.05)}\sqrt{\frac{s_{22}}{n}}$$

From Example 6.3, we get

$$n = 100,\ \bar{x}_1 = 20,\ \bar{x}_2 = 30,\ s_{11} = 9,\ s_{22} = 4 \text{ and } \chi_2^2(0.05) = 5.99$$

Putting appropriate values for the 95% SCI for μ_1, we get

$$20-\sqrt{5.99}\sqrt{\frac{9}{100}} \le \mu_1 \le 20+\sqrt{5.99}\sqrt{\frac{9}{100}}$$

$$\text{or} \quad 19.27 \le \mu_1 \le 20.73$$

Similarly, putting appropriate values for the 95% SCI for μ_2, we get

$$30-\sqrt{5.99}\sqrt{\frac{4}{100}} \le \mu_2 \le 30+\sqrt{5.99}\sqrt{\frac{4}{100}}$$

$$\text{or} \quad 29.51 \le \mu_2 \le 30.49$$

Using the Bonferroni Approach

The required intervals are

$$\bar{x}_1 - z_{(\alpha_1/2)}\sqrt{\frac{s_{11}}{n}} \le \mu_1 \le \bar{x}_1 + z_{(\alpha_1/2)}\sqrt{\frac{s_{11}}{n}}$$

$$\bar{x}_2 - z_{(\alpha_2/2)}\sqrt{\frac{s_{22}}{n}} \le \mu_2 \le \bar{x}_2 + z_{(\alpha_2/2)}\sqrt{\frac{s_{22}}{n}}$$

Here, $\alpha = 0.05$. Considering equal weights for μ_1 and μ_2, we get

$\alpha_1 = \alpha_2 = \alpha / 2 = 0.05 / 2 = 0.025$

From Z table $z(0.025 / 2) = z(0.0125) = 2.24$

Now, putting appropriate values for the 95% SCI for μ_1, we get

$$20-2.24\sqrt{\frac{9}{100}} \le \mu_1 \le 20+2.24\sqrt{\frac{9}{100}}$$

$$\text{or} \quad 19.33 \le \mu_1 \le 20.67$$

Similarly, for the 95% SCI for μ_2 is

$$30-2.24\sqrt{\frac{4}{100}} \le \mu_2 \le 30+2.24\sqrt{\frac{4}{100}}$$

$$\text{or} \quad 29.55 \le \mu_2 \le 30.15.$$

Scenario 3: Σ is Unknown and *n* is Small to Medium

When Σ is unknown and n is small to medium, the sampling distribution of interest is $F_{p,n-p}$.

Using the linear combination approach, the 100(1 – α)% SCI for μ_j is

$$\bar{x}_j - \sqrt{\frac{(n-1)p}{n-p}F^{(\alpha)}_{p,n-p}}\sqrt{s_{jj}/n} \le \mu_j \le \bar{x}_j + \sqrt{\frac{(n-1)p}{n-p}F^{(\alpha)}_{p,n-p}}\sqrt{s_{jj}/n};\ j =,1,2,\ldots,p \quad (6.32)$$

For the Bonferroni interval, we use t-distribution. The SCI is

$$\bar{x}_j - t^{(\alpha_j/2)}_{n-1}\sqrt{s_{jj}/n} \le \mu_j \le \bar{x}_j + t^{(\alpha_j/2)}_{n-1}\sqrt{s_{jj}/n};\ j = 1,2,\ldots,p \quad (6.33)$$

Example 6.9

Consider Example 6.4. Obtain 95% SCI for each of process means μ_1 and μ_2 using (a) linear combination approach and (b) the Bonferroni approach.

Solution:

Here, the population is bivariate normal, population covariance Σ is unknown and sample size is small. So, we need to use Equations 6.32 and 6.33 for linear combination and Bonferroni approaches, respectively.

Using Linear Combination Approach

The required intervals are

$$\bar{x}_1 - \sqrt{\frac{(n-1)p}{n-p}F^{(\alpha)}_{p,n-p}}\sqrt{\frac{s_{11}}{n}} \le \mu_1 \le \bar{x}_1 + \sqrt{\frac{(n-1)p}{n-p}F^{(\alpha)}_{p,n-p}}\sqrt{\frac{s_{11}}{n}}$$

$$\bar{x}_2 - \sqrt{\frac{(n-1)p}{n-p}F^{(\alpha)}_{p,n-p}}\sqrt{\frac{s_{22}}{n}} \le \mu_2 \le \bar{x}_2 + \sqrt{\frac{(n-1)p}{n-p}F^{(\alpha)}_{p,n-p}}\sqrt{\frac{s_{22}}{n}}$$

From Example 6.4, we get

$$n = 20,\ \bar{x}_1 = 10,\ \bar{x}_2 = 20,\ s_{11} = 40,\ s_{22} = 100 \text{ and } \frac{(n-1)p}{n-p}F^{(\alpha)}_{p,n-p} = \frac{38}{18}F^{(0.05)}_{2,18} = 7.51$$

Now, putting appropriate values for the 95% SCI for μ_1, we get

$$10 - \sqrt{7.51}\sqrt{\frac{40}{20}} \le \mu_1 \le 10 + \sqrt{7.51}\sqrt{\frac{40}{20}}$$

$$\text{or} \quad 6.14 \le \mu_1 \le 13.86$$

Similarly, putting appropriate values for the 95% SCI for μ_2, we get

$$20-\sqrt{7.51}\sqrt{\frac{100}{20}} \le \mu_2 \le 20+\sqrt{7.51}\sqrt{\frac{100}{20}}$$

$$\text{or} \quad 13.87 \le \mu_2 \le 26.13$$

Using the Bonferroni Approach

The required intervals are

$$\bar{x}_1 - t_{n-1}^{(\alpha_1/2)}\sqrt{\frac{s_{11}}{n}} \le \mu_1 \le \bar{x}_1 + t_{n-1}^{(\alpha_1/2)}\sqrt{\frac{s_{11}}{n}}$$

$$\bar{x}_2 - t_{n-1}^{(\alpha_2/2)}\sqrt{\frac{s_{22}}{n}} \le \mu_2 \le \bar{x}_2 + t_{n-1}^{(\alpha_2/2)}\sqrt{\frac{s_{22}}{n}}$$

Again $\alpha = 0.05$. Considering equal weights for μ_1 and μ_2,

$$\alpha_1 = \alpha_2 = \alpha/2 = 0.05/2 = 0.025$$

From t table $t_{20-1}^{0.025/2} = t_{19}^{0.0125} = 2.46$

Now, putting appropriate values for the 95% SCI for μ_1, we get

$$10-2.46\sqrt{\frac{40}{20}} \le \mu_1 \le 10+2.46\sqrt{\frac{40}{20}}$$

$$\text{or} \quad 6.52 \le \mu_1 \le 13.48$$

Similarly, putting appropriate values for the 95% SCI for μ_2, we get

$$20-2.46\sqrt{\frac{100}{20}} \le \mu_2 \le 20+2.46\sqrt{\frac{100}{20}}$$

$$\text{or} \quad 14.50 \le \mu_2 \le 25.50$$

Example 6.10

Construct 95% simultaneous confidence intervals for the problem given in Example 6.1.

Solution:

Here, the population is bivariate normal, Σ is unknown and sample size is small. So, it comes under Scenario 3.

Accordingly, $\bar{x}_j - \sqrt{\frac{(n-1)p}{n-p}F_{p,n-p}^{(\alpha)}}\sqrt{\frac{s_{jj}}{n}} \le \mu_j \le \bar{x}_j + \sqrt{\frac{(n-1)p}{n-p}F_{p,n-p}^{(\alpha)}}\sqrt{\frac{s_{jj}}{n}}$

The 95% SCI for μ_1 is

$$\bar{x}_1 - \sqrt{7.51}\sqrt{\frac{40}{20}} \le \mu_1 \le \bar{x}_1 + \sqrt{7.51}\sqrt{\frac{40}{20}}$$
$$\text{or} \quad 10 - 3.88 \le \mu_1 \le 10 + 3.88$$
$$\text{or } 6.12 < \mu_1 < 13.88$$

Similarly, the 95% SCI for μ_2 is

$$20 - \sqrt{7.51 \times \frac{100}{20}} \le \mu_2 \le 20\sqrt{7.51 \times \frac{100}{20}}$$
$$\text{or} \quad 13.88 \le \mu_2 \le 26.12$$

The readers may compare it with Bonferroni approach.

6.5.3 Hypothesis Testing

The concept of hypothesis testing is discussed in Chapter 2 for the univariate case. The same holds true for the multivariate case too. We will describe them below for three different scenarios. The scenarios are explained in Section 6.5.1. For all the scenarios, the steps to be followed for hypothesis testing are given below.

Step 1: Set null (H_0) and alternative (H_1) hypotheses.
Step 2: Obtain appropriate test statistics.
Step 3: Identify appropriate distribution for the test statistic when H_0 is true.
Step 4: Set cut off value for the test static.
Step 5: Make decision about H_0 by comparing the test statistic value with its cut-off.

Setting of null hypothesis (H_0) is governed by the purpose of the test (with due consideration to practical purpose) and selection of appropriate hypothesis, as there could be a large number of hypotheses concerning the parameters of a distribution. So, Step 1 is crucial. Following Step 1, the task in Step 2 is to select the appropriate test statistics from a set of available test statistics. The two most commonly used approaches are likelihood ratio test (LRT) and union intersection test (UIT).

Both the approaches may lead to the same test statistic or different test statistics. It is better to choose a statistic for which sampling distribution is available (see Step 3). When there are multiple test statistics available for H_0, the best one could be one for which a plausible easy-to-handle sampling distribution is available. The selection of cut-off value (Step 4) depends on the errors of the test (known as type-I and type-II errors). The type-I error gives the probability (α) of accepting a false H_0, also known as level of significance (see Chapter 2, Section 2.6.1 for further discussion).

Scenario 1: $\mathbf{X} \sim N_p(\mu, \Sigma)$ and Σ is Known

- Hypotheses: H_0: $\boldsymbol{\mu} = \boldsymbol{\mu}_0$
  H_1: $\boldsymbol{\mu} \neq \boldsymbol{\mu}_0$
- Test statistic: $T^2 = n\left(\bar{\mathbf{x}} - \boldsymbol{\mu}_0\right)^T \boldsymbol{\Sigma}^{-1}\left(\bar{\mathbf{x}} - \boldsymbol{\mu}_0\right)$
- Sampling distribution: If H_0 is true T^2 follows χ_p^2
- Cut-off value: $\chi_p^{(\alpha)}$ where α is the level of significance
- Decision: Reject H_0, if $T^2 > \chi_p^2(\alpha)$

Example 6.11

Consider Example 6.2. Suppose, based on past information, the mean vector of the process is $\boldsymbol{\mu}_0^T = [8 \quad 16]$. Test the hypothesis that $\mu = \mu_0$ considering $\alpha = 0.01$.

Solution:

From Example 6.2, we get

$$\boldsymbol{\Sigma} = \begin{bmatrix} 10 & 3 \\ 3 & 2 \end{bmatrix}, \ \bar{\mathbf{x}} = \begin{bmatrix} 10 \\ 15 \end{bmatrix}, \ n = 30$$

As the process is multivariate normal with known Σ, it falls under scenario 1. The step wise computations are given below.

Step 1: Set null and alternative hypothesis:

H_0: $\boldsymbol{\mu} = \boldsymbol{\mu}_0$, where $\boldsymbol{\mu} = \begin{bmatrix} \mu_1 \\ \mu_2 \end{bmatrix}$ and $\boldsymbol{\mu}_0 = \begin{bmatrix} 8 \\ 16 \end{bmatrix}$.

H_1: $\boldsymbol{\mu} \neq \boldsymbol{\mu}_0$.

Step 2: The appropriate test statistic is

$$T^2 = n(\bar{\mathbf{x}} - \boldsymbol{\mu}_0)^T \boldsymbol{\Sigma}^{-1} (\bar{\mathbf{x}} - \boldsymbol{\mu}_0)$$

From Example 6.2, we see that

$$\begin{aligned} T^2 &= n\left(\bar{\mathbf{x}} - \boldsymbol{\mu}_0\right)^T \boldsymbol{\Sigma}^{-1} \left(\bar{\mathbf{x}} - \boldsymbol{\mu}_0\right) \\ &= 5.45(10-\mu_1)^2 - 16.36(10-\mu_1)(15-\mu_2) + 27.27(15-\mu_2)^2 \end{aligned}$$

If H_0 is true, then $\boldsymbol{\mu} = \boldsymbol{\mu}_0$. So, $\mu_1 = \mu_{10} = 8$ and $\mu_2 = \mu_{20} = 16$.

Putting these values, we get

$$\begin{aligned} T^2 &= 5.45(10-8)^2 - 16.36(10-8)(15-16) + 27.27(15-16)^2 \\ &= 21.80 + 32.72 + 27.27 \\ &= 81.79 \end{aligned}$$

Step 3: As the case belongs to scenario 1, the sampling distribution χ_2^2.

Step 4: For $\alpha = 0.01$, the cut-off value is $\chi_2^2(0.01) = 9.21$.

Step 5: As computed $T^2 = 81.79 > 9.21 =$ the cut-off value, we reject H_0, (i.e., $\boldsymbol{\mu} = \boldsymbol{\mu}_0$).

The result so obtained can be compared with the 99% simultaneous confidence intervals for μ_1 and μ_2 obtained in Example 6.6. For evaluation purpose, the 99% SCIs are shown below.

Using linear combination approach: $8.25 \leq \mu_1 \leq 11.75$ and $14.22 \leq \mu_2 \leq 15.78$.

Using the Bonferroni approach (from Example 6.7) $8.38 \leq \mu_1 \leq 11.62$ and $14.27 \leq \mu_2 \leq 15.173$.

For both the above cases, the 99% SCI for μ_1 doesn't contain 8 and the 99% SCI for μ_2 also doesn't contain 16. Hence, the null hypothesis H_0 is rejected.

Note: It should be noted here that any contradictory results for SCI and hypothesis testing might be attributed to the violations of assumptions and sample size limitation.

Scenario 2: $X \sim N_p(\boldsymbol{\mu}, \boldsymbol{\Sigma})$ and $\boldsymbol{\Sigma}$ is Unknown but n is Large

- Hypotheses: H_0: $\boldsymbol{\mu} = \boldsymbol{\mu}_0$
  H_1: $\boldsymbol{\mu} \neq \boldsymbol{\mu}_0$
- Test statistics: $T^2 = n(\bar{\mathbf{x}} - \boldsymbol{\mu}_0)^T \mathbf{S}^{-1}(\bar{\mathbf{x}} - \boldsymbol{\mu}_0)$
- Sampling distribution: If H_0 is true, then $T^2 \sim \chi_p^2$
- Cut-off value: $\chi_p^2(\alpha)$
- Decision: Reject H_0 if $T^2 > \chi_p^2(\alpha)$

Example 6.12

Consider Example 6.3. Suppose, based on past information, the mean vector of the process is $\boldsymbol{\mu}_0^T = [18 \quad 25]$. Test the hypothesis that $\boldsymbol{\mu} = \boldsymbol{\mu}_0$ considering $\alpha = 0.05$.

Solution:

From Example 6.3, we get

$$\bar{\mathbf{x}} = \begin{bmatrix} \bar{x}_1 \\ \bar{x}_2 \end{bmatrix} = \begin{bmatrix} 20 \\ 30 \end{bmatrix}, \ \mathbf{S} = \begin{bmatrix} s_1^2 & s_{12} \\ s_{12} & s_2^2 \end{bmatrix} = \begin{bmatrix} 9 & 0 \\ 0 & 4 \end{bmatrix} \quad \text{and} \quad n = 100.$$

As the process is multivariate normal with unknown $\boldsymbol{\Sigma}$, but the sample size is large, it falls under Scenario 2. The step-wise computations are given below.

Step 1: The null and alternate hypotheses are

$$H_0: \boldsymbol{\mu} = \boldsymbol{\mu}_0, \text{ where } \boldsymbol{\mu} = \begin{bmatrix} \mu_1 \\ \mu_2 \end{bmatrix} \text{ and } \boldsymbol{\mu}_0 = \begin{bmatrix} \mu_{10} \\ \mu_{20} \end{bmatrix} = \begin{bmatrix} 18 \\ 25 \end{bmatrix}$$

$$H_1: \boldsymbol{\mu} \neq \boldsymbol{\mu}_0$$

Step 2: The appropriate test statistic is

$$T^2 = n(\bar{\mathbf{x}} - \boldsymbol{\mu}_0)^T \mathbf{S}^{-1}(\bar{\mathbf{x}} - \boldsymbol{\mu}_0)$$

From Example 6.3, we see that

$$T^2 = 100\left[\left(\frac{20 - \mu_1}{3}\right)^2 + \left(\frac{30 - \mu_2}{2}\right)^2\right]$$

If H_0 is true, then $\boldsymbol{\mu} = \boldsymbol{\mu}_0$ or $\mu_1 = \mu_{10} = 18$ and $\mu_2 = \mu_{20} = 25$. Putting these values, we get

$$T^2 = 100\left[\left(\frac{20-18}{3}\right)^2 + \left(\frac{30-25}{2}\right)^2\right] = 669.44$$

Step 3: As the case belongs to Scenario 2, the sampling distribution is χ_2^2.

Step 4: For $\alpha = 0.05$, the cut off value is $\chi_2^2(0.05) = 5.99$.

Step 5: As computed $T^2 = 669.44 > 5.99 =$ the cut-off value, we reject H_0, (i.e., $\boldsymbol{\mu} = \boldsymbol{\mu}_0$).

The results so obtained, when compared with the SCIs obtained in Example 6.8, are compatible.

Scenario 3: $X \sim N_p(\boldsymbol{\mu}, \boldsymbol{\Sigma})$, and $\boldsymbol{\Sigma}$ is Unknown but n is Small to Medium

- Hypotheses: H_0: $\boldsymbol{\mu} = \boldsymbol{\mu}_0$

  H_1: $\boldsymbol{\mu} \neq \boldsymbol{\mu}_0$
- Test statistics: $T^2 = n\left(\bar{\mathbf{x}} - \boldsymbol{\mu}_0\right)^T \mathbf{S}^{-1}\left(\bar{\mathbf{x}} - \boldsymbol{\mu}_0\right)$
- Sampling distribution: If H_0 is true, then $T^2 \sim \frac{(n-1)p}{n-p} F_{p,n-p}$
- Cut-off value: $\frac{(n-1)p}{n-p} F_{p,n-p}^{(\alpha)}$
- Decision: Reject H_0 if $T^2 > \frac{(n-1)p}{n-p} F_{p,n-p}^{\alpha}$

Example 6.13

Consider Example 6.4. Suppose, based on past information, the mean vector of the process is $\boldsymbol{\mu}_0^T = [9, \quad 18]$. Test the hypothesis that $\boldsymbol{\mu} = \boldsymbol{\mu}_0$ considering $\alpha = 0.05$.

Solution:

From Example 6.4, we get

$$\bar{\mathbf{x}} = \begin{bmatrix} \bar{x}_1 \\ \bar{x}_2 \end{bmatrix} = \begin{bmatrix} 10 \\ 20 \end{bmatrix}, \mathbf{S} = \begin{bmatrix} s_1^2 & s_{12} \\ s_{12} & s_2^2 \end{bmatrix} = \begin{bmatrix} 40 & -50 \\ -50 & 100 \end{bmatrix} \quad \text{and} \quad n = 20.$$

As the process is multivariate normal with unknown $\boldsymbol{\Sigma}$ and small sample size, it falls under Scenario 3. The step-wise computations are:

Step 1: The null and alternate hypotheses are

$$H_0: \boldsymbol{\mu} = \boldsymbol{\mu}_0, \text{ where } \boldsymbol{\mu} = \begin{bmatrix} \mu_1 \\ \mu_2 \end{bmatrix} \text{ and } \boldsymbol{\mu}_0 = \begin{bmatrix} \mu_{10} \\ \mu_{20} \end{bmatrix} = \begin{bmatrix} 9 \\ 18 \end{bmatrix}$$

$H_1: \boldsymbol{\mu} \neq \boldsymbol{\mu}_0$.

Step 2: The appropriate test statistic is

$$T^2 = n\left(\bar{\mathbf{x}} - \boldsymbol{\mu}_0\right)^T \mathbf{S}^{-1}\left(\bar{\mathbf{x}} - \boldsymbol{\mu}_0\right)$$

From Example 6.4, we get

$$T^2 = 1.33\left(10-\mu_1\right)^2 + 1.33\left(10-\mu_1\right)\left(20-\mu_2\right) + 0.53\left(20-\mu_2\right)^2$$

If H_0 is true, then $\boldsymbol{\mu} = \boldsymbol{\mu}_0$ or $\mu_1 = \mu_{10} = 9$ and $\mu_2 = \mu_{20} = 18$. Putting these values, we get

$$T^2 = 1.33\left(10-9\right)^2 + 1.33\left(10-9\right)\left(20-18\right) + 0.53\left(20-18\right)^2$$

$$= 1.33 + 2.66 + 2.12$$
$$= 6.11.$$

Step 3: As the case belongs to Scenario 3, the sampling distribution is $\frac{(n-1)p}{n-p}F_{p,n-p} = \frac{38}{18}F_{2,18}$.

Step 4: For $\alpha = 0.05$, the cut off value is $\frac{38}{18}F_{2,18}^{(0.05)} = 7.51$.

Step 5: As computed $T^2 = 6.11 < 7.51 =$ the cut-off value, H_0 (i.e. $\boldsymbol{\mu} = \boldsymbol{\mu}_0$) cannot be rejected.

The result so obtained, when is compared with the SCIs obtained in Example 6.9, is compatible. The intervals contain $\mu_{10} = 9$ and $\mu_{20} = 18$.

6.6 INFERENCE ABOUT EQUALITY OF TWO-POPULATION MEAN VECTORS

6.6.1 Confidence Region

The development of a confidence region (CR) for the difference between two-population mean vectors, i.e., $\boldsymbol{\mu}_A - \boldsymbol{\mu}_B$, follows the same principles applied to single population mean vectors (see Sections 6.5.1 and 6.5.2). The difference lies in the selection of appropriate random variables and computation. The necessary steps pertaining to the developments are given below for Scenarios 1, 2, and 3.

Scenario 1: Sampling from Multivariate Normal Populations with Known $\boldsymbol{\Sigma}_A$ and $\boldsymbol{\Sigma}_B$

Populations: $X_A \sim N_p\left(\boldsymbol{\mu}_A, \boldsymbol{\Sigma}_A\right); X_B \sim N_p\left(\boldsymbol{\mu}_B, \boldsymbol{\Sigma}_B\right)$ and $\boldsymbol{\Sigma}_A$ and $\boldsymbol{\Sigma}_B$ are known.

Samples: $X_A, \bar{\mathbf{x}}_A, \mathbf{S}_A, n_A$ from population A

$X_B, \bar{\mathbf{x}}_B, \mathbf{S}_B, n_B$ from population B

X_A and X_B are independent samples.

Random vector (RV) : $\bar{\mathbf{x}}_A - \bar{\mathbf{x}}_B$

$$E\left(\bar{\mathbf{x}}_A - \bar{\mathbf{x}}_B\right) = \boldsymbol{\mu}_A - \boldsymbol{\mu}_B \tag{6.34}$$

$$Cov\left(\bar{\mathbf{x}}_A - \bar{\mathbf{x}}_B\right) = Cov\left(\bar{\mathbf{x}}_A\right) + Cov\left(\bar{\mathbf{x}}_B\right)$$
$$= \frac{1}{n_A}\boldsymbol{\Sigma}_A + \frac{1}{n_B}\boldsymbol{\Sigma}_B \tag{6.35}$$

Sample statistic :

$$T^2 = \left[(\bar{\mathbf{x}}_A - \bar{\mathbf{x}}_B) - (\boldsymbol{\mu}_A - \boldsymbol{\mu}_B)\right]^T \left[\frac{\boldsymbol{\Sigma}_A}{n_A} + \frac{\boldsymbol{\Sigma}_B}{n_B}\right]^{-1} \left[(\bar{\mathbf{x}}_A - \bar{\mathbf{x}}_B) - (\boldsymbol{\mu}_A - \boldsymbol{\mu}_B)\right] \quad (6.36)$$

Sampling distribution: $T^2 \sim \chi_p^2$

The $100(1-\alpha)\%$ CR is characterized by

$$\left[(\bar{\mathbf{x}}_A - \bar{\mathbf{x}}_B) - (\boldsymbol{\mu}_A - \boldsymbol{\mu}_B)\right]^T \left[\frac{\boldsymbol{\Sigma}_A}{n_A} + \frac{\boldsymbol{\Sigma}_B}{n_B}\right]^{-1} \left[(\bar{\mathbf{x}}_A - \bar{\mathbf{x}}_B) - (\boldsymbol{\mu}_A - \boldsymbol{\mu}_B)\right] \le \chi_p^2(\alpha) \quad (6.37)$$

The CR is an ellipsoid, centered at $\bar{\mathbf{x}}_A - \bar{\mathbf{x}}_B$ and axes are defined with $\sqrt{\chi_p^2(\alpha)}$.

Example 6.14

Electric overhead trolley (EOT) crane is extensively used in all kinds of manufacturing industries. The EOT cranes can be equipped with movable or static cabins. As crane operation is a risky job, the management is interested to find out whether the operators with static cabins perceive different job risks and management support requirements than the operators operating cranes with static cabins. A study was conducted involving 31 operators using movable cabins and 45 operators using static cabins. The summary statistics obtained are given below.

Population A: Movable cabin

i) $\bar{\mathbf{x}}_A = \begin{bmatrix} \bar{x}_{A1} \\ \bar{x}_{A2} \end{bmatrix} = \begin{bmatrix} 30.77 \\ 33.97 \end{bmatrix}$

ii) $\boldsymbol{\Sigma}_A = \begin{bmatrix} \sigma_{A1}^2 & \sigma_{A12} \\ \sigma_{A12} & \sigma_{A2}^2 \end{bmatrix} = \begin{bmatrix} 59.92 & -8.49 \\ -8.49 & 47.84 \end{bmatrix}$

Population B: Static cabin

i) $\bar{\mathbf{x}}_B = \begin{bmatrix} \bar{x}_{B1} \\ \bar{x}_{B2} \end{bmatrix} = \begin{bmatrix} 28.78 \\ 37.09 \end{bmatrix}$

ii) $\boldsymbol{\Sigma}_B = \begin{bmatrix} \sigma_{B1}^2 & \sigma_{B12} \\ \sigma_{B12} & \sigma_{B2}^2 \end{bmatrix} = \begin{bmatrix} 32.26 & -11.40 \\ -11.40 & 68.79 \end{bmatrix}$

Obtain 95% confidence region (CR) for the difference of the two-population mean vectors $(\boldsymbol{\mu}_A - \boldsymbol{\mu}_B)$.

Solution:

In this situation, $\boldsymbol{\Sigma}_A$ and $\boldsymbol{\Sigma}_B$ are known and different. So, it falls under Scenario 1. Accordingly,

(i) Populations: $\boldsymbol{X}_A \sim N_2(\boldsymbol{\mu}_A, \boldsymbol{\Sigma}_A) \quad (\because p = 2)$

$\boldsymbol{X}_B \sim N_2(\boldsymbol{\mu}_B, \boldsymbol{\Sigma}_B)$

(ii) Random vector (RV): $(\bar{\mathbf{x}}_A - \bar{\mathbf{x}}_B)$

Expected value of RV: $E\left(\bar{\mathbf{x}}_A - \bar{\mathbf{x}}_B\right) = \left(\boldsymbol{\mu}_A - \boldsymbol{\mu}_B\right)$

Covariance of RV: $\text{Cov}\left(\bar{\mathbf{x}}_A - \bar{\mathbf{x}}_B\right) = \frac{1}{n_A}\boldsymbol{\Sigma}_A + \frac{1}{n_B}\boldsymbol{\Sigma}_B$

$$= \frac{1}{31}\begin{bmatrix} 59.92 & -8.49 \\ -8.49 & 47.84 \end{bmatrix} + \frac{1}{45}\begin{bmatrix} 32.26 & -11.40 \\ -11.40 & 68.79 \end{bmatrix} \qquad \begin{pmatrix} \because n_A = 31, n_B = 45 \\ \boldsymbol{\Sigma}_A \,\&\, \boldsymbol{\Sigma}_B \text{ given} \end{pmatrix}$$

$$= \begin{bmatrix} 1.93 & -0.27 \\ -0.27 & 1.54 \end{bmatrix} + \begin{bmatrix} 0.72 & -0.25 \\ -0.25 & 1.53 \end{bmatrix}$$

$$= \begin{bmatrix} 2.65 & -0.52 \\ -0.52 & 3.07 \end{bmatrix}$$

(iii) Sample statistic:

$$T^2 = \left[(\bar{\mathbf{x}}_A - \bar{\mathbf{x}}_B) - (\boldsymbol{\mu}_A - \boldsymbol{\mu}_B)\right]^T \left[\frac{\boldsymbol{\Sigma}_A}{n_A} + \frac{\boldsymbol{\Sigma}_B}{n_B}\right]^{-1} \left[(\bar{\mathbf{x}}_A - \bar{\mathbf{x}}_B) - (\boldsymbol{\mu}_A - \boldsymbol{\mu}_B)\right]$$

$$\text{Now, let, } \boldsymbol{\mu}_A = \begin{bmatrix} \mu_{A1} \\ \mu_{A2} \end{bmatrix}, \boldsymbol{\mu}_B = \begin{bmatrix} \mu_{B1} \\ \mu_{B2} \end{bmatrix}$$

$$\therefore \left(\boldsymbol{\mu}_A - \boldsymbol{\mu}_B\right) = \begin{bmatrix} (\mu_{A1} - \mu_{B1}) \\ (\mu_{A2} - \mu_{B2}) \end{bmatrix}_{2\times 1} = \begin{bmatrix} \theta_1 \\ \theta_2 \end{bmatrix} = \boldsymbol{\theta} \text{ [say]}$$

Now,

$$\left[(\bar{\mathbf{x}}_A - \bar{\mathbf{x}}_B) - \boldsymbol{\theta}\right] = \left[\left\{\begin{bmatrix} 30.77 \\ 33.97 \end{bmatrix} - \begin{bmatrix} 28.78 \\ 37.09 \end{bmatrix}\right\} - \begin{bmatrix} \theta_1 \\ \theta_2 \end{bmatrix}\right] = \left[\begin{bmatrix} 1.99 \\ -3.12 \end{bmatrix} - \begin{bmatrix} \theta_1 \\ \theta_2 \end{bmatrix}\right] = \begin{bmatrix} (1.99 - \theta_1) \\ (-3.12 - \theta_2) \end{bmatrix}$$

So,

$$T^2 = \left[(1.99 - \theta_1) \quad (-3.12 - \theta_2)\right] \begin{bmatrix} 2.65 & -0.52 \\ -0.52 & 3.07 \end{bmatrix}^{-1} \begin{bmatrix} (1.99 - \theta_1) \\ (-3.12 - \theta_2) \end{bmatrix}$$

$$= \left[(1.99 - \theta_1) \quad (-3.12 - \theta_2)\right]_{1\times 2} \begin{bmatrix} 0.39 & 0.07 \\ 0.07 & 0.34 \end{bmatrix}_{2\times 2} \begin{bmatrix} (1.99 - \theta_1) \\ (-3.12 - \theta_2) \end{bmatrix}_{2\times 1}$$

$$= \left[(1.99 - \theta_1)\times .39 - (3.12 + \theta_2)\times .07 \qquad (1.99 - \theta_1)\times .07 - (3.12 + \theta_2)\times .34\right]_{1\times 2} \begin{bmatrix} (1.99 - \theta_1) \\ (-3.12 - \theta_2) \end{bmatrix}_{2\times 1}$$

$$= \left[(1.99 - \theta_1)^2 \times .39 - (3.12 + \theta_2)(1.99 - \theta_1)\times 0.07 - (3.12 + \theta_2)(1.99 - \theta_1)\times 0.07 + (3.12 + \theta_2)^2 \times 0.34\right]_{1\times 1}$$

(iv) Sample distribution: $T^2 \sim \chi_2^2$

(v) $100(1-\alpha)\%\text{CR}$:

$$T^2 \le \chi_p^2(\alpha) \le \chi_2^2(0.05) \le 5.99 \left[\because p=2, \alpha=0.05, \text{ and } \chi_2^2(0.05)=5.99 \text{ from } \chi^2 \text{table}\right]$$

So, $0.39(1.99-\theta_1)^2 - 0.14(3.12+\theta_2)(1.99-\theta_1) + 0.34(3.12+\theta_2)^2 \le 5.99$

where $\theta_1 = \mu_{A1} - \mu_{B1}$ and $\theta_2 = \mu_{A2} - \mu_{B2}$

Scenario 2: Sampling from Multivariate Normal Populations with Unknown but Equal Σ

Populations: $\boldsymbol{X}_A \sim N_p(\boldsymbol{\mu}_A, \boldsymbol{\Sigma}_A)$; $\boldsymbol{X}_B \sim N_p(\boldsymbol{\mu}_B, \boldsymbol{\Sigma}_B)$

$\boldsymbol{\Sigma}_A$ and $\boldsymbol{\Sigma}_B$ arc unknown but $\boldsymbol{\Sigma}_A = \boldsymbol{\Sigma}_B = \boldsymbol{\Sigma}$.

Samples: $\boldsymbol{X}_A, \overline{\mathbf{x}}_A, \mathbf{S}_A, n_A$ from population A

$\boldsymbol{X}_B, \overline{\mathbf{x}}_B, \mathbf{S}_B, n_B$ from population B

$\boldsymbol{X}_A$ *and* $\boldsymbol{X}_B$ are independent samples.

Random vector (RV): $\overline{\mathbf{x}}_A - \overline{\mathbf{x}}_B$

$$E(\overline{\mathbf{x}}_A - \overline{\mathbf{x}}_B) = \mu_A - \mu_B \tag{6.38}$$

$$Cov(\overline{\mathbf{x}}_A - \overline{\mathbf{x}}_B) = \left(\frac{1}{n_A} + \frac{1}{n_B}\right)\text{S} \tag{6.39}$$

Now, estimate of $\boldsymbol{\Sigma}$ is $\mathbf{S}_{\text{pooled}}$, where,

$$\mathbf{S}_{pooled} = \frac{(n_A - 1)\mathbf{S}_A + (n_B - 1)\mathbf{S}_B}{n_A + n_B - 2} \tag{6.40}$$

Sample statistic:

$$T^2 = \left[(\overline{\mathbf{x}}_A - \overline{\mathbf{x}}_B) - (\boldsymbol{\mu}_A - \boldsymbol{\mu}_B)\right]^T \left[\left(\frac{1}{n_A} + \frac{1}{n_B}\right)\mathbf{S}_{\text{pooled}}\right]^{-1} \left[(\overline{\mathbf{x}}_A - \overline{\mathbf{x}}_B) - (\boldsymbol{\mu}_A - \boldsymbol{\mu}_B)\right] \tag{6.41}$$

Sampling distribution: T^2 follows χ_p^2 when n_A and n_B are large, and follows $\dfrac{(n_A + n_B - 2)p}{n_A + n_B - p - 1} F_{p,\, n_A + n_B - p - 1}$, when n_A and n_B are small to medium.

Confidence region: The $100(1-\alpha)\%$ CR is characterized by

$$T^2 \le \chi_p^2(\alpha) \text{ when } n_A \text{ and } n_B \text{ are large} \tag{6.42}$$

$$\text{and } \ T^2 \le \frac{(n_A + n_B - 2)p}{n_A + n_B - p - 1} F^{(\alpha)}_{p,\, n_A + n_B - p - 1}, \text{ when } n_A \text{ and } n_B \text{ is small to medium.} \tag{6.43}$$

For both Equations 6.42 and 6.43, T^2 is obtained by Equation 6.41.

Example 6.15

Age and risky behavior are the two important factors that make differences in safety performance between accident group (AG) and non-accident group (NAG) of workers. Random samples of 20 individuals from AG and 50 individuals from NAG were collected. The sample mean vectors and sample covariance matrices are given below.

$$\text{Population A: AG } \bar{\mathbf{x}}_A = \begin{bmatrix} 50 \\ 6 \end{bmatrix}; \ \mathbf{S}_A = \begin{bmatrix} 16 & -5 \\ -5 & 4 \end{bmatrix}$$

$$\text{Population B: NAG } \bar{\mathbf{x}}_B = \begin{bmatrix} 40 \\ 8 \end{bmatrix}; \mathbf{S}_B = \begin{bmatrix} 25 & -6 \\ -6 & 9 \end{bmatrix}$$

Obtain 95% CR for the difference between the two-group mean vectors.

Solution:

$n_A = 20$; $n_B = 50$ (given).

Here, nothing is known about population covariance matrices. So, we may consider $\boldsymbol{\Sigma}_A = \boldsymbol{\Sigma}_B = \boldsymbol{\Sigma}$ (as Scenario 2).

Here, $\boldsymbol{\Sigma}_A$ = Covariance matrix for population A (for AG group)

$\boldsymbol{\Sigma}_B$ = Covariance matrix for population B (for NAG group)

i) Populations:

$$X_A \sim N_p(\boldsymbol{\mu}_A, \boldsymbol{\Sigma}_A) \sim N_2(\boldsymbol{\mu}_A, \boldsymbol{\Sigma}_A) \quad (\because p = 2)$$
$$X_B \sim N_p(\boldsymbol{\mu}_B, \boldsymbol{\Sigma}_B)$$

$$\boldsymbol{\Sigma}_A = \boldsymbol{\Sigma}_B = \boldsymbol{\Sigma} \text{ [say]}$$

ii) Random vector: $(\bar{\mathbf{x}}_A - \bar{\mathbf{x}}_B)$

$$E(\bar{\mathbf{x}}_A - \bar{\mathbf{x}}_B) = (\boldsymbol{\mu}_A - \boldsymbol{\mu}_B)$$
$$Cov(\bar{\mathbf{x}}_A - \bar{\mathbf{x}}_B) = \left(\frac{\boldsymbol{\Sigma}_A}{n_A} + \frac{\boldsymbol{\Sigma}_B}{n_B}\right) = \left(\frac{1}{n_A} + \frac{1}{n_B}\right)\boldsymbol{\Sigma} \quad (\because \boldsymbol{\Sigma}_A = \boldsymbol{\Sigma}_B)$$

Now, $\hat{\boldsymbol{\Sigma}} = \mathbf{S}_{\text{pooled}}$, where,

$$\mathbf{S}_{\text{pooled}} = \left\{\frac{(n_A - 1)\mathbf{S}_A + (n_B - 1)\mathbf{S}_B}{(n_A + n_B - 2)}\right\}$$

$$= \left\{\frac{(20-1)\begin{bmatrix} 16 & -5 \\ -5 & 4 \end{bmatrix} + (50-1)\begin{bmatrix} 25 & -6 \\ -6 & 9 \end{bmatrix}}{(20+50-2)}\right\} \quad [\because n_A = 20,\ n_B = 50,\ \mathbf{S}_A,\ \mathbf{S}_B \text{ given}]$$

$$= \left[\frac{\begin{bmatrix} 304 & -95 \\ -95 & 76 \end{bmatrix} + \begin{bmatrix} 1225 & -294 \\ -294 & 441 \end{bmatrix}}{68}\right]$$

$$= \left[\frac{\begin{bmatrix} 1529 & -389 \\ -389 & 517 \end{bmatrix}}{68}\right]$$

$$= \begin{bmatrix} 22.49 & -5.72 \\ -5.72 & 7.60 \end{bmatrix}$$

iii) Sample statistic:

$$T^2 = \left[(\bar{\mathbf{x}}_A - \bar{\mathbf{x}}_B) - (\boldsymbol{\mu}_A - \boldsymbol{\mu}_B)\right]^T \left[\left(\frac{1}{n_A} + \frac{1}{n_B}\right)\mathbf{S}_{\text{pooled}}\right]^{-1} \left[(\bar{\mathbf{x}}_A - \bar{\mathbf{x}}_B) - (\boldsymbol{\mu}_A - \boldsymbol{\mu}_B)\right]$$

Now,

$$\left[(\bar{\mathbf{x}}_A - \bar{\mathbf{x}}_B) - (\boldsymbol{\mu}_A - \boldsymbol{\mu}_B)\right]$$

$$= \left[\begin{pmatrix} 10 \\ -2 \end{pmatrix} - \begin{pmatrix} \mu_{A1} - \mu_{B1} \\ \mu_{A2} - \mu_{B2} \end{pmatrix}\right]$$

$$= \begin{bmatrix} 10 - (\mu_{A1} - \mu_{B1}) \\ -2 - (\mu_{A2} - \mu_{B2}) \end{bmatrix}$$

$$= \begin{bmatrix} 10 - \theta_1 \\ -2 - \theta_2 \end{bmatrix} \quad \begin{bmatrix} \because \theta_1 = \mu_{A1} - \mu_{B1} \\ \theta_1 = \mu_{A2} - \mu_{B2} \end{bmatrix}$$

Let $\mathbf{B} = \left(\frac{1}{n_A} + \frac{1}{n_B}\right)\mathbf{S}_{\text{pooled}}$

$$= \left(\frac{1}{20} + \frac{1}{50}\right)\begin{bmatrix} 22.49 & -5.72 \\ -5.72 & 7.60 \end{bmatrix}$$

$$= 0.07\begin{bmatrix} 22.49 & -5.72 \\ -5.72 & 7.60 \end{bmatrix}$$

$$=\begin{bmatrix} 1.57 & -0.4 \\ -0.4 & 0.53 \end{bmatrix}$$

$$\text{So, } \mathbf{B}^{-1} = \begin{bmatrix} 0.79 & 0.60 \\ 0.60 & 2.34 \end{bmatrix}$$

$$T^2 = \begin{bmatrix} (10-\theta_1) & -(2+\theta_2) \end{bmatrix}_{1\times 2} \begin{bmatrix} .79 & 0.60 \\ 0.60 & 2.34 \end{bmatrix}_{2\times 2} \begin{bmatrix} (10-\theta_1) \\ -(2+\theta_2) \end{bmatrix}_{2\times 1}$$

$$= \begin{bmatrix} .79(10-\theta_1)-0.60(2+\theta_2) & 0.60(10-\theta_1)-2.34(2+\theta_2) \end{bmatrix}_{1\times 2} \begin{bmatrix} (10-\theta_1) \\ -(2+\theta_2) \end{bmatrix}_{2\times 1}$$

$$= 0.79(10-\theta_1)^2 - 0.60(2+\theta_2)(10-\theta_1) - 0.60(2+\theta_2)(10-\theta_1) + 2.34(2+\theta_2)^2$$

$$= 0.79(10-\theta_1)^2 - 1.2(10-\theta_1)(2+\theta_2) + 2.34(2+\theta_2)^2$$

iv) Sample distribution

$$T^2 \sim \frac{(n_A+n_B-2)p}{n_A+n_B-p-1} F^{\alpha}_{p,n_A+n_B-p-1}$$

$$= \frac{(20+50-2)\times 2}{20+50-2-1} F^{\alpha}_{2,67} = 2.03 F^{\alpha}_{2,67}$$

[as $n_A = 20$, (i.e., small)]

v) $100(1-\alpha)\%$CR

$$T^2 \le 2.03 F^{\alpha}_{2,67} = 2.03 F^{\alpha}_{2,67} = 2.03 \times 3.13 = 6.35$$

$[\because p = 2,\ \alpha = 0.05\text{(taken)}]$

$\therefore$ The 95% CR is

$$0.79(10-\theta_1)^2 + 2.34(2+\theta_2)^2 - 1.2(10-\theta_1)(2+\theta_2) \le 6.35.$$

Scenario 3: Sampling from Multivariate Normal Populations with Unknown and Unequal Σ and Large Samples

Populations: $X_A \sim N_p(\mu_A, \Sigma_A)$
$X_B \sim N_p(\mu_B, \Sigma_B)$

$\Sigma_A \neq \Sigma_B$ and unknown

Samples: $X_A, \bar{\mathbf{x}}_A, \mathbf{S}_A, n_A$ from population A

$X_B, \bar{\mathbf{x}}_B, \mathbf{S}_B, n_B$ from population B

n_A and n_B is large

X_A *and* X_B are independent samples.

Random vector (RV): $\bar{\mathbf{x}}_A - \bar{\mathbf{x}}_B$

$$E\left(\bar{\mathbf{x}}_A - \bar{\mathbf{x}}_B\right) = \boldsymbol{\mu}_A - \boldsymbol{\mu}_B \tag{6.44}$$

$$Cov\left(\bar{\mathbf{x}}_A - \bar{\mathbf{x}}_B\right) = \frac{1}{n_A}\boldsymbol{\Sigma}_A + \frac{1}{n_B}\boldsymbol{\Sigma}_B = \frac{1}{n_A}\mathbf{S}_A + \frac{1}{n_B}\mathbf{S}_B \tag{6.45}$$

$\mathbf{S}_A$ and $\mathbf{S}_B$ are estimates of $\boldsymbol{\Sigma}_A$ and $\boldsymbol{\Sigma}_B$, respectively.

Sample statistic:

$$T^2 = \left[\left(\bar{\mathbf{x}}_A - \bar{\mathbf{x}}_B\right) - \left(\boldsymbol{\mu}_A - \boldsymbol{\mu}_B\right)\right]^T \left[\frac{\mathbf{S}_A}{n_A} + \frac{\mathbf{S}_B}{n_B}\right]^{-1} \left[\left(\bar{\mathbf{x}}_A - \bar{\mathbf{x}}_B\right) - \left(\boldsymbol{\mu}_A - \boldsymbol{\mu}_B\right)\right] \tag{6.46}$$

$$\text{Sampling distribution}: T^2 \sim \chi_p^2 \tag{6.47}$$

Confidence region: The 100 (1-α) % CR is characterized by $T^2 \le \chi_p^2(\alpha)$ where the CR is an ellipsoid centered at $\bar{\mathbf{x}}_A - \bar{\mathbf{x}}_B$ and axes are defined with $\sqrt{\chi_p^2(\alpha)}$.

Example 6.16

Inoculant is used to improve the quality of cast iron in a foundry. So, in order to evaluate the effect of the amount of inoculant, a study was conducted considering 2 kg and 4kg inoculant, separately. The effect on the quality of cast iron produced is evaluated considering two quality variables namely hardness (X_1) and tensile strength (X_2). The relevant data is given below.

Population A: Amount of inoculant = 2kg

No. of observation = $n_A = 56$

$$\bar{\mathbf{x}}_A = \begin{bmatrix} 198.25 \\ 25.61 \end{bmatrix}$$

$$\mathbf{S}_A = \begin{bmatrix} 125.68 & -0.54 \\ -0.54 & 0.52 \end{bmatrix}$$

Population B: The amount of inoculant = 4kg

No. of observation = $n_B = 137$

$$\bar{\mathbf{x}}_B = \begin{bmatrix} 201.12 \\ 25.99 \end{bmatrix}$$

$$\mathbf{S}_B = \begin{bmatrix} 127.90 & -0.98 \\ -0.98 & 1.39 \end{bmatrix}$$

Obtain 95% CR for the difference between the mean vectors for use of 2kg and 4kg inoculants.

Solution:

In this case, $n_A = 56$, $n_B = 137$ and both $\mathbf{\Sigma}_A$ and $\mathbf{\Sigma}_B$ are unknown. This situation falls under Scenario 3.

(i) Population:

$$X_A \sim N_2(\boldsymbol{\mu}_A, \mathbf{\Sigma}_A)$$
$$X_B \sim N_2(\boldsymbol{\mu}_B, \mathbf{\Sigma}_B)$$

(ii) Random vector: $(\overline{\mathbf{x}}_A - \overline{\mathbf{x}}_B)$

$$E(\overline{\mathbf{x}}_A - \overline{\mathbf{x}}_B) = (\boldsymbol{\mu}_A - \boldsymbol{\mu}_B)$$

$$\text{Cov}(\overline{\mathbf{x}}_A - \overline{\mathbf{x}}_B) = \frac{1}{n_A}\mathbf{\Sigma}_A + \frac{1}{n_B}\mathbf{\Sigma}_B$$

$$= \frac{1}{n_A}\mathbf{S}_A + \frac{1}{n_B}\mathbf{S}_B \qquad \begin{bmatrix} \text{Both } \mathbf{\Sigma}_A \text{ and } \mathbf{\Sigma}_B \text{ are unknown. } \mathbf{S}_A \text{and } \mathbf{S}_B \\ \text{are taken as their estimates, respectively} \end{bmatrix}$$

$$= \frac{1}{56}\begin{bmatrix} 125.68 & -0.54 \\ -0.54 & 0.52 \end{bmatrix} + \frac{1}{137}\begin{bmatrix} 127.90 & -0.98 \\ -0.98 & 1.39 \end{bmatrix} \left[\because n_A = 56, n_B = 137, \mathbf{S}_A, \mathbf{S}_B = \text{given}\right]$$

$$= \begin{bmatrix} 2.24 & -0.0096 \\ -0.0096 & 0.0093 \end{bmatrix} + \begin{bmatrix} 0.934 & -.00715 \\ -.00715 & 0.01 \end{bmatrix}$$

$$= \begin{bmatrix} 3.174 & -0.01675 \\ -0.01675 & 0.0193 \end{bmatrix} \approx \begin{bmatrix} 3.17 & -0.02 \\ -0.02 & 0.02 \end{bmatrix}$$

(iii) Sample statistic

$$T^2 = \left[(\overline{\mathbf{x}}_A - \overline{\mathbf{x}}_B) - (\boldsymbol{\mu}_A - \boldsymbol{\mu}_B)\right]^T \left[\frac{\mathbf{S}_A}{n_A} + \frac{\mathbf{S}_B}{n_B}\right]^{-1} \left[(\overline{\mathbf{x}}_A - \overline{\mathbf{x}}_B) - (\boldsymbol{\mu}_A - \boldsymbol{\mu}_B)\right]$$

$$\text{Now, } \left[\frac{\mathbf{S}_A}{n_A} + \frac{\mathbf{S}_B}{n_B}\right]^{-1} = \begin{bmatrix} 3.17 & -0.02 \\ -0.02 & 0.02 \end{bmatrix}^{-1} = \begin{bmatrix} 0.32 & 0.32 \\ 0.32 & 50.32 \end{bmatrix}$$

Again

$$\left[(\overline{\mathbf{x}}_A - \overline{\mathbf{x}}_B) - (\boldsymbol{\mu}_A - \boldsymbol{\mu}_B)\right] = \left[\begin{bmatrix} -2.87 \\ -0.38 \end{bmatrix} - \begin{bmatrix} (\mu_{A1} - \mu_{B1}) \\ (\mu_{A2} - \mu_{B2}) \end{bmatrix}\right] \qquad \begin{bmatrix} \because \text{say, } \boldsymbol{\mu}_A = \begin{bmatrix} \mu_{A1} \\ \mu_{A2} \end{bmatrix}, \boldsymbol{\mu}_B = \begin{bmatrix} \mu_{B1} \\ \mu_{B2} \end{bmatrix} \\ \overline{\mathbf{x}}_A, \overline{\mathbf{x}}_B \text{ as given} \end{bmatrix}$$

$$= \begin{bmatrix} -2.87 - (\mu_{A1} - \mu_{B1}) \\ -0.38 - (\mu_{A2} - \mu_{B2}) \end{bmatrix}_{2\times1}$$

$$= \begin{bmatrix} -2.87 - \theta_1 \\ -0.38 - \theta_2 \end{bmatrix}, \quad (\text{say}) \begin{bmatrix} \theta_1 = \mu_{A1} - \mu_{B1} \\ \theta_2 = \mu_{A2} - \mu_{B2} \end{bmatrix}$$

$$T^2 = \left[\left(-2.87-\theta_1\right) \ \left(-0.38-\theta_2\right) \right]_{1\times2} \begin{bmatrix} 0.32 & 0.32 \\ 0.32 & 50.32 \end{bmatrix}_{2\times2} \begin{bmatrix} \left(-2.87-\theta_1\right) \\ \left(-0.38-\theta_2\right) \end{bmatrix}_{2\times1}$$

$$= \left[0.32\left(-2.87-\theta_1\right)+0.32\left(-0.38-\theta_2\right) \quad 0.32\left(-2.87-\theta_1\right)+50.32\left(-0.38-\theta_2\right) \right]_{1\times2} \begin{bmatrix} \left(-2.87-\theta_1\right) \\ \left(-0.38-\theta_2\right) \end{bmatrix}_{2\times1}$$

$$= 0.32\left(-2.87-\theta_1\right)^2 + 0.32\left(-0.38-\theta_2\right)\left(-2.87-\theta_1\right) + 0.32\left(-2.87-\theta_1\right)\left(-0.38-\theta_2\right) + 50.32\left(-0.38-\theta_2\right)^2_{1\times2}$$

(iv) Sample distribution

$$T^2 \sim \chi_2^2$$

v) $100(1-\alpha)\%\text{CR}$

$$T^2 \le \chi_p^2(\alpha)$$
$$\le \chi_2^2(0.05) \quad \left[\because p=2; \alpha=0.05(\text{taken})\right]$$

$$\le 5.99[\text{from } \chi^2 \text{ table}]$$

$\because$ We can write

$$+0.32\left(2.87+\theta_1\right)^2 + 0.64\left(0.38+\theta_2\right)\left(2.87+\theta_1\right) + 50.32\left(0.38+\theta_2\right)^2 \le 5.99$$

6.6.2 Simultaneous Confidence Intervals

The development of simultaneous confidence intervals for the difference between two population mean vectors, i.e., $\mathbf{\mu}_A - \mathbf{\mu}_B$ follows the same principles applied for single population mean vector (Section 6.5.2). In the following section, we discuss them for $\mathbf{\mu}_A - \mathbf{\mu}_B$ for the three scenarios mentioned earlier.

Scenario 1: Sampling from Multivariate Normal Populations with Known Σ_A and Σ_B

The simultaneous confidence intervals can be computed using linear combination approach and Bonferroni approach.

Using linear combination approach, the $100(1-\alpha)\%$ SCI for $\mathbf{\mu}_{A_j} - \mathbf{\mu}_{B_j}$ is

$$\left(\bar{x}_{Aj}-\bar{x}_{Bj}\right) - \sqrt{\chi_p^2(\alpha)}\sqrt{\frac{\sigma_{Ajj}}{n_A}+\frac{\sigma_{Bjj}}{n_B}} \le \mu_{Aj}-\mu_{Bj} \le \left(\bar{x}_{Aj}-\bar{x}_{Bj}\right) + \sqrt{\chi_p^2(\alpha)}\sqrt{\frac{\sigma_{Ajj}}{n_A}+\frac{\sigma_{Bjj}}{n_B}} \quad (6.48)$$

for $j = 1,2,\ldots\ldots,p$.

Using Bonferroni approach, the $100(1-\alpha)\%$ SCI for $\mathbf{\mu}_{Aj} - \mathbf{\mu}_{Bj}$ is

$$\left(\bar{x}_{Aj}-\bar{x}_{Bj}\right) - Z_{\alpha_j/2}\sqrt{\frac{\sigma_{Ajj}}{n_A}+\frac{\sigma_{Bjj}}{n_B}} \le \mu_{Aj}-\mu_{Bj} \le \left(\bar{x}_{Aj}-\bar{x}_{Bj}\right) + Z_{\alpha_j/2}\sqrt{\frac{\sigma_{Ajj}}{n_A}+\frac{\sigma_{Bjj}}{n_B}} \quad (6.49)$$

where $\sum_{j=1}^{p} \alpha_j = 1$ and $j = 1, 2, \ldots\ldots, p$.

Example 6.17

Consider Example 6.14. Obtain 95% SCI for each of the mean differences for movable (μ_A) and static crane operators (μ_B) using (a) linear combination approach and (b) the Bonferroni approach.

Solution:

Considering Example 6.14, the situation described here falls under Scenario 1. $\left[\mathbf{\Sigma}_A \neq \mathbf{\Sigma}_B \text{ but known}\right]$.

Now, we have to calculate 95% simultaneous CI for each of the mean differences, i.e., $(\mu_{A1} - \mu_{B1})$ and $(\mu_{A2} - \mu_{B2})$ for movable and static crane operators.

Using Linear Combination Approach

$$95\% \text{ SCI for } (\mu_{A1} - \mu_{B1}) = \left\{(\bar{x}_{A1} - \bar{x}_{B1}) \pm \sqrt{\chi_p^2(\alpha)}\sqrt{\frac{\sigma_{Ajj}}{n_A} + \frac{\sigma_{Bjj}}{n_B}}\right\}$$

$$= \left\{(30.77 - 28.78) \pm \sqrt{5.99}\sqrt{\frac{59.92}{31} + \frac{32.26}{45}}\right\}$$

$$\left[\because \sigma_{A1}^2 = 59.92, \sigma_{B1}^2 = 32.26\right]$$

$$= 1.99 \pm 2.45\sqrt{1.93 + 0.72} = 1.99 \pm 2.45\sqrt{2.65}$$

$$= 1.99 \pm 2.45 \times 1.63$$

$$= 1.99 \pm 3.99$$

$$= (-2.00, 5.98)$$

$$95\% \text{ SCI for } (\mu_{A2} - \mu_{B2}) = \left\{(\bar{x}_{A2} - \bar{x}_{B2}) \pm \sqrt{\chi_p^2(\alpha)}\sqrt{\frac{\sigma_{Ajj}}{n_A} + \frac{\sigma_{Bjj}}{n_B}}\right\}$$

$$= \left\{(33.97 - 37.09) \pm 2.45\sqrt{\frac{47.84}{31} + \frac{68.79}{45}}\right\}$$

$$= -3.12 \pm 2.45 \times 1.75 = (-7.40,\ 1.17)$$

Using Bonferroni Approach

$$95\% \text{ SCI for } (\mu_{A1} - \mu_{B1}) = \left\{(\bar{x}_{A1} - \bar{x}_{B1}) \pm z_{\alpha/2p}\sqrt{\frac{\sigma_{Ajj}}{n_A} + \frac{\sigma_{Bjj}}{n_B}}\right\} \quad \left[\text{Here, } Z_{0.0125} = 2.24\right]$$

$$= \left\{1.99 \pm Z_{0.0125} \times 1.63\right\}$$

$$= (1.99 \pm 2.24 \times 1.63) \quad \left[\because Z_{0.0125} \text{ from table } Z = 2.24\right]$$

$$= (-1.66,\ 5.64)$$

$$95\%\ \text{SCI for } (\mu_{A2}-\mu_{B2}) = \left\{(\bar{x}_{A2}-\bar{x}_{B2}) \pm Z_{\alpha/2p}\sqrt{\frac{\sigma_{Ajj}}{n_A}+\frac{\sigma_{Bjj}}{n_B}}\right\}$$

$$=\{-3.12 \pm Z_{0.0125} \times 1.75\} \quad [\because \sigma^2_{A2}=47.84, \sigma^2_{B2}=68.79,\ n_A=31,\ n_B=45]$$
$$=(-3.12 \pm 2.24 \times 1.75)$$
$$=(-7.04, 0.8)$$

Scenario 2: Sampling from Multivariate Normal Populations with Unknown but Equal Σ

Here, $\boldsymbol{\Sigma}_A=\boldsymbol{\Sigma}_B=\boldsymbol{\Sigma}$ but unknown.

Using linear combination approach, the SCI is

$$(\bar{x}_{Aj}-\bar{x}_{Bj})-c\sqrt{\left(\frac{1}{n_A}+\frac{1}{n_B}\right)s_{jj,pooled}} \le \mu_{Aj}-\mu_{Bj} \le (\bar{x}_{Aj}-\bar{x}_{Bj})+c\sqrt{\left(\frac{1}{n_A}+\frac{1}{n_B}\right)s_{jj,pooled}} \quad (6.50)$$

for $j = 1, 2, \ldots\ldots, p$

where, $c=\sqrt{\frac{(n_A+n_B-2)p}{n_A+n_B-p-1}F^{\alpha}_{p,n_A+n_B-p-1}}$ for sample size small to medium or $c=\sqrt{\chi^2_p(\alpha)}$ for large sample.

Using the Bonferroni approach, the SCI is

$$(\bar{x}_{Aj}-\bar{x}_{Bj})-d\sqrt{\left(\frac{1}{n_A}+\frac{1}{n_B}\right)s_{jj,pooled}} \le \mu_{Aj}-\mu_{Bj} \le (\bar{x}_{Aj}-\bar{x}_{Bj})+d\sqrt{\left(\frac{1}{n_A}+\frac{1}{n_B}\right)s_{jj,pooled}}$$

where $\sum_{j=1}^{p}\alpha_j=1$ and $j = 1, 2, \ldots\ldots, p,$ (6.51)

$d=t^{\alpha_j/2}_{n_A+n_B-2}$ for sample size small to medium

or $d=Z_{\alpha_j/2}$ for large sample.

Example 6.18

Consider example 6.15. Obtain 95% SCI for each of the mean for accident group (AG) and non-accident group (NAG) of workers using (a) linear combination approach and (b) the Bonferroni approach.

Solution:

As per the information given in Example 6.15, this case falls under Scenario 2. Here $n_A=20, n_B=50$. We consider the sample size is small to medium. Further, $\boldsymbol{\Sigma}_A=\boldsymbol{\Sigma}_B=\boldsymbol{\Sigma}$.

Using Linear Combination Approach (see Equation 6.50)

$$95\%\ \text{SCI for } (\mu_{A1}-\mu_{B1}) = (\bar{x}_{A1}-\bar{x}_{B1}) \pm c\sqrt{\left(\frac{1}{n_A}+\frac{1}{n_B}\right)s^2_{1,\text{pooled}}}$$

$$= (50-40) \pm 2.52\sqrt{\left(\frac{1}{20}+\frac{1}{50}\right)\times 22.49}$$
$$= 10 \pm 2.52\sqrt{1.57}$$
$$= 10 \pm 3.16$$
$$= (6.84, 13.16)$$

Note:

$$c = \sqrt{\frac{(n_A+n_B-2)p}{(n_A+n_B-p-1)}F^{(\alpha)}_{(p,n_A+n_B-p-1)}}$$
$$= \sqrt{\frac{(20+50-2)2}{(20+50-2-1)}F^{(0.05)}_{(2,67)}}$$
$$= \sqrt{2.0298\times 3.13} = \sqrt{6.35}$$
$$= 2.52 \qquad \left(\because F^{0.05}_{2,67} = 3.13\right)$$

$$95\% \text{ SCI for } (\mu_{A2}-\mu_{B2}) = (\bar{x}_{A2}-\bar{x}_{B2}) \pm c\sqrt{\left(\frac{1}{n_A}+\frac{1}{n_B}\right)s^2_{2,\text{pooled}}}$$

$$= (6-8) \pm 2.52\sqrt{\left(\frac{1}{20}+\frac{1}{50}\right)\times 7.60}$$
$$= -2 \pm 2.52\sqrt{0.532}$$
$$= -2 \pm 1.84$$
$$= (-3.84, -0.16)$$

Using the Bonferroni Approach (see Equation 6.51)

$$95\% \text{ SCI for } (\mu_{A1}-\mu_{B1}) = (\bar{x}_{A1}-\bar{x}_{B1}) \pm t^{\alpha/2p}_{(n_A+n_B-2)}\sqrt{\left(\frac{1}{n_A}+\frac{1}{n_B}\right)s^2_{1,\text{pooled}}}$$

$$= 10 \pm t^{0.0125}_{68}\sqrt{1.57} = 10 \pm 2.386\sqrt{1.57} \qquad \left[\because t^{0.0125}_{68} = 2.386 \text{ from table}\right]$$
$$= 10 \pm 2.99$$
$$= (7.01, 12.99)$$

Similarly, 95% SCI for

$$(\mu_{A2}-\mu_{B2}) = (\bar{x}_{A2}-\bar{x}_{B2}) \pm t^{\alpha/2p}_{(n_A+n_B-2)}\sqrt{\left(\frac{1}{n_A}+\frac{1}{n_B}\right)s^2_{2,\text{pooled}}}$$
$$= -2 \pm 2.386 \times \sqrt{.532}$$
$$= (-3.740, -0.259)$$

Scenario 3: Sampling from Multivariate Normal Populations with Unknown, Unequal Σ and Large Samples

Here, $\mathbf{\Sigma}_A \neq \mathbf{\Sigma}_B$ and unknown. As sample size is large, $\mathbf{S}_A$ and $\mathbf{S}_B$ can be used in place of $\mathbf{\Sigma}_A$ and $\mathbf{\Sigma}_B$, respectively.

Using linear combination approach, the 100(1 – α)% SCI is

$$\left(\bar{x}_{Aj}-\bar{x}_{Bj}\right)-\sqrt{\chi_p^2(\alpha)}\sqrt{\frac{s_{Ajj}}{n_A}+\frac{s_{Bjj}}{n_B}} \le \mu_{Aj}-\mu_{Bj} \le \left(\bar{x}_{Aj}-\bar{x}_{Bj}\right)+\sqrt{\chi_p^2(\alpha)}\sqrt{\frac{s_{Ajj}}{n_A}+\frac{s_{Bjj}}{n_B}} \quad (6.52)$$

for $j = 1, 2, \ldots\ldots, p$.

Using Bonferroni approach, the 100(1 – α)% SCI is

$$\left(\bar{x}_{Aj}-\bar{x}_{Bj}\right)-Z_{\alpha_j/2}\sqrt{\frac{s_{Ajj}}{n_A}+\frac{s_{Bjj}}{n_B}} \le \mu_{Aj}-\mu_{Bj} \le \left(\bar{x}_{Aj}-\bar{x}_{Bj}\right)+Z_{\alpha_j/2}\sqrt{\frac{s_{Ajj}}{n_A}+\frac{s_{Bjj}}{n_B}} \quad (6.53)$$

where $\sum_{j=1}^{p}\alpha_j = 1$ and $j = 1, 2, \ldots\ldots, p$.

Example 6.19

Consider Example 6.16. Obtain 95% SCI for each of the mean differences for CI production considering 2kg inoculant (μ_A) and 4kg inoculant (μ_B) using (a) linear combination approach and (b) the Bonferroni approach.

Solution:

According to the Example 6.16, the problem falls under Scenario 3.

$n_A = 56,\ n_B = 137$

$\boldsymbol{\Sigma}_A \neq \boldsymbol{\Sigma}_B$ and unknown, and n is large

Using Linear Combination Approach

$$95\%\ \text{SCI for}\left(\mu_{A1}-\mu_{B1}\right)=\left\{\left(\bar{x}_{A1}-\bar{x}_{B1}\right)\pm\sqrt{\chi_p^2(\alpha)}\sqrt{\frac{s_{A1}^2}{n_A}+\frac{s_{B1}^2}{n_B}}\right\}$$
$$= -2.87 \pm 4.36 = (-7.23, 1.49)$$

$$95\%\ \text{SCI for}\left(\mu_{A2}-\mu_{B2}\right)=\left\{\left(\bar{x}_{A2}-\bar{x}_{B2}\right)\pm\sqrt{\chi_p^2(\alpha)}\sqrt{\frac{s_{A2}^2}{n_A}+\frac{s_{B2}^2}{n_B}}\right\}$$
$$= -0.38 \pm 0.35 = (-0.73, -0.03)$$

Using the Bonferroni Approach

$$95\%\ \text{SCI for}\left(\mu_{A1}-\mu_{B1}\right)=\left\{\left(\bar{x}_{A1}-\bar{x}_{B1}\right)\pm Z_{\alpha/2p}\sqrt{\frac{s_{A1}^2}{n_A}+\frac{s_{B1}^2}{n_B}}\right\}$$
$$= -2.87 \pm 2.24 \times 1.78 = (-6.86, 1.12)$$

$$95\%\ \text{SCI for}\left(\mu_{A2}-\mu_{B2}\right)=\left\{\left(\bar{x}_{A2}-\bar{x}_{B2}\right)\pm Z_{\alpha/2p}\sqrt{\frac{s_{A2}^2}{n_A}+\frac{s_{B2}^2}{n_B}}\right\}$$

$$= -0.38 \pm 2.24 \times 0.14 = (-0.69, -0.07)$$

Note:

$$\left(\bar{x}_{A1}-\bar{x}_{B1}\right)=(198.25-201.12)=-2.87, \quad \frac{s_{A1}^2}{n_A}+\frac{s_{B1}^2}{n_B}=\frac{125.68}{56}+\frac{127.90}{137}=2.24+.93=3.17.$$

$$\left(\bar{x}_{A2}-\bar{x}_{B2}\right)=(25.61-25.99)=-0.38, \quad \frac{s_{A2}^2}{n_A}+\frac{s_{B2}^2}{n_B}=\frac{.52}{56}+\frac{1.39}{137}=0.02.$$

$$\sqrt{\chi_p^2(\alpha)}=\sqrt{\chi_2^2(0.05)}=\sqrt{5.99}=2.45.$$

$$Z_{\alpha/2p}=Z_{0.0125}=2.24.$$

6.6.3 Hypothesis Testing

As discussed in Section 6.5.3, the steps to be followed for hypothesis testing are:

Step 1: Set null (H_0) and alternative (H_1) hypothesis.
Step 2: Obtain appropriate test statistics.
Step 3: Identify appropriate distribution for the test statistic when H_0 is true.
Step 4: Set cut off value for the test static.
Step 5: Make decision about H_0 by comparing test statistic value with its cut-off.

The following section describes the hypothesis testing of equality of two-population mean vectors under the three scenarios mentioned earlier in Section 6.6.2.

Scenario 1: Sampling from Multivariate Normal Populations with Known Σ_A and Σ_B

- Null hypothesis $H_0: \mu_A = \mu_B$
- Alternate hypothesis $H_1: \mu_A \neq \mu_B$
- Test statistic: $T^2=\left[\left(\bar{\mathbf{x}}_A-\bar{\mathbf{x}}_B\right)-\left(\mu_{A0}-\mu_{B0}\right)\right]^T\left[\frac{\Sigma_A}{n_A}+\frac{\Sigma_B}{n_B}\right]^{-1}\left[\left(\bar{\mathbf{x}}_A-\bar{\mathbf{x}}_B\right)-\left(\mu_{A0}-\mu_{B0}\right)\right]$ given that $\mu_A=\mu_{A0}$ and $\mu_B=\mu_{B0}$.
- Sampling distribution: If H_0 is true, T^2 follows χ_p^2
- Cut-off value: $\chi_p^2(\alpha)$ where α is the level of significance
- Decision: Reject H_0, if $T^2>\chi_p^2(\alpha)$

Example 6.20

Consider Example 6.14. In addition to the data given in Example 6.14, suppose the hypothesized population mean vector for movable and static cabins are as given below.

$$\mu_{Ao}=\begin{bmatrix}30\\25\end{bmatrix}_{2\times1} \quad \mu_{Bo}=\begin{bmatrix}35\\40\end{bmatrix}_{2\times1}$$

Test the hypothesis that $\mu_A=\mu_B$ at $\alpha=0.05$.

Solution: According to the information given in example 6.14, the situation falls under scenario 1. $\Sigma_A \neq \Sigma_B =$ known, $n_A=31$, $n_B=45$. From example 6.14,

$$\bar{\mathbf{x}}_A - \bar{\mathbf{x}}_B = \begin{bmatrix} 1.99 \\ -3.12 \end{bmatrix} \text{ and } \left[\frac{\boldsymbol{\Sigma}_A}{n_A} + \frac{\boldsymbol{\Sigma}_B}{n_B}\right] = \begin{bmatrix} 2.65 & -0.52 \\ -0.52 & 3.07 \end{bmatrix}$$

i) Null hypothesis: $H_0 : \boldsymbol{\mu}_A = \boldsymbol{\mu}_B$.

Alternate hypothesis: $H_1 : \boldsymbol{\mu}_A \neq \boldsymbol{\mu}_B$.

ii) Test statistic

$$T^2 = \left[(\bar{\mathbf{x}}_A - \bar{\mathbf{x}}_B) - (\boldsymbol{\mu}_{A0} - \boldsymbol{\mu}_{B0})\right]^T \left[\frac{\boldsymbol{\Sigma}_A}{n_A} + \frac{\boldsymbol{\Sigma}_B}{n_B}\right]^{-1} \left[(\bar{\mathbf{x}}_A - \bar{\mathbf{x}}_B) - (\boldsymbol{\mu}_{A0} - \boldsymbol{\mu}_{B0})\right]$$

$$= \begin{bmatrix} 6.99 & 11.88 \end{bmatrix} \begin{bmatrix} 0.39 & 0.07 \\ 0.07 & 0.34 \end{bmatrix} \begin{bmatrix} 6.99 \\ 11.88 \end{bmatrix}$$

$$= 78.67$$

iii) Sampling distribution

$$T^2 \sim \chi^2_p$$
$$\sim \chi^2_2 \qquad (\because p = 2).$$

iv) Cut-off value

$$\chi^2_p(\alpha) \text{ at } \alpha = 0.05$$
$$= \chi^2_2(0.05)$$
$$= 5.99 \text{ (from table)}$$

v) Decision

$$T^2 > \chi^2_2(0.05)$$

∴ Reject H_0 (null) hypothesis.

Compare the result with SCIs in example 6.17. The 95% SCIs are as follows (taken from example 6.17, using Bonferroni approach).

$$95\% \text{ SCI for } (\mu_{A1} - \mu_{B1}) : -1.66 \text{ to } 5.64$$

$$95\% \text{ SCI for } (\mu_{A2} - \mu_{B2}) : -7.04 \text{ to } 0.80$$

$$\text{Here, } \mu_{A1} - \mu_{B1} = -5$$
$$\mu_{B1} - \mu_{B2} = -15$$

Neither of the differences (i.e., –5 or –15) falls within 95% SCI, obtained in example 6.17. So, the rejection of H_0 is supported by the 95% SCIs.

Scenario 2: Sampling from Multivariate Normal Population with Unknown but Equal Σ

- Null hypothesis $H_0 : \mu_A = \mu_B$
- Alternate hypothesis $H_1 : \mu_A \neq \mu_B$
- Test statistic:

$$T^2 = \left[\left(\bar{\mathbf{x}}_A - \bar{\mathbf{x}}_B\right) - \left(\mu_{A0} - \mu_{B0}\right)\right]^T \left[\left(\frac{1}{n_A} + \frac{1}{n_B}\right)\mathbf{S}_{\text{pooled}}\right]^{-1} \left[\left(\bar{\mathbf{x}}_A - \bar{\mathbf{x}}_B\right) - \left(\mu_{A0} - \mu_{B0}\right)\right]$$

given that $\mu_A = \mu_{A0}$ and $\mu_B = \mu_{B0}$.

- Sampling distribution: If H_0 is true T^2 follows $\dfrac{\left(n_A + n_B - 2\right)p}{n_A + n_B - p - 1} F_{p, n_A + n_B - p - 1}$ if sample size is small to medium and T^2 follows χ_p^2 for large samples.
- Cut-off value (c^2): $\dfrac{\left(n_A + n_B - 2\right)p}{n_A + n_B - p - 1} F^{\alpha}_{p, n_A + n_B - p - 1}$, where α is the level of significance for small to medium samples, and $\chi_p^2(\alpha)$ for large samples.
- Decision: Reject H_0, if $T^2 > c^2$

Example 6.21
Consider Example 6.15. In addition to the data given in Example 6.15, suppose the hypothesized population mean vector for AG and NAG are as given below:

$$\mu_{A0} = \begin{bmatrix} 55 \\ 5 \end{bmatrix} ; \mu_{B0} = \begin{bmatrix} 40 \\ 10 \end{bmatrix}$$

Test the hypothesis that $\mu_A = \mu_B$ at $\alpha = 0.05$.

Solution:

According to the information given in Example 6.15, this situation falls under Scenario 2.

(i) Null hypothesis: H_0: $\mu_A = \mu_B \quad \therefore (\mu_A - \mu_B) = 0$

Alternate hypothesis: H_1: $\mu_A \neq \mu_B \quad \therefore (\mu_A - \mu_B) \neq 0$

(ii) Test statistics

$$T^2 = \left[(\bar{\mathbf{x}}_A - \bar{\mathbf{x}}_B) - (\mu_{A0} - \mu_{B0})\right]^T \left[\left(\frac{1}{n_A} + \frac{1}{n_B}\right)\mathbf{S}_{\text{pooled}}\right]^{-1} \left[(\bar{\mathbf{x}}_A - \bar{\mathbf{x}}_B) - (\mu_{A0} - \mu_{B0})\right]$$

$$= \begin{bmatrix} -5 & 3 \end{bmatrix} \begin{bmatrix} 0.79 & 0.60 \\ 0.60 & 2.34 \end{bmatrix} \begin{bmatrix} -5 \\ 3 \end{bmatrix}$$

$$= 22.81$$

(iii) Sampling distribution (considering small to medium sample as $n_A = 20$)

$$T^2 \sim \frac{(n_A + n_B - 2)p}{(n_A + n_B - p - 1)} F_{(p, n_A + n_B - p - 1)}$$

$$\sim \frac{(20 + 50 - 2)2}{(20 + 50 - 2 - 1)} F_{(2,67)}$$

$$\sim 2.02985 \times F_{2,67} \quad \left(\because n_A = 20, n_B = 50, p = 2, n_A + n_B - p - 1 = 67\right)$$

(iv) Cut-off value

$$c^2 = \frac{(n_A + n_B - 2)p}{(n_A + n_B - p - 1)} F^{(\alpha)}_{(p, n_A + n_B - p - 1)}$$

$$= 2.02985 \times F^{0.05}_{2,67}$$

$$= 2.02985 \times 3.13$$

$$= 6.35 \qquad \left(\because F^{0.05}_{2,67} = 3.13\right)$$

v) Decision

$T^2_{calculated} = 22.81,\ T^2_{tabulated} = c^2 = 6.35$

$\therefore T^2_{calculated} > T^2_{tabulated}.$

$\therefore H_0$ is rejected.

Note: Compare the results with the SCIs in Example 6.18.The 95% SCIs are as follows (from Example 6.18, using the Bonferroni approach) are as follows::

95% SCI for $\left(\mu_{A1} - \mu_{B1}\right)$: +7.01 to 12.99

95% SCI for $\left(\mu_{A2} - \mu_{B2}\right)$: –3.74 to – 0.259

Here, $\mu_{A1} - \mu_{B1} = 55 - 40 = 15$

$\mu_{A2} - \mu_{B2} = 5 - 10 = -5$

Neither of the SCIs contains their respective mean differences (i.e., 15 or –5). So, the rejection of H_0 is supported by the SCIs.

Scenario 3: Sampling from Multivariate Normal Population with Unknown, Unequal Σ and Large Samples

- Null hypothesis H_0: $\mu_A = \mu_B$
- Alternate hypothesis H_1: $\mu_A \neq \mu_B$
- Test statistic:

$$T^2 = \left[\left(\bar{\mathbf{x}}_A - \bar{\mathbf{x}}_B\right) - \left(\mu_{A0} - \mu_{B0}\right)\right]^T \left[\frac{\mathbf{S}_A}{n_A} + \frac{\mathbf{S}_B}{n_B}\right]^{-1} \left[\left(\bar{\mathbf{x}}_A - \bar{\mathbf{x}}_B\right) - \left(\mu_{A0} - \mu_{B0}\right)\right]$$ given that

$\mu_A = \mu_{A0}$ and $\mu_B = \mu_{B0}$.

- Sampling distribution: If H_0 is true T^2 follows χ^2_p.
- Cut-off value: $\chi^2_p(\alpha)$ where α is the level of significance
- Decision: Reject H_0, if $T^2 > \chi^2_p(\alpha)$

Example 6.22

Consider Example 6.16. In addition to the data given in Example 6.16, suppose the hypothesized mean vector for population inoculant 2kg and 4kg are as given below.

$$\boldsymbol{\mu}_{A0} = \begin{bmatrix} 195 \\ 30 \end{bmatrix}; \quad \boldsymbol{\mu}_{B0} = \begin{bmatrix} 200 \\ 20 \end{bmatrix}.$$

Test the hypothesis $H_0: \boldsymbol{\mu}_A = \boldsymbol{\mu}_B$ at $\alpha = 0.05$.

Solution:

From Example 6.16,

$$\bar{\mathbf{x}}_A - \bar{\mathbf{x}}_B = \begin{bmatrix} -2.87 \\ -0.38 \end{bmatrix}; \quad \left[\frac{\mathbf{S}_A}{n_A} + \frac{\mathbf{S}_B}{n_B}\right]^{-1} = \begin{bmatrix} 0.32 & 0.32 \\ 0.32 & 50.32 \end{bmatrix}$$

(i) Null hypothesis:

$H_0: \boldsymbol{\mu}_A = \boldsymbol{\mu}_B$, i.e, $(\boldsymbol{\mu}_A - \boldsymbol{\mu}_B) = 0$

Alternate hypothesis:

$H_1: \boldsymbol{\mu}_A \neq \boldsymbol{\mu}_B$, i.e, $(\boldsymbol{\mu}_A - \boldsymbol{\mu}_B) \neq 0$

(ii) Test statistic:

$$T^2 = \left[(\bar{\mathbf{x}}_A - \bar{\mathbf{x}}_B) - (\boldsymbol{\mu}_{A0} - \boldsymbol{\mu}_{B0})\right]^T \left[\frac{\mathbf{S}_A}{n_A} + \frac{\mathbf{S}_B}{n_B}\right]^{-1} \left[(\bar{\mathbf{x}}_A - \bar{\mathbf{x}}_B) - (\boldsymbol{\mu}_{A0} - \boldsymbol{\mu}_{B0})\right]$$

$$= \begin{bmatrix} 2.13 & -10.38 \end{bmatrix}_{1\times2} \begin{bmatrix} 0.32 & 0.32 \\ 0.32 & 50.32 \end{bmatrix}_{2\times2} \begin{bmatrix} 2.13 \\ -10.38 \end{bmatrix}$$

$$= 5409.$$

(iii) Sampling distribution (considering large sample)

$$T^2 \sim \chi_p^2$$
$$\sim \chi_2^2 \qquad (\because p = 2)$$

(iv) Cut-off value

$\chi_p^2(\alpha) = \chi_2^2(0.05) = 5.99$ (from table)

(v) Decision

$T^2_{\text{calculated}} = 5409$
$T^2_{\text{tabulated}} = \chi_2^2(0.05) = 5.99.$

$\therefore T^2_{\text{calculated}} > T^2_{\text{tabulated}}$

$\therefore$ H_0 is rejected.

Note: Comparison of the results with the results of example 6.19 are as follows: The 95% SCIs for Example 6.19 (using the Bonferroni approach) are as follows:

95% SCI for $(\mu_{A1} - \mu_{B1})$: –6.86 to 1.12

95% SCI for $(\mu_{A2} - \mu_{B2})$: –0.69 to – 0.07

Here, $\mu_{A1} - \mu_{B1} = -5$

$\mu_{A2} - \mu_{B2} = 10$

The SCI for $\mu_{A1} - \mu_{B1}$ contains it (the population mean difference) but the SCI for $\mu_{A2} - \mu_{B2}$ doesn't. As one of the SCIs doesn't contain its respective mean difference, the rejection of H_0 is supported overall.

6.7 CONFIDENCE REGION AND HYPOTHESIS TESTING FOR COVARIANCE MATRIX Σ

6.7.1 Confidence Region for Σ

In Section 6.5.1, we described confidence region (CR) for population mean vector ($\boldsymbol{\mu}_{p\times 1}$) using one sample $\mathbf{X}_{n\times p} = [\mathbf{x}_1 \ \mathbf{x}_2 \ \cdots \ \mathbf{x}_n]^T$, where $\mathbf{x}_1, \mathbf{x}_2, \ldots\ldots, \mathbf{x}_n$ are the n number of multivariate observations, each is measured on p variables. In Equation 6.13, we have defined the CR, $R(\overline{\mathbf{x}})$ that might contain $\boldsymbol{\mu}$ with probability $(1-\alpha)$. Analogously one may be interested to know what could be the CR for the population covariance matrix $\boldsymbol{\Sigma}_{p\times p}$ using the sample $\mathbf{X}_{n\times p}$. If $\mathbf{S}_{p\times p}$ is the sample covariance matrix, then $\hat{\boldsymbol{\Sigma}} = \frac{n-1}{n}\mathbf{S} \sim \mathbf{S}$ (see Section 6.1). So, the key question is what is the desired region $R(\mathbf{S})$ that contains $\boldsymbol{\Sigma}_{p\times p}$ with a confidence of $100(1-\alpha)\%$. Krzanowski and Marriott (2014a) stated that the corresponding confidence region theory is not available. The possible reason, as stated by Krzanowski and Marriott (2014a), is that it is intractable to have all possible positive definite $\boldsymbol{\Sigma}_{p\times p}$ within an interpretable region. Bootstrapping-based methods were proposed by researchers. For example, interested readers may refer to Beran and Srivastava (1985).

6.7.2 Hypothesis Testing for Σ

The test of covariance matrices can be grouped as (i) one-sample, (ii) two-sample, and (iii) multi-sample tests. In one-sample test, the popular tests are (a) sphericity test, i.e., H_0: is $\boldsymbol{\Sigma} = \sigma^2\mathbf{I}$, (b) identity test, i.e., $\mathbf{H}_0 : \boldsymbol{\Sigma} = \mathbf{I} = \boldsymbol{\rho}$ ($\boldsymbol{\rho}$ = correlation matrix), (c) equality test with pre-specified covariance matrix, i.e., $\mathbf{H}_0 : \Sigma = \Sigma_0$, and (d) test of proportionality to a pre-specified covariance matrix, i.e., $\mathbf{H}_0 : \Sigma = m\,\Sigma_0$ (m is a constant). In two-sample test, usually (c) and (d) above are modified as (e) $\mathbf{H}_0 : \Sigma_A = \Sigma_B$ and (f) $\mathbf{H}_0 : \Sigma_A = m\,\Sigma_B$. In multi-sample test, (e) is generalized to L population as $\mathbf{H}_0 : \Sigma_1 = \Sigma_2 = \ldots\ldots = \Sigma_L$, where L is the number of populations to be considered. Note that for two-sample test, L = 2. We describe below the likelihood ratio based tests for (e), above. For tests in (a) and (b), see Chapter 11, and for (e) and its generalization see Chapter 7. For proof and more details of the above tests including (d) and (f), readers may refer to Krzanowski and Marriott (2014a), Anderson (2003) and Mardia et al. (1979).

We consider single population case.

Following the steps of hypothesis testing as mentioned in Section 6.6.3, we find the following:

Step 1: Set null and alternate hypothesis

$$H_0 : \Sigma = \Sigma_0$$

$$H_1 : \Sigma \neq \Sigma_0$$

Step 2: Obtain appropriate test statistic

In likelihood ratio test (LRT), the test statistic, say λ is derived using the ratio of the maximum value of the likelihood function in equation 6.1 for H_0 and H_1. If we denote the maximum likelihood value for H_0 and H_1 as $L_{\max 0}$ and $L_{\max 1}$, then

$$\lambda = \frac{L_{\max 0}}{L_{\max 1}} \tag{6.54}$$

Taking –2log of the above, we get

$$\lambda_{LR} = -2\ln\lambda = 2\left(\ln L_{\max 1} - \ln L_{\max 0}\right) \tag{6.55}$$

Now, the likelihood function (Equation 6.1) or its log likelihood (Equation 6.2) depends on $\boldsymbol{\mu}$ and $\boldsymbol{\Sigma}$. So, to get $L_{\max 0}$ and $L_{\max 1}$, we need to know $\boldsymbol{\mu}$ and $\boldsymbol{\Sigma}$ for both H_0 and H_1. As the maximum likelihood value for $\boldsymbol{\mu}$ is always $\bar{\mathbf{x}}$, the unknown $\boldsymbol{\mu}$ can be substituted by $\bar{\mathbf{x}}$ in the Equations 6.1 and 6.2. Further, the $\boldsymbol{\Sigma}$ value under H_0 is $\boldsymbol{\Sigma}_0$. As $\boldsymbol{\Sigma}$ under H_1 is not known and we are interested to test $H_0 : \boldsymbol{\Sigma} = \boldsymbol{\Sigma}_0$, any value other than $\boldsymbol{\Sigma}_0$ will be plausible to use. The MLE of $\boldsymbol{\Sigma}$ under H_1 is $\mathbf{S}$ (see Section 6.1). Putting these values in Equation 6.2, we get

$$\begin{aligned}
\lambda_{LR} &= -2\ln\lambda = -n\ln|\mathbf{S}| - \sum_{i=1}^{n}\left(\mathbf{x}_i - \bar{\mathbf{x}}\right)^T \mathbf{S}^{-1}\left(\mathbf{x}_i - \bar{\mathbf{x}}\right) + n\ln|\boldsymbol{\Sigma}_0| + \sum_{i=1}^{n}\left(\mathbf{x}_i - \bar{\mathbf{x}}\right)^T \boldsymbol{\Sigma}_0^{-1}\left(\mathbf{x}_i - \bar{\mathbf{x}}\right) \\
&= -n\ln\left|\boldsymbol{\Sigma}_0^{-1}\mathbf{S}\right| + n\, trace\left(\boldsymbol{\Sigma}_0^{-1}\mathbf{S}\right) - n\, trace\left(\mathbf{S}^{-1}\mathbf{S}\right) \qquad \text{(from equation 6.3)} \\
&= -n\ln\left|\boldsymbol{\Sigma}_0^{-1}\mathbf{S}\right| + n\, trace\left(\boldsymbol{\Sigma}_0^{-1}\mathbf{S}\right) - np
\end{aligned} \tag{6.56}$$

The matrix $\boldsymbol{\Sigma}_0^{-1}\mathbf{S}$ in equation 6.56 is very important matrix and used extensively in multivariate tests of hypothesis. λ_{LR} is the appropriate statistic to test $H_0 : \boldsymbol{\Sigma} = \boldsymbol{\Sigma}_0$.

Step 3: Identify the sampling distribution of the test statistic under H_0.

The exact distribution of λ_{LR} (in Equation 6.56) is not readily available, but asymptotically, under H_0, it follows χ^2 distribution with $\frac{1}{2}p(p+1)$ degrees of freedom (Krzanowski and Marriott, 2014a). Using the eigenvalues of $\boldsymbol{\Sigma}^{-1}\mathbf{S}$, λ_{LR} in Equation 6.56 can be written as

$$\lambda_{LR} = -n\ln\left|\boldsymbol{\Sigma}_0^{-1}\mathbf{S}\right| + n\,trace\left(\boldsymbol{\Sigma}_0^{-1}\mathbf{S}\right) - np = np\left(a - \ln g - 1\right) \tag{6.57}$$

Where a = arithmetic mean of the eigenvalues of $\boldsymbol{\Sigma}_0^{-1}\mathbf{S}$
and g = geometric mean of the eigenvalues of $\boldsymbol{\Sigma}_0^{-1}\mathbf{S}$.

Step 4: Set the cut-off value of λ_{LR}

As $\lambda_{LR} \sim \chi^2_{p(p+1)/2}$, the cut-off value for rejection of H_0 is $\chi^2_{p(p+1)/2}(\alpha)$, where α is the level of significance.

Step 5: Make decision on the rejection or acceptance of H_0

If λ_{LR} computed using Equation 6.56 or 6.57 exceeds $\chi^2_{p(p+1)/2}(\alpha)$, H_0 is rejected.

Example 6.23

Consider Example 6.1. Suppose, based on past information, the covariance matrix of the process is assumed to be $\boldsymbol{\Sigma}_0$ as below.

$$\boldsymbol{\Sigma}_0 = \begin{bmatrix} 36 & -45 \\ -45 & 81 \end{bmatrix}$$

Test the null hypothesis: $H_0 : \boldsymbol{\Sigma} = \boldsymbol{\Sigma}_0$, $\alpha = 0.05$.

Solution:

From Example 6.1, the following data are obtained.

$$X \sim N_2(\mu, \boldsymbol{\Sigma}), \quad \bar{\mathbf{x}} = \begin{bmatrix} 10 \\ 20 \end{bmatrix}, \quad \mathbf{S} = \begin{bmatrix} 40 & -50 \\ -50 & 100 \end{bmatrix}, \quad \text{and } n = 20.$$

(i) Null hypothesis $H_0 : \boldsymbol{\Sigma} = \boldsymbol{\Sigma}_0 = \begin{bmatrix} 36 & -45 \\ -45 & 81 \end{bmatrix}$

Alternate hypothesis $H_1 : \boldsymbol{\Sigma} \neq \boldsymbol{\Sigma}_0$; here $\boldsymbol{\Sigma} = \mathbf{S} = \begin{bmatrix} 40 & -50 \\ -50 & 100 \end{bmatrix}$

(ii) Test statistic

From Equation 6.57, we get the test statistic λ_{LR} is

$$\lambda_{LR} = -n \ln\left|\boldsymbol{\Sigma}_0^{-1}\mathbf{S}\right| + n\,trace\left(\boldsymbol{\Sigma}_0^{-1}\mathbf{S}\right) - np$$
$$= np(a - \ln g - 1)$$

$$\text{Now, } \boldsymbol{\Sigma}_0^{-1} = \frac{1}{|\boldsymbol{\Sigma}_0|} Adj(\boldsymbol{\Sigma}_0) = \frac{1}{36 \times 81 - 45^2} \begin{bmatrix} 81 & 45 \\ 45 & 36 \end{bmatrix}$$

$$= \begin{bmatrix} 0.091 & 0.051 \\ 0.051 & 0.040 \end{bmatrix}$$

$$\text{So, } \boldsymbol{\Sigma}_0^{-1}\mathbf{S} = \begin{bmatrix} 0.091 & 0.051 \\ 0.051 & 0.040 \end{bmatrix} \begin{bmatrix} 40 & -50 \\ -50 & 100 \end{bmatrix} = \begin{bmatrix} 1.09 & 0.55 \\ 0.04 & 1.45 \end{bmatrix}$$

$$\text{So, } \left|\boldsymbol{\Sigma}_0^{-1}\mathbf{S}\right| = \begin{vmatrix} 1.09 & 0.55 \\ 0.04 & 1.45 \end{vmatrix} = 1.5585 \text{ and } trace\left(\boldsymbol{\Sigma}_0^{-1}\mathbf{S}\right) = 2.54$$

$$\text{So, } \lambda_{LR} = -20 \ln(1.5585) + 20 \times 2.54 - 20 \times 2$$
$$= -8.874 + 50.8 - 40 = 1.926$$

Alternative computation using eigenvalues of $\boldsymbol{\Sigma}_0^{-1}\mathbf{S}$, is shown below.

Let λ_1 and λ_2 are the eigenvalues of $\mathbf{\Sigma}_0^{-1}\mathbf{S}$.

$$\text{So, } \begin{bmatrix} 1.09-\lambda & 0.55 \\ 0.04 & 1.45-\lambda \end{bmatrix} = 0 \quad \text{gives } \lambda_1 = 1.503238 \quad \text{and} \quad \lambda_2 = 1.036762$$

The arithmetic mean (a) of λ_1 and λ_2 is $\frac{1}{2}(\lambda_1+\lambda_2) = 1.27$ and the geometric mean (g) of λ_1 and λ_2 is $(\lambda_1\lambda_2)^{1/2} = (1.5585)^{1/2} = 1.2484$.

So, using eigenvalues, we get

$$\begin{aligned} \lambda_{LR} &= np(a - \ln g - 1) \\ &= 20 \times 2(1.27 - \ln(1.2484) - 1) \\ &= 40(1.27 - 0.221863 - 1) \\ &= 1.925 \text{ (Hence, verified)} \end{aligned}$$

(iii) Asymptotic distribution of λ_{LR}

$$\lambda_{LR} \sim \chi^2_{\frac{p(p+1)}{2}} = \chi^2_3$$

(iv) The cut-off value = $\chi^2_3(0.05) = 7.81$ (from χ^2 table)

(v) Decision: As $\lambda_{LR} = 1.925 < \chi^2_3(0.05) = 7.81$, we fail to reject H_0.

In the next chapter (Chapter 7), we will describe Box M tests for equality of multiple covariance matrices, each coming from different populations.

6.8 SAMPLING FROM NON-NORMAL POPULATION

Krzanowski and Marriott (2014a) stated that for large samples from multivariate non-normal populations, the procedures described under large sample can be used. Multivariate CLT allows this. For small to medium samples, they recommended bootstrapping. The general procedures of bootstrapping (for univariate cases) are given in Chapter 3 (Section 3.5.2). The bootstrapping procedure for multivariate cases is given in Krzanowski and Marriott (2014a). While the steps are similar to univariate bootstrapping, the mathematical complexity increases in a multivariate situation. Readers may refer to Davison and Hinkley (1997), Efron and Tibshirani (1997), and Mandel and Betensky (2008). Alternatively, non-parametric methods could be used. For example, Puri and Sen (1968) provided confidence regions for location vectors for one- and two-sample cases.

The emergence of the industrial Internet of Things and other Industry 4.0 technologies enables organizations to collect big data. The major problem in applying the classical statistics in big data is the assumption of large sample size (n) as compared to the number of variables/features (p). For classical statistics, $n>>p$. But big data often involves high dimensional data with large number of variables/features (p), and there might be situations when $p>n$. Big data are a mixture of different data types (such as text, categories, and topics). For statistical modeling, these data require transformation to numeric data. The transformation often leads to non-normal data. Recent statistics-based research focuses on high-dimensional data analysis using statistics. For a theoretical treatment point of view, readers may refer to Giraud (2021). For multivariate tests of mean vectors, readers may consult Srivastava and Du (2008) and Wang et al. (2015), and for covariance structures, Cai (2017) provides an excellent review.

6.9 LEARNING SUMMARY

Multivariate statistical inference deals with estimation of parameters, defining 100(1-α)% confidence region (CR), dissecting CR to simultaneous confidence intervals (SCI), and finally conducting hypothesis testing for parameter vectors involving multiple variables. The inferencing can be done under different situations of interest (scenarios). One may deal with single population mean vector, differences between two-population mean vectors, or comparing several-population mean vectors. In this chapter, multivariate statistical inferences concerning single and two-population mean vectors were discussed. We primarily relied on the most commonly used concepts, applicable to real life problem-solving.

In summary,

- If $\boldsymbol{X} \sim N_p(\boldsymbol{\mu}, \boldsymbol{\Sigma})$, then sample statistics $\overline{\mathbf{x}} \sim N_p(\boldsymbol{\mu}, \boldsymbol{\Sigma}/n)$, and $(n-1)\mathbf{S}$ is distributed as Wishart random matrix with $n-1$ degrees of freedom.
- The central limit theorem says that for large sample, $\sqrt{n}\left(\overline{\mathbf{x}}-\boldsymbol{\mu}\right) \sim N_p\left(0, \boldsymbol{\Sigma}\right)$ and the Hotelling's T^2 is $n\left(\overline{\mathbf{x}}-\boldsymbol{\mu}\right)^T \mathbf{S}^{-1}\left(\overline{\mathbf{x}}-\boldsymbol{\mu}\right) \sim \chi_p^2$.
- For small to medium sample, the Hotelling's $T^2 = n\left(\overline{\mathbf{x}}-\boldsymbol{\mu}\right)^T \mathbf{S}^{-1}\left(\overline{\mathbf{x}}-\boldsymbol{\mu}\right) \sim \frac{(n-1)p}{n-p} F_{p,n\,p}$.
- For a multivariate sample, $\mathbf{X} = \left[\mathbf{x}_1, \mathbf{x}_2, \cdots \mathbf{x}_n\right]^T$, the $100(1-\alpha)\%$ confidence region for $\boldsymbol{\mu}$ vector, defined by the sample mean vector $(\overline{\mathbf{x}})$ is $R(\overline{\mathbf{x}})$, where $P\left[R\left(\overline{\mathbf{x}}\right) \text{ will contain } \boldsymbol{\mu}\right] = 1-\alpha$.
- There could be different scenarios for confidence regions for single population mean vector as (i) sampling from $\boldsymbol{X} \sim N_p\left(\boldsymbol{\mu}, \boldsymbol{\Sigma}\right)$ when $\boldsymbol{\Sigma}$ is known, (ii) sampling from $X \sim N_p\left(\boldsymbol{\mu}, \boldsymbol{\Sigma}\right)$ when $\boldsymbol{\Sigma}$ is unknown and sample size is large, and (iii) sampling from $\boldsymbol{X} \sim N_p\left(\boldsymbol{\mu}, \boldsymbol{\Sigma}\right)$ when $\boldsymbol{\Sigma}$ is unknown and sample size is small. For (i) and (ii), Hotelling's T^2 follows χ_p^2 and for (iii) Hotelling's T^2 follows $\frac{(n-1)p}{n-p} F_{p,n\,p}$.
- The CR forms an ellipse for bivariate populations and ellipsoid for multivariate populations (involving three or more variables).
- The SCI is used to determine the CI for each of the variables separately without compromising the overall 100(1-α)% confidence. Two popular approaches for determining SCI are linear combination approach and the Bonferroni approach.
- Following the appropriate distribution for Hotelling's T^2, the null hypothesis (H_0: $\boldsymbol{\mu} = \boldsymbol{\mu}_0$) is tested and decision on its rejection or acceptance is determined.
- For the difference between two-population mean vectors, CR and SCI are computed and the hypothesis is tested. The approach remains the same as that for single populations with a difference of the random variable vector to be utilized. In single populations, the random variable vector is sample average vector, whereas for two populations, the random variable vector is the difference between the sample average vectors obtained from the two populations. In addition, when the covariance matrices for both the populations are same, the sample pooled covariance matrix is used as the estimate of the population covariance matrices.
- Multivariate statistical inferencing is a vast subject. Readers may consult advanced books in this area. The next logical extension of this chapter could be the statistical inferencing concerning more than two multivariate populations.

EXERCISES

(A) Short conceptual questions

6.1 How does multivariate statistical inferencing differ from univariate statistical inferencing?

6.2 Define multivariate central limit theorem. How does it differ from univariate central limit theorem?

6.3 What is Hotelling's T^2? Derive that $T^2 = n(\bar{\mathbf{x}} - \boldsymbol{\mu})^T \mathbf{S}^{-1}(\bar{\mathbf{x}} - \boldsymbol{\mu})$.

6.4 What would be the distribution of Hotelling's T^2 when population covariance matrix $\boldsymbol{\Sigma}$ is unknown and the sample collected for inferencing is small?

6.5 What is a confidence region (CR)? Show that for a bivariate population, CR represents an ellipse.

6.6 The CR for a p-variable is an ellipsoid. How do you define its axes?

6.7 Define simultaneous confidence intervals (SCI). What is their use?

6.8 Consider $\boldsymbol{X} \sim N_p(\boldsymbol{\mu}, \boldsymbol{\Sigma})$. Using linear combination approach show that the j-th SCI is $\bar{x}_j - \sqrt{\chi_p^2(\alpha)}\sqrt{\frac{\sigma_{jj}}{n}} \le \mu_j \le \bar{x}_j + \sqrt{\chi_p^2(\alpha)}\sqrt{\frac{\sigma_{jj}}{n}}, j = 1, 2, \ldots p.$

6.9 How does the Bonferroni approach define SCI? Explain advantages and disadvantages of the Bonferroni approach.

6.10 Explain the steps to be followed while conducting hypothesis testing in relation to multivariate observations.

6.11 Show that when sampling from two multivariate normal populations with known covariance matrices $\boldsymbol{\Sigma}_A$ and $\boldsymbol{\Sigma}_B$, the $100(1-\alpha)\%$CR is

$$\left[\bar{\mathbf{x}}_A - \bar{\mathbf{x}}_B - (\boldsymbol{\mu}_A - \boldsymbol{\mu}_B)\right]^T \left[\frac{\boldsymbol{\Sigma}_A}{n_A} + \frac{\boldsymbol{\Sigma}_B}{n_B}\right]^{-1} \left[\bar{\mathbf{x}}_A - \bar{\mathbf{x}}_B - (\boldsymbol{\mu}_A - \boldsymbol{\mu}_B)\right] \le \chi_p^2(\alpha)$$

6.12 Show that when sampling from two multivariate normal population with unknown but equal covariance matrices, i.e. $\boldsymbol{\Sigma}_A = \boldsymbol{\Sigma}_B = \boldsymbol{\Sigma}$, the $100(1-\alpha)\%$CR considering small samples is

$$\left[\bar{\mathbf{x}}_A - \bar{\mathbf{x}}_B - (\boldsymbol{\mu}_A - \boldsymbol{\mu}_B)\right]^T \left[\left(\frac{1}{n_A} + \frac{1}{n_B}\right)\mathbf{S}_{\text{pooled}}\right]^{-1} \left[\bar{\mathbf{x}}_A - \bar{\mathbf{x}}_B - (\boldsymbol{\mu}_A - \boldsymbol{\mu}_B)\right] \le C,$$

where $C = \frac{(n_A + n_B - 2)p}{n_A + n_B - p - 1} F^{\alpha}_{p, n_A + n_B - p - 1}$.

6.13 What will happen to the $100(1-\alpha)\%$CR as given in question 6.12, when n_A and n_B are large?

6.14 Given simultaneous confidence intervals (SCI), is it possible to conduct hypothesis testing?

(B) Long conceptual and/or numerical questions: see web resources (Multivariate Statistical Modeling in Engineering and Management – 1st (routledge.com))

N.B.: R and MS Excel software are used for computation, as and when required.

Part III

Multivariate Models

7 Multivariate Analysis of Variance

In Chapter 6, we described multivariate hypothesis testing for equality of two-population means. When the number of populations in consideration is more than two (>2), the procedures described in Chapter 6 are not sufficient. We require more advanced techniques to test the hypothesis of equality of population means. Depending on the number of populations to be compared and the number of response variables that characterizes the populations of interest, the scenarios for comparisons can be categorized as (i) two populations, each characterized by one variable of interest, (ii) two populations, each characterized by two or more variables of interest, (iii) more than two populations, each characterized by one variable of interest, and (iv) more than two populations, each characterized by two or more variables of interest. For case (i), t-test is commonly used (see Chapter 2) and for case (ii), Hotelling's T^2 is commonly used (see Chapter 6). For case (iii), analysis of variance (ANOVA) is commonly used and for case (iv), multivariate variance of analysis (MANOVA) is used. The scenarios (iii) and (iv) described above is called one-way ANOVA and one-way MANOVA, respectively and the populations are treatments/levels of one particular kind (e.g., a factor with different levels; each level represent a population). When there are two or more factors, each with two or more treatments/levels, the corresponding models of ANOVA (with one response variable) or MANOVA (with two or more response variables) are called two-way and multi-way ANOVA or two-way and multi-way MANOVA, respectively. In this chapter, both one and two-way ANOVA and MANOVA are described.

7.1 ANALYSIS OF VARIANCE (ANOVA)

7.1.1 Conceptual Model

A manufacturer produces steel washers of some quality. Let the quality variable is outer diameter (OD). Three processes, A, B, and C, are in operations. A sample ($n = 10$) was randomly collected from each of the three processes A, B, and C separately. The purpose of this sampling is to test if there are mean differences in OD of the washers produced by the three processes. The data set is given in Table 7.1. Here, the response variable is outer diameter. The factor is the process and the three treatments/levels are process A, process B, and process C. Note that we have one factor with three treatments/levels (A, B, C). Each treatment/level is considered as a population.

The quality control engineer of the plant initially uses graphical plots such as box plots to extract the pattern in the data. The box plot is shown in Figure 7.1. A box plot is a powerful tool that depicts the central tendency, spread, and skewness of a given univariate data set. It also shows outliers, if any, in the data set. A box plot contains a box in the middle and two whiskers at the two ends. The box length, i.e., the distance between the top and bottom horizontal lines of the box, represents the inter-quartile range (IQR) and the horizontal line in the middle of the box represents median of the data set. The two vertical lines at the bottom and top of the box are whiskers. The whiskers' lengths are the lines emanating from the top and bottom of the box and terminating at the points of maximum and minimum, respectively. Alternatively, a whisker's length is 1.5 IQR. Any observation beyond the whisker's length (1.5 IQR) is suspected as being an outlier. The mean of the observed data can also be added in the box plot (see the small circle within the box). For a symmetric data, both mean and median coincide. From Figure 7.1, it can be seen that the steel washer quality varies

DOI: 10.1201/9781003303060-10

TABLE 7.1
Data Set for Production of Steel Washers (Response Variable OD)

Sl. No.	Process A	Process B	Process C
1	20	17	20
2	21	17	20
3	20	19	21
4	21	17	20
5	23	16	21
6	19	19	21
7	20	18	22
8	19	18	19
9	19	18	22
10	20	20	20

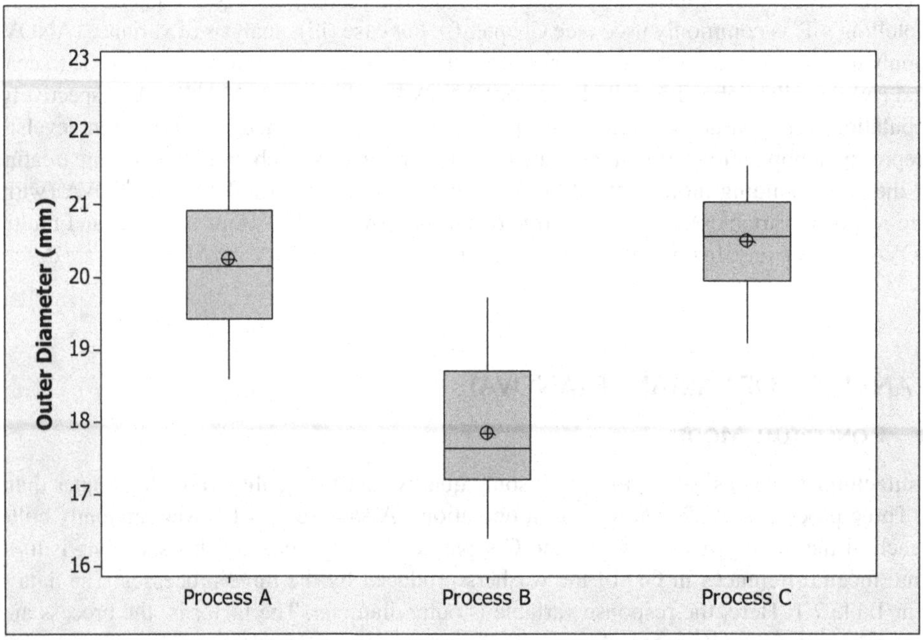

FIGURE 7.1 Box plot for steel washer outer diameter (in mm). Each box visually represents the central tendencies namely, median (the horizontal line within the box) and mean (circle circumscribed a + symbol), the variability (dispersion) in terms of interquartile range (IQR, represented by the box), and minimum and maximum values (the lower and upper whiskers of the box) of the outer diameter. The mean outer diameter of the steel washers, produced by three processes A, B, and C seem to be different (based on this visual plot).

from process to process. The mean OD for the three processes differs. We will conceptualize this as an ANOVA model below.

Suppose the population mean OD for the three processes A, B, and C is μ_1, μ_2, and μ_3, respectively. We want to test whether there are differences in the process means, i.e., between μ_1 and μ_2 or between μ_2 and μ_3 or between μ_3, and μ_1. Alternatively, we can set hypothesis as:

$$H_0 : \mu_1 = \mu_2 = \mu_3$$
$$H_1 : \mu_\ell \neq \mu_m \text{ for at least one pair of } \ell \text{ and m, } \ell \neq \text{m, } \ell = 1,2,3 \text{ and m} = 1,2,3.$$

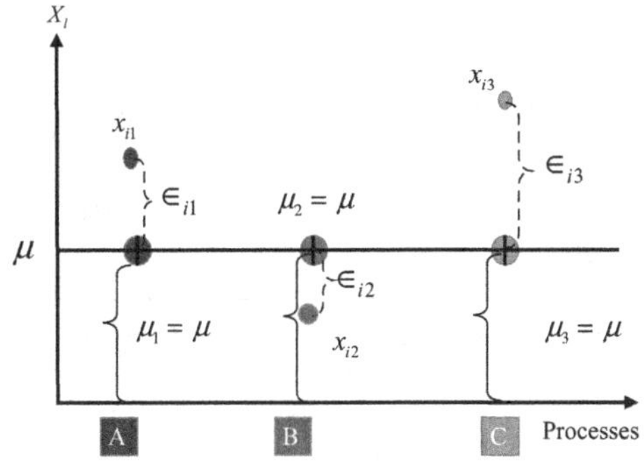

FIGURE 7.2 H_0 is true (Scenario 1). The horizontal line represents the grand mean (μ). As H_0 is true (i.e., $\mu_1 = \mu_2 = \mu_3 = \mu$), all the process means fall on the horizontal line representing the grand mean (μ). The error terms are represented by $\in_{il} = x_{il} - \mu_l, l = 1,2,3$.

Now let us define grand mean (μ) for the three processes. It is

$$\mu = \frac{1}{3}\sum_{\ell=1}^{3}\mu_\ell = \frac{1}{3}(\mu_1 + \mu_2 + \mu_3) \tag{7.1}$$

If H_0 is true, then each of the mean is equal to μ (i.e., $\mu_1 = \mu_2 = \mu_3 = \mu$). Now, if we consider a general observation $x_{i\ell}$ representing the $i-th$ observation from the $\ell - th$ process, then we can write from Figure 7.2 that

$$x_{i\ell} = \mu_\ell + \in_{i\ell} = \mu + \in_{i\ell}, i = 1,2,\ldots,n, \text{ and } l = 1,2,3 \text{ for the three processes A, B, and C, respectively.} \tag{7.2}$$

The term $\in_{i\ell}$ relates to random error for the i-th observation from the l-th process, which is inevitable for every natural process.

When H_0 is false, we accept H_1, i.e., at least one pair of (μ_ℓ, μ_m) is different. The scenario is depicted in Figure 7.3.

From Figure 7.3, it is seen that if H_0 is false, we need to rewrite Equation 7.2 by considering the process effect. The resultant equation becomes

$$x_{i\ell} = \mu + (\mu_l - \mu) + (x_{il} - \mu_l) \text{ or } x_{il} = \mu + \tau_l + \in_{il},\ l = 1,2,3 \text{ and } i = 1,2,\ldots,n \tag{7.3}$$

Where $\tau_\ell = \mu_\ell - \mu$, represents the effect of the l-th process

$\in_{i\ell} = x_{i\ell} - \mu_\ell$, represents the error

Now,

$$\sum_{l=1}^{3}\tau_l = \tau_1 + \tau_2 + \tau_3 = (\mu_1 - \mu) + (\mu_2 - \mu) + (\mu_3 - \mu) = \mu_1 + \mu_2 + \mu_3 - 3\mu$$
$$= 3\mu - 3\mu = 0 \quad \text{(from equation 7.1)}$$

So, sum of the process effects is zero. This is obvious as the process effects are nothing but the deviation of the process mean (μ_l) from the grand mean (μ).

The three-process case can be generalized to L processes. We now use the word *population* in place of *processes* as population is the common word used by statisticians to represent such situation. For a general case consisting of L populations, Equation 7.3 holds where ℓ ranges from 1 to L and i ranges from 1 to n. The ANOVA model can be specified as below.

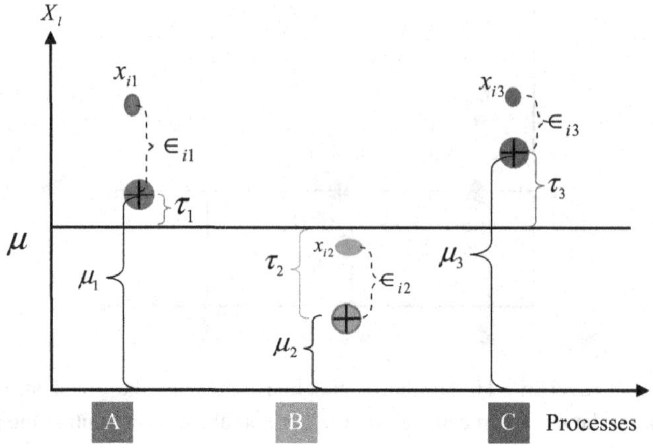

FIGURE 7.3 H_0 is false (Scenario 2). As H_0 is false (favoring H_1 to be true), the process means differ from the grand mean and these do not fall on the horizontal line representing the grand mean (μ). The effects of processes are represented by τ_l and the error terms are represented by $\in_{il} = x_{il} - \mu, l = 1,2,3$.

$$x_{i\ell} = \mu + \tau_\ell + \in_{i\ell},\ l = 1,2,....,L \text{ and } i = 1,2,...,n \tag{7.4}$$

$$\tau_\ell = \mu_\ell - \mu, \text{and}$$

$$\in_{i\ell} = x_{i\ell} - \mu_\ell$$

$\in_{i\ell}$ is assumed to be normally distributed with mean zero and variance σ^2.

In addition, the following condition is satisfied:

$$\left.\begin{array}{l}\sum_{\ell=1}^{L} \tau_\ell = 0, \text{ when number of observations (n) are equal across all the L populations and} \\ \sum_{\ell=1}^{L} n_\ell \tau_\ell = 0, \text{ when the number of observations } (n_l) \text{ differ across populations}\end{array}\right\} \tag{7.5}$$

Accordingly, in ANOVA the null and alternative hypothesis can be as follows:

$$H_0 = \tau_\ell = 0, \text{ for all } \ell = 1,2,...\text{L}$$
$$H_1 = \tau_\ell \neq 0, \text{ for at least one } \ell.$$

7.1.2 Assumptions

The ANOVA requires the following assumptions to be satisfied.

- Independence: the observations $x_{i\ell}$, $i = 1,2,...n_\ell$ & $\ell = 1,2,...L$ must be independent. The assumption can be tested by auto correlation or serial correlation plot of the errors (residuals), $\in_{i\ell}, \ell = 1,2,...n_\ell$ and $\ell = 1,2,...L$.
- Homogeneity of variance: The variance of the response variable for all the L populations must be equal. That is, if σ_ℓ^2 is the variance of $x_{i\ell}$, then $\sigma_1^2 = \sigma_2^2 = ... = \sigma_\ell^2 = ... = \sigma_L^2 = \sigma^2$. This can be tested by modified Levene test or Bartlett test.
- Normality: $x_{i\ell}$ must be normal with mean μ_ℓ and variance $\sigma_\ell^2 (= \sigma^2)$. So, $x_{i\ell} \sim N(\mu_\ell, \sigma^2)$. This can be tested by testing the normality of the errors, i.e., $\in_{i\ell} \sim N(0, \sigma^2)$.

The abovementioned assumptions are critical for the success of ANOVA. The independent observation assumption is a must. Violation of this assumption affects the decision taken using F-statistic. The type-I error (α error) in hypothesis testing gets inflated (Scariano and Davenport, 1987). As an example, Field (2009) mentioned that with Pearson correlation coefficient of 0.50 between observations across three groups involving ten observations in each group, the type-I error rate (α) becomes 0.74 instead of 0.05. This is a gross inflation.

If the "homogeneity of variance" assumption is violated, it is recommended to use equal sample size (n), preferably large in number. ANOVA is fairly robust to violation of "homogeneity of variance" so long the sample size is equal (balanced ANOVA). For unequal sample sizes (unbalanced ANOVA), the "homogeneity of variance" assumption needs to be satisfied. Field (2009) categorizes the problem with three scenarios. First, when populations with larger sample size have larger variances than the populations with smaller sample size, the F-ratio becomes conservative. That means the F-ratio underestimates the differences in means of the population. Second, when populations with larger samples have smaller variances than the populations with smaller samples, the F-ratio becomes liberal. That means the F-ratio overestimates the mean differences. Third, if variances are proportional to the means, F-ratio is unaffected by heterogeneity of variance.

There are several tests of homogeneity of variance, namely the Bartlett test (Bartlett, 1937a), Cochran test (Cochran, 1947), Scheffé test (Scheffé, 1959), Levene test (Levene, 1960), and modified Levene test (Brown and Forsythe, 1974). The tests proposed by Bartlett and Cochran require normality assumptions to be satisfied whereas the tests proposed by Levene and Scheffé, including the modified Levene test, are robust against normality. We describe the Bartlett test below. The modified Levene test will be described later as it requires ANOVA procedure to be known.

Bartlett Test

Suppose there are L number of populations and the variable of interest is X. Assume that the means of the populations are $\mu_1, \mu_2, \ldots \mu_\ell, \ldots \mu_L$ and variances are $\sigma_1^2, \sigma_2^2, \ldots \sigma_\ell^2, \ldots \sigma_L^2$. It is to be tested that $\sigma_1^2 = \sigma_2^2 = \ldots = \sigma_\ell^2 = \ldots = \sigma_L^2 = \sigma^2$ or not. The steps to be followed are given below.

Step 1: Set null (H_0) and alternate (H_1) hypothesis

$$H_O : \sigma_1^2 = \sigma_2^2 = \ldots = \sigma_\ell^2 = \ldots = \sigma_L^2 = \sigma^2.$$

$$H_1 : \sigma_\ell^2 \neq \sigma_m^2 \text{ for at least one pair, } l = 1,2,\ldots L,\ m = 1,2,\ldots L, l \neq m.$$

Step 2: Define test statistic. The test statistic is $Q = 2.3026\dfrac{q}{C}$ (7.6)

Where

$$q = (N-L)\log_{10} s_p^2 - \sum_{\ell=1}^{L}(n_\ell - 1)\log_{10} s_\ell^2$$

$$C = 1 + \frac{1}{3(L-1)}\left[\sum_{\ell=1}^{L}(n_\ell - 1)^{-1} - (N-L)^{-1}\right]$$

$$N = \sum_{\ell=1}^{L} n_\ell,\ s_\ell^2 = \ell - th \text{ sample variance.}$$

s_p^2 = Pooled variance

$$= \frac{(n_1-1)s_1^2+(n_2-1)s_2^2+\cdots+(n_L-1)s_L^2}{n_1+n_2+\cdots+n_L-L}$$

$$= \frac{(n_1-1)s_1^2+(n_2-1)s_2^2+\cdots+(n_L-1)s_L^2}{(N-L)}, \text{ where } N=\sum_{\ell=1}^{L} n_\ell .$$

Step 3: Define sampling distribution

Q follows χ^2 distribution with L-1 degree of freedom, i.e., $Q \sim \chi^2_{L-1}$

Step 4: Obtain cut-off value

For α level of significance, the cut-off value is $\chi^2_{L-1}(\alpha)$.

Step 5: Decision

Reject H_0, if $Q > \chi^2_{L-1}(\alpha)$.

Example 7.1

Consider the data given in Table 7.1. Test whether the equality of the population variances is satisfied or not.

Solution:

There are three processes, A, B, and C, producing steel washers. The response variable of interest is outer diameter (OD). The task at hand is to test whether the three processes have equal variances or not with respect to OD. We use the Bartlett test and the steps are described below.

(i) Hypothesis

H_0 (null hypothesis): $\sigma_1^2 = \sigma_2^2 = \sigma_3^2$
H_1 (alternative hypothesis): $\sigma_l^2 \neq \sigma_m^2$ for at least one pair (l, m), $l = 1,2,3$ and $m = 1,2,3$, $l \neq m$.

(ii) Test statistic

$$Q = 2.3026\frac{q}{C}$$

Now,

$$q = (N-L)\log_{10} s_p^2 - \sum_{l=1}^{L}(n_\ell - 1)\log_{10} s_\ell^2$$
$$= (30-3)\log_{10} s_p^2 - \left[(10-1)\left\{\log_{10} s_1^2 + \log_{10} s_2^2 + \log_{10} s_3^2\right\}\right]$$
$$(\because N = n_1 + n_2 + n_3 = 10+10+10 = 30, L = 3;)$$

s_1^2, s_2^2, s_3^2 & s_p^2 are the sample variances for process A, B, C and pooled variance, respectively which are to be calculated from the data given in Table 7.1. These are

Sample variance for $A: s_1^2 = 1.51$
Sample variance for $B: s_2^2 = 1.43$
Sample variance for $C: s_3^2 = 0.93$

$$s_p^2 = \text{Pooled variance}$$
$$= \sum_{\ell-1}^{L}(n_\ell - 1)\, s_l^2 / (N-L)$$
$$= \frac{(n_1-1)s_1^2 + (n_2-1)s_2^2 + (n_3-1)s_3^2}{(N-L)}$$
$$= \frac{(10-1)1.51 + (10-1)1.43 + (10-1)0.93}{(30-3)}$$
$$= 1.29$$

Now,

$$q = 27\log_{10}(1.29) - \left[9\left\{\log_{10}1.51 + \log_{10}1.43 + \log_{10}0.93\right\}\right]$$
$$= 2.99–2.73$$
$$= 0.26$$

Again,

$$C = 1 + \frac{1}{3(L-1)}\left(\sum_{\ell=1}^{L}(n_\ell-1)^{-1} - (N-L)^{-1}\right)$$
$$= 1 + \frac{1}{3(3-1)}\left[\left\{(n_1-1)^{-1} + (n_2-1)^{-1} + (n_3-1)^{-1}\right\} - (30-3)^{-1}\right] \quad [\because N = 30,\ L = 3]$$
$$= 1 + 0.17\left[\left\{(10-1)^{-1} + (10-1)^{-1} + (10-1)^{-1}\right\} - 0.04\right]$$
$$= 1.05$$

Therefore, $Q = 2.3026\dfrac{q}{C}$

$$= 2.3026 \times \frac{0.26}{1.05} \quad (\because q = 0.26, C = 1.05).$$
$$= 0.57$$

(iii) Sampling distribution: Q follows $\chi^2_{(L-1)} = \chi^2_2(\alpha)$ as L = 3.
(iv) Cut-off value: $\chi^2_2(0.05) = 5.99$ considering $\alpha = 0.05$.
(v) Decision: $Q = 0.57 << \chi^2_2(0.05) = 5.99$. So, we fail to reject the null hypothesis. Therefore, there are no differences among the variances of the three processes A, B, and C.

7.1.3 Total Sum Squares Decomposition

For a general case, suppose that a sample of size n is collected from each of the L populations. As stated earlier, x_{il} is a data point representing the i-th observation on the l-th population. x_{il} can be written as

$$x_{il} - \bar{x} = (\bar{x}_l - \bar{x}) + (x_{il} - \bar{x}_l) \tag{7.7}$$

where $\bar{x}$ = grand mean of the sample data, and $\bar{x}_l$ = mean of the l-th sample.

Equation 7.7 is known as partitioning of an observation into different deviation values. The left hand side of Equation 7.7 represents mean (grand) subtracted i-th observation for the l-th sample. The first term of the right hand side of Equation 7.7 shows the deviation of the l-th sample mean from the grand mean and the second term represents the deviation of the i-th observation of the l-th sample from the l-th sample mean.

The partitioning of an observation in ANOVA is shown in Table 7.2. The partitioning of an observation will enable us to derive the sum squares total (SST) of variances contained in the data set. In order to do so, we first square equation 7.7 and then take sum over $i = 1, 2, \ldots, n$ and then over $l = 1, 2, \ldots, L$.

Squaring Equation 7.7, we get

$$\begin{aligned}(x_{i\ell}-\bar{x})^2 &= \{(\bar{x}_\ell-\bar{x})+(x_{i\ell}-\bar{x}_\ell)\}^2\\ &= (\bar{x}_\ell-\bar{x})^2+(x_{i\ell}-\bar{x}_\ell)^2+2(\bar{x}_\ell-\bar{x})(x_{i\ell}-\bar{x}_\ell)\end{aligned}$$

Taking sum over $i = 1, 2, \ldots, n$, we get

$$\begin{aligned}\sum_{i=1}^{n}(x_{i\ell}-\bar{x})^2 &= \sum_{i=1}^{n}(x_\ell-\bar{x})^2+\sum_{i=1}^{n}(x_{i\ell}-\bar{x}_\ell)^2+2\sum_{i=1}^{n}(\bar{x}_\ell-\bar{x})(x_{i\ell}-\bar{x}_\ell)\\ &= n(x_\ell-\bar{x})^2+\sum_{i=1}^{n}(x_{i\ell}-\bar{x}_\ell)^2+2(\bar{x}_\ell-\bar{x})\sum_{i=1}^{n}(x_{i\ell}-\bar{x}_\ell)\\ &= n(x_\ell-\bar{x})^2+\sum_{i=1}^{n}(x_{i\ell}-\bar{x}_\ell)^2+2(\bar{x}_\ell-\bar{x})\left(\sum_{i=1}^{n}x_{i\ell}-\sum_{i=1}^{n}\bar{x}_\ell\right)\\ &= n(x_\ell-\bar{x})^2+\sum_{i=1}^{n}(x_{i\ell}-\bar{x}_\ell)^2+2(\bar{x}_\ell-\bar{x})(n\bar{x}_\ell-n\bar{x}_\ell)\\ &= n(x_\ell-\bar{x})^2+\sum_{i=1}^{n}(x_{i\ell}-\bar{x}_\ell)^2\end{aligned}$$

Again taking sum over $l = 1, 2, \ldots, L$, we get

$$\sum_{\ell=1}^{L}\sum_{i=1}^{n}(x_{i\ell}-\bar{x})^2 = n\sum_{\ell=1}^{L}(x_\ell-\bar{x})^2+\sum_{\ell=1}^{L}\sum_{i=1}^{n}(x_{i\ell}-\bar{x}_\ell)^2 \tag{7.8}$$

TABLE 7.2
Partitioning of Observations

Population	$i = 1,2,..n$	$\bar{x}_l = \sum_{i=1}^{n}\frac{x_{il}}{n}$	Partitioning of Observations (x_{il})
1	$x_{11}, x_{21}, \ldots x_{i1}, \ldots x_{n1}$	$\bar{x}_1 = \frac{1}{n}\sum_{i=1}^{n}x_{i1}$	$x_{i\ell} = \bar{x}+(\bar{x}_\ell-\bar{x})+(x_{i\ell}-\bar{x}_\ell)$ $i = 1,2,\ldots,n$ and $l = 1,2,\ldots,L$
2	$x_{12}, x_{22}, \ldots x_{i2}, \ldots x_{n2}$	$\bar{x}_2 = \frac{1}{n}\sum_{i=1}^{n}x_{i2}$	
⋮		⋮	
ℓ	$x_{1\ell}, x_{2\ell}, \ldots x_{i\ell}, \ldots x_{n\ell}$	$\bar{x}_\ell = \frac{1}{n}\sum_{i=1}^{n}x_{i\ell}$	
⋮		⋮	
L	$x_{1L}, x_{2L}, \ldots x_{iL}, \ldots x_{nL}$	$\bar{x}_L = \frac{1}{n}\sum_{i=1}^{n}x_{iL}$	
	Grand mean	$\bar{x} = \frac{1}{nL}\sum_{\ell=1}^{L}\sum_{i=1}^{n}x_{i\ell}$	

The three important statistics that can be derived using Equation 7.8 are as follows:

$$\text{Sum squares total (SST)} = \sum_{\ell=1}^{L}\sum_{i=1}^{n}\left(x_{i\ell} - \bar{x}\right)^2 \tag{7.9}$$

$$\text{Sum squares between (SSB) populations} = n\sum_{\ell=1}^{L}\left(\bar{x}_{\ell} - \bar{x}\right)^2 \tag{7.10}$$

$$\text{Sum squares errors (SSE)} = \sum_{\ell=1}^{L}\sum_{i=1}^{n}\left(x_{i\ell} - \bar{x}_{\ell}\right)^2 \tag{7.11}$$

SSE is also known as sum squares within (SSW) populations. Equation 7.8 also represents that the total variation (SST) of the sample data is portioned into two components as SSB and SSE.

The computation of SSB, SSE, and SST following Equations 7.9–7.11 is tedious. A comparatively easier scheme is given below (Montgomery, 2012).

Let the total number of observations be $N = nL$. Further, let $G = \sum_{\ell=1}^{L}\sum_{i=1}^{n} x_{i\ell}$, and $A_{\ell} = \sum_{\ell=1}^{n} x_{i\ell}$. Then,

$$SST = \sum_{\ell=1}^{L}\sum_{i=1}^{n} x_{i\ell}^2 - \frac{G^2}{N} \tag{7.12}$$

$$SSB = \sum_{\ell=1}^{L}\frac{A_l^2}{n} - \frac{G^2}{N} \tag{7.13}$$

$$SSE = SST - SSB \tag{7.14}$$

The above equations hold when the sample size for each of the L populations is same (n). If the sample size varies from population to population, computation of SST and SSB can be done using Equations 7.15 and 7.16 given below.

$$SST = \sum_{\ell=1}^{L}\sum_{i=1}^{n_\ell} x_{i\ell}^2 - \frac{G^2}{N} \tag{7.15}$$

$$SSB = \sum_{\ell=1}^{L}\frac{A_\ell^2}{n_\ell} - \frac{G^2}{N} \tag{7.16}$$

Where $A_l = \sum_{\ell=1}^{n_\ell} x_{i\ell}, G = \sum_{\ell=1}^{L} A_\ell$ and $N = \sum_{\ell=1}^{L} n_\ell$. The equation of SSE will be same as Equation 7.14.

Example 7.2

Compute SST, SSB and SSE for the data set given in Table 7.1.

Solution: The data with preliminary computations are given in Table 7.3.

So, $G = \sum_{\ell=1}^{3}\sum_{i=1}^{10} x_{i\ell} = 587$ and $\sum_{\ell=1}^{3}\sum_{i=1}^{10} x_{i\ell}^2 = 11563$.

Now, $SST = \sum_{\ell=1}^{3}\sum_{i=1}^{10} x_{i\ell}^2 - \frac{G^2}{N} = 11563 - (587)^2/30 = 11563 - 11485.63 = 77.37$.

TABLE 7.3
Data With Preliminary Computations for Example 7.2

	Observations												
Process	**1**	**2**	**3**	**4**	**5**	**6**	**7**	**8**	**9**	**10**	$A_l = \sum_{l=1}^{n} x_{il}$	$\sum_{l=1}^{n} x_{il}^2$	$A_l^2 = \left(\sum_{l=1}^{n} x_{il}\right)^2$
A	20	21	20	21	23	19	20	19	19	20	202	4094	40804
B	17	17	19	17	16	19	18	18	18	20	179	3217	32041
C	20	20	21	20	21	21	22	19	22	20	206	4252	42436
					Grand total						587	11563	

TABLE 7.4
Resultant Table

Sources of Variation	Sum Squares (SS)
Process	SSB = 42.47
Error	SSE = 34.90
Total	SST = 77.37

Again, $SSB = \sum_{\ell=1}^{3} \frac{A_\ell^2}{n_\ell} - \frac{G^2}{N} = \left(\frac{40804}{10} + \frac{32041}{10} + \frac{42436}{10}\right) - \frac{587^2}{30} = 11528.10 - 11485.63 = 42.47.$

Finally, SSE = SST – SSB = 77.37 – 42.47 = 34.90.

The resultant table is shown in Table 7.4.

7.1.4 Hypothesis Testing

ANOVA tests whether SSB is sufficiently larger than SSE when each of them is weighted (divided) by their respective degrees of freedom. Table 7.5 explains the development. By computing Table 7.5, we test the following hypothesis as described before.

$$H_0 : \mu_1 = \mu_2 = ... = \mu_L$$
$$H_1 : \mu_\ell \neq \mu_m, \text{ for at least one pair of } (\ell, m),\ l = 1,2,...,L \text{ and } m = 1,,2,...,L.$$

Alternatively,

$$H_0 : \tau_l = 0,\ l = 1,2,...,L$$
$$H_1 : \tau_l \neq 0, \text{ for at least one population.}$$

It is customary to present ANOVA results in a table (Table 7.5) and the table contains a lot of useful information. The first column defines the sources of variation which comprises population and error. As we are interested to test the equality of several population (treatment) means, every population (treatment) is a source of variation. For example, for the production of steel washers, the three processes, A, B, and C, represent three populations. In ANOVA, populations are also called treatments (usually in medical treatment), groups (such as age groups), or levels (usually in laboratory-based experiments like pressure at three levels, low, medium, and high). Now, the capability may differ from process to process (population to population); hence, the processes are the source of variability. The error variability is induced by sources within a process beyond its control and is random in nature.

TABLE 7.5
ANOVA Table

Sources of Variation	Sums Square (SS)	Degrees of Freedom*	Mean Ssquare (MS)	F
Population (treatment)	SSB	L-1	$MSB = \frac{SSB}{L-1}$	$F = \frac{MSB}{MSE}$
Error (random component)	SSE	N-L	$MSE = \frac{SSE}{N-L}$	
Total	SST	N-1		

***N = nL**

These two sources (i.e., population and error) collectively provide the information related to the total variability (third row of first column). The second column provides the information on the sum squares (SS), a measure of variability, for the two sources and the total, represented by SSB, SSE and SST, respectively. We have seen earlier that SST=SSB+SSE. Pictorially,

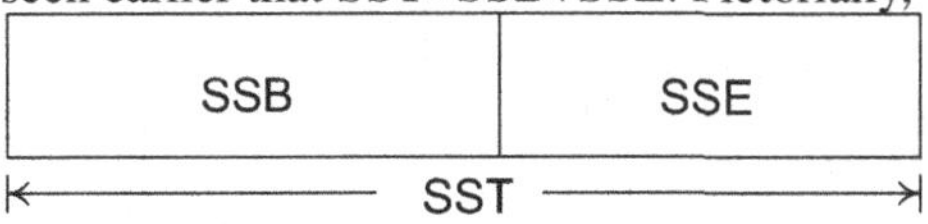

SSB is an estimate of variability due to the differences among population means, whereas SSE is the estimate of variability of the differences in observations within a given population. SSE is a result of randomness inherent to a population. This randomness can be attributed, for example, for the three processes considered, to uncontrollable environmental and operational factors affecting the response variable, OD.

The third column of Table 7.5 defines degrees of freedom (DOF) available while computing SST, SSB, and SSE (shown in column 2). For computation of SST, we have N (=nL) observations but $\bar{x}$ needs to be computed first (see Equation 7.9). So, the DOF available to compute SST is N-1. Next, we compute SSB using L population means, i.e., $\bar{x}_1$, $\bar{x}_2$,...$\bar{x}_\ell$,...$\bar{x}_L$. Further one DOF is lost to compute $\bar{x}$ (see Equation 7.10). So, SSB has L-1 DOF. Finally, SSE defines the sum square errors within population. As every population has n observations and we require computing $\bar{x}_\ell$ before SSE, there are n-1 DOF available per population. So, for L populations, SSE has L(n–1) = nL–L = N–L DOF. If we add the DOFs of SSB and SSE, [i.e., (L–1) + (N–L) = N–1], we get the DOF of SST. Therefore, like SST, the DOF of SST is also decomposed into DOF for SSB and DOF for SSE. *So, the important concept of ANOVA is that total variability is the sum total of individual source variabilities and total DOF is the sum total of individual source DOFs.*

The fourth column of Table 7.5 shows two very important statistics used in ANOVA namely mean square between populations (MSB) and mean square errors (MSE) within population. The computation of MSB and MSE is explained below.

From Equation 7.10, we see that

$$SSB = n\sum_{l=1}^{L}\left(\bar{x}_l - \bar{x}\right)^2.$$

Dividing SSB by its DOF we get MSB.

$$MSB = \frac{SSB}{L-1} = \frac{n\sum_{l=1}^{L}\left(\bar{x}_l - \bar{x}\right)^2}{L-1} \tag{7.17}$$

What does MSB represent? Let H_0 is true, i.e., $\mu_1 = \mu_2 = \ldots = \mu_L = \mu$. Further assume that $\bar{x}$ is an estimate of μ. So, $E(\bar{x}_l) = \mu_l = \mu$ when H_0 is true. Therefore,

$$\frac{1}{L-1}\sum_{l=1}^{L}\left(\bar{x}_l - \bar{x}\right)^2 = \frac{1}{L-1}\sum_{l=1}^{L}\left(\bar{x}_l - \mu\right)^2 = \hat{\sigma}_l^2 = \frac{\hat{\sigma}^2}{n}$$

Then, $\text{MSB} = n\hat{\sigma}_l^2 = \hat{\sigma}^2$ when H_0 is true. That is MSB is an estimate of σ^2 when there is no differences in population means.

Similarly, MSE is obtained by dividing SSE by its DOF.

$$MSE = \frac{SSE}{N-L} = \frac{\sum_{l=1}^{L}\sum_{i=1}^{n}\left(x_{il} - \bar{x}_l\right)^2}{N-L} \tag{7.18}$$

What does MSE represent? Expanding Equation 7.18, we get

$$\begin{aligned}
MSE &= \frac{1}{N-L}\left[\sum_{l=1}^{L}\sum_{i=1}^{n}\left(x_{il} - \bar{x}_l\right)^2\right] = \frac{1}{N-L}\sum_{l=1}^{L}\left[\sum_{i=1}^{n}\left(x_{il} - \bar{x}_l\right)^2\right] \\
&= \frac{1}{N-L}\sum_{l=1}^{L}\left[(n-1)\times\frac{1}{n-1}\sum_{i=1}^{n}\left(x_{il} - \bar{x}_l\right)^2\right] \\
&= \frac{1}{N-L}\sum_{l=1}^{L}\left[(n-1)\times s_l^2\right], \text{ as } s_l^2 = \frac{1}{n-1}\sum_{i=1}^{n}\left(x_{il} - \bar{x}_l\right)^2 \\
&= \frac{(n-1)s_1^2 + (n-1)s_2^2 + \cdots + (n-1)s_L^2}{nL - L}, \quad as\ N = nL \\
&= \frac{(n-1)s_1^2 + (n-1)s_2^2 + \cdots + (n-1)s_L^2}{(n-1) + (n-1) + \cdots + (n-1)} \\
&= s^2_{\text{pooled}}
\end{aligned}$$

So, MSE is the pooled estimate of the common variance of each of the L populations. Assuming homogeneity of population variances, MSE is an estimate of the common variance σ^2.

Now, we have two estimates of the common variance σ^2. They are MSB when H_0 is true and MSE. So, from an unbiased estimation point of view, the expected values of MSB and MSE must be equal to σ^2. It can be proven that the expected values of MSB and MSE are (Montgomery, 2012)

$$E(MSE) = \sigma^2 \quad \text{and}$$

$$E(MSB) = \sigma^2 + \frac{n\sum_{l=1}^{L}\tau_l^2}{L-1}$$

Now, if there are no differences in population means, τ_ℓ becomes zero. Accordingly, E (MSB) becomes σ^2. Therefore, if H_0 is true, i.e., there are no differences among the means of the L populations, MSB will be more or less equal to MSE (random error). Alternatively, if H_0 is false, MSB should be sufficiently larger than MSE.

The final (fifth) column of Table 7.5 represents a statistic called F_0 which is the ratio of MSB to MSE. So,

$$F_0 = \frac{MSB}{MSE} = \frac{SSB/L-1}{SSE/N-L} \tag{7.19}$$

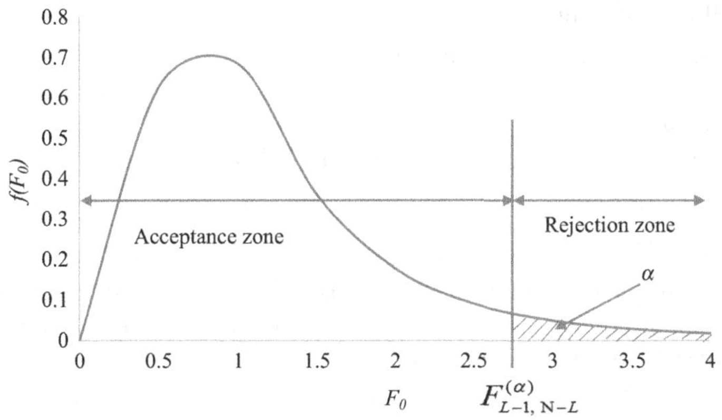

FIGURE 7.4 Decision-making using F-test. The tabulated (theoretical) F value ($F^{(\alpha)}_{L-1,N-L}$) defines the acceptance and rejection zones for H_0. If computed F value (F_0) is greater than $F^{(\alpha)}_{L-1,N-L}$, H_0 will be rejected. The hatched area represents the value of α (the probability level of significance).

TABLE 7.6
ANOVA Table for Example 7.3

Sources of Variation	Sum Squares (SS)	Degrees of Freedom (DOF)	Mean Squares (MS)	F_0-Value
Process	SSB = 42.47	L-1 = 3-1=2	MSB = SSB/L-1 = 42.47/2 = 21.235	F = MSB/MSE = 21.235/ 1.2926 = 16.43
Error	SSE = 34.90	N-L = 29-2=27	MSE = SSE/N-L = 34.90/27 = 1.2926	
Total	SST = 77.37	N-1 = 30-1=29		

If H_0 is true, F_0 follows F distribution with L–1 numerator and N–L denominator degree of freedom. In order to test H_0, F_0 is compared with $F^{(\alpha)}_{L-1,N-L}$ where α is the level of significance. If F_0 is greater than $F^{(\alpha)}_{L-1,N-L}$, we reject H_0 (see Figure 7.4).

Note that in the discussion above, we have considered equal sample size across population (balanced ANOVA). If sample size varies, necessary modifications in the equations are to be made but all the concepts and interpretations remain the same.

Example 7.3

Conduct ANOVA for the data set given in Table 7.1.

Solution:

The ANOVA table is given in Table 7.6.

Using ANOVA table given above, the steps of hypothesis testing are

(i) Hypothesis

$H_0 : \mu_1 = \mu_2 = \mu_3$

$H_1 : \mu_l \neq \mu_m$ for at least one of the three pairs, $l = 1, 2, 3$ and $m = 1, 2, 3$.

(ii) Test statistics

$$F_0 = \frac{MSB}{MSE} = \frac{SSB / L-1}{SSE / N-L} = \frac{42.47 / 2}{34.90 / 27} = 16.43.$$

(iii) Sampling distribution

F_0 follows $F_{2,27}$ as L–1 = 2 and N–L = 27.

(iv) Cut-off value

At α = 0.05, the cut-off value $F_{(2,27)}(0.05) = 3.3541$.

(v) Decision

Since the computed $F_0 = 16.43 >$ tabulated $F_{(2,27)}(0.05) = 3.3541$, we can reject the null hypothesis (H_0). So, we can say that there are differences in the means of the three processes A, B, and C with respect to the outer diameter (OD).

7.1.5 Estimation of Parameters

Consider the ANOVA model given in Equation 7.4. The parameters of ANOVA are

μ = Grand mean
μ_ℓ = Mean of population ℓ, ℓ =1, 2, …., L
τ_ℓ = Effect of population ℓ, ℓ =1, 2, …., L
$\varepsilon_{i\ell}$ = Random error, ℓ = 1, 2, …., L and i = 1, 2, …, n_l

The point estimates of the above parameters are

$$\hat{\mu} = \overline{x}$$
$$\hat{\mu}_l = \overline{x}_l$$
$$\hat{\tau}_l = \overline{x}_l - \overline{x}$$
$$\hat{\epsilon}_{il} = x_{il} - \overline{x}_l$$

Example 7.4

Consider Example 7.2. Obtain point estimates of grand mean, population means, population effects, and errors (residuals).

Solution:

The point estimates of grand mean, population means, and population effects are given in Table 7.7.

TABLE 7.7
Point Estimates of Grand Mean, Population Means, and Population Effects for Example 7.4

	Observations											Population Means	Population Effects
Process	**1**	**2**	**3**	**4**	**5**	**6**	**7**	**8**	**9**	**10**	$A_l = \sum_{l=1}^{n} x_{il}$	$\hat{\mu}_l = \overline{x}_l = A_l / 10$	$\hat{\tau}_l = \overline{x}_l - \overline{x}$
A	20	21	20	21	23	19	20	19	19	20	202	20.20	0.63
B	17	17	19	17	16	19	18	18	18	20	179	17.90	−1.67
C	20	20	21	20	21	21	22	19	22	20	206	20.60	1.04
	Grand total										587		
	Grand mean ($\hat{\mu} = \overline{x}$)											19.57	

TABLE 7.8
The Residuals for Example 7.4

	Residuals									
Process	**1**	**2**	**3**	**4**	**5**	**6**	**7**	**8**	**9**	**10**
A	−0.2	0.8	−0.2	0.8	2.8	−1.2	−0.2	−1.2	−1.2	−0.2
B	−0.9	−0.9	1.1	−0.9	−1.9	1.1	0.1	0.1	0.1	2.1
C	−0.6	−0.6	0.4	−0.6	0.4	0.4	1.4	−1.6	1.4	−0.6

Using $\hat{\epsilon}_{il} = x_{il} - \bar{x}_l$, we can compute the error estimates (residuals). The residuals are given in Table 7.8.

100(1–α)% Confidence Interval (CI) and Simultaneous Confidence Interval (SCI) for μ_l

To compute 100(1–α)% confidence interval (CI) for μ_l, the random variable of interest is $\bar{x}_l$, which is normally distributed with mean μ_l and variance $\frac{\sigma^2}{n_l}$, i.e., $\bar{x}_l \sim N\left(\mu_l, \frac{\sigma^2}{n_l}\right)$. If σ^2 is known in advance, we can use z distribution to find out $100(1-\alpha)\%$ CI for μ_l. However, σ^2 is seldom known. We use the estimate of σ^2. We have seen in Section 7.1.4 that MSE is the estimator of σ^2. For small *n*, we use t-distribution to compute 100(1–α)% CI for μ_l as

$$\bar{x}_l - t_{N-L}^{(\alpha/2)}\sqrt{\frac{MSE}{n_l}} \le \mu_l \le \bar{x}_l + t_{N-L}^{(\alpha/2)}\sqrt{\frac{MSE}{n_l}} \tag{7.20}$$

Where l ranges from 1 to L and $N = \sum_{l=1}^{L} n_l$. For balanced ANOVA, $n_l = n$ and $N = nL$.

The above equation will not simultaneously satisfy 100(1 – α)% CI for all the L populations. Therefore, it is recommended to use Bonferroni simultaneous CI. For the Bonferroni approach, see Chapter 6. Using the Bonferroni approach, the simultaneous CI (SCI) for $\mu_l, l = 1,2,...L$ will be

$$\bar{x}_l - t_{N-L}^{(\alpha_l/2)}\sqrt{\frac{MSE}{n_l}} \le \mu_l \le \bar{x}_l + t_{N-L}^{(\alpha_l/2)}\sqrt{\frac{MSE}{n_l}} \tag{7.21}$$

Where $\sum_{l=1}^{L} \alpha_l = \alpha$ and $N = \sum_{l=1}^{L} n_l$. For balanced ANOVA, $n_l = n$ and $N = nL$.

For large sample, we can use z-distribution irrespective of σ is known or not.

Example 7.5

Consider Example 7.4. Obtain 100(1 – α)% CI and 100(1 – α)% SCI for the population means. What difference do you notice in 100(1 – α)% CI and 100(1 – α)% SCI?

Solution:

There are three population means. So, there will be three 100(1 – α)% CIs and three 100(1 – α)% SCIs. The three 100(1 – α)% CIs can be computed using Equation 7.20 and other information obtained in Example 7.4. Similarly, the three 100(1 – α)% SCIs can be computed using Equation 7.21 and other information obtained in Example 7.4.

(a) 100(1–α)% CI for mean using Equation 7.20

(i) Process A (μ_1)

$$\bar{x}_1 - t_{30-3}^{(0.05/2)}\sqrt{\frac{MSE}{n_1}} \le \mu_1 \le \bar{x}_1 + t_{30-3}^{(0.05/2)}\sqrt{\frac{MSE}{n_1}}$$

$$\text{or, } 20.20 - 2.052\sqrt{\frac{1.2926}{10}} \le \mu_1 \le 20.20 + 2.052\sqrt{\frac{1.2926}{10}}$$

$$\text{or, } 20.20 - 0.738 \le \mu_1 \le 20.20 + 0.738$$

$$\text{or, } 19.46 \le \mu_1 \le 20.94$$

(ii) Process B (μ_2)

$$\bar{x}_2 - t_{30-3}^{(0.05/2)}\sqrt{\frac{MSE}{n_2}} \le \mu_2 \le \bar{x}_2 + t_{30-3}^{(0.05/2)}\sqrt{\frac{MSE}{n_2}}$$

$$\text{or, } 17.90 - 2.052\sqrt{\frac{1.2926}{10}} \le \mu_2 \le 17.90 + 2.052\sqrt{\frac{1.2926}{10}}$$

$$\text{or, } 17.90 - 0.738 \le \mu_2 \le 17.90 + 0.738$$

$$\text{or, } 17.16 \le \mu_2 \le 18.64$$

(iii) Process C (μ_3)

$$\bar{x}_3 - t_{30-3}^{(0.05/2)}\sqrt{\frac{MSE}{n_3}} \le \mu_3 \le \bar{x}_3 + t_{30-3}^{(0.05/2)}\sqrt{\frac{MSE}{n_3}}$$

$$\text{or, } 20.60 - 2.052\sqrt{\frac{1.2926}{10}} \le \mu_3 \le 20.60 + 2.052\sqrt{\frac{1.2926}{10}}$$

$$\text{or, } 20.60 - 0.738 \le \mu_3 \le 20.60 + 0.738$$

$$\text{or, } 19.86 \le \mu_3 \le 21.34$$

(b) 100(1–α)% SCI for mean using Equation 7.21

We have 3 means. Considering $\sum_{\ell=1}^{3}\alpha_\ell = \alpha = 0.05$ and $\alpha_1 = \alpha_2 = \alpha_3$, we can write $\alpha_1 = \alpha_2 = \alpha_3 = 0.05/3 = 0.0167$.

(i) Process A (μ_1)

$$\bar{x}_1 - t_{30-3}^{(0.0167/2)}\sqrt{\frac{MSE}{n_1}} \le \mu_1 \le \bar{x}_1 + t_{30-3}^{(0.0167/2)}\sqrt{\frac{MSE}{n_1}}$$

$$\text{or, } 20.20 - 2.552\sqrt{\frac{1.2926}{10}} \le \mu_1 \le 20.20 + 2.552\sqrt{\frac{1.2926}{10}}$$

$$\text{or, } 20.20 - 0.918 \le \mu_1 \le 20.20 + 0.918$$

$$\text{or, } 19.28 \le \mu_1 \le 21.12$$

(ii) Process B (μ_2)

$$\bar{x}_2 - t_{30-3}^{(0.0167/2)}\sqrt{\frac{MSE}{n_2}} \le \mu_2 \le \bar{x}_2 + t_{30-3}^{(0.0167/2)}\sqrt{\frac{MSE}{n_2}}$$

$$\text{or, } 17.90 - 2.552\sqrt{\frac{1.2926}{10}} \le \mu_2 \le 17.90 + 2.552\sqrt{\frac{1.2926}{10}}$$

$$\text{or, } 17.90 - 0.918 \le \mu_2 \le 17.90 + 0.918$$

$$\text{or, } 16.98 \le \mu_2 \le 18.82$$

(iii) Process C (μ_3)

$$\bar{x}_3 - t_{30-3}^{(0.0167/2)}\sqrt{\frac{MSE}{n_3}} \le \mu_3 \le \bar{x}_3 + t_{30-3}^{(0.0167/2)}\sqrt{\frac{MSE}{n_3}}$$

$$\text{or, } 20.60 - 2.552\sqrt{\frac{1.2926}{10}} \le \mu_3 \le 20.60 + 2.552\sqrt{\frac{1.2926}{10}}$$

$$\text{or, } 20.60 - 0.918 \le \mu_3 \le 20.60 + 0.918$$

$$\text{or, } 19.68 \le \mu_2 \le 21.52$$

From (a) and (b) above, it is seen that the 100(1–α)% SCI for mean becomes wider than the 100(1–α)% CI for mean.

100(1–α)% CI and Simultaneous Confidence Interval (SCI) for $\tau_l - \tau_m$

We know

$$\hat{\tau}_l - \hat{\tau}_m = (\bar{x}_l - \bar{x}) - (\bar{x}_m - \bar{x}) = \bar{x}_l - \bar{x}_m$$

Here, our random variable is $\bar{x}_l - \bar{x}_m$. So, it is seen that finding out 100(1 – α)% CI for $\tau_l - \tau_m$ is equivalent to finding out 100(1 – α)% CI for $\mu_l - \mu_m$.

Now,

$$E(\bar{x}_l - \bar{x}_m) = \mu_l - \mu_m$$

$$V(\bar{x}_l - \bar{x}_m) = V(\bar{x}_l) + V(\bar{x}_m)$$

$$= \frac{\sigma_l^2}{n_l} + \frac{\sigma_m^2}{n_m}, \quad \text{as } \bar{x}_l \text{ and } \bar{x}_m \text{ are independent.}$$

Again, we assume that $\sigma_l^2 = \sigma_m^2 = \sigma^2$. The estimate of σ^2 in ANOVA is MSE. Therefore, the 100(1 – α)% confidence interval for $\tau_l - \tau_m$ or $\mu_l - \mu_m$ is

$$(\bar{x}_l - \bar{x}_m) - t_{N-L}^{(\alpha/2)}\sqrt{MSE\left(\frac{1}{n_l} + \frac{1}{n_m}\right)} \le \mu_l - \mu_m \le (\bar{x}_l - \bar{x}_m) + t_{N-L}^{(\alpha/2)}\sqrt{MSE\left(\frac{1}{n_l} + \frac{1}{n_m}\right)} \quad (7.22)$$

For balanced ANOVA $n_l = n_m = n$. So, Equation 7.22 yields to

$$(\bar{x}_l - \bar{x}_m) - t_{N-L}^{(\alpha/2)}\sqrt{\frac{2MSE}{n}} \le \mu_l - \mu_m \le (\bar{x}_l - \bar{x}_m) + t_{N-L}^{(\alpha/2)}\sqrt{\frac{2MSE}{n}} \quad (7.23)$$

Equations 7.22 and 7.23 do not simultaneously satisfy 100(1 – α)% CI. We use the Bonferroni approach to compute 100(1 – α)% SCI. When there are L populations, we can have ${}^{L}C_2 = \frac{L(L-1)}{2}$ comparisons. Let $g = {}^{L}C_2 = \frac{L(L-1)}{2}$.

So, using the Bonferroni approach with equal probability levels of significance for each of the g comparisons, the 100(1–α)% of SCI is

$$(\bar{x}_l - \bar{x}_m) + t_{N-L}^{(\alpha/2g)}\sqrt{MSE\left(\frac{1}{n_l} + \frac{1}{n_m}\right)} \le \mu_l - \mu_m \le (\bar{x}_l - \bar{x}_m) + t_{N-L}^{(\alpha/2g)}\sqrt{MSE\left(\frac{1}{n_l} + \frac{1}{n_m}\right)} \quad (7.24)$$

For balanced ANOVA, Equation 7.24 becomes

$$\left(\bar{x}_l - \bar{x}_m\right) - t_{N-L}^{(\alpha/2g)}\sqrt{\frac{2MSE}{n}} \le \mu_l - \mu_m \le \left(\bar{x}_l - \bar{x}_m\right) + t_{N-L}^{(\alpha/2g)}\sqrt{\frac{2MSE}{n}} \quad (7.25)$$

Example 7.6

Consider Example 7.4, and develop 100(1 – α)% CI and 100(1 – α)% SCI for pair-wise mean differences.

Solution:

We have three processes A, B and C. Thus there are g = L(L–1)/2 = 3(3 – 1)/2 = 3 pairs of comparison. These are A vs B, B vs C, and C vs A. So, there will be three 100(1 – α)% CIs and three 100(1 – α)% SCIs. The three 100(1 – α)% CIs can be computed using Equation 7.23, as the problem is a balanced ANOVA problem, and other information obtained in Example 7.4. Similarly, three 100(1 – α)% SCIs can be computed using Equation 7.25 (balanced ANOVA) and other information obtained in Example 7.4.

(a) 100(1 – α)% CI for mean difference using Equation 7.23

(i) Process A vs B ($\mu_1 - \mu_2$)

$$\left(\bar{x}_1 - \bar{x}_2\right) - t_{N-L}^{(\alpha/2)}\sqrt{\frac{2MSE}{n}} \le \mu_1 - \mu_2 \le \left(\bar{x}_1 - \bar{x}_2\right) + t_{N-L}^{(\alpha/2)}\sqrt{\frac{2MSE}{n}}$$

$$\text{or, } (20.20 - 17.90) - t_{30-3}^{(0.05/2)}\sqrt{\frac{2\times 1.2926}{10}} \le \mu_1 - \mu_2 \le (20.20 - 17.90) + t_{30-3}^{(0.05/2)}\sqrt{\frac{2\times 1.2926}{10}}$$

$$\text{or, } 2.30 - 2.052\times 0.5084 \le \mu_1 - \mu_2 \le 2.30 + 2.052\times 0.5084$$

$$\text{or, } 1.257 \le \mu_1 - \mu_2 \le 3.343$$

(ii) Process B vs C ($\mu_2 - \mu_3$)

$$\left(\bar{x}_2 - \bar{x}_3\right) - t_{N-L}^{(\alpha/2)}\sqrt{\frac{2MSE}{n}} \le \mu_2 - \mu_3 \le \left(\bar{x}_2 - \bar{x}_3\right) + t_{N-L}^{(\alpha/2)}\sqrt{\frac{2MSE}{n}}$$

$$\text{or, } (17.90 - 20.60) - t_{30-3}^{(0.05/2)}\sqrt{\frac{2\times 1.2926}{10}} \le \mu_2 - \mu_3 \le (17.90 - 20.60) + t_{30-3}^{(0.05/2)}\sqrt{\frac{2\times 1.2926}{10}}$$

$$\text{or, } -2.70 - 2.052\times 0.5084 \le \mu_2 - \mu_3 \le 2.70 + 2.052\times 0.5084$$

$$\text{or, } -3.743 \le \mu_2 - \mu_3 \le -1.657$$

(iii) Process C vs A ($\mu_3 - \mu_1$)

$$\left(\bar{x}_3 - \bar{x}_1\right) - t_{N-L}^{(\alpha/2)}\sqrt{\frac{2MSE}{n}} \le \mu_3 - \mu_1 \le \left(\bar{x}_3 - \bar{x}_1\right) + t_{N-L}^{(\alpha/2)}\sqrt{\frac{2MSE}{n}}$$

$$\text{or, } (20.60 - 20.20) - t_{30-3}^{(0.05/2)}\sqrt{\frac{2\times 1.2926}{10}} \le \mu_3 - \mu_1 \le (20.60 - 20.20) + t_{30-3}^{(0.05/2)}\sqrt{\frac{2\times 1.2926}{10}}$$

$$\text{or, } 0.40 - 2.052\times 0.5084 \le \mu_3 - \mu_1 \le 0.40 + 2.052\times 0.5084$$

$$\text{or, } -0.643 \le \mu_3 - \mu_1 \le 1.443$$

From the 100(1 – α)% CI for pair-wise mean differences it is observed that the means of the processes A and C do not differ but process B's mean is different from that of the processes A and C as only the 100(1 – α)% CI for $\mu_3 - \mu_1$ does contain zero value but other two pairs do not contain zero.

(b) 100(1–α)% SCI for mean difference using Equation 7.25

We have three pairs of comparisons. Considering $\sum_{l=1}^{3} \alpha_l = \alpha = 0.05$ and $\alpha_1 = \alpha_2 = \alpha_3$, we can write $\alpha_1 = \alpha_2 = \alpha_3 = 0.05/3 = 0.0167$.

(i) Process A vs B ($\mu_1 - \mu_2$)

$$(\bar{x}_1 - \bar{x}_2) - t_{N-L}^{(\alpha/2g)}\sqrt{\frac{2MSE}{n}} \le \mu_1 - \mu_2 \le (\bar{x}_1 - \bar{x}_2) + t_{N-L}^{(\alpha/2g)}\sqrt{\frac{2MSE}{n}}$$

$$\text{or, } (20.20 - 17.90) - t_{30-3}^{(0.05/2\times3)}\sqrt{\frac{2\times1.2926}{10}} \le \mu_1 - \mu_2 \le (20.20 - 17.90) + t_{30-3}^{(0.05/2\times3)}\sqrt{\frac{2\times1.2926}{10}}$$

$$\text{or, } 2.30 - 2.552\times0.5084 \le \mu_1 - \mu_2 \le 2.30 + 2.552\times0.5084$$

$$\text{or, } 1.003 \le \mu_1 - \mu_2 \le 3.597$$

(ii) Process B vs C ($\mu_2 - \mu_3$)

$$(\bar{x}_2 - \bar{x}_3) - t_{N-L}^{(\alpha/2g)}\sqrt{\frac{2MSE}{n}} \le \mu_2 - \mu_3 \le (\bar{x}_2 - \bar{x}_3) + t_{N-L}^{(\alpha/2g)}\sqrt{\frac{2MSE}{n}}$$

$$\text{or, } (17.90 - 20.60) - t_{30-3}^{(0.05/2\times3)}\sqrt{\frac{2\times1.2926}{10}} \le \mu_2 - \mu_3 \le (17.90 - 20.60) + t_{30-3}^{(0.05/2\times3)}\sqrt{\frac{2\times1.2926}{10}}$$

$$\text{or, } -2.70 - 2.552\times0.5084 \le \mu_2 - \mu_3 \le 2.70 + 2.552\times0.5084$$

$$\text{or, } -3.997 \le \mu_2 - \mu_3 \le -1.403$$

(iii) Process A vs C ($\mu_3 - \mu_1$)

$$(\bar{x}_3 - \bar{x}_1) - t_{N-L}^{(\alpha/2g)}\sqrt{\frac{2MSE}{n}} \le \mu_3 - \mu_1 \le (\bar{x}_3 - \bar{x}_1) + t_{N-L}^{(\alpha/2g)}\sqrt{\frac{2MSE}{n}}$$

$$\text{or, } (20.60 - 20.20) - t_{30-3}^{(0.05/2\times3)}\sqrt{\frac{2\times1.2926}{10}} \le \mu_3 - \mu_1 \le (20.60 - 20.20) + t_{30-3}^{(0.05/2\times3)}\sqrt{\frac{2\times1.2926}{10}}$$

$$\text{or, } 0.40 - 2.552\times0.5084 \le \mu_3 - \mu_1 \le 0.40 + 2.552\times0.5084$$

$$\text{or, } -0.897 \le \mu_3 - \mu_1 \le 1.697$$

From the 100(1 – α)% SCI for pair-wise mean differences also shows the same trend as is seen in (a) but the difference lies only in wider spread in case of (b) when compared to (a).

7.1.6 Model Adequacy Tests

By model adequacy we mean whether the developed model is able to explain the given situation for the purpose it is made. The quantity, SSB/SST, known as Fisher's correlation ratio, can be used to compute the percentage of variance explained by ANOVA. In addition, to judge the model adequacy we analyze the residuals (errors) and test the assumptions underlying ANOVA. In this section, we describe the tests of assumptions in ANOVA.

Test of Normality

The residuals, $\hat{\epsilon}_{il}$, $i = 1,2,...n$ and $l = 1,2,...L$ should be normally distributed with mean 0 and variance σ^2. The estimate of σ^2 is MSE. There are both plots and absolute measures of checking normality. The most frequently used plots are probability-probability (P-P) plot and quantile-quantile (Q-Q) plot. In P-P plot, cumulative probability is plotted against residuals and in Q-Q plot, normal quantile values (z_{il}) are plotted against residuals ($\hat{\epsilon}_{il}$). For clarity, see the procedures discussed in Chapter 5. If ϵ_{il} is normal, both P-P and Q-Q plot will resemble a straight line. Montgomery (2012) suggested that while reading the straight line, we should emphasize the central values of the plot, rather than the extreme values.

Table 7.9 reproduces the residuals, computed in Example 7.4, based on ANOVA model for the steel washers' data (Table 7.1). The P-P and Q-Q plots are shown in Figures 7.5 and 7.6, respectively. The residual values almost lie along the desired straight line which indicates that the normality assumption is satisfied.

It may so happen that one or more residuals lie far away from the straight line. Such observations may be indicative of outliers. We can use standardized residuals to identify outliers. Let e_{il} be the standardized residual for the i-th observation of the l-th population. Then,

TABLE 7.9
Residuals Based on ANOVA Model for Steel Washers' Data

Process	Residuals									
Process A	–0.2	0.8	–0.2	0.8	2.8	–1.2	–0.2	–1.2	–1.2	–0.2
Process B	–0.9	–0.9	1.1	–0.9	–1.9	1.1	0.1	0.1	0.1	2.1
Process C	–0.6	–0.6	0.4	–0.6	0.4	0.4	1.4	–1.6	1.4	–0.6

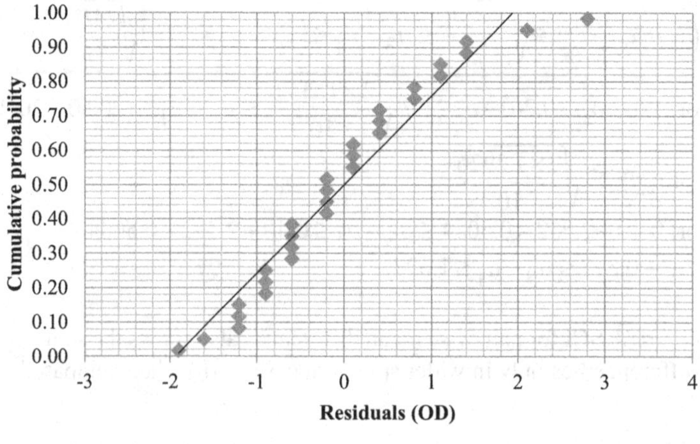

FIGURE 7.5 Residual P-P plot for steel washers' outer diameter (OD) data. The residuals fall along the straight line satisfying the normality of the error terms.

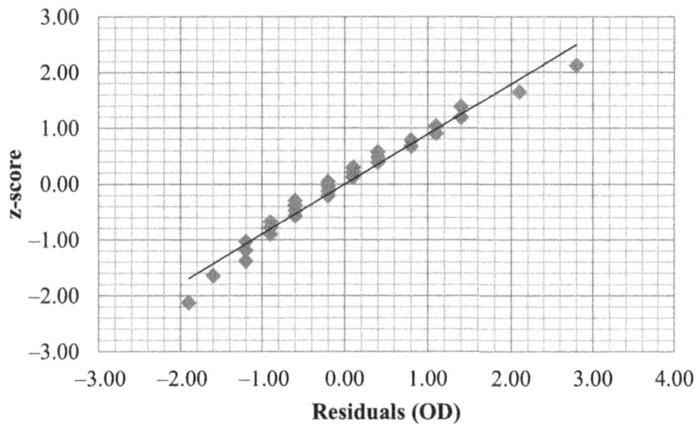

FIGURE 7.6 Residual Q-Q plot for steel washers' outer diameter (OD) data. The residuals fall along the straight line passing through the origin satisfying the normality of the error terms.

TABLE 7.10
Standardized Residuals for Steel Washers' ata

Process	Standardized Residuals									
Process A	−0.18	0.70	−0.18	0.70	2.46	−1.06	−0.18	−1.06	−1.06	−0.18
Process B	−0.79	−0.79	0.97	−0.79	−1.67	0.97	0.09	0.09	0.09	1.85
Process C	−0.53	−0.53	0.35	−0.53	0.35	0.35	1.23	−1.41	1.23	−0.53

$$e_{il} = \frac{\hat{\epsilon}_{il} - E(\hat{\epsilon}_{il})}{\sqrt{\text{var}(\hat{\epsilon}_{il})}} = \frac{\hat{\epsilon}_{il}}{\sqrt{MSE}} \tag{7.26}$$

As $\hat{\epsilon}_{il} \sim N(0, \text{MSE})$, e_{il} follows unit normal distribution, i.e., $e_{il} \sim N(0,1)$. The value of e_{il} can be compared with cut-off value to detect outliers. It is customary to use $\pm 3\sigma (= \pm 3 \times 1 = \pm 3)$ limit as cut-off value. So, e_{il} values falling beyond the range from −3 to +3 are potential outliers. Once outliers are detected, it is advisable to remove those outliers and conduct a fresh ANOVA by repeating the entire process. But there is no guarantee that we will not encounter outliers again. If this happens, the normality assumption cannot be justified. However, the F-ratio is robust against departure from normality if equal sample sizes are considered from each of the populations.

The standardized residuals for the steel washers' data are shown in Table 7.10. From Table 7.10, it is seen that none of e_{il} values fall beyond ±3. So, there are no outliers encountered for the production of steel washers' data set.

Test of Independence

It is a test of whether each of the n data points is collected independently or there is bias in data collection. If we use cross-sectional data where n numbers of individuals or items are observed or measured at a point or fixed period of time, the data should be so collected that measurement on one individual (or item) does not affect the other. If the data are observed in a time (or space) sequence, measurement on the subsequent time periods (or spaces) should not affect the next. In general sense, the order of observation (order in which the data were collected) should not have an influence on the observed values. Residual versus order of observation plot, namely serial correlation plot, is used

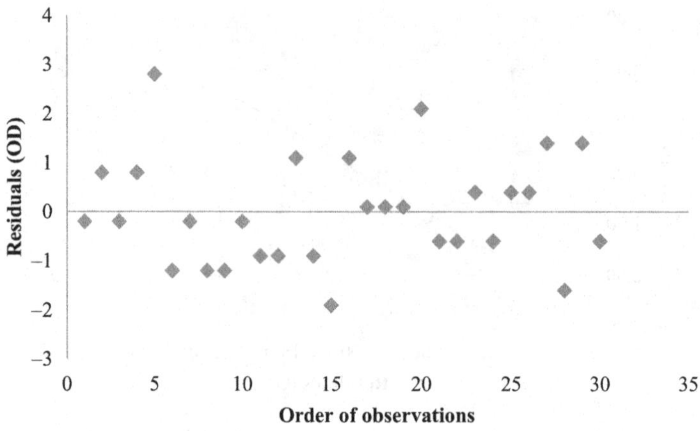

FIGURE 7.7 Serial correlation plot for steel washers' outer diameter (OD) data. The plot exhibits randomness indicating that the observations are independent.

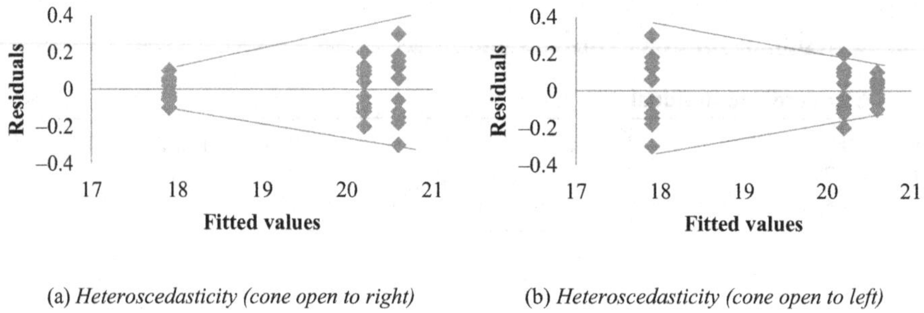

FIGURE 7.8 Residuals vs fitted values indicating heteroscedasticity. The plot of residual versus fitted values exhibits cone pattern (open to left or open to right) when error variances are not equal across fitted values.

to test this independence. The plot should not show any systematic patterns; rather the observations should exhibit randomness.

Figure 7.7 shows the serial correlation plot of the steel washers' data. The plot exhibits randomness. So, the N(=30) observations are independent.

Test of Homogeneity of Variances

In Section 7.1.2, we described the Bartlett test to test the homogeneity of population variances (homoscedasticity). Here, we will first see residual plots to visually identify heterogeneity (lack of homogeneity or heteroscedasticity) of population variances. We will then discuss the modified Levene test.

In one-way ANOVA, the plot of residuals ($\hat{\epsilon}_{il}$) versus fitted values ($\hat{x}_{il} = \hat{\mu}_l = \bar{x}_l$) show no systematic patterns if the homogeneity of variance assumption is met. In the presence of heterogeneity of variance, the plot usually portrays a conical structure, open to right (Figure 7.8a) or open to left (Figure 7.8b). For the steel washers' data, the plot of residuals versus fitted values (Figure 7.9) exhibits random patterns, thereby indicating constant variance across populations.

Modified Leven Test

If x_{il} doesn't follow normal distribution, the Bartlett test cannot be used for testing the homogeneity of variances across population. Levene (1960) proposed a robust test for homogeneity of variances following one way ANOVA of $d_{il} = |x_{il} - \bar{x}_l|, i = 1,2,...,n_l$ and $l = 1,2,....,L$, $\bar{x}_l$ is the sample mean

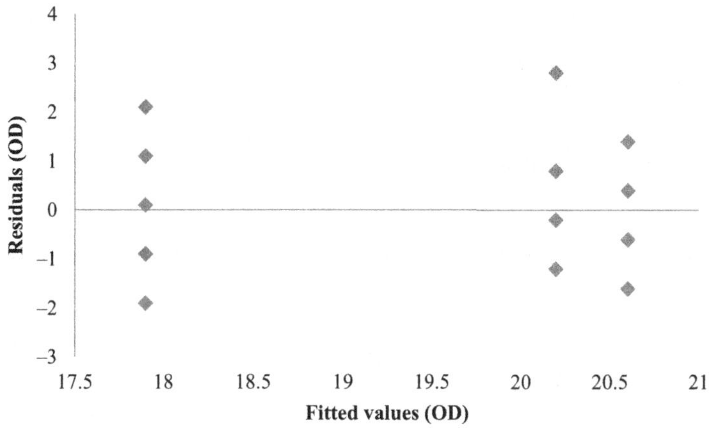

FIGURE 7.9 Residuals versus fitted values for the steel washers' outer diameter (OD) data. The plot exhibits no appreciable differences of residual variance across the fitted values, which indicates homogeneity of error variances.

collected from the l-th population. In the modified Levene test, while computing d_{il}, $\bar{x}_l$ is replaced by the median of the n_l data points, say x'_l. The steps to be followed for the modified Levene test are given below.

Step 1: Set null (H_0) and alternate (H_1) hypothesis

$$H_0 : \sigma_1^2 = \sigma_2^2 = \ldots = \sigma_l^2 = \ldots = \sigma_L^2 = \sigma^2$$
$$H_1 : \sigma_l^2 \neq \sigma_m^2 \text{ for at least one pair, } \ell = 1,2,\ldots L,\ m = 1,2,\ldots L,\ \ell \neq m$$

Step 2: Define variable of interest

$$d_{il} = \left|x_{il} - x'_l\right|, i = 1,2,\ldots,n_l \text{ and } l = 1,2,\ldots.,L$$

x'_l is the l-th population median based on the sample collected.

Step 3: Compute SST, SSB, SSE, MSB and MSE for d_{il}

Let $D_l = \sum_{l=1}^{n_l} d_{il}, T = \sum_{l=1}^{L} D_l$ and $N = \sum_{l=1}^{L} n_l$. Then,

$$SST_d = \sum_{l=1}^{L}\sum_{i=1}^{n_l} d_{il}^2 - \frac{T^2}{N}$$
$$SSB_d = \sum_{l=1}^{L}\frac{D_l^2}{n_l} - \frac{T^2}{N}$$
$$SSE_d = SST_d - SSB_d$$

It should be noted here that the DOFs for SST_d, SSB_d and SSE_d are N–1, L–1, and N–L, respectively. So, MSB and MSE for d_{il} are

$$MSB_d = \frac{SSB_d}{L-1}$$
$$MSE_d = \frac{SSE_d}{N-L}$$

Step 4: Define test statistic

The test statistic is $F_0 = \frac{MSB_d}{MSE_d}$.

Step 5: Define sampling distribution

$F_0 = \frac{MSB_d}{MSE_d}$ follows F distribution with L–1 numerator DOF and N–L denominator DOF, i.e., $F_0 \sim F_{L-1,N-L}$.

Step 6: Obtain cut-off value

For α level of significance, the cut-off value of F_0 is $F^{\alpha}_{L-1,N-L}$.

Step 7: Decision

Reject H_0 if $F_0 > F^{\alpha}_{L-1,N-L}$.
Rejection of H_0 is not desirable as it indicates that the homoscedasticity assumption is violated.

Example 7.7

Conduct modified Levene test for the steel washers' data given in Table 7.1.

Solution:

The step-wise solutions are given below.

Step 1: Set null (H_0) and alternate (H_1) hypothesis
As there are three processes A, B and C, H_0 and H_1 are:
$H_0 : \sigma_1^2 = \sigma_2^2 = \sigma_3^2 = \sigma^2$
$H_1 : \sigma_l^2 \neq \sigma_m^2$ for at least one pair, $l = 1,2,3,\ m = 1,2,3,\ 1 \neq m$

Step 2: Define variable of interest

$d_{il} = \left| x_{il} - x_l' \right|, i = 1,2,...,10$ and $l = 1,2,3$
x_l' is the l-th population median based on the sample collected.
The median (OD) data for the three processes A, B, and C are 20, 18, and 20.5 respectively. The d_{il} data table is shown in Table 7.11.

Step 3: Compute SST_d, SSB_d, SSE_d, MSB_d and MSE_d for d_{il}

Now, from the d_{il} data given above, we get

TABLE 7.11
Data Table for d_{il}

Sl. No.	Process A	Process B	Process C
1	0	1	0.5
2	1	1	0.5
3	0	1	0.5
4	1	1	0.5
5	3	2	0.5
6	1	1	0.5
7	0	0	1.5
8	1	0	1.5
9	1	0	1.5
10	0	2	0.5

$$D_1 = \sum_{i=1}^{n_1} d_{i1} = \sum_{i=1}^{10} d_{i1} = (0+1+0+.....+1+1+0) = 8$$

$$D_2 = \sum_{i=1}^{n_2} d_{i2} = \sum_{i=1}^{10} d_{i2} = (1+1+1+.....+0+0+2) = 9$$

$$D_3 = \sum_{i=1}^{n_3} d_{i3} = \sum_{i=1}^{10} d_{i3} = (0.5+0.5+0.5+.....+1.5+1.5+0.5) = 8$$

$$T = \sum_{l=1}^{L} D_l = \sum_{l=1}^{3} D_l = (D_1 + D_2 + D_3) = (8+9+8) = 25$$

$$N = \sum_{l=1}^{L} n_l = \sum_{l=1}^{3} n_l = (n_1 + n_2 + n_3) = (10+10+10) = 30$$

$$SST_d = \sum_{l=1}^{3}\sum_{i=1}^{10} d_{il}^2 - \frac{T^2}{N} = (0+1+0+.....+2.25+2.25+0.25) - \frac{25^2}{30} = 14.667$$

$$SSB_d = \sum_{l=1}^{3} \frac{D_l^2}{n_l} - \frac{T^2}{N} = \left(\frac{D_1^2}{10} + \frac{D_2^2}{10} + \frac{D_3^2}{10}\right) - \frac{T^2}{N} = \frac{1}{10}(64+81+64) - \frac{25^2}{30} = 0.067$$

$$SSE_d = SST_d - SSB_d = 14.667 - 0.067 = 14.60$$

The DOFs for SST_d, SSB_d and SSE_d are 29 (=N–1), 2 (=L–1) and 27 (=N–L), respectively.

So, MSB_d and MSE_d for d_{il} are:

$$MSB_d = \frac{SSB_d}{L-1} = \frac{0.0667}{2} = 0.033$$

$$MSE_d = \frac{SSE_d}{N-L} = \frac{14.60}{27} = 0.541$$

Step 4: The test statistic is $F_0 = \dfrac{MSB_d}{MSE_d} = \dfrac{0.033}{0.541} = 0.062.$

Step 5: Define sampling distribution

$F_0 = \frac{MSB_d}{MSE_d}$ follows F distribution with 2 numerator DOF and 27 denominator DOF, i.e., $F_0 \sim F_{2,27}$.

Step 6: Obtain cut-off value

Let α=0.05, the cut-off value of F_0 is $F_{2,27}^{0.05} = 3.35$.

Step 7: Decision

Accept H_0 as $F_0 = 0.062 < F_{2,27}^{0.05} = 3.35$. So, the three processes A, B and C do not differ in their process variance (with respect to OD).

7.1.7 Interpretation of Results

As described in Chapter 1, the statistical solutions (obtained through statistical modeling) must be realistic for implementable decision-making in practice. In ANOVA, if the group means (population means) differ, then it is clear to see which groups are different so that similar groups can be collapsed into a single larger group (provided homoscedasticity assumption is satisfied). Thereby implementable decisions can be applied to them. Dissimilar groups will be treated differently. For example, ANOVA of the steel washers' data reveals that processes A and C do not differ from the mean and variability of outer diameter (OD) point of view, while process B is different than processes A and C from the mean OD point of view. A closer look into process B may reveal that process B is comparatively newer and also handled by a less experienced operator. The lower mean OD (of process B) may not fit the customer requirements. But the management might have mixed all steel washers during sale, irrespective of the processes from which the washers are produced. This might have led to frequent rejection of lots by the customers. Under such situations, the ANOVA results are extremely useful to the management. The management can assign an experienced operator to process B, and maintain process B at a lesser interval as it is going through the burn-in (new) phase. It is also advisable not to mix the steel washers produced by the three processes.

The pair-wise comparisons[1] of means (see SCI of differences between two-population means given in Section 7.1.5) are therefore extremely useful. If H_0 is accepted, all population means are equal and they can be considered as a single group (population). If H_0 is rejected, it is important to know which pairs are different. The 100(1–α)% SCI developed in Section 7.1.5 can be used for this purpose. If any of the 100(1–α)% SCI contains zero, it indicates that the corresponding pair is indifferent from their means. For example, the 100(1–α)% SCI for the pair A and C contains zero whereas that of the pairs B and C, and C and A does not.

On the other hand, the similarity or dissimilarity in the means of different populations is often known in advance. Planned contrasts can be used to test hypotheses where the contrast is a linear combination of different population parameters (here means). Planned contrasts help make multiple comparisons possible in single hypothesis testing. For more details, see Montgomery (2012).

The ANOVA procedures described in this section are known as one-way ANOVA. We assumed randomized design. There are two-way and multi-way ANOVA with full and fractional factorial designs. Interested readers may consult Morris (2011) and Montgomery (2012).

7.2 MULTIVARIATE ANALYSIS OF VARIANCE (MANOVA)

Table 7.12 describes several scenarios for comparison of population means. These are: (i) two populations ($l = 2$), each characterized by one variable of interest (say, X1); (ii) two populations ($l = 2$), each characterized by two or more variables of interest (say, $\boldsymbol{X} = [X_1, X_2, \cdots, X_p]^T$);

(iii) more than two populations ($l > 2$), each characterized by one variable of interest (say, X_1); and (iv) more than two populations ($l > 2$), each characterized by two or more variables of interest (say, $\boldsymbol{X} = [X_1, X_2, \cdots, X_p]^T$). For univariate cases ($p = 1$) comprising several ($l > 2$) population means, we have described analysis of variance (ANOVA) in Section 7.1. The multivariate counterpart of ANOVA is multivariate analysis of variance (MANOVA), which is described in the following sections.

7.2.1 Conceptual Model

As shown in Table 7.12, MANOVA is the extension of ANOVA when the number of response variables is two or more. So, in MANOVA, from analysis point of view one more dimension ($j = 1, 2, \ldots, p$) is added. Pictorially, Figure 7.10 depicts the dimensions in balanced MANOVA. For unbalanced MANOVA, n will be replaced by n_l.

In ANOVA, an observation, represented by x_{il}, is a scalar quantity. But in MANOVA, an observation $\mathbf{x}_{il}$ is a multivariate observation and is a vector quantity. Accordingly, the following notations are used in MANOVA.

$$\mathbf{x}_{il} = \begin{bmatrix} x_{i1l} \\ x_{i2l} \\ \vdots \\ x_{ipl} \end{bmatrix}_{p\times 1} = i-\text{th multivariate observation vector on } p\text{-variables for } l\text{-th population.}$$

$$\boldsymbol{\mu}_l = \begin{bmatrix} \mu_{1l} \\ \mu_{2l} \\ \vdots \\ \mu_{pl} \end{bmatrix}_{p\times 1} = \text{vector of } l\text{-th population mean on } p\text{-variables}$$

$$\boldsymbol{\mu} = \begin{bmatrix} \mu_1 \\ \mu_2 \\ \vdots \\ \mu_p \end{bmatrix}_{p\times 1} = \text{vector of grand means on } p\text{-variables}$$

TABLE 7.12
Scenarios for Comparisons of Population Means

No. of Populations (l)	No of Variables (p)	Hypothesis	Technique Applicable
$l = 2$	$p = 1$	$H_0 : \mu_1 = \mu_2$ $H_1 : \mu_1 \neq \mu_2$	t-test
	$p \geq 2$	$H_0 : \mu_1 = \mu_2$ $H_1 : \mu_1 \neq \mu_2$	Hotelling's T^2
$l > 2$	$p = 1$	$H_0 : \mu_1 = \mu_2 = ... = \mu_l$ H_1 : at least one pair of means are not equal	ANOVA
	$p \geq 2$	$H_0 : \mu_1 = \mu_2 = ... = \mu_l$ H_1 : at least one pair of mean vectors are not equal	MANOVA

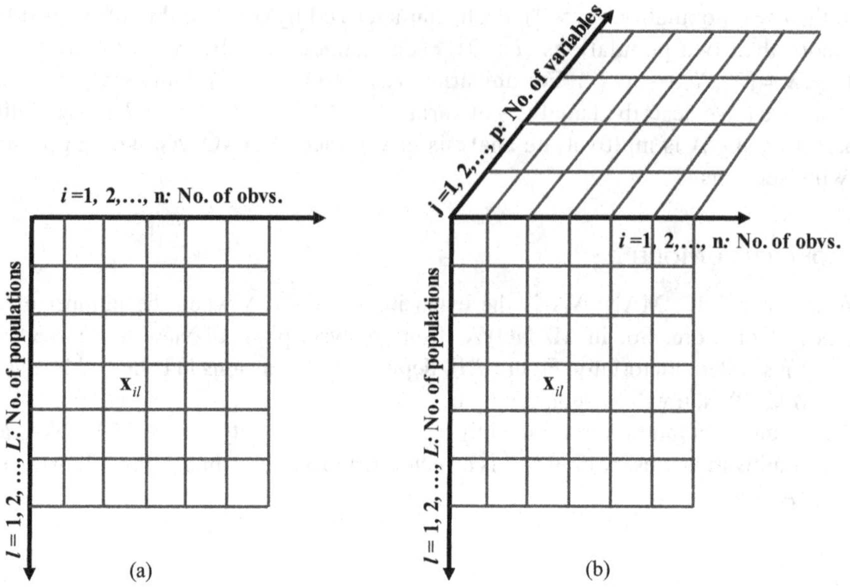

FIGURE 7.10 Data dimensions in (a) ANOVA and (b) MANOVA (balanced). While data for ANOVA has two dimensions namely, population number (l) and observation number (i) per population, in MANOVA, another dimension, the variable number (j) is added (see Figure (b)).

The MANOVA model is given below.

$$\begin{aligned}\mathbf{x}_{il} &= \boldsymbol{\mu} + (\boldsymbol{\mu}_l - \boldsymbol{\mu}) + (\mathbf{x}_{il} - \boldsymbol{\mu}_l) \\ &= \boldsymbol{\mu} + \tau_l + \epsilon_{il}\end{aligned} \tag{7.27}$$

Where $\mathbf{x}_{il}$, $\boldsymbol{\mu}$, and $\boldsymbol{\mu}_\ell$ are i-th multivariate observation vector, grand mean vector and l-th population mean vector as shown above. $\boldsymbol{\tau}_l$ is also a vector of the order $p \times 1$ representing the effect of p-variables for the l-th population. So,

$$\boldsymbol{\tau}_l = \begin{bmatrix} \tau_{1l} \\ \tau_{2l} \\ \vdots \\ \tau_{pl} \end{bmatrix}_{p\times 1} \text{ and } \boldsymbol{\epsilon}_{il} = \begin{bmatrix} x_{i1l} - \mu_{1l} \\ x_{i2l} - \mu_{2l} \\ \vdots \\ x_{ipl} - \mu_{pl} \end{bmatrix}.$$

In addition, $\sum_{l=1}^{L} \boldsymbol{\tau}_l = \mathbf{0}$ when equal observations are considered from all populations, and $\sum_{l=1}^{L} n_l \boldsymbol{\tau}_l = \mathbf{0}$ when different number of observations are considered from different populations.

MANOVA tests the following hypothesis

$$H_0 : \boldsymbol{\mu}_1 = \boldsymbol{\mu}_2 = \cdots = \boldsymbol{\mu}_L$$

$$H_1 : \boldsymbol{\mu}_l \neq \boldsymbol{\mu}_m \text{ for at least one pair } (l, m),\ l = 1, 2, \ldots \text{L and } m = 1, 2, \ldots L,\ l \neq m.$$

Alternatively,

$$H_0 : \boldsymbol{\tau}_l = 0,\ l = 1,2,\ldots,L$$

$$H_1 : \boldsymbol{\tau}_l \neq 0, \text{ for at least one } l,\ l = 1,2,\ldots,L.$$

Consider the steel washers' production case described in Section 7.1.1. Let us consider that in addition to the outer diameter (OD), the inner diameter (ID) of the steel washers is also an important quality variable. So, the response variables of interest are now OD and ID. Thereby, each of the processes (populations) is now bivariate in nature. The univariate ANOVA is not applicable to test the hypothesis of equality of population mean vectors. MANOVA is used for this purpose. Ten bivariate observations from each of the three processes A, B, and C were collected. The observations are shown in Table 7.13. In addition, the box plot for OD and ID is shown in Figure 7.11. Please see that the box plot of OD is same as shown in Figure 7.1 (as OD values remain the same). From Figure 7.11, it is seen that the three processes differ in terms of mean OD and mean ID.

The MANOVA model for the steel washers' data for two variables (p=2) namely OD and ID is shown below.

$$\mathbf{x}_{il} = \begin{bmatrix} x_{i1l} \\ x_{i2l} \end{bmatrix}_{2\times1},\ \boldsymbol{\mu}_l = \begin{bmatrix} \mu_{1l} \\ \mu_{2l} \end{bmatrix}_{2\times1} \text{ and } \boldsymbol{\mu} = \begin{bmatrix} \mu_1 \\ \mu_2 \end{bmatrix}_{2\times1}.$$

As there are three processes, A, B, and C, and ten observations are recorded, l ranges from 1 to 3 and i ranges from 1 to 10.

$$\begin{aligned} \mathbf{x}_{il} &= \boldsymbol{\mu} + (\boldsymbol{\mu}_l - \boldsymbol{\mu}) + (\mathbf{x}_{il} - \boldsymbol{\mu}_l) \\ &= \boldsymbol{\mu} + \boldsymbol{\tau}_l + \boldsymbol{\epsilon}_{il} \end{aligned}$$

where

$$\boldsymbol{\tau}_l = \begin{bmatrix} \tau_{1l} \\ \tau_{2l} \end{bmatrix}_{2\times1} \text{ and } \boldsymbol{\epsilon}_{il} = \begin{bmatrix} x_{i1l} - \mu_{1l} \\ x_{i2l} - \mu_{2l} \end{bmatrix}.$$

$$i = 1,2,\ldots 10 \text{ and } l = 1,2,3.$$

TABLE 7.13
Steel Washers' Data with Response Variables OD and ID (p = 2)

	Process A		Process B		Process C	
No of Observations	**OD**	**ID**	**OD**	**ID**	**OD**	**ID**
1	20	6	17	6	20	8
2	21	6	17	6	20	7
3	20	9	19	7	21	8
4	21	6	17	8	20	7
5	23	7	16	6	21	8
6	19	7	19	7	21	9
7	20	6	18	7	22	7
8	19	7	18	6	19	7
9	19	5	18	6	22	6
10	20	6	20	8	20	8

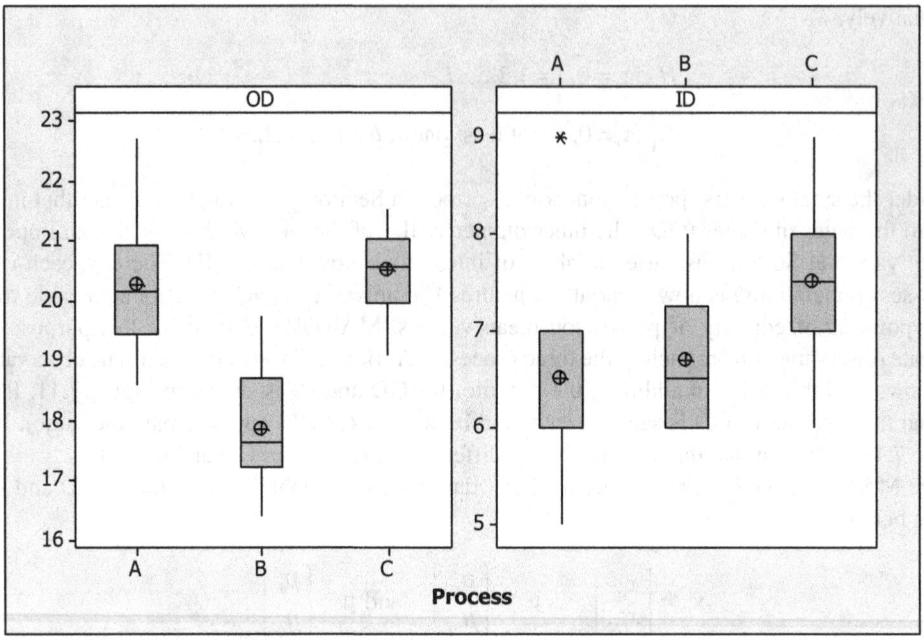

FIGURE 7.11 Box plot of OD and ID.

In addition, $\sum_{l=1}^{3} \boldsymbol{\tau}_l = \mathbf{0}$ as equal number of (n =10) observations are considered from each of the three processes, A, B, and C.

The hypothesis to be tested is

$$H_0 : \boldsymbol{\mu}_1 = \boldsymbol{\mu}_2 = \boldsymbol{\mu}_3$$
$$H_1 : \boldsymbol{\mu}_l \neq \boldsymbol{\mu}_m \text{ for at least one pair } (l, m),\ l = 1, 2, 3,\ m = 1, 2, 3 \text{ and } l \neq m.$$

Alternatively,

$$H_0 : \boldsymbol{\tau}_l = \mathbf{0},\ \ l = 1,2,3$$
$$H_1 : \boldsymbol{\tau}_l \neq \mathbf{0}, \text{ for at least one } l,\ l = 1, 2, 3.$$

7.2.2 Assumptions

MANOVA requires the following three assumptions to be met.

- Independence: The multivariate observations $\mathbf{x}_{il}$, $i = 1, 2, \ldots, n_l$ and $l = 1, 2, \ldots, L$ must be independent. The common violation is that responses from one population are not independent of the responses from the other populations (Hair et al., 2010). The other sources are time-ordered effects (serial correlation) and experience bias. This assumption can be tested by serial correlation plot of the residuals.
- Homogeneity of covariance matrices: The population covariance matrices must be equal. That is $\boldsymbol{\Sigma}_1 = \boldsymbol{\Sigma}_2 = \ldots = \boldsymbol{\Sigma}_\ell = \ldots = \boldsymbol{\Sigma}_L$, where $\boldsymbol{\Sigma}_l$, $l = 1,2,\ldots,L$ is a $p \times p$ matrix and p is the number of response variables of interest. This assumption can be tested using Box's M method.

- Normality: The response variables must follow multivariate normal distribution. That is $\mathbf{x}_{il} \sim N_p(\boldsymbol{\mu}_l, \boldsymbol{\Sigma}_l)$. This assumption can be tested by assessing the multivariate normality of the errors ($\boldsymbol{\epsilon}_{il}$). χ^2 Q-Q plot can be used for this purpose.

Box's M Test

This test was first introduced by Box(1949). Consider L populations, each one is characterized by p response variables. The population mean vectors and covariance matrices are $\boldsymbol{\mu}_l$ and $\boldsymbol{\Sigma}_l$, respectively where l varies from 1 to L. Suppose n_p $l = 1, 2, \ldots$ L observations are obtained from the l-th population. All are interested to test whether the covariances of the L populations are different or not. This can be accomplished using the following steps (Johnson and Wichern, 2013):

Step 1: Set null (H_0) and alternate (H_1) hypothesis

$$H_0 : \boldsymbol{\Sigma}_1 = \boldsymbol{\Sigma}_2 = \ldots = \boldsymbol{\Sigma}_l = \ldots = \boldsymbol{\Sigma}_L$$

$H_1 : \boldsymbol{\Sigma}_l \neq \boldsymbol{\Sigma}_m$ for at least one pair (l, m), $l \neq m$

Step 2: Define test statistic.

The test statistic is $D = (1-u)M$ where

$$M = -2ln\left[\prod_{l=1}^{L}\left[\frac{|\mathbf{S}_l|}{|\mathbf{S}_{\text{pooled}}|}\right]^{\frac{n_l-1}{2}}\right].$$

$$= \sum_{l=1}^{L}(n_l - 1)\left[\ell n|\mathbf{S}_{\text{pooled}}|\right] - \sum_{l=1}^{L}\left[(n_l-1)\ell n|\mathbf{S}_l|\right]. \tag{7.28}$$

$$u = \left[\sum_{l=1}^{L}\frac{1}{(n_l-1)} - \frac{1}{\sum_{l=1}^{L}(n_l-1)}\right]\left[\frac{2p^2+3p-1}{6(p+1)(L-1)}\right]. \tag{7.29}$$

$$\text{and } \mathbf{S}_{\text{pooled}} = \frac{(n_1-1)\mathbf{S}_1 + (n_2-1)\mathbf{S}_2 + \cdots + (n_L-1)\mathbf{S}_L}{\sum_{l=1}^{L} n_l - L}. \tag{7.30}$$

$|\mathbf{S}_l|$ and $|\mathbf{S}_{\text{pooled}}|$ denote the determinant of $\mathbf{S}_l$ and $\mathbf{S}_{\text{pooled}}$.

Step 3: Define sampling distribution

The statistic D follows χ^2 distribution with $v = \frac{1}{2}p(p+1)(L-1)$ degrees of freedom.

Step 4: Obtain cut-off value

For level of significance α, the cut-off value is $\chi^2_v(\alpha)$.

Step 5: Decision

Reject H_0 if $D > \chi^2_v(\alpha)$.

Example 7.8

Consider the data set given in Table 7.13. Test equality of covariance matrices using Box's M test. Consider $\alpha = 0.05$.

Solution:

Let the population covariance matrices for the three processes A, B and C are $\mathbf{\Sigma}_1, \mathbf{\Sigma}_2$ and $\mathbf{\Sigma}_3$, respectively. The task at hand is to test where the three covariance matrices are equal or not. This can be done using Box's M test. The step-wise computations are given below.

Step 1: Set null (H_0) and alternate (H_1) hypothesis

$$H_0 : \mathbf{\Sigma}_1 = \mathbf{\Sigma}_2 = \mathbf{\Sigma}_3$$
$$H_1 : \mathbf{\Sigma}_l \neq \mathbf{\Sigma}_m \text{ for at least one pair } (l, m),\ l = 1, 2, 3 \text{ and } m = 1, 2, 3;\ 1 \neq m$$

Step 2: Define test statistic

The test statistic is $D = (1-u)M$.

Let the sample covariance matrices of three processes A, B and C are $\mathbf{S}_1$, $\mathbf{S}_2$, and $\mathbf{S}_3$, respectively. Using data given in Table 7.13 and the sample covariance formula given in Chapter 4 (Equation 4.14), $\mathbf{S}_1$, $\mathbf{S}_2$, and $\mathbf{S}_3$ are

$$\mathbf{S}_1 = \begin{pmatrix} 1.51 & 0.11 \\ 0.11 & 1.17 \end{pmatrix},\ \mathbf{S}_2 = \begin{pmatrix} 1.43 & 0.52 \\ 0.52 & 0.68 \end{pmatrix},\ \mathbf{S}_3 = \begin{pmatrix} 0.93 & -0.11 \\ -0.11 & 0.72 \end{pmatrix}$$

Using Equation 7.30, $\mathbf{S}_{\text{pooled}}$ is

$$\mathbf{S}_{\text{pooled}} = \frac{(10-1)S_1 + (10-1)S_2 + (10-1)S_3}{(30-3)}$$
$$= \frac{(10-1)\begin{pmatrix} 1.51 & 0.11 \\ 0.11 & 1.17 \end{pmatrix} + (10-1)\begin{pmatrix} 1.43 & 0.52 \\ 0.52 & 0.68 \end{pmatrix} + (10-1)\begin{pmatrix} 0.93 & -0.11 \\ -0.11 & 0.72 \end{pmatrix}}{(30-3)}$$
$$= \begin{pmatrix} 1.29 & 0.17 \\ 0.17 & 0.86 \end{pmatrix}$$

Now, using Equation 7.28, M is

$$M = \sum_{l=1}^{3}(n_l - 1)\left[\ell n\left|\mathbf{S}_{\text{pooled}}\right|\right] - \sum_{l=1}^{3}\left[(n_l - 1)\ell n\left|\mathbf{S}_l\right|\right]$$
$$= (3\times 9)\left[\ell n\begin{vmatrix} 1.29 & 0.17 \\ 0.17 & 0.86 \end{vmatrix}\right] - \left[\left[9\times \ell n\begin{vmatrix} 1.51 & 0.11 \\ 0.11 & 1.17 \end{vmatrix}\right] + \left[9\times \ell n\begin{vmatrix} 1.43 & 0.52 \\ 0.52 & 0.68 \end{vmatrix}\right] + \left[9\times \ell n\begin{vmatrix} 0.93 & -0.11 \\ -0.11 & 0.72 \end{vmatrix}\right]\right]$$
$$= 3\times 9\times 0.0774 - \left[[9\times 0.562] + [9\times(-0.354)] + [9\times(-0.419)]\right]$$
$$= 2.09 - (-1.90) = 3.99$$

Further, using Equation 7.29, u is

$$u=\left[\sum_{l=1}^{3}\frac{1}{(n_l-1)}-\frac{1}{\sum_{l=1}^{3}(n_l-1)}\right]\left[\frac{2p^2+3p-1}{6(p+1)(L-1)}\right], \text{ where } L=3,\ n_1=n_2=n_3=10 \text{ and } p=2.$$

$$=\left[\sum_{l=1}^{3}\frac{1}{(10-1)}-\frac{1}{\sum_{l=1}^{3}(10-1)}\right]\left[\frac{2\times 2^2+3\times 2-1}{6(2+1)(3-1)}\right]=0.107$$

Finally, the test statistic D is

$D=(1-u)M=(1-0.107)\times 3.99=3.56$

Step 3: Define sampling distribution

The statistic D follows χ^2 distribution with $v=\frac{1}{2}p(p+1)(L-1)=\frac{1}{2}\times 2(2+1)(3-1)=6$ degrees of freedom.

Step 4: Obtain cut-off value

For level of significance $\alpha=0.05$, the cut-off value is $\chi^2_v(0.05)=\chi^2_6(0.05)=12.59$.

Step 5: Decision

As $D=3.56<\chi^2_6(0.05)=12.59$, we fail to reject H_0. So, the process covariance matrices are equal.

Box's M test is sensitive to departure from normality. But if sample size is large, Box's M works well. Johnson and Wichern (2013) mentioned that for $n_l>20$, $p<5$ and $L<5$, Box's M provides good results. When data violates normality and sample sizes are unequal, one has to be cautious in using MANOVA. It is recommended to use equal size samples from all the populations. If data exhibits heterogeneity, the test statistics should be carefully chosen (see Section 7.2.4). Readers may consult Box (1949) and Hair et al. (2010). Hair et al. (2010) provide several guidelines to deal with unequal variances.

7.2.3 Total Sum Squares and Cross Product (SSCP) Decomposition

The data structure used in MANOVA is shown in Figure 7.10. Suppose a sample of n_l observations was collected from l-th population. Now, consider the following notations.

$$\mathbf{x}_{il}=\begin{bmatrix}x_{i1l}\\x_{i2l}\\\vdots\\x_{ipl}\end{bmatrix}_{p\times 1}$$ = i – th multivariate observation vector collected on p-variables from the l-th population (Note: It is reproduced here for convienece)

$$\bar{\mathbf{x}}_l=\begin{bmatrix}\bar{x}_{1l}\\\bar{x}_{2l}\\\vdots\\\bar{x}_{pl}\end{bmatrix}_{p\times 1}$$ = vector of sample means on p-variables for the l-th population

$$\bar{\mathbf{x}} = \begin{bmatrix} \bar{x}_1 \\ \bar{x}_2 \\ \vdots \\ \bar{x}_p \end{bmatrix}_{p\times 1}$$ = vector of sample grand means on p-variables considering all L populations together

We will now partition the multivariate observation x_{il}. Following the partitioning procedure used in ANOVA, we can partition x_{il} as follows:

$$\mathbf{x}_{il} - \bar{\mathbf{x}} = \left(\bar{\mathbf{x}}_l - \bar{\mathbf{x}}\right) + \left(\mathbf{x}_{il} - \bar{\mathbf{x}}_l\right) \tag{7.31}$$

What is the difference between Equation 7.7 and Equation 7.31? In Equation 7.7, a scalar quantity $\mathrm{x}_{il} - \bar{\mathrm{x}}$ is partitioned into set of two other scalar quantities. But in Equation 7.31, the partitioning is done for a vector $\mathbf{x}_{il} - \bar{\mathbf{x}}$ of the order of p × 1. Equation 7.31 involves matrix operations.

Now, squaring Equation 7.31, we get

$$\begin{aligned}\left(\mathbf{x}_{il} - \bar{\mathbf{x}}\right)\left(\mathbf{x}_{il} - \bar{\mathbf{x}}\right)^T &= \left[\left(\bar{\mathbf{x}}_l - \bar{\mathbf{x}}\right) + \left(\mathbf{x}_{il} - \bar{\mathbf{x}}_l\right)\right]\left[\left(\bar{\mathbf{x}}_l - \bar{\mathbf{x}}\right) + \left(\mathbf{x}_{il} - \bar{\mathbf{x}}_l\right)\right]^T \\ &= \left(\bar{\mathbf{x}}_l - \bar{\mathbf{x}}\right)\left(\bar{\mathbf{x}}_l - \bar{\mathbf{x}}\right)^T + \left(\bar{\mathbf{x}}_l - \bar{\mathbf{x}}\right)\left(\mathbf{x}_{il} - \bar{\mathbf{x}}_l\right)^T + \left(\mathbf{x}_{il} - \bar{\mathbf{x}}_l\right)\left(\bar{\mathbf{x}}_l - \bar{\mathbf{x}}\right)^T + \left(\mathbf{x}_{il} - \bar{\mathbf{x}}_l\right)\left(\mathbf{x}_{il} - \bar{\mathbf{x}}_l\right)^T\end{aligned}$$

Taking sum over $i = 1, 2, \ldots, n_l$, we get

$$\sum_{i=1}^{n_l}\left(\mathbf{x}_{il} - \bar{\mathbf{x}}\right)\left(\mathbf{x}_{il} - \bar{\mathbf{x}}\right)^T = \sum_{i=1}^{n_l}\left(\bar{\mathbf{x}}_l - \bar{\mathbf{x}}\right)\left(\bar{\mathbf{x}}_l - \bar{\mathbf{x}}\right)^T + \sum_{i=1}^{n_l}\left(\mathbf{x}_{il} - \bar{\mathbf{x}}_l\right)\left(\mathbf{x}_{il} - \bar{\mathbf{x}}_l\right)^T$$

as $\sum_{i=1}^{n_l}\left(\mathbf{x}_{il} - \bar{\mathbf{x}}_l\right) = \sum_{i=1}^{n_l}\left(\mathbf{x}_{il} - \bar{\mathbf{x}}_l\right)^T = \mathbf{0}$ and $\sum_{i=1}^{n_l}\left(\bar{\mathbf{x}}_l - \bar{\mathbf{x}}\right)$ is free from the i-th term, the term

$$\sum_{i=1}^{n_l}\left(\mathbf{x}_{il} - \bar{\mathbf{x}}_l\right)\left(\bar{\mathbf{x}}_l - \bar{\mathbf{x}}\right)^T = \sum_{i=1}^{n_l}\left(\bar{\mathbf{x}}_l - \bar{\mathbf{x}}\right)\left(\mathbf{x}_{il} - \bar{\mathbf{x}}_l\right)^T = \mathbf{0}$$

Finally, taking sum over $l = 1, 2, \ldots, L$, we get,

$$\begin{aligned}&\sum_{l=1}^{L}\sum_{i=1}^{n_l}\left(\mathbf{x}_{il} - \bar{\mathbf{x}}\right)\left(\mathbf{x}_{il} - \bar{\mathbf{x}}\right)^T \\ &= \sum_{l=1}^{L} n_l\left(\bar{\mathbf{x}}_l - \bar{\mathbf{x}}\right)\left(\bar{\mathbf{x}}_l - \bar{\mathbf{x}}\right)^T + \sum_{l=1}^{L}\sum_{i=1}^{n_l}\left(\mathbf{x}_{il} - \bar{\mathbf{x}}_l\right)\left(\mathbf{x}_{il} - \bar{\mathbf{x}}_l\right)^T\end{aligned} \tag{7.32}$$

Equation 7.32 partitions the total sum squares and cross product (SSCP) matrix into two SSCP matrices. Like ANOVA, the LHS of Equation 7.32 is known as total SSCP or $\mathrm{SSCP_T}$ and the first term of the RHS of Equation 7.32 as SSCP between populations or $\mathrm{SSCP_B}$ and second term as SSCP for errors or $\mathrm{SSCP_E}$, or SSCP within populations or $\mathrm{SSCP_W}$, where T stands for total, B stands for between, E stands for error and W stands for within.

For computation, it is recommended to compute $\mathrm{SSCP_B}$ and $\mathrm{SSCP_E}$, as $\mathrm{SSCP_B}$ has single summation over $l = 1, 2, \ldots, L$, and $\mathrm{SSCP_E}$ can be expressed in a simpler form.

$$SSCP_B = \sum_{l=1}^{L} n_l\left(\bar{\mathbf{x}}_l - \bar{\mathbf{x}}\right)\left(\bar{\mathbf{x}}_l - \bar{\mathbf{x}}\right)^T \tag{7.33}$$

$$SSCP_E = \left(n_1 - 1\right)\mathbf{S}_1 + \left(n_2 - 1\right)\mathbf{S}_2 + \ldots + \left(n_L - 1\right)\mathbf{S}_L \tag{7.34}$$

$$SSCP_T = SSCP_B + SSCP_E \tag{7.35}$$

Where, $\mathbf{S}_l$, $l = 1, 2, \ldots L$ is the sample covariance matrix for the l-th population.

Example 7.9

Consider the data set given in Table 7.13. Compute $SSCP_B$, $SSCP_E$, and $SSCP_T$.

Solution:

There are three processes, A, B, and C (i.e., L = 3), two variables, OD and ID (i.e., p = 2), and ten observations from each of the three processes (i.e., $n_1 = n_2 = n_3 = 10$). To compute $SSCP_B$, $SSCP_E$, and $SSCP_T$, we require first computing $\bar{\mathbf{x}}$, $\bar{\mathbf{x}}_\ell$, and $\mathbf{S}_\ell$, for $\ell = 1,2,3$. Using formulae discussed in Chapter 4, the computed values of $\bar{\mathbf{x}}$, $\bar{\mathbf{x}}_\ell$, and $\mathbf{S}_\ell$, for $\ell = 1,2,3$ are given below.

Process A: $\bar{\mathbf{x}}_1 = \begin{pmatrix} 20.20 \\ 6.50 \end{pmatrix}$, $\mathbf{S}_1 = \begin{pmatrix} 1.51 & 0.11 \\ 0.11 & 1.17 \end{pmatrix}$, $n_1 = 10$

Process B: $\bar{\mathbf{x}}_2 = \begin{pmatrix} 17.90 \\ 6.70 \end{pmatrix}$, $\mathbf{S}_2 = \begin{pmatrix} 1.43 & 0.52 \\ 0.52 & 0.68 \end{pmatrix}$, $n_2 = 10$

Process C: $\bar{\mathbf{x}}_3 = \begin{pmatrix} 20.60 \\ 7.50 \end{pmatrix}$, $\mathbf{S}_3 = \begin{pmatrix} 0.93 & -0.11 \\ -0.11 & 0.72 \end{pmatrix}$, $n_3 = 10$

Finally, the grand average is $\bar{\mathbf{x}} = \begin{pmatrix} 19.57 \\ 6.90 \end{pmatrix}$.

(i) Computation of $SSCP_B$

$$SSCP_B = \sum_{\ell=1}^{3} n_l \left(\bar{\mathbf{x}}_l - \bar{\mathbf{x}}\right)\left(\bar{\mathbf{x}}_l - \bar{\mathbf{x}}\right)^T$$

$$= 10\left(\begin{bmatrix} 20.20 \\ 6.50 \end{bmatrix} - \begin{bmatrix} 19.57 \\ 6.90 \end{bmatrix}\right)\left(\begin{bmatrix} 20.20 \\ 6.50 \end{bmatrix} - \begin{bmatrix} 19.57 \\ 6.90 \end{bmatrix}\right)^T + 10\left(\begin{bmatrix} 17.90 \\ 6.70 \end{bmatrix} - \begin{bmatrix} 19.57 \\ 6.90 \end{bmatrix}\right)\left(\begin{bmatrix} 17.90 \\ 6.70 \end{bmatrix} - \begin{bmatrix} 19.57 \\ 6.90 \end{bmatrix}\right)^T$$

$$+ 10\left(\begin{bmatrix} 20.60 \\ 7.50 \end{bmatrix} - \begin{bmatrix} 19.57 \\ 6.90 \end{bmatrix}\right)\left(\begin{bmatrix} 20.60 \\ 7.50 \end{bmatrix} - \begin{bmatrix} 19.57 \\ 6.90 \end{bmatrix}\right)^T$$

$$= \begin{bmatrix} 42.47 & 7.00 \\ 7.00 & 5.60 \end{bmatrix}$$

(ii) Computation of $SSCP_E$

$$SSCP_E = (n_1 - 1)\mathbf{S}_1 + (n_2 - 1)\mathbf{S}_2 + (n_3 - 1)\mathbf{S}_3$$

$$= (10-1)\begin{pmatrix} 1.51 & 0.11 \\ 0.11 & 1.17 \end{pmatrix} + (10-1)\begin{pmatrix} 1.43 & 0.52 \\ 0.52 & 0.68 \end{pmatrix} + (10-1)\begin{pmatrix} 0.93 & -0.11 \\ -0.11 & 0.72 \end{pmatrix}$$

$$= \begin{bmatrix} 34.83 & 4.68 \\ 4.68 & 23.22 \end{bmatrix}$$

(iii) Computation of $SSCP_T$

$$SSCP_T = SSCP_B + SSCP_E = \begin{bmatrix} 42.47 & 7.00 \\ 7.00 & 5.60 \end{bmatrix} + \begin{bmatrix} 34.83 & 4.68 \\ 4.68 & 23.22 \end{bmatrix} = \begin{bmatrix} 77.30 & 11.68 \\ 11.68 & 28.82 \end{bmatrix}$$

What do these three matrices $SSCP_B$, $SSCP_E$, and $SSCP_T$ mean? Do these relate to SSB, SSE, and SST developed in ANOVA? Yes, these do. When there are p response variables, all the three SSCP

TABLE 7.14
One Way MANOVA Table

Sources of Variation	SSCP	Degrees of Freedom
Population	$SSCP_B$	L–1
Error	$SSCP_E$	$\sum_{\ell=1}^{L} n_\ell - L = N - L$
Total	$SSCP_T$	$\sum_{\ell=1}^{L} n_\ell - 1 = N - 1$

matrices are of the order of p × p. The p diagonal elements are the sum of squares for the p response variables, respectively and the off-diagonal elements represent sum of cross products for each pair of the p variables, respectively. Each of the p diagonal elements of the $SSCP_B$ represents SSB for the corresponding response variable. That means the *j*-th diagonal element of $SSCP_B$ is the SSB for the *j*-th response variable (X_j), $j = 1, 2, \ldots$, p. Similarly, the *j*-th diagonal element of $SSCP_E$ is the SSE for the *j*-th response variable (X_j), $j = 1, 2, \ldots$, p and the *j*-th diagonal element of $SSCP_T$ is the SST for the *j*-th response variable (X_j), $j = 1, 2, \ldots$, p.

Recall that in ANOVA, SSB, SSE and SST have L–1, N–L, and N–1 DOF, respectively where N is the total number of observations ($N = \sum_{l=1}^{L} n_l$) and L is the number of populations. Similarly, in MANOVA, $SSCP_B$ has L–1 DOF, $SSCP_E$ has N–L DOF, and $SSCP_T$ has N–1 DOF. Accordingly, one way a MANOVA table can be prepared as shown in Table 7.14.

The next question of interest is the rank of these three SSCP matrices. If there is no linear dependency in the p response variables, the rank (r) of each of these three SSCP matrices is the minimum of p and v, where v stands for DOF. So, if r_B is the rank of $SSCP_B$, then $r_B = \min(p, L-1)$. Similarly, the rank of $SSCP_E$ and $SSCP_T$ is $r_E = \min(p, N-L)$ and $r_T = \min(p, N-1)$, respectively. However, it is unlikely that p will be greater than N–L and N–1. So, usually the rank of both the $SSCP_E$ and $SSCP_T$ is p (assuming no or insignificant linear dependency).

In addition, it should be noted that there is relationship between $SSCP_E$ and the population covariance matrix $\boldsymbol{\Sigma}$ because of homogeneity of population variances (see that $\boldsymbol{\Sigma}_1 = \boldsymbol{\Sigma}_2 = \ldots = \boldsymbol{\Sigma}_L = \boldsymbol{\Sigma}$) as

$$E\left(\frac{SSCP_E}{N-L}\right) = \boldsymbol{\Sigma} \tag{7.36}$$

So, $\dfrac{SSCP_E}{N-L}$ can be used as the estimate of $\boldsymbol{\Sigma}$.

7.2.4 Hypothesis Testing

The procedures followed for hypothesis testing are the same as mentioned in one-way ANOVA except the test statistic and its distribution. These are outlined below.

Step 1: Set null (H_0) and alternate (H_1) hypothesis

$H_0 : \boldsymbol{\mu}_1 = \boldsymbol{\mu}_2 = \ldots. = \boldsymbol{\mu}_L$

$H_1 : \boldsymbol{\mu}_l \neq \boldsymbol{\mu}_m$ *for* at least one pair of mean vectors (l,m), $l \neq m$

Step 2: Define test statistic

There are four popular multivariate statistics for hypothesis testing in MANOVA. They are (i) Wilks' lambda, (ii) Pillai's trace, (iii) Hotelling-Lawley's trace, and (iv) Roy's largest root. All the four tests make use of the $SSCP_B$ and $SSCP_E$ matrices. We will describe them in the next sub-section.

Step 3: Define sampling distribution of the test statistic

Defining the sampling distribution of the multivariate test statistic is a tedious job. Statisticians have researched over the years and have produced a handful of solutions. The most popular two distributions used are χ^2 and F. We will describe them in the next sub-section.

Step 4: Obtain cut-off value

This is a straightforward case as described under any hypothesis testing conditions. Once the theoretical distribution of the test statistic is known, the cut-off value is the $100(1 - \alpha)\%$ value of the corresponding sampling distribution, where α is the significance level (usually a value of 0.05 or 0.01).

Step 5: Decision

Reject H_0 if the computed value of the test statistic exceeds its theoretical value.

Multivariate Test Statistics Used in MANOVA

Wilks' lambda (Λ)

This is the first MANOVA test statistic developed by Wilks (1932). It is a likelihood ratio-based statistic and computed as the ratio of the determinants of $SSCP_E$ and $SSCP_T$.

$$\Lambda = \frac{|SSCP_E|}{|SSCP_T|} = \frac{|SSCP_E|}{|SSCP_B + SSCP_E|} = \prod_{j=1}^{s} \frac{1}{1+\lambda_j} \tag{7.37}$$

Where λ_j denotes the j-th eigenvalue of the matrix $(SSCP_E)^{-1} SSCP_B$, and $s = \min(p, L-1)$, p being the number of response variables and L the number of populations. Under H_0, $SSCP_E$ and $SSCP_B$ follow Wishart distribution as $W_p(\Sigma, N-L)$ and $W_p(\Sigma, L-1)$, respectively, where $SSCP_E$ and $SSCP_B$ are independent (Mardia, 1970). Wilks' Λ, as given in Equation 7.37, is three-parameter Wilks' lambda distribution with parameter p, $N - L$, and $L - 1$ and denoted by $\Lambda(p, N-L, L-1)$.

Wilks' Λ varies from 0 to 1. If H_0 is true, the ideal value of Λ is 1. Where the population means differ widely making $SSCP_B >> SSCP_E$, Λ approaches 0. The quantity $(1-\Lambda)$ can be interpreted as the proportion of total variances of the p response variables explained by the population effects. So, the higher the Λ value the lower the population effects. Therefore, if H_0 is true (population mean vectors are equal), Λ value should be very low.

Wilks' Λ depends on the number of variables p, population DOF, L–1 (say df_h), and error DOF, N–L (say df_e). Accordingly, tables were prepared (Wall, 1967). Interestingly, it is possible to convert Λ to equivalent F-distribution. For smaller values of p or L, exact F-distributions are available, but for higher p and L values, there is no exact F-distribution. An approximation of F-distribution of Wilks' Λ, F_Λ is given by Rao (1952) as shown below.

$$F_\Lambda = \frac{1-\Lambda^{1/t}}{\Lambda^{1/t}} \times \frac{\nu_e}{\nu_h} \tag{7.38}$$

Where, $\nu_h = p(df_h) = p(L-1)$,

$\nu_e = 1 + mt - \nu_h / 2$,

$t = \sqrt{\frac{p^2(df_h)^2 - 4}{p^2 + (df_h)^2 - 5}} = \sqrt{\frac{p^2(L-1)^2 - 4]}{p^2 + (L-1)^2 - 5}}$, or t=1 if $p(df_h)=2$, and

$m = df_e - \frac{(p+1-df_h)}{2}$

F_Λ follows F distribution with ν_h numerator and ν_e denominator degrees of freedom. It may so happen that ν_e can be fractional. Then, ν_e value can be rounded to the next lowest integer.

For a theoretical treatment on the exact F-distribution of Wilks' Λ when p or L are small, reader may consult Mardia (1970). It is to be noted here that t becomes undefined when $p(df_h)=2$ and under such situations t = 1 will give the exact test statistic.

In this connection, Bartlett (1947) gave a less accurate test for Wilks' Λ. If H_0 is true and $\sum n_l = N$ is large, the statistic $-m\, \ell n\, \Lambda$ follows $\chi^2_{\nu_h}$. For one-way MANOVA, m becomes $\left(N - 1 - \frac{p+L}{2}\right)$ and $\nu_h = p(df_h) = p(L-1)$. It is to be noted here that for smaller sample size, F approximation is preferable than the χ^2 approximation.

Pillai Trace (V)

Pillai trace V is (Pillai, 1967):

$$V = trace\left[\left(SSCP_B + SSCP_E\right)^{-1} SSCP_B\right] = \sum_{j=1}^{s} \frac{\lambda_j}{1+\lambda_j}. \tag{7.39}$$

where $s = \min(p, L-1)$, and $\lambda_j, j = 1,2,....s$ are the eigenvalues of $\left(SSCP_E\right)^{-1} SSCP_B$.

An approximation of F-distribution of Pillai's trace V, F_V is shown below (Haase and Ellis, 1987).

$$F_V = \frac{V}{s-V} \times \frac{\nu_e}{\nu_h} \tag{7.40}$$

Where, $\nu_h = p(L-1)$, and $\nu_e = s(N - L + s - p)$. F_V follows F distribution with ν_h numerator and ν_e denominator degrees of freedom.

Hotelling-Lawley's trace (U)

$$U = trace\left(SSCP_E^{-1} SSCP_B\right) = \sum_{j=1}^{s} \lambda_j \tag{7.41}$$

An approximation of F-distribution of Hotelling-Lawley's trace U, F_U is shown below (Haase and Ellis, 1987).

$$F_V = \frac{U}{S} \times \frac{\nu_e}{\nu_h} \tag{7.42}$$

Where, $\nu_h = p(L-1)$, and $\nu_e = s(N - L - p - 1)$. F_U follows F distribution with ν_h numerator and ν_e denominator degrees of freedom.

Roy's largest root (λ_1)

If $\lambda_1 \geq \lambda_2 \geq .... \geq \lambda_s$ are the eigenvalues of $SSCP_E^{-1}SSCP_B$, then Roy's largest root is λ_1. The test statistic is

$$\theta = \frac{\lambda_1}{1+\lambda_1} \tag{7.43}$$

An approximation of F-distribution of Roy's statistic θ, F_θ is shown below (Haase and Ellis, 1987).

$$F_V = \frac{\theta}{1-\theta} \times \frac{v_e}{v_h} \tag{7.44}$$

Where, $v_h = p(L-1)$, and $v_e = N - L$.. F_6 follows F distribution with v_h numerator and v_e denominator degrees of freedom.

The Pillai trace (V), Hotelling-Lawley's trace (U), and Roy's largest root statistic (θ) are dependent on the number of response variables (p), the population DOF (L–1) and the error DOF (N–L). Accordingly, statistical tables were developed. Rencher (2002) has given the upper critical values of θ, V, and U (see Rencher, 2002, Appendix A10–A12, pp. 574–586). There are several approximate F-statistics for Λ, V, U, and θ that are quite complex. Software like SAS, SPSS, etc. provide confidence bounds for all the four test statistics.

The four statistics may differ in overall decision-making as they are computed differently. Wilks' Λ uses multiplication operator for all s eigenvalues, Pillai trace (V) and Hotelling-Lawley's trace (U) use summation operator for all s eigenvalues and Roy's largest root statistic (θ) uses only the largest single eigenvalue. Then, which one should be preferred? Two frequently used criteria are power and robustness of the test, where power is defined as the probability of rejecting a false null hypothesis and robustness is the ability to provide correct inferences when one or more assumptions are violated. Roy's test (θ) is the most powerful if the population mean vectors are collinear, i.e., the aggregated variance of the response variables is concentrated on a single eigenvalue (Haase and Ellis, 1987). When the aggregated variance of the response variables is diffused across several eigenvalues, the other three statistics are preferred. From robustness point of view, Pillai's trace (V) is the most preferred. Wilks' Λ is very popular but fails to give accurate results if the homogeneity of variances is grossly violated. For detailed discussion, reader may refer to Rencher (2002, pp. 176–178) and Olson (1976).

Example 7.10

Consider the data set given in Table 7.13. Do the mean vectors of the three processes differ?

Solution:

There are three processes, A, B, and C (i.e., L = 3), two variables OD and ID (i.e., p = 2), and ten observations from each of the three processes (i.e., $n_1 = n_2 = n_3 = 3$). The matrices $SSCP_B$, $SSCP_E$, and $SSCP_T$ were computed in example 7.9. The MANOVA table is given in Table 7.15.

Further,

$$SSCP_E^{-1} = \begin{bmatrix} 0.030 & -0.006 \\ -0.006 & 0.044 \end{bmatrix}$$

$$\left(SSCP_B + SSCP_E\right)^{-1} = SSCP_T^{-1} = \begin{bmatrix} 0.014 & -0.006 \\ -0.006 & 0.037 \end{bmatrix}$$

TABLE 7.15
MANOVA Table for Example 7.10

Sources of Variation	SSCP	Degrees of Freedom
Population	$SSCP_B = \begin{bmatrix} 42.47 & 7.00 \\ 7.00 & 5.60 \end{bmatrix}$	L–1 = 2
Error	$SSCP_E = \begin{bmatrix} 34.83 & 4.68 \\ 4.68 & 23.22 \end{bmatrix}$	$\sum_{\ell=1}^{L} n_\ell - L = N - L = 30 - 3 = 27$
Total	$SSCP_T = \begin{bmatrix} 77.30 & 11.68 \\ 11.68 & 28.82 \end{bmatrix}$	$\sum_{\ell=1}^{L} n_\ell - 1 = N - 1 = 30 - 1 = 29$

$$SSCP_E^{-1} SSCP_B = \begin{bmatrix} 1.212 & 0.173 \\ 0.057 & 0.206 \end{bmatrix}$$

So, Wilks' Λ

$$\Lambda = \frac{|SSCP_E|}{|SSCP_T|} = \frac{\begin{vmatrix} 34.83 & 4.68 \\ 4.68 & 23.22 \end{vmatrix}}{\begin{vmatrix} 77.3 & 11.68 \\ 11.68 & 28.82 \end{vmatrix}} = \frac{786.85}{2091.36} = 0.376$$

Pillai's trace V is

$$V = trace\left[\left(SSCP_B + SSCP_E\right)^{-1} SSCP_B\right] = trace\left[\left(SSCP_T\right)^{-1} SSCP_B\right]$$

$$= trace\left[\begin{pmatrix} 0.014 & -0.006 \\ -0.006 & 0.037 \end{pmatrix}\begin{pmatrix} 42.47 & 7 \\ 7 & 5.6 \end{pmatrix}\right] = trace\begin{pmatrix} 0.552 & 0.065 \\ 0.00418 & 0.165 \end{pmatrix} = 0.717$$

Hotelling Lawley's trace (U)

$$U = trace(SSCP_E^{-1} SSCP_B) = trace\begin{bmatrix} 1.212 & 0.173 \\ 0.057 & 0.206 \end{bmatrix} = 1.418$$

Roy's test statistic (θ)

The matrix $SSCP_E^{-1} SSCP_B = \begin{bmatrix} 1.212 & 0.173 \\ 0.057 & 0.206 \end{bmatrix}$ is of the order of 2×2 and its determinant is 0.24 (positive). Let $\lambda_1 \geq \lambda_2$ are the two eigenvalues of $SSCP_E^{-1} SSCP_B$. Using the procedure given in Section 3.1.12 (see Chapter 3), the two eigenvalues are $\lambda_1 = 1.22$ and $\lambda_2 = 0.20$. So, the Roy's test statistic is

$$\theta = \frac{\lambda_1}{1+\lambda_1} = \frac{1.22}{2.22} = 0.55.$$

As the data set exhibit homogeneity of variances, of all the four statistics Wilks' Λ is the best statistic to test $H_0 : \mu_1 = \mu_2 = \mu_3$. Wilks' Λ value is 0.376 which is comparatively a low value. So, we can conclude that the population mean vectors differ. To confirm statistically, we test the statistic $-\left(N - 1 - \frac{p+L}{2}\right) \ell n\ \Lambda$ that follows $\chi^2_{P(L-1)}$.

Test statistic: $-\left(N - 1 - \frac{p+L}{2}\right) \ell n\ \Lambda = -\left(30 - 1 - \frac{2+3}{2}\right) \ell n\ (0.376) = 25.92$

Cut-off value: $\chi^2_{p(L-1),\alpha} = \chi^2_{2(3-1),0.05} = 9.49$

Decision: As the computed test statistic is (25.92) greater than the cut-off value (9.49), we can safely reject $H_0 : \mu_1 = \mu_2 = \mu_3$.

The hypothesis test using the F approximation for all the four multivariate test statistics discussed in this section could also be followed.

7.2.5 Estimation of Parameters

Consider the MANOVA model given in Section 7.2.1. The parameters of MANOVA are

μ = Grand mean vector
μ_ℓ = Population mean vectors l, $l = 1, 2, \ldots, L$
τ_ℓ = Population effect vectors l, $l = 1, 2, \ldots, L$
$\epsilon_{i\ell}$ = Random error vectors, $l = 1, 2, \ldots, L$ and $i = 1, 2, \ldots, n_l$

The point estimates of the above parameters are

$$\begin{aligned} \hat{\mu} &= \bar{\mathbf{x}} \\ \hat{\mu}_\ell &= \bar{\mathbf{x}}_\ell \\ \hat{\tau}_\ell &= \bar{\mathbf{x}}_\ell - \bar{\mathbf{x}} \\ \hat{\epsilon}_{i\ell} &= \mathbf{x}_{i\ell} - \bar{\mathbf{x}}_\ell \end{aligned} \tag{7.45}$$

In addition, $\dfrac{SSCP_E}{N-L}$ can be used as the estimate of $\hat{\Sigma}$, (the population covariance matrix) assuming homogeneity of variances (see Equation 7.36).

Example 7.11

Consider Example 7.10. Obtain point estimates of grand mean, population means, population effects, and errors (residuals).

Solution:

Using Equation 7.45, we compute the following tables, where Table 7.16 shows the estimates of grand mean ($\hat{\mu} = \bar{\mathbf{x}}$), population mean vectors ($\hat{\mu}_\ell = \bar{\mathbf{x}}_\ell$), and population effect ($\hat{\tau}_\ell = \bar{\mathbf{x}}_\ell - \bar{\mathbf{x}}$) vectors under the column (estimates) and Table 7.17 shows the residuals (errors) ($\hat{\epsilon}_{i\ell} = \mathbf{x}_{i\ell} - \bar{\mathbf{x}}_\ell$).

TABLE 7.16
Estimation of Parameters

		Observations										Estimates	
Population	**Variables**	**1**	**2**	**3**	**4**	**5**	**6**	**7**	**8**	**9**	**10**	**Average**	**Effect**
Process A	OD	20	21	20	21	23	19	20	19	19	20	20.20	0.63
	ID	6	6	9	6	7	7	6	7	5	6	6.50	–0.40
Process B	OD	17	17	19	17	16	19	18	18	18	20	17.90	–1.67
	ID	6	6	7	8	6	7	7	6	6	8	6.70	–0.20
Process C	OD	20	20	21	20	21	21	22	19	22	20	20.60	1.03
	ID	8	7	8	7	8	9	7	7	6	8	7.50	0.60
Grand average											OD	19.57	
											ID	6.90	

TABLE 7.17
Residuals

		Residuals									
Population	**Variables**	**1**	**2**	**3**	**4**	**5**	**6**	**7**	**8**	**9**	**10**
Process A	OD	–0.20	0.80	–0.20	0.80	2.80	–1.20	–0.20	–1.20	–1.20	–0.20
	ID	–0.50	–0.50	2.50	–0.50	0.50	0.50	–0.50	0.50	–1.50	–0.50
Process B	OD	–0.90	–0.90	1.10	–0.90	–1.90	1.10	0.10	0.10	0.10	2.10
	ID	–0.70	–0.70	0.30	1.30	–0.70	0.30	0.30	–0.70	–0.70	1.30
Process C	OD	–0.60	–0.60	0.40	–0.60	0.40	0.40	1.40	–1.60	1.40	–0.60
	ID	0.50	–0.50	0.50	–0.50	0.50	1.50	–0.50	–0.50	–1.50	0.50

100(1 – α)% Simultaneous CI

When H_0 is rejected (as seen in Example 7.10), the follow up questions are (i) which pair(s) of the population mean vectors differ and (ii) which of the response variables contribute to the decision (of rejecting H_0). Here we rely on the simultaneous confidence interval (SCI) approach.

For one-way MANOVA there are L populations (treatments), so we can have ${}^{L}C_2 = L(L-1)/2$ pairs of populations to compare. The methodology for the difference between two-population mean vectors described in Chapter 6 can be used. Each comparison gives 100(1 – α)% confidence region (CR). So, there will be ${}^{L}C_2 = L(L-1)/2$ CRs. The problem of inflation of type-I error rate can be overcome by using the Bonferroni approach. The CRs that reject H_0 can be further developed to obtain the 100(1 – α)% SCI for each of the p-variables.

In order to get 100(1 – α)% SCI for each of the p response variables, we require considering, in addition to ${}^{L}C_2$ pairs of comparisons, the p response variables. This makes the total number of comparisons equal to $p \times {}^{L}C_2 = pL(L-1)/2$. Considering the approach presented in Section 6.5.2 of Chapter 6, the Bonferroni 100(1 – α)% SCI for $\mu_{lj} - \mu_{mj}$, $l = 1,2,\ldots,L;\ m = 1,2,\ldots,L;\ l \neq m$ and $j = 1,2,\ldots,p$ is given below.

$$\left(\bar{x}_{lj} - \bar{x}_{mj}\right) - t_{N-L}^{\alpha_j/2}\sqrt{\left(\frac{1}{n_l}+\frac{1}{n_m}\right)s_{jj-pooled}} \le \mu_{lj} - \mu_{mj} \le \left(\bar{x}_{lj} - \bar{x}_{mj}\right) + t_{N-L}^{\alpha_j/2}\sqrt{\left(\frac{1}{n_l}+\frac{1}{n_m}\right)s_{jj-pooled}} \quad (7.46)$$

where $\alpha_j = \dfrac{2\alpha}{pL(L-1)}$ and $s_{jj-pooled}$ is the *j*-th diagonal element of the matrix $\dfrac{SSCP_E}{N-L}$.

It should be noted here that knowing the 100(1 – α)% SCI for mean difference $\mu_{lj} - \mu_{mj}$, $l = 1,2,\ldots,L;\ m = 1,2,\ldots,L;\ l \neq m$ and $j = 1,2,\ldots,p$ is equivalent to knowing the 100(1 – α)% SCI for effect difference $\tau_{lj} - \tau_{mj}$, $l = 1,2,\ldots,L;\ m = 1,2,\ldots,L;\ l \neq m$ and $j = 1,2,\ldots,p$ as shown below.

$$\tau_{lj} - \tau_{mj} = (\mu_{lj} - \mu_j) - (\mu_{mj} - \mu_j) = (\mu_{lj} - \mu_{mj})$$

When the 100(1 – α)% SCI for mean difference $\mu_{lj} - \mu_{mj}$ does not contain zero, we can conclude that the *j*-th response variable is responsible for the mean differences in the *l*-th and *m*-th populations.

Example 7.12

Consider the data set given in Table 7.13. Obtain 100(1–α)% SCI for OD and ID for all possible comparisons.

Solution:

From Example 7.9, we get the following

$$\text{Process A: } \bar{\mathbf{x}}_1 = \begin{bmatrix} 20.20 \\ 6.50 \end{bmatrix}, n_1 = 10$$

$$\text{Process B: } \bar{\mathbf{x}}_2 = \begin{bmatrix} 17.90 \\ 6.70 \end{bmatrix}, n_2 = 10$$

$$\text{Process C: } \bar{\mathbf{x}}_3 = \begin{bmatrix} 20.60 \\ 7.50 \end{bmatrix}, n_3 = 10$$

$$SSCP_E = \begin{bmatrix} 34.83 & 4.68 \\ 4.68 & 23.22 \end{bmatrix}$$

Further N = 30, L = 3, and P = 2. So, using Equation 7.46, there will be $P(L-1)L/2 = \frac{2\times2\times3}{2} = 6$ comparisons.

Let's say α = 0.05 and all pairs of comparisons are equally weighted. Then, $\alpha_j = \frac{2\alpha}{PL(L-1)} = \frac{0.05}{6} = 0.0083.$

Now, using Equation 7.46, we can write that the 100(1 – α)% = 95% SCI for all the pairs of comparisons, $\mu_{\ell j} - \mu_{mj}$ is $\left(\bar{x}_{\ell j} - \bar{x}_{mj}\right) \pm t_{N-L}^{(\alpha_j/2)} \sqrt{\left(\frac{1}{n_\ell} + \frac{1}{n_m}\right) s_{jj-pooled}}$

where, $t_{N-L}^{(\alpha_j/2)} = t_{30-3}^{\left(\frac{0.0083}{2}\right)} = t_{27}^{(0.0042)} = 2.844$ and

$$\mathbf{S}_{pooled} = \frac{SSCP_E}{N-L} = \frac{1}{27}\begin{bmatrix} 34.83 & 4.68 \\ 4.68 & 23.22 \end{bmatrix} = \begin{bmatrix} 1.29 & 0.16 \\ 0.16 & 0.86 \end{bmatrix}.$$

(i) For OD, 100(1–α)% SCI for mean difference between processes A and B $(\mu_{11} - \mu_{21})$ is

$$\left(\bar{x}_{11} - \bar{x}_{21}\right) \pm t_{27}^{(0.0042)} \sqrt{\left(\frac{1}{n_\ell} + \frac{1}{n_m}\right) s_{11-pooled}}$$

$$= \left(20.20 - 17.90\right) \pm 2.844 \sqrt{\left(\frac{1}{10} + \frac{1}{10}\right) 1.29}$$

$$= 2.3 \pm 2.844 \times 0.508 = 2.3 \pm 1.44$$

So, the SCI is (0.89, 3.74) which does not contain zero. Hence, the mean difference is significant.

(ii) For OD, 100(1 – α)% SCI for mean difference between processes A and C $(\mu_{11} - \mu_{31})$

$$\left(\bar{x}_{11} - \bar{x}_{31}\right) \pm t_{27}^{(0.0042)} \sqrt{\left(\frac{1}{n_1} + \frac{1}{n_3}\right) s_{11-pooled}}$$

$$= \left(20.20 - 20.60\right) \pm 2.844 \sqrt{\left(\frac{1}{10} + \frac{1}{10}\right) 1.29}$$

$$= -0.40 \pm 1.44$$

So, the SCI is (–1.84, 1.04) which contains zero. Hence, the mean difference is not significant.

(iii) For OD, 100(1 – α)% SCI for the mean difference between processes B and C $(\mu_{21} - \mu_{31})$

$$\left(\bar{x}_{21} - \bar{x}_{31}\right) \pm t_{27}^{(0.0042)} \sqrt{\left(\frac{1}{n_2} + \frac{1}{n_3}\right) s_{11-pooled}}$$

$$= \left(17.90 - 20.60\right) \pm 2.844 \sqrt{\left(\frac{1}{10} + \frac{1}{10}\right) \times 1.29}$$

$$= -2.7 \pm 1.44$$

So, the SCI is (–4.14, –1.26) which does not contain zero. Hence, the mean difference is significant.

(iv) For ID, 100(1 – α)% SCI for the mean difference between processes A and B $(\mu_{12} - \mu_{22})$

$$\left(\bar{x}_{12} - \bar{x}_{22}\right) \pm t_{27}^{(0.0042)} \sqrt{\left(\frac{1}{n_1} + \frac{1}{n_2}\right) s_{22-pooled}}$$

$$= \left(6.50 - 6.70\right) \pm 2.844 \sqrt{\left(\frac{1}{10} + \frac{1}{10}\right) \times 0.86}$$

$$= 0.20 \pm 2.844 \times 0.412$$

$$= -0.20 \pm 1.18$$

So, the SCI is (–1.38, 0.98) which contains zero. Hence, the mean difference is not significant.

(v) For ID, 100(1 – α)% SCI for the mean difference between processes A and C $(\mu_{12} - \mu_{32})$

$$\left(\bar{x}_{12} - \bar{x}_{32}\right) \pm t_{27}^{(0.0042)} \sqrt{\left(\frac{1}{n_1} + \frac{1}{n_3}\right) s_{22-pooled}}$$

$$= \left(6.50 - 7.50\right) \pm 2.844 \sqrt{\left(\frac{1}{10} + \frac{1}{10}\right) \times 0.86}$$

$$= -1.00 \pm 1.18$$

So, the SCI is (–2.18, 0.18) which contain zero. Hence, the mean difference is not significant.

(vi) For ID, 100(1 – α)% SCI for the mean difference between processes B and C $(\mu_{22} - \mu_{32})$

$$\left(\bar{x}_{22} - \bar{x}_{32}\right) \pm t_{27}^{(0.0042)} \sqrt{\left(\frac{1}{n_2} + \frac{1}{n_3}\right) s_{22-pooled}}$$

$$= \left(6.70 - 7.50\right) \pm 2.844 \sqrt{\left(\frac{1}{10} + \frac{1}{10}\right) \times 0.86}$$

$$= -0.80 \pm 1.18$$

So, the SCI is (–1.98, 0.38) which contain zero. Hence, the mean difference is not significant.

The use of multiple ANOVAs were also suggested where each ANOVA considers one response variable. The response variable with the largest univariate F ratio has the largest between population (treatment) differences with respect to within population (treatment) differences, and vice versa (Finn, 1974). These F ratios, however, do not control the correlation between the p response variables. For criticism of this approach, please see Hummel and Sligo (1971), and Muller and Peterson (1984).

In search of better methods, Tatsuoka (1969) used discriminant analysis. Discriminant analysis creates s = min(p, L–1) discriminant functions, where each function is a linear combination of the p response variables. The functions are extracted in such a manner that the ratio between population variance to within population variance is maximized for each of the discriminant functions.

The other important concept in ANOVA and MANOVA is the use of contrasts to test different groups of comparisons, popularly known as patterns of comparisons. Contrast is used to manipulate comparisons among the populations (treatments). For a review on MANOVA contrasts, see Bray and Maxwell (1982) and for mathematical treatment see Rencher (2002), Anderson (2003), and Mardia (1970).

7.2.6 Model Adequacy Tests

Model adequacy test refers to the ability of a model to explain the purpose for which it is built. For ANOVA or MANOVA, the purpose is to test whether the population means or mean vectors are different or not. As seen earlier, both ANOVA and MANOVA do it by partitioning the total variability into model and error variability. If there is a difference, the model variability must be adequately large in comparison to error variability, which is captured through F-ratio. Another measure (similar to squared multiple correlation in regression, R^2, see Chapter 8) is Fisher's correlation ratio η^2. For ANOVA, η^2 is

$$\eta^2 = \frac{\text{Sum squares between populations}}{\text{Sum squares total}} = \frac{SSB}{SST} \tag{7.47}$$

η^2 represents the proportion of variance in the response variable that can be attributed to the differences among the population means.

The MANOVA counterpart of η^2, called multivariate association, is given below.

(i) Using Wilks' Λ: $\eta^2 = 1 - \Lambda^{1/s}$

(ii) Using Pillai's trance (V): $\eta^2 = \dfrac{V}{S}$

(iii) Using Hotelling-Lawley's trace (U): $\eta^2 = \dfrac{U/S}{1+U/S}$

(iv) Using Roy's statistic (θ): $\eta^2 = \theta$

Example 7.13

Consider the data set given in Table 7.13. Compute multivariate association η^2 using the four MANOVA test statistics.

Solution:

From Example 7.10, we get

Wilks' lambda, $\Lambda = 0.376$,

Pillai's trace, $V = 0.714$,

Hotelling Lawley's trace, $U = 1.418$, and

Roy's statistic, $\theta = 0.55$.

Now, using the concept given in Equation 7.47 for multivariate association η^2 and its MANOVA counterparts, we get the results as in Table 7.18.

TABLE 7.18
MANOVA Table for Example 7.13

MANOVA Test Statistic	Multivariate Association η^2	s Value for The Data	η^2-Value for The Data
Wilks' lambda Λ	$\eta^2 = 1 - \Lambda^{1/s}$	s = min (P, L-1) = min (2,2) = 2	$\eta^2 = 1 - 0.376^{1/2} = 0.39$
Pillai's trace V	$\eta^2 = \frac{V}{s}$	s = 2	$\eta^2 = \frac{0.714}{2} = 0.36$
Hotelling-Lawley's trace U	$\eta^2 = \frac{U/s}{1+U/s}$	s = 2	$\eta^2 = \frac{1.418/2}{1+1.418/2} = 0.42$
Roy's statistic, θ	$\eta^2 = \theta$	-	$\eta^2 = 0.55$

The η^2 values are different for different test statistics. Considering the problem of small samples, the most conservative estimate, 0.36, is given by Pillai's trace (V) which is also close to Hotelling-Lawley's trace (U) (0.42) and Wilks' lambda Λ (0.39). However, Roy's statistic (θ) gives an association of 0.55, which is appreciable. Which one is the best estimate is difficult to judge but can be approximated with a particular test statistic, if the conditions of the use of that statistic are satisfied.

7.2.7 Test of Assumptions

Test of Normality

The usual procedure follows the test the multivariate normality of the residuals using χ^2 Q-Q plot of Mahalanobis distances of the residual vectors. If the multivariate normality of residuals is violated, then the univariate normality test of residuals is performed for each of the response variables separately. The multiple univariate tests help in identifying which of the response variables contributes to the departure from multivariate normality of the residuals.

Following the procedures given in Section 5.6.2 that uses the Mahalanobis distance formula reproduced below (Equation 7.48) and using the data in Table 7.13, the χ^2 Q-Q plot (Figure 7.12) was obtained.

$$MD_i^2 = (\mathbf{x}_i - \bar{\mathbf{x}})^T \mathbf{S}^{-1} (\mathbf{x}_i - \bar{\mathbf{x}}), \quad i = 1,2,\ldots n \tag{7.48}$$

From Figure 7.12, it is seen that there is no apparent departure from multivariate normality as the MD^2 observations more or less lie along a straight line with slope 1 passing through the origin.

Test of Independence

As stated earlier, the test of independence is the test of whether each of the n data vectors is collected independently or there is bias in data collection. If there is a bias, the order of observation influences on the observed values. Residual versus order of observation plot, namely serial correlation plot, is used to test this independence. We assume that if the squared MDs of the residual vectors are not related to the orders of observations, the observed vectors are collected independently. So, the plot of the MD^2 versus the "order of observation" should not show any systematic patterns, rather it exhibits randomness. Figure 7.13 shows the serial correlation plot of MD^2 versus the "order of observation." The plot exhibits randomness. So, the observed data vectors are independent.

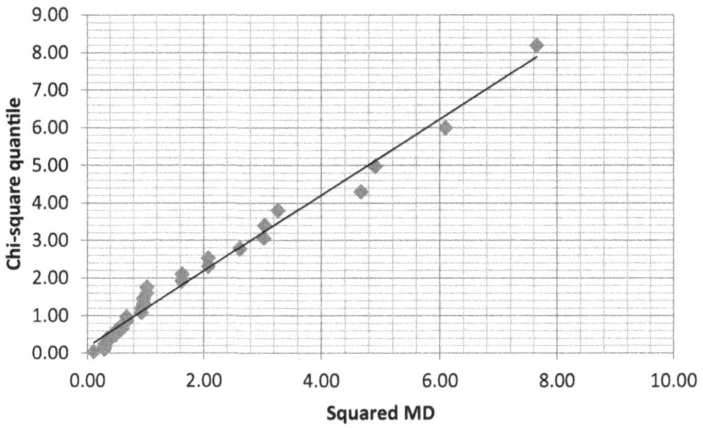

FIGURE 7.12 χ^2 Q-Q plot of squared Mahalanobis distances (MD^2) of the residual vectors.

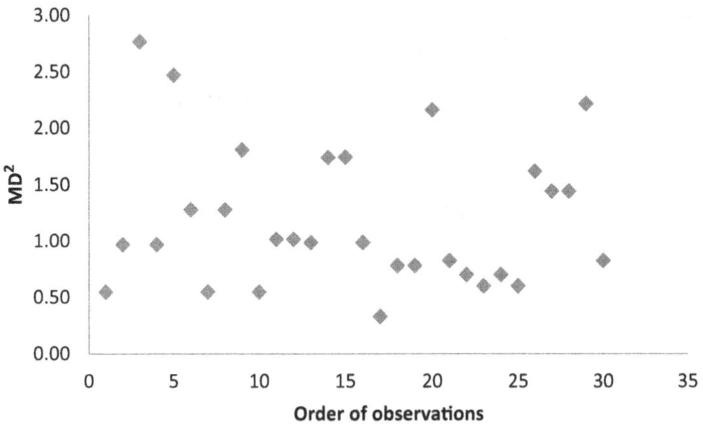

FIGURE 7.13 Serial correlation plot of MD^2 of the residual vectors and its order of observation.

Test of Homogeneity of Covariance Matrices

The Box M test in Section 7.2.2 shows the covariance matrices of the processes A, B, and C do not differ in satisfying the homogeneity of the covariance matrices. For visualization purposes, a plot of log of determinants of the components of the Box M test is shown in Figure 7.14. For construction of the plot, readers may refer to Friendly and Sigal (2020).

The plot compares the log determinants of the covariance matrices of Processes A, B, and C together with that of their pooled covariance matrix (see Example 7.8 for computation). Figure 7.14 can be used to throw insights on (i) the extent that the covariance matrices are all equal and (ii) if they differ, how the groups differ from each other and from the pooled covariance matrix. The bars represent the 95% confidence intervals of the log determinants of the covariance matrices of Processes A, B, C, and the log determinant of their pooled covariance matrix. The approximate central limit theorem applies to the log determinant of a covariance matrix (Cai et al., 2015). All the three process intervals overlap with that for the log determinant of the pooled covariance. This signifies that the covariance matrices across the three processes do not differ from each other. The Box M test (see Example 7.8) also provides same conclusion. Further, we check the plots residuals versus predicted values for both the response variables outer diameter (OD) and inner diameter (ID)

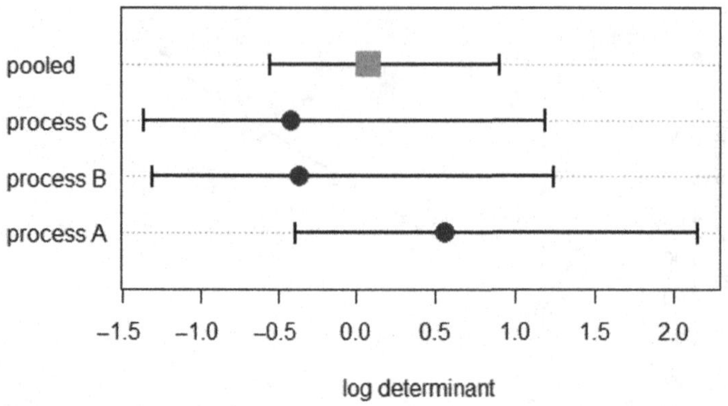

FIGURE 7.14 Plot of log of determinants of covariance matrices (for processes A, B, C, and pooled).

TABLE 7.19
Modified Steel Washers' Data with Supplier Information

Factors		Observations									
Supplier	**Process**	**OD**					**ID**				
I	A	20	21	20	21	23	6	6	9	6	7
	B	17	17	19	17	16	6	6	7	8	6
	C	20	20	21	20	21	8	7	8	7	8
II	A	19	20	19	19	20	7	6	7	5	6
	B	19	18	18	18	20	7	7	6	6	8
	C	21	22	19	22	20	9	7	7	6	8

(Figure 7.15). The plots do not show any appreciable funnel structure (open to right or to left), satisfying the assumption of homogeneity.

7.3 TWO-WAY MANOVA

Consider the steel washers' production data given in Table 7.13. Let us add that the raw materials for production of steel washers are supplied by two suppliers, supplier I and supplier II. Accordingly, Table 7.13 is modified and the resultant data is given in Table 7.19.

We introduced one more terminology "factor" in Table 7.19, under which "supplier" with two levels (categories) and "process" with three levels (categories) are mentioned. So, there are two factors, namely supplier and process. The levels or categories against each of the (two) factors are equivalent to the populations or treatments mentioned under one-way ANOVA and one-way MANOVA. The term factor represents the categorical independent variable, which affects the response (dependent) variable(s); for example, in steel washers' data, the response variable(s) is (are) either OD or ID or both. It should be noted here that in ANOVA or MANOVA we test the equality of mean or mean vectors at each of the levels (categories) of the factors considered.

With respect to the data given in Table 7.19, the following questions are of immediate interest:

1. Of the three processes, A, B, and C, which of the process means with respect to OD differs from the others?

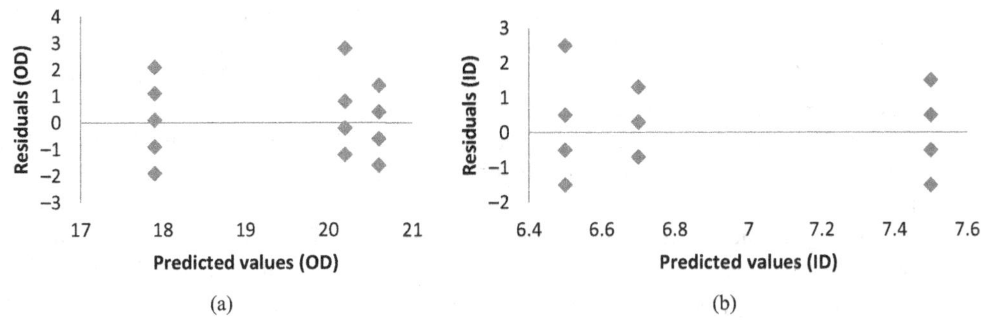

FIGURE 7.15 Plot of residuals versus predicted values for (a) OD and (b) ID.

2. Of the three processes, A, B, and C, which of the process means with respect to ID differs from the others?
3. Of the two suppliers, I and II, which of the supplier means with respect to OD differs from the other?
4. Of the two suppliers, I and II, which of the supplier means with respect to ID differs from the other?
5. Of the three processes, A, B, and C, which of the process mean vectors with respect to OD and ID differs from the others?
6. Of the two suppliers, I and II, which of the supplier mean vectors with respect to OD and ID differs from the other?
7. If we consider the two factors process and supplier simultaneously, which of the process means or which of the supplier means with respect to OD differs from the others?
8. If we consider the two factors process and supplier simultaneously, which of the process means or which of the supplier means with respect to ID differs from the others?
9. If we consider the two factors process and supplier simultaneously, which of the process mean vectors or which of the suppliers mean vector with respect to OD and ID differs from the others?
10. Do the two factors process and supplier interact with respect to OD or ID or both?

Answers to questions 1 and 2 can be obtained using one-way ANOVA (see Section 7.1), and to questions 3 and 4 by using t-test (see Chapter 2). Question 5 can be answered by using one-way MANOVA (see Section 7.2) and Hoteling's T^2 gives answer to question 6 (see Chapter 6). Questions 7 and 8 can be answered by using two-way ANOVA while for question 9 we employ two-way MANOVA. In addition, to answer the question 10, we require using two-way ANOVA if we consider the response variables OD and ID separately, and two-way MANOVA if the effects of the two factors on both OD and ID are evaluated simultaneously. The following sections entail the answers to questions 7 to 10.

7.3.1 Two-Way ANOVA

We briefly describe the essence of two-way ANOVA so that we can draw an analogy to two-way MANOVA that will be described in the next section. In addition, it is to be noted here that the assumptions underlined in Section 7.1 are equally applicable here. To get into the model of two-way ANOVA, let us assume the following notations.

- There are two factors F_1 and F_2 with levels L and M, respectively. For steel washers' data, let F_1 be process with L (=3) levels and F_2 is supplier with M (=2) levels.

- x_{ilm} represents the i-th observation to be collected on a response variable attributable to the l-th level of F_1 and m-th level of F_2, where $i = 1, 2, \ldots, n$, $l = 1, 2, \ldots, L$, and $m = 1, 2, \ldots, M$. For steel washers' data, the response variable could be the OD or ID, the l-th level of F_1 could be process B ($l = 2$) and the m-th level of F_2 could be supplier I ($m = 1$).
- μ, μ_l, μ_m and μ_{lm} are the grand mean, *l-th* level (population) mean for F_1, m-th level (population) mean for F_2, and the mean at the l-th and m-th level combination of the factors F_1 and F_2, respectively.
- τ_l, β_m and γ_{lm} are the l-th level (population) fixed effect for F_1, m-th level (population) fixed effect for F_2, and the fixed interaction effect at the l-th and m-th level combination of the factors F_1 and F_2, respectively. τ_l and β_m are known as main effects of the factors F_1 and F_2, respectively, and γ_{lm} is known as the interaction effect between the two factors.
- ϵ_{ilm} represents the i-th error to be observed on a response variable attributable to the l-th level of F_1 and the m-th level of F_2, where $l = 1, 2, \ldots, L$ and $m = 1, 2, \ldots, M$, that cannot be captured by a two-way ANOVA model.

The population two-way ANOVA model can be written as

$$x_{ilm} = \mu + \tau_l + \beta_m + \gamma_{lm} + \epsilon_{ilm} \tag{7.49}$$

where

$$\begin{aligned} \tau_l &= \mu_l - \mu \\ \beta_m &= \mu_m - \mu \\ \gamma_{lm} &= \mu_{lm} - \mu_l - \mu_m + \mu \\ \epsilon_{ilm} &= x_{ilm} - \mu_{lm} \end{aligned}$$

and

$$\sum_{l=1}^{L} \tau_l = \sum_{m=1}^{M} \beta_m = \sum_{l=1}^{L} \gamma_{lm} = \sum_{m=1}^{M} \gamma_{lm} = 0$$

ϵ_{ilm} is an independent random variable normally distributed with mean 0 and variance σ^2.

The four parameters shown in Equation 7.49 can be estimated from sample data. To estimate the parameters, assume the following notations attributable to the sample data.

- x_{ilm} represents the i-th observation collected on a response variable attributable to the l-th level of F_1 and the m-th level of F_2 where $i = 1, 2, \ldots, n$, $l = 1, 2, \ldots, L$, and $m = 1, 2, \ldots, M$. Here the total observations are $N = nLM$.
- G, A_l, B_m and $(AB)_{lm}$ are the grand total, l-th level (sample) total for F_1, m-th level (sample) total for F_2, and the cell (sample) total at the l-th and m-th level combination of the factors F_1 and F_2, respectively. See Table 7.20 under Example 7.14 for further clarity.
- $\bar{x}, \bar{x}_{l.}, \bar{x}_{.m}$ and $\bar{x}_{lm}$ are the grand average, l-th level (sample) average for F_1, m-th level (sample) average for F_2, and the sample average at the l-th and m-th level combination of the factors F_1 and F_2, respectively. See Table 7.21 for further clarity.

The estimates of the totals and means are,

$$\begin{aligned} N &= nLM \\ G &= \sum_{m=1}^{M} \sum_{\ell=1}^{L} \sum_{i=1}^{n} x_{i\ell m} \\ A_\ell &= \sum_{m=1}^{M} \sum_{i=1}^{n} x_{i\ell m} \\ B_m &= \sum_{l=1}^{L} \sum_{i=1}^{n} x_{i\ell m} \\ (AB)_{lm} &= \sum_{i=1}^{n} x_{i\ell m} \end{aligned} \tag{7.50}$$

$$\begin{aligned} \hat{\mu} &= \bar{x} = \frac{G}{N} \\ \hat{\mu}_l &= \bar{x}_l = \frac{A_l}{nM} \\ \hat{\mu}_m &= \bar{x}_m = \frac{B_m}{nL} \\ \hat{\mu}_{lm} &= \bar{x}_{lm} = \frac{(AB)_{lm}}{n} \end{aligned} \tag{7.51}$$

And the estimates of the main and interaction effects and error are

$$\begin{aligned} \hat{\tau}_l &= \hat{\mu}_l - \hat{\mu} = \bar{x}_l - \bar{x} \\ \hat{\beta}_m &= \hat{\mu}_m - \hat{\mu} = \bar{x}_m - \bar{x} \\ \hat{\gamma}_{lm} &= \hat{\mu}_{lm} - \hat{\mu}_l - \hat{\mu}_m + \hat{\mu} == \bar{x}_{lm} - \bar{x}_l - \bar{x}_m + \bar{x} \\ \hat{\epsilon}_{ilm} &= x_{ilm} - \hat{\mu}_{lm} = x_{ilm} - \bar{x}_{lm} \end{aligned} \tag{7.52}$$

Example 7.14

Consider the OD data set given in Table 7.13. Compute the main effects, interaction effects, and residuals using two-way ANOVA.

Solution:

Consider the OD data set given in Table 7.13 and perform the following steps:

Step 1: Compute grand total (G), row total (A_l), column total (B_m) and cell total ($(AB)_{lm}$) using Equation 7.50. The results are given in Table 7.20.

Step 2: Compute grand average ($\bar{x}$), row average ($\bar{x}_l$), column average ($\bar{x}_m$), and cell average ($\bar{x}_{lm}$). These can be obtained by dividing the overall total, row total, column total, and cell total by the respective number of observations (see Equation 7.51). The results are given in Table 7.21.

Step 3: Compute main ($\hat{\tau}_l$, $\hat{\beta}_m$) and interaction effects ($\hat{\gamma}_{lm}$) using Equation 7.52. The results are given in Table 7.22.

Step 4: Compute residuals ($\hat{\epsilon}_{ilm}$) using Equation 7.52. The residuals are given in Table 7.23.

TABLE 7.20
Grand Total, Row Total, Column Total and Cell Total for Example 7.14

Response Variable (OD)	Supplier		
Process	**S-I**	**S-II**	**Row Total (A_l)**
A	20, 21, 20, 21, 23 Cell total ($(AB)_{11}$): 105	19, 20, 19, 19, 20 Cell total ($(AB)_{12}$): 97	**202**
B	17, 17, 19, 17, 16 Cell total ($(AB)_{21}$): 86	19, 18, 18, 18, 20 Cell total ($(AB)_{22}$): 93	**179**
C	20, 20, 21, 20, 21 Cell total ($(AB)_{31}$): 102	21, 22, 19, 22, 20 Cell total ($(AB)_{32}$): 104	**206**
Column Total (B_m)	**293**	**294**	**Grand total (G): 587**

TABLE 7.21
Grand Average, Row Average, Column Average and Cell Average for Example 7.14

Response Variable (OD)	Supplier		
Process	**S-I**	**S-II**	**Row Average ($\bar{x}_l$)**
A	Cell average ($\bar{x}_{11}$): 21.00	Cell average ($\bar{x}_{12}$): 19.40	**20.20**
B	Cell average ($\bar{x}_{21}$): 17.20	Cell average ($\bar{x}_{22}$): 18.60	**17.90**
C	Cell average ($\bar{x}_{31}$): 20.40	Cell average ($\bar{x}_{32}$): 20.80	**20.60**
Column average ($\bar{x}_m$)	**19.53**	**19.60**	**Grand average ($\bar{x}$): 19.57**

TABLE 7.22
Main and Interaction Effects for Example 7.14

Response Variable (OD)	Supplier		
Process	**S-I**	**S-II**	**Main Effect (ME) for Process ($\hat{\tau}_l$)**
A	IE ($\hat{\gamma}_{11}$): 0.83	IE ($\hat{\gamma}_{12}$): –0.83	**0.63**
B	IE ($\hat{\gamma}_{21}$): –0.67	IE ($\hat{\gamma}_{22}$): 0.67	**–1.67**
C	IE ($\hat{\gamma}_{31}$): –0.17	IE ($\hat{\gamma}_{32}$): 0.17	**1.03**
Main Effect (ME) for Supplier ($\hat{\beta}_m$)	**–0.03**	**0.03**	IE stands for interaction effects

TABLE 7.23
The Residuals for Example 7.14

Response Variable (OD)	Supplier									
Process	**S-I**						**S-II**			
A	−1.00	0.00	−1.00	0.00	2.00	−0.40	0.60	−0.40	−0.40	0.60
B	−0.20	−0.20	1.80	−0.20	−1.20	0.40	−0.60	−0.60	−0.60	1.40
C	−0.40	−0.40	0.60	−0.40	0.60	0.20	1.20	−1.80	1.20	−0.80

Two-way ANOVA tests the following hypotheses:

(i) Hypothesis concerning main effects (ME) of factor 1

$$H_0: \ \tau_1 = \tau_2 = ... = \tau_l = ... = \tau_L = 0$$
$$H_1: \ \tau_l \neq 0, \quad \text{for at least one } l, \ l = 1,2,....,L$$

(ii) Hypothesis concerning main effects (ME) of factor 2

$$H_0: \ \beta_1 = \beta_2 = ... = \beta_m = ... = \beta_M = 0$$
$$H_1: \ \beta_m \neq 0, \quad \text{for at least one } m, \ m = 1,2,....,M$$

(iii) Hypothesis concerning interaction effects (IE) between factors 1 and 2

$$H_0: \ \gamma_{11} = \gamma_{12} = ... = \gamma_{lm} = ... = \gamma_{LM} = 0$$
$$H_1: \ \gamma_{lm} \neq 0, \quad \text{for at least one combination of } (l, m), \ l = 1,2,....,L \text{ and } m = 1,2,....,M$$

In order to test the abovementioned hypotheses, partitioning of sum squares total (SST) (as seen in ANOVA) into individual sources of variation is done. In two-way ANOVA, the sources of variations are (i) factor 1 (F_1), (ii) factor 2 (F_2), (iii) their interaction, and (iv) error, those can be termed as SSF_1, SSF_2, SSF_{12}, and SSE, respectively. Accordingly,

$$SST = SSF_1 + SSF_2 + SSF_{12} + SSE \tag{7.53}$$

Where,

$$\begin{aligned}
SST &= \sum_{m=1}^{M}\sum_{l=1}^{L}\sum_{i=1}^{n}(x_{ilm} - \bar{x})^2 \\
SSF_1 &= nM\sum_{l=1}^{L}(\bar{x}_l - \bar{x})^2 \\
SSF_2 &= nL\sum_{m=1}^{M}(\bar{x}_m - \bar{x})^2 \\
SSF_{12} &= n\sum_{l=1}^{L}\sum_{m=1}^{M}(\bar{x}_{lm} - \bar{x}_l - \bar{x}_m + \bar{x})^2 \\
SSE &= \sum_{m=1}^{M}\sum_{l=1}^{L}\sum_{i=1}^{n}(x_{ilm} - \bar{x}_{lm})^2
\end{aligned} \tag{7.54}$$

Equation 7.54 is computationally tedious. An easier formula can be described as below (Montgomery, 2012).

TABLE 7.24
Two-Way ANOVA Table

Sources of Variation	Sum Squares (*SS*)	Degrees of Freedom (DOF)	Mean Squares (*MS*)	F_0-Value
Process	SSF_1	$L-1$	$MSF_1 = SSF_1/L-1$	$F_1 = MSF_1/MSE$
Supplier	SSF_2	$M-1$	$MSF_2 = SSF_2/M-1$	$F_2 = MSF_2/MSE$
Interaction	SSF_{12}	$(L-1)(M-1)$	$MSF_{12} = SSF_{12}/(L-1)(M-1)$	$F_{12} = SSF_{12}/MSE$
Error	SSE	$LM(n-1)$	$MSE = SSE/LM(n-1)$	
Total	SST	$N-1$		

$$
\begin{aligned}
SST &= \sum_{m=1}^{M}\sum_{\ell=1}^{L}\sum_{i=1}^{n} x_{i\ell m}^2 - \frac{G^2}{N} \\
SSF_1 &= \sum_{\ell=1}^{L}\frac{A_l^2}{nM} - \frac{G^2}{N} \\
SSF_2 &= \sum_{m=1}^{M}\frac{B_m^2}{nL} - \frac{G^2}{N} \\
SS_{subtotal} &= \sum_{m=1}^{M}\sum_{\ell=1}^{L}\frac{(AB)_{lm}^2}{n} - \frac{G^2}{N} \\
SSF_{12} &= SS_{subtotal} - SSF_1 - SSF_2 \\
SSE &= SST - SSF_{12} - SSF_1 - SSF_2
\end{aligned}
\tag{7.55}
$$

The DOF for SST, SSF_1, SSF_2, SSF_{12}, and SSE are $N-1$, $L-1$, $M-1$, $(L-1)(M-1)$, and $LM(n-1)$, respectively. The two-way ANOVA table is given in Table 7.24.

It should be noted here that in the abovementioned equations for two-way ANOVA we consider equal sample size (n) across all the combinations of the two factors F_1 and F_2. For unequal sample sizes, necessary changes must be made in the formulae. However, it is recommended to collect equal sample size (n) across all the combinations of the factors (balanced two-way ANOVA).

Example 7.15

Consider the OD data set given in Table 7.13. Develop a two-way ANOVA table and test the hypotheses concerning main and interactions effects.

Solution:

Using Equation 7.55, the following results are obtained.

$$
\begin{aligned}
SST &= \sum_{m=1}^{M}\sum_{\ell=1}^{L}\sum_{i=1}^{n} x_{i\ell m}^2 - \frac{G^2}{N} = (20^2 + 21^2 + \dots + 22^2 + 20^2) - \frac{587^2}{(5\times 3\times 2)} \\
&= 11563 - 11485.63 = 77.37 \\
SSF_1 &= \sum_{\ell=1}^{L}\frac{A_l^2}{nM} - \frac{G^2}{N} = \frac{(202^2 + 179^2 + 206^2)}{5\times 2} - \frac{587^2}{(5\times 3\times 2)} \\
&= 11528.10 - 11485.63 = 42.47 \\
SSF_2 &= \sum_{m=1}^{M}\frac{B_m^2}{nL} - \frac{G^2}{N} = \frac{(293^2 + 294^2)}{5\times 3} - \frac{587^2}{(5\times 3\times 2)} = 11485.67 - 11485.63 = 0.04
\end{aligned}
$$

TABLE 7.25
Two-Way ANOVA Table for Example 7.15

Sources of Variation	Sum Squares (*SS*)	Degrees of Freedom (DOF)	Mean Squares (*MS*)	F_0-Value
Process	$SSF_1 = 42.47$	$L - 1 = 3 - 1 = 2$	$MSF_1 = SSF_1/L - 1 = 42.47/2 = 21.24$	$F_1 = MSF_1/MSE = 21.24/0.967 = 21.96$
Supplier	$SSF_2 = 0.04$	$M - 1 = 2 - 1 = 1$	$MSF_2 = SSF_2/M - 1 = 0.04/1 = 0.04$	$F_2 = MSF_2/MSE = 0.04/0.967 = 0.04$
Interaction	$SSF_{12} = 11.66$	$(L - 1)(M - 1) = 2 \times 1 = 2$	$MSF_{12} = SSF_{12}/(L - 1)(M-1) = 11.66/2 = 5.83$	$F_{12} = MSF_{12}/MSE = 5.83/0.967 = 6.03$
Error	$SSE = 23.20$	$LM(n - 1) = 3 \times 2 \times (5 - 1) = 24$	$MSE = SSE/LM(n - 1) = 23.20/24 = 0.967$	
Total	$SST = 77.37$	$N-1=30-1=29$		

$$SS_{subtotal} = \sum_{m=1}^{M} \sum_{\ell=1}^{L} \frac{(AB)^2_{lm}}{n} - \frac{G^2}{N} = \frac{(105^2 + 97^2 + 86^2 + 93^2 + 102^2 + 104^2)}{5} - \frac{587^2}{(5 \times 3 \times 2)}$$
$$= 11539.80 - 11485.63 = 54.17$$
$$SSF_{12} = SS_{subtotal} - SSF_1 - SSF_2 = 54.17 - 42.47 - 0.04 = 11.63$$
$$SSE = SST - SSF_{12} - SSF_1 - SSF_2 = 77.37 - 11.66 - 42.47 - 0.04 = 23.20$$

The two-way ANOVA table is given in Table 7.25. The main effect of process and the interaction effect between process and supplier are significant at 0.05 probability level.

7.3.2 Two-Way MANOVA

Like in one-way MANOVA, in two-way MANOVA, an observation $\mathbf{x}_{ilm}$ is a multivariate observation and is a vector quantity. $\mathbf{x}_{ilm}$ represents the i-th multivariate observation vector on p-variables attributable to the l-th level of F_1 and m-th level of F_2, where $i = 1, 2, \ldots, n$, $l = 1, 2, \ldots, L$, and $m = 1, 2, \ldots, M$. Accordingly, the following notations are used in two-way MANOVA.

$$\mathbf{x}_{ilm} = \begin{bmatrix} x_{i1lm} \\ x_{i2lm} \\ \vdots \\ x_{iplm} \end{bmatrix}_{p\times 1}$$

$$\boldsymbol{\mu}_l = \begin{bmatrix} \mu_{1l} \\ \mu_{2l} \\ \vdots \\ \mu_{pl} \end{bmatrix}_{p\times 1} = \text{vector of } l - \text{th level means on } p - \text{variables for factor } F_1$$

$$\boldsymbol{\mu}_m = \begin{bmatrix} \mu_{1m} \\ \mu_{2m} \\ \vdots \\ \mu_{pm} \end{bmatrix}_{p\times 1} = \text{vector of } m - \text{th level means on } p - \text{variables for factor } F_2$$

$$\boldsymbol{\mu}_{lm} = \begin{bmatrix} \mu_{1lm} \\ \mu_{2lm} \\ \vdots \\ \mu_{plm} \end{bmatrix}_{p\times 1}$$ = vector of means on p-variables for the combination of the l – th level of factor F_1 nad the m-th level of factor F_2.

$$\boldsymbol{\mu} = \begin{bmatrix} \mu_1 \\ \mu_2 \\ \vdots \\ \mu_p \end{bmatrix}_{p\times 1}$$ = vector of grand eans on p – variables

The two-way MANOVA model is given below.

$$\begin{aligned} \mathbf{x}_{ilm} &= \boldsymbol{\mu} + \left(\boldsymbol{\mu}_l - \boldsymbol{\mu}\right) + \left(\boldsymbol{\mu}_m - \boldsymbol{\mu}\right) + \left(\boldsymbol{\mu}_{lm} - \boldsymbol{\mu}_l - \boldsymbol{\mu}_m + \boldsymbol{\mu}\right) + \left(\mathbf{x}_{ilm} - \boldsymbol{\mu}_{lm}\right) \\ &= \boldsymbol{\mu} + \boldsymbol{\tau}_l + \boldsymbol{\beta}_m + \boldsymbol{\gamma}_{lm} + \boldsymbol{\epsilon}_{ilm} \end{aligned} \tag{7.56}$$

Where the effects, $\boldsymbol{\tau}_l$, $\boldsymbol{\beta}_m$ and $\boldsymbol{\gamma}_{lm}$ are each a vector of the order $p\times 1$ representing the effect of p-variables for the l-th level of factor F_1, for the m-th level of factor F_2, and for their interactions respectively. ξ_{ilm} represents error vector of size $p\times 1$. So,

$$\boldsymbol{\tau}_l = \begin{bmatrix} \tau_{1l} \\ \tau_{2l} \\ \vdots \\ \tau_{pl} \end{bmatrix}_{p\times 1}$$

$$\boldsymbol{\beta}_m = \begin{bmatrix} \beta_{1m} \\ \beta_{2m} \\ \vdots \\ \beta_{pm} \end{bmatrix}_{p\times 1}$$

$$\boldsymbol{\gamma}_{lm} = \begin{bmatrix} \gamma_{1\ell m} \\ \gamma_{2\ell m} \\ \vdots \\ \gamma_{p\ell m} \end{bmatrix}_{p\times 1}$$

$$\text{and, } \boldsymbol{\epsilon}_{ilm} = \begin{bmatrix} x_{i1lm} - \mu_{1lm} \\ x_{i2lm} - \mu_{2lm} \\ \vdots \\ x_{iplm} - \mu_{plm} \end{bmatrix}$$

In addition, $\sum_{l=1}^{L} \boldsymbol{\tau}_l = \sum_{m=1}^{M} \boldsymbol{\beta}_m = \sum_{l=1}^{L} \boldsymbol{\gamma}_{lm} = \sum_{m=1}^{M} \boldsymbol{\gamma}_{lm} = \mathbf{0}.$

Two-way MANOVA tests the following hypotheses:

(i) Hypothesis concerning main effect (ME) vector of factor F_1

$$H_0: \boldsymbol{\tau}_1 = \boldsymbol{\tau}_2 = \ldots\ldots = \boldsymbol{\tau}_l = \ldots\ldots = \boldsymbol{\tau}_L = \mathbf{0}$$
$$H_1: \boldsymbol{\tau}_l \neq \mathbf{0}, \quad \text{for at least one } l, \; l = 1,2,\ldots\ldots,L$$

(ii) Hypothesis concerning main effect (ME) vector of factor F_2

$$H_0: \boldsymbol{\beta}_1 = \boldsymbol{\beta}_2 = \ldots\ldots = \boldsymbol{\beta}_m = \ldots\ldots\ldots = \boldsymbol{\beta}_M = \mathbf{0}$$
$$H_1: \boldsymbol{\beta}_m \neq \mathbf{0}, \quad \text{for at least one } m, \; m = 1,2,\ldots\ldots,M$$

(iii) Hypothesis concerning interaction effect (IE) vector between factors F_1 and F_2

$$H_0: \boldsymbol{\gamma}_{11} = \boldsymbol{\gamma}_{12} = \ldots\ldots = \boldsymbol{\gamma}_{lm} = \ldots\ldots\ldots = \boldsymbol{\gamma}_{LM} = \mathbf{0}$$
$$H_1: \boldsymbol{\gamma}_{lm} \neq \mathbf{0}, \quad \text{for at least one combination of } (l,\ m), \; l = 1,2,\ldots\ldots,L \text{ and } m = 1,2,\ldots\ldots,M$$

In order to test the abovementioned hypotheses, partitioning of sum squares and cross product total ($SSCP_T$) (as seen in one-way MANOVA) into individual sources of variation is done. In two-way MANOVA, the sources of variations are (i) factor 1 (F_1), (ii) factor 2 (F_2), (iii) their interaction, and (iv) error, those can be termed as $SSCP_{F1}$, $SSCP_{F2}$, $SSCP_{F12}$, and $SSCP_E$, respectively. These are computed using the following equations:

$$\begin{aligned}
SSCP_{F1} &= \sum_{\ell=1}^{L} Mn\left(\bar{\mathbf{x}}_l - \bar{\mathbf{x}}\right)\left(\bar{\mathbf{x}}_l - \bar{\mathbf{x}}\right)^T \\
SSCP_{F2} &= \sum_{m=1}^{M} Ln\left(\bar{\mathbf{x}}_m - \bar{\mathbf{x}}\right)\left(\bar{\mathbf{x}}_m - \bar{\mathbf{x}}\right)^T \\
SSCP_{F12} &= \sum_{m=1}^{M}\sum_{l=1}^{L} n\left(\bar{\mathbf{x}}_{\ell m} - \bar{\mathbf{x}}_l - \bar{\mathbf{x}}_m + \bar{\mathbf{x}}\right)\left(\bar{\mathbf{x}}_{\ell m} - \bar{\mathbf{x}}_l - \bar{\mathbf{x}}_m + \bar{\mathbf{x}}\right)^T \\
SSCP_E &= \sum_{m=1}^{M}\sum_{l=1}^{L}\sum_{i=1}^{n} \left(\mathbf{x}_{ilm} - \bar{\mathbf{x}}_{lm}\right)\left(\mathbf{x}_{ilm} - \bar{\mathbf{x}}_{lm}\right)^T \\
SSCP_T &= \sum_{m=1}^{M}\sum_{l=1}^{L}\sum_{i=1}^{n} \left(\mathbf{x}_{ilm} - \bar{\mathbf{x}}\right)\left(\mathbf{x}_{ilm} - \bar{\mathbf{x}}\right)^T
\end{aligned} \tag{7.57}$$

Each of the sources of variation is attributed to their corresponding DOF. Table 7.26 summarizes the information.

TABLE 7.26
Two-Way MANOVA Table

Sources of Variation	SSCP Matrix	Degrees of Freedom (DOF)
Factor F_1	$SSCP_{F1}$	$L - 1$
Factor F_2	$SSCP_{F2}$	$M - 1$
Interaction	$SSCP_{F12}$	$(L - 1)(M - 1)$
Residual	$SSCP_E$	$LM(n - 1)$
Total	$SSCP_T$	$LMn - 1$

Example 7.16

Consider the OD and ID data set given in Table 7.13. Compute SSCP matrices and develop two-way MANOVA table.

Solution:

The computation of SSCP matrices as given in Equation 7.57 is very tedious. We need to have programs using R, MATLAB, or SAS, or we can use software modules available in SAS or SPSS to compute. Using SPSS, we get the following results:

$$SSCP_{F1} = \begin{bmatrix} 42.47 & 7.00 \\ 7.00 & 5.60 \end{bmatrix}$$

$$SSCP_{F2} = \begin{bmatrix} 0.033 & -0.10 \\ -0.10 & 0.30 \end{bmatrix}$$

$$SSCP_{F12} = \begin{bmatrix} 11.67 & 3.00 \\ 3.00 & 0.80 \end{bmatrix}$$

$$SSCP_{E} = \begin{bmatrix} 23.2 & 1.80 \\ 1.80 & 22.00 \end{bmatrix}$$

$$SSCP_{T} = \begin{bmatrix} 77.37 & 11.70 \\ 11.70 & 28.70 \end{bmatrix}$$

The results are given in Table 7.27.

7.3.3 Hypothesis Testing

As stated earlier, there are four popular multivariate statistics for hypothesis testing in MANOVA. They are (i) Wilks' lambda, (ii) Pillai's trace, (iii) Hotelling-Lawley's trace, and (iv) Roy's largest root. We will explain below Wilks' lambda for the main and interactions effects. For other tests, follow Section 7.2.4 with appropriate inputs for the effects to be tested. It is to be noted here that some of the equations below are similar to that explained in Section 7.2.4 but used here for clarity and explanation.

TABLE 7.27
Two-Way MANOVA Table for Example 7.16

Sources of Variation	SSCP Matrix	Degrees of Freedom (DOF)
Process (Factor F_1)	$SSCP_{F1} = \begin{bmatrix} 42.47 & 7.00 \\ 7.00 & 5.60 \end{bmatrix}$	$L-1 = 3-1 = 2$
Supplier (Factor F_2)	$SSCP_{F2} = \begin{bmatrix} 0.033 & -0.10 \\ -0.10 & 0.30 \end{bmatrix}$	$M-1 = 2-1 = 1$
Interaction F_{12}	$SSCP_{F12} = \begin{bmatrix} 11.67 & 3.00 \\ 3.00 & 0.80 \end{bmatrix}$	$(L-1)(M-1) = (2\times 1) = 2$
Residual	$SSCP_{E} = \begin{bmatrix} 23.2 & 1.80 \\ 1.80 & 22.0 \end{bmatrix}$	$LM(n-1) = 3\times 2\times (10-1) = 54$
Total	$SSCP_{T} = \begin{bmatrix} 77.37 & 11.70 \\ 11.70 & 28.70 \end{bmatrix}$	$LMn - 1 = 3\times 2\times 10-1 = 59$

Factor F_1 (Wilks' lambda, Λ_{F1})

The hypothesis to be tested is

$$H_0: \boldsymbol{\tau}_1 = \boldsymbol{\tau}_2 = ... = \boldsymbol{\tau}_l = ... = \boldsymbol{\tau}_L = \mathbf{0}$$
$$H_1: \boldsymbol{\tau}_l \neq \mathbf{0}, \quad \text{for at least one } l, \; l = 1, 2, ..., L$$

The test statistics is Wilks' lambda, Λ_{F1}, where

$$\Lambda_{F1} = \frac{|SSCP_E|}{|SSCP_{F1} + SSCP_E|} = \prod_{j=1}^{s_1} \frac{1}{1+\lambda_j^{F1}} \tag{7.58}$$

Where λ_j^{F1} denotes the j-th eigenvalue of the matrix $(SSCP_E)^{-1} SSCP_{F1}$, and $s_1 = \min(p, L-1)$, p being the number of response variables and L the number of levels of factor F_1.

An approximation of F-distribution of Wilks' Λ_{F1}, $F_{\Lambda-F1}$ following Rao (1952) is shown below.

$$F_{\Lambda-F1} = \frac{1-(\Lambda_{F1})^{1/t_1}}{(\Lambda_{F1})^{1/t_1}} \times \frac{v_{e1}}{v_{h1}} \tag{7.59}$$

Where, $v_{h1} = p(df_{h1}) = p(L-1)$,

$v_{e1} = 1 + m_1 t_1 - v_{h1}/2$,

$t_1 = \sqrt{\dfrac{p^2(df_{h1})^2 - 4}{p^2 + (df_{h1})^2 - 5}} = \sqrt{\dfrac{p^2(L-1)^2 - 4]}{p^2 + (L-1)^2 - 5}}$, or $t_1=1$ if $p(df_{h1})=2$, and

$m_1 = df_{e1} - \dfrac{(p+1-df_{h1})}{2}$

$F_{\Lambda-F1}$ follows F distribution with v_{h1} numerator and v_{e1} denominator degrees of freedom when H_0 is true. It may so happen that v_{e1} can be fractional. Then, v_{e1} value can be rounded to the next lowest integer. It is to be noted here that t_1 becomes undefined when $p(df_{h1})=2$ and under such situations $t_1=1$ will give the exact test statistic (Finn, 1974). See the similarity of Equation 7.59 with Equation 7.38.

In this connection, Bartlett's (1954) test for Wilks' Λ_{F1} is

$-\left(LM(n-1) - \dfrac{p+1-(L-1)}{2}\right) \ell n\, \Lambda_{F1}$ follows $\chi^2_{p(L-1)}$ when H_0 is true and sample size is large.

Factor F_2 (Wilks' lambda, Λ_{F2})

The hypothesis to be tested is

$$H_0: \boldsymbol{\beta}_1 = \boldsymbol{\beta}_2 = ... = \boldsymbol{\beta}_m = ... = \boldsymbol{\beta}_M = \mathbf{0}$$
$$H_1: \boldsymbol{\beta}_m \neq \mathbf{0}, \quad \text{for at least one } m, \; m = 1, 2, ..., M$$

The test statistics is Wilks' lambda, Λ_{F2}, where

$$\Lambda_{F2} = \frac{|SSCP_E|}{|SSCP_{F2} + SSCP_E|} = \prod_{j=1}^{s_2} \frac{1}{1+\lambda_j^{F2}} \tag{7.60}$$

Where λ_j^{F2} denotes the j-th eigenvalue of the matrix $(SSCP_E)^{-1} SSCP_{F2}$, and $s_2 = \min(p, M-1)$, p being the number of response variables and M the number of levels of factor F_2.

An approximation of F-distribution of Wilks' Λ_{F2}, $F_{\Lambda-F2}$ following Rao (1952) is shown below.

$$F_{\Lambda-F2} = \frac{1-(\Lambda_{F2})^{1/t_2}}{(\Lambda_{F2})^{1/t_2}} \times \frac{v_{e2}}{v_{h2}} \tag{7.61}$$

Where, $v_{h2} = p(df_{h2}) = p(M-1)$,

$v_{e2} = 1 + m_2 t_2 - v_{h2}/2$,

$t_2 = \sqrt{\frac{p^2(df_{h2})^2 - 4}{p^2 + (df_{h2})^2 - 5}} = \sqrt{\frac{p^2(M-1)^2 - 4]}{p^2 + (M-1)^2 - 5}}$, or $t_2=1$ if $p(df_{h2})=2$, and

$m_2 = df_{e2} - \frac{(p+1-df_{h2})}{2}$

$F_{\Lambda-F2}$ follows F distribution with v_{h2} numerator and v_{e2} denominator degrees of freedom when H_0 is true. If v_{e2} becomes fractional, v_{e2} value can be rounded to the next lowest integer, and for undefined t_2 (when $p(df_{h2})=2$), consider $t_2 = 1$ (Finn, 1974).

In this connection, Bartlett's (1954) test for Wilks' Λ_{F2} is

$-\left(LM(n-1) - \frac{p+1-(M-1)}{2}\right) \ell n\, \Lambda_{F2}$ follows $\chi^2_{p(M-1)}$ when H_0 is true and sample size is large.

Interaction F_{12} (Wilks' lambda, Λ_{F12})

The hypothesis to be tested is

$$H_0: \boldsymbol{\gamma}_{11} = \boldsymbol{\gamma}_{12} = \ldots\ldots = \boldsymbol{\gamma}_{lm} = \ldots\ldots\ldots = \boldsymbol{\gamma}_{LM} = \mathbf{0}$$
$$H_1: \boldsymbol{\gamma}_{lm} \neq \mathbf{0}, \quad \text{for at least one combination of } (l,\ m),\ \ l = 1,2,\ldots, L \text{ and } m = 1,2,\ldots, M$$

The test statistics is Wilks' lambda, Λ_{F12}, where

$$\Lambda_{F12} = \frac{|SSCP_E|}{|SSCP_{F12} + SSCP_E|} = \prod_{j=1}^{s_{12}} \frac{1}{1+\lambda_j^{F12}} \tag{7.62}$$

Where λ_j^{F12} denotes the j-th eigenvalue of the matrix $(SSCP_E)^{-1} SSCP_{F12}$, and $s_{12} = \min\,[p, (L-1)(M-1)]$, p being the number of response variables, L the number of levels of factor F_1 and M the number of levels of factor F_2.

An approximation of F-distribution of Wilks' Λ_{F12}, $F_{\Lambda-F12}$ following Rao (1952) is shown below.

$$F_{\Lambda-F12} = \frac{1-(\Lambda_{F12})^{1/t_{12}}}{(\Lambda_{F12})^{1/t_{12}}} \times \frac{v_{e12}}{v_{h12}} \tag{7.63}$$

Where, $v_{h12} = p(df_{h12}) = p(L-1)(M-1)$,

$$v_{e2} = 1 + m_{12}t_{12} - v_{h12}/2,$$

$$t_{12} = \sqrt{\frac{p^2(df_{h12})^2 - 4}{p^2 + (df_{h12})^2 - 5}} = \sqrt{\frac{p^2[L-1)(M-1)]^2 - 4]}{p^2 + [(L-1)(M-1)]^2 - 5}},$$ or $t_{12}=1$ if $p(df_{h12})=2$, and

$$m_{12} = df_{e12} - \frac{(p+1-df_{h12})}{2}$$

$F_{\Lambda-F12}$ follows F distribution with v_{h12} numerator and v_{e12} denominator degrees of freedom when H_0 is true. It may so happen that v_{e12} can be fractional. Then, v_{e12} value can be rounded to the next lowest integer. It is to be noted here that t_{12} becomes undefined when $p(df_{h12})=2$ and under such situations $t_{12} = 1$ will give the exact test statistic (Finn, 1974).

In this connection, Bartlett's (1954) test for Wilks' Λ_{F12} is

$-\left(LM(n-1) - \frac{p+1-(L-1)(M-1)}{2}\right) \ell n\, \Lambda_{F12}$ follows $\chi^2_{p(L-1)(M-1)}$ when H_0 is true and sample size is large.

Example 7.17

Consider Example 7.16. Conduct hypothesis testing for the main and interaction effects.

Solution:

Using the results of Example 7.16 and Equations 7.58–7.63, the following results are given in Table 7.28.

From the multivariate test results, it is found that the main effect of process and interaction effect between process and supplier are significant at 0.05 probability level, whereas the main effect of supplier is insignificant.

TABLE 7.28
Main and Interaction Effects for Example 7.16

Multivariate Test	Sources of Variation	DF	Test Statistic	Computed F	Numerator/ Denominator DF	Pr(>F)
Pillai trace	Process	2	0.82	8.28	4 /48	3.67E-05
	Supplier	1	0.02	0.18	2 / 23	**0.83**
	Process × supplier	2	0.34	2.49	4 / 48	0.06
Wilks' lambda	Process	2	0.29	9.77	4 / 46	8.33E-06
	Supplier	1	0.98	0.18	2 / 23	**0.83**
	Process × Supplier	2	0.66	2.69	4 / 46	0.04
Hotelling-Lawley	Process	2	2.05	11.27	4 / 44	2.22E-06
	Supplier	1	0.02	0.18	2 / 23	**0.83**
	Process × Supplier	2	0.52	2.87	4 / 44	0.03
Roy's largest root	Process	2	1.85	22.16	2 / 24	3.53E-06
	Supplier	1	0.02	0.18	2 / 23	**0.83**
	Process × Supplier	2	0.52	6.24	2 / 24	0.01

7.4 CASE STUDY

Operations in coke plants are hazardous in nature and the employees are subjected to several job-related stressors. The health and safety executive (HSE) provides seven job stress risk factors, namely demand (D), control (C), management support (MS), peer support (PS), relationship (RE), role (RO), and change (CH). Knowing how different groups of employees differ in experiencing job stress is important for the management to develop and implement interventions for job stress management. In this study, three groups of employees—namely officers, maintenance workers, and operation workers—are considered. MANOVA is used to assess how different groups of employees differ in experiencing job stress.

The details of the case study are available in the web resources of this book (Multivariate Statistical Modeling in Engineering and Management – 1st (routledge.com)).

7.5 LEARNING SUMMARY

Multivariate analysis of variance (MANOVA) is a well-developed and widely used technique for comparing the multivariate mean vectors of several groups (populations) when there are two or more response variables of interest. In this chapter, we have discussed both one-way and two-way MANOVA. However, to make the learning process easy and sequential, we described one-way ANOVA first, followed by one-way MANOVA and then two-way ANOVA followed by two-way MANOVA. This chapter is equipped with the important concepts and derivations of ANOVA and MANOVA, along with a large number of examples. Finally, a case study on job stress is presented. In summary,

- In one-way MANOVA, there are L groups (populations) for comparison; each is measured by p number of response variables (metric in nature). The hypotheses of interest are equality of group mean vectors (H_0) versus different mean vectors (H_1) as $H_0 : \mu_1 = \mu_2 = ... = \mu_L$ versus $H_1 : \mu_l \neq \mu_m$ for at least one pair (ℓ, m), $\ell = 1,2,...L$ and $m = 1,2,...L$, $\ell \neq m$.
- The important assumptions are (i) multivariate normality of data, (ii) homogeneity of covariance matrices of the groups (populations), and (iii) independent observations. Slight violation of assumptions (i) and (ii) may not cause problems in estimation of parameters, tests, and interpretation, but care must be taken in drawing insights from the results.
- The fundamental treatment in MANOVA includes decomposition of the total sum squares and cross-products matrix ($SSCP_T$) into its counterparts across the sources of variations such as between sum squares and cross-products matrix ($SSCP_B$) and within (error) sum squares and cross-products matrix ($SSCP_E$). Accordingly, the total degrees of freedom is also partitioned into sources of variation.
- The MANOVA design varies based on sources of variation. For example, in one-way MANOVA, the sources of variation are the L groups of one-type (say employee designations), and errors; whereas in two-way MANOVA, there are two or more types of nested groups to be compared (e.g., employee designations and income levels). Hence, the sources of variation are L groups of first-type (say employee designations), M groups of second-type (say income levels), their interactions, and errors.
- The four most widely used multivariate statistics are Wilks' lambda (see Equations 7.37–7.38), Pillai trace (see Equations 7.39–7.40), Hotelling-Lowley's trace (see Equations 7.41–7.42), and Roy's largest root (see Equations 7.43–7.44). Care must be taken in identifying appropriate F distribution for each of the test statistics and their degrees of freedom.
- Parameter estimation is simple. Excel-based tabular calculations would give the desired results. But the computation of the multivariate tests is cumbersome and so a suitable computer

program or software is recommended. In addition, this enables the analyst to easily obtain the values of test and diagnostic statistics.
- Although MANOVA gives robust findings related to the overall hypothesis testing, the post hoc analysis often depends on its univariate counterparts (e.g., ANOVA, pairwise comparison, etc.).
- Recently, many visualization-based techniques have been developed such as plot of log of determinants of covariance matrices together with the log determinant of pooled covariance matrix, HE plot, etc. Readers may use the recently available literature for this (e.g., Friendly and Sigal, 2020).

Being a well-developed field in statistics, there are many good books and numerous case applications on MANOVA in the literature. Interested readers may use them for further information.

EXERCISES

(A) Short conceptual questions

7.1 What is ANOVA? In what way it is different from multiple univariate *t* tests?

7.2 (a) Set the hypothesis tested in ANOVA.
(b) Write down the assumptions of ANOVA.

7.3 (a) Define the ANOVA model.
(b) Identify the parameters with their physical significance.
(c) State the conditions applicable for estimation of parameter in ANOVA.

7.4 Explain Bartlett's test for testing homogeneity of population variances for ANOVA.

7.5 (a) What are the sources of variation considered in ANOVA?
(b) What is sum squares decomposition?
(c) How is it accomplished in one-way ANOVA?
(d) What is "degrees of freedom" (DOF)? How do you compute DOF for one-way ANOVA?

7.6 Conduct test of hypothesis for ANOVA. Develop an ANOVA table.

7.7 (a) How do you estimate the parameters of ANOVA?
(b) Derive 100(1 – α)% CI for $\mu_\ell \left(\ell = 1,2,...L\right)$ in ANOVA.
(c) Derive 100(1 – α)% SCI for pairwise mean difference in ANOVA.

7.8 How do you test model adequacy in ANOVA? Test the following using residuals $\epsilon_{i\ell} \left(i = 1,2,...n\right)$ and $\left(\ell = 1,2,...L\right)$:
(a) Test of normality.
(b) Test of independence.
(c) Test of homogeneity of variances.

7.9 What is the modified Leven test? How is it done?

7.10 What is MANOVA? Explain the one-way MANOVA model. In what ways is one-way MANOVA different from one-way ANOVA?

7.11 What will happen if multiple one-way ANOVAs are conducted instead of one-way MANOVA?

7.12 State the assumptions of MANOVA.

7.13 What is the Box M test? How is it conducted?

7.14 Show that in one-way MANOVA, $SSCP_T = SSCP_B + SSCP_E$. Also mention the DOF associated with each of the terms stated above.

7.15 What is the physical significance of $SSCP_B$ and $SSCP_E$?

7.16 Develop a one-way MANOVA table and compare with the one-way ANOVA table.

7.17 (a) What hypotheses are tested in one-way MANOVA?
(b) Explain the following with reference to hypothesis testing in one-way MANOVA:
(i) Wilks' lambda (Λ)
(ii) Pillai's trace (V)
(iii) Hotelling-Lawley's trace (U)
(iv) Roy's largest root (λ_1)
(c) Compare the merits and demerits of the above mentioned test statistics with reference to hypothesis testing in one-way MANOVA.

7.18 (a) How do you estimate the parameters in one-way MANOVA?
(b) Derive 100(1–α)% SCI for pair-wise mean differences in MANOVA.

(B) Long conceptual and/or numerical questions: see web resources (Multivariate Statistical Modeling in Engineering and Management – 1st (routledge.com))

N.B.: R, MINITAB and MS Excel software are used for computation, as and when required.

NOTE

1 Tukey's test and Fisher's least significant difference (LSD) test are popular in pairwise comparisons. Readers may refer to Montgomery (2012) for these tests.

8 Multiple Linear Regression

In this chapter, we will describe multiple linear regression (MLR) which quantifies the relationship between a dependent variable (say, Y) and a set of independent variables (say, X). The relationship is considered to be linear and the dependent variable is assumed to follow normal distribution. MLR not only describes the pattern of relationship between Y and X, but also infers about the strength of the relationship. In addition, MLR is also used for prediction. In this chapter, MLR is conceptualized with reference to the CityCan problem and then, after describing the assumptions of MLR, parameter estimation is described. Several model-building issues, including the model adequacy test, test of assumptions and model diagnostics, are discussed. Finally, a case study is presented. MLR is a very widely used technique and its application areas are vast. How MLR could be used for prediction of dependent/response variables is also discussed. The explanations are kept as lucid as possible and explained all through with a single data set (CityCan).

8.1 CONCEPTUAL MODEL

Recall the CityCan problem discussed in Chapter 1. CityCan has a good recording scheme for its business performance as well as related causes. The primary task that CityCan is interested in is to monitor sales volumes and profit over time and quantify their relationships with operation, maintenance, and marketing performances, so as to explain the variation in business performances, if any. Another issue that interests CityCan will be to predict future sales volume, in order to make its production processes ready to produce the required amount of items. In a nutshell, we can say that CityCan is interested to explain the changes in business performance measures, namely sales volume and profit, as well as to predict the future sales and profit with the help of operation, maintenance, and marketing performances, measured through absenteeism (%), machine breakdown in hours, and M-ratio, respectively. Now, we require a statistical model that serves the purpose of CityCan. Regression analysis is used to serve such purpose.

In the above discussion, we have considered two types of variables. One is related to business performance, namely sales volumes and profit, and the other one is related to operation, maintenance, and marketing, namely absenteeism, breakdown hours, and M-ratio. The first type of variables depends on the second type of variables. Hence, we can say that sales volume and profit depend on absenteeism, breakdown hours and M-ratio. In regression terminology, the first kind of variables is termed as *dependent variables* (DVs) and the second kind is termed as *independent variables* (IVs). Other names used for DVs are response variables, criterion variables, effect variables, or endogenous variables. Similarly, IVs are also known as explanatory variables, causal variables, predictor variables, or exogenous variables. Whatever the terminology used, the basic purpose is to build a relationship model between a DV or multiple DVs with the help of one or more IVs and to predict future values for the DVs using the relationship obtained among the DVs and IVs. When the relationship contains one DV and several IVs, the statistical method used is known as multiple regression. If the relationship is linear, it is termed as *multiple linear regression* (MLR), otherwise *non-linear regression*. A special case of MLR, involving one DV and one IV is known as *simple linear regression*. On the other hand, simultaneous relationships involving multiple DVs with multiple IVs are termed as *multivariate regression*. The multivariate regression can be of linear and non-linear types. MLR is a special case of multivariate multiple linear regression. In the following sections, we will discuss multiple linear regression.

DOI: 10.1201/9781003303060-11

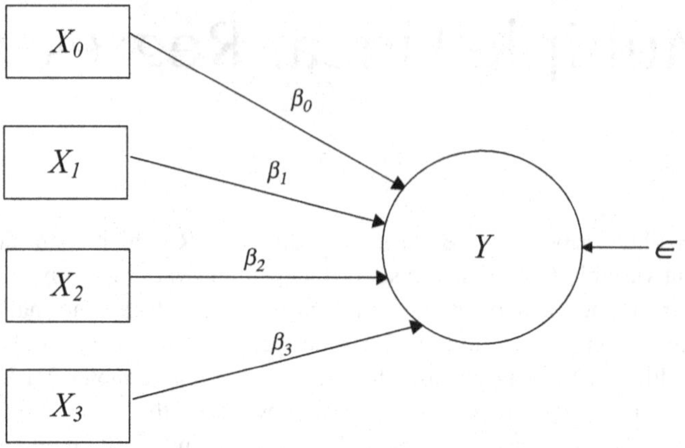

FIGURE 8.1 The diagrammatic representation of MLR involving one Y (the DV) and 3 IVs, X_1, X_2, and X_3. β_0 represents the intercept term and β_1, β_2, and β_3 represent the regression coefficients for the IVs, X_1, X_2 and X_3, respectively. $\in$ represents the error term.

Let us define pictorially the relationship between the DV, sales volume (*Y*) and the IVs, absenteeism (X_1), breakdown hours (X_2), and M-ratio (X_3). Figure 8.1 depicts this. We define the following terms with respect to Figure 8.1.

- X_1, X_2 and X_3 are IVs and represented by rectangles (□). They are independent as there are no variables in the MLR model (Figure 8.1) that affect them. It should be noted here that sufficient care must be taken while identifying IVs for building MLR. We will discuss this issue in a later stage. X_0 assumes a value of 1 only and its inclusion facilitates to quantify the value of *Y* when all IVs assume 0 (zero) values.
- *Y* is the DV and represented by a circle (○) or oval (⬭). The purpose of MLR is either to explain *Y*'s relationships with the IVs or to predict *Y*'s future values or both.
- β_1, β_2, and β_3 are constant terms that represent the amount of contribution of X_1, X_2, and X_3, respectively in explaining or predicting *Y*. They are represented by solid arrows (→) where an arrow emanates from an IV and terminates at the DV (*Y*). β_0, a constant term is known as the value of *Y* when all the IVs, X_1, X_2, and X_3 assume 0 (zero) value. β_0 is also called regression intercept.
- $\in$ is a random error term which represents the amount (variance) of *Y* that cannot be explained or predicted by the IVs, X_1, X_2, and X_3.

Mathematically, we can write

$$\begin{aligned} Y &= \beta_0 X_0 + \beta_1 X_1 + \beta_2 X_2 + \beta_3 X_3 + \in \\ &= \beta_0 + \beta_1 X_1 + \beta_2 X_2 + \beta_3 X_3 + \in \end{aligned} \tag{8.1}$$

(as X_0 takes value of 1 only).

For a general case when there are *p* numbers of IVs, Equation 8.1 is

$$Y = \beta_0 + \beta_1 X_1 + \cdots + \beta_p X_p + \in \tag{8.2}$$

where β_0 is known as regression intercept and β_j, $j = 1, 2, \cdots p$ represents the regression coefficient for the *j*-th IV, i.e., X_j. The linear combination $\beta_1 X_1 + \beta_2 X_2 + \cdots + \beta_p X_p$ is termed as variate (Hair et al., 2010).

Being variables, X_1, X_2, …, X_p can take many values, collectively termed as multivariate observations of IVs. If, we consider n such observations to be collected, then the observation matrix[1] is

$$\mathbf{X}_{nxp} = \begin{bmatrix} x_{11} & x_{12} & \cdots & x_{1j} & \cdots & x_{1p} \\ x_{21} & x_{22} & \cdots & x_{2j} & \cdots & x_{2p} \\ \vdots & \vdots & & & & \\ x_{i1} & x_{i2} & \cdots & x_{ij} & \cdots & x_{ip} \\ \vdots & \vdots & & & & \\ x_{n1} & x_{n2} & \cdots & x_{nj} & \cdots & x_{np} \end{bmatrix}$$

and the corresponding $\mathbf{y}$ and $\in$ values are

$$\mathbf{y}_{n\times 1} = \begin{bmatrix} y_1 \\ y_2 \\ \vdots \\ y_i \\ \vdots \\ y_n \end{bmatrix} \text{ and } \boldsymbol{\in}_{n\times 1} = \begin{bmatrix} \in_1 \\ \in_2 \\ \vdots \\ \in_i \\ \vdots \\ \in_n \end{bmatrix}$$

Therefore, Equation 8.2 for the i-th observation is

$$y_i = \beta_0 + \beta_1 x_{i1} + \beta_2 x_{i2} + \cdots + \beta_p x_{ip} + \in_i \tag{8.3}$$

The linear combination $\beta_0 + \beta_1 x_{i1} + \beta_2 x_{i2} + \cdots + \beta_p x_{ip}$ is known as the expected value of y_i conditional to the information contained in $x_{i1}, x_{i2}, \ldots, x_{ip}$ and is termed as $E\left(y_i \middle| x_{i1}, x_{i2}, \cdots, x_{ip}\right)$.

Therefore,

$$\text{or, } y_i = E\left(y_i \middle| x_{i1}, x_{i2}, \cdots, x_{ip}\right) + \in_i \text{, where}$$

$$E\left(y_i \middle| x_{i1}, x_{i2}, \ldots, x_{ip}\right) = \begin{cases} = E(\beta_0 + \beta_1 x_{i1} + \beta_2 x_{i2} + \cdots + \beta_p x_{ip} + \in_i) \\ = \beta_0 + \beta_1 x_{i1} + \beta_2 x_{i2} + \cdots + \beta_p x_{ip} \qquad (as\, E(\in_i) = 0) \\ = \hat{y}_i \\ = \text{Predicted value for the } i\text{-}th \text{ observation of } Y. \end{cases} \tag{8.4}$$

Therefore, from Equations 8.3 and 8.4, we can write

$$\in_i = y_i - \hat{y}_i \tag{8.5}$$

$\in_i$ is normally distributed with mean 0, and variance equal to that of y_i, i.e., $\sigma^2_{y_i}$.

Further, if we expand Equation 8.3 for all n observations, i = 1, 2, …, n, we get the following linear equations:

$$\left.\begin{aligned} y_1 &= \beta_0 + \beta_1 x_{11} + \beta_2 x_{12} + \cdots + \beta_p x_{1p} + \in_1 \\ y_2 &= \beta_0 + \beta_1 x_{21} + \beta_2 x_{22} + \cdots + \beta_p x_{2p} + \in_2 \\ &\vdots \\ y_n &= \beta_0 + \beta_1 x_{n1} + \beta_2 x_{n2} + \cdots + \beta_p x_{np} + \in_n \end{aligned}\right\} \tag{8.6}$$

In matrix notation

$$\begin{bmatrix} y_1 \\ y_2 \\ \vdots \\ y_n \end{bmatrix}_{n\times1} = \begin{bmatrix} 1 & x_{11} & x_{12} & \cdots & x_{1p} \\ 1 & x_{21} & x_{22} & \cdots & x_{2p} \\ \vdots & \vdots & \vdots & \cdots & \vdots \\ 1 & x_{n1} & x_{n2} & \cdots & x_{np} \end{bmatrix}_{n\times(p+1)} \begin{bmatrix} \beta_0 \\ \beta_1 \\ \vdots \\ \beta_p \end{bmatrix}_{(p+1)\times1} + \begin{bmatrix} \epsilon_1 \\ \epsilon_2 \\ \vdots \\ \epsilon_n \end{bmatrix}_{n\times1} \quad (8.7)$$

or,

$$\mathbf{y}_{n\times1} = \mathbf{X}_{n\times(p+1)}\boldsymbol{\beta}_{(p+1)\times1} + \boldsymbol{\epsilon}_{n\times1} \quad (8.8)$$

In Equation 8.8, **y** is the observation matrix of the order $n \times 1$ for the DV, Y and **X** is known as design matrix of the order $n \times (p + 1)$, and is usually considered as fixed values. $\boldsymbol{\beta}$ is known as the regression coefficient vector of the order of $(p + 1) \times 1$, and is usually considered as having fixed values. The regression model is then called a *fixed effect* regression model. If we assume the β_j, $j = 1, 2, \ldots, p$, are random, then the regression model is called a *random effect* regression model. $\boldsymbol{\epsilon}$ is the vector of order $n \times 1$ for the errors which are *random* in nature. The individual error term ϵ_i, $i = 1,2,\ldots n$ has the following properties:

$$E(\epsilon_i) = 0, \mathrm{Var}(\epsilon_i) = \sigma^2_{y_i} = \sigma^2_y \text{ and } \mathrm{Cov}(\epsilon_i, \epsilon_k) = 0,\ i \neq k.$$

So, $E(\boldsymbol{\epsilon}) = \mathbf{0}$ and $\mathrm{Cov}(\boldsymbol{\epsilon}) = \sigma^2_y \mathbf{I}$ where I is the identity matrix. The property $\sigma^2_{y_i} = \sigma^2_y$, $i = 1, 2, \ldots, n$ denotes equal y variance across n sets of values on p number of IVs (*i.e.*, $\mathbf{X}_{p\times1}$), which is known as *homoscedasticity*. The assumption of homoscedasticity will be discussed later.

A special case of MLR is simple linear regression where there is only one IV. Equation 8.4 for simple linear regression is

$$E(y_i|x_{i1}) = \beta_0 + \beta_1 x_{i1} \quad (8.9)$$

This relationship can be graphically depicted as shown in Figure 8.2.

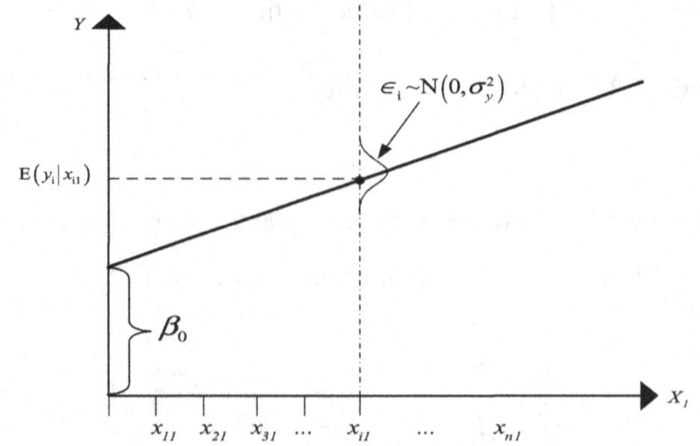

FIGURE 8.2 The regression line involving the DV (Y) and one IV (X_1). $E(y_i|x_{i1})$ represents the expected value of Y when X_1 assumes the value X_{i1}. The ϵ_i represents the error variable which is normally distributed $[N(0,\sigma^2_y)]$. β_0 is the intercept and the slope of the line is β_1.

A closer look at Figure 8.2 reveals that for the observation x_{i1}, the corresponding value on the regression line is the expected value of y_i given x_{i1}, i.e., $E(y_i|x_{i1})$. A normal distribution around this point is the distribution of the random error ϵ_i which has zero mean and a standard deviation equal to the standard deviation of the dependent observation y_i (i.e., $\sigma_{y_i} = \sigma_y$). MLR requires same error distribution for all ϵ_i, $i = 1,2 \ldots n$. This feature, as said earlier, is known as homoscedasticity or equal variance of Y across different values of X (here X_1).

8.2 ASSUMPTIONS

A model is a representation of the real-world phenomenon and is established under certain assumptions. MLR is also subjected to certain assumptions. Hair et al. (2010) stated that the following assumptions must be examined for developing a MLR:

- Linearity of the phenomenon measured, i.e., Y is linearly related with IVs (see Figure 8.2).
- Constant error variance across observations of *IVs*, known as homoscedasticity.
- Uncorrelated error terms, i.e., $Cov(\epsilon_i, \epsilon_k) = 0,\ i \neq k$.
- Normal distribution of the error terms, i.e., $\epsilon_i \sim N(0, \sigma_y^2)$.

Hair et al. (2010) also pointed out that these tests can be done once the model parameters are estimated and the analyst is in a position to compute residuals (the estimated errors). As most of the assumptions are tested using residuals, we will discuss these issues in a later stage. Nevertheless, the analyst should be well aware of these assumptions before doing costly data collection and should be in a position to handle the data collected if some of the assumptions are violated.

8.3 ESTIMATION OF MODEL PARAMETERS

The regression model given in Equations 8.7 and 8.8 contains (p + 1) parameters that are required to be estimated for MLR to specify, explain, and predict the dependence relationship between a DV and IVs. The ordinary least square (OLS) method is used to estimate the parameters.

The basic concept behind OLS is given in Chapter 3. Rearranging Equation 8.3, we get

$$\begin{aligned} \epsilon_i &= y_i - \left(\beta_0 + \beta_1 x_{i1} + \beta_2 x_{i2} + \cdots + \beta_p x_{ip}\right) \\ &= y_i - \sum_{j=0}^{p} \beta_j x_{ij} \end{aligned} \tag{8.10}$$

Squaring Equation 8.10, we get squared errors,

$$\epsilon_i^2 = \left(y_i - \sum_{j=0}^{p} \beta_j x_{ij}\right)^2$$

Taking sum over $i = 1, 2, \ldots, n$, we get sum squared errors (SSE),

$$SSE = \sum_{i=1}^{n} \epsilon_i^2 = \sum_{i=1}^{n} \left(y_i - \sum_{j=0}^{p} \beta_j x_{ij}\right)^2 \tag{8.11}$$

The OLS estimates of β_j, $j = 0, 1, 2, \ldots, p$ are those values of β_j for which *SSE* is the minimum. Taking derivatives of β_j, $j = 0, 1, 2, \ldots, p$, we get a system of $(p + 1)$ linear equations.

$$\frac{\partial SSE}{\partial \beta_j} = 0, \quad j = 0, 1, 2, \cdots p \tag{8.12}$$

subject to the condition that the Hessian matrix containing the several derivatives $\left[\frac{\partial^2 SSE}{\partial \beta_j \cdot \partial \beta_k}\right]$ for $j = 0, 1, 2, \ldots, p$ and $k = 0, 1, 2, \ldots, p$ is positive definite.

Equations 8.11 and 8.12 can be written in matrix form using Equation 8.8. In matrix notation *SSE* is

$$SSE = \epsilon^T \epsilon = (\mathbf{y} - \mathbf{X}\boldsymbol{\beta})^T (\mathbf{y} - \mathbf{X}\boldsymbol{\beta}) \tag{8.13}$$

Now, taking derivative of *SSE* over $\boldsymbol{\beta}$ we get,

$$\frac{\partial SSE}{\partial \boldsymbol{\beta}} = -\mathbf{X}^T (\mathbf{y} - \mathbf{X}\boldsymbol{\beta}) - \mathbf{X}^T (\mathbf{y} - \mathbf{X}\boldsymbol{\beta}) = -2\mathbf{X}^T (\mathbf{y} - \mathbf{X}\boldsymbol{\beta}) \tag{8.14}$$

Putting $\frac{\partial SSE}{\partial \boldsymbol{\beta}} = 0$ gives $\hat{\boldsymbol{\beta}}$. So, using $\hat{\boldsymbol{\beta}}$ in place of $\boldsymbol{\beta}$, we get

$$\begin{aligned} -\mathbf{X}^T \mathbf{y} + \mathbf{X}^T \mathbf{X}\hat{\boldsymbol{\beta}} &= 0 \\ \text{or } \hat{\boldsymbol{\beta}} &= (\mathbf{X}^T \mathbf{X})^{-1} \mathbf{X}^T \mathbf{y} \end{aligned} \tag{8.15}$$

where $\hat{\boldsymbol{\beta}}$ is the estimate of $\boldsymbol{\beta}$.

Example 8.1

Obtain multiple regression model using sales volume (Y), absenteeism (X_1) and machine breakdown in hours (X_2) for CityCan data. For data, see Chapter 1.

Solution:

The following steps will be performed.

Step 1: Identity variables of interest and divide them into a DV and IVs.

In this example, the dependent variable is sales volume (Y) and IVs are absenteeism in % (X_1) and machine breakdown in hours (X_2).

Step 2: Develop design matrix **X**.

The numbers of IVs are 2 and the numbers of observations are 12. So, the design matrix **X** is a $12 \times (2+1)$ matrix as shown below.

$$\mathbf{X}_{12\times 3} = \begin{bmatrix} 1 & 9 & 62 \\ 1 & 8 & 58 \\ 1 & 7 & 64 \\ 1 & 14 & 60 \\ 1 & 12 & 63 \\ 1 & 10 & 57 \\ 1 & 7 & 55 \\ 1 & 4 & 56 \\ 1 & 6 & 59 \\ 1 & 5 & 61 \\ 1 & 7 & 57 \\ 1 & 6 & 60 \end{bmatrix}$$

Step 3: Obtain $\mathbf{X}^{\mathbf{T}}\mathbf{X}$

$$\mathbf{X}^{\mathbf{T}}\mathbf{X} = \begin{bmatrix} 1 & 1 & 1 & 1 & 1 & 1 & 1 & 1 & 1 & 1 & 1 & 1 \\ 9 & 8 & 7 & 14 & 12 & 10 & 7 & 4 & 6 & 5 & 7 & 6 \\ 62 & 58 & 64 & 60 & 63 & 57 & 55 & 56 & 59 & 61 & 57 & 60 \end{bmatrix}_{(3\times 12)} \begin{bmatrix} 1 & 9 & 62 \\ 1 & 8 & 58 \\ 1 & 7 & 64 \\ 1 & 14 & 60 \\ 1 & 12 & 63 \\ 1 & 10 & 57 \\ 1 & 7 & 55 \\ 1 & 4 & 56 \\ 1 & 6 & 59 \\ 1 & 5 & 61 \\ 1 & 7 & 57 \\ 1 & 6 & 60 \end{bmatrix}_{(12\times 3)}$$

$$= \begin{bmatrix} 12 & 95 & 712 \\ 95 & 845 & 5663 \\ 712 & 5663 & 42334 \end{bmatrix}_{(3\times 3)}$$

Step 4: Obtain $\left(\mathbf{X}^{\mathbf{T}}\mathbf{X}\right)^{-1}$

$$\left(\mathbf{X}^{\mathbf{T}}\mathbf{X}\right)^{-1} = \frac{1}{\left|\mathbf{X}^{\mathbf{T}}\mathbf{X}\right|}\, adj\left(\mathbf{X}^{\mathbf{T}}\mathbf{X}\right)$$

$$= \begin{bmatrix} 40.894402598 & 0.114046520 & -0.703043891 \\ 0.114046520 & 0.011751452 & -0.003490093 \\ -0.703043891 & -0.003490093 & 0.012314727 \end{bmatrix}$$

N.B: To avoid large rounding off errors in computation, up to nine decimal places are considered.

Determinant of $\left(\mathbf{X}^{\mathbf{T}}\mathbf{X}\right) = 90542$.

Step 5: Obtain $\mathbf{X}^{\mathbf{T}}\mathbf{y}$

$$\mathbf{X}^{\mathbf{T}}\mathbf{y} = \begin{bmatrix} 1 & 1 & 1 & 1 & 1 & 1 & 1 & 1 & 1 & 1 & 1 & 1 \\ 9 & 8 & 7 & 14 & 12 & 10 & 7 & 4 & 6 & 5 & 7 & 6 \\ 62 & 58 & 64 & 60 & 63 & 57 & 55 & 56 & 59 & 61 & 57 & 60 \end{bmatrix}_{(3\times 12)} \begin{bmatrix} 100 \\ 110 \\ 105 \\ 94 \\ 95 \\ 99 \\ 104 \\ 108 \\ 105 \\ 98 \\ 103 \\ 110 \end{bmatrix}_{(12\times 1)}$$

$$= \begin{bmatrix} 1231 \\ 9622 \\ 72980 \end{bmatrix}_{3\times 1}$$

Step 6: Obtain $\hat{\boldsymbol{\beta}}$

$$\hat{\boldsymbol{\beta}} = \left(\mathbf{X}^{\mathbf{T}}\mathbf{X}\right)^{-1}\mathbf{X}^{\mathbf{T}}\mathbf{y}$$

$$= \begin{bmatrix} 130.22 \\ -1.24 \\ -0.30 \end{bmatrix}_{3\times 1}$$

So, the regression equation is

$$Y = 130.22 - 1.24X_1 - 0.30X_2 + \epsilon$$

or, sales volume = 130.22–1.24* absenteeism –0.30* breakdown hours + ϵ.

Once $\boldsymbol{\beta}$ is estimated, i.e., $\hat{\boldsymbol{\beta}}$ is known, the regression Equation 8.8 can be written as

$$\mathbf{y} = \mathbf{X}\hat{\boldsymbol{\beta}} + \hat{\boldsymbol{\epsilon}} \tag{8.16}$$

$$\text{and } \hat{\mathbf{y}} = \mathbf{X}\hat{\boldsymbol{\beta}} \tag{8.17}$$

Equation 8.17 is used to predict y values. The error term can be estimated as

$$\hat{\boldsymbol{\epsilon}} = \mathbf{y} - \mathbf{X}\hat{\boldsymbol{\beta}} = \mathbf{y} - \hat{\mathbf{y}} \tag{8.18}$$

Again, we can expand $\hat{\boldsymbol{\epsilon}}$ as below.

$$\begin{aligned} \hat{\boldsymbol{\epsilon}} &= \mathbf{y} - \mathbf{X}\hat{\boldsymbol{\beta}} = \mathbf{y} - \mathbf{X}\left[\left(\mathbf{X}^T\mathbf{X}\right)^{-1}\mathbf{X}^T\mathbf{y}\right] \\ &= \mathbf{y} - \left[\mathbf{X}\left(\mathbf{X}^T\mathbf{X}\right)^{-1}\mathbf{X}^T\right]\mathbf{y} = \mathbf{y} - \mathbf{H}\mathbf{y} = (\mathbf{I} - \mathbf{H})\mathbf{y} \end{aligned} \tag{8.19}$$

The matrix $\mathbf{H} = \mathbf{X}\left(\mathbf{X}^{\mathbf{T}}\mathbf{X}\right)^{-1}\mathbf{X}^{\mathbf{T}}$ is known as *hat matrix* and is used for several diagnostic tests in MLR.

The other estimate of importance is the value of sum squared errors (*SSE*). The estimated value of *SSE* is $\hat{\boldsymbol{\epsilon}}^T\hat{\boldsymbol{\epsilon}}$.

$$SSE = \hat{\boldsymbol{\epsilon}}^T\hat{\boldsymbol{\epsilon}} = \mathbf{y}^T(\mathbf{I} - \mathbf{H})^T(\mathbf{I} - \mathbf{H})\mathbf{y} = \mathbf{y}^T(\mathbf{I} - \mathbf{H})\mathbf{y} = \mathbf{y}^T\mathbf{y} - \mathbf{y}^T\mathbf{H}\mathbf{y} \tag{8.20}$$

[**I**–**H** is symmetric and idempotent matrix]

Example 8.2

Obtain *SSE* for Example 8.1.

Solution:

Step 1: Obtain hat matrix, **H**

$$\mathbf{H} = \mathbf{X}\left(\mathbf{X}^{\mathbf{T}}\mathbf{X}\right)^{-1}\mathbf{X}^{\mathbf{T}}$$

From Example 8.1, step 4, we get

$$\left(\mathbf{X}^{\mathbf{T}}\mathbf{X}\right)^{-1} = \begin{bmatrix} 40.894402598 & 0.114046520 & -0.703043891 \\ 0.114046520 & 0.011751452 & -0.003490093 \\ -0.703043891 & -0.003490093 & 0.012314727 \end{bmatrix}_{3\times 3},$$

So, $\mathbf{H} = \mathbf{X}\left(\mathbf{X}^{\mathbf{T}}\mathbf{X}\right)^{-1}\mathbf{X}^{\mathbf{T}} =$

$$
\begin{bmatrix} 1 & 9 & 62 \\ 1 & 8 & 58 \\ 1 & 7 & 64 \\ 1 & 14 & 60 \\ 1 & 12 & 63 \\ 1 & 10 & 57 \\ 1 & 7 & 55 \\ 1 & 4 & 56 \\ 1 & 6 & 59 \\ 1 & 5 & 61 \\ 1 & 7 & 57 \\ 1 & 6 & 60 \end{bmatrix}
\begin{bmatrix} 40.894402598 & 0.114046520 & -0.703043891 \\ 0.114046520 & 0.011751452 & -0.003490093 \\ -0.703043891 & -0.003490093 & 0.012314727 \end{bmatrix}
\begin{bmatrix} 1 & 1 & 1 & 1 & 1 & 1 & 1 & 1 & 1 & 1 & 1 & 1 \\ 9 & 8 & 7 & 14 & 12 & 10 & 7 & 4 & 6 & 5 & 7 & 6 \\ 62 & 58 & 64 & 60 & 63 & 57 & 55 & 56 & 59 & 61 & 57 & 60 \end{bmatrix}
$$

$$
= \begin{bmatrix}
0.16 & 0.04 & 0.22 & 0.12 & 0.20 & 0.02 & -0.05 & -0.03 & 0.07 & 0.12 & 0.01 & 0.10 \\
0.04 & 0.11 & 0.00 & 0.11 & 0.05 & 0.13 & 0.15 & 0.12 & 0.08 & 0.04 & 0.12 & 0.06 \\
0.22 & 0.00 & 0.39 & -0.04 & 0.20 & -0.11 & -0.15 & -0.01 & 0.11 & 0.26 & -0.03 & 0.18 \\
0.12 & 0.11 & -0.04 & 0.50 & 0.32 & 0.26 & 0.08 & -0.14 & -0.04 & -0.14 & 0.05 & -0.06 \\
0.20 & 0.05 & 0.20 & 0.32 & 0.34 & 0.08 & -0.08 & -0.16 & 0.01 & 0.03 & -0.02 & 0.04 \\
0.02 & 0.13 & -0.11 & 0.26 & 0.08 & 0.24 & 0.21 & 0.08 & 0.03 & -0.07 & 0.14 & 0.00 \\
-0.05 & 0.15 & -0.15 & 0.08 & -0.08 & 0.21 & 0.30 & 0.23 & 0.09 & -0.01 & 0.20 & 0.04 \\
-0.03 & 0.12 & -0.01 & -0.14 & -0.16 & 0.08 & 0.23 & 0.31 & 0.16 & 0.14 & 0.18 & 0.13 \\
0.07 & 0.08 & 0.11 & -0.04 & 0.01 & 0.03 & 0.09 & 0.16 & 0.12 & 0.15 & 0.10 & 0.13 \\
0.12 & 0.04 & 0.26 & -0.14 & 0.03 & -0.07 & -0.01 & 0.14 & 0.15 & 0.25 & 0.05 & 0.18 \\
0.01 & 0.12 & -0.03 & 0.05 & -0.02 & 0.14 & 0.20 & 0.18 & 0.10 & 0.05 & 0.15 & 0.07 \\
0.10 & 0.06 & 0.18 & -0.06 & 0.04 & 0.00 & 0.04 & 0.13 & 0.13 & 0.18 & 0.07 & 0.14
\end{bmatrix}
$$

N.B.: The figures in the hat matrix (**H**) up to two decimal places are shown above. However, for further computations, we have used up to 9 decimal places (as usually done in Excel).

Step 2: Obtain $\mathbf{y}^T\mathbf{y}$

$$
\mathbf{y}^T\mathbf{y} = \begin{bmatrix} 100 & 110 & 105 & 94 & 95 & 99 & 104 & 108 & 105 & 98 & 103 & 110 \end{bmatrix}
\begin{bmatrix} 100 \\ 110 \\ 105 \\ 94 \\ 95 \\ 99 \\ 104 \\ 108 \\ 105 \\ 98 \\ 103 \\ 110 \end{bmatrix}
$$

$$= 126605$$

Step 3: Obtain $\mathbf{y}^T\mathbf{H}\mathbf{y}$

Using data obtained in Steps 1 and 2, we get

$$\mathbf{y}^T\mathbf{H}\mathbf{y} = 126451.30$$

Step 4: Obtain $SSE = \hat{\epsilon}^T \hat{\epsilon} = \mathbf{y}^T\mathbf{y} - \mathbf{y}^T\mathbf{H}\mathbf{y}$

$$\therefore SSE = \hat{\epsilon}^T \hat{\epsilon} = 126605 - 126451.30 = 153.70$$

Check the result:

$$\hat{\epsilon} = \mathbf{y} - \hat{\mathbf{y}} = \mathbf{y} - \mathbf{X}\hat{\boldsymbol{\beta}}$$

$$= \begin{bmatrix} 100 \\ 110 \\ 105 \\ 94 \\ 95 \\ 99 \\ 104 \\ 108 \\ 105 \\ 98 \\ 103 \\ 110 \end{bmatrix} - \begin{bmatrix} 100.44 \\ 102.88 \\ 102.32 \\ 94.82 \\ 96.41 \\ 100.69 \\ 105.02 \\ 108.45 \\ 105.07 \\ 105.71 \\ 104.42 \\ 104.77 \end{bmatrix} = \begin{bmatrix} -0.44 \\ 7.12 \\ 2.68 \\ -0.82 \\ -1.41 \\ -1.69 \\ -1.02 \\ -0.45 \\ -0.07 \\ -7.71 \\ -1.42 \\ 5.23 \end{bmatrix}$$

$$\text{So}, \hat{\epsilon}^T \hat{\epsilon} = \begin{bmatrix} -0.44 & 7.12 & 2.68 & -0.82 & -1.41 & -1.69 & -1.02 & -0.45 & -0.07 & -7.71 & -1.42 & 5.23 \end{bmatrix} \begin{bmatrix} -0.44 \\ 7.12 \\ 2.68 \\ -0.82 \\ -1.41 \\ -1.69 \\ -1.02 \\ -0.45 \\ -0.07 \\ -7.71 \\ -1.42 \\ 5.23 \end{bmatrix}_{12\times1}$$

$$= 153.70$$

8.4 SAMPLING DISTRIBUTION OF $\hat{\boldsymbol{\beta}}$

The estimate $\hat{\boldsymbol{\beta}}$ is obtained using Equation 8.15 and data on X and Y collected through sampling. Usually, a sample of size n is collected. But if we attempt collecting several samples, the estimate $\hat{\boldsymbol{\beta}}$ will change from sample to sample and so for $\hat{\epsilon}$. As a result, $\hat{\boldsymbol{\beta}}$, like $\hat{\epsilon}$, is a random variable and possesses certain properties, known as sampling properties of $\hat{\boldsymbol{\beta}}$ (Johnson and Wichern, 2002).

In order to define the sampling distribution of $\hat{\boldsymbol{\beta}}$, we need to define the expected value, $E(\hat{\boldsymbol{\beta}})$ and covariance matrix, $Cov(\hat{\boldsymbol{\beta}})$. As $\hat{\boldsymbol{\beta}}$ is the unbiased estimate of $\boldsymbol{\beta}$, the expected value of $\hat{\boldsymbol{\beta}}$ will be $\boldsymbol{\beta}$, i.e., $E(\hat{\boldsymbol{\beta}}) = \boldsymbol{\beta}$ and is proven below.

$$
\begin{aligned}
E(\hat{\boldsymbol{\beta}}) &= E\left[(\mathbf{X}^T\mathbf{X})^{-1}\mathbf{X}^T\mathbf{y}\right] \\
&= E\left[(\mathbf{X}^T\mathbf{X})^{-1}\mathbf{X}^T(\mathbf{X}\boldsymbol{\beta}+\boldsymbol{\epsilon})\right], \text{ as } \mathbf{y}=\mathbf{X}\boldsymbol{\beta}+\boldsymbol{\epsilon} \\
&= E\left[(\mathbf{X}^T\mathbf{X})^{-1}\mathbf{X}^T\mathbf{X}\boldsymbol{\beta}+(\mathbf{X}^T\mathbf{X})^{-1}\mathbf{X}^T\boldsymbol{\epsilon}\right] \\
&= E\left[\boldsymbol{\beta}+(\mathbf{X}^T\mathbf{X})^{-1}\mathbf{X}^T\boldsymbol{\epsilon}\right] \\
&= E(\boldsymbol{\beta})+(\mathbf{X}^T\mathbf{X})^{-1}\mathbf{X}^T E(\boldsymbol{\epsilon}) \\
&= \boldsymbol{\beta}+\mathbf{0}, \text{ as } E(\boldsymbol{\epsilon})=\mathbf{0} \\
&= \boldsymbol{\beta}
\end{aligned}
\tag{8.21}
$$

The $Cov(\hat{\boldsymbol{\beta}})$ can be estimated as below.

$$
Cov(\hat{\boldsymbol{\beta}}) = E\left[(\hat{\boldsymbol{\beta}}-\boldsymbol{\beta})(\hat{\boldsymbol{\beta}}-\boldsymbol{\beta})^T\right]
$$

$$
\begin{aligned}
\text{Now, } \underset{[(p+1)\times 1]}{\hat{\boldsymbol{\beta}}-\boldsymbol{\beta}} &= (\mathbf{X}^T\mathbf{X})^{-1}\mathbf{X}^T Y-\boldsymbol{\beta} \\
&= -\boldsymbol{\beta}+(\mathbf{X}^T\mathbf{X})^{-1}\mathbf{X}^T(\mathbf{X}\boldsymbol{\beta}+\boldsymbol{\epsilon}) \\
&= -\boldsymbol{\beta}+(\mathbf{X}^T\mathbf{X})^{-1}\mathbf{X}^T\mathbf{X}\boldsymbol{\beta}+(\mathbf{X}^T\mathbf{X})^{-1}\mathbf{X}^T\boldsymbol{\epsilon} \\
&= -\boldsymbol{\beta}+\boldsymbol{\beta}+(\mathbf{X}^T\mathbf{X})^{-1}\mathbf{X}^T\boldsymbol{\epsilon} \\
&= (\mathbf{X}^T\mathbf{X})^{-1}\mathbf{X}^T\boldsymbol{\epsilon}
\end{aligned}
$$

$$
\begin{aligned}
\because \underset{[1\times(p+1)]}{(\hat{\boldsymbol{\beta}}-\boldsymbol{\beta})^T} &= \left[(\mathbf{X}^T\mathbf{X})^{-1}\mathbf{X}^T\boldsymbol{\epsilon}\right]^T \\
&= \boldsymbol{\epsilon}^T\mathbf{X}(\mathbf{X}^T\mathbf{X})^{-1}, as\,(\mathbf{X}^T\mathbf{X})^{-1} \text{ is symmetric}
\end{aligned}
$$

$$
\begin{aligned}
&\therefore E\left[(\hat{\boldsymbol{\beta}}-\boldsymbol{\beta})(\hat{\boldsymbol{\beta}}-\boldsymbol{\beta})^T\right] \\
&= E\left[(\mathbf{X}^T\mathbf{X})^{-1}\mathbf{X}^T\boldsymbol{\epsilon}\boldsymbol{\epsilon}^T\mathbf{X}(\mathbf{X}^T\mathbf{X})^{-1}\right] \\
&= (\mathbf{X}^T\mathbf{X})^{-1}\mathbf{X}^T E(\boldsymbol{\epsilon}\boldsymbol{\epsilon}^T)\mathbf{X}(\mathbf{X}^T\mathbf{X})^{-1} \\
&= (\mathbf{X}^T\mathbf{X})^{-1}\mathbf{X}^T(\sigma^2 I)\mathbf{X}(\mathbf{X}^T\mathbf{X})^{-1} \\
&= \sigma^2(\mathbf{X}^T\mathbf{X})^{-1}(\mathbf{X}^T\mathbf{X})(\mathbf{X}^T\mathbf{X})^{-1} \\
&= \sigma^2(\mathbf{X}^T\mathbf{X})^{-1}
\end{aligned}
$$

$$
\therefore Cov(\hat{\boldsymbol{\beta}}) = \sigma^2(\mathbf{X}^T\mathbf{X})^{-1} \tag{8.22}
$$

Now, in order to estimate Cov ($\hat{\boldsymbol{\beta}}$), we require to estimate σ^2, the error variance. If the variance of the estimated errors is s_e^2, then

$$
E(s_e^2) = \sigma^2 \tag{8.23}
$$

That is s_e^2 is the unbiased estimate of σ^2. Now, the value of s_e^2 can be computed using the following equation:

$$s_e^2 = \frac{SSE}{n-p-1} = \frac{\hat{\boldsymbol{\epsilon}}^T \hat{\boldsymbol{\epsilon}}}{n-p-1} = \frac{\mathbf{y}^T(\mathbf{I}-\mathbf{H})\mathbf{y}}{n-p-1} \tag{8.24}$$

(using Equation 8.20)

Therefore,

$$Cov(\hat{\boldsymbol{\beta}}) = s_e^2 (\mathbf{X}^T\mathbf{X})^{-1} \tag{8.25}$$

If we denote $(\mathbf{X}^T\mathbf{X})^{-1}$ as $\mathbf{C}$, a $(p+1)\times(p+1)$ matrix, then,

$$Cov(\hat{\boldsymbol{\beta}}) = s_e^2 \mathbf{C}$$
$$= s_e^2 \begin{bmatrix} c_{00} & c_{01} & c_{02} & \cdots & c_{0p} \\ c_{01} & c_{11} & c_{12} & \cdots & c_{1p} \\ \vdots & \vdots & \vdots & \cdots & \vdots \\ c_{0j} & c_{1j} & c_{jj} & \cdots & c_{jp} \\ \vdots & \vdots & \vdots & \cdots & \vdots \\ c_{0p} & c_{1p} & c_{2p} & \cdots & c_{pp} \end{bmatrix} \tag{8.26}$$

The diagonal elements of the matrix $\mathbf{C}$, i.e. c_{jj} for $j = 0, 1, 2, \ldots, p$, when multiplied by s_e^2, represent the variance components of the estimates $\hat{\beta}_j$, for $j = 0, 1, 2, \ldots, p$, respectively. In other words, for the variable X_j, the estimate of the regression coefficient β_j, i.e., $\hat{\beta}_j$ has a mean value β_j and variance value $s_e^2 c_{jj}$, where c_{jj} is the j-th diagonal element of $\mathbf{C}$, i.e., $(\mathbf{X}^T\mathbf{X})^{-1}$.

Now, we can define the distribution of $\hat{\boldsymbol{\beta}}$. As we have seen, $\hat{\boldsymbol{\beta}}$ is a $(p + 1) \times 1$ vector of random variables having a mean vector of $\boldsymbol{\beta} = [\beta_0, \beta_1, \ldots, \beta_p]^T$ and covariance matrix of $\sigma^2\mathbf{C}$ (Equatiom 8.22) of the order $(p + 1) \times (p + 1)$, $\hat{\boldsymbol{\beta}}$ is multivariate in nature and is distributed as $N_{p+1}(\boldsymbol{\beta}, \sigma^2\mathbf{C})$. For practical purposes with large n, we can use the estimates of $\boldsymbol{\beta}$ and σ^2, to get the sampling distribution of $\hat{\boldsymbol{\beta}}$. The estimate of σ^2 is s_e^2 (Equation 8.23) and $\boldsymbol{\beta}$ is $\hat{\boldsymbol{\beta}}$. So, there could be a situation when the approximate sampling distribution[2] of $\hat{\boldsymbol{\beta}}$ is $N_{p+1}(\hat{\boldsymbol{\beta}}, s_e^2\mathbf{C})$. [Note that expected value of **estimated** $\boldsymbol{\beta}$ is equal to $\boldsymbol{\beta}$].

Example 8.3

Obtain sampling distribution of $\hat{\boldsymbol{\beta}}$ for Example 8.1

Solution:

Here, $\hat{\boldsymbol{\beta}} = \begin{bmatrix} \hat{\beta}_0 \\ \hat{\beta}_1 \\ \hat{\beta}_2 \end{bmatrix}$, a three-variable vector.[2]

The mean value of $\hat{\boldsymbol{\beta}}$ is $\boldsymbol{\beta}$ which is not known. If we take the estimate, then

$$E\left(\hat{\boldsymbol{\beta}}\right)=\boldsymbol{\beta}=\begin{bmatrix}130.22\\-1.24\\-0.30\end{bmatrix}\text{ (assumed)}$$

Note: This assumption is theoretically not correct.

Similarly, the $Cov\left(\hat{\boldsymbol{\beta}}\right)=\sigma^2\mathbf{C}=s_e^2\mathbf{C}$.

Here,

$$\mathbf{C} = \left(\mathbf{X}^T\mathbf{X}\right)^{-1}$$

$$\mathbf{C} = \begin{bmatrix}40.894402598 & 0.114046520 & -0.703043891\\ 0.114046520 & 0.011751452 & -0.003490093\\ -0.703043891 & -0.003490093 & 0.012314727\end{bmatrix}\text{, from Example (8.1).}$$

Further, $s_e^2=\dfrac{\mathbf{y}^T\left(\mathbf{I}-\mathbf{H}\right)\mathbf{y}}{n-p-1}=\dfrac{SSE}{n-p-1}=\dfrac{153.7}{12-3}=17.08$

So, the sampling distribution of $\hat{\boldsymbol{\beta}}$ is $N_3\left(\boldsymbol{\beta},\ \sigma^2\mathbf{C}\right)$.

8.5 CONFIDENCE REGION (CR) AND SIMULTANEOUS CONFIDENCE INTERVALS (SCI) FOR β

As $\hat{\boldsymbol{\beta}}$ is $N_{p+1}(\boldsymbol{\beta},\ \sigma^2\mathbf{C})$, a multivariate normal, we can have the CR and SCIs for $\boldsymbol{\beta}$. The 100(1–α)% CR for $\boldsymbol{\beta}$ is given by

$$\left(\hat{\boldsymbol{\beta}}-\boldsymbol{\beta}\right)^T\left(\mathbf{X}^T\mathbf{X}\right)\left(\hat{\boldsymbol{\beta}}-\boldsymbol{\beta}\right)\le(p+1)s_e^2\,F_{p+1,n-p-1}^{(\alpha)} \tag{8.27}$$

And 100(1–α)% SCI for β_j is given by

$$\hat{\beta}_j\pm\sqrt{s_e^2c_{jj}}\;\sqrt{(p+1)F_{p+1,n-p-1}^{(\alpha)}},\ j=0,1,2,\cdots,p \tag{8.28}$$

Where $s_e^2c_{jj}$ can be obtained from Equation 8.26. The derivations for Equations 8.27 and 8.28 are not straightforward. Interested readers may consult Johnson and Wichern (2013, p. 371).

However, in practice, SCIs for β_j are seldom used. Even most of the software packages by default do not generate SCIs for regression parameters. The packages report *t*-statistic values considering each of the parameters separately. It is assumed that $(\hat{\beta}_j-\beta_j)/SE(\hat{\beta}_j)$ follows *t*-distribution with $n–p–1$degrees of freedom.

Therefore, as shown in Figure 8.3 the 100(1–α)% confidence interval for β_j is computed as below.

$$-t_{n-p-1}^{(\alpha/2)}<\frac{\hat{\beta}_j-\beta_j}{SE\left(\beta_j\right)}<t_{n-p-1}^{(\alpha/2)}$$

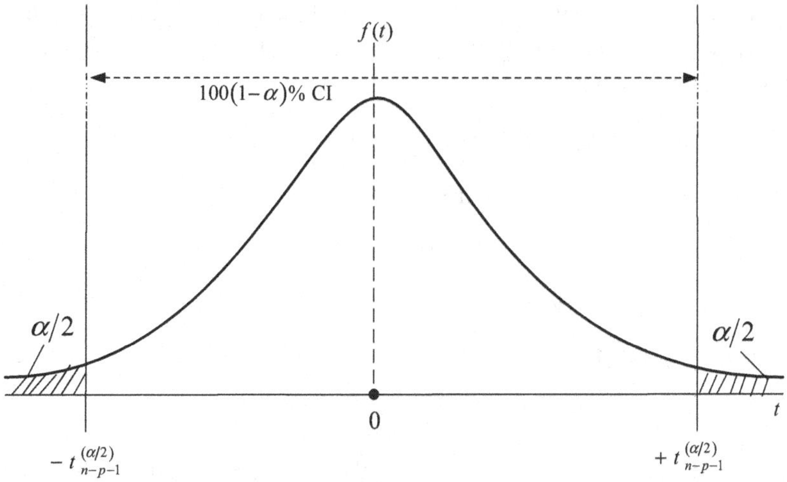

FIGURE 8.3 Distribution of $(\hat{\beta}_j - \beta_j)/\mathrm{SE}(\hat{\beta}_j)$. The hatched areas represent the rejection region for $H_0: \beta_j = 0$. $\left|t_{n-p-1}^{(\alpha/2)}\right|$ is the threshold value for rejection.

$$\text{or,} -t_{n-p-1}^{(\alpha/2)} < \frac{\hat{\beta}_j - \beta_j}{s_e\sqrt{c_{jj}}} < t_{n-p-1}^{(\alpha/2)}$$

which yields

$$\hat{\beta}_j - t_{n-p-1}^{(\alpha/2)} \cdot s_e\sqrt{c_{jj}} < \beta_j < \hat{\beta}_j + t_{n-p-1}^{(\alpha/2)} \cdot s_e\sqrt{c_{jj}} \tag{8.29}$$

The value of $s_e\sqrt{c_{jj}}$ will be obtained from Equation 8.26.

Example 8.4

Obtain 100(1 – α)% SCI and individual CI for β_j for Example 8.1. Take α = 0.05.

Solution:

The 100(1 – α)% SCI for β_j is (Equation 8.28)

$$\hat{\beta}_j \pm \sqrt{s_e^2 c_{jj}} \cdot \sqrt{(p+1)F_{p+1,n-p-1}^{\alpha}}$$

Here, $p = 2$ and $n = 12$.

So, 100(1 – α)% SCI for β_j is

$$\hat{\beta}_j \pm \sqrt{s_e^2 c_{jj}} \cdot \sqrt{3F_{3,9}^{(0.05)}} = \hat{\beta}_j \pm \sqrt{17.08c_{jj}}\sqrt{3\times 3.86} \text{ as } F_{3,9}^{0.05} = 3.86 (\text{from F-table})$$

$$= \hat{\beta}_j \pm 14.06\sqrt{c_{jj}}$$

From Example 8.3, $c_{00} = 40.89$, $c_{11} = 0.012$, $c_{22} = 0.012$ and from Example 8.1, $\hat{\beta}_0 = 130.22$ $\hat{\beta}_1 = -1.24$, and $\hat{\beta}_2 = -0.30$. Putting these values in the above interval, we get the derived 95% SCIs as given below.

95% SCI for β_0 : $130.22 \pm 14.06\sqrt{40.89} = 130.22 \pm 89.91$

i.e., the 95% SCI of β_0 is (40.31 to 220.13)

95% SCI for β_1 : $-1.24 \pm 14.06\sqrt{0.012} = -1.24 \pm 1.55$

i.e., the 95% SCI of β_1 is (–2.79 to 0.31)

95% SCI for β_2 : -0.30 ± 1.55

i.e., the 95% SCI of β_2 is (–1.85 to 1.25).

Using Equation 8.29, the 100(1–α)% CI for β_j is $\hat{\beta}_j \pm t_{n-p-1}^{(\alpha/2)} \sqrt{s_e^2 c_{jj}}$. Now, $t_9(0.025) = 2.262$.

So, the 95% CI for β_j is $\beta_j \pm 2.262\sqrt{17.08\, c_{jj}}$

$= \beta_j \pm 9.35\sqrt{c_{jj}}$

Accordingly,

95% CI for β_0 : $130.22 \pm 9.35\sqrt{40.89} = 130.22 \pm 59.79$

i.e., the 95% CI of β_0 is (70.43 to 190.01)

95% CI for β_1 : $-1.24 \pm 9.35\sqrt{0.012} = -1.24 \pm 1.03$

i.e., the 95% CI of β_1 is (–2.27 to –0.21)

95% CI for β_2 : -0.30 ± 1.03

i.e., the 95% SCI of β_2 is (–1.33 to 0.73).

There is reasonable difference in the results. The SCI is wider than the individual CI. This is because the individual CI is not tightened enough to satisfy the overall (joint) 100(1–α)% (= 95%) confidence for all the parameters together.

What will happen if we use the Bonferroni approach?

We need to calculate $-t_{n-p-1}^{(\alpha/2(p+1))} = t_9\left(\frac{0.05}{6}\right) = t_9(0.0083) = 2.936.$

$\therefore$ 95% CIs are $\hat{\beta}_j \pm 2.936\sqrt{17.08\, c_{jj}}$

$= \hat{\beta}_j \pm 12.13\sqrt{c_{jj}}$

95% CIs for

$\beta_0 : 130.22 \pm 12.13\sqrt{40.89} = 130.22 \pm 77.57$

i.e., the 95% SCI of β_0 is (52.65 to 207.79)

$\beta_1 : -1.24 \pm 12.13\sqrt{0.012} = -1.24 \pm 1.33$

i.e., the 95% SCI of β_1 is (–2.57 to 0.09)

$\beta_2 : -0.30 \pm 1.33$

i.e., the 95% SCI of β_2 is (–1.63 to 1.03).

The findings indicate that confidence intervals are very large when we use SCI formula (Equation 8.28) compared to individual CIs using Equation 8.29. Applying the Bonferroni approach, the intervals get closer to that of the SCI.

8.6 SAMPLING DISTRIBUTION OF $\hat{\epsilon}$

First, we will define the mean vector and covariance matrix of $\hat{\epsilon}$. We know,

$$\begin{aligned} \hat{\epsilon} &= (\mathbf{I}-\mathbf{H})\mathbf{y} \\ &= (\mathbf{I}-\mathbf{H})(\mathbf{X}\boldsymbol{\beta}+\epsilon) \\ &= (\mathbf{I}-\mathbf{H})\mathbf{X}\boldsymbol{\beta}+(\mathbf{I}-\mathbf{H})\epsilon \\ \therefore \quad E(\hat{\epsilon}) &= (\mathbf{I}-\mathbf{H})\mathbf{X}\boldsymbol{\beta}+(\mathbf{I}-\mathbf{H})E(\epsilon) \\ &= \mathbf{0}+(\mathbf{I}-\mathbf{H})\times\mathbf{0} = \mathbf{0} \end{aligned} \tag{8.30}$$

As $\mathbf{I}-\mathbf{H}$ and $\mathbf{X}\boldsymbol{\beta}$ are fixed, $(\mathbf{I}-\mathbf{H})\mathbf{X}\boldsymbol{\beta} = \mathbf{0}$ and $E(\epsilon) = \mathbf{0}$.

The covariance matrix of $\hat{\epsilon}$ is

$$\begin{aligned} Cov(\hat{\epsilon}) &= E\left[\left\{\hat{\epsilon}-E(\hat{\epsilon})\right\}\left\{\hat{\epsilon}-E(\hat{\epsilon})\right\}^T\right] = E\left(\hat{\epsilon}\,\hat{\epsilon}^T\right) \text{ as } E(\hat{\epsilon}) = \mathbf{0} \\ &= Cov\left[(\mathbf{I}-\mathbf{H})\mathbf{y}\right] \\ &= Cov\left[(\mathbf{I}-\mathbf{H})(\mathbf{X}\boldsymbol{\beta}+\epsilon)\right] \\ &= Cov\left[(\mathbf{I}-\mathbf{H})\mathbf{X}\boldsymbol{\beta}+(\mathbf{I}-\mathbf{H})\epsilon\right] \\ &= Cov\left[(\mathbf{I}-\mathbf{H})\mathbf{X}\boldsymbol{\beta}\right]+Cov\left[(\mathbf{I}-\mathbf{H})\epsilon\right] \\ &= Cov\left[(\mathbf{I}-\mathbf{H})\epsilon\right] \text{ as } (\mathbf{I}-\mathbf{H})\mathbf{X}\boldsymbol{\beta}=\mathbf{0} \\ \therefore \quad Cov(\hat{\epsilon}) &= Cov\left[(\mathbf{I}-\mathbf{H})\epsilon\right] \\ &= (\mathbf{I}-\mathbf{H})^T(\mathbf{I}-\mathbf{H})\,Cov(\epsilon) \\ &= (\mathbf{I}-\mathbf{H})\sigma^2\mathbf{I} \\ &= \sigma^2(\mathbf{I}-\mathbf{H}), \text{ as } \mathbf{I}-\mathbf{H} \text{ is a symmetric and idempotent matrix.} \end{aligned} \tag{8.31}$$

What is the sampling distribution of $\hat{\epsilon}$? We know that one of the assumptions of MLR is normality of the error terms. Therefore, $\hat{\epsilon}_{n\times 1}$ being a vector of n error terms follows multivariate normal as $N_n(0, \sigma^2(\mathbf{I}-\mathbf{H}))$. The individual error variance σ^2 will be estimated using s_e^2 given in Equation 8.24.

Therefore, $Cov\left(\hat{\in}\right) = s_e^2\left(\mathbf{I}-\mathbf{H}\right) = s_e^2\mathbf{E}(\text{say})$

$$= s_e^2 \begin{bmatrix} e_{11} & e_{12} & \cdots & e_{1n} \\ e_{21} & e_{22} & \cdots & e_{2n} \\ \vdots & \vdots & \vdots & \vdots \\ e_{i1} & e_{i2} & e_{ii} & e_{in} \\ \vdots & \vdots & \vdots & \vdots \\ e_{n1} & e_{n2} & \cdots & e_{nn} \end{bmatrix}_{n\times n} \tag{8.32}$$

So, the variance of the *i-th* residual $\left(\hat{\in}_i\right)$ is $s_e^2\, e_{ii}$.

Similarly, we are interested to know the distribution of *SSE*,(i.e., $\hat{\in}^T\hat{\in}$). $\hat{\in}^T\hat{\in}$ is distributed as $\sigma^2\chi^2_{n-p-1}$, where the estimate of σ^2 is s_e^2.

Example 8.5

Obtain the covariance matrix for the residuals $\left(\hat{\in}\right)$ for Example 8.1.

Solution:

$$Cov\left(\hat{\in}\right) = s_e^2\left(\mathbf{I}-\mathbf{H}\right) = 17.08\left(\mathbf{I}-\mathbf{H}\right)$$

Now, $\mathbf{H} = \mathbf{X}\left(\mathbf{X}^T\mathbf{X}\right)^{-1}\mathbf{X}^T$ is a 12×12 matrix and $\mathbf{I}$ is a 12×12 identify matrix. See Example 8.2 for the values of H matrix. Subtracting $\mathbf{H}$ from $\mathbf{I}$, we get

$$\mathbf{I}-\mathbf{H} = \begin{bmatrix}
0.84 & -0.04 & -0.22 & -0.12 & -0.20 & -0.02 & 0.05 & 0.03 & -0.07 & -0.12 & -0.01 & -0.10 \\
-0.04 & 0.89 & 0.00 & -0.11 & -0.05 & -0.13 & -0.15 & -0.12 & -0.08 & -0.04 & -0.12 & -0.06 \\
-0.22 & 0.00 & 0.61 & 0.04 & -0.20 & 0.11 & 0.15 & 0.01 & -0.11 & -0.26 & 0.03 & -0.18 \\
-0.12 & -0.11 & 0.04 & 0.50 & -0.32 & -0.26 & -0.08 & 0.14 & 0.04 & 0.14 & -0.05 & 0.06 \\
-0.20 & -0.05 & -0.20 & -0.32 & 0.66 & -0.08 & 0.08 & 0.16 & -0.01 & -0.03 & 0.02 & -0.04 \\
-0.02 & -0.13 & 0.11 & -0.26 & -0.08 & 0.76 & -0.21 & -0.08 & -0.03 & 0.07 & -0.14 & 0.00 \\
0.05 & -0.15 & 0.15 & -0.08 & 0.08 & -0.21 & 0.70 & -0.23 & -0.09 & 0.01 & -0.20 & -0.04 \\
0.03 & -0.12 & 0.01 & 0.14 & 0.16 & -0.08 & -0.23 & 0.69 & -0.16 & -0.14 & -0.18 & -0.13 \\
-0.07 & -0.08 & -0.11 & 0.04 & -0.01 & -0.03 & -0.09 & -0.16 & 0.88 & -0.15 & -0.10 & -0.13 \\
-0.12 & -0.04 & -0.26 & 0.14 & -0.03 & 0.07 & 0.01 & -0.14 & -0.15 & 0.75 & -0.05 & -0.18 \\
-0.01 & -0.12 & 0.03 & -0.05 & 0.02 & -0.14 & -0.20 & -0.18 & -0.10 & -0.05 & 0.85 & -0.07 \\
-0.10 & -0.06 & -0.18 & 0.06 & -0.04 & 0.00 & -0.04 & -0.13 & -0.13 & -0.18 & -0.07 & 0.86
\end{bmatrix}$$

Then, $Cov(\hat{\boldsymbol{\epsilon}}) = s_e^2(\mathbf{I} - \mathbf{H}) = 17.08(\mathbf{I} - \mathbf{H})$

$$= \begin{bmatrix} 14.27 & -0.77 & -3.69 & -2.11 & -3.48 & -0.39 & 0.78 & 0.46 & -1.15 & -2.08 & -0.21 & -1.64 \\ -0.77 & 15.27 & 0.00 & -1.82 & -0.77 & -2.29 & -2.57 & -2.00 & -1.33 & -0.67 & -2.00 & -1.05 \\ -3.69 & 0.00 & 10.40 & 0.70 & -3.34 & 1.96 & 2.64 & 0.22 & -1.96 & -4.50 & 0.57 & -3.00 \\ -2.11 & -1.82 & 0.70 & 8.62 & -5.43 & -4.40 & -1.30 & 2.46 & 0.77 & 2.39 & -0.86 & 0.99 \\ -3.48 & -0.77 & -3.34 & -5.43 & 11.27 & -1.44 & 1.41 & 2.69 & -0.10 & -0.55 & 0.36 & -0.62 \\ -0.39 & -2.29 & 1.96 & -4.40 & -1.44 & 13.06 & -3.58 & -1.29 & -0.56 & 1.23 & -2.35 & 0.05 \\ 0.78 & -2.57 & 2.64 & -1.30 & 1.41 & -3.58 & 12.01 & -3.99 & -1.57 & 0.22 & -3.35 & -0.71 \\ 0.46 & -2.00 & 0.22 & 2.46 & 2.69 & -1.29 & -3.99 & 11.80 & -2.71 & -2.36 & -3.05 & -2.24 \\ -1.15 & -1.33 & -1.96 & 0.77 & -0.10 & -0.56 & -1.57 & -2.71 & 14.97 & -2.56 & -1.65 & -2.15 \\ -2.08 & -0.67 & -4.50 & 2.39 & -0.55 & 1.23 & 0.22 & -2.36 & -2.56 & 12.79 & -0.83 & -3.09 \\ -0.21 & -2.00 & 0.57 & -0.86 & 0.36 & -2.35 & -3.35 & -3.05 & -1.65 & -0.83 & 14.60 & -1.22 \\ -1.64 & -1.05 & -3.00 & 0.99 & -0.62 & 0.05 & -0.71 & -2.24 & -2.15 & -3.09 & -1.22 & 14.67 \end{bmatrix}$$

8.7 ASSESSMENT OF OVERALL FIT OF THE MODEL

Once a regression model is estimated, we need to know whether the model is adequate for the purpose for which it was built. As we outlined earlier, a multiple regression model is a dependence model and it helps to explain the variation in the dependent variable Y. The amount of variance of Y that is explained by a MLR is a measure of its adequacy, which is also known as goodness of fit. If the variance explained by the model is acceptable, then we say that the model is fit for the purpose to be served.

In order to assess the variance explained by the model, we demonstrate partitioning the total variation of Y into that explained by the model and that accounted for by the error terms. The latter part is known as variance not explained by the model. The ratio of variance of Y explained by the model (explained variance) to the total variance of Y is known as coefficient of determination (R^2).

$$\therefore R^2 = \frac{\text{Explained variance of } Y}{\text{Total variance of } Y} \tag{8.33}$$

How do we compute the value of explained and total variance? Let us explain this with the help of Figure 8.4.

From Figure 8.4, we see that for the i-th observation of Y, i.e. y_i,

Total variation = Explained variation + Unexplained variation

$$y_i - \bar{y} = \left(\hat{y}_i - \bar{y}\right) + \left(y_i - \hat{y}_i\right) \tag{8.34}$$

where $\bar{y}$ is the average of all y_i, $i = 1, 2, \ldots, n$.

Squaring and taking sum over $i = 1, 2, \ldots, n$, Equation 8.34 becomes

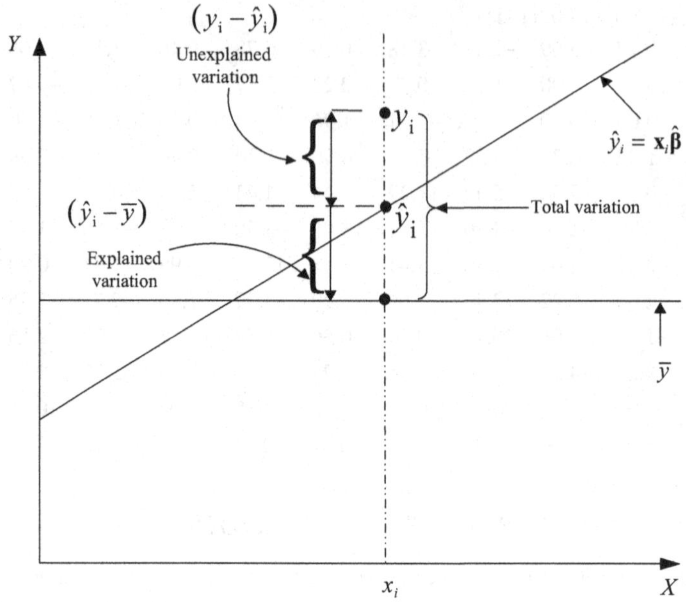

FIGURE 8.4 Partitioning of variation of Y in MLR. The total variation $(y_i - \bar{y})$ is divided into two parts: explained variation $(\hat{y}_i - \bar{y})$ and unexplained variation $(y_i - \hat{y}_i)$, where $\bar{y}$ is the average of y_i, $i = 1,2,\ldots n$ and $\hat{y}_i = \mathbf{x}_i\hat{\boldsymbol{\beta}}$ represents the fitted regression line.

$$
\begin{aligned}
\sum_{i=1}^{n}(y_i-\bar{y})^2 &= \sum_{i=1}^{n}(\hat{y}_i-\bar{y})^2+\sum_{i=1}^{n}(y_i-\hat{y}_i)^2+2\sum_{i=1}^{n}(\hat{y}_i-\bar{y})(y_i-\hat{y}_i)\\
&= \sum_{i=1}^{n}(\hat{y}_i-\bar{y})^2+\sum_{i=1}^{n}(y_i-\hat{y}_i)^2 \text{ as} \sum_{i=1}^{n}(\hat{y}_i-\bar{y})(y_i-\hat{y}_i)=0 \qquad (8.35)\\
\text{So, } \sum_{i=1}^{n}(y_i-\bar{y})^2 &= \sum_{i=1}^{n}(\hat{y}_i-\bar{y})^2+\sum_{i=1}^{n}(y_i-\hat{y}_i)^2
\end{aligned}
$$

The left-hand side of Equation 8.35 is known as sum squared total (*SST*) while the two terms of the right-hand side of Equation 8.35 are called sum squared regression (*SSR*) and sum squared errors (*SSE*).

$$\text{So, } SST = \sum_{i=1}^{n}(y_i-\bar{y})^2 = (\mathbf{y}-\bar{y}\mathbf{1})^T(\mathbf{y}-\bar{y}\mathbf{1}) \qquad (8.36)$$

$$SSR = \sum_{i=1}^{n}(\hat{y}_i-\bar{y})^2 = (\hat{\mathbf{y}}-\bar{y}\mathbf{1})^T(\hat{\mathbf{y}}-\bar{y}\mathbf{1}) \qquad (8.37)$$

$$SSE = \sum_{i=1}^{n}(y_i-\hat{y}_i)^2 = \sum_{i=1}^{n}\hat{\epsilon}_i^2 = \hat{\epsilon}^T\hat{\epsilon} \qquad (8.38)$$

Here, $\mathbf{y}$ is $n\times1$ vector, $\bar{y}$ is a scalar, $\hat{\mathbf{y}}$ is $n\times1$ vector and $\mathbf{1}$ is a $n\times1$ vector of 1 only.

Now,

$$R^2 = \frac{SSR}{SST} = \frac{SST - SSE}{SST} = 1 - \frac{SSE}{SST} \tag{8.39}$$

R^2 varies from 0 to 1 with 1 representing perfect fit and 0 representing no fit. Usually, $R^2 \geq 0.90$ is considered acceptable for controlled experimentations. However, for the management, administrative, social, and socio-technical applications, R^2 lower than 0.90 is also acceptable. The rule of thumb is that the higher the R^2 value, the better the model fit. What does $R^2 = 0.90$ mean? It says that 90% of the Y's (dependent variable) variance is explained or accounted for by the regression model.

Again, we know that $SSE = (n-p-1)s_e^2$ (from Equation 8.24) and $SST = (n-1)s_y^2$.

So,

$$R^2 = 1 - \frac{(n-p-1)s_e^2}{(n-1)s_y^2} \tag{8.40}$$

Therefore, R^2 is sensitive to the number of parameters to be estimated $(p + 1)$ and sample size (n). For $n = (p + 1)$, R^2 is always 1 representing a perfect fit, which is misleading. In order to overcome this, R^2 is adjusted for $p + 1$ and n. The adjusted measure is called adjusted R^2 or R_a^2.

$$R_a^2 = 1 - \frac{SSE/n-p-1}{SST/n-1} = 1 - \frac{s_e^2}{s_y^2} \tag{8.41}$$

R_a^2 also varies from 0 to 1 and a value of 0.90 or more is desirable for controlled experimentations. For management, administrative, social, and socio-technical applications, lower R_a^2 values are also acceptable depending on the situations being modeled. However, $R_a^2 \leq R^2$.

Example 8.6

Assess the overall fit of MLR for Example 8.1.

Solution:

MLR is,

$Y = 130.22 - 1.24X_1 - 0.30X_2 + \epsilon$ (from Example 8.1)

Here, n = 12, and p + 1 = 3

SSE = 153.70 (from Example 8.2)

From data

$$\mathbf{y} = [100 \quad 110 \quad 105 \quad 94 \quad 95 \quad 99 \quad 104 \quad 108 \quad 105 \quad 98 \quad 103 \quad 110]$$

$$\bar{y} = 102.58$$

$$\mathbf{y} - \bar{y}\mathbf{1} = [-2.58 \quad 7.42 \quad 2.42 \quad -8.58 \quad -7.58 \quad -3.58 \quad 1.42 \quad 5.42 \quad 2.42 \quad -4.58 \quad 0.42 \quad 7.42]^T$$

TABLE 8.1
ANOVA Table for MLR

Source of Variation	Sum of Squares (SS)	Degrees of Freedom	Mean Squares (MS)	F-statistic
Regression (model)	*SSR*	p	*SSR/p* = MSR	$F=\dfrac{MSR}{MSE}$
Error	*SSE*	n–p–1	*MSE* = *SSE*/n–p–1	
Total	*SST*	n–1		

So, $SST=\sum_{i=1}^{12}(y_i-\bar{y})^2=(\mathbf{y}-\bar{y}\mathbf{1})^T(\mathbf{y}-\bar{y}\mathbf{1})=324.92$

$$R^2=1-\frac{SSE}{SST}=1-\frac{153.7}{324.92}=0.53$$

$$R_a^2=1-\frac{s_e^2}{s_y^2}=1-\frac{153.70/9}{324.92/11}=0.42$$

So, the model doesn't fit to the data. Even if the value of both R^2 and R_a^2 comes out to be more than 0.90, a desirable value, the model fit value cannot be considered blindly. In this example, the sample size n is very small (only 12) which does not permit to develop MLR. We should therefore first determine the sample size based on a systematic sampling method.

Like other statistics, R^2 is also a statistic (in MLR) and as such follows certain distribution. This property is used to test hypothesis for goodness of fit. We can develop an ANOVA table for MLR as given in Table 8.1.

In MLR, the null and alternate hypothesis are

$$\left.\begin{aligned} H_0 &: \beta_1=\beta_2\cdots=\beta_p=0 \\ H_1 &: \beta_j\neq 0 \text{ for at least one } j,\ j=1,2,\ldots p \end{aligned}\right\} \tag{8.42}$$

Now, under the null hypothesis H_0, the quantity $\dfrac{MSR}{MSE}=\dfrac{SSR/p}{SSE/n-p-1}$ follows F distribution with p and n–p–1 numerator and denominator degrees of freedom, respectively. If the computed $F\left(=\dfrac{MSR}{MSE}\right)$ is greater than tabulated $F_{p,\,n-p-1}^{(\alpha)}$ where α is the level of significance, we reject H_0.

Example 8.7

Compute an ANOVA table for Example 8.1 and comment on the overall fit of the model for Example 8.1.

TABLE 8.2
ANOVA Table for Example 8.1

Source of Variation	SS	DOF	MS	F	Comments
1. Regression	171.22	2	85.61	$F = \frac{85.61}{17.08}$	As computed F = 5.01 >
2. Error	153.70	9	17.08		$F_{2,9}^{(0.05)} = 4.26$
3. Total	324.92	11		= 5.01	Reject H_0.

Solution:

The null hypothesis (H_0) is rejected. So, there is at least one regression parameter (i.e., β_j, j = 0, 1, and 2) that is statistically significant in explaining the variability of Y.

8.8 TEST OF INDIVIDUAL REGRESSION PARAMETERS

Once the overall fit of a regression model (through R^2 and/or F-test) is adequate, we will test the fit of the individual parameters. By "fit of the individual parameter," we mean to test whether each of the parameters contributes significantly to the overall fit of the model. In order to do so we do hypothesis testing as below.

$$\left.\begin{aligned} H_0 &: \beta_j = 0 \\ H_1 &: \beta_j \neq 0 \end{aligned}\right\} \tag{8.43}$$

As we have discussed in Section 8.5, traditionally t- test is done for individual β_j although $\boldsymbol{\beta} = [\beta_0 \ \beta_1 \ \cdots \ \beta_p]^T$ is multivariate normal. Under the assumption of H_0: β_j = 0, the quantity $(\hat{\beta}_j - \beta_j)/SE(\hat{\beta}_j)$ follows t_{n-p-1} degrees of freedom.

Therefore, the test-statistic t is

$$\begin{aligned} t &= \frac{\hat{\beta}_j - \beta_j}{SE(\hat{\beta}_j)} = \frac{\hat{\beta}_j}{SE(\hat{\beta}_j)} \text{ as } \beta_j = 0 \left(\text{under } H_0\right) \\ &= \frac{\hat{\beta}_j}{s_e\sqrt{c_{jj}}} \end{aligned} \tag{8.44}$$

The value of c_{jj} will be obtained from Equation 8.26.

If the computed t is greater than $\left|t_{n-p-1}^{(\alpha/2)}\right|$, we reject H_0. Table 8.3 explains the test.

The parameters with probability value (p-value) less than α (a predetermined value) are said to be statistically significant. Alternatively, for a predetermined probability (α), the cut-off t-value, $t_{n-p-1}^{(\alpha/2)}$ is obtained from t-table, and if the computed $|t| > t_{n-p-1}^{(\alpha/2)}$, the corresponding parameter is considered to be statistically significant at α probability level of significance.

What does the significance test of parameters physically mean for? A parameter, β_j is significant, meaning that the null hypothesis H_0: $\beta_j = 0$ is rejected. So, we can claim that a unit change in the X_j value will increase or decrease Y (the dependent variable) by β_j units depending on the positive or negative relationship between Y and X_j, respectively.

TABLE 8.3
Test of Individual Regression Parameters

Variables	Parameter	Estimate	Standard Error	Computed t-value (t_c)	p-value	Cut-off t and Decision
Intercept	β_0	$\hat{\beta}_0$	$s_e\sqrt{c_{00}}$	$\hat{\beta}_0/s_e\sqrt{c_{00}}$	Obtained from t-table for $n-p-1$ degrees of freedom.	Cut-off $= t_{n-p-1}^{(\alpha/2)}$
X_1	β_1	$\hat{\beta}_1$	$s_e\sqrt{c_{11}}$	$\hat{\beta}_1/s_e\sqrt{c_{11}}$		If $\lvert t_c \rvert > t_{n-p-1}^{(\alpha/2)}$, the corresponding parameter is significant.
X_2	β_2	$\hat{\beta}_2$	$s_e\sqrt{c_{22}}$	$\hat{\beta}_2/s_e\sqrt{c_{22}}$		
⋮	⋮	⋮	⋮	⋮		
X_p	β_p	$\hat{\beta}_p$	$s_e\sqrt{c_{pp}}$	$\hat{\beta}_p/s_e\sqrt{c_{pp}}$		

TABLE 8.4
Test Results of Regression Coefficients

Variables	Parameter	Estimate	Standard Error	Computed t-value (t_c)	Cut-off $t_9^{(0.05)}$	Comments
Intercept	β_0	130.22	26.43	4.93	2.262	Significant
X_1	β_1	−1.24	0.45	− 2.76		Significant
X_2	β_2	− 0.30	0.45	− 0.67		Not significant

Example 8.8

Test the individual parameters for MLR given in Example 8.1.

Solution:

$$s_e^2 = 17.08 \text{ and } V\left(\hat{\boldsymbol{\beta}}\right) = \begin{bmatrix} 40.89 \\ 0.012 \\ 0.012 \end{bmatrix}$$

$$\therefore SE\left(\hat{\boldsymbol{\beta}}\right) = \sqrt{17.08}\begin{bmatrix} \sqrt{40.89} \\ \sqrt{0.012} \\ \sqrt{0.012} \end{bmatrix} = \begin{bmatrix} 26.43 \\ 0.45 \\ 0.45 \end{bmatrix}$$

The individual parameters test results are shown in Table 8.4.

So, of the two independent variables absenteeism (X_1) and breakdown hours (X_2), absenteeism (X_1) has a significant negative relationship with sales volume (Y). However, for breakdown hours (X_2), we cannot claim this at 0.05 probability level of significance.

8.9 INTERPRETATION OF REGRESSION PARAMETERS

By interpretation of regression parameters, we usually mean that for a fitted (accepted) MLR how the regression coefficients (including intercept) be interpreted statistically and for practical use.

Consider the regression equation obtained in Example 8.1. For interpretation purposes, we have reproduced it below.

$$Y = 130.22 - 1.24X_1 - 0.30X_2 + \epsilon$$

or, sales volume = 130.22–1.24 absenteeism–0.30 breakdown hours + ϵ.

Here, the regression coefficients for X_1 and X_2 are –1.24 and –0.30, and the intercept is 130.22. In Example 8.8, we saw that the coefficient for X_1(–1.24) is significantly different from zero but the coefficient for X_2(–0.30) is not. So, we should remove X_2 from the model and a fresh regression must be done using X_1 only. If X_1 and X_2 are truly independent, the coefficient for X_1 will remain significant in the new regression model (except minor changes in the absolute value of the coefficient). In the model given above, the β coefficient for X_1 is –1.24. This implies, if we change X_1 by one unit, the change in Y value will be –1.24 units. However, if both the coefficients for X_1 and X_2 would become significant, then the interpretation would have been different as follows. If we change X_1 by one unit keeping X_2 constant, the change in Y value will be –1.24 units. Similarly, if we change X_2 by one unit keeping X_1 constant, the change in Y value will be –0.30 units (as the β coefficient for X_2 is –0.30)

Note that the above interpretation of regression coefficients is valid for continuous IVs and is measured in their original units. If one standardizes the IVs to overcome the scaling effect, the above interpretation is not valid. Please see the path model (Chapter 10) for such interpretation. If an IV is categorical (say, having K categories), K-1 dummy IVs are used. A one unit change in this case represents switching from the reference category (assumed to have 0 values) to the other while keeping other continuous and categorical IVs (if used together) constant.

On the other hand, the intercept has different meaning altogether. The intercept is the y-value when all the IVs assume 0 values. For example, in the above regression model the intercept (130.22) is the Y-value when both X_1 and X_2 assume 0 values. However, if one or more IVs cannot assume 0 values, the intercept has no physical meaning. It is only used to position the regression line in the right place. To avoid the intercept, the regression can be passed through the origin. However, this may create problems in interpretation of residuals and regression coefficients.

8.10 TEST OF ASSUMPTIONS

When a MLR passes the overall fit test (see R^2) and some of the regression parameters emerge as significant, one may ignore the test of assumptions using errors. This happens because we generally examine the observed data for MLR assumptions before fitting the model. But we also need to test the assumptions using errors. Fortunately, if there are any violations of assumptions, the error terms should capture this and we need to carefully analyze the error terms to reveal those violations. Similarly, if a model is rejected because of poor fit, it may be due to the violation of one or more of the assumptions. Test of assumptions, therefore, has utmost importance in MLR modeling. In the following sections, we test the MLR assumptions outlined in Section 8.2.

8.10.1 Test of Linearity

Let us consider the non-linear relationship between a dependent variable Y and an independent variable X as shown in Figure 8.5. Suppose we fit a linear model $\hat{y}_i = \mathbf{x}_i\hat{\boldsymbol{\beta}}$. The model $\hat{y}_i = \mathbf{x}_i\hat{\boldsymbol{\beta}}$ captures the linearity of the data between Y and X. Where does the non-linearity relationship go? It is captured by the residuals, $\hat{\epsilon}_i$, $i = 1, 2, \ldots n$. Therefore, if we plot $\hat{\boldsymbol{\epsilon}}$ versus $\hat{Y}$, we will get a plot similar to Figure 8.5. Figure 8.6 represents this.

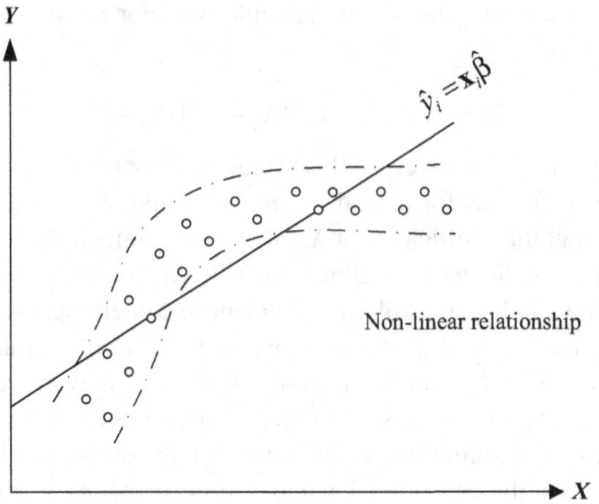

FIGURE 8.5 Non-linear relationship between Y and X. As expected, the fitted regression line, $\hat{y}_i = \mathbf{x}_i\hat{\boldsymbol{\beta}}$ could not capture the non-linear relationship.

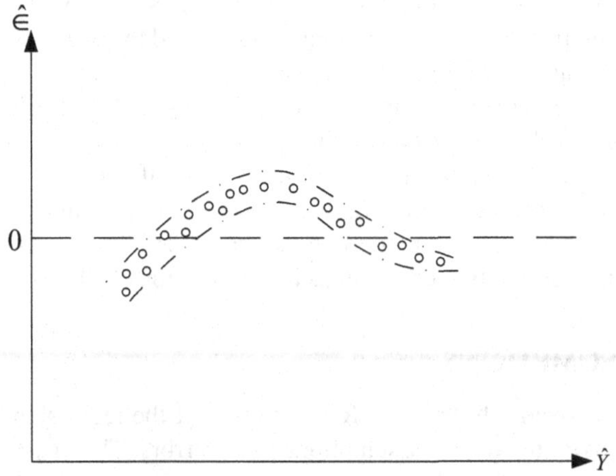

FIGURE 8.6 Residual plot showing non-linearity relationship between Y and X. It is clear from this plot that if a linear regression is fit to model having a non-linear relationship, the non-linearity will be transferred to its residuals.

Example 8.9

Check the assumptions of linearity for MLR developed in Example 8.1.

Solution:

The regression equation is $Y = 130.22 - 1.24\, X_1 - 0.30X_2 + \epsilon$. Using this equation the predicted values $\hat{y}_i, i = 1,2,...n$ and corresponding $\hat{\boldsymbol{\epsilon}}_i$ are computed as shown below.

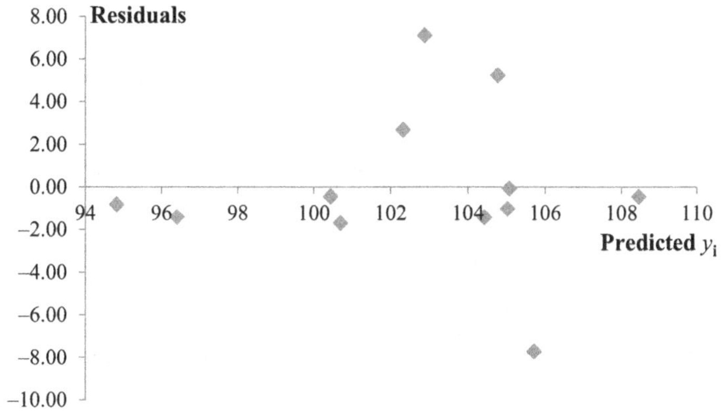

FIGURE 8.7 Residual plot for Example 8.9. The scatter plot does not show any apparent trends that violate the linear relationship between Y and X.

$$
\begin{matrix} \text{Predicted } y_i \text{ (i.e., } \hat{y}_i\text{)} & \text{Residuals } \left(\hat{\epsilon}_i\right) \\ \begin{bmatrix} 100.44 \\ 102.88 \\ 102.32 \\ 94.82 \\ 96.41 \\ 100.69 \\ 105.02 \\ 108.45 \\ 105.07 \\ 105.71 \\ 104.42 \\ 104.77 \end{bmatrix} & \begin{bmatrix} -0.44 \\ \mathbf{7.12} \\ 2.68 \\ -0.82 \\ -1.41 \\ -1.69 \\ -1.02 \\ -0.45 \\ -0.07 \\ \mathbf{-7.71} \\ -1.42 \\ \mathbf{5.23} \end{bmatrix} \end{matrix}
$$

The plot (Figure 8.7) shows no apparent trend. Hence, we can conclude that the relationship is linear. But there are a few high value residuals like 7.12, –7.71, and 5.23.

In case the residual plot shows non-linearity as shown in Figure 8.6, we often want to know which of the p-independent variables are responsible for such non-linearity. Figure 8.6 cannot tell this. The answer can be obtained by using partial residual plot. The partial residual plot is a plot of $\hat{\epsilon} + \mathbf{x}_j\hat{\beta}_j$ versus X_j. The ordinate of the plot is composed of two components, $\hat{\epsilon}$, representing randomness and $\mathbf{x}_j\hat{\beta}_j$, representing systematic trend (Cook and Weisberg, 1982). A hypothetical partial residual plot is shown in Figure 8.8.

What is the difference between residual plot (Figure 8.6) and partial residual plot (Figure 8.8)? Apart from the abscissa and ordinate difference, the non-linearity, if it exists, is centered around a horizontal line in the residual plot and centered along the trend line in the partial residual plot (Hair et al., 2010). Although both the plots can determine non-linearity, the partial residual plot can also determine the responsible IVs for which the non-linearity is introduced.

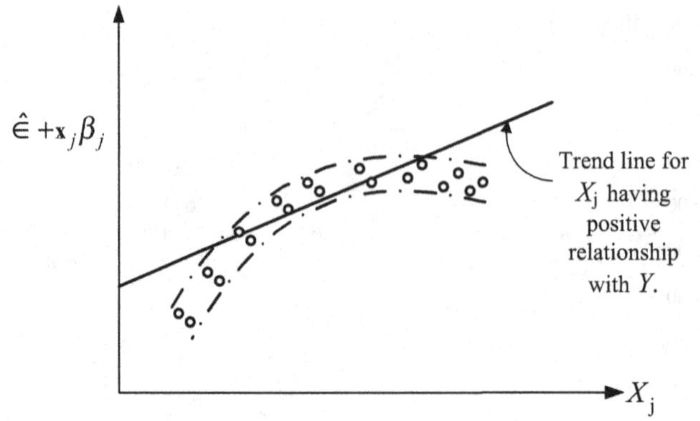

FIGURE 8.8 Partial residual plot for X_j. The ordinate of the plot is composed of two components $\hat{\in}$, representing randomness and $X_j\hat{\beta}_j$, representing systematic trend. The plot shows non-linear relationship and the non-linearity (represented by the dotted lines) is centered along the trend line.

Example 8.10

Obtain partial residual plots for X_1 and X_2 for MLR for Example 8.1.

Solution:

Step 1: Find out $\mathbf{x}_1\hat{\beta}_1$, $\mathbf{x}_2\hat{\beta}_2$, $\hat{\in}$. They represent a vector of 12 × 1 as sample size is 12.

$$\mathbf{x}_1\hat{\beta}_1 = [-11.2 \quad -9.92 \quad -8.68 \quad -17.4 \quad -14.9 \quad -12.4 \quad -8.68 \quad -4.96 \quad -7.44 \quad -6.2 \quad -8.68 \quad -7.44]^T$$

$$\mathbf{x}_2\hat{\beta}_2 = [-18.6 \quad -17.4 \quad -19.2 \quad -18 \quad -18.9 \quad -17.1 \quad -16.5 \quad -16.8 \quad -17.7 \quad -18.3 \quad -17.1 \quad -18]^T$$

and, $\hat{\in} = [-0.44 \quad 7.12 \quad 2.68 \quad -0.82 \quad -1.41 \quad -1.69 \quad -1.02 \quad -0.45 \quad -0.07 \quad -7.71 \quad -1.42 \quad 5.23]^T$

Step 2: Compute $\hat{\in} + \mathbf{x}_1\hat{\beta}_1$ and $\hat{\in} + \mathbf{x}_2\hat{\beta}_2$.

$$\hat{\in} + \mathbf{x}_1\hat{\beta}_1 = [-11.64 \quad -2.8 \quad -6 \quad -18.2 \quad -16.3 \quad -14.1 \quad -9.7 \quad -5.41 \quad -7.51 \quad -13.91 \quad -10.10 \quad -2.21]^T$$

$$\hat{\in} + \mathbf{x}_2\hat{\beta}_2 = [-19.04 \quad -10.3 \quad -16.5 \quad -18.8 \quad -20.3 \quad -18.8 \quad -17.5 \quad -17.25 \quad -17.77 \quad -26.01 \quad -18.52 \quad -12.77]^T$$

Step 3: Plot $\hat{\in} + \mathbf{x}_1\hat{\beta}_1$ versus X_1 (Figure 8.9) and $\hat{\in} + \mathbf{x}_2\hat{\beta}_2$ versus X_2 (Figure 8.10).

The partial residual plots (See Figures 8.9 and 8.10) show that both the independent variables X_1 and X_2 show linear relationship with Y, the dependent variable.

8.10.2 Homoscedasticity or Constant Error Variance

The homoscedasticity or constant error variance can be tested through the residual plot. We examine two such plots below (Figures 8.11a, 8.11b, and 8.11 c).

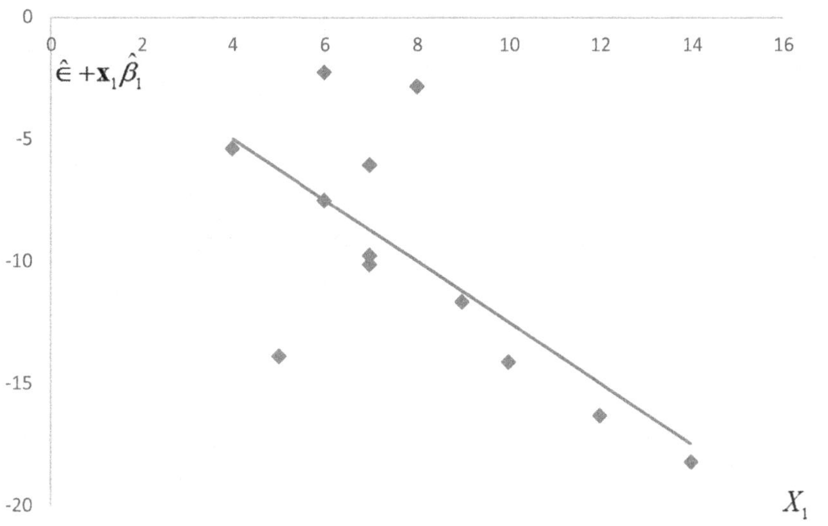

FIGURE 8.9 Partial residual plots between $\mathbf{x}_1\hat{\beta}_1$ and $\mathbf{x}_1$. The plot shows a linear relationship between Y and $\mathbf{x}_1$.

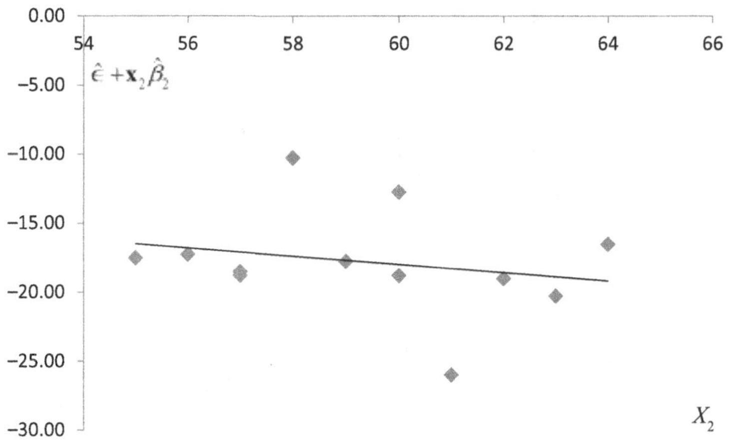

FIGURE 8.10 Partial residual plots between $\mathbf{x}_2\hat{\beta}_2$ and X_2. The plot shows linear relationship between Y and X_2.

In Figure 8.11a, the error terms spread uniformly across different values of $\hat{Y}$ and indicate that the assumption of constant error terms is satisfied. On the other hand, Figure 8.11b shows that the variability in $\hat{\epsilon}$ increases with increase in $\hat{Y}$. In Figure 8.11b, the residual plot funnels out to the right. It can be funneled out to the left also (Figure 8.11c). This type of funneling shape of residuals confirms heteroscedastic behavior, which is undesirable in MLR.

Under the influence of heteroscedasticity, the OLS estimates of the parameters are not efficient (do not pose minimum variance) and the test statistics become large (Jobson, 1991). Therefore, heteroscedasticity must be eliminated before accepting the MLR model. The remedial measures will be discussed in Section 8.11.

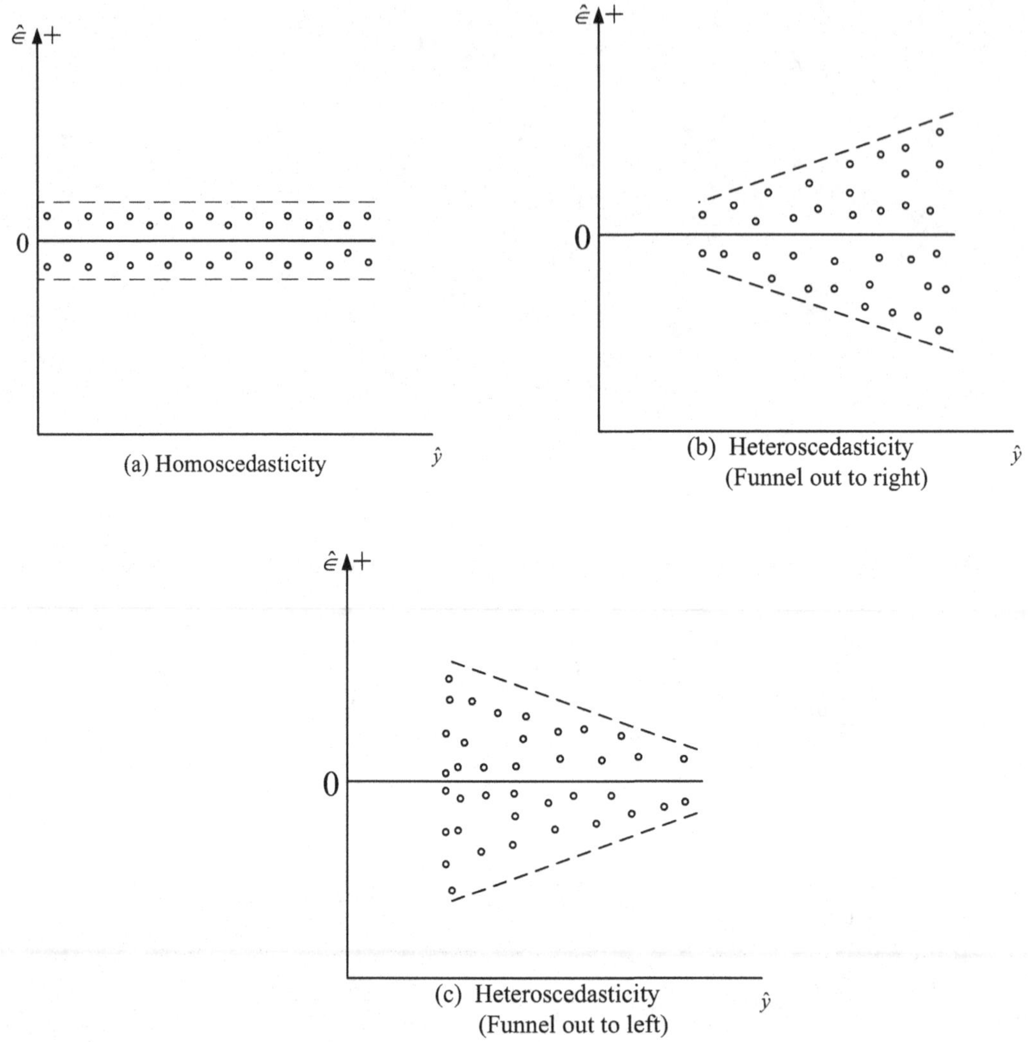

FIGURE 8.11 Residual plot showing (a) homoscedasticity (equal error variance), (b) and (c) heteroscedasticity (unequal error variance) behavior.

8.10.3 Uncorrelated Error Terms

The error terms must be uncorrelated, i.e., $Cov\left(\epsilon_i, \epsilon_k\right) = 0, i \neq k$. In case of time series data (like in the CityCan example), this issue is called *autocorrelation*. The error terms should not be autocorrelated. If we plot the error values over time in a chronological order (order of observations), the plot should resemble randomness for uncorrelated error terms. Figures 8.12a and 8.12b depict the situation.

Alternatively, the Durbin-Watson test can be used to detect autocorrelation (Durbin and Watson, 1951). The Durbin-Watson Statistic (DW) is

$$DW = \frac{\sum_{i=1}^{n-1}\left(\hat{\epsilon}_{i+1} - \hat{\epsilon}_i\right)^2}{\sum_{i=1}^{n}\left(\hat{\epsilon}_i^2\right)} \tag{8.45}$$

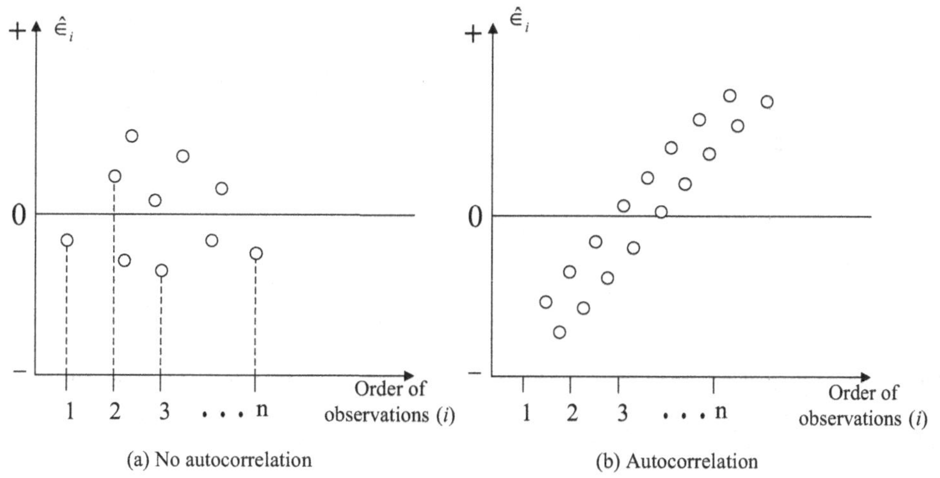

FIGURE 8.12 Residual plots showing (a) uncorrelated (random scatter) and (b) correlated (scatter with a trend or pattern) behaviors.

Further, $DW = 2(1-r)$ where r is the first order autocorrelation of $\hat{\epsilon}$ and is explained below.

$$r = \frac{\sum_{i=1}^{n-1} \hat{\epsilon}_{i+1}\, \hat{\epsilon}_i}{\sum_{i=1}^{n} \hat{\epsilon}_i^2} \tag{8.46}$$

- For no autocorrelation, $r = 0$ yields DW = 2
- For positive autocorrelation, $r > 0$ yields DW < 2
- For negative autocorrelation, $r < 0$ yields DW > 2

For MLR, it is desirable that DW value should be around 2.

Example 8.11

Obtain residual plot for autocorrelation and DW statistic value for the CityCan case (Example 8.1).

Solution:

The residuals given below when plotted against observation number resemble the following plot (Figure 8.13).

$$\hat{\epsilon} = [-0.44 \quad 7.12 \quad 2.68 \quad -0.82 \quad -1.41 \quad -1.69 \quad -1.02 \quad -0.45 \quad -0.07 \quad -7.71 \quad -1.42 \quad 5.23]^T$$

The plot shows no clear trend for residuals but three high values are observed. More conclusive evidences are needed to ascertain autocorrelation.

The DW test results are shown in Table 8.5.

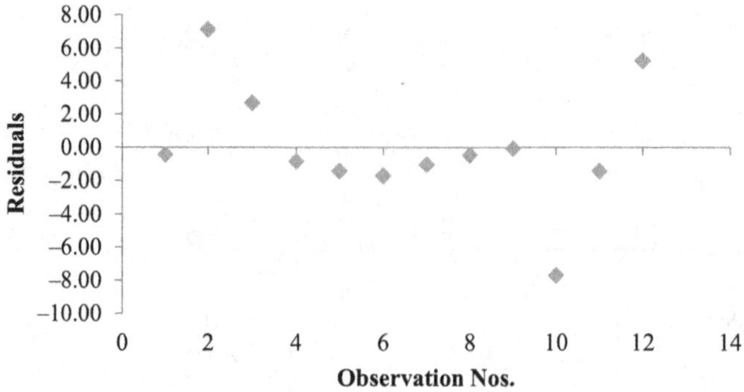

FIGURE 8.13 Residual vs observation number for Example 8.11. Scatter does not show any systematic pattern. The errors could be considered as uncorrelated. But the presence of three high value residuals prohibits to make a clear conclusion.

TABLE 8.5
DW Test for Example 8.11

$\hat{\epsilon}_i$	$\hat{\epsilon}_{i+1}$	$\hat{\epsilon}_{i+1}\hat{\epsilon}_i$	$\hat{\epsilon}_i^2$	$(\hat{\epsilon}_{i+1}-\hat{\epsilon}_i)^2$
−0.44	7.12	−3.11	0.19	57.11
7.12	2.68	19.06	50.70	19.75
2.68	−0.82	−2.20	7.16	12.23
−0.82	−1.41	1.15	0.67	0.34
−1.41	−1.69	2.38	1.98	0.08
−1.69	−1.02	1.73	2.87	0.45
−1.02	−0.45	0.46	1.05	0.33
−0.45	−0.07	0.03	0.20	0.15
−0.07	−7.71	0.51	0.00	58.42
−7.71	−1.42	10.97	59.44	39.52
−1.42	5.23	−7.45	2.02	44.31
5.23			27.39	
Sum(Σ)		23.55	153.68	232.69
r		0.15		
DW		1.69		1.51

$$DW = \sum_{i=1}^{n-1}\left(\hat{\epsilon}_{i+1} - \hat{\epsilon}_i\right)^2 \Big/ \sum_{i=1}^{n}\hat{\epsilon}_i^2 = \frac{232.69}{153.68} = 1.51 \text{ using Equation 8.45}$$

$$r = \sum_{i=1}^{n-1}\hat{\epsilon}_{i+1}\hat{\epsilon}_i \Big/ \sum_{i=1}^{n}\hat{\epsilon}_i^2 = \frac{23.55}{153.68} = 0.15 \text{ using Equation 8.46}$$

$$DW = 2(1-r) = 2\times\cdot 85 = 1.70$$

The value of DW is less than 2, indicating little positive correlation but the presence of three high residuals should be removed before making such a conclusion. The difference in DW values obtained through Equation 8.45 and using r can be attributed to low sample size and rounding-off errors.

8.10.4 Normality of Error Terms

The most basic tool for assessing normality is drawing a histogram. A bell-shaped histogram resembles normality. Testing normality through histogram plots is a visual technique and subject to misinterpretation. The widely used plots are P-P and Q-Q plots.

The probability-probability plot or P-P plot compares the empirical cumulative distribution of the actual data to the theoretical cumulative normal distribution (Hair et al., 2010). For normal error terms, the plot will be a straight line. The scheme for plotting Q-Q plots developed by Lloyd (1952) and Blom (1958) are given below (Cook and Weisberg, 1982).

Step 1: Sort the residuals $\left(\hat{\epsilon}_i\right)$ in ascending order and let the ordered residuals be $u_1, u_2, \ldots, u_n$, where $u_n = \max(\hat{\epsilon}_i)$.

If $\hat{\epsilon}_1, \hat{\epsilon}_2, \cdots \hat{\epsilon}_n$ comes from a normal distribution with mean 0 and standard deviation σ^2, then, let $a_i = E\left(u_i/\sigma\right)$. The value of a_i can be obtained as

$$a_i = \phi^{-1}\left\{(i-3/8)\Big/\left(n+\frac{1}{4}\right)\right\} \tag{8.47}$$

where ϕ is standard normal *cdf.* If the errors are normal, then the plot between u and a will be linear and pass through the origin as shown in Figure 8.14.

Step 2: Obtain the value of $\left(i-\frac{3}{8}\right)\Big/\left(n+\frac{1}{4}\right)$ for $i = 1, 2, \cdots n$.

Step 3: Find out $\phi^{-1}\left(a_i\right),\ i = 1, 2, \cdots n$

Step 4: Plot u versus a. The plot should exhibit a straight line with slope 1passing through the origin like Figure 8.14.

Example 8.12

Obtain histogram and Q-Q plot for the residuals obtained in Example 8.2.

Solution:

The histogram of the residuals with fitted normal distribution using MINITAB is given in Figure 8.15. The data does not show perfect normality. From Figure 8.16, it is seen that the presence of three high value residuals has grossly affected the normality. The issue of small sample size (n =12) also can't be ruled out.

The computation details (Table 8.6) as well as the Q-Q plot (Figure 8.16) are given below. The Q-Q plot shows that the data do not exhibit normality.

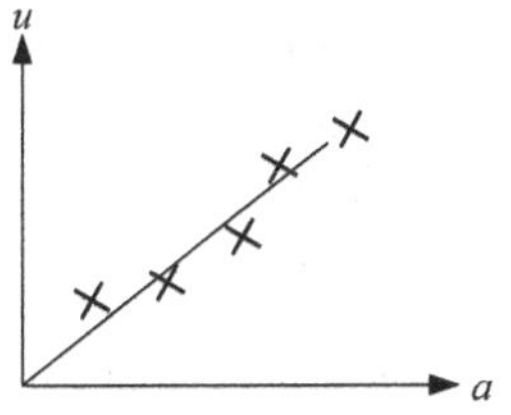

FIGURE 8.14 Q-Q plot for residuals. For normality of error terms, the scatter should follow a straight line with slope 1 passing through the origin.

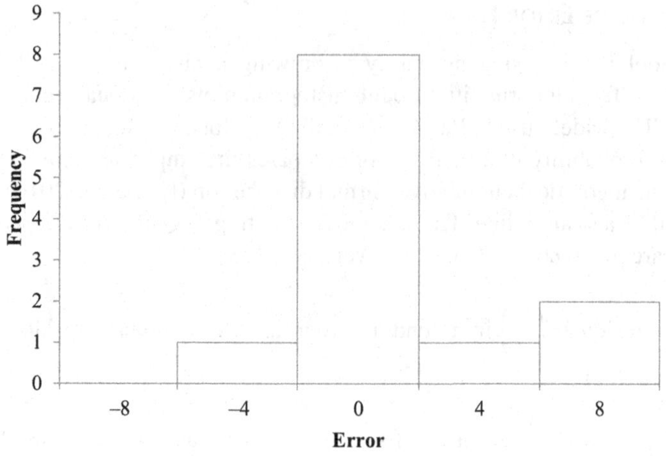

FIGURE 8.15 Histogram of residuals for sales volume.

TABLE 8.6
Computation Details of Q-Q plot

I	(i-3/8)/(n + 1/4)	a_i	u_i
1	0.05	−1.64	-7.71
2	0.13	−1.13	-1.69
3	0.21	–0.81	-1.42
4	0.30	–0.52	-1.41
5	0.38	–0.31	-1.02
6	0.46	–0.10	–0.82
7	0.54	0.10	–0.45
8	0.62	0.31	–0.44
9	0.70	0.52	–0.07
10	0.79	0.81	2.68
11	0.87	1.13	5.23
12	0.95	1.64	7.12

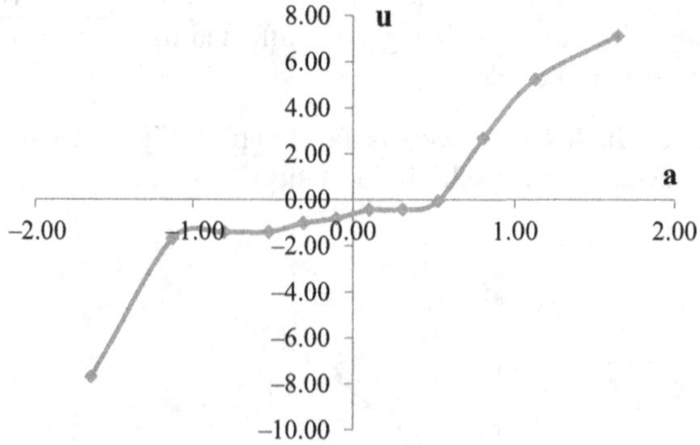

FIGURE 8.16 Q-Q plot for residuals. The scatter does not follow a straight line passing through the origin indicating violation of normality assumption.

8.11 REMEDY AGAINST VIOLATION OF ASSUMPTIONS

If one or more of the assumptions are violated, MLR results cannot be used for decision-making or proper cushion should be made for such decisions. Therefore, it is better to find out the remedy for each of the violations. In brief, we will describe them below.

8.11.1 Remedies Against Linearity

Hair et al. (1998) documented two popular remedies against linearity as (i) transforming Y or X or both variables and (ii) creation of new variables to represent the non-linear part of the relationship. They proposed four basic non-linear relationships (Hair et al., 1998, Figure 2.9, p. 77), and the possible transformation to achieve linearity. Montgomery et al. (2003), based on Daniel and Wood (1980), have shown several linearizable functions, their transformation and the corresponding linear forms (Montgomery et al., 2003, Figure 5.4 and Table 5.4, pp. 179–180). The possible transformations to achieve linearity are as follows:

- Squaring X or Y or both.
- Considering square root of X or Y or both.
- Considering log of X or Y or both.
- Considering negative reciprocal of X (i.e., $-\frac{1}{X}$) or $Y\left(\text{i.e.,} -\frac{1}{Y}\right)$ or both.
- Any other combinations using the above mentioned transformations.

It should be noted here that for noticeable effect of transformation of a variable, its mean to standard deviation ratio should be less than 4 (Hair et al., 1998). Further, it is also advisable to prefer transforming X (IVs) instead of Y (DVs) when either of the transformation serves the purpose.

8.11.2 Remedies Against Non-Normality and Heteroscedasticity

The most plausible remedy against non-normality as well as heteroscedasticity is data transformation. Out of many ways of transforming data, the Box-Cox method serves the purpose efficiently. Hence, in this section we will describe the Box-Cox method of data transformation (Box and Cox, 1964). It is to be noted here that in multiple linear regression, normality and homoscedasticity are attributed to the dependent variable, Y, not to the independent variables X.

In the Box-Cox method, the dependent variable's observations $(y_1, y_2, \ldots, y_n)$ are raised to a power of λ (i.e., y_i^{λ}, $i = 1,2,\ldots n$).The Box-Cox transformation of y_i is

$$y_i^{(\lambda)} = \begin{cases} \dfrac{y_i^{\lambda} - 1}{\lambda} & if\, \lambda \neq 0 \\ \log\, y_i & if\, \lambda = 0 \end{cases} \tag{8.48}$$

and $\mathbf{y}^{(\lambda)} = \left[y_1^{(\lambda)}, y_2^{(\lambda)}, \ldots, y_n^{(\lambda)}\right]$, a vector of size $n \times 1$.

Montgomery et al. (2003, pp. 186–187) stated that Equation 8.48 suffers from dramatic changes of $\left(y_i^{\lambda} - 1\right)/\lambda$ with respect to changes in λ for lower values of λ (tends to zero).

A better procedure for solving this problem is to use the following formulation (Montgomery et al., 2003):

$$y_i^{(\lambda)} = \begin{cases} \dfrac{y_i^{\lambda} - 1}{\lambda \dot{y}^{\lambda - 1}} & , \lambda \neq 0 \\ \dot{y} \ln y_i & , \lambda = 0 \end{cases} \tag{8.49}$$

$$\text{where } \dot{y} = \ell n^{-1}\left[\frac{1}{n}\sum_{i=1}^{n} \ell n y_i\right]$$

In order to find out the best value of λ, a variance minimization approach is preferred. The steps to be followed are given below.

Step 1: Set a range of values of λ. For most of the cases, λ ranges from –3 to 3. Let the minimum value of λ be λ_1. Also set an increment, δ by which λ values will be changed. That is $\lambda_2 = \lambda_1 + \delta, \lambda_3 = \lambda_1 + 2\delta$, and so on. Alternatively, set the number of increments (iterations). Usually 10–20 values of λ are sufficient.

Step 2: Consider λ_1, compute $\mathbf{y}^{(\lambda_1)}$ using Equation 8.49 and obtain the MLR model as given below.

$$\mathbf{y}^{(\lambda_1)} = \mathbf{X}\boldsymbol{\beta} + \epsilon$$

Compute SSE_{λ_1}.

Step 3: Change λ from λ_1 to $\lambda_2 = \lambda_1 + \delta$. Repeat Step 2. This will give SSE_{λ_2}. Repeat this process until the required number of iterations are completed.

Step 4: Plot SSE_{λ_k} versus λ_k, for k = 1, 2, …, K, where K = number of λ values considered (or iterations performed).

Step 5: Choose the best value of λ, i.e. λ^* for which SSE_λ is the minimum.

Once the appropriate λ (i.e., λ^*) value is chosen, we will use the following equation for MLR.

$$\mathbf{y}^{(\lambda^*)} = \mathbf{X}\boldsymbol{\beta} + \epsilon \tag{8.50}$$

It is quite likely that Equation 8.50 will satisfy MLR assumptions, particularly the assumption of normality and homoscedasticity. However, care should be taken in interpreting the results of Equation 8.50 as the *Y* variable is transformed. It is therefore recommended to make simple transformation with λ values of 0.5 (square root) or 2 (square) or log even if we require to sacrifice a little from minimum SSE_λ point of view.

The Box-Cox transformation, being a widely used technique, has been studied by many statisticians after it was first proposed. Interested readers may consult Manly (1976), John and Draper (1980), and Bickel and Docksum (1981) for modifications of the Box-Cox transformation and Sakia (1992) for a review.

Example 8.13

Consider Example 8.12. It was seen that the sales data do not exhibit normality. Transform the sales data that exhibits normality.

Solution:

Using Equation 8.49 and the procedure given above, we obtain the following graph (Figure 8.17).

It is seen that the minimum *SSE* value is 144.2495 for λ^* equal to –4.25. It is expected that if the sales data is transformed, it exhibits normality and homoscedasticity. In this example, we demonstrate the procedure with hypothetical data with a small sample (size = 12). It is unlikely that with such a small data set, we will get a satisfactory result. In addition, data transformation results in the problem of interpretability of the findings. Nevertheless, the procedure is adopted by researchers and analyst to make the data analyzable with linear regression.

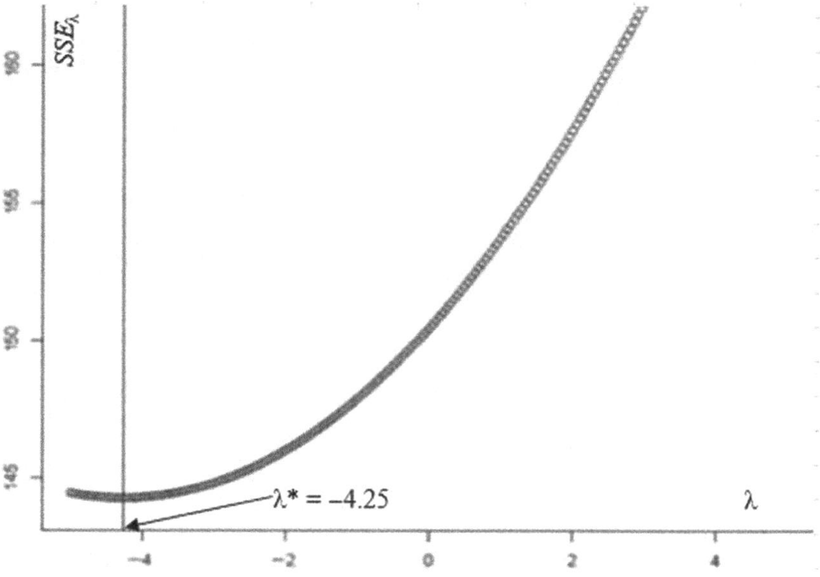

FIGURE 8.17 Demonstration of Box-Cox transformation procedure. At the best value of $\lambda(\lambda^* = -4.25)$, the SSE value is the minimum.

8.12 DIAGNOSTIC ISSUES IN MLR

The diagnostic issues that are important in MLR are identifying influential observations and detecting multicollinearity. Hair et al. (1998) stated that influential observations are of three types namely, outliers, leverage points, and influential observations. Belsley et al. (2004) stated that influential observations are those observations that do not conform to the pattern set by other data points or those that strongly influence the results of the regression. They also pointed out that influential observations may contain important information related to the sample collected, which the analyst must not overlook. Therefore, they cannot be always treated as bad data points. Multicollinearity refers to the presence of collinear relationships among the independent variables which may grossly affect the least squares estimation. Detecting and removing multicollinearity is therefore extremely important before using MLR results for practical purpose. We discuss these issues in the subsequent sections.

8.12.1 Outliers

Outliers are those observations which do not belong to the general mass of observations. For one variable, a dot plot may be used to detect outliers. For two variables, a scatter plot can determine outliers. For more than two variables, we rely on squared Mahalanobis distances of the observations.

Figure 8.18 depicts situations for outliers in single variable and two variable cases. For more than two variables, the squared Mahalanobis distances for the observations are computed as below[3].

$$d_i^2 = \left(\mathbf{x}_i - \overline{\mathbf{x}}\right)^T \mathbf{S}^{-1} \left(\mathbf{x}_i - \overline{\mathbf{x}}\right), \; i = 1,2,\ldots n \tag{8.51}$$

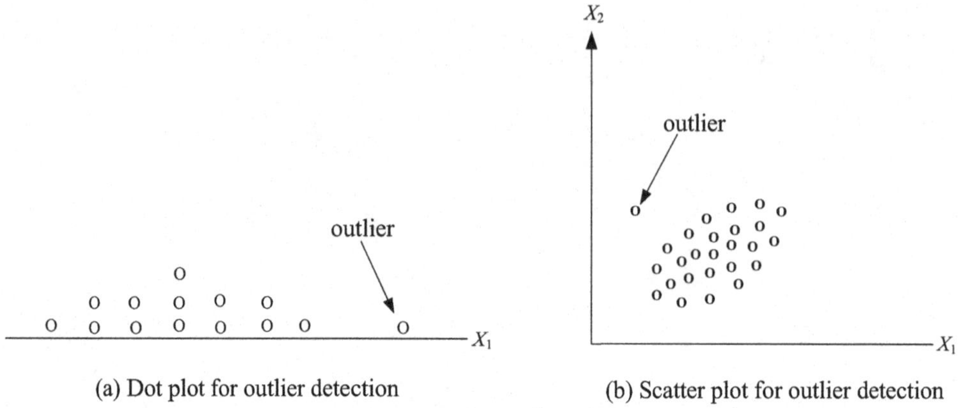
outlier
X_1
(a) Dot plot for outlier detection
X_2
outlier
X_1
(b) Scatter plot for outlier detection

FIGURE 8.18 Graphical methods for outlier detection.

where, $\mathbf{x}_i = \begin{bmatrix} x_{i1} & x_{i2} & \cdots & x_{ip} \end{bmatrix}^T$, i-th multivariate observation p-variables, $\bar{\mathbf{x}}$ is the $p \times 1$ vector of sample means and $\mathbf{S}^{-1}$ is the inverse of the sample covariance matrix. The dot plot of $d_i^2, i = 1, 2, \ldots n$ detects large values of d_i^2, which are treated as outliers. If we draw a χ^2-quantile plot, the larger values will be the points far away from the origin (Johnson and Wichern, 2002).

The most frequently used remedy against outliers is their deletion, but that may be limited. If the analyst finds that deletion of outlier is not due, then the outlier can be properly weighted in MLR. Hair et al. (1998) pointed out that a large number of outliers may represent a segment of a population and hence they may be retained. The regression analysis can take care of these outliers by incorporating a dummy variable that will represent the segments of the population studied.

8.12.2 Leverage Points

In MLR, the outliers have large residual values but there may be other observations having not so large residuals. Such observations may also affect the estimates of one or more regression coefficients. These observations are known as leverage points. Identification of leverage points is not as easy as outliers. In order to identify leverage points, we rely on one of the most important matrix in MLR namely the *hat matrix* (**H**). Let us now understand the physical meaning of the elements of **H**.

$\mathbf{H} = \mathbf{X}\left(\mathbf{X}^T\mathbf{X}\right)^{-1}\mathbf{X}^T$ is a $n \times n$ matrix and can be written as

$$\mathbf{H} = \begin{bmatrix} h_{11} & h_{12} & \cdots & h_{1n} \\ h_{21} & h_{22} & \cdots & h_{2n} \\ h_{i1} & h_{i2} & \cdots & h_{in} \\ h_{n1} & h_{n2} & \cdots & h_{nn} \end{bmatrix}_{n \times n}$$

The diagonal elements of the **H** matrix, i.e., h_{ii}, $i = 1, 2, \ldots, n$, are called leverage values of the observations $i = 1, 2, \ldots, n$. It can be shown that (Jobson, 1991):

$$\frac{1}{n} \le h_{ii} \le 1 \tag{8.52}$$

and for full rank $\mathbf{X}_{n\times(p+1)}$,

$$\sum_{i=1}^{n} h_{ii} = p+1 \tag{8.53}$$

Before data collection, $x_1, x_2, \ldots, x_n$ are random observations. It is likely that they contribute equally to the regression estimates and hence, should have h_{ii}, $i = 1, 2, \ldots, n$ roughly equal to $(p+1)/n$.

The statistic $\dfrac{\left(h_{ii}-\frac{1}{n}\right)/p}{\left(1-h_{ii}\right)/n-p-1}$ follows F-distribution with p-numerator and n–p–1 denominator degrees of freedom (Rao, 1973 and Belsley et al. 2004). For $p>10$ and $n-p-1>50$, the value of F for $\alpha = 0.05$ is less than 2. Therefore, $2(p+1)/n$ is usually considered to be large and considered as a cut-off for identifying leverage points.

Example 8.14

Obtain leverage points for MLR for the CityCan data (Example 8.1)

Solution:

From Example 8.2, we get the hat matrix (**H**) whose diagonal elements are h_{ii}, $i = 1,2,\ldots,12$. The h_{ii} values are given below.

$\mathbf{h}^T = [0.16 \quad 0.11 \quad 0.39 \quad 0.50 \quad 0.34 \quad 0.24 \quad 0.30 \quad 0.31 \quad 0.12 \quad 0.25 \quad 0.15 \quad 0.14]$

From $\mathbf{h}^T$, we observe the followings:

All h_{ii} values are within 1/n and 1, i.e., 0.083 and 1 (from Equation 8.52).
Sum of h_{ii} values are 3 (i.e., p + 1 = 2 + 1) (from Equation 8.53).

$$\frac{\left(h_{ii}-\frac{1}{n}\right)/p}{\left(1-h_{ii}\right)/n-p-1} \text{ follows } F_{p,n-p-1} \quad \text{and} \quad F^{\alpha=0.05}_{p>10,n-p-1>50} < 2.$$

So, it can be concluded that there is no leverage point. Readers may note that there are three high value residuals. But h values for them are low. Why?

8.12.3 Influential Observations

Hair et al. (1998) stated that influential observations comprise the superset of unusual observations like outliers, leverage points, and other observations, if any, which influence the regression results. The sources of influential observations are many. Some of them are improperly recorded data, observational errors, extreme observations, generated data from models not specified (Belsley et al., 2004), and too few observations for too many variables (Jobson, 1991).

The other measures of influential observations are PRESS statistic, Cook's D, standardized residuals, deleted residuals, DF Fit, DF Beta, and covariance ratio (CVR). Excellent discussions on these measures are given by Belsley et al. (2004).

8.12.4 Multicollinearity

Multicollinearity refers to a situation when one or more of the independent variables are not truly independent. There may be a high or significant degree of correlation between the collinear variables. In such a situation, OLS fails to estimate the parameters ($\boldsymbol{\beta}$) (in case of perfect correlation) or the estimates will be distorted (in case of high correlation). Therefore, multicollinearity should be identified and removed before fitting an acceptable MLR.

By identifying multicollinearity we mean the extent of collinearity and its impact on the estimated coefficients. The first and rudimentary method of identifying multicollinearity is examining the correlation matrix $R_{p\times p}$ of the predictor variables vector $\boldsymbol{X}_{p\times 1}$. The correlation coefficient values above 0.80 are indicative of multicollinearity. However, values below 0.80 do not rule out multicollinearity, hence more sophisticated methods are required to identify multicollinearity. The measures that can be used are given below.

- Variance inflation factor (VIF)
- Tolerance statistic
- Eigen value structure
- Multicollinearity condition number

Variance Inflation Factor (VIF)

In case of multicollinearity, there may be one or more independent variables X_j, $j = 1, 2, \ldots, p$ that are dependent on the other independent variables. Let X_j be dependent on the remaining p–1 independent variables and we develop a regression model considering X_j as dependent and remaining p–1 variables as independent. Let the design matrix without the variable X_j be $\mathbf{X}_{(j)}$ which is of the order of $n \times p$. Pictorially the regression situation is shown in Figure 8.19.

Now, let's say the multiple coefficient of determination for the MLR model (Figure 8.19) is R_j^2. Then VIF for X_j is

$$VIF_j = \frac{1}{1-R_j^2} \tag{8.54}$$

If X_j is dependent on the other IVs, R_j^2 value will be large and vice versa. For different R_j^2, the VIF is given in Table 8.7.

A large value of VIF is indicative of multicollinearity. Regarding what constitutes "large," Myers (1990) recommended the cut-off for multicollinearity as $VIF \geq 10$.

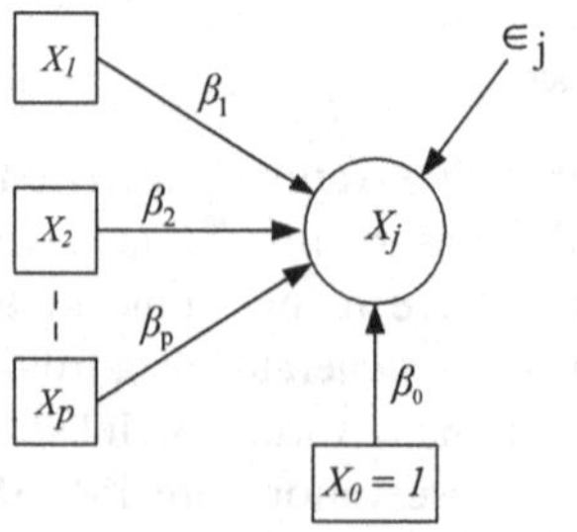

FIGURE 8.19 The MLR model for X_j as DV and remaining (p–1) X variables as IVs.

TABLE 8.7
***VIF* for Different Values of R_j^2**

R_j^2	0	0.2	0.4	0.5	0.6	0.8	0.9	1.0
VIF_j	1	1.25	1.67	2	2.5	5	10	∞
$1/VIF_j$	1	0.8	0.60	0.5	0.4	0.2	0.1	0.0

Tolerance Statistic

Table 8.7 also shows the reciprocal of VIF_j, i.e., $\frac{1}{VIF_j}$, which is known as the tolerance statistic. The tolerance statistic corresponds to $VIF = 10$ is 0.10. Menard (1995) has given the using limit for multicollinearity as tolerance = 0.2 or $VIF = 5$.

Eigenvalue Structure of the Correlation Matrix, R

We pointed out earlier that the correlation coefficient between $\boldsymbol{X}_{p\times1}$ variables hints the presence of multicollinearity. If $\boldsymbol{X}_{p\times1}$ is multicollinear, then the rank of the correlation matrix **R** will be less than p. In other words, we can reduce the dimensions of **R**. This can be done by obtaining the eigenvalues and eigenvectors of **R**.

Let $\lambda_1 \le \lambda_2 \le \cdots \le \lambda_p$ be the eigenvalues of **R** and the corresponding eigenvectors are $v_1, v_2, \cdots, v_p$. We can reproduce **R** using λ_j and v_j, $j = 1,2,\ldots p$ based on spectral decomposition method of a matrix.

$$\mathbf{R} = \sum_{j=1}^{p} v_j \lambda_j v_j^T \tag{8.55}$$

The eigenvectors are orthogonal (independent) to each other. If $\boldsymbol{X}_{p\times1}$ is truly independent, we will get all λ_j, $j=1,2,\ldots p$ non-zero and nearly equal. Otherwise, some of the λ_j values will become zero or close to zero. Therefore, one or more zero or near zero eigenvalues indicate multicollinearity.

Once we know that multicollinearity exists, we will be interested to know which variables are contributing to the multicollinearity. This can be tested using the eigenvectors corresponding to the very small eigenvalues.

If λ_j is close to zero, the v_j can be used to find out responsible variables for multicollinearity. The elements of v_j are $\left[v_{j1},\ v_{j2}, \cdots v_{jp}\right]^T$. Relatively large values of v_{jk}, $k=1,2, \cdots m$ $(m \le p)$, determine the responsible variable (Jobson, 1991).

Multicollinearity Condition Number (MCN)

The multicollinearity condition number (MCN) is the ratio of the largest eigenvalue of **R**, i.e., λ_1 to the smallest eigenvalue of **R**, i.e., λ_p. So,

$$MCN = \frac{\lambda_1}{\lambda_p} \tag{8.56}$$

If λ_p is zero, *MCN* is undefined. If all eigenvalues are same *MCN* is 1. The range for *MCN* is (Berk, 1977 and Jobson, 1991)

$$VIF_m \leq MCN \leq p\sum_{j=1}^{p} VIF_j \tag{8.57}$$

where, $VIF_m = \max_j VIF_j$.

$MCN < 100$ is not considered as serious for multicollinearity but $MCN > 1000$ indicates a severe multicollinearity problem (Jobson, 1991).

Example 8.15

Obtain eigenvalue structure of the correlation matrix and *MCN* for the MLR model obtained in Example 8.1.

Solution:

From Example 8.1, we can extract the correlation matrix of the two predictor (*X*) variables, absenteeism in % and breakdown in hours. The **R** matrix is given below.

$$\mathbf{R} = \begin{pmatrix} 1.00 & 0.29 \\ 0.29 & 1.00 \end{pmatrix}$$

The two eigenvalues of **R** are 1.29 and 0.71, respectively. Using Equation 8.56, we get $MCN = (1.29/.71) = 1.82$, which is much less than 100, the threshold value for multicollinearity. So, there is no multicollinearity. This example, with only two IVs is shown just to demonstrate the computation on MCN. For a better feel of this concept, we need higher order $\mathbf{R}_{p \times p}$, involving more IVs.

If we want to develop a MLR model, multicollinearity should be removed. But the researcher or analyst should not develop a statistical model by scarifying relevant information. For example, if the predictors variables are inherently related, it is advisable to keep the correlated structure inherent during model-building and accordingly appropriate statistical model should be used. For example, the path model (Chapter 10) may be used in such cases. Nevertheless, under many situations, a MLR is preferable and multicollinearity, if exists, should be removed. The most frequently used techniques are ridge regression and principal component regression. As multicollinearity increases the variance of the estimated parameters, in ridge regression a weight is added to the parameter variance while estimating to adjust the inflated variance value of the estimates. For discussion on ridge regression, see Vinod (1978). The principal component regression transforms the predictor variables into orthogonal (independent) dimensions and the regression is carried on between *Y* and the orthogonal principal components (PCs). Principal component analysis (PCA) is discussed in Chapter 11. The difficulty with principal component regression is in interpreting the regression coefficients. However, if one is interested in predicting the future *y* values, principal component regression is very useful.

8.13 PREDICTION WITH MLR

So far we have discussed MLR as a tool in explaining the variance of the dependent variables with the help of a set of independent explanatory variables. The explanatory variables in MLR are also known as predictor variables as they can be used (i.e., MLR) in predicting the *Y* (the DV) variable's values in future. In this section, we will describe the predictive ability of MLR.

The fitted regression equation is $\hat{y}_i = \mu_{y_i/\mathbf{x}_i} = \mathbf{x}_i\hat{\boldsymbol{\beta}}$, where, $\hat{y}_i$ is the fitted y_i, $\mu_{y_i/\mathbf{x}_i}$ is the mean of y_i at $\mathbf{x}_i$, $\hat{\boldsymbol{\beta}}$ is the estimated parameter vector and $\mathbf{x}_i$ is the i-th row of the design matrix $\mathbf{X}_{n\times(p+1)}$. If MLR is adequate $\left(i.e., R^2 \geq 0.90\right)$, we can use it to find out $\hat{y}_i$.

Now, the source of uncertainty in $\hat{y}_i$ is that of $\hat{\boldsymbol{\beta}}$ which can be estimated as below.

$$V(\hat{y}_i) = V(\mathbf{x}_i\hat{\boldsymbol{\beta}}) = \mathbf{x}_i^T V(\hat{\boldsymbol{\beta}})\mathbf{x}_i = \mathbf{x}_i^T \sigma^2 (\mathbf{X}^T\mathbf{X})^{-1}\mathbf{x}_i = \sigma^2 \mathbf{x}_i^T (\mathbf{X}^T\mathbf{X})^{-1}\mathbf{x}_i = s_e^2 \mathbf{x}_i^T (\mathbf{X}^T\mathbf{X})^{-1}\mathbf{x}_i \quad (8.58)$$

where s_e^2 is the estimate of σ^2.

The 100(1–α)% confidence interval for $y_i / \mathbf{x}_i$ is

$$\hat{y}_i - t_{n-p-1}^{\alpha/2}\sqrt{s_e^2 \mathbf{x}_i^T (\mathbf{X}^T\mathbf{X})^{-1}\mathbf{x}_i} \leq y_i / \mathbf{x}_i \leq \hat{y}_i + t_{n-p-1}^{\alpha/2}\sqrt{s_e^2 \mathbf{x}_i^T (\mathbf{X}^T\mathbf{X})^{-1}\mathbf{x}_i} \quad (8.59)$$

The new value for Y (say $\hat{y}_{new}$) can be predicted if a new multivariate observation on $\mathbf{X}$ (say $\mathbf{x}_{new}$) is available. The relationship between $\hat{y}_{new}$ and $\mathbf{x}_{new}$ is

$$\hat{y}_{new} = \mu_{y_{new}/\mathbf{x}_{new}} = \mathbf{x}_{new}\hat{\boldsymbol{\beta}} \quad (8.60)$$

Now, the source of uncertainty in $\hat{y}_{new}$ is that of $\hat{\boldsymbol{\beta}}$ and the error for $\mathbf{x}_{new}$ (i.e., $\in_{new}$) which is given below.

$$\begin{aligned} V(\hat{y}_{new}) &= V(\mathbf{x}_{new}\hat{\boldsymbol{\beta}}) + V(\in_{new}) = \mathbf{x}_{new}^T V(\hat{\boldsymbol{\beta}})\mathbf{x}_{new} + \sigma^2 = \mathbf{x}_{new}^T \sigma^2 (\mathbf{X}^T\mathbf{X})^{-1}\mathbf{x}_{new} + \sigma^2 \\ &= \sigma^2 (1 + \mathbf{x}_{new}^T (\mathbf{X}^T\mathbf{X})^{-1}\mathbf{x}_{new}) = s_e^2 (1 + \mathbf{x}_{new}^T (\mathbf{X}^T\mathbf{X})^{-1}\mathbf{x}_{new}) \end{aligned} \quad (8.61)$$

So, the 100(1–α)% confidence interval for $y_{new} / \mathbf{x}_{new}$ is

$$\hat{y}_{new} - t_{n-p-1}^{\alpha/2}\sqrt{s_e^2[1 + \mathbf{x}_{new}^T (\mathbf{X}^T\mathbf{X})^{-1}\mathbf{x}_{new}]} \leq y_{new} / \mathbf{x}_{new} \leq \hat{y}_{new} + t_{n-p-1}^{\alpha/2}\sqrt{s_e^2[1 + \mathbf{x}_{new}^T (\mathbf{X}^T\mathbf{X})^{-1}\mathbf{x}_{new}]} \quad (8.62)$$

Example 8.16

(a) Consider the CityCan data and Example 8.1. Obtain 95% CI for the observed sales values and (b) suppose a new observation ($\mathrm{x}_{\mathrm{new}}$) was obtained as [11, 60]. Find out 95% CI for y_{new}.

Solution:

(a) Using Equation 8.59, 95% CI values ($\hat{y}_-$ and $\hat{y}_+$) for the sales data are shown in Table 8.8.
The 95% CI plot is shown in Figure 8.20. It is seen that the variability in prediction is less and constant across the observations. This again proves that the sales data exhibit homoscedasticity.

(b) Using Equation 8.60, $\hat{y}_{new}$ is 98.55 and using Equation 8.62, the 95% CI for y_{new} is 104.47 to 96.40.

TABLE 8.8
95% CI Values ($\hat{y}_-$ and $\hat{y}_+$) for The Sales Data

i	1	2	3	4	5	6	7	8	9	10	11	12
$\hat{y}_+$	104.23	105.92	108.17	101.40	101.86	105.23	110.11	113.65	108.35	110.40	107.99	108.28
$\hat{y}_-$	96.64	99.83	96.48	88.24	90.95	96.16	99.93	103.25	101.78	101.02	100.86	101.26
$\hat{y}$	100.44	102.88	102.32	94.82	96.41	100.69	105.02	108.45	105.07	105.71	104.42	104.77

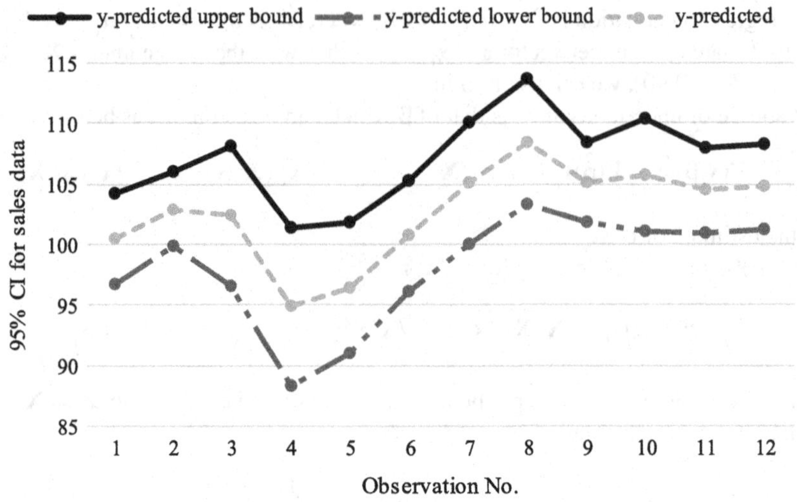

FIGURE 8.20 95% CI for sales data. The 95% CI band indicates that the variance of sales data is constant across the observations. This also proves that the sales data exhibit homoscedasticity.

8.14 MODEL VALIDATION

The regression model so built needs to be validated for its generalizability. Two methods are described by Lattin et al. (2003) as (i) cross-validation when the sample size is large and can be split into calibration and holdout sample, and (ii) jackknife validation when the sample size is too small to split. In cross-validation, usually a large sample (say, of size n) is split into two samples with two-thirds (2n/3) and one-third (n/3) observations of the original sample, respectively known as calibration sample and holdout observations. The calibration sample is used to fit the regression model and the holdout observations are used to validate the model by assessing the accuracy of prediction for the holdout observations by the regression model developed using the calibration sample. The accuracy can be measured by measuring the correlation coefficient between the holdout observations and the predicted values. The correlation coefficient should be positive and statistically significant, preferably a high value.

For small sample size, the jackknife method, described in Section 3.5.1, can be used. The basic idea is to use n–1 observations to build the regression model and the remaining one is predicted and the process is repeated for all the observations, separately. The correlation coefficient between the actual observations and the predicted values is used to measure the validity of the regression model.

8.15 CASE STUDY

The study was conducted in a worm gear manufacturing plant of India (Mondal et al., 2013). Manufacturing process comprises (i) heating of ingots in crucible furnace, (ii) casting of molten metal in centrifugal casting machine, and (iii) gear cutting in a hobbing machine. The purpose of this study is to model relationship of worm wheel quality with input and hobbing process variables. The input variables are hardness, Cu, Si, Ni, and P, and hobbing process variables are speed, feed, and depth of cut, etc. The quality variables of interest are backlash and contact %. In this study, multiple liner regression (MLR) is demonstrated using backlash as the dependent variable.

The details of the case study are available in the web resources of this book (Multivariate Statistical Modeling in Engineering and Management – 1st (routledge.com)).

8.16 LEARNING SUMMARY

Multiple linear regression (MLR) is a widely used statistical model for developing the dependence relationship of a DV with multiple IVs. It is also used to predict future values of the DV. This chapter gives the reader a through overview of MLR with the necessary skills for model-building, its verification and prediction.

In summary,

- The regression model of Y (DV) and a set of independent variables (X) is $Y = \beta_0 + \beta_1 X_1 + \cdots + \beta_p X_p + \epsilon$, where $\boldsymbol{\beta} = [\beta_0, \beta_1, \ldots, \beta_p]^T$ is the parameter vector.
- The ordinary least square regression (OLS) is used to estimate the $\boldsymbol{\beta}$ parameters. The estimated $\boldsymbol{\beta}$ (i.e., $\hat{\boldsymbol{\beta}}$) is $\hat{\boldsymbol{\beta}} = \left(\mathbf{X}^T\mathbf{X}\right)^{-1}\mathbf{X}^T\mathbf{y}$, where $\mathbf{X}$ is the design matrix.
- The fitted values of $\mathbf{y}_{n\times1}$ and the residual, $\boldsymbol{\epsilon}_{n\times1}$ are computed as $\hat{\mathbf{y}} = \mathbf{X}\hat{\boldsymbol{\beta}}$ and $\hat{\boldsymbol{\epsilon}} = \mathbf{y} - \mathbf{X}\hat{\boldsymbol{\beta}} = \mathbf{y} - \hat{\mathbf{y}}$, respectively.
- The expected value of $\hat{\boldsymbol{\beta}}$ is $\boldsymbol{\beta}$ and the covariance matrix of $\hat{\boldsymbol{\beta}}$ is $Cov\left(\hat{\boldsymbol{\beta}}\right) = \sigma^2\left(\mathbf{X}^T\mathbf{X}\right)^{-1}$.
- The confidence interval for β_j is computed as $\hat{\beta}_j - t_{n-p-1}^{(\alpha/2)} \cdot s_e\sqrt{c_{jj}} < \beta_j < \hat{\beta}_j + t_{n-p-1}^{(\alpha/2)} \cdot s_e\sqrt{c_{jj}}$ (see Section 8.5 for details).
- The sampling distribution of $\hat{\boldsymbol{\epsilon}}$ is $N_n\left(0, \sigma^2\left(\mathbf{I} - \mathbf{H}\right)\right)$.
- The assumptions in MLR include linearity, homoscedasticity, normality, and independent and identically distributed (iid) observations.
- Violation of assumptions in MLR may lead to imprecise results and wrong inferences. Therefore, it is better to apply the remedial actions against each of the violations.
- The model adequacy (fit) of MLR is tested using R^2 and adjusted R^2 (R_a^2). The formulas for computations are as follows: $R^2 = 1 - \dfrac{(n-p-1)s_e^2}{(n-1)s_y^2}$ and $R_a^2 = 1 - \dfrac{SSE/n-p-1}{SST/n-1} = 1 - \dfrac{s_e^2}{s_y^2}$.

  Both R^2 and R_a^2 range between 0 and 1. The higher the value the better is the fit.
- The t-statistic is used to test the individual regression coefficient (β_j, $j = 1, 2, \ldots, p$). The test confirms whether a regression coefficient is different from 0 or not.
- Diagnostic issues in MLR includes the detection of outliers, leverage points, influential observations, and multicollinearity (see Section 8.12 for details).
- MLR is also used for prediction. The new value for Y (say $\hat{y}_{new}$) can be predicted using the formula $\hat{y}_{new} = \mu_{y_{new}/\mathbf{x}_{new}} = \mathbf{x}_{new}\hat{\boldsymbol{\beta}}$. The 100(1–α)% confidence interval for $y_{new}/\mathbf{x}_{new}$ is

$$\hat{y}_{new} - t_{n-p-1}^{\alpha/2} s_e^2[1 + \mathbf{x}_{new}^T(\mathbf{X}^T\mathbf{X})^{-1}\mathbf{x}_{new}] \le y_{new}/\mathbf{x}_{new} \le \hat{y}_{new} + t_{n-p-1}^{\alpha/2} s_e^2[1 + \mathbf{x}_{new}^T(\mathbf{X}^T\mathbf{X})^{-1}\mathbf{x}_{new}]$$

- Some topics such as step-wise regression and ridge regression are not included. However, there are many good books available on this topic, such as Cook and Weisberg (1982) and Belsley et al. (2004).

EXERCISES

(A) Short conceptual questions

8.1 Define multiple linear regression (MLR). Give a pictorial representation of MLR.

8.2 State the assumptions of MLR.

8.3 Define design matrix in MLR. State the properties of design matrix.

8.4 Prove that the OLS estimate of regression coefficient vectors ($\hat{\boldsymbol{\beta}}$) is $\hat{\boldsymbol{\beta}} = \left(\mathbf{X}^T\mathbf{X}\right)^{-1}\mathbf{X}^T\mathbf{y}$.

8.5 Prove that
(a) $\hat{\boldsymbol{\epsilon}} = (\mathbf{I} - \mathbf{H})\mathbf{y}$ [where $\mathbf{H}_{n\times n}$ = Hat matrix]
(b) $SSE = \hat{\boldsymbol{\epsilon}}^T \hat{\boldsymbol{\epsilon}} = \mathbf{y}^T (\mathbf{I} - \mathbf{H})\mathbf{y}$, where $(\mathbf{I} - \mathbf{H})$ is symmetric and idempotent matrix
(c) $E(\hat{\boldsymbol{\beta}}) = \boldsymbol{\beta}$
(d) $Cov(\hat{\boldsymbol{\beta}}) = \sigma^2 (\mathbf{X}^T\mathbf{X})^{-1}$
(e) $S_e^2 = (\mathbf{y}^T (\mathbf{I} - \mathbf{H})\mathbf{y}) / (n - p - 1)$

8.6 If $\hat{\boldsymbol{\beta}}$ is $N_{p+1}(\boldsymbol{\beta}, \sigma^2\mathbf{C})$, a multivariate normal, then prove that
(a) $100(1-\alpha)\%$ CR for $\boldsymbol{\beta}$ is

$$\left[(\hat{\boldsymbol{\beta}}-\boldsymbol{\beta})^T (\mathbf{X}^T\mathbf{X})(\hat{\boldsymbol{\beta}}-\boldsymbol{\beta}) \le (p+1) S_e^2 \, F^{(\alpha)}_{p+1,n-p-1}\right], \text{where } \mathbf{C} = (\mathbf{X}^T\mathbf{X})^{-1}.$$

(b) $100(1-\alpha)\%$ SCI for β_j is $\hat{\beta}_j \pm \sqrt{S_e^2 c_{jj}}\sqrt{(p+1) F^{(\alpha)}_{p+1,(n-p-1)}}$, $(j = 0,1,\ldots p)$

8.7 If $(\hat{\beta}_j - \beta_j) / SE(\hat{\beta}_j)$ follows t-distribution with $(n-p-1)$ dof then prove that

$$\left(\hat{\beta}_j - t^{(\alpha/2)}_{n-p-1} S_e \sqrt{c_{jj}}\right) \le \beta_j \le \left(\hat{\beta}_j + t^{\alpha/2}_{n-p-1} S_e \sqrt{c_{jj}}\right).$$

8.8 Show that: $E(\hat{\boldsymbol{\epsilon}}) = 0$ and $Cov(\hat{\boldsymbol{\epsilon}}) = \sigma^2(\mathbf{I} - \mathbf{H})$ [here $\mathbf{H}$ = Hat matrix], [Hint: take $\hat{\boldsymbol{\epsilon}} = (\mathbf{I} - \mathbf{H})\mathbf{y}$].

8.9 What is R^2? Show that $R^2 = \left(1 - \frac{SSE}{SST}\right)$. State the limitation of R^2 as a measure of model adequacy in MLR. How does R_a^2 overcome the limitation of R^2?

8.10 Develop ANOVA table for MLR. What hypotheses are tested in MLR? How is individual regression parameter tested in MLR?

8.11 Test the following using residuals of MLR:
- Test of linearity
- Test of homoscedasticity
- Test of uncorrelated error terms
- Normality of error terms

8.12 Explain the remedial measures against the violation of assumptions of MLR.

8.13 What are the diagnostic issues involved in MLR? Define outliers and leverage points. Explain influential observations. What are good and bad leverage points?

8.14 What are the measures of influence of observations in MLR?

8.15 Define multicollinearity. Explain the following with reference to multicollinearity.
- Variance inflation factor (VIF)
- Tolerance statistics
- Eigen value structure
- Multicollinearity condition number

(B) Long conceptual and/or numerical questions: see web resources (Multivariate Statistical Modeling in Engineering and Management – 1st (routledge.com)).

NOTES

1 Note that the DV and IVs are denoted by upper-case italic alphabets (such as Y, X_1, X_2, etc.), the data matrix in bold upper-case regular $\mathbf{X}$, the observation vector for Y in bold lower-case regular $\mathbf{y}$ and the error vector in bold lower-case regular $\boldsymbol{\epsilon}$.

2 The consideration, $\hat{\boldsymbol{\beta}} \sim N_{p+1}(\hat{\boldsymbol{\beta}}, s_e^2\mathbf{C})$ is to facilitate characterization of $\hat{\boldsymbol{\beta}}$ in terms of its probability distribution, only from practical use point of view.

3 We used X notation here for computation. Note that in MLR both Y and X variables are used.

N.B.: R, MINITAB and MS Excel software are used for computation, as and when required.

9 Multivariate Multiple Linear Regression

In Chapter 8, we described a dependence model called multiple linear regression (MLR) where one dependent variable (DV) is regressed over multiple independent variables (IVs). But in real-life problems, we may encounter several situations where the relationships of more than one DV need to be modeled with a set of independent variables (IVs). Multivariate multiple linear regression (MMLR) is used to model the linear relationships in such situations. In this chapter, we describe MMLR along with its subtle differences with MLR in terms of model conceptualization, model assumptions, parameter estimation, sampling distribution of parameter estimates, test of model fit, and test of model parameters. Finally, a case study is presented.

9.1 CONCEPTUAL MODEL

Consider a set of p number of independent variables ($\boldsymbol{X}_{p\times 1}$) that could influence a set of q number of dependent variables ($\boldsymbol{Y}_{q\times 1}$) simultaneously. The pictorial representation of the dependence model is shown in Figure 9.1. Figure 9.1 is the graphical representation of multivariate multiple linear regression (MMLR).

The difference between multiple linear regression (MLR) (Figure 8.1) and multivariate multiple linear regression (MMLR) (Figure 9.1) is that the later involves more than one DV and hence there is more than one error variable representing the model.

Analogous to MLR, the equations for MMLR are:

$$\left.\begin{aligned} Y_1 &= \beta_{10} + \beta_{11}X_1 + \beta_{12}X_2 + \cdots + \beta_{1p}X_p + \epsilon_1 \\ Y_2 &= \beta_{20} + \beta_{21}X_1 + \beta_{22}X_2 + \cdots + \beta_{2p}X_p + \epsilon_2 \\ &\vdots \\ Y_q &= \beta_{q0} + \beta_{q1}X_1 + \beta_{q2}X_2 + \cdots + \beta_{qp}X_p + \epsilon_p \end{aligned}\right\} \tag{9.1}$$

Therefore,

$$\boldsymbol{Y} = \begin{bmatrix} Y_1 \\ Y_2 \\ \vdots \\ Y_q \end{bmatrix}_{q\times 1} \quad \text{represents q dependent variables (DVs)}$$

$$\boldsymbol{X} = \begin{bmatrix} X_1 \\ X_2 \\ \vdots \\ X_p \end{bmatrix}_{p\times 1} \quad \text{represents } p \text{ independent variables (IVs)}$$

DOI: 10.1201/9781003303060-12

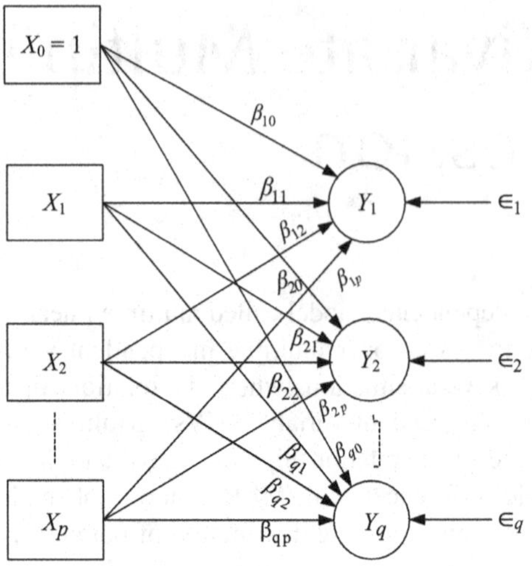

FIGURE 9.1 Linear dependence model involving p number of IVs ($\boldsymbol{X}_{p\times1}$) and q number of DVs ($\boldsymbol{Y}_{q\times1}$). Each β represents the regression coefficient and each ∈ represents the error term.

$$\boldsymbol{\epsilon} = \begin{bmatrix} \epsilon_1 \\ \epsilon_2 \\ \vdots \\ \epsilon_q \end{bmatrix}_{q\times1} \quad \text{represents } q \text{ error variables}$$

and

$$\boldsymbol{\beta} = \begin{bmatrix} \beta_{10} & \beta_{20} & \cdots & \beta_{q0} \\ \beta_{11} & \beta_{21} & \cdots & \beta_{q1} \\ \vdots & \vdots & \vdots & \vdots \\ \beta_{1p} & \beta_{2p} & \cdots & \beta_{qp} \end{bmatrix}_{(p+1)\times q} \quad \text{is } (p+1)\times q \text{ matrix of multivariate regression parameters}$$

Now, let n multivariate observations are proposed to be collected. Then,

$$\mathbf{Y}_{n\times q} = \begin{bmatrix} y_{11} & y_{12} & \cdots & y_{1q} \\ y_{21} & y_{22} & \cdots & y_{2q} \\ \vdots & \vdots & \vdots & \vdots \\ y_{n1} & y_{n2} & \cdots & y_{nq} \end{bmatrix}_{n\times q} \quad \text{is the dependent data matrix}$$

$$\mathbf{X}_{n\times(p+1)} = \begin{bmatrix} 1 & x_{11} & x_{12} & \cdots & x_{1p} \\ 1 & x_{21} & x_{22} & \cdots & x_{2p} \\ 1 & x_{31} & x_{32} & \cdots & x_{3p} \\ \vdots & \vdots & \vdots & \vdots & \vdots \\ 1 & x_{n1} & x_{n2} & \cdots & x_{np} \end{bmatrix}_{n\times(p+1)} \quad \text{is the design matrix}$$

The design matrix is the multivariate data matrix for the p number of IVs with constant term 1 in the first column.

Finally,

$$\epsilon = \begin{bmatrix} \epsilon_{11} & \epsilon_{12} & \cdots & \epsilon_{1q} \\ \epsilon_{21} & \epsilon_{22} & \cdots & \epsilon_{2q} \\ \vdots & \vdots & & \vdots \\ \epsilon_{n1} & \epsilon_{n2} & \cdots & \epsilon_{nq} \end{bmatrix} \text{ is the error data matrix}$$

Now, for the i-th multivariate observation, Equation 9.1 becomes

$$\left.\begin{aligned} y_{i1} &= \beta_{10} + \beta_{11}x_{i1} + \beta_{12}x_{i2} + \cdots + \beta_{1p}x_{ip} + \epsilon_{i1} \\ y_{i2} &= \beta_{20} + \beta_{21}x_{i1} + \beta_{22}x_{i2} + \cdots + \beta_{2p}x_{ip} + \epsilon_{i2} \\ &\vdots \\ y_{ik} &= \beta_{k0} + \beta_{k1}x_{i1} + \beta_{k2}x_{i2} + \cdots + \beta_{kp}x_{ip} + \epsilon_{ik} \\ &\vdots \\ y_{iq} &= \beta_{q0} + \beta_{q1}x_{i1} + \beta_{q2}x_{i2} + \cdots + \beta_{qp}x_{ip} + \epsilon_{iq} \end{aligned}\right\} \tag{9.2}$$

Again, considering $i = 1, 2, \ldots, n$, we will get n sets of linear equations where the i-th set is shown in Equation 9.2. In matrix form, we can write,

$$\mathbf{Y}_{n\times q} = \mathbf{X}_{n\times(p+1)}\, \boldsymbol{\beta}_{(p+1)\times q} + \boldsymbol{\epsilon}_{n\times q} \tag{9.3}$$

Equation 9.3 is known as multivariate multiple linear regression (MMLR).

9.2 ASSUMPTIONS

Apart from linearity assumption, there are four important assumptions of MMLR. These are listed below. From Equation 9.2, the i-th error term $\boldsymbol{\epsilon}_i$ is $\begin{bmatrix} \epsilon_{i1} & \epsilon_{i2} & \cdots\cdots & \epsilon_{iq} \end{bmatrix}^T$. There are n such error vectors, resulting into the error matrix of the order $n \times q$ (i.e., $\boldsymbol{\epsilon}_{n\times q}$ in Equation 9.3).

- Each error vector $\boldsymbol{\epsilon}_i$ (the i-th row of $\boldsymbol{\epsilon}_{n\times q}$) is multivariate normal with zero mean vector and covariance matrix $\Sigma_{q\times q}$, where $\text{cov}(\epsilon_k, \epsilon_m) = \sigma_{km} I,\ k, m = 1, 2, \ldots\ldots, q$.
- Errors have common covariance structure ($\Sigma_{q\times q}$) across observations.
- Error variances are equal (homogenous) across observations conditional on predictors, $\boldsymbol{X}_{p\times 1}$.
- The error vectors $\boldsymbol{\epsilon}_i$ ((the i-th row of $\boldsymbol{\epsilon}_{n\times q}$, $\text{i} = 1, 2, \ldots., n$) are independent and identically distributed. So, $\boldsymbol{\epsilon}_i \sim N_q(0, \Sigma_{q\times q})$ and $\text{cov}(\boldsymbol{\epsilon}_i, \boldsymbol{\epsilon}_{i'}) = 0,\ i \neq i',\ i, i' = 1, 2, \ldots., n$.

These assumptions can be tested after the MMLR is fit with sample data and errors are computed. However, one can examine the $\mathbf{Y}_{\text{nxq}}$ data for the above assumptions, as any departures from the above assumptions in the $\mathbf{Y}_{\text{nxq}}$ data will be captured through error (residual) analysis. A chi-square Q-Q plot can be used to test multivariate normality. Matrix scatter plots of $\mathbf{Y}_{\text{nxq}}$ and $\mathbf{X}_{\text{nx(p+1)}}$ data can be used to examine heteroscedasticity. Essentially, we require examining each DV separately with all IVs. For independent observations, a scatter plot of a DV versus observation orders can be useful. The test of variance and covariance is possible if we collect data through a study design where multiple observations of the DVs on every set of values of the IVs are available. However, this is seldom possible when data are obtained using cross-sectional or correlational study design. It is very difficult to get equal covariance structures across observations and, therefore, it is suggested in the literature that an equal number of observations across all possible groups of IVs should be collected. It is also noticed that regression estimators are very robust and the multivariate central limit theorem ensures that a small departure from the multivariate normality can be tolerable.

9.3 ESTIMATION OF PARAMETERS

Rearranging Equation 9.3, we can write

$$\boldsymbol{\epsilon} = \mathbf{Y} - \mathbf{X}\boldsymbol{\beta} \tag{9.4}$$

Squaring Equation 9.4 we get

$$\boldsymbol{\epsilon}^T \boldsymbol{\epsilon} = \left(\mathbf{Y} - \mathbf{X}\boldsymbol{\beta}\right)^T \left(\mathbf{Y} - \mathbf{X}\boldsymbol{\beta}\right) \tag{9.5}$$

$\boldsymbol{\epsilon}^T \boldsymbol{\epsilon}$ is a $q \times q$ matrix as derived below.

$$\boldsymbol{\epsilon}^T \boldsymbol{\epsilon} = \begin{bmatrix} \epsilon_{11} & \epsilon_{21} & \epsilon_{31} & \cdots & \epsilon_{n1} \\ \epsilon_{12} & \epsilon_{22} & \epsilon_{32} & \cdots & \epsilon_{n2} \\ \vdots & \vdots & \vdots & & \vdots \\ \epsilon_{1q} & \epsilon_{2q} & \epsilon_{3q} & & \epsilon_{nq} \end{bmatrix}_{q\times n} \begin{bmatrix} \epsilon_{11} & \epsilon_{12} & \cdots & \epsilon_{1q} \\ \epsilon_{21} & \epsilon_{22} & \cdots & \epsilon_{2q} \\ \vdots & \vdots & & \vdots \\ \vdots & \vdots & & \vdots \\ \epsilon_{n1} & \epsilon_{n2} & \cdots & \epsilon_{nq} \end{bmatrix}_{n\times q}$$

$$= \begin{bmatrix} \sum_{i=1}^{n} \epsilon_{i1}^2 & \sum_{i=1}^{n} \epsilon_{i1}\epsilon_{i2} & \cdots & \sum_{i=1}^{n} \epsilon_{i1}\epsilon_{iq} \\ \sum_{i=1}^{n} \epsilon_{i1}\epsilon_{i2} & \sum_{i=1}^{n} \epsilon_{i2}^2 & \cdots & \sum_{i=1}^{n} \epsilon_{i2}\epsilon_{iq} \\ \vdots & \vdots & \vdots & \vdots \\ \sum_{i=1}^{n} \epsilon_{i1}\epsilon_{iq} & \sum_{i=1}^{n} \epsilon_{i2}\epsilon_{iq} & \cdots & \sum_{i=1}^{n} \epsilon_{iq}^2 \end{bmatrix}_{q\times q}$$

Each element on the diagonal of the matrix $\boldsymbol{\epsilon}^T \boldsymbol{\epsilon}$ is the sum squares errors (*SSE*) for one of the q error variables. Therefore, the sum of *SSE* for all the q error variables is the trace of the matrix $\boldsymbol{\epsilon}^T \boldsymbol{\epsilon}$, which is required to be minimized while estimating $\boldsymbol{\beta}_{(p+1)\times q}$. If we denote *SSE* for the k-th error variable as SSE_k and sum of *SSE* for all error variables as SSE_T, then,

$$SSE_T = \sum_{k=1}^{q} SSE_k = trace\left(\boldsymbol{\in}^T \boldsymbol{\in}\right)$$
$$= trace\left\{\left(\mathbf{Y} - \mathbf{X\beta}\right)^T \left(\mathbf{Y} - \mathbf{X\beta}\right)\right\} \quad (9.6)$$

Now to estimate the elements of $\boldsymbol{\beta}_{(p+1)\times q}$, we take partial derivatives of SSE_T with respect to each element of $\boldsymbol{\beta}$ and equate them to zero. This gives the following equations:

$$\left(\mathbf{X}^T\mathbf{X}\right)\boldsymbol{\beta} = \mathbf{X}^T\mathbf{Y} \quad (9.7)$$

Now solving for $\boldsymbol{\beta}$, we get

$$\hat{\boldsymbol{\beta}} = \left(\mathbf{X}^T\mathbf{X}\right)^{-1}\mathbf{X}^T\mathbf{Y} \quad (9.8)$$

Alternatively, the maximum likelihood estimation (MLE) for $\boldsymbol{\beta}_{(p+1)\times q}$ could be made using Equation 9.4. MLE also gives Equation 9.8.

Equation 9.8 looks similar to Equation 8.15 (Chapter 8) but differs in dimensions. In MLR (Chapter 8) $\hat{\boldsymbol{\beta}}$ is a $(p+1)\times 1$ vector while in MMLR (in this Chapter) $\hat{\boldsymbol{\beta}}$ is a $(p+1)\times q$ matrix. So, we can write,

$$\hat{\boldsymbol{\beta}} = \left[\hat{\boldsymbol{\beta}}_1 : \hat{\boldsymbol{\beta}}_2 : \cdots\cdots\cdots : \hat{\boldsymbol{\beta}}_{q-1} : \hat{\boldsymbol{\beta}}_q\right].$$

What would have happened if we have estimated $\hat{\boldsymbol{\beta}}_1, \hat{\boldsymbol{\beta}}_2, \ldots, \hat{\boldsymbol{\beta}}_q$, each having a dimension of $(p+1)\times 1$, considering each of the dependent variables separately? In other words, would we arrive at the same estimate of $\hat{\boldsymbol{\beta}}_1, \hat{\boldsymbol{\beta}}_2, \ldots, \hat{\boldsymbol{\beta}}_q$ if we develop q separate multiple regressions for $Y_1, Y_2, \ldots Y_q$, respectively? The answer is yes. Then why should we go for multivariate regression? The answer will be obvious when we describe later the covariance structure of the error variables, model adequacy test, and test of regression coefficients. It is to be noted here that in MMLR there is non-zero correlation among the error terms for different DVs. That is,

$$Cov(\in_j, \in_k) = \sigma_{jk}\mathbf{I} \text{ and } E(\in_j) = 0, j = 1,2,3,\ldots,q, \ \mathrm{k} = 1,2,3,\ldots,q, \ j \neq k.$$

Example 9.1

Develop a multivariate regression model for the CityCan data considering Y_1 = Profit, Y_2 = Sales and X_1 = absenteeism (%), and X_2 = breakdown hours.

Solution:

The following steps will be performed.

Step 1: Identity variables of interest and divide them into dependent variables (DVs) and independent variables (IVs).

In this example, the DVs are profit (Y_1) and sales volume (Y_2) and IVs are absenteeism in % (X_1) and machine breakdown in hours (X_2).

Step 2: Develop design matrix **X**.

The numbers of *IVs* are 2 and the numbers of observations are 12. So, the design matrix **X** is a 12 × (2+1) matrix as shown below.

$$\mathbf{X}_{12\times 3} = \begin{bmatrix} 1 & 9 & 62 \\ 1 & 8 & 58 \\ 1 & 7 & 64 \\ 1 & 14 & 60 \\ 1 & 12 & 63 \\ 1 & 10 & 57 \\ 1 & 7 & 55 \\ 1 & 4 & 56 \\ 1 & 6 & 59 \\ 1 & 5 & 61 \\ 1 & 7 & 57 \\ 1 & 6 & 60 \end{bmatrix}$$

Step 3: Obtain $\mathbf{X}^T\mathbf{X}$

$$\mathbf{X}^T\mathbf{X} = \begin{bmatrix} 1 & 1 & 1 & 1 & 1 & 1 & 1 & 1 & 1 & 1 & 1 & 1 \\ 9 & 8 & 7 & 14 & 12 & 10 & 7 & 4 & 6 & 5 & 7 & 6 \\ 62 & 58 & 64 & 60 & 63 & 57 & 55 & 56 & 59 & 61 & 57 & 60 \end{bmatrix}_{(3\times 12)} \begin{bmatrix} 1 & 9 & 62 \\ 1 & 8 & 58 \\ 1 & 7 & 64 \\ 1 & 14 & 60 \\ 1 & 12 & 63 \\ 1 & 10 & 57 \\ 1 & 7 & 55 \\ 1 & 4 & 56 \\ 1 & 6 & 59 \\ 1 & 5 & 61 \\ 1 & 7 & 57 \\ 1 & 6 & 60 \end{bmatrix}_{(12\times 3)}$$

$$= \begin{bmatrix} 12 & 95 & 712 \\ 95 & 845 & 5663 \\ 712 & 5663 & 42334 \end{bmatrix}_{(3\times 3)}$$

Step 4: Obtain $\left(\mathbf{X}^T\mathbf{X}\right)^{-1}$

$$\left(\mathbf{X}^T\mathbf{X}\right)^{-1} = \frac{1}{\left|\mathbf{X}^T\mathbf{X}\right|} adj\left(\mathbf{X}^T\mathbf{X}\right)$$

$$= \begin{bmatrix} 40.894402598 & 0.114046520 & -0.703043891 \\ 0.114046520 & 0.011751452 & -0.003490093 \\ -0.703043891 & -0.003490093 & 0.012314727 \end{bmatrix}$$

Determinant of $\left(\mathbf{X}^T\mathbf{X}\right) = 90542$.

Step 5: Obtain $\mathbf{X}^T\mathbf{Y}$

$$\mathbf{X}^T\mathbf{Y} = \begin{bmatrix} 1 & 1 & 1 & 1 & 1 & 1 & 1 & 1 & 1 & 1 & 1 & 1 \\ 9 & 8 & 7 & 14 & 12 & 10 & 7 & 4 & 6 & 5 & 7 & 6 \\ 62 & 58 & 64 & 60 & 63 & 57 & 55 & 56 & 59 & 61 & 57 & 60 \end{bmatrix}_{(3\times 12)} \begin{bmatrix} 10 & 100 \\ 12 & 110 \\ 11 & 105 \\ 9 & 94 \\ 9 & 95 \\ 10 & 99 \\ 11 & 104 \\ 12 & 108 \\ 11 & 105 \\ 10 & 98 \\ 11 & 103 \\ 12 & 110 \end{bmatrix}_{(12\times 2)}$$

$$= \begin{bmatrix} 128 & 1231 \\ 987 & 9622 \\ 7580 & 72980 \end{bmatrix}_{3\times 2}$$

Step 6: Obtain $\hat{\boldsymbol{\beta}}$

$$\begin{aligned} \hat{\boldsymbol{\beta}} &= \left(\mathbf{X}^T\mathbf{X}\right)^{-1}\mathbf{X}^T\mathbf{Y} \\ &= \begin{bmatrix} 40.894402598 & 0.114046520 & -0.703043891 \\ 0.114046520 & 0.011751452 & -0.003490093 \\ -0.703043891 & -0.003490093 & 0.012314727 \end{bmatrix}_{3\times 3} \begin{bmatrix} 128 & 1231 \\ 987 & 9622 \\ 7580 & 72980 \end{bmatrix}_{3\times 2} \\ &= \begin{bmatrix} 17.97 & 130.22 \\ -0.26 & -1.24 \\ -0.09 & -0.30 \end{bmatrix}_{3\times 2} \end{aligned}$$

So, the multivariate regression equations are

$$\begin{aligned} Y_1 &= 17.97-0.26X_1-0.09X_2+\epsilon_1 \\ Y_2 &= 130.22-1.24X_1-0.30X_2+\epsilon_2 \end{aligned}$$

or, Profit = 17.97 – 0.26 × absenteeism – 0.09 × breakdown-hrs + ϵ_1

and Sales volume = 130.22 – 1.24 × absenteeism – 0.30 × breakdown-hrs + ϵ_2.

Example 9.2

Consider Example 9.1. If the independent variable M-ratio (X_3) is considered along with absenteeism (%) (X_1) and breakdown in hours (X_2), what will happen to the regression equations?

Solution:

In this example, the DVs remain same as Example 9.1, i.e., profit (Y_1) and sales volume (Y_2); whereas the IVs are now three in number (instead of two) and they are absenteeism in % (X_1), machine breakdown in hours (X_2), and M-ratio (X_3). Accordingly, the matrix $\mathbf{X}^T\mathbf{X}$ will become a 4 × 4 matrix as shown below.

$$\mathbf{X}^T\mathbf{X} = \begin{bmatrix} 1 & 1 & 1 & 1 & 1 & 1 & 1 & 1 & 1 & 1 & 1 & 1 \\ 9 & 8 & 7 & 14 & 12 & 10 & 7 & 4 & 6 & 5 & 7 & 6 \\ 62 & 58 & 64 & 60 & 63 & 57 & 55 & 56 & 59 & 61 & 57 & 60 \\ 1 & 1.3 & 1.2 & 0.8 & 0.8 & 0.9 & 1 & 1.2 & 1.1 & 1 & 1.2 & 1.2 \end{bmatrix}_{(4\times12)} \begin{bmatrix} 1 & 9 & 62 & 1 \\ 1 & 8 & 58 & 1.3 \\ 1 & 7 & 64 & 1.2 \\ 1 & 14 & 60 & 0.8 \\ 1 & 12 & 63 & 0.8 \\ 1 & 10 & 57 & 0.9 \\ 1 & 7 & 55 & 1 \\ 1 & 4 & 56 & 1.2 \\ 1 & 6 & 59 & 1.1 \\ 1 & 5 & 61 & 1 \\ 1 & 7 & 57 & 1.2 \\ 1 & 6 & 60 & 1.2 \end{bmatrix}_{(12\times4)}$$

$$= \begin{bmatrix} 12 & 95 & 712 & 12.7 \\ 95 & 845 & 5663 & 96.6 \\ 712 & 5663 & 42334 & 752.4 \\ 12.7 & 96.6 & 752.4 & 13.75 \end{bmatrix}_{(4\times4)}$$

Then, $\left(\mathbf{X}^T\mathbf{X}\right)^{-1} = \dfrac{1}{\left|\mathbf{X}^T\mathbf{X}\right|}\, adj\left(\mathbf{X}^T\mathbf{X}\right)$

$$= \begin{bmatrix} 54.81224 & -0.30544 & -0.70502 & -9.90183 \\ -0.30544 & 0.024395 & -0.00343 & 0.298445 \\ -0.70502 & -0.00343 & 0.012315 & 0.001408 \\ -9.90183 & 0.298445 & 0.001408 & 7.044639 \end{bmatrix}$$

Determinant of $\left(\mathbf{X}^T\mathbf{X}\right) = 12852.61$, and

$$\mathbf{X}^T\mathbf{Y} = \begin{bmatrix} 1 & 1 & 1 & 1 & 1 & 1 & 1 & 1 & 1 & 1 & 1 & 1 \\ 9 & 8 & 7 & 14 & 12 & 10 & 7 & 4 & 6 & 5 & 7 & 6 \\ 62 & 58 & 64 & 60 & 63 & 57 & 55 & 56 & 59 & 61 & 57 & 60 \\ 1 & 1.3 & 1.2 & 0.8 & 0.8 & 0.9 & 1 & 1.2 & 1.1 & 1 & 1.2 & 1.2 \end{bmatrix}_{(4\times12)} \begin{bmatrix} 10 & 100 \\ 12 & 110 \\ 11 & 105 \\ 9 & 94 \\ 9 & 95 \\ 10 & 99 \\ 11 & 104 \\ 12 & 108 \\ 11 & 105 \\ 10 & 98 \\ 11 & 103 \\ 12 & 110 \end{bmatrix}_{(12\times2)}$$

$$= \begin{bmatrix} 128 & 1231 \\ 987 & 9622 \\ 7580 & 72980 \\ 137.3 & 1312 \end{bmatrix}_{4\times2}$$

$$\text{Now, } \hat{\boldsymbol{\beta}} = \left(\mathbf{X}^T\mathbf{X}\right)^{-1}\mathbf{X}^T\mathbf{Y}$$

$$= \begin{bmatrix} 54.81224 & -0.30544 & -0.70502 & -9.90183 \\ -0.30544 & 0.024395 & -0.00343 & 0.298445 \\ -0.70502 & -0.00343 & 0.012315 & 0.001408 \\ -9.90183 & 0.298445 & 0.001408 & 7.044639 \end{bmatrix}_{4\times4} \begin{bmatrix} 128 & 1231 \\ 987 & 9622 \\ 7580 & 72980 \\ 137.3 & 1312 \end{bmatrix}_{4\times2}$$

$$= \begin{bmatrix} 10.90 & 91.10 \\ -0.04 & -0.06 \\ -0.09 & -0.29 \\ 5.04 & 27.84 \end{bmatrix}_{4\times2}$$

So, the multivariate regression equations are

$$Y_1 = 10.90 - 0.04X_1 - 0.09X_2 + 5.04X_3 + \epsilon_1$$
$$Y_2 = 91.10 - 0.06X_1 - 0.29X_2 + 27.84X_3 + \epsilon_2$$

or, Profit = 10.90 – 0.04 × absenteeism – 0.09 × breakdown-hrs + 5.04 × M-ratio + ϵ_1, and Sales volume = 91.10 – 0.06 × absenteeism – 0.29 × breakdown-hrs + 27.84 × M-ratio + ϵ_2.

So, the changes in regression equations are as follows:

- M-ratio (X_3) dominates the relationship, both with profit ($\beta_{13} = 5.04$) and sales volume ($\beta_{23} = 27.84$), as compared to absenteeism (X_1, $\beta_{11} = -0.04$ and $\beta_{21} = -0.06$) and breakdown hours (X_2, $\beta_{12} = -0.09$ and $\beta_{22} = -0.29$), respectively.
- Inclusion of M-ratio in the IVs has an effect on the relationship between absenteeism (X_1) with profit (Y_1) and sales volume (Y_2). It reduces the strength of relationship between absenteeism (X_1) with profit (Y_1) from $\beta_{11} = -0.26$ to -0.04 and between absenteeism (X_1) with sales volume (Y_2) from $\beta_{21} = -1.24$ to -0.06. This may be because of the presence of high correlation (-0.74) between absenteesism (X_1) and M-ratio (X_3). Such an influence may indicate the violation of the independence assumption of the independent (X) variables.
- However, the relationship between breakdown hours (X_2) with profit (Y_1) and sales volume (Y_2) almost remains unchanged. This is desirable in regression.

Once $\hat{\boldsymbol{\beta}}$, the estimates of $\boldsymbol{\beta}$, is obtained, the fitted values of $\mathbf{Y}$ can be written as

$$\hat{\mathbf{Y}} = \mathbf{X}\hat{\boldsymbol{\beta}} \tag{9.9}$$

and estimated errors are

$$\begin{aligned} \hat{\boldsymbol{\epsilon}} &= \mathbf{Y} - \hat{\mathbf{Y}} = \mathbf{Y} - \mathbf{X}\hat{\boldsymbol{\beta}} \\ &= \mathbf{Y} - \mathbf{X}\left(\mathbf{X}^T\mathbf{X}\right)^{-1}\mathbf{X}^T\mathbf{Y} \\ &= (\mathbf{I} - \mathbf{H})\mathbf{Y} \end{aligned} \tag{9.10}$$

$\mathbf{H}$ is the hat matrix. The errors for the k-th DV, Y_k are

$$\hat{\boldsymbol{\epsilon}}_k = (\mathbf{I} - \mathbf{H})\mathbf{Y}_k, \quad k = 1, 2, \ldots q \tag{9.11}$$

The SSE for the k-th DV, Y_k is

$$SSE_{\hat{k}} = \hat{\boldsymbol{\epsilon}}_k^T \hat{\boldsymbol{\epsilon}}_k = \sum_{i=1}^{n} \hat{\epsilon}_{ik}^2 \tag{9.12}$$

The covariance matrix of errors, $SSCP_E$ is

$$SSCP_E = \hat{\boldsymbol{\epsilon}}^T \hat{\boldsymbol{\epsilon}} \tag{9.13}$$

The $SSCP_E$ is a $q \times q$ matrix and its diagonals are SSE_k, $k = 1, 2, \ldots, q$.

Example 9.3

Obtain fitted values, estimated errors and SSE for Example 9.2.

Solution:

We use Equation 9.9 to estimate fitted values ($\hat{\mathbf{Y}}$), Equation 9.10 to estimate errors ($\hat{\boldsymbol{\epsilon}}$) and Equation 9.12 or 9.13 to estimate SSE. The computations (using R-software) are shown below.

$$\hat{\mathbf{Y}} = \mathbf{X}\hat{\boldsymbol{\beta}} = \begin{bmatrix} 1 & 9 & 62 & 1 \\ 1 & 8 & 58 & 1.3 \\ 1 & 7 & 64 & 1.2 \\ 1 & 14 & 60 & 0.8 \\ 1 & 12 & 63 & 0.8 \\ 1 & 10 & 57 & 0.9 \\ 1 & 7 & 55 & 1 \\ 1 & 4 & 56 & 1.2 \\ 1 & 6 & 59 & 1.1 \\ 1 & 5 & 61 & 1 \\ 1 & 7 & 57 & 1.2 \\ 1 & 6 & 60 & 1.2 \end{bmatrix} \begin{bmatrix} 10.90 & 91.10 \\ -0.04 & -0.06 \\ -0.09 & -0.29 \\ 5.04 & 27.84 \end{bmatrix} = \begin{bmatrix} 10.090 & 100.105 \\ 11.997 & 109.697 \\ 11.012 & 105.212 \\ 9.034 & 94.807 \\ 8.861 & 94.052 \\ 9.980 & 98.730 \\ 10.794 & 102.294 \\ 11.848 & 107.759 \\ 10.992 & 103.964 \\ 10.358 & 100.656 \\ 11.626 & 107.272 \\ 11.408 & 106.453 \end{bmatrix}$$

Now, the errors are (using Equation 9.10)

$$\hat{\boldsymbol{\epsilon}} = \mathbf{Y} - \hat{\mathbf{Y}} = \begin{bmatrix} 10 & 100 \\ 12 & 110 \\ 11 & 105 \\ 9 & 94 \\ 9 & 95 \\ 10 & 99 \\ 11 & 104 \\ 12 & 108 \\ 11 & 105 \\ 10 & 98 \\ 11 & 103 \\ 12 & 110 \end{bmatrix} - \begin{bmatrix} 10.090 & 100.105 \\ 11.997 & 109.697 \\ 11.012 & 105.212 \\ 9.034 & 94.807 \\ 8.861 & 94.052 \\ 9.980 & 98.730 \\ 10.794 & 102.294 \\ 11.848 & 107.759 \\ 10.992 & 103.964 \\ 10.358 & 100.656 \\ 11.626 & 107.272 \\ 11.408 & 106.453 \end{bmatrix} = \begin{bmatrix} -0.090 & -0.105 \\ 0.003 & 0.303 \\ -0.012 & -0.212 \\ -0.034 & -0.807 \\ 0.139 & 0.948 \\ 0.020 & 0.270 \\ 0.206 & 1.706 \\ 0.152 & 0.241 \\ 0.008 & 1.036 \\ -0.358 & -2.656 \\ -0.626 & -4.272 \\ 0.592 & 3.547 \end{bmatrix}$$

And the $SSCP_E$ is (using Equation 9.13)

$$SSCP_E = \hat{\epsilon}^T \hat{\epsilon} =$$

$$= \begin{bmatrix} -0.090 & 0.003 & -0.012 & -0.034 & 0.139 & 0.020 & 0.206 & 0.152 & 0.008 & -0.358 & -0.626 & 0.592 \\ -0.105 & 0.303 & -0.212 & -0.807 & 0.948 & 0.270 & 1.706 & 0.241 & 1.036 & -2.656 & -4.272 & 3.547 \end{bmatrix}$$

$$* \begin{bmatrix} -0.090 & -0.105 \\ 0.003 & 0.303 \\ -0.012 & -0.212 \\ -0.034 & -0.807 \\ 0.139 & 0.948 \\ 0.020 & 0.270 \\ 0.206 & 1.706 \\ 0.152 & 0.241 \\ 0.008 & 1.036 \\ -0.358 & -2.656 \\ -0.626 & -4.272 \\ 0.592 & 3.547 \end{bmatrix}$$

$$= \begin{bmatrix} 0.97 & 6.30 \\ 6.30 & 43.70 \end{bmatrix}$$

The SSE_1 and SSE_2 are the first and second elements in the diagonal of $SSCP_E$, which are 0.97 and 43.70, respectively.

9.4 SAMPLING DISTRIBUTION OF $\hat{\beta}$

Let's first describe the expected value and covariance of $\hat{\boldsymbol{\beta}}$, denoted by $E(\hat{\boldsymbol{\beta}})$ and $Cov(\hat{\boldsymbol{\beta}})$, respectively.

$$\begin{aligned} E(\hat{\boldsymbol{\beta}}) &= E\left[(\mathbf{X}^T\mathbf{X})^{-1}\mathbf{X}^T\mathbf{Y}\right] \\ &= (\mathbf{X}^T\mathbf{X})^{-1}\mathbf{X}^T E(\mathbf{Y}) \\ &= (\mathbf{X}^T\mathbf{X})^{-1}\mathbf{X}^T\mathbf{X}\boldsymbol{\beta} \\ &= \boldsymbol{\beta} \end{aligned}$$

$$Cov(\hat{\boldsymbol{\beta}}) = E\left[(\hat{\boldsymbol{\beta}} - \boldsymbol{\beta})(\hat{\boldsymbol{\beta}} - \boldsymbol{\beta})^T\right]$$

Now

$$\begin{aligned} \hat{\boldsymbol{\beta}} - \boldsymbol{\beta} &= (\mathbf{X}^T\mathbf{X})^{-1}\mathbf{X}^T\mathbf{Y} - \boldsymbol{\beta} \\ &= (\mathbf{X}^T\mathbf{X})^{-1}\mathbf{X}^T(\mathbf{X}\boldsymbol{\beta} + \boldsymbol{\epsilon}) - \boldsymbol{\beta} \\ &= (\mathbf{X}^T\mathbf{X})^{-1}\mathbf{X}^T\mathbf{X}\boldsymbol{\beta} + (\mathbf{X}^T\mathbf{X})^{-1}\mathbf{X}^T\boldsymbol{\epsilon} - \boldsymbol{\beta} \\ &= \boldsymbol{\beta} + (\mathbf{X}^T\mathbf{X})^{-1}\mathbf{X}^T\boldsymbol{\epsilon} - \boldsymbol{\beta} \\ &= (\mathbf{X}^T\mathbf{X})^{-1}\mathbf{X}^T\boldsymbol{\epsilon} \end{aligned}$$

So,

$$\begin{aligned} Cov(\hat{\boldsymbol{\beta}}) &= E\left[\{(\mathbf{X}^T\mathbf{X})^{-1}\mathbf{X}^T \in\}\{(\mathbf{X}^T\mathbf{X})^{-1}\mathbf{X}^T \in\}^T\right] \\ &= E\left[(\mathbf{X}^T\mathbf{X})^{-1}\mathbf{X}^T \in \in^T \mathbf{X}(\mathbf{X}^T\mathbf{X})^{-1}\right] \\ &= (\mathbf{X}^T\mathbf{X})^{-1}\mathbf{X}^T E(\in \in^T)\mathbf{X}(\mathbf{X}^T\mathbf{X})^{-1} \\ &= (\mathbf{X}^T\mathbf{X})^{-1}\mathbf{X}^T \mathbf{I} \otimes \boldsymbol{\Sigma}\mathbf{X}(\mathbf{X}^T\mathbf{X})^{-1} \\ &= (\mathbf{X}^T\mathbf{X})^{-1}(\mathbf{X}^T\mathbf{X})(\mathbf{X}^T\mathbf{X})^{-1} \otimes \boldsymbol{\Sigma} \\ &= (\mathbf{X}^T\mathbf{X})^{-1} \otimes \boldsymbol{\Sigma} \\ &= \text{The kronecker product of } (\mathbf{X}^T\mathbf{X})^{-1} \text{ and } \boldsymbol{\Sigma} \end{aligned}$$

If we denote $(\mathbf{X}^T\mathbf{X})^{-1} = \mathbf{C}$, then

$$Cov(\hat{\boldsymbol{\beta}}) = \mathbf{C} \otimes \boldsymbol{\Sigma}$$

$$\text{Where } \hat{\boldsymbol{\beta}} = \begin{bmatrix} \hat{\beta}_{10} & \hat{\beta}_{20} \cdots \hat{\beta}_{q0} \\ \hat{\beta}_{11} & \hat{\beta}_{21} \cdots \hat{\beta}_{qi} \\ \vdots & \vdots \quad \vdots \\ \hat{\beta}_{1p} & \hat{\beta}_{2p} \cdots \hat{\beta}_{qp} \end{bmatrix}_{(p+1)\times q}$$

$$\mathbf{C} = \begin{bmatrix} c_{00} & c_{01} \cdots & c_{0p} \\ c_{01} & c_{11} \cdots & c_{1p} \\ \vdots & \vdots & \vdots \\ c_{0p} & c_{1p} \cdots & c_{pp} \end{bmatrix}_{(p+1)\times(p+1)}$$

$$\boldsymbol{\Sigma} = \begin{bmatrix} \sigma_1^2 & \sigma_{12} & \cdots & \sigma_{1q} \\ \sigma_{12} & \sigma_2^2 & \cdots & \sigma_{2q} \\ \vdots & \vdots & & \vdots \\ \sigma_{1q} & \sigma_{2q} & \cdots & \sigma_q^2 \end{bmatrix}_{q\times q}.$$

In MLR, we have seen that the sampling distribution of $\hat{\boldsymbol{\beta}}$ is $N_{p+1}(\boldsymbol{\beta}, \sigma^2(\mathbf{X}^T\mathbf{X})^{-1})$ where σ^2 is the variability of Y across X satisfying homoscedasticity assumption. In MMLR, we have q number of Y variables and the k-th Y variable can be termed as Y_k, $k = 1, 2, \ldots, q$ and the corresponding regression coefficient vector is $\hat{\boldsymbol{\beta}}_k$, $k = 1, 2, \ldots, q$, which is the k-th column of the $\hat{\boldsymbol{\beta}}_{(p+1)\times q}$ matrix given above. For $\mathbf{Y}_k$, the following holds: $\mathbf{E}(\hat{\boldsymbol{\beta}}_k) = \boldsymbol{\beta}_k$ and $Cov(\hat{\boldsymbol{\beta}}_k, \hat{\boldsymbol{\beta}}_m) = \sigma_{km}(\mathbf{X}^T\mathbf{X})^{-1}$, $k, m = 1, 2, \ldots, q$. The sampling distribution of $\hat{\boldsymbol{\beta}}_k$ is $N_{p+1}(\boldsymbol{\beta}_k, (\mathbf{X}^T\mathbf{X})^{-1}\sigma_k^2)$ where σ_k^2 is the k-th diagonal element of $\boldsymbol{\Sigma}$ given above.

Now, we need to find the value of σ_k^2. In MLR, the estimate of σ^2 is s_e^2 where s_e^2 is $SSE/(n-p-1)$. We will follow similar approach here. As there are q number of DVs; SSE, instead of being a scalar, will be a $SSCP$ matrix. So, we call it $SSCP_E$. $SSCP_E$ is computed as below.

$$SSCP_E = \hat{\in}^T \hat{\in} = (\mathbf{Y} - \mathbf{X}\hat{\boldsymbol{\beta}})^T(\mathbf{Y} - \mathbf{X}\hat{\boldsymbol{\beta}}) = (n-p-1)\hat{\boldsymbol{\Sigma}} \tag{9.14}$$

Further, $\hat{\in}$ and $\hat{\boldsymbol{\beta}}$ are uncorrelated.

If we represent $SSCP_E$ as

$$SSCP_E = \begin{bmatrix} s_1^2 & s_{12} & \cdots & s_{1q} \\ s_{12} & s_2^2 & \cdots & s_{2q} \\ \vdots & \vdots & & \vdots \\ s_{1q} & s_{2q} & \cdots & s_q^2 \end{bmatrix}_{q \times q}.$$

Then, $\hat{\sigma}_k^2 = \dfrac{s_k^2}{n-p-1}, k = 1,2,...q.$

Example 9.4

Consider Example 9.2. Obtain sampling distribution of $\hat{\boldsymbol{\beta}}$.

Solution:

$$\text{Here, } \hat{\boldsymbol{\beta}} = \begin{bmatrix} \hat{\beta}_{10} & \hat{\beta}_{20} \\ \hat{\beta}_{11} & \hat{\beta}_{21} \\ \hat{\beta}_{12} & \hat{\beta}_{22} \\ \hat{\beta}_{13} & \hat{\beta}_{23} \end{bmatrix} = \begin{bmatrix} 10.90 & 91.10 \\ -0.04 & -0.06 \\ -0.09 & -0.29 \\ 5.04 & 27.84 \end{bmatrix}, \text{ from Example 9.2.}$$

Now, using the explanation given in Section 9.4, the sampling distribution of $\hat{\boldsymbol{\beta}}_k$ (k = 0, 1, 2) is $N_3(\boldsymbol{\beta}_k, (\mathbf{X}^T\mathbf{X})^{-1}\sigma_k^2)$ with $(\mathbf{X}^T\mathbf{X})^{-1}$ and σ_k^2 as given below.

$$(\mathbf{X}^T\mathbf{X})^{-1} = \begin{bmatrix} 54.81224 & -0.30544 & -0.70502 & -9.90183 \\ -0.30544 & 0.024395 & -0.00343 & 0.298445 \\ -0.70502 & -0.00343 & 0.012315 & 0.001408 \\ -9.90183 & 0.298445 & 0.001408 & 7.044639 \end{bmatrix}, \text{ from Example 9.2.}$$

$$\hat{\sigma}_1^2 = \frac{s_1^2}{n-p-1} = \frac{0.97}{12-3-1} = 0.12$$
$$\hat{\sigma}_2^2 = \frac{s_2^2}{n-p-1} = \frac{43.70}{12-3-1} = 5.46$$

Note that the value of $\boldsymbol{\beta}_k$ is unknown, and $\hat{\boldsymbol{\beta}}_k$ as given here, is the estimate from one sample. To get the closer approximate value of $\boldsymbol{\beta}_k$, several samples might be chosen and the average of all the estimates may be considered as close to $\boldsymbol{\beta}_k$.

9.5 ASSESSMENT OF OVERALL FIT OF THE MODEL

Let us recall the general MMLR in matrix form (Equation 9.3)

$$\mathbf{Y} = \mathbf{X}\boldsymbol{\beta} + \in$$

The total sum squares and cross products ($SSCP$) matrix of $\mathbf{Y}_{n \times q}$ observations is

$$SSCP_{\mathbf{Y}} = (\mathbf{Y} - \bar{\mathbf{Y}})^T(\mathbf{Y} - \bar{\mathbf{Y}}) \quad (9.15)$$

Where, $\bar{\mathbf{Y}} = \frac{1}{n}\mathbf{1}\mathbf{1}^T\mathbf{Y}$, $\mathbf{1}$ is a n × 1 vector of ones and $\bar{\mathbf{Y}}$ is the n × q matrix, where every element of a column is equal to the mean of the corresponding *Y* variable.

When $\boldsymbol{\beta}$ is estimated, the fitted values of $\mathbf{Y}_{n\times q}$ can be estimated using $\hat{\mathbf{Y}} = \mathbf{X}\hat{\boldsymbol{\beta}}$ (Equation 9.9), and the residuals are $\hat{\boldsymbol{\epsilon}} = \mathbf{Y} - \hat{\mathbf{Y}}$.

So, the fitted *SSCP* matrix, i.e., $SSCP_{\hat{\mathbf{Y}}}$ is

$$SSCP_{\hat{\mathbf{Y}}} = (\hat{\mathbf{Y}} - \bar{\mathbf{Y}})^T(\hat{\mathbf{Y}} - \bar{\mathbf{Y}}) \quad (9.16)$$

Similarly, the *SSCP* for $\hat{\boldsymbol{\epsilon}}$ is

$$SSCP_{\hat{\boldsymbol{\epsilon}}} = \hat{\boldsymbol{\epsilon}}^T\hat{\boldsymbol{\epsilon}} \quad (9.17)$$

Similar to MLR, in MMLR we can write,

$$SSCP_{\mathbf{Y}} = SSCP_{\hat{\mathbf{Y}}} + SSCP_{\hat{\boldsymbol{\epsilon}}} \quad (9.18)$$

Equation 9.18 shows the partition of Y's variability ($SSCP_{\mathbf{Y}}$) into variability explained by MMLR ($SSCP_{\hat{\mathbf{Y}}}$) and variability unexplained ($SSCP_{\hat{\boldsymbol{\epsilon}}}$). However, unlike MLR, it is not possible to have the coefficient of determination (R^2) in MMLR. The likelihood ratio-based approach as followed in MANOVA is used here.

Let us consider the following hypothesis:

$H_0 : \boldsymbol{\beta} = \mathbf{0}$

$H_1 : \boldsymbol{\beta} \neq \mathbf{0}$ for at least one β_{jk}, $j = 0, 1, \ldots, p$ and $k = 1, 2, \ldots, q$.

Following the above, the commonly used multivariate statistics to test H_0 against H_1 are Wilks' lambda, Pillai trace, Hotelling-Lawley's trace, and Roy's largest root. The formulas for their computation and distributions they follow are given below. Please note that these test statistics are similar as used in multivariate analysis of variance (Chapter 7).

Wilks' lambda (Λ)

The quantity Wilks' lambda (Λ) is the ratio of log-likelihood functions under H_0 (l_0) and H_1 (l_l) and the corresponding test is known as the likelihood ratio test. The likelihood ratio Λ is (Wilks, 1932).

$$\Lambda = \frac{\left|SSCP_{\hat{\boldsymbol{\epsilon}}}\right|}{\left|SSCP_{\hat{\boldsymbol{\epsilon}}} + SSCP_{\hat{\mathbf{Y}}}\right|} = \prod_{j=1}^{s}\frac{1}{1+\lambda_j} \quad (9.19)$$

Here, λ_j are the *j*-th eigenvalues of the matrix $[(SSCP_{\hat{\boldsymbol{\epsilon}}})^{-1}SSCP_{\hat{\mathbf{Y}}}]$.

Λ varies from 0 to 1. For sufficiently low value of Λ (close to zero) we can reject H_0. For Λ value close to 1, H_0 is accepted. Λ in Equation 9.19 is a point estimate and based on this value it is difficult to conclude objectively about H_0. So, we need to know the distribution of Λ as well as the confidence interval for which H_0 can be accepted. However, because of complexity of the likelihood ratio (Finn 2003), exact distribution of Λ across all situations is not amenable. Several alternative distributions were given by Bartlett (1938) and Rao (1952).

Bartlett's Test

Let us define a multiplier M as shown below.

$$M = (n - p - 1) - (q + 1 - m) / 2$$

Where n = number of observations
p = number of IVs
q = number of DVs
m = number IVs tested under H_0.

Then the quantity $-M\ell n(\Lambda)$ follows χ^2 distribution with mq degrees of freedom (Bartlett, 1938). The 100(1–α)% confidence interval is

$$P\left\{-M\ell n(\Lambda) \le \chi^{2(\alpha)}_{mq}\right\} = 1 - \alpha$$

The χ^2 approximation works well when sample size is large. Schatzoff (1966) pointed out that the threshold value of n as $n>p+q+30$.

Rao's Test

Rao (1952) has given approximate F-distribution for Λ. Let's define Q as below.

$$Q = \left(\frac{q^2m^2 - 4}{q^2 + m^2 - 5}\right)^{\frac{1}{2}}, qm \neq 2.$$

Then the statistic

$\left(\frac{1-\Lambda^{\frac{1}{Q}}}{\Lambda^{\frac{1}{Q}}}\right)\left(\frac{MQ+1-mq/2}{mq}\right)$ follows F-distribution with mq and $MQ+1-mq/2$ numerator and denominator degrees of freedom, respectively. Recall that Q becomes zero when $qm = 2$. Under such condition Q can be approximated to unity (Finn, 2003).

Example 9.5

Demonstrate Wilks' lambda and its interpretation for Example 9.2.

Solution:

The $SSCP_{\hat{e}}$ matrix can be obtained from Example 9.3, which is

$$SSCP_{\hat{e}} = \begin{bmatrix} 0.97 & 6.30 \\ 6.30 & 43.70 \end{bmatrix}$$

Using Equations 9.16 and 9.18, we

$$SSCP_{\mathbf{Y}} = (\mathbf{Y}-\bar{\mathbf{Y}})^T(\mathbf{Y}-\bar{\mathbf{Y}})$$

$$= \begin{bmatrix} 10-10.67 & 100-102.58 \\ 12-10.67 & 110-102.58 \\ 11-10.67 & 105-102.58 \\ 9-10.67 & 94-102.58 \\ 9-10.67 & 95-102.58 \\ 10-10.67 & 99-102.58 \\ 11-10.67 & 104-102.58 \\ 12-10.67 & 108-102.58 \\ 11-10.67 & 105-102.58 \\ 10-10.67 & 98-102.58 \\ 11-10.67 & 103-102.58 \\ 12-10.67 & 110-102.58 \end{bmatrix}^T \begin{bmatrix} 10-10.67 & 100-102.58 \\ 12-10.67 & 110-102.58 \\ 11-10.67 & 105-102.58 \\ 9-10.67 & 94-102.58 \\ 9-10.67 & 95-102.58 \\ 10-10.67 & 99-102.58 \\ 11-10.67 & 104-102.58 \\ 12-10.67 & 108-102.58 \\ 11-10.67 & 105-102.58 \\ 10-10.67 & 98-102.58 \\ 11-10.67 & 103-102.58 \\ 12-10.67 & 110-102.58 \end{bmatrix} = \begin{bmatrix} 12.67 & 63.33 \\ 63.33 & 324.92 \end{bmatrix}$$

Again,

$$SSCP_{\hat{\mathbf{Y}}} = SSCP_{\mathbf{Y}} - SSCP_{\hat{\mathbf{e}}}$$

$$= \begin{bmatrix} 12.67 & 63.33 \\ 63.33 & 324.92 \end{bmatrix} - \begin{bmatrix} 0.97 & 6.30 \\ 6.30 & 43.70 \end{bmatrix} = \begin{bmatrix} 11.70 & 57.03 \\ 57.03 & 281.22 \end{bmatrix}$$

To compute Wilks' Λ, we use Equation 9.19, as shown below.

$$\Lambda = \frac{|SSCP_{\hat{\mathbf{e}}}|}{|SSCP_{\hat{\mathbf{e}}} + SSCP_{\hat{\mathbf{Y}}}|} = \frac{|SSCP_{\hat{\mathbf{e}}}|}{|SSCP_{\mathbf{Y}}|} = \frac{2.699}{106.05} = 0.025$$

The Λ value (0.025) is very close to zero. So, we reject H_0.

Now, we use Bartlett's test to objectively make a decision.

$$M = (n-p-1)-(q+1-m)/2 = (12-3-1)-(2+1-3)/2 = 8$$

So, $-M\ell n(\Lambda) = -8\ell n(0.025) = -8 \times (-3.689) = 29.512$

Again $-M\ell n(\Lambda)$ follows χ^2 distribution with mq degrees of freedom. Here, $mq = 3 \times 2 = 6$. The tabulated $\chi^2_6(0.05)$ value is 12.59. As the computed statistic (29.512) is more than the tabulated theoretical value (12.59), we reject H_0. Hence, we can say the at least one of the three IVs is significantly influencing the DVs. Which of the three X variables has effect on Y will be shown in the next section (see Section 9.6). Note that the required sample size for the Bartlett's test is $n > p+q+30 = 35$, but we have concluded with only 12 observations, which may be wrong due to small sample size.

Following Rao's test, Q = 2 and $mq = 6$, the computed value for Rao's test statistic is

$$\left(\frac{1-\Lambda^{1/Q}}{\Lambda^{1/Q}}\right)\left(\frac{MQ+1-mq/2}{mq}\right)=\left(\frac{1-\Lambda^{1/2}}{\Lambda^{1/2}}\right)\left(\frac{2M+1-mq/2}{mq}\right)=\left(\frac{1-\sqrt{0.025}}{\sqrt{0.025}}\right)\left(\frac{16+1-6/2}{6}\right)$$
$$=(5.25)(2.33)=12.23$$

As $\left(\frac{1-\Lambda^{1/Q}}{\Lambda^{1/Q}}\right)\left(\frac{MQ+1-mq/2}{mq}\right)$ follows F-distribution with mq and $MQ+1-mq/2$, the tabulated theoretical value for the Rao's statistic (α = 0.05) is $F_{6,6}(0.05) = 4.28$. Again, as the computed statistic (12.23) is more than the tabulated theoretical value (4.28), we reject H_0.

Pillai trace (V) can be computed using Equation 9.20

$$V = trace\left[SSCP_{\hat{Y}}(SSCP_{Y})^{-1}\right] = \sum_{j=1}^{s}\frac{\lambda_j}{1+\lambda_j} \tag{9.20}$$

where s is the number of non-zero eigenvalues (s ≤ q), of $(SSCP_{\hat{e}})^{-1}SSCP_{\hat{Y}}$.

Hotelling-Lawley's trace (U) is

$$U = trace\left[(SSCP_{\hat{e}})^{-1}SSCP_{\hat{Y}}\right] = \sum_{j=1}^{s}\lambda_j \tag{9.21}$$

Finally, the Roy's statistic (θ) is computed as

$$\theta = \frac{\lambda_{max}}{1+\lambda_{max}} \tag{9.22}$$

Separate charts and tables are available for Roy's test (Pillai, 1967). Wilks' Λ, Roy's θ, and Hotelling-Lawley's trace tests are almost equivalent for large sample size. Compare the similarity of Equations 9.19–9.22 with the same statistics given in Chapter 7 (Section 7.2.4). It is reproduced here for the sake of clarity and ease of use.

Which of the above four test statistics is the best? There is no clear answer. These tests answer how large is $SSCP_{\hat{Y}}$ relative to $SSCP_{\hat{e}}$. None of them is the best under all situations. The extreme situations could be (i) when there is only one non-zero eigenvalue $[(SSCP_{\hat{e}})^{-1}SSCP_{\hat{Y}}]$, and (ii) when all eigenvalues are almost equal. For case (i), Roy's largest root and for case (ii) Pillai's trace are preferred. In other situations, such as more than one non-zero unequal eigenvalues, Wilks' Λ or Hotelling-Lawley's trace can be used. But researchers prefer Wilks' Λ as it has better distributional treatment. However, for large sample size, all four tests give nearly equivalent results. There are several computer packages such as SAS and SPSS that provide all four test results with approximate p-values for hypothesis test. The readers may see Chapter 7, Section 7.2.4.

Example 9.6

Consider Example 9.5. Test goodness of fit using Pillai trace, Hotelling-Lawley's trace, and Roy's largest root.

Solution:

We will compute these fit measures using the eigenvalues of $(SSCP_{\hat{e}})^{-1}SSCP_{\hat{Y}}$. So, using Equation 9.24, Pillai trace (V) can be computed as follows:

$$V = \sum_{j=1}^{s} \frac{\lambda_j}{1+\lambda_j}$$

Using R software, the following results (rounded up to two decimal points) are obtained.

$$SSCP_{\hat{Y}} = \begin{bmatrix} 11.70 & 57.03 \\ 57.03 & 281.22 \end{bmatrix}, \text{ and } SSCP_{\hat{e}} = \begin{bmatrix} 0.97 & 6.30 \\ 6.30 & 43.70 \end{bmatrix}$$

$$\text{So, } \left(SSCP_{\hat{e}}\right)^{-1} = \begin{bmatrix} 17.46 & -2.52 \\ -2.52 & 0.39 \end{bmatrix}, \text{ and } \left[(SSCP_{\hat{e}})^{-1}SSCP_{\hat{Y}}\right] = \begin{bmatrix} 60.76 & 288.06 \\ -7.45 & -35.09 \end{bmatrix}.$$

$$\left|(SSCP_{\hat{e}})^{-1}SSCP_{\hat{Y}} - \lambda\mathbf{I}\right| = \left\| \begin{bmatrix} 60.76 & 288.06 \\ -7.45 & -35.09 \end{bmatrix} - \lambda \begin{bmatrix} 1 & 0 \\ 0 & 1 \end{bmatrix} \right\| = \begin{bmatrix} (60.76-\lambda) & 288.06 \\ -7.45 & (-35.09-\lambda) \end{bmatrix}$$

The characteristic equation is

$$\begin{vmatrix} (60.76-\lambda) & 288.06 \\ -7.45 & (-35.09-\lambda) \end{vmatrix} = 0, \text{ which yields}$$

$$\begin{aligned} &(60.76-\lambda)(-35.09-\lambda)+(288.06\times 7.45)=0 \\ &\text{or } -2132.07-60.76\lambda+\lambda^2+35.09\lambda+2146.05=0 \\ &\text{or } \lambda^2-25.67\lambda+13.98=0 \end{aligned}$$

The roots of the characteristic equation (eigenvalues) are as given below.

$\lambda_1 = 25.08$, and $\lambda_2 = 0.60$.

So, Pillai trace is

$$V = \sum_{j=1}^{s} \frac{\lambda_j}{1+\lambda_j} = \left(\frac{25.08}{1+25.08} + \frac{0.60}{1+0.60}\right) = 1.337.$$

Hotelling-Lawley's is

$$U = \sum_{j=1}^{s} \lambda_j = 25.08 + 0.60 = 25.68.$$

Finally, the Roy's largest root is

$$\theta = \frac{\lambda_{max}}{1+\lambda_{max}} = \frac{25.08}{1+25.08} = 0.962$$

As $\lambda_1 = 25.08 >> 0.60 = \lambda_2$, we may consider s = 1. As the aggregated variance of the dependent variables is concentrated in one eigenvalue, Roy's largest root is the preferred statistic. The high value of Roy's statistic suggests that H_0 can be rejected.

TABLE 9.1
Measures of Multivariate Association for Example 9.6

Multivariate Tests	Statistic Value	Measure of Multivariate Association (η^2) (for s = 1)	Comments
Wilks' Λ	0.025	0.98	• As $\lambda_1 = 25.08 >> 0.60 = \lambda_2$, we may consider s = 1.
Pillai's trace (V)	1.337	1.337	• When there is only one non-zero eigenvalue, Roy's largest root is preferred.
Hotelling-Lawley's trace (U)	25.68	0.96	• When all eigenvalues are almost equal, Pillai's trace is preferred. As s = 1, Pillai's trace value cannot be considered.
Roy's statistic (θ)	0.962	0.96	• As Roy's largest root value is high, H_0 can be rejected.

Further, one may be interested to use R^2 like measure used in regression. In MANOVA (Chapter 7), we have shown the measures of multivariate association using the above four test statistics. The measures of multivariate association are bounded between 0 and 1. The measures are also used in MMLR.

Example 9.7

Consider Example 9.6. Compute measures of multivariate association and comment on the results.

Solution:

The measures of multivariate association (η^2) along with the comments are given in Table 9.1. For formula, see Chapter 7. Pillai's trace can not be used here and Roy's statistic is a preferred statistic under one eigenvalue situation. The η^2 value of 0.96 shows a very good fit to the data.

9.6 TEST OF SUBSET OF REGRESSION PARAMETERS

In the discussion on multivariate tests (Equations 9.19–9.22), $H_0 : \boldsymbol{\beta} = \mathbf{0}$, i.e., none of the $\boldsymbol{X}$ variables are contributing in explaining the variability of $\boldsymbol{Y}$ variables, against $H_1 : \boldsymbol{\beta} \neq \mathbf{0}$ for at least one β_{jk}, $j = 0, 1, \ldots, p$ and $k = 1, 2, \ldots, q$, contributes. Alternatively, one may be interested to test the effect of a subset of $\boldsymbol{X}$ variables. The null and alternative hypotheses as well as the modification of the multivariate test statistics (in Equations 9.19–9.22) are presented below.

Let us now partition the β matrix into two halves as follows:

$$\underset{(p+1)\times q}{\boldsymbol{\beta}} = \begin{bmatrix} \boldsymbol{\beta}_{(0)} \\ \cdots \\ \boldsymbol{\beta}_{(m)} \end{bmatrix} \cdots$$

$\beta_{(0)}$ corresponds to (p+1−m) regression coefficients including constant and $\beta_{(m)}$ corresponds to the remaining m regression coefficients. Following Equation 9.3 we can write the expected value of Y as

$$E(Y) = \mathbf{X\beta} = \begin{bmatrix} \mathbf{X}_{(0)} & \vdots & \mathbf{X}_{(m)} \\ \uparrow & & \uparrow \\ n\times(p+1-m) & & n\times m \end{bmatrix} \begin{bmatrix} \boldsymbol{\beta}_{(0)} \\ \cdots\cdots \\ \boldsymbol{\beta}_{(m)} \end{bmatrix} = \underset{n\times(p+1-m)}{\mathbf{X}_{(0)}} \underset{(p+1-m)\times q}{\boldsymbol{\beta}_{(0)}} + \underset{n\times m}{\mathbf{X}_{(m)}} \underset{m\times q}{\boldsymbol{\beta}_{(m)}} \tag{9.23}$$

If we want to assess whether the subset of independent variables X_m contributes or not, then the null and alternative hypotheses are

$$H_0 : \boldsymbol{\beta}_{(m)} = \mathbf{0}$$

$$H_1 : \boldsymbol{\beta}_{(m)} \neq \mathbf{0} \text{ for at least one element of } \boldsymbol{\beta}_{(m)}.$$

If H_0 is true, the MMLR equation becomes

$$\mathbf{Y} = \mathbf{X}_{(0)}\boldsymbol{\beta}_{(0)} + \boldsymbol{\epsilon}_{(0)} \tag{9.24}$$

Upon estimation of $\hat{\boldsymbol{\beta}}_{(0)}$, we get the fitted values of $\mathbf{Y}$, i.e., $\hat{\mathbf{Y}}_0$ and residual $\boldsymbol{\epsilon}_{(0)}$ as

$$\hat{\mathbf{Y}}_{(0)} = \mathbf{X}_{(0)}\hat{\boldsymbol{\beta}}_{(0)} \text{ and } \boldsymbol{\epsilon}_{(0)} = \mathbf{Y} - \hat{\mathbf{Y}}_{(0)}.$$

The corresponding regression and error SSCP matrices are

$$SSCP_{\hat{\mathbf{Y}}_{(0)}} = \hat{\mathbf{Y}}_{(0)}^T \hat{\mathbf{Y}}_{(0)} \tag{9.25}$$

and

$$SSCP_{\hat{\boldsymbol{\epsilon}}_{(0)}} = \hat{\boldsymbol{\epsilon}}_{(0)}^T \boldsymbol{\epsilon}_{(0)} \tag{9.26}$$

So, based on partitioning of $\mathbf{Y}$'s variability, we can write

$$SSCP_{\mathbf{Y}} = SSCP_{\hat{\mathbf{Y}}_{(0)}} + SSCP_{\hat{\boldsymbol{\epsilon}}_{(0)}} \tag{9.27}$$

Now, to simplify the abovementioned notations (in Equation 9.27), let us denote that $SSCP_{\hat{\boldsymbol{\epsilon}}} = \hat{\boldsymbol{\Sigma}}$ and $SSCP_{\hat{\boldsymbol{\epsilon}}_{(0)}} = \hat{\boldsymbol{\Sigma}}_0$. So, $\hat{\boldsymbol{\Sigma}}_0$ is the estimate of the $SSCP_{\hat{\boldsymbol{\epsilon}}_{(0)}}$ when H_0 is true and suppose $\hat{\boldsymbol{\Sigma}}$ is the estimate of the $SSCP_{\hat{\boldsymbol{\epsilon}}}$ when H_1 is true. The difference $\hat{\boldsymbol{\Sigma}}_0 - \hat{\boldsymbol{\Sigma}}$ is the increase in error covariance due to non-inclusion of $X_{(m)}$ under H_0. Let, $\hat{\boldsymbol{\Sigma}}_0 - \hat{\boldsymbol{\Sigma}} = \hat{\boldsymbol{\Sigma}}_d$, for convenience. If $X_{(m)}$, the drop out variables under H_0 has no influence on $\mathbf{Y}$, then $\hat{\boldsymbol{\Sigma}}_d$ will be negligible.

Under the above situation, the multivariate tests are as below (Johnson and Wichern, 2002). Note, θ_j is the j-th eigenvalue of $\hat{\boldsymbol{\Sigma}}_d(\hat{\boldsymbol{\Sigma}})^{-1}$.

$$\textbf{Wilk's lambda} = \frac{|\hat{\boldsymbol{\Sigma}}|}{|\hat{\boldsymbol{\Sigma}}_d + \hat{\boldsymbol{\Sigma}}|} = \frac{|\hat{\boldsymbol{\Sigma}}|}{|\hat{\boldsymbol{\Sigma}}_0|} = \sum_{j=1}^{s} \frac{\theta_j}{1+\theta_j}$$

$$\textbf{Pillai Trace}(V) = trace\left[\hat{\boldsymbol{\Sigma}}_d(\hat{\boldsymbol{\Sigma}}_d + \hat{\boldsymbol{\Sigma}})^{-1}\right] = trace\left[\hat{\boldsymbol{\Sigma}}_d(\hat{\boldsymbol{\Sigma}}_0)^{-1}\right] = \sum_{j=1}^{s} \frac{\theta_j}{1+\theta_j}$$

$$\textbf{Hotelling – Lowley's trace}(U) = trace\left[\hat{\boldsymbol{\Sigma}}_d(\hat{\boldsymbol{\Sigma}})^{-1}\right] = \sum_{j=1}^{s} \theta_j$$

$$\textbf{Roy's statistic}(\theta) = \frac{\theta_1}{1+\theta_1} \tag{9.28}$$

The distributions and interpretations of these four multivariate statistics are similar, as discussed in Section 9.5.

Now, consider Example 9.2. There are three ***X*** variables: absenteeism (%) (X_1), breakdown hours (X_2), and M-ratio (X_3). From Examples 9.5 and 9.6, it is seen that the null hypothesis $H_0 : \boldsymbol{\beta} = \mathbf{0}$ is rejected, indicating that at least one of the three ***X*** variables is significantly contributing to explain the variability of Y_1 (sales) and Y_2 (profit). The subsets of ***X*** are $\{X_1\}$, $\{X_2\}$, $\{X_3\}$, $\{X_1, X_2\}$, $\{X_1, X_3\}$, and $\{X_2, X_3\}$. Following the above discussion, the hypothesis test scenarios are given in Table 9.2.

From Table 9.2, it can be concluded that the variable X_1 could have negligible or no influence on both Y_1 and Y_2.

9.7 TEST OF INDIVIDUAL REGRESSION PARAMETERS

In Section 9.6, we discussed the impact of subset of ***X*** on ***Y***. In some of the subsets (see Table 9.2), only one *X* is included in the model. For example, in rows 4–6 of Table 9.2, the test procedure examines the influence of only one *X* (e.g., in row 4, the effect of X_3 is tested). Does the test procedure (in Section 9.6) test the individual regression parameter, like in MLR? The answer is no. The individual regression parameter in MMLR is β_{jk}, which tells the influence of the *j*-th *X* on the *k*-th *Y*. Whereas, even when we exclude only one *X* variable from the model, such as in row 4 of Table 9.2, $H_0 : \boldsymbol{\beta}_3 = \mathbf{0}$ tests whether X_3 (M-ratio) affects either Y_1 (sales) or Y_2 (profit) or both. But by the test of individual regression parameter, we test H_0: $\beta_{jk} = 0$ against H_1: $\beta_{jk} \neq 0$. We will discuss below.

The point estimation and sampling distribution of multivariate regression parameters $\boldsymbol{\beta}_k$ are discussed in Sections 9.3 and 9.4. The confidence region (CR) of $\boldsymbol{\beta}_k$ and the simultaneous confidence interval (SCI) of β_{jk} are discussed in Chapter 8 (MLR). The same is applied in MMLR. The estimation of the confidence interval for individual regression coefficient β_{jk}, $j = 1, 2, \ldots, p$ and $k = 1, 2, \ldots, q$ is extremely important and *t*-statistic is used for this purpose.

Following Chapter 8, the 100(1–α)% for β_{jk} is

$$\hat{\beta}_{jk} - t_{n-p-1}^{(\alpha/2)} \sqrt{s_k^2 c_{jj} / n-p-1} \le \beta_{jk} \le \hat{\beta}_{jk} + t_{n-p-1}^{(\alpha/2)} \sqrt{s_k^2 c_{jj} / n-p-1} \qquad (9.29)$$

TABLE 9.2
Hypothesis Test Scenarios and Wilks' Λ Test Results for The Test of Parameters (with α = 0.05) for Example 9.2

Contributing Sub-Set	$H_0 : \boldsymbol{\beta}_{(m)} = 0$	m	Wilks' Λ	*M*	$-M\ell n(\Lambda)$	Cut-off $\chi^2_{2m}(0.05)$	Decision About H_0
$\{X_1\}$	$\boldsymbol{\beta}_2 = \boldsymbol{\beta}_3 = \mathbf{0}$	2	0.074	7.5	19.49	9.49	Reject H_0
$\{X_2\}$	$\boldsymbol{\beta}_1 = \boldsymbol{\beta}_3 = \mathbf{0}$	2	0.045	7.5	23.34	9.49	Reject H_0
$\{X_3\}$	$\boldsymbol{\beta}_1 = \boldsymbol{\beta}_2 = \mathbf{0}$	2	0.169	7.5	13.32	9.49	Reject H_0
$\{X_1, X_2\}$	$\boldsymbol{\beta}_3 = \mathbf{0}$	1	0.164	7	12.67	5.99	Reject H_0
$\{X_1, X_3\}$	$\boldsymbol{\beta}_2 = \mathbf{0}$	1	0.246	7	9.82	5.99	Reject H_0
$\{X_2, X_3\}$	$\boldsymbol{\beta}_1 = \mathbf{0}$	1	0.522	7	4.55	5.99	Fail to reject H_0
$\{X_1, X_2, X_3\}$	$\boldsymbol{\beta}_1 = \boldsymbol{\beta}_2 = \boldsymbol{\beta}_3 = \mathbf{0}$	3	0.025	8	29.512	12.59	Reject H_0

Note: m = number of ***X*** variables under test, $M = [(n-p-1) - (q+1-m)/2] = [8 - (3-m)/2]$

It is to be noted here that Equation 9.29 doesn't simultaneously satisfy 100(1–α)% confidence interval (CI) for all regression parameters in β_k. As stated in Chapter 8, alternatively, Bonferroni 100(1 – α)% simultaneous confidence intervals (SCI) can also be computed for the parameter vector in MMLR. Bonferroni 100(1 – α)% SCI adjusts $t_{n-p-1}^{\alpha/2}$ to $t_{n-p-1}^{\alpha_j/2}$, where $\sum_{j=1}^{p}\alpha_j = \alpha$. Further to be noted here is that in MLR there is one Y but in MMLR, there are q number of $\boldsymbol{Y}$. It is not clear whether Bonferroni adjustment should consider both pq (in line with MANOVA) or only p (as in MLR). For simultaneous testing of regression coefficients in complex survey, based on the comparison of Wald procedure based χ^2 and F statistics with Bonferroni adjustment of t-statistic, Korn and Graubard (1990) recommended, in addition, the use of the Bonferroni approach. They stated that although researchers suggest that Bonferroni adjustment works well for lower number of variables, their study found it suitable for large number of variables also. The other interesting point to observe is that for all practical purposes, the researchers provide t-distribution-based individual CI. The next step that one may be interested is to conduct hypothesis test of β_{jk}. See Chapter 8 for the procedure.

Example 9.8

Consider Example 9.1. Obtain 95% CI for the multivariate regression parameters and conduct hypothesis test at α = 0.05. Also compare the results with the Bonferroni approach.

Solution:

In Example 9.1, the multivariate linear regression equations are

$$
\begin{aligned}
Y_1 &= 17.97 - 0.26X_1 - 0.09X_2 + \epsilon_1 \\
Y_2 &= 130.22 - 1.24X_1 - 0.30X_2 + \epsilon_2
\end{aligned}
$$

We use Equation 9.9 to estimate fitted values ($\hat{\mathbf{Y}}$), Equation 9.10 to estimate errors ($\hat{\boldsymbol{\epsilon}}$) and Equation 9.12 or 9.13 to estimate SSE. The computations are shown below.

$$
\hat{\mathbf{Y}} = \mathbf{X}\hat{\boldsymbol{\beta}} = \begin{bmatrix} 1 & 9 & 62 \\ 1 & 8 & 58 \\ 1 & 7 & 64 \\ 1 & 14 & 60 \\ 1 & 12 & 63 \\ 1 & 10 & 57 \\ 1 & 7 & 55 \\ 1 & 4 & 56 \\ 1 & 6 & 59 \\ 1 & 5 & 61 \\ 1 & 7 & 57 \\ 1 & 6 & 60 \end{bmatrix} \begin{bmatrix} 17.97 & 130.22 \\ -0.26 & -1.24 \\ -0.09 & -0.30 \end{bmatrix} = \begin{bmatrix} 10.15 & 100.44 \\ 10.76 & 102.88 \\ 10.49 & 102.32 \\ 9.04 & 94.82 \\ 9.29 & 96.41 \\ 10.34 & 100.69 \\ 11.29 & 105.02 \\ 11.97 & 108.45 \\ 11.19 & 105.07 \\ 11.27 & 105.71 \\ 11.11 & 104.42 \\ 11.10 & 104.77 \end{bmatrix}
$$

Now, the errors are (using Equation 9.10)

$$\hat{\epsilon} = \mathbf{Y} - \hat{\mathbf{Y}} = \begin{bmatrix} 10 & 100 \\ 12 & 110 \\ 11 & 105 \\ 9 & 94 \\ 9 & 95 \\ 10 & 99 \\ 11 & 104 \\ 12 & 108 \\ 11 & 105 \\ 10 & 98 \\ 11 & 103 \\ 12 & 110 \end{bmatrix} - \begin{bmatrix} 10.15 & 100.44 \\ 10.76 & 102.88 \\ 10.49 & 102.32 \\ 9.04 & 94.82 \\ 9.29 & 96.41 \\ 10.34 & 100.69 \\ 11.29 & 105.02 \\ 11.97 & 108.45 \\ 11.19 & 105.07 \\ 11.27 & 105.71 \\ 11.11 & 104.42 \\ 11.10 & 104.77 \end{bmatrix} = \begin{bmatrix} -0.15 & -0.44 \\ 1.24 & 7.12 \\ 0.51 & 2.68 \\ -0.04 & -0.82 \\ -0.29 & -1.41 \\ -0.34 & -1.69 \\ -0.29 & -1.02 \\ 0.03 & -0.45 \\ -0.19 & -0.07 \\ -1.27 & -7.71 \\ -0.11 & -1.42 \\ 0.90 & 5.23 \end{bmatrix}$$

And the $SSCP_{\hat{\epsilon}}$ is (using Equation 9.13)

$$SSCP_{\epsilon} = \hat{\epsilon}^T \hat{\epsilon} =$$

$$= \begin{bmatrix} -0.15 & 1.24 & 0.51 & -0.04 & -0.29 & -0.34 & -0.29 & 0.03 & -0.19 & -1.27 & -0.11 & 0.90 \\ -0.44 & 7.12 & 2.68 & -0.82 & -1.41 & -1.69 & -1.02 & -0.45 & -0.07 & -7.71 & -1.42 & 5.23 \end{bmatrix} \begin{bmatrix} -0.15 & -0.44 \\ 1.24 & 7.12 \\ 0.51 & 2.68 \\ -0.04 & -0.82 \\ -0.29 & -1.41 \\ -0.34 & -1.69 \\ -0.29 & -1.02 \\ 0.03 & -0.45 \\ -0.19 & -0.07 \\ -1.27 & -7.71 \\ -0.11 & -1.42 \\ 0.90 & 5.23 \end{bmatrix}$$

$$= \begin{bmatrix} 4.56 & 26.20 \\ 26.20 & 153.68 \end{bmatrix}$$

The SSE_1 and SSE_2 are the first and second elements in the diagonal of $SSCP_{\hat{\epsilon}}$, which are 4.56 and 153.68, respectively. See that the SSE_2 value matches with the result obtained in Example 8.2 (Chapter 8).

The error variances are

$$\hat{\sigma}_1^2 = \frac{s_1^2}{n-p-1} = \frac{4.56}{12-2-1} = 0.51$$

$$\hat{\sigma}_2^2 = \frac{s_2^2}{n-p-1} = \frac{153.68}{12-2-1} = 17.08$$

Now, using Equation 9.29, the 95% CI for the regression parameters are given in Table 9.3.

TABLE 9.3
Regression Parameter Estimates for Example 9.1

Dependent Variable	Independent Variable	Parameter	Estimate	Standard Error (SE)	95% CI	95% SCI
Y_1	Intercept	β_{10}	17.97	√(0.51 × 40.894) = 4.57	(+7.67) – (+28.28)	(+5.30) – (+30.64)
	X_1	β_{11}	–0.26	√(0.51 × 0.012) = 0.078	(–0.44) – (–0.08)	(–0.48) – (–0.04)
	X_2	β_{12}	–0.09	√(0.51 × 0.012) = 0.078	(–0.27) – (+0.09)#	(-0.31) – (+0.13)#
Y_2	Intercept	β_{20}	130.22	√(17.08 × 40.894) = 26.43	(+70.44) – (+190.00)	(+56.96) – (+203.48)
	X_1	β_{21}	–1.24	√(17.08 × 0.012) = 0.452	(–2.26) – (–0.23)	(–2.49) – (+0.01)#
	X_2	β_{22}	–0.30	√(17.08 × 0.012) = 0.452	(–1.34) – (+0.74)#	(–1.55) – (+0.95)#

$SE = \sqrt{s_k^2 c_{jj} / n - p - 1}$, $95\% SCI : \hat{\beta}_{jk} \pm t_{n-p-1}^{(\alpha/2)} \sqrt{s_k^2 c_{jj} / n - p - 1}$, $t_{n-p-1}^{(\alpha/2)} = t_9^{0.05/2} = 2.262$, $t_{n-p-1}^{(\alpha/2p)} = t_9^{0.05/4} = 2.772$,

# indicates that the interval contains zero.

TABLE 9.4
Hypothesis Testing for Example 9.1

Dependent Variable	Independent Variable	Parameter	Estimate (SE)	t-computed	Tabulated t = $t_{n-p-1}^{(\alpha/2)}$
Y_1	Intercept	β_{10}	17.97 (4.57)	3.94*	$t_{n-p-1}^{(\alpha/2)} = t_9^{0.05/2} = 2.262$
	X_1	β_{11}	–0.26 (0.078)	–3.34*	
	X_2	β_{12}	–0.09 (0.078)	–1.12	
Y_2	Intercept	β_{20}	130.22 (26.43)	4.93*	
	X_1	β_{21}	–1.24 (0.452)	–2.78*	
	X_2	β_{22}	–0.30 (0.452)	0.65	

* indicates significant at 0.05 probability level.

Note that the 95% CI and 95% SCI provide similar results (except for β_{21}). 95% CI and 95% SCI contain zero for both β_{12} and β_{22}. So, these two parameters may not be different from zero with 5% error. Hence, we can conclude that the IV X_2 doesn't influence the DVs Y_1 and Y_2. In addition, 95% SCI for β_{21} also contain zero. This reveals that if we consider Bonferroni adjustment, none of the X variables has influence on Y_2. This is interesting and such situations warrant expert knowledge for decision-making.

The ANOVA table below shows the results of hypothesis testing (see Table 9.4). Note, we showed the traditional t-test-based decisions.

From the above results, we can conclude that the variable X_1 affects both Y_1 and Y_2, whereas the effect of X_2 on both Y_1 and Y_2 is statistically insignificant.

Example 9.9

Consider Example 9.2. Obtain 95% CI for the multivariate regression parameters and conduct hypothesis test at $\alpha = 0.05$. Also compare the results with the Bonferroni approach.

Solution:

In Example 9.2, the multivariate linear regression equations are

$$\begin{aligned} Y_1 &= 10.90 - 0.04X_1 - 0.09X_2 + 5.04X_3 + \epsilon_1 \\ Y_2 &= 91.10 - 0.06X_1 - 0.29X_2 + 27.84X_3 + \epsilon_2 \end{aligned}$$

Now, using Equation 9.27, the 95% CI for the regression parameters are given in Table 9.5.

Note that the 95% CI and 95% SCI provide similar results. 95% CI and 95% SCI contain zero for $\beta_{11}, \beta_{12}, \beta_{21}$, and β_{22}. So, these parameters may not be different from zero with 5% error. So, it can be inferred that X_1 and X_2 doesn't influence Y_1 and Y_2. Note that any errors in these interpretations may be attributed to the low sample size (n = 12).

The ANOVA table below shows the results of hypothesis testing (Table 9.6). From the results, we can conclude that only X_3 (M-ratio) affects both Y_1 and Y_2. Note, we showed the traditional t-test based decisions.

9.8 INTERPRETATION AND DIAGNOSTICS

For interpretation of multivariate regression results, we consider Example 9.2. The multivariate goodness of fit tests are explained in Example 9.7 and the test of regression parameters is shown in Example 9.9. It should be considered that all the multivariate goodness of fit tests are not applicable under all situations and are influenced by eigenvalue structure of the $(SSCP_{\hat{\epsilon}})^{-1} SSCP_{\hat{Y}}$ matrix, sample size, and the theoretical distributions of each of the tests. As stated earlier, each of the

TABLE 9.5
Regression Parameter Estimates and 95% CI for Example 9.2

Dependent Variable	Independent Variable	Parameter	Estimate	SE	95% CI	95% SCI
Y_1	Intercept	β_{10}	10.897	2.572	(+4.97) – (+16.83)	(+3.14) – (+18.65)
	X_1	β_{11}	–0.045	0.0543	(-0.17) – (+0.08)#	(-0.21) – (+0.12)#
	X_2	β_{12}	–0.088	0.039	(-0.18) – (+0.001)#	(-0.21) – (+0.03)#
	X_3	β_{13}	5.035	0.922	(+2.91) – (+7.16)	(+2.25) – (+7.82)
Y_2	Intercept	β_{20}	91.097	17.303	(+51.20) – (+131.00)	(+38.91) – (+143.28)
	X_1	β_{21}	–0.064	0.365	(-0.91) – (+0.78)#	(-1.16) – (+1.04)#
	X_2	β_{22}	–0.294	0.259	(-0.89) – (+0.30)#	(-1.08) – (+0.49)#
	X_3	β_{23}	27.835	6.203	(+13.53) – (+42.14)	(+9.13) – (+46.54)

$SE = \sqrt{s_k^2 c_{jj} / n-p-1}$, $95\% SCI : \hat{\beta}_{jk} \pm t_{n-p-1}^{(\alpha/2)} \sqrt{s_k^2 c_{jj} / n-p-1}$, $t_{n-p-1}^{(\alpha/2)} = t_8^{0.05/2} = 2.306$, $t_{n-p-1}^{(\alpha/2p)} = t_8^{0.05/6} = 3.016$,

# indicates that the interval contains zero.

TABLE 9.6
Hypothesis Testing for Example 9.2

Dependent Variable	Independent Variable	Parameter	Estimate (SE)	t-computed	Tabulated t = $t_{n-p-1}^{(\alpha/2)}$
Y_1	Intercept	β_{10}	10.897 (2.57)	4.237*	$t_{n-p-1}^{(\alpha/2)} = t_8^{0.05/2} = 2.306$
	X_1	β_{11}	−0.045 (0.05)	−0.828	
	X_2	β_{12}	−0.088 (0.04)	−2.275	
	X_3	β_{13}	5.035 (0.92)	5.462*	
Y_2	Intercept	β_{20}	91.097 (17.30)	5.265*	
	X_1	β_{21}	−0.064 (0.37)	−0.175	
	X_2	β_{22}	−0.294 (2.59)	−1.135	
	X_3	β_{23}	27.835 (6.20)	4.487*	

* indicates significant at 0.05 probability level.

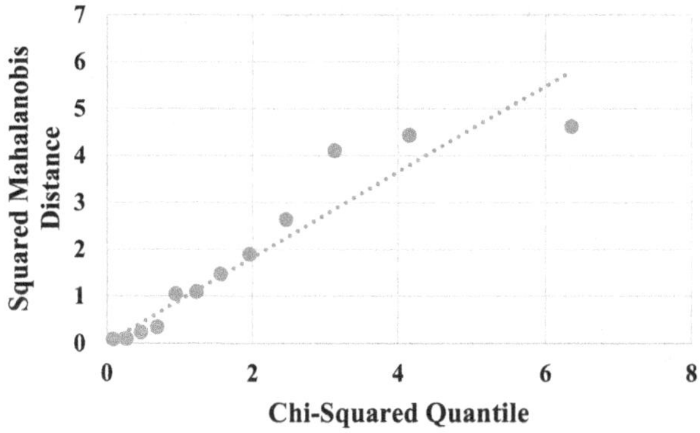

FIGURE 9.2 Chi-square Q-Q plot. The Mahalanobis distances and corresponding Chi-square quantiles for the error vectors follow a straight line with slope 1 passing through the origin. This indicates that the errors are multivariate normal.

regression equations may be treated as a separate multiple linear regression (MLR, see Chapter 8) and the practical meaning of the regression coefficients including intercept can be interpreted similarly as given in Section 8.9 of Chapter 8. The graphical tests of assumptions for Example 9.2 are explained below.

The multivariate normality test is performed using Chi-square Q-Q plot of the residual vectors of the dependent variables, profit and sales (Figure 9.2). It is seen from the figure that the error vectors follow a straight line with slope 1 passing through the origin which indicates that the errors are multivariate normal. The confidence interval increases along the error vectors, a possible indication of violation of the assumption of equal error variances (homoscedasticity).

For the visual test of homoscedasticity, the plot of residuals vs fitted values for each of the dependent variables, profit and sales, is shown in Figure 9.3. The residuals vs fitted values plots for profit (P) and sales volume (S) resemble a funnel shape opening to the right. So, it can be inferred that for profit, the homoscedasticity assumption is violated. However, the funnel for sales volume(S) shows lower degree of violation.

To test the independence of observations, the serial correlation plot of MD^2 (squared Mahalanobis distance) of the residual vectors against the order of observations is shown in Figure 9.4. If the MD^2 values of the residual vectors are not related to the order of observations, the observations are collected independently. The MD^2 plot shows two clusters of observations, one with lower MD^2 (observations 1 to 7) and the other with comparatively higher MD^2 (observations 8 to 12). So, the observed data vectors might not be independent. Further, the residuals vs order of observations plots for both profit (P) and sales (S) separately are shown in Figure 9.5. The figures do not show any conclusive patterns.

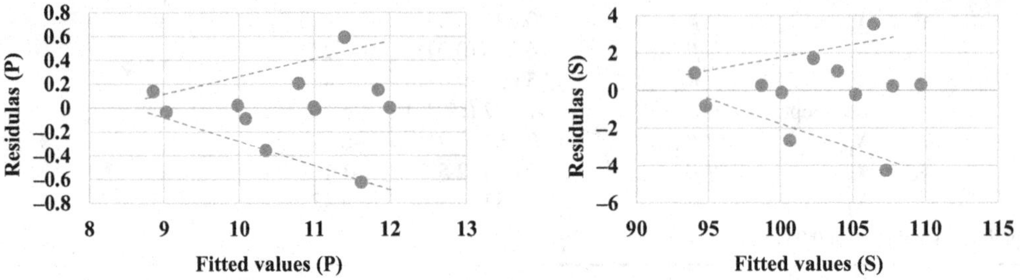

FIGURE 9.3 Residual vs fitted values for (a) profit (P) and (b) sales (S). Both the scatter plots show funneling effects (open to right) indicating that the homoscedasticity assumption is violated. The funnel for sales volume(S) shows lower degree of violation. However, an objective test is needed for making final conclusions.

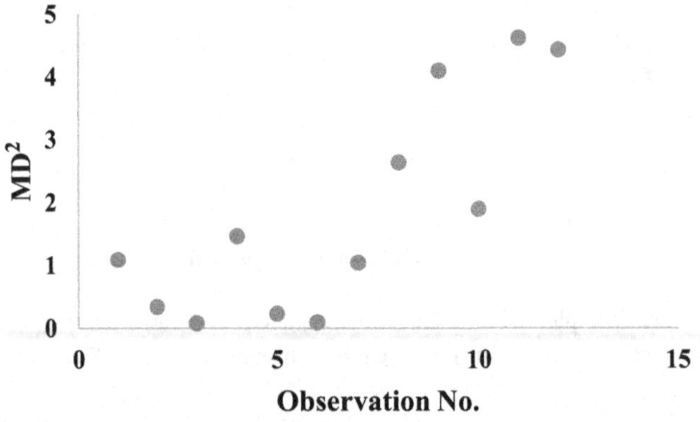

FIGURE 9.4 Serial correlation plot of MD^2 of the residual vectors and its order of observations. The plot shows two groups of observations with low and moderate to high MD^2 values, indicating that the observed data vectors may not be independent.

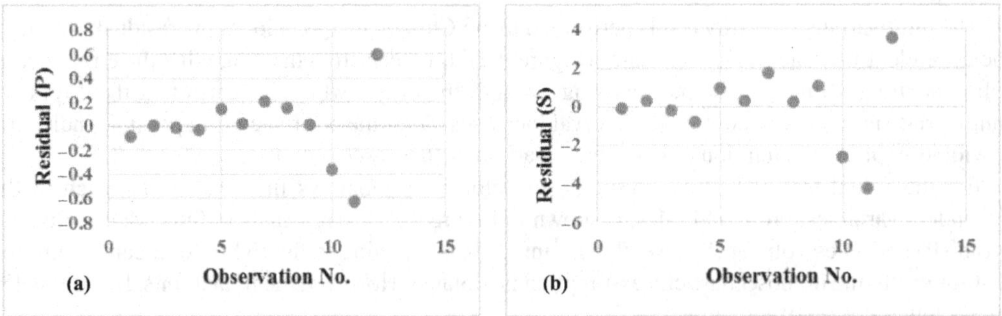

FIGURE 9.5 Serial correlation plot of residual vectors and its order of observations for (a) profit (P) and (b) sales (S). Both the figures do not conclusively confirm the independence of the error vectors.

9.9 CASE STUDY

The case study is in continuation with the one given in Chapter 8 (Section 8.15), where multiple linear regression (MLR) was demonstrated using one DV (backlash). Here, we use the case to demonstrate multivariate linear regression (MMLR) using two DVs backlash (Y_1) and contact % (Y_2).

The details of the case study are available in the web resources of this book (Multivariate Statistical Modeling in Engineering and Management – 1st (routledge.com)).

9.10 LEARNING SUMMARY

Multivariate multiple regression (MMLR) is an extension of the multiple linear regression (MLR) that accommodates more than one dependent variable simultaneously. In this chapter, we have discussed the conceptual model, assumptions, estimation of parameters, and assessment of overall fit of the model. A large number of solved examples are provided. Finally, a case study on the hobbing process in a manufacturing plant is presented.

In summary,

- The difference between MMLR and MLR is explained in the chapter.
- The important assumptions are (i) errors are independent and identically distributed, (ii) errors are multivariate normal, (iii) error variances are equal (homogenous) across observations conditional on predictors, and (iv) errors have common covariance structure across observations.
- Slight violation of these assumptions may not distort the model adequacy as well as the estimation of parameters, but if the violation is significant, the variables may be transformed to satisfy the assumptions or care must be taken in drawing insights from the results, in case the variables are used in their original form for ease in interpretation of the relationships.
- Parameter estimation is simple and Excel-based tabular calculations would give the desired results. But the computation of the multivariate tests, like in case of MANOVA, is cumbersome and hence, suitable computer programs or software are recommended.
- The four most widely used goodness of fit multivariate statistics for MMLR are Wilks' lambda (see Equation 9.19), Pillai trace (see Equation 9.24), Hotelling-Lawley's trace (see Equation 9.25), and Roy's largest root (see Equation 9.26). All four tests will not yield identical results in all situations. Care must be taken in identifying appropriate goodness of fit multivariate statistics under different situations. However, the distribution of Wilks' lambda is well developed and hence, favored in most of the situations, if the distributions assumptions are satisfied.
- The ease of interpretability of R^2 in MLR encourages statisticians to find out its equivalent for MMLR. One such measure is multivariate association which has been developed for each of the multivariate test statistics, Wilks' lambda, Pillai trace, Hotelling-Lawley's trace, and Roy's largest root.
- The test of regression parameters is similar to that of MLR.
- Finally, a case study was conducted for hobbing process in manufacturing plant to demonstrate the usefulness of MMLR in real-world situation.

EXERCISES

(A) Short conceptual questions

9.1 Define multivariate multiple linear regression (MMLR) with a pictorial representation of it.

9.2 State the assumptions of MMLR.

9.3 Define design matrix in MMLR. Is it different from the design matrix used in multiple linear regression (MLR), if the independent variables remain same both in MLR and MMLR?

9.4 Prove that the OLS estimates of MMLR regression coefficient matrix ($\hat{\boldsymbol{\beta}}$) is $\hat{\boldsymbol{\beta}} = \left(\mathbf{X}^T\mathbf{X}\right)^{-1}\mathbf{X}^T\mathbf{Y}$. In what way is it different from the OLS estimates of regression coefficient vector in MLR?

9.5 Prove that

(a) $\hat{\boldsymbol{\epsilon}} = \left(\mathbf{I} - \mathbf{H}\right)\mathbf{Y}$ [where $\mathbf{H}_{n \times n}$ = Hat matrix]

(b) $SSCP_E = (n - p - 1)\hat{\boldsymbol{\Sigma}}$ [where $\hat{\boldsymbol{\Sigma}}$ is estimated covariance matrix of the errors]

(c) $Cov(\hat{\boldsymbol{\beta}}) = (\mathbf{X}^T\mathbf{X})^{-1} \otimes \boldsymbol{\Sigma}$

9.6 What are the multivariate test statistics used in MMLR? Do all tests provide equivalent results?

9.7 Define the measures of multivariate association for MMLR. How do you interpret these measures for model adequacy?

9.8 What is the significance of $(SSCP_{\hat{\boldsymbol{\epsilon}}})^{-1} SSCP_{\hat{\mathbf{Y}}}$ matrix in MMLR? Which of the multivariate test statistics will be used under the following conditions?

(a) The matrix has one dominant non-zero eigenvalue.

(b) All eigenvalues of the matrix are more or less equal.

(c) The matrix has more than one non-zero unequal eigenvalues.

9.9 How is individual regression parameter tested in MMLR?

9.10 Test the following assumptions using residuals of MMLR:

(a) Test of linearity.

(b) Test of homoscedasticity.

(c) Test of uncorrelated error terms.

(d) Multivariate normality of error terms.

(B) Long conceptual and/or numerical questions: see web resources (Multivariate Statistical Modeling in Engineering and Management – 1st (routledge.com))

N.B.: R, MINITAB and MS Excel software are used for computation, as and when required.

10 Path Model

In Chapter 9, we described the multivariate multiple linear regression model (MMLR), where more than one dependent variable is modeled with one or more independent variables. In MMLR, it is assumed that the dependent variables are not affected by other dependent variables present in the model. In addition, the independent variables are considered uncorrelated with each other. In this chapter, we will describe the path model, which is capable of relaxing the above two requirements of the MMLR. So, the path model is more flexible as compared to the MMLR in exploring the dependence relationships of a set of dependent variables with other dependent and independent variables present in the model. A complete path model comprises a system of simultaneous equations.

The chapter begins with the conceptual path model followed by assumptions, types of model, model identification, parameter estimation, overall fit, and diagnostic tests for path modeling. The path model is known by different names, such as causal model, and structural model. It is applied in different domains such as sociology, psychology, management science, econometrics, and engineering applications. There have been many methods developed over the years in estimating path model parameters, particularly for the non-recursive path model. In this chapter, we will focus on the estimation of both recursive and non-recursive path models, primarily using the casual model and the simultaneous equation modeling approaches followed in econometrics. We will describe the structural modeling approaches in Chapter 14.

10.1 CONCEPTUAL MODEL

In multiple linear regression (MLR), one dependent variable (DV) is affected by a set of independent variables (IVs). Figure 10.1a shows the MLR pictorially (please note that the intercept term is removed).

Let's assume that the arrow(s) in Figure 10.1a indicate(s) causal relationships, i.e., cause and effect relationship(s), where the arrow emanates from a cause (say, X) and terminates at the effect (say, Y). So, $X \rightarrow Y$ indicates that X causes Y. The MLR equation for Figure 10.1a is $Y = \beta_0 + \beta_1 X_1 + \beta_2 X_2 + \beta_3 X_3 + \epsilon$, where ϵ represents error. One of the assumptions of the MLR is that the X variables are independent. Now, let X_1, X_2 and X_3 covary with each other, and let the covariances between X_1 and X_2, X_2 and X_3, and X_1 and X_3 are $Cov(X_1X_2) = \phi_{21}, Cov(X_2X_3) = \phi_{32}, Cov(X_1X_3) = \phi_{31}$, respectively. The pictorial representation of this situation is shown in Figure 10.1b, which is an example of a simple path model. On the other hand, the complex path model accommodates multiple X and multiple Y (in single or multi-layers, see Figure 10.2b).

Now, let draw some insights from Figure 10.2. In Figure 10.2a, the problem at hand (situation) is that there are three DVs and three IVs, neither the DVs among themselves, nor the IVs among themselves are correlated, but the IVs (X) affect the DVs (Y). So, the researcher (or analyst) will use multivariate multiple linear regression (MMLR). On the contrary, in addition that the DVs Y_1, Y_2, and Y_3 are affected by all the three IVs, suppose the DV (Y_3) is affected by Y_1 and Y_2 and the DV (Y_2) is affected by Y_1. This situation is complicated. The MMLR (Figure 10.2a) cannot handle this. The path model (Figure 10.2b) is appropriate in such situations.

In the path models described above, we have used three kinds of variables, Y, X, and ϵ. In path analysis terminology, the X variables are called *exogenous variables*, the Y variables are called *endogenous variables*, and the ϵ variables are called *residual* or *error variables*. The exogenous variables are determined outside the model and are assumed not to be affected by any other variables in the path model, but may be correlated among them. The endogenous variables (Y) are completely

DOI: 10.1201/9781003303060-13

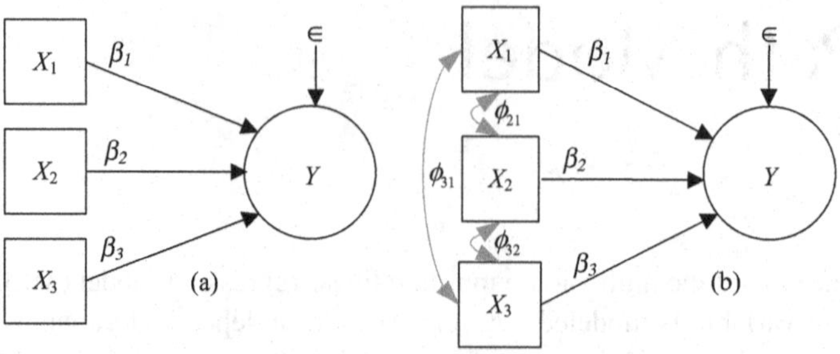

FIGURE 10.1 Pictorial representation of (a) multiple linear regression and (b) path model, with one DV (Y) and three IVs (X). β represents the regression coefficients and ϕ represents the correlation coefficient between two IVs. Note that no intercept term (β_0) is used as in path model, the mean subtracted DV (Y) is considered.

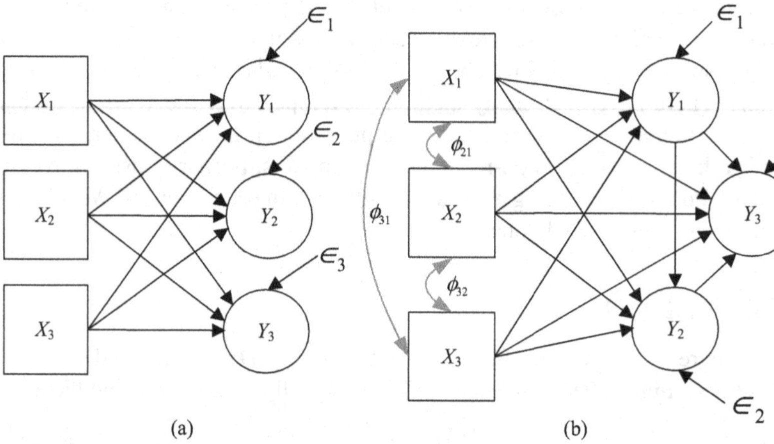

FIGURE 10.2 Pictorial representation of (a) multivariate multiple linear regression (MMLR) and (b) path model, with three DVs (Y) and three IVs (X) [the notation for regression coefficients are omitted for clarity in presentation]. The Y variables are mean subtracted and hence, no intercept term is included.

determined by the exogenous (X) and residual ($\in$) variables. The residual ($\in$) variables represent the effects of all other variables not included in the model.

The path diagram graphically depicts the linear, additive relationships among the variables included in the model. There are two kinds of relationships among the variables: (i) *causal relationships* or cause and effect relationships, represented by single-headed arrows (e.g., between X and Y), and (ii) *correlation relationships*, represented by double-headed curved arrows (e.g., between X variables). The benefits of the path model over MLR and MMLR are many. Apart from estimating the regression coefficients, it helps in (i) identifying and/or estimating moderating and confounding variables, (ii) estimating direct, indirect, and correlated causal relationships as well as spurious (joint) relationships among the variables of interest, and (iii) tracing the causal paths (sequence) involving two or more variables along the path, which may be intelligently used in generating causation rules.

Causation is defined as the cause and effect relationship. Establishing the existence cause and effect ($X \rightarrow Y$) relationship between two variables is not an easy task. As the minimum requirements, the following need to be satisfied (Yang and Trewn, 2004; Cohen et al., 2003):

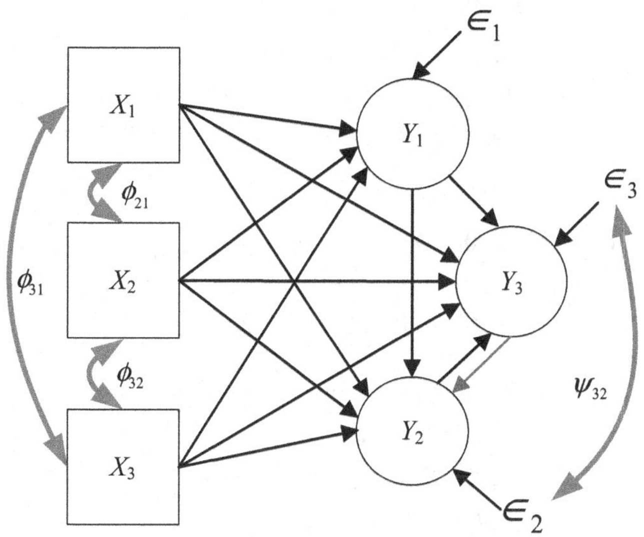

FIGURE 10.3 Non-recursive path model. In the non-recursive path model, one or more reciprocal causations among the DVs are involved. In this figure, Y_2 and Y_3 have reciprocal causations.

1. Existence of a sufficient degree of correlation between two variables, *X* and *Y*.
2. The cause variable (*X*) occurs before the effect variable (*Y*) on time (temporal precedence).
3. The effect variable (*Y*) is clearly the outcome of the cause (*X*), i.e., the *X*-*Y* relationship holds when the influences of other variables are eliminated.

Based on the types of causal relations, path models are categorized as *recursive* and *non-recursive* path models. In recursive path model, all the causal links are one-way. For example, the path models in Figures 10.1b and 10.2b are recursive path models. Anderson (1978) pointed out that "such models involve a priori assumptions that no reciprocal links are involved, nor is there indirect feedback in which a variable that appears at one point in the causal sequence directly or indirectly affect a variable that appears earlier in the sequence." In addition, a fully recursive model is one where each pair of disturbances are uncorrelated, i.e., $E(\epsilon_j \epsilon_k) = 0, j \neq k$. On the other hand, Figure 10.3 depicts a path model where reciprocal causation lies between Y_2 and Y_3. Further, the error variables ϵ_2 and ϵ_3 are correlated (i.e., $E(\epsilon_2 \epsilon_3) \neq 0$). Such path models are called non-recursive path models. Non-recursive path models often involve tedious mathematical computations for parameter estimation. Whether a path model will be recursive or non-recursive depends on the cause–effect relationships among the variables of the path model.

The steps that can be followed in developing a path model and conducting path analysis are summarized below.

- Define the system and its associated problems, and identify the variables of interest.
- Develop the theoretically based conceptual path model (e.g., recursive or non-recursive) and respective path diagram (e.g., Figure 10.2b or Figure 10.3).
- Construct path equations representing the causal structure (also called structural equations).
- Collect data that is representative of the system's behavior of interest (e.g., the sample size should be adequate).
- Estimate the parameters.
- Evaluate adequacy (goodness of fit) of the model.
- Interpret the results and make decisions.

The path equations for Figure 10.2b (recursive model) are

$$\begin{aligned} Y_1 &= \gamma_{11}X_1 + \gamma_{12}X_2 + \gamma_{13}X_3 + \epsilon_1 \\ Y_2 &= \beta_{21}Y_1 + \gamma_{21}X_1 + \gamma_{22}X_2 + \gamma_{23}X_3 + \epsilon_2 \\ Y_3 &= \beta_{31}Y_1 + \beta_{32}Y_2 + \gamma_{31}X_1 + \gamma_{32}X_2 + \gamma_{33}X_3 + \epsilon_3 \end{aligned} \tag{10.1}$$

In matrix form

$$\begin{pmatrix} Y_1 \\ Y_2 \\ Y_3 \end{pmatrix} = \begin{pmatrix} 0 & 0 & 0 \\ \beta_{21} & 0 & 0 \\ \beta_{31} & \beta_{32} & 0 \end{pmatrix} \begin{pmatrix} Y_1 \\ Y_2 \\ Y_3 \end{pmatrix} + \begin{pmatrix} \gamma_{11} & \gamma_{12} & \gamma_{13} \\ \gamma_{21} & \gamma_{22} & \gamma_{23} \\ \gamma_{31} & \gamma_{32} & \gamma_{33} \end{pmatrix} \begin{pmatrix} X_1 \\ X_2 \\ X_3 \end{pmatrix} + \begin{pmatrix} \epsilon_1 \\ \epsilon_2 \\ \epsilon_3 \end{pmatrix} \tag{10.2}$$

Similarly, the path equations for Figure 10.3 (non-recursive model) are

$$\begin{aligned} Y_1 &= \gamma_{11}X_1 + \gamma_{12}X_2 + \gamma_{13}X_3 + \epsilon_1 \\ Y_2 &= \beta_{21}Y_1 + \beta_{23}Y_3 + \gamma_{21}X_1 + \gamma_{22}X_2 + \gamma_{23}X_3 + \epsilon_2 \\ Y_3 &= \beta_{31}Y_1 + \beta_{32}Y_2 + \gamma_{31}X_1 + \gamma_{32}X_2 + \gamma_{33}X_3 + \epsilon_3 \end{aligned} \tag{10.3}$$

In matrix form

$$\begin{pmatrix} Y_1 \\ Y_2 \\ Y_3 \end{pmatrix} = \begin{pmatrix} 0 & 0 & 0 \\ \beta_{21} & 0 & \beta_{23} \\ \beta_{31} & \beta_{32} & 0 \end{pmatrix} \begin{pmatrix} Y_1 \\ Y_2 \\ Y_3 \end{pmatrix} + \begin{pmatrix} \gamma_{11} & \gamma_{12} & \gamma_{13} \\ \gamma_{21} & \gamma_{22} & \gamma_{23} \\ \gamma_{31} & \gamma_{32} & \gamma_{33} \end{pmatrix} \begin{pmatrix} X_1 \\ X_2 \\ X_3 \end{pmatrix} + \begin{pmatrix} \epsilon_1 \\ \epsilon_2 \\ \epsilon_3 \end{pmatrix}$$

$$\text{Or, } \mathbf{Y}_{q\times 1} = \boldsymbol{\beta}_{q\times q}\mathbf{Y}_{q\times 1} + \boldsymbol{\Gamma}_{q\times p}\mathbf{X}_{p\times 1} + \boldsymbol{\epsilon}_{q\times 1} \tag{10.4}$$

Thus, $\boldsymbol{\beta}$ is the matrix of coefficients of the endogenous variables and $\boldsymbol{\Gamma}$ is the matrix of coefficients of the exogenous variables. In the example above (Figure 10.2b), $p = q = 3$. Also note that Equation 10.4 is the general equation of a path model. For recursive path model $\boldsymbol{\beta}_{q\times q}$ (in Equation 10.4) will be lower triangular, which is not true for a non-recursive path model. For example, the difference between the path models in Figures 10.2b (recursive) and 10.3 (non-recursive) is reflected in the Y_2 and Y_3 Equations 10.1 and 10.3. Note that the term $\beta_{23}Y_3$ and $\beta_{32}Y_2$ are added in Equation 10.3 for Y_2 and Y_3, respectively, which makes the $\boldsymbol{\beta}$ matrix non-triangular.

In addition, the path model allows covariation among the exogenous (X) variables. For both the recursive and non-recursive models, if there are p number of X variables, the covariance ($\boldsymbol{\phi}$) and correlation ($\boldsymbol{\rho}$) matrices are

$$\boldsymbol{\Phi} = \begin{pmatrix} \phi_{11} & \phi_{21} & \cdots & \phi_{p1} \\ \phi_{21} & \phi_{22} & \cdots & \phi_{p2} \\ \cdots & \cdots & \cdots & \cdots \\ \phi_{p1} & \phi_{p2} & \cdots & \phi_{pp} \end{pmatrix}, \quad \boldsymbol{\rho} = \begin{pmatrix} \rho_{11} & \rho_{21} & \cdots & \rho_{p1} \\ \rho_{21} & \rho_{22} & \cdots & \rho_{p2} \\ \cdots & \cdots & \cdots & \cdots \\ \rho_{p1} & \rho_{p2} & \cdots & \rho_{pp} \end{pmatrix}, \text{ where } \rho_{jk} = \frac{\phi_{jk}}{\sqrt{\phi_{jj}\phi_{kk}}}.$$

Another important matrix for path model is the covariance matrix ($\boldsymbol{\Psi}$) of the error (disturbance) variables, given below. Note that for recursive path model, $\boldsymbol{\Psi}$ is diagonal.

$$\boldsymbol{\Psi} = \begin{pmatrix} \psi_{11} & \psi_{21} & \cdots & \psi_{q1} \\ \psi_{21} & \psi_{22} & \cdots & \psi_{q2} \\ \cdots & \cdots & \cdots & \cdots \\ \psi_{q1} & \psi_{q2} & \cdots & \psi_{qq} \end{pmatrix}$$

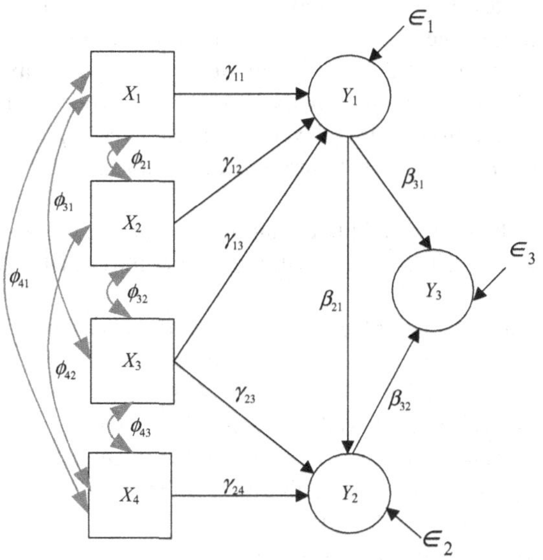

FIGURE 10.4 Example 10.1 path diagram. There is no reciprocal causation among the DVs indicating that the path model is a recursive model.

It is to be further mentioned that in traditional path equations, only X terms are used for both exogenous and endogenous variables. That means, the endogenous variables Y_1, Y_2, and Y_3 may be denoted by X_4, X_5, and X_6, respectively. We are using both X and Y terms to keep in the parity with regression modeling (see Chapter 9).

Example 10.1

Consider the following causal relations. Develop path diagram and path equations. Also comment on the path model.

Causal relationships

Explanatory variables	Relations	Dependent variables
X_1, X_2, and X_3	⟶	Y_1
X_3, X_4, and Y_1	⟶	Y_2
Y_1 and Y_2	⟶	Y_3

Solution:

The path diagram is shown in Figure 10.4.
The path equations are

$$Y_1 = \gamma_{11}X_1 + \gamma_{12}X_2 + \gamma_{13}X_3 + \epsilon_1$$
$$Y_2 = \beta_{21}Y_1 + \gamma_{23}X_3 + \gamma_{24}X_4 + \epsilon_2$$
$$Y_3 = \beta_{31}Y_1 + \beta_{32}Y_2 + \epsilon_3$$

The path model is recursive as no reciprocal causation is involved in the modeling.

10.2 ASSUMPTIONS

A path model involves several assumptions. The assumptions of multivariate linear models such as multivariate multiple linear regression (assuming multiple equations simultaneously) and multiple linear regression (with single equation) are applicable to path equations. Apart from the methodological assumptions of multivariate analyses, the special structure of path models explicitly requires some more assumptions that enables the parameters of the model estimable. We will first describe below the assumptions of a *recursive path model* having q number of endogenous variables (Y) (Heise, 1969) and the assumptions that are relaxed in a *non-recursive path model*. Readers may also consult Duncan (1975) and Berry (1984) for further explanations.

1. The path model involves linear, additive relationships among the set of variables.
2. The causal laws are established scientifically to specify the causal ordering (priorities) among the variables.
3. There is no reciprocal causation or feedback loops (this is relaxed in non-recursive path models).
4. Disturbance (residual) variables ($\in$) are uncorrelated with each other and with the predetermined variables (PV), i.e., $E(\in_j \in_k) = 0 \, and \, E(\in_j \, PV) = 0, j = 1,2,...,q$. We will define predetermined variables in Section 10.3. In non-recursive models, this assumption is weakened. Some of the Y variables in the right-hand side (RHS) of the path equation may be correlated with the disturbance (residual) variables ($\in$), i.e., $E(\in_j \, Y_k) \neq 0, j = 1,2,...,q, \, k = 1,2,\ldots,q$ and $j \neq k$.
5. Each of the disturbance (residual) variable ($\in_j$) is normally distributed with mean zero and variance σ_j^2, i.e., $E(\in_j) = 0$ and $V(\in_j) = \sigma_j^2$, where V stands for variance.
6. Each of the disturbance (residual) variable ($\in_j$) has equal variance (σ_j^2) across all settings of causal/explanatory variables. This is the homoscedasticity assumption in linear multiple regression.
7. Other multivariate assumptions like sample observations are independent and identically distributed are to be met.

10.3 ESTIMATION OF PARAMETERS FOR RECURSIVE MODEL

10.3.1 Normal Equation Approach

From Equation 10.4, we see that the parameters $\boldsymbol{\beta}$, $\boldsymbol{\Gamma}$, and error $\in$ are unknown, which are to be estimated from sample data. For a recursive path model, $\boldsymbol{\beta}$ is triangular. The steps to estimate the parameters are given below.

1. For each of the path equations, find out the predetermined variables.
2. For each of the path equations, obtain the normal equations.
3. Check the identification conditions.
4. Estimate the parameters using a suitable method (e.g., ordinary least squares).

Let us consider the recursive model in Equation 10.1. There are three endogenous (Y) and three exogenous (X) variables. We will consider the third equation, given below, for demonstration.

$$Y_3 = \beta_{31}Y_1 + \beta_{32}Y_2 + \gamma_{31}X_1 + \gamma_{32}X_2 + \gamma_{33}X_3 + \in_3 \tag{10.5}$$

Here, we further assume that all the variables are standardized, i.e., each of the X and Y variables has zero mean and one variance. So, $E(X_j) = 0$, $E(X_j^2) = 1$, and $E(Y_j) = 0$, $E(Y_j^2) = 1$, $j = 1,2,3$.

Now, we define predetermined variable (PV). For any path equation (say Equation 10.5), there could be a set of variables which are determined independently and prior to the determination of the endogenous variable (say, Y_3) of that equation. These variables are called predetermined variables (PVs). So, in Equation 10.5, the variables X_1, X_2, X_3, Y_1, and Y_2 are the PVs for Y_3. Please note that the exogenous variables are always the PV for any endogenous variables. Further, some of the endogenous variables, which occur prior to the causal ordering with respect to the endogenous variable of interest, are also predetermined variables (PVs) for that endogenous variable. In Equation

10.5, the endogenous variables Y_1 and Y_2 do therefore qualify as the PV for Y_3. Note that for Y_2 in Equation 10.1, Y_3 is not a PV of Y_2, as Y_3 does not precede the causal ordering of Y_2. Using the above logic, for Y_1 in Equation 10.1, only X_1, X_2, and X_3 are the PV of Y_1. Note that the PVs are uncorrelated with the error variable, i.e., $E(PV \times \epsilon_j) = 0$.

Once the PVs of a path equation are known, the normal equations can be obtained by multiplying each of its PVs with the equation and then by taking the expectation (i.e., method of moments). So, if there are K number of PVs for a path equation, we will get K number of normal equations, which is detailed below for Equation 10.5.

Multiplying Equation 10.5 with X_1 and then taking expectation, we get

$$
\begin{aligned}
X_1Y_3 &= \beta_{31}X_1Y_1 + \beta_{32}X_1Y_2 + \gamma_{31}X_1^2 + \gamma_{32}X_1X_2 + \gamma_{33}X_1X_3 + X_1\,\epsilon_3 \\
E(X_1Y_3) &= E(\beta_{31}X_1Y_1) + E(\beta_{32}X_1Y_2) + E(\gamma_{31}X_1^2) + E(\gamma_{32}X_1X_2) + E(\gamma_{33}X_1X_3) + E(X_1\,\epsilon_3) \\
&= \beta_{31}E(X_1Y_1) + \beta_{32}E(X_1Y_2) + \gamma_{31}E(X_1^2) + \gamma_{32}E(X_1X_2) + \gamma_{33}E(X_1X_3) + E(X_1\,\epsilon_3)
\end{aligned}
$$

Denoting $E(X_jY_k) = \rho_{X_jY_k} = \rho_{Y_kX_j}$, $E(X_jX_k) = \rho_{jk} = \rho_{kj}$, and as $E(X_j\,\epsilon_k) = 0$, we get

$$\rho_{Y_3X_1} = \beta_{31}\rho_{Y_1X_1} + \beta_{32}\rho_{Y_2X_1} + \gamma_{31} + \gamma_{32}\rho_{21} + \gamma_{33}\rho_{31} \tag{10.6}$$

Similarly, multiplying Equation 10.5 with X_2, X_3, Y_1, and Y_2, respectively and then taking expectation, we get

$$\rho_{Y_3X_2} = \beta_{31}\rho_{Y_1X_2} + \beta_{32}\rho_{Y_2X_2} + \gamma_{31}\rho_{21} + \gamma_{32} + \gamma_{33}\rho_{32} \tag{10.7}$$

$$\rho_{Y_3X_3} = \beta_{31}\rho_{Y_1X_3} + \beta_{32}\rho_{Y_2X_3} + \gamma_{31}\rho_{31} + \gamma_{32}\rho_{32} + \gamma_{33} \tag{10.8}$$

$$\rho_{Y_3Y_1} = \beta_{31} + \beta_{32}\rho_{Y_2Y_1} + \gamma_{31}\rho_{Y_1X_1} + \gamma_{32}\rho_{Y_1X_2} + \gamma_{33}\rho_{Y_1X_3} \tag{10.9}$$

$$\rho_{Y_3Y_2} = \beta_{31}\rho_{Y_2Y_1} + \beta_{32} + \gamma_{31}\rho_{Y_2X_1} + \gamma_{32}\rho_{Y_2X_2} + \gamma_{33}\rho_{Y_2X_3} \tag{10.10}$$

Equations 10.6–10.10 are the five normal equations for the path equation 10.5. Here, the number of PVs is five (K = 5) and hence, we get five normal equations. In matrix form, Equations 10.6–10.10, after little rearranging, can be written as

$$
\begin{pmatrix} \rho_{Y_3Y_1} \\ \rho_{Y_3Y_2} \\ \rho_{Y_3X_1} \\ \rho_{Y_3X_2} \\ \rho_{Y_3X_3} \end{pmatrix}
=
\begin{pmatrix}
1 & \rho_{Y_2Y_1} & \rho_{Y_1X_1} & \rho_{Y_1X_2} & \rho_{Y_1X_3} \\
\rho_{Y_2Y_1} & 1 & \rho_{Y_2X_1} & \rho_{Y_2X_2} & \rho_{Y_2X_3} \\
\rho_{Y_1X_1} & \rho_{Y_2X_1} & 1 & \rho_{21} & \rho_{31} \\
\rho_{Y_1X_2} & \rho_{Y_2X_2} & \rho_{21} & 1 & \rho_{32} \\
\rho_{Y_1X_3} & \rho_{Y_2X_3} & \rho_{31} & \rho_{32} & 1
\end{pmatrix}
\begin{pmatrix} \beta_{31} \\ \beta_{32} \\ \gamma_{31} \\ \gamma_{32} \\ \gamma_{33} \end{pmatrix}
\tag{10.11}
$$

$$\text{Or, } \boldsymbol{\rho} \quad = \quad \boldsymbol{\Omega} \quad \times \quad \boldsymbol{\Pi}$$

Note that $\boldsymbol{\Omega}$ in Equation 10.11 is symmetric. The correlation coefficients $\boldsymbol{\rho}, \boldsymbol{\Omega}$ and the path coefficients $\boldsymbol{\Pi}$ are population parameters and hence, unknown. $\boldsymbol{\rho}, \boldsymbol{\Omega}$ can be estimated using sample data and then, $\boldsymbol{\Pi}$ can be estimated solving Equation 10.11.

However, before solving Equation 10.11, we need to test whether it is identifiable or not. The counting rules for identification are as follows (Wonnacott and Wonnacott, 1970). Let K_1 (<K) be the number of predetermined variables (PVs) excluded from a path equation and q_1 is the number of endogenous variables in the right-hand side (RHS) of that equation. Then,

1. If $K_1 = q_1$, the path equation is just-identified.
2. If $K_1 > q_1$, the path equation is over-identified.
3. If $K_1 < q_1$, the path equation is under-identified.

A similar rule is presented in Maddala (2007), where the *a priori* identification of the PVs for an equation is not needed. Let q be the number of endogenous variables (*Y*) in the path model and m be the number of variables (both endogenous and exogenous) missing from the equation under consideration. Then,

1. If $m = q - 1$, the path equation is just-identified.
2. If $m > q - 1$, the path equation is over-identified.
3. If $m < q - 1$, the path equation is under-identified.

The above condition is called order condition, which is only necessary. The sufficient condition is known as rank condition (we will discuss under non-recursive model). Further, if condition 3 satisfies, the path parameters cannot be estimated. Interestingly, a fully recursive model is either just-identified or over-identified.

Now, as Equation 10.11 is just identified, the unknown path parameters, $\mathbf{\Pi}$ can be obtained as

$$\hat{\mathbf{\Pi}} = \hat{\mathbf{\Omega}}^{-1}\hat{\boldsymbol{\rho}} \tag{10.12}$$

The hat symbol (∧) in Equation 10.12 indicates estimate. The rank condition satisfies that $\hat{\mathbf{\Omega}}^{-1}$ exists. The solution obtained in Equation 10.12 will exactly be the same as derived from using OLS estimate of Equation 10.5. The OLS estimate is

$$\hat{\mathbf{\Pi}} = (\mathbf{Z}^T\mathbf{Z})^{-1}\mathbf{Z}^T\mathbf{y} \tag{10.13}$$

In Equation 10.13, $\mathbf{Z}$ is a $n \times K$ data matrix comprising n observations on the K number of PVs, and $\mathbf{y}$ is a $n \times 1$ data vector on the endogenous dependent variable in the *RHS*. For Equation 10.5, $\mathbf{Z}$ in Equation 10.13 comprises n observations on Y_1, Y_2, X_1, X_2 and X_3.

Example 10.2

Consider the path model in Figure 10.2b. The sample correlation matrix is given below. Estimate the path coefficients for Y_3.

	X_1	X_2	X_3	Y_1	Y_2	Y_3
X_1	1					
X_2	0.87	1				
X_3	−0.26	−0.30	1			
Y_1	−0.50	−0.65	0.28	1		
Y_2	−0.20	−0.20	0.08	0.70	1	
Y_3	0.45	0.25	−0.30	−0.50	−0.40	1

Solution:

The path equation for Y_3 is given in Equation 10.5 and its normal equations are given in Equations 10.6–10.10. Further, the relationships among correlation coefficients and path coefficients are given in Equation 10.11. Now, using the relation for just identified path equation in Equation 10.12, the path coefficients for Y_3 are computed below.

$$\Pi = \begin{bmatrix} 1 & 0.70 & -0.50 & -0.65 & 0.28 \\ 0.70 & 1 & -0.20 & -0.20 & 0.08 \\ -0.50 & -0.20 & 1 & 0.87 & -0.26 \\ -0.65 & -0.20 & 0.87 & 1 & -0.30 \\ 0.28 & 0.08 & -0.26 & -0.30 & 1 \end{bmatrix}^{-1} \begin{bmatrix} -0.50 \\ -0.40 \\ 0.45 \\ 0.25 \\ -0.30 \end{bmatrix}$$

$$= \begin{bmatrix} 4.85 & -2.91 & -1.62 & 3.86 & -0.39 \\ -2.91 & 2.8 & 1.08 & -2.21 & 0.21 \\ -1.62 & 1.08 & 4.66 & -4.86 & 0.12 \\ 3.86 & -2.21 & -4.86 & 7.3 & 0.02 \\ -0.39 & 0.21 & 0.12 & 0.02 & 1.13 \end{bmatrix} \begin{bmatrix} -0.50 \\ -0.40 \\ 0.45 \\ 0.25 \\ -0.30 \end{bmatrix} = \begin{bmatrix} -0.91 \\ 0.21 \\ 1.22 \\ -1.41 \\ -0.17 \end{bmatrix}$$

If the over-identified condition satisfies, the abovementioned normal equation procedure yields multiple solutions as there will be K!/(K-P)! choices to estimate the parameters, where K is the number of normal equations and P is the number of parameters to be estimated. As pointed out in Anderson (1978), each solution will generally differ in their values from the other and under such conditions the best parameter estimates could be obtained by averaging the individual estimates, a procedure that was proposed by Boudon (1968). However, this results in inefficient estimates.

For example, suppose the effects of the variables X_2 (γ_{32}) and Y_1 (β_{31}) in Y_3 in Equation 10.5 are negligible or not supported by literature or intuitively less obvious. Hence, the researcher wants to exclude those variables from Y_3 equation, but not from the path model considering their relations with other variables in the path model. Essentially then, the path arrows (in Figure 10.2b) linking X_2 and Y_3 (effect coefficient γ_{32}) and Y_1 and Y_3 (effect coefficient β_{31}) will be removed. The resulting path equation is

$$Y_3 = \beta_{32}Y_2 + \gamma_{31}X_1 + \gamma_{33}X_3 + \epsilon_3 \tag{10.14}$$

Equation 10.14 is over-identified with three unknown path coefficients and five estimating equations (as it has five PVs). So, we can generate 5!/(5–3)!=10 different estimates for each the three path coefficients and the residual coefficient. Which one do you chose? By Boudon's procedure, the best estimate is the average of the ten estimates (although it is inefficient). Goldberger (1970) proposed to use those PVs that appear in the equation. Then, the number of PVs to use for Equation 10.14 is three, which results into three normal equations. Interestingly, the results will match with that of the OLS estimates. The general rule for a fully recursive model, as pointed out by Duncan (1975), is "estimate the coefficients in each equation by OLS regression of the dependent variable on the predetermined variables included in that equation."

Example 10.3

Consider Equation 10.14 and the sample correlation matrix given in Example 10.2. Estimate the path coefficients for Y_3 by (i) Boudon's approach, and (ii) Goldberger's approach.

Solution:

Here, based on the normal equations (10.6–10.10) for Y_3 and sample correlation matrix given in Example 10.2, we can obtain the following set of normal equations for the path equation 10.14.

$$\rho_{Y_3X_1} = \beta_{32}\rho_{Y_2X_1} + \gamma_{31} + \gamma_{33}\rho_{31}$$
$$\rho_{Y_3X_2} = \beta_{32}\rho_{Y_2X_2} + \gamma_{31}\rho_{21} + \gamma_{33}\rho_{32}$$

TABLE 10.1
Solutions for The Unknown Parameters for Example 10.3

Comb. of Equations	β_{32}	γ_{31}	γ_{33}
(i, ii, iii)	6.56	1.66	–0.39
(i, ii, iv)	–1.03	0.84	2.28
(i, ii, v)	–0.38	0.91	2.05
(i, iii, iv)	–0.41	0.32	–0.18
(i, iii, v)	–0.32	0.34	–0.19
(i, iv, v)	–0.31	0.23	–0.61
(ii, iii, iv)	–0.57	0.08	–0.23
(ii, iii, v)	–0.36	0.12	–0.24
(ii, iv, v)	–0.34	–0.29	–1.45
(iii, iv, v)	–0.28	0.52	–0.14

$$\rho_{Y_3X_3} = \beta_{32}\rho_{Y_2X_3} + \gamma_{31}\rho_{31} + \gamma_{33}$$
$$\rho_{Y_3Y_1} = \beta_{32}\rho_{Y_2Y_1} + \gamma_{31}\rho_{Y_1X_1} + \gamma_{33}\rho_{Y_1X_3}$$
$$\rho_{Y_3Y_2} = \beta_{32} + \gamma_{31}\rho_{Y_2X_1} + \gamma_{33}\rho_{Y_2X_3}$$

Now by putting the estimated correlation coefficients from Example 10.2 into the above equations, the following equations are obtained.

$$(-0.20)\beta_{32} + \gamma_{31} + (-0.26)\gamma_{33} = 0.45 \ldots\ldots\ldots\ldots (i)$$
$$(-0.20)\beta_{32} + (0.87)\gamma_{31} + (-0.30)\gamma_{33} = 0.25 \ldots\ldots\ldots (ii)$$
$$(0.08)\beta_{32} + (-0.26)\gamma_{31} + \gamma_{33} = -0.30 \ldots\ldots\ldots\ldots (iii)$$
$$(0.70)\beta_{32} + (-0.50)\gamma_{31} + (0.28)\gamma_{33} = -0.50 \ldots\ldots\ldots (iv)$$
$$\beta_{32} + (-0.20)\gamma_{31} + (0.08)\gamma_{33} = -0.40 \ldots\ldots\ldots\ldots (v)$$

There are five equations and three unknowns. So, we can obtain ${}^5C_3 = 10$ combinations of equations. By solving each of the combinations, we will obtain the values of the unknowns. As there are ten such combinations, we get ten solutions for the unknown parameters. The solutions are given in Table 10.1.

Using Boudon's approach, the solutions are obtained based on average and these are

$$\beta_{32} = 0.26;\ \gamma_{31} = 0.47;\ \gamma_{33} = 0.09$$

Note that the solutions differ from that obtained in Example 10.2.

Now, the solution using Goldberg's approach is given below.

The path equation is

$$Y_3 = \beta_{32}Y_2 + \gamma_{31}X_1 + \gamma_{33}X_3 + \epsilon_3$$

So the normal equations can be obtained as

$$\rho_{Y_3X_1} = \beta_{32}\rho_{Y_2X_1} + \gamma_{31} + \gamma_{33}\rho_{31}$$
$$\rho_{Y_3X_3} = \beta_{32}\rho_{Y_2X_3} + \gamma_{31}\rho_{31} + \gamma_{33}$$
$$\rho_{Y_3Y_2} = \beta_{32} + \gamma_{31}\rho_{X_1Y_2} + \gamma_{33}\rho_{X_3Y_2}$$

In matrix form,

Therefore,

$$\Pi = \begin{bmatrix} \beta_{32} \\ \gamma_{31} \\ \gamma_{33} \end{bmatrix} = \begin{bmatrix} 1 & \rho_{Y_2X_1} & \rho_{Y_2X_3} \\ \rho_{Y_2X_1} & 1 & \rho_{31} \\ \rho_{Y_2X_3} & \rho_{31} & 1 \end{bmatrix}^{-1} \begin{bmatrix} \rho_{Y_3Y_2} \\ \rho_{Y_3X_1} \\ \rho_{Y_3X_3} \end{bmatrix}$$

$$\therefore \begin{bmatrix} \beta_{32} \\ \gamma_{31} \\ \gamma_{33} \end{bmatrix} = \begin{bmatrix} 1 & -0.20 & 0.08 \\ -0.20 & 1 & -0.26 \\ 0.08 & -0.26 & 1 \end{bmatrix}^{-1} \begin{bmatrix} -0.40 \\ 0.45 \\ -0.30 \end{bmatrix}$$

$$= \begin{bmatrix} 1.04 & 0.2 & -0.03 \\ 0.2 & 1.11 & 0.27 \\ -0.03 & 0.27 & 1.07 \end{bmatrix} \begin{bmatrix} -0.40 \\ 0.45 \\ -0.30 \end{bmatrix} = \begin{bmatrix} -0.32 \\ 0.34 \\ -0.19 \end{bmatrix}$$

Again, note that the solutions differ from that obtained in example 10.2.

10.3.2 Reduced Form Approach

Let us consider again the path equations (Equation 10.1) for the recursive path model in Figure 10.2b. The Y_1 equation is a function of exogenous variables (X) only, whereas, Y_2 and Y_3 are not. The reduced form approach converts each of the endogenous variables (Y) as a function of exogenous variables (X) only. With little algebraic manipulation we can do the same.

As Y_l is already a function of exogenous variables only, the reduced form is (note: we use different symbols for reduced form coefficients and error term).

$$Y_1 = a_{11}X_1 + a_{12}X_2 + a_{13}X_3 + \theta_1$$

where, $a_{11} = \gamma_{11}$, $a_{12} = \gamma_{12}$, $a_{13} = \gamma_{13}$, and $\theta_1 = \epsilon_1$.

The reduced form equation for Y_2 is

$$\begin{aligned} Y_2 &= \beta_{21}Y_1 + \gamma_{21}X_1 + \gamma_{22}X_2 + \gamma_{23}X_3 + \epsilon_2 \\ &= \beta_{21}(\gamma_{11}X_1 + \gamma_{12}X_2 + \gamma_{13}X_3 + \epsilon_1) + \gamma_{21}X_1 + \gamma_{22}X_2 + \gamma_{23}X_3 + \epsilon_2 \\ &= \gamma_{21}X_1 + \beta_{21}\gamma_{11}X_1 + \gamma_{22}X_2 + \beta_{21}\gamma_{12}X_2 + \gamma_{23}X_3 + \beta_{21}\gamma_{13}X_3 + \beta_{21}\epsilon_1 + \epsilon_2 \\ &= (\gamma_{21} + \beta_{21}\gamma_{11})X_1 + (\gamma_{22} + \beta_{21}\gamma_{12})X_2 + (\gamma_{23} + \beta_{21}\gamma_{13})X_3 + \beta_{21}\epsilon_1 + \epsilon_2 \\ &= a_{21}X_1 + a_{22}X_2 + a_{23}X_3 + \theta_2 \end{aligned}$$

where,

$a_{21} = \gamma_{21} + \beta_{21}\gamma_{11}$

$a_{22} = \gamma_{22} + \beta_{21}\gamma_{12}$

$a_{23} = \gamma_{23} + \beta_{21}\gamma_{13}$

$\theta_2 = \beta_{21}\epsilon_1 + \epsilon_2$

Similarly, the reduced form equation for Y_3 is

$$\begin{aligned}
Y_3 &= \beta_{31}Y_1 + \beta_{32}Y_2 + \gamma_{31}X_1 + \gamma_{32}X_2 + \gamma_{33}X_3 + \epsilon_3 \\
&= \beta_{31}(\gamma_{11}X_1 + \gamma_{12}X_2 + \gamma_{13}X_3 + \epsilon_1) \\
&\quad + \beta_{32}((\gamma_{21} + \beta_{21}\gamma_{11})X_1 + (\gamma_{22} + \beta_{21}\gamma_{12})X_2 + (\gamma_{23} + \beta_{21}\gamma_{13})X_3 + \beta_{21}\epsilon_1 + \epsilon_2)) \\
&\quad + \gamma_{31}X_1 + \gamma_{32}X_2 + \gamma_{33}X_3 + \epsilon_3 \\
&= (\gamma_{31} + \beta_{31}\gamma_{11} + \beta_{32}\gamma_{21} + \beta_{32}\beta_{21}\gamma_{11})X_1 + (\gamma_{32} + \beta_{31}\gamma_{12} + \beta_{32}\gamma_{22} + \beta_{32}\beta_{21}\gamma_{12})X_2 \\
&\quad + (\gamma_{33} + \beta_{31}\gamma_{13} + \beta_{32}\gamma_{23} + \beta_{32}\beta_{21}\gamma_{13})X_3 + (\beta_{32}\beta_{21}\epsilon_1 + \beta_{32}\epsilon_2 + \epsilon_3) \\
&= a_{31}X_1 + a_{32}X_2 + a_{33}X_3 + \theta_3
\end{aligned}$$

where,

$$\begin{aligned}
a_{31} &= \gamma_{31} + \beta_{31}\gamma_{11} + \beta_{32}\gamma_{21} + \beta_{32}\beta_{21}\gamma_{11} \\
a_{32} &= \gamma_{32} + \beta_{31}\gamma_{12} + \beta_{32}\gamma_{22} + \beta_{32}\beta_{21}\gamma_{12} \\
a_{33} &= \gamma_{33} + \beta_{31}\gamma_{13} + \beta_{32}\gamma_{23} + \beta_{32}\beta_{21}\gamma_{13} \\
\theta_3 &= \beta_{32}\beta_{21}\epsilon_1 + \beta_{32}\epsilon_2 + \epsilon_3
\end{aligned}$$

In matrix form, these three reduced form of equations can be written as

$$\begin{pmatrix} Y_1 \\ Y_2 \\ Y_3 \end{pmatrix} = \begin{pmatrix} a_{11} & a_{12} & a_{13} \\ a_{21} & a_{22} & a_{23} \\ a_{31} & a_{32} & a_{33} \end{pmatrix} \begin{pmatrix} X_1 \\ X_2 \\ X_3 \end{pmatrix} + \begin{pmatrix} \theta_1 \\ \theta_2 \\ \theta_3 \end{pmatrix} \tag{10.15}$$

For q number of endogenous (Y) and p number of exogenous (X) variables, the generic reduced form of equations 10.15, in matrix form, is

$$\mathbf{Y}_{q\times 1} = \mathbf{A}_{q\times p}\,\mathbf{X}_{p\times 1} + \boldsymbol{\theta}_{q\times 1} \tag{10.16}$$

It can be proven that the exogenous variables (X) are uncorrelated with the reduced form errors (θ), i.e., $E(X_j\,\theta_k) = 0, j = 1, 2, \ldots, p, k = 1, 2, \ldots, q$. For example, let us prove $E(X_1\,\theta_k) = 0, k = 1, 2, 3$ for Equation 10.16. Using assumption 4 discussed earlier,

$$\begin{aligned}
E(X_1\theta_1) &= E(X_1\,\epsilon_1) = 0 \\
E(X_1\theta_2) &= E(X_1(\beta_{21}\epsilon_1 + \epsilon_2)) = \beta_{21}E(X_1\,\epsilon_1) + E(X_1\,\epsilon_2) = 0 \\
E(X_1\theta_3) &= E(X_1(\beta_{32}\beta_{21}\epsilon_1 + \beta_{32}\epsilon_2 + \epsilon_3)) = \beta_{32}\beta_{21}E(X_1\,\epsilon_1) + \beta_{32}E(X_1\,\epsilon_2) + E(X_1\,\epsilon_3) = 0
\end{aligned}$$

Further to note that the reduced form errors (θ), unlike the path form errors (ϵ), are not uncorrelated among themselves (Duncan, 1975).

Again, let us rewrite Equation 10.4 in the form below.

$$\begin{aligned}
&(\mathbf{I} - \boldsymbol{\beta})\boldsymbol{Y} = \boldsymbol{\Gamma X} + \epsilon, \text{ or considering } \mathbf{B} = (\mathbf{I} - \boldsymbol{\beta}) \\
&\mathbf{B}\boldsymbol{Y} = \boldsymbol{\Gamma X} + \epsilon
\end{aligned} \tag{10.17}$$

For example, for Equation 10.1, the $\mathbf{\Gamma}$ and $\mathbf{B}$ matrices are

$$\mathbf{\Gamma} = \begin{pmatrix} \gamma_{11} & \gamma_{12} & \gamma_{13} \\ \gamma_{21} & \gamma_{22} & \gamma_{23} \\ \gamma_{31} & \gamma_{32} & \gamma_{33} \end{pmatrix}, \quad \mathbf{B} = \begin{pmatrix} 1 & 0 & 0 \\ -\beta_{21} & 1 & 0 \\ -\beta_{31} & -\beta_{32} & 1 \end{pmatrix}$$

If $\mathbf{B}^{-1}$ exists, post-multiplying $\mathbf{B}^{-1}$ with Equation 10.17, we get

$$\mathbf{B}^{-1}\mathbf{B}Y = \mathbf{B}^{-1}\boldsymbol{\Gamma}X + \mathbf{B}^{-1}\epsilon, \text{ or}$$

$$Y = \mathbf{B}^{-1}\boldsymbol{\Gamma}X + \mathbf{B}^{-1}\epsilon \tag{10.18}$$

Comparing Equations 10.16 and 10.18, we get

$$\mathbf{A} = \mathbf{B}^{-1}\boldsymbol{\Gamma} \tag{10.19}$$

$$\boldsymbol{\theta} = \mathbf{B}^{-1}\epsilon \tag{10.20}$$

and,

$$E(\boldsymbol{\theta}) = E(\mathbf{B}^{-1}\epsilon) = 0$$

$$V(\boldsymbol{\theta}) = V(\mathbf{B}^{-1}\epsilon) = (\mathbf{B}^{-1})^T E(\epsilon\epsilon^T)(\mathbf{B}^{-1}) = (\mathbf{B}^{-1})^T \boldsymbol{\Psi}(\mathbf{B}^{-1}) = \boldsymbol{\Psi}_\theta \tag{10.21}$$

Now, if we collect n observations on X and Y, and the data matrices are denoted as $\mathbf{X}_{n\times p}$ and $\mathbf{Y}_{n\times q}$, then, the OLS estimate of $\mathbf{A}$ is

$$\hat{\mathbf{A}} = (\mathbf{X}^T\mathbf{X})^{-1}\mathbf{X}^T\mathbf{Y}$$

$$\hat{\boldsymbol{\Psi}}_\theta = \frac{1}{n-p}\hat{\boldsymbol{\theta}}^T\hat{\boldsymbol{\theta}} \tag{10.22}$$

However, the estimated $\hat{\mathbf{A}}$ will not enable us to uniquely estimate $\mathbf{B}$ and $\boldsymbol{\Gamma}$. Instead, any general solution for $\boldsymbol{\Gamma}$ will involve elements $\mathbf{B}$ which are unknown, and vice versa. Carmines (1990) pointed out that any arbitrary non-singular transformation of the reduced form (Equation 10.18) will generate an observationally equivalent system of equations. So, given $\mathbf{X}_{n\times p}$ and $\mathbf{Y}_{n\times p}$ data, there could be an unlimited number of path systems, resulting in an under-identified system of equations. Fortunately, in the recursive path model, the transformation is restricted to identity matrix (I) owing to the two restrictions (derived from its assumptions), (i) the diagonal elements of $\mathbf{B}$ are one, and (ii) the error covariance matrix ($\boldsymbol{\Psi}$) is diagonal. Hence, every recursive model is either just-identified or over-identified. The reduced form approach is also known as *indirect least squares* (ILS) method, as the path coefficients are estimated indirectly from OLS estimates of the reduced form coefficients.

Duncan (1975) provided the following steps to estimate $\mathbf{B}$ and $\boldsymbol{\Gamma}$ using $\boldsymbol{A}$.

1. Multiply through each reduced form equation by each exogenous variable (X) and take expectation.
2. Multiply through each reduced form equation by each endogenous variable (Y) and take expectation.
3. Multiply through each reduced form equation by each reduced form disturbance variable (θ) and take expectation.
4. Using the results from steps 1–3, algebraically derive the equation for each of the variances of the reduced form disturbances as a function of the variances and covariances of the observed variables (i.e., X and Y), and the elements of $\mathbf{A}$.
5. Using sample estimates of the variances and covariances of the observed variables (i.e., X and Y) and the elements of $\mathbf{A}$, estimate the variances of the reduced form disturbances.
6. Compute path coefficients from the information obtained above in steps 4 and 5.

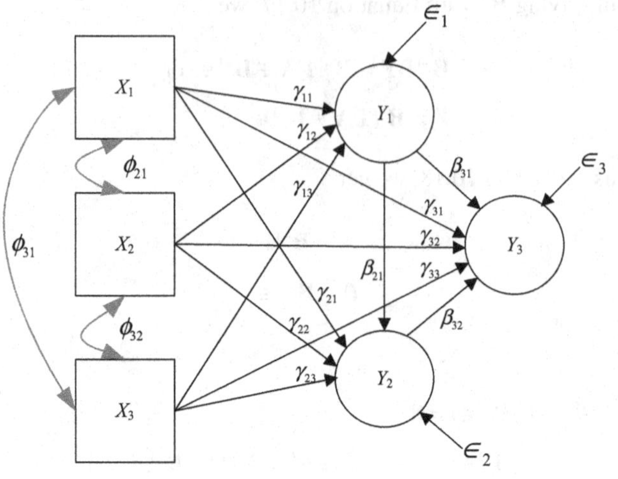

FIGURE 10.5 Example 10.4 original form path model.

Example 10.4

Consider the recursive path diagram in Figure 10.2b. Develop the reduced form path diagram using the reduced form coefficients (A) given in Equation 10.16. Show that the number of path parameters is equal to the number of reduced form parameters. Where do the parameters differ?

Solution:

The original form of the path model is shown in Figure 10.5.

The path equations for the original path diagram in matrix form (from Equation 10.2) are shown below.

$$\begin{pmatrix} Y_1 \\ Y_2 \\ Y_3 \end{pmatrix} = \begin{pmatrix} 0 & 0 & 0 \\ \beta_{21} & 0 & 0 \\ \beta_{31} & \beta_{32} & 0 \end{pmatrix} \begin{pmatrix} Y_1 \\ Y_2 \\ Y_3 \end{pmatrix} + \begin{pmatrix} \gamma_{11} & \gamma_{12} & \gamma_{13} \\ \gamma_{21} & \gamma_{22} & \gamma_{23} \\ \gamma_{31} & \gamma_{32} & \gamma_{33} \end{pmatrix} \begin{pmatrix} X_1 \\ X_2 \\ X_3 \end{pmatrix} + \begin{pmatrix} \epsilon_1 \\ \epsilon_2 \\ \epsilon_3 \end{pmatrix}$$

The number of parameters to be estimated are

$$\boldsymbol{\Gamma} = [\gamma_{11}, \gamma_{12}, \ldots\ldots\ldots, \gamma_{33}] = 9$$
$$\boldsymbol{\beta} = [\beta_{21}, \beta_{31}, \beta_{32}] = 3$$
$$\boldsymbol{\Psi} = [\psi_{11}, \psi_{22}, \psi_{33}] = 3$$
$$\boldsymbol{\Phi} = [\phi_{11}, \phi_{22}, \phi_{33}, \phi_{21}, \phi_{31}, \phi_{32}] = 6$$

$$\text{Total} = 9 + 3 + 3 + 6 = 21$$

The reduced form model is shown in Figure 10.6.

The reduced model path equation in matrix form can be written as,

$$\begin{pmatrix} Y_1 \\ Y_2 \\ Y_3 \end{pmatrix} = \begin{pmatrix} a_{11} & a_{12} & a_{13} \\ a_{21} & a_{22} & a_{23} \\ a_{31} & a_{32} & a_{33} \end{pmatrix} \begin{pmatrix} X_1 \\ X_2 \\ X_3 \end{pmatrix} + \begin{pmatrix} \theta_1 \\ \theta_2 \\ \theta_3 \end{pmatrix}$$

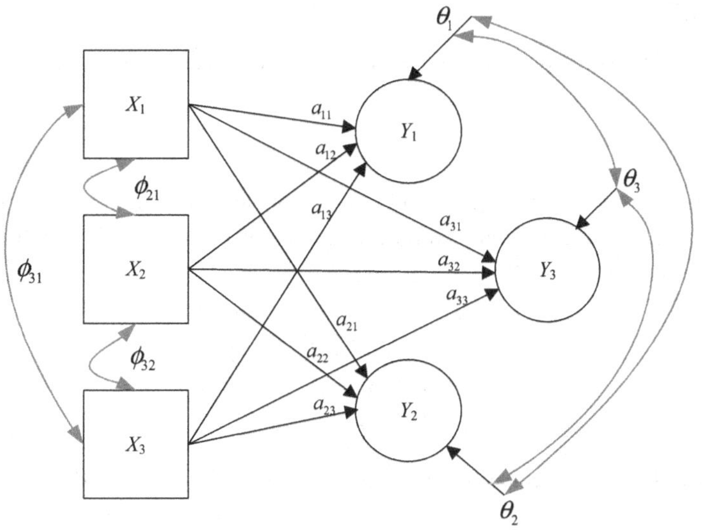

FIGURE 10.6 Example 10.4 reduced form model. In the reduced form path model, the error terms are correlated.

The number of parameters to be estimated are

$$\begin{aligned}
&\mathbf{A} = [a_{11}, a_{12}, \ldots\ldots\ldots, a_{33}] = 9 \\
&\boldsymbol{\beta} = 0 \\
&\boldsymbol{\Psi}_\theta = [\psi_{\theta_{11}}, \psi_{\theta_{22}}, \psi_{\theta_{33}}, \psi_{\theta_{21}}, \psi_{\theta_{31}}, \psi_{\theta_{32}}] = 6 \\
&\boldsymbol{\Phi} = [\phi_{11}, \phi_{22}, \phi_{33}, \phi_{21}, \phi_{31}, \phi_{32}] = 6
\end{aligned}$$

$$\text{Total} = 9 + 0 + 6 + 6 = 21$$

So, the number of parameters are equal (=21) in the original and reduced form path models. The number of parameters differ in $\boldsymbol{\beta}$ and $\boldsymbol{\Psi}$. In the original form path model, there are three $\boldsymbol{\beta}$ coefficients, which are removed in the reduced form path model. On the other hand, in the original form path model, the $\boldsymbol{\Psi}$ matrix is diagonal with three error variances in the diagonal, whereas in the reduced form path model the $\boldsymbol{\Psi}$ matrix is full with three error variances and three unique covariances (having six parameters to be estimated).

The reduced form coefficients give an important interpretation of the causal relationships of a path model. Duncan (1975) stated that "the reduced form coefficients sum up the several direct and indirect paths through which the exogenous variable exerts its effects on each dependent/endogenous variable." That means the reduced form coefficient represents the total effect of an exogenous variable (X) on an endogenous variable (Y). In addition, the reduced form coefficients help in identifying correlated causal relationships and spurious (joint) relationships among the variables of interest. However, causation, being a difficult issue, should not treated lightly and the reduced form coefficient-based decomposition may not provide the precise total effect of X on Y, particularly when correlational relationship is present. In this context, it is to be mentioned that the correlation between X and Y can be directly read from the path diagram using Wright's multiplication rule (Wright, 1921). The reduced form approach also helps in diagnosing an under-identification problem. Duncan (1975) stated that "if there are not enough reduced-form coefficients to define solutions for the path coefficients, at least one of the equations of the model is under-identified."

Example 10.5

Consider Example 10.4. Show that the reduced form coefficients sum up the several direct and indirect paths.

Solution:

To demonstrate the direct and indirect paths, we use the relationships of Y_3 with X_1, X_2 and X_3. Following the derivation of the reduced form equation for Y_3 (given earlier in this section), we have the following relationships:

$$a_{31} = \gamma_{31} + \beta_{31}\gamma_{11} + \beta_{32}\gamma_{21} + \beta_{32}\beta_{21}\gamma_{11}$$
$$a_{32} = \gamma_{32} + \beta_{31}\gamma_{12} + \beta_{32}\gamma_{22} + \beta_{32}\beta_{21}\gamma_{12}$$
$$a_{33} = \gamma_{33} + \beta_{31}\gamma_{13} + \beta_{32}\gamma_{23} + \beta_{32}\beta_{21}\gamma_{13}$$
$$\theta_3 = \beta_{32}\beta_{21}\in_1 + \beta_{32}\in_2 + \in_3$$

First check the RHS of a_{31} equation. a_{31} is summed up of γ_{31}, $\beta_{31}\gamma_{11}$, $\beta_{32}\gamma_{21}$ and $\beta_{32}\beta_{21}\gamma_{11}$., where γ_{31} is the direct effect of X_1 on Y_3, and other terms represent the indirect effects. The following figures (Figures 10.7 and 10.8) pictorially represent this. Similar explanations can be derived for the relationships between X_2 and Y_3 as well as X_3 and Y_3 using a_{32} equation and a_{33} equation, respectively. In the same manner, the relationships between Y_2 and each of the exogenous variables (X_1, X_2, and X_3) can be explained. Note that Y_1 has only direct relationships with X_1, X_2, and X_3.

10.4 ESTIMATION OF PARAMETERS FOR NON-RECURSIVE MODEL

A non-recursive path model entails reciprocal causation. In Figure 10.3, Y_1 involves one-way causation and Y_2 and Y_3 are reciprocally caused, i.e., Y_2 causes Y_3 and vice versa. So, Y_1 equation is recursive and Y_2 and Y_3 equations are non-recursive. In addition, this path model is block-recursive. Anderson (1978) stated that a block-recursive model has reciprocal causation within individual blocks but recursive between blocks.

As Y_2 and Y_3 are reciprocally caused, the error (disturbance) variables $\in_2$ and $\in_3$ are no longer uncorrelated. The double headed curved arrow linking $\in_2$ and $\in_3$ (Figure 10.3) shows this relation.

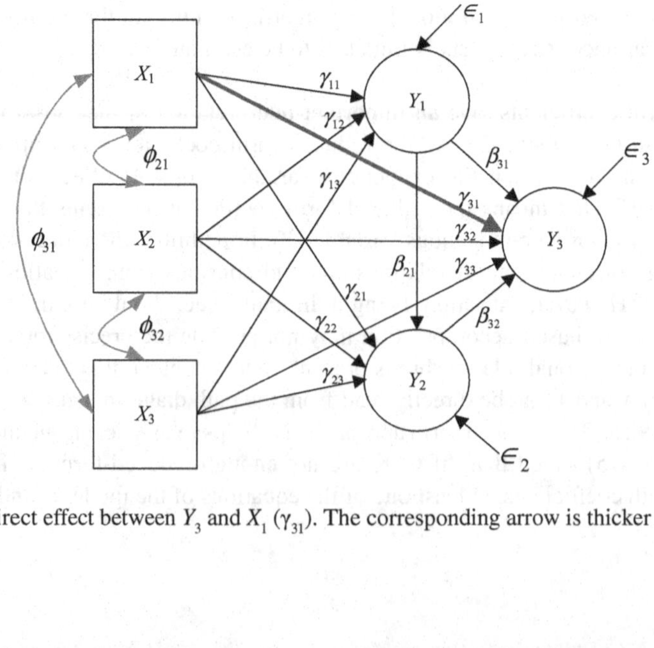

FIGURE 10.7 Direct effect between Y_3 and X_1 (γ_{31}). The corresponding arrow is thicker in size.

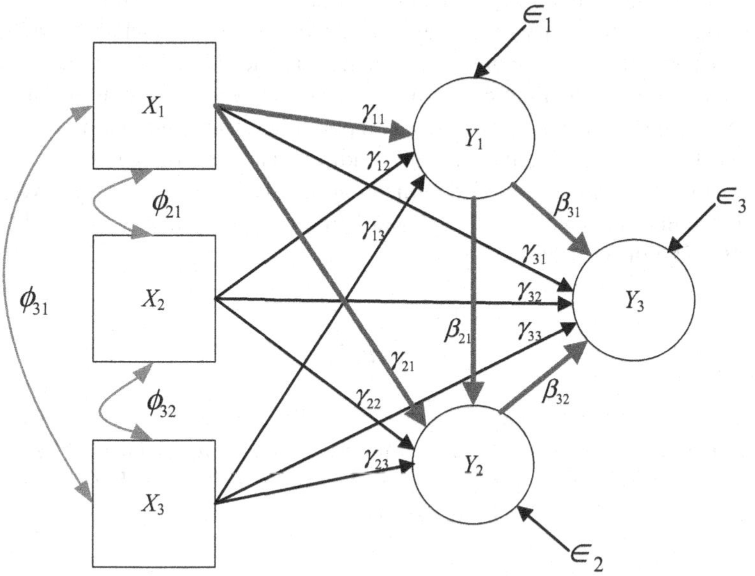

FIGURE 10.8 Indirect effect components between Y_3 and X_1 ($\beta_{31}\gamma_{11}$, $\beta_{32}\gamma_{21}$ and $\beta_{32}\beta_{21}\gamma_{11}$). The corresponding arrows are thicker in size. Note that there could be multiple paths for indirect effects.

This essentially violates assumption 4 of the recursive model (see Section 10.2). Under this condition, the $\boldsymbol{\beta}$ matrix in Equation 10.4 and error covariance matrix ($\boldsymbol{\Psi}$) become non-triangular and non-diagonal, respectively. These make estimation of parameters for the non-recursive path model a difficult task and the traditional OLS fails. In the following subsections, we describe identification issues and estimation methods for non-recursive path model.

There are two popular approaches to estimate the parameters of a path model (for both recursive and non-recursive models). These are (i) regression analysis approach and (ii) covariance structure analysis approach. The former uses ordinary least squares (OLS), instrumental variable (IV) methods and its variants such as two stage least squares (2SLS), three stage least squares (3SLS) and maximum likelihood estimation (MLE). These techniques are popular in econometrics, particularly in simultaneous equations modeling. In covariance structure analysis approach, the most popular method is MLE. Other methods are generalized least squares (GLS), unweighted least squares (ULS), etc. In this chapter, we primarily focus on the first approach. For the second approach, readers may see Chapter 14 (Section 14.6, estimation of structural model parameters. It is to be noted here that for directly observed path variables, the structural model in Section 14.3.1 is nothing but the path model).

10.4.1 Model Identification

Let consider Figure 10.3 and its path equations given in Equation 10.3. There are three path equations for three endogenous variables Y_1, Y_2, and Y_3. Following the order condition for identification in Section 10.3.1, we can establish that the path equation for Y_1 is *just-identified* and Y_2 and Y_3 equations are *under-identified*. The under-identification problem may be solved by assuming *a priori* that certain coefficients are equal to zero or introducing other exogenous variables in the model (Anderson, 1978).

Now, let assume that the path model in Figure 10.3 is wrongly specified. None of the three exogenous variables directly cause Y_3 to happen. Then, the Y_3 equation becomes

$$Y_3 = \beta_{31}Y_1 + \beta_{32}Y_2 + \epsilon_3 \tag{10.23}$$

The order condition suggests that this non-recursive equation is over-identified.

The *order condition* is necessary for model identification but not sufficient. A path equation, that satisfies the order condition may not always be identified. The sufficient condition for model identification is *rank condition*. The rank condition satisfies that "each equation of a model is distinct from every other equation in the model and from all possible linear combinations of equations in the model" (Duncan, 1975). We can define the rank condition using matrix algebra.

In order to test the rank condition of identification for the j-th path equation, we present below the procedure given in Maddala (2007). Let $\mathbf{B}_j$ and $\mathbf{\Gamma}_j$ represent the j-th row of $\mathbf{B}$ and $\mathbf{\Gamma}$, respectively (see Equation 10.17). Further, divide $\mathbf{B}_j$ and $\mathbf{\Gamma}_j$ as follows:

$$\mathbf{B}_j = [\boldsymbol{\beta}_j, \boldsymbol{\beta}_j^*]$$
$$\mathbf{\Gamma}_j = [\boldsymbol{\gamma}_j, \boldsymbol{\gamma}_j^*]$$

$\boldsymbol{\beta}_j$ corresponds to q_1 included and $\boldsymbol{\beta}_j^*$ corresponds q_2 excluded endogenous (Y) variables ($q = q_1 + q_2$) in the j-th path equation. Similarly, $\boldsymbol{\gamma}_j$ corresponds to p_1 included and $\boldsymbol{\gamma}_j^*$ corresponds p_2 excluded exogenous (X) variables ($p = p_1 + p_2$) in the j-th path equation. Further, conforming with the partitioning of $\mathbf{B}_j$ and $\mathbf{\Gamma}_j$, $\mathbf{B}$ and $\mathbf{\Gamma}$ are also portioned and we have the following partitioning matrices.

$$\mathbf{B} = \begin{pmatrix} \boldsymbol{\beta}_j & \mathbf{0} \\ \mathbf{B}_j & \mathbf{B}_j^* \end{pmatrix}, \text{ and } \mathbf{\Gamma} = \begin{pmatrix} \boldsymbol{\gamma}_j & \mathbf{0} \\ \mathbf{\Gamma}_j & \mathbf{\Gamma}_j^* \end{pmatrix}$$

Note that any parameter or parameter matrix marked with an asterisk (*) denotes it exclusion. Now, we define the exclusion-variable matrix (M_j) for the j-th path equation as

$$\mathbf{M}_j = \begin{pmatrix} \mathbf{0} & \mathbf{0} \\ \mathbf{B}_j^* & \mathbf{\Gamma}_j^* \end{pmatrix}, \ \mathbf{M} = [\mathbf{B} - \mathbf{\Gamma}] \text{ and } \mathbf{B} = \mathbf{I} - \boldsymbol{\beta}$$

The columns in $\mathbf{M}_j$ are the columns of $\mathbf{B}$ and $\mathbf{\Gamma}$ corresponding to the excluded endogenous and excluded exogenous variables in the j-th path equation. The rank condition for identification is as follows:

> The j-th path equation is identified if rank ($\mathbf{M}_j$)=q–1, where q is the total number of endogenous (Y) variables in the path model.

So, the general guidelines for identification of any non-recursive path equation can be summarized as below.

- If the order condition is satisfied for a non-recursive path equation, the equation may be *identified*. This is the necessary condition for identification. Otherwise, the equation is *under-identified* and necessary actions such as re-specifying the model (obviously based on sound theory) can be taken.
- If a non-recursive path equation satisfies the order condition, it should be tested through the rank condition. If rank condition is also satisfied, then the equation is identified.
- Note that the rank condition states whether a non-recursive path equation is identified or not. It does not state whether the equation is just-identified or over-identified, which is obtained using the order condition.

Following the above guidelines, we can test the non-recursive path model in Figure 10.3. The order condition states that the Y_1 equation is identified and the Y_2 and Y_3 equations are under-identified. Now, the exclusion-variable matrix ($\mathbf{M}_1$) for the Y_1 path equation and its rank are

TABLE 10.2
Exclusion Variable Matrix for Y_1

Equation	Y_1	Y_2	Y_3	X_1	X_2	X_3
1	1	0	0	1	1	1
2	1	1	1	1	1	1
3	1	1	1	1	1	1

$$\mathbf{M}_1 = \begin{pmatrix} 0 & 0 \\ 1 & -\beta_{23} \\ -\beta_{32} & 1 \end{pmatrix}, \text{ and}$$

$\text{rank}(\mathbf{M}_1) = 2$, assuming $\beta_{23} \neq \alpha\beta_{32}$, α is proportionality constant.

The number of endogenous variables in the path model is three. The rank condition, *rank* $(\mathbf{M}_1) = q - 1 = 3 - 1 = 2$ is satisfied. So, the Y_1 equation is identified. Interestingly, for both the Y_2 and Y_3 equations, the exclusion-variable matrices, $\mathbf{M}_2$ and $\mathbf{M}_3$, are null matrix. So, both the Y_2 and Y_3 equations are under-identified.

Alternatively, a simple procedure is presented in Maddala (2007) to test whether the path equation under consideration can be obtained as a linear combination of the other equations in the model. The steps are as follows:

1. Prepare an equation-variable cross-table, where each row represents a path equation, each column represents a variable (either endogenous or exogenous), and each cell contains either 0 or 1. A 1 denotes that the corresponding variable is present in the equation and 0 denotes its exclusion (missing).
2. Consider the *j*-th equation for identification and delete that particular row. Usually, the *j*-th equation represents the *j*-th row.
3. Pick up the columns which have 0 entries in the deleted *j*-th row, and form the exclusion-variable matrix ($\mathbf{M}_j$) comprising the picked-up columns.
4. If q–1 rows and columns of $\mathbf{M}_j$, where q is the number of endogenous variables, are not all zeros, and no column (or row) is proportional to other column (or row) for all coefficient values, then the equation is identified.

We illustrate this for Y_1 equation with reference to the non-recursive path model in Figure 10.3.

From Table 10.2, it is seen that $\mathbf{M}_1$ contains two rows and columns whose all entries are not zero. So, Y_1 equation is identified. For the Y_2 and Y_3 equations, there is no $\mathbf{M}$ matrix or $\mathbf{M}$ matrix is null. This demonstrates the procedure.

Example 10.6

Consider the following non-recursive path model as in Figure 10.9. Test the identification of the model.

Solution:

The path equations are

$$\begin{aligned} Y_1 &= \beta_{12}Y_2 + \gamma_{11}X_1 + \epsilon_1 \\ Y_2 &= \beta_{21}Y_1 + \gamma_{22}X_2 + \gamma_{23}X_3 + \epsilon_2 \\ Y_3 &= \beta_{31}Y_1 + \beta_{32}Y_2 + \epsilon_3 \end{aligned}$$

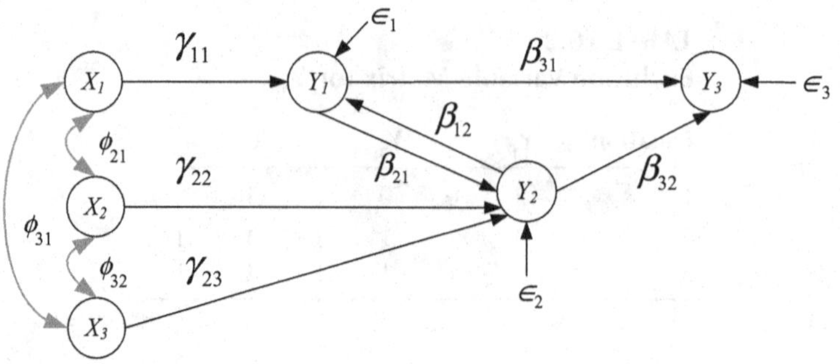

FIGURE 10.9 Example 10.6 non-recursive path model.

So,

$$\boldsymbol{\beta} = \begin{bmatrix} 0 & \beta_{12} & 0 \\ \beta_{21} & 0 & 0 \\ \beta_{31} & \beta_{32} & 0 \end{bmatrix}; \boldsymbol{\Gamma} = \begin{bmatrix} \gamma_{11} & 0 & 0 \\ 0 & \gamma_{22} & \gamma_{23} \\ 0 & 0 & 0 \end{bmatrix}$$

Hence, the matrix **M** will be,

$$\mathbf{M} = \begin{bmatrix} 1 & -\beta_{12} & 0 & -\gamma_{11} & 0 & 0 \\ -\beta_{21} & 1 & 0 & 0 & -\gamma_{22} & -\gamma_{23} \\ -\beta_{31} & -\beta_{32} & 1 & 0 & 0 & 0 \end{bmatrix}$$

Based on this matrix, the following observations can be made on the order conditions:

1. For Y_1, the first row of matrix **M** has three zeros, which is greater than $(q - 1) = 2$. The path equation Y_1 is over-identifiable.
2. For Y_2, the second row of matrix **M** has two zeros, which is equal to $(q - 1) = 2$. The path equation Y_2 is just-identifiable.
3. For Y_3, the third row of matrix **M** has three zeros, which is greater than $(q - 1) = 2$. The path equation Y_3 is over-identifiable.

So, based on the order condition for model identification, it can be said that the model is identifiable. For sufficiency, the model identification should pass through rank condition, which is shown below.

To test the rank condition for Y_1, we have the following **M** matrix (see theory described earlier).

$$\mathbf{M}_1 = \begin{bmatrix} 0 & 0 & 0 \\ 0 & -\gamma_{22} & -\gamma_{23} \\ 1 & 0 & 0 \end{bmatrix}$$

Rank $(\mathbf{M}_1) = 2$, which is equal to $(q - 1) = 2$. So, the path equation Y_1 is identifiable.

Similarly, for Y_2, we get

$$\mathbf{M}_2 = \begin{bmatrix} 0 & -\gamma_{11} \\ 0 & 0 \\ 1 & 0 \end{bmatrix}$$

Rank $(\mathbf{M}_2) = 2$, which is equal to $(q - 1) = 2$. So, the path equation Y_2 is identifiable.

and for Y_3,

$$\mathbf{M}_3 = \begin{bmatrix} -\gamma_{11} & 0 & 0 \\ 0 & -\gamma_{22} & -\gamma_{23} \\ 0 & 0 & 0 \end{bmatrix}$$

Rank $(\mathbf{M}_3) = 2$, which is again equal to $(q - 1) = 2$. So, the path equation Y_3 is also identifiable. Finally, based on order and rank conditions, all the three model equations are identifiable. Hence, the model is identifiable.

10.4.2 Estimation of Parameters

There are several methods for estimation of parameters in non-recursive model. In indirect least squares (ILS), the reduced form coefficients are first estimated through OLS and then the path coefficients are estimated. The ILS method is not preferred for large path model. The two better known methods are (i) instrumental variable (IV[1]) method, preferably for just-identified non-recursive equations and (ii) two stage least squares (2SLS) for over-identified non-recursive equations. The IV and 2SLS methods consider one equation at a time from the system of equations (representing a path model) and do not consider cross-equation correlation effects. That is why these are called limited information estimators. The other widely used method in this category is limited information maximum likelihood (LIML). However, LIML additionally assumes normality of the reduced form disturbance matrix of each equation. To accommodate cross-equation correlation effects, full information estimators such as three stage least squares (3SLS) and full information maximum likelihood (FIML) are used. Note that the estimation methods described here are widely popular in econometric literature for solving system of linear equations.

Instrumental Variable (IV) Method

In order to make a non-recursive path equation estimable, we need to find out some variables that satisfy the following two conditions for the equation considered:

1. Uncorrelated with disturbance (error) variable ($\in$), and
2. Correlated with the explanatory variables, also called relevance.

The variables that satisfy the above two conditions are called instrumental variables (IV). Condition 1 confirms the *exogeneity* requirement but cannot be tested *a priori*. Condition 2 ensures *relevance* of the IV and can be tested by regressing the concerned explanatory variable on the IV as well as other exogenous variables in the system. We will discuss below the following three situations to demonstrate IV.

1. Simple regression context involving one endogenous variable as the sole explanatory variable in the *RHS* of the equation.
2. Multiple regression context with one endogenous variable and multiple exogenous variables in the *RHS* of the equation.
3. Multiple regression context involving multiple endogenous and multiple exogenous variables in the *RHS* of the equation.

In the *simple regression context*, there will be one dependent variable (say, Y_j) and one explanatory variable (say, Y_k or X_j) in the *j*-th equation. As we consider notation X for exogenous variables (that does not depend on any other variables including the disturbance in the model) and IV is needed only for those explanatory variables that violate assumption 4, we use Y_k as the required explanatory variable. The equation of interest is $Y_j = \beta_{jk} Y_k + \in_j$ with $E(Y_k \in_j) \neq 0$. We need an IV for Y_k and let Z_k be the IV satisfying the two conditions mentioned above. Now, multiplying Z_k with the equation and then taking expectation, we will get the required normal equation to estimate the path coefficient (β_{jk}).

$$Cov(Z_k, Y_j) = \beta_{jk} Cov(Z_k, Y_k) + Cov(Z_k, \epsilon_j)$$

Since, $Cov(Z_k, \epsilon_j) = 0$

$$\beta_{jk} = \frac{Cov(Z_k, Y_j)}{Cov(Z_k, Y_k)}$$

Now, if a sample of size n on Z_k, Y_k, and Y_j is collected, the estimated β_{jk} is

$$\hat{\beta}_{jk} = \frac{Cov(Z_k, Y_j)}{Cov(Z_k, Y_k)} = \frac{\sum_{i=1}^{n}(y_{ji} - \bar{y}_j)(z_{ki} - \bar{z}_k)}{\sum_{i=1}^{n}(y_{ki} - \bar{y}_k)(z_{ki} - \bar{z}_k)}, \quad Cov(Z_k, Y_k) \neq 0, \text{ and}$$

$$\bar{y}_j = \bar{y}_k = \bar{z}_k = 0 \text{ for mean subtracted variables.}$$

Pictorially, the IV for simple regression context is given in Figure 10.10a. The IV Z_k affects both Y_j (i.e., *relevant*) and Y_k (i.e., *regression requirement*) but is uncorrelated with ϵ_k.

For multiple regression context with single endogenous variable and multiple exogenous variables in the RHS of the path equation, the IV is demonstrated in Figure 10.10b. The equation of interest is $Y_j = \beta_{jk} Y_k + \mathbf{\Gamma}_j \mathbf{X}_j + \epsilon_j$ with $E(Y_k\ \epsilon_j) \neq 0$. As there is only one explanatory variable (Y_k) in the RHS that violates assumption 4, we need one IV and it is Z_k for Y_k. So, $E(Z_k\ \epsilon_j) = 0$. Here, the X_j variables (say p_j in number), being exogenous in nature, satisfy the two conditions needed for an IV. So, each of the X_j variables is its own IV. Now, multiplying Z_k and each of the X_j variables through the equation, separately and then taking expectation, we will get the required normal equations to estimate path coefficients β_{jk} and $\mathbf{\Gamma}_j$.

For multiple regression context involving multiple endogenous and multiple exogenous variables in the RHS of the equation, we need to identify appropriate IVs for the endogenous variables in the RHS, which will meet the IV conditions. We will elaborate it below as a generic non-recursive equation.

Let consider the general path model in matrix form (Equation 10.4). Further, consider that n observations will be collected. Then, Equation 10.4 for the n observations together can be represented as

$$\begin{aligned} \mathbf{Y}_{n\times q} &= \mathbf{Y}_{n\times q}\boldsymbol{\beta}_{q\times q} + \mathbf{X}_{n\times p}\boldsymbol{\Gamma}_{p\times q} + \boldsymbol{\epsilon}_{n\times q} \\ &= \mathbf{Z}_{n\times(p+q)}\boldsymbol{\Pi}_{(p+q)\times q} + \boldsymbol{\epsilon}_{n\times q} \end{aligned} \tag{10.24}$$

Where, $\mathbf{Z} = [\mathbf{Y}\ \ \mathbf{X}]$ and $\boldsymbol{\Pi}^T = [\boldsymbol{\beta}\ \ \boldsymbol{\Gamma}]$.

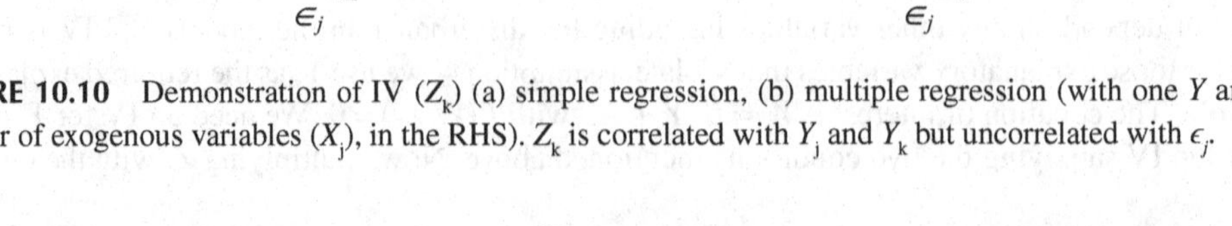

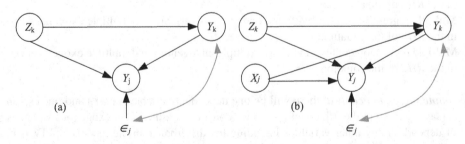

FIGURE 10.10 Demonstration of IV (Z_k) (a) simple regression, (b) multiple regression (with one Y and p_j number of exogenous variables (X_j), in the RHS). Z_k is correlated with Y_j and Y_k but uncorrelated with ϵ_j.

The j-th equation of 10.24 is

$$\mathbf{y}_j = \mathbf{Y}_j \boldsymbol{\beta}_j + \mathbf{X}_j \boldsymbol{\Gamma}_j + \boldsymbol{\epsilon}_j \tag{10.25}$$

Where,

$\mathbf{y}_j = n \times 1$ vector of observations on the j-th Y
$\mathbf{Y}_j = n \times q_j$ matrix of observations on the Y variables included in the RHS of Equation 10.25 $(q_j \le q - 1)$
$\mathbf{X}_j = n \times p_j$ matrix of observations on the X variables included in the RHS of Equation 10.25
$\boldsymbol{\beta}_j = q_j \times 1$ vector of endogenous coefficients in Equation 10.25
$\boldsymbol{\Gamma}_j = p_j \times 1$ vector of exogenous coefficients in Equation 10.25
$\boldsymbol{\epsilon}_j = n \times 1$ vector of error values

It is to be noted here that Equation 10.25 may contain intercept terms also. To avoid the intercept, as described earlier, the variable Y_j needs to be mean subtracted. It is observed that a path model, in practice, considers only the causal effects (links) that are conceptually viable and so all the path equations may not comprise all the Y and X variables included in the path model. So, with reference to the general Equation 10.25, we can write the following:

$$\mathbf{X} = [\mathbf{X}_j \;\; \mathbf{X}_j^*]$$
$$\mathbf{Y} = [\mathbf{Y}_j \;\; \mathbf{Y}_j^*]$$
$$\boldsymbol{\beta}^T = [\boldsymbol{\beta}_j \;\; \boldsymbol{\beta}_j^*]$$
$$\boldsymbol{\Gamma}^T = [\boldsymbol{\Gamma}_j \;\; \boldsymbol{\Gamma}_j^*]$$

The variables or parameters with asterisk (*) represent those that are excluded from the j-th equation. Therefore, $\boldsymbol{\beta}_j^*$ and $\boldsymbol{\Gamma}_j^*$ are null vectors as they are the coefficient vectors for $\mathbf{Y}_j^*$ and $\mathbf{X}_j^*$, the endogenous and exogenous variables excluded from the j-th equation, respectively. If q_j^* and p_j^* represent the number of variables excluded from the j-th equation, then $q = q_j + q_j^*$ and $p = p_j + p_j^*$, respectively.

Further, Equation 10.25 can be written as

$$\mathbf{y}_j = \mathbf{Z}_j \boldsymbol{\Pi}_j + \boldsymbol{\epsilon}_j$$

Where, $\mathbf{Z}_j = [\mathbf{Y}_j \;\; \mathbf{X}_j]$ and $\boldsymbol{\Pi}_j^T = [\boldsymbol{\beta}_j \;\; \boldsymbol{\Gamma}_j]$. (10.26)

Note that all the variables are mean-subtracted and hence, Equations 10.24–10.26 do not contain the intercept term.

Equation 10.26 satisfies all the assumptions in Section 10.2 except $E(\mathbf{Y}_j \boldsymbol{\epsilon}_j) = \mathbf{0}$. Hence, OLS provides biased and inconsistent estimates of $\boldsymbol{\Pi}_j$. The OLS estimate is $\hat{\boldsymbol{\Pi}}_j = (\mathbf{Z}_j^T \mathbf{Z}_j)^{-1} \mathbf{Z}_j^T \mathbf{y}_j$ with expected value as

$$E[\hat{\boldsymbol{\Pi}}_j] = E[(\mathbf{Z}_j^T \mathbf{Z}_j)^{-1} \mathbf{Z}_j^T \mathbf{y}_j] = E[(\mathbf{Z}_j^T \mathbf{Z}_j)^{-1} \mathbf{Z}_j^T (\mathbf{Z}_j \boldsymbol{\Pi}_j + \boldsymbol{\epsilon}_j)]$$
$$= E[(\mathbf{Z}_j^T \mathbf{Z}_j)^{-1} \mathbf{Z}_j^T \mathbf{Z}_j \boldsymbol{\Pi}_j + (\mathbf{Z}_j^T \mathbf{Z}_j)^{-1Z} \mathbf{Z}_j^T \boldsymbol{\epsilon}_j)] = \boldsymbol{\Pi}_j + E[(\mathbf{Z}_j^T \mathbf{Z}_j)^{-1} \mathbf{Z}_j^T \boldsymbol{\epsilon}_j]$$

Expanding $\mathbf{Z}_j^T \boldsymbol{\epsilon}_j$, we get, $\mathbf{Z}_j^T \boldsymbol{\epsilon}_j = \begin{bmatrix} \mathbf{Y}_j^T \boldsymbol{\epsilon}_j \\ \mathbf{X}_j^T \boldsymbol{\epsilon}_j \end{bmatrix}$. So, $E[\mathbf{Z}_j^T \boldsymbol{\epsilon}_j] \neq 0$, as $E[\mathbf{Y}_j^T \boldsymbol{\epsilon}_j] \neq 0$, and $E[\hat{\boldsymbol{\Pi}}_{j(OLS)}] \neq \boldsymbol{\Pi}_j$. Hence, $\hat{\boldsymbol{\Pi}}_{j(OLS)}$ is biased and inconsistent.

Fortunately, the instrumental variable (IV) method provides a consistent estimator for $\boldsymbol{\Pi}_j$. However, the critical issue in the IV method is finding out the required number of IVs. For example, for the multiple regression context with one endogenous variable in the RHS, we require one IV for the explanatory variable y_k. For the general path equation in Equation 10.25, we require q_j number of IVs for the q_j number of Y_j variables. There are three ways the problem can be handled. First, the IVs can be obtained externally, but this is difficult. Particularly, considering the relevance of the IVs, the obvious question is why these IVs did not take part in the proposed path model? Is it not a case of omitted variables, which perhaps introduces endogeneity? Are these IVs the proxy variables? Note that an IV is not a proxy variable, as a good proxy will always be well correlated with the disturbance terms. Second, some of the missing exogenous variables, X_j^*, can be used, as these X_j^* satisfy both the IV conditions. Being exogenous, X_j^* are uncorrelated with the disturbance variable (ϵ_j), and it is quite likely that these will also be significantly correlated with the Y_j (the variables included in the RHS of Equation 10.25). It is recommended to test whether the IVs are weak or strong instruments (this will be discussed later). Third, the IVs can be constructed by deriving the reduced form path equations and then regressing each of the endogenous variables on all the exogenous variables in the reduced form equation using OLS. The predicted endogenous variables, $\hat{Y}_j$, $j = 1,2,...,q_j$, will correspondingly qualify for the IV of Y_j.

Now, we will explain the single equation-based limited information method using IV-based estimation. Let $\mathbf{W}_j$ be $n \times k_j$ data matrix that satisfy the IV conditions, and let $k_j = q_j + p_j$. Note that $\mathbf{W}_j$ and $\mathbf{Z}_j$ are the same size. Then, the IV estimator is (Greene, 2012)

$$\hat{\mathbf{\Pi}}_{j(IV)} = (\mathbf{W}_j^T\mathbf{Z}_j)^{-1}\mathbf{W}_j^T\mathbf{y}_j \tag{10.27}$$

It can be proven that the expected value and covariance matrix of $\hat{\mathbf{\Pi}}_{j(IV)}$ are

$$\begin{aligned} &E[\hat{\mathbf{\Pi}}_{j(IV)}] = \mathbf{\Pi}_j, \text{ and} \\ &Cov[\hat{\mathbf{\Pi}}_{j(IV)}] = \sigma_{jj}(\mathbf{W}_j^T\mathbf{Z}_j)^{-1}(\mathbf{W}_j^T\mathbf{W}_j)(\mathbf{Z}_j^T\mathbf{W}_j)^{-1} \end{aligned} \tag{10.28}$$

Where,

$$\hat{\sigma}_{jj} = E[\hat{\epsilon}_j^T\hat{\epsilon}_j] = \frac{1}{n - p_j - q_j}(\mathbf{y}_j - \mathbf{Z}_j\hat{\mathbf{\Pi}}_{j(IV)})^T(\mathbf{y}_j - \mathbf{Z}_j\hat{\mathbf{\Pi}}_{j(IV)}) \tag{10.29}$$

Further,

$$\begin{aligned} \mathbf{S}_{\mathbf{W}_j\mathbf{W}_j} &= \frac{1}{n-1}(\mathbf{W}_j^T\mathbf{W}_j) \\ \mathbf{S}_{\mathbf{W}_j\mathbf{Z}_j} &= \frac{1}{n-1}(\mathbf{W}_j^T\mathbf{Z}_j) \\ \mathbf{S}_{\mathbf{Z}_j\mathbf{W}_j} &= \frac{1}{n-1}(\mathbf{Z}_j^T\mathbf{W}_j) \end{aligned} \tag{10.30}$$

Note that for large n, the denominator in Equations 10.29 and 10.30 could be n, instead of $n - p_j - q_j$ and $n - 1$, respectively.

Using multivariate central limit theorem, it can be proved that $\hat{\mathbf{\Pi}}_{j(IV)}$ is asymptotically multivariate normal, i.e.,

$$\hat{\mathbf{\Pi}}_{j(IV)} \sim N_{p_j+q_j}\left(\mathbf{\Pi}_j, \frac{\hat{\sigma}_{jj}}{n}\mathbf{S}_{\mathbf{W}_j\mathbf{Z}_j}^{-1}\mathbf{S}_{\mathbf{W}_j\mathbf{W}_j}\mathbf{S}_{\mathbf{Z}_j\mathbf{W}_j}^{-1}\right)$$

For further details and mathematical treatments, readers may refer Green (2012). Note that in the IV method, both $\mathbf{Z}_j$ and $\mathbf{W}_j$ are of same size $n \times (p_j + q_j)$. This is the case for just-identified equation. If there could be more IVs than required, i.e., $K_j > k_j = p_j + q_j$, known as an over-identifying condition, the IV-based procedure (Equations 10.27–10.30) described above will be unusable. $\mathbf{W}_j^T\mathbf{Z}_j$ will be a rectangular matrix of size $K_j \times (p_j + q_j)$, and hence, $(\mathbf{W}_j^T\mathbf{Z}_j)^{-1}$ cannot be computed (Greene, 2012). Two stage least squares (2SLS) is preferred for estimating over-identified non-recursive equations.

Two Stage Least Squares (2SLS)

Two stage least squares (2SLS) comprises two stages given below. Let the number of IVs is $K_j > k_j = p_j + q_j$.

Stage 1: Regress $\mathbf{Z}_j$ (all explanatory variables including exogenous ($\mathbf{X}_j$) and endogenous ($\mathbf{Y}_j$) variables in the RHS of the equation considered) on the $\mathbf{W}_j$ (i.e., the IVs), and then predict $\mathbf{Z}_j$, (say $\hat{\mathbf{Z}}_j$). Note that the size of $\mathbf{W}_j$ and $\mathbf{Z}_j$ is different. Further, $\mathbf{Z}_j = [\mathbf{Y}_j \ \mathbf{X}_j]$ and we need IVs for the $\mathbf{Y}_j$ variables only. $\mathbf{X}_j$ variables can act as their own IVs. So, we could proceed by predicting $\mathbf{Y}_j$ by regressing $\mathbf{Y}_j$ on the entire $\mathbf{W}_j$. However, a better choice is projecting $\mathbf{Z}_j$ in the column space of $\mathbf{W}_j$ (Greene, 2012, Johnston and DiNardo, 1996).

$$\hat{\mathbf{Z}}_j = \mathbf{W}_j(\mathbf{W}_j^T\mathbf{W}_j)^{-1}\mathbf{W}_j^T\mathbf{Z}_j = \mathbf{H}_j\mathbf{Z}_j, \text{ so, } \mathbf{H}_j = \mathbf{W}_j(\mathbf{W}_j^T\mathbf{W}_j)^{-1}\mathbf{W}_j^T. \quad (10.31)$$

It is to be noted here that if any regressors in $\mathbf{Z}$ are valid instruments (e.g., $\mathbf{X}$ variables in $\mathbf{Z}$), no first stage regression is needed for them as they serve as their own instruments (Davidson and MacKinnon, 2004).

Stage 2: Regress $\mathbf{y}_j$ on $\hat{\mathbf{Z}}_j$, the predicted $\mathbf{Z}_j$.

$$\begin{aligned}\hat{\Pi}_{j(2SLS)} &= (\hat{\mathbf{Z}}_j^T\hat{\mathbf{Z}}_j)^{-1}\hat{\mathbf{Z}}_j^T\mathbf{y}_j = [(\mathbf{H}_j\mathbf{Z}_j)^T(\mathbf{H}_j\mathbf{Z}_j)]^{-1}(\mathbf{H}_j\mathbf{Z}_j)^T\mathbf{y}_j \\ &= (\mathbf{Z}_j^T\mathbf{H}\mathbf{Z}_j)^{-1}\mathbf{Z}_j^T\mathbf{H}_j\mathbf{y}_j \\ &\text{as } \mathbf{H} = \mathbf{Z}_j(\mathbf{Z}_j^T\mathbf{Z}_j)^{-1}\mathbf{Z}_j^T \text{ is symetric and idempotent.}\end{aligned} \quad (10.32)$$

Note that in, $\hat{\mathbf{Z}}_j$, $\mathbf{X}$ variables take their actual values.
Now, the expected value and covariance of $\hat{\Pi}_{j(2SLS)}$ are

$$\begin{aligned} E(\hat{\Pi}_{j(2SLS)}) &= \Pi_{j(2SLS)}, \text{ and} \\ Cov(\hat{\Pi}_{j(2SLS)}) &= E(\hat{\Pi}_{j(2SLS)} - \Pi_{j(2SLS)})(\hat{\Pi}_{j(2SLS)} - \Pi_{j(2SLS)})^T \\ &= \hat{\sigma}_{jj}(\mathbf{Z}_j^T\mathbf{H}_j\mathbf{Z}_j)^{-1}, \\ E(\epsilon_j^T\epsilon_j) &= \sigma_{jj}\mathbf{I}, \text{ and} \\ \hat{\sigma}_{jj} &= \frac{(\mathbf{y}_j - \mathbf{Z}_j\hat{\Pi}_{j(2SLS)})^T(\mathbf{y}_j - \mathbf{Z}_j\hat{\Pi}_{j(2SLS)})}{n - k_j}. \end{aligned} \quad (10.33)$$

Example 10.7

Consider Example 10.6. Assume a sample of size n (=100) and the covariance matrix of the mean subtracted variables as given in Table 10.3. Estimate the parameters of the model by (i) OLS method, (ii) IV method, and (iii) 2SLS. Compare the results.

Solution:

The correlation matrix is shown in Table 10.4.

Interestingly, the correlations among the exogenous variables (X) are low but among the endogenous variables (Y) are moderate to high. In addition, there are several moderate to high correlations among X and Y.

TABLE 10.3
Example 10.7 Covariance Matrix

	X_1	X_2	X_3	Y_1	Y_2	Y_3
X_1	0.691					
X_2	−0.379	10.353				
X_3	−1.417	10.785	1964.273			
Y_1	−1.566	1.363	2.294	4.185		
Y_2	−4.671	34.773	−844.292	14.414	539.999	
Y_3	−3.808	33.818	−318.952	11.197	292.862	321.052

TABLE 10.4
Correlation Matrix

	X_1	X_2	X_3	Y_1	Y_2	Y_3
X_1	1.000					
X_2	−0.142	1.000				
X_3	−0.038	0.076	1.000			
Y_1	−0.921	0.207	0.025	1.000		
Y_2	−0.242	0.465	−0.820	0.303	1.000	
Y_3	−0.256	0.587	−0.402	0.305	0.703	1.000

TABLE 10.5
Parameter Estimates of The Regressors and Their t-values Based on OLS Regression, IVMethod, and 2SLS Method for Y_1

			IV Estimates (t-value)			
Y Equation	RHS Variables In The Model	OLS Estimates (t-value)	X_2 and X_3 as IV	X_2 as IV	X_3 as IV	2 SLS Estimates (t-value)
Y_1	X_1	−2.224 (−23.12)	−2.244 (−23.22)	−2.179 (−20.25)	−2.265 (−22.93)	−2.243 (−23.222)
	Y_2	0.007 (2.119)	0.004 (1.23)	0.014 (1.86)	0.00093 (0.227)	0.0045 (1.281)
	R^2 (R_a^2)	0.86 (0.86)	0.86 (0.85)	0.85 (0.85)	0.85 (0.85)	0.86 (0.85)

The parameter estimates of the regressors and their t-values based on OLS regression, IV method, and 2SLS method for equation Y_1 are shown in Table 10.5.[2]

The equation of Y_1 comprises X_1 and Y_2 in its RHS, where Y_2 is endogenous. Therefore, OLS regression is not preferred. We need to use the instrumental variable (IV) method or two-stage least square (2SLS) method. Y_1 equation is over-identified and its RHS contains one endogenous variable Y_2. So, we need at least one IV for its estimation. Further, we have two potential IVs, X_2 and X_3. So, the possible combinations are (i) both X_2 and X_3 may be considered as IVs, (ii) only X_2 as IV, and (iii) only X_3 as IV. As there are more IVs than required (over-identified equation), 2SLS is preferable. But, for comparison purposes all the possible models are used including the OLS estimation. The results are given in Table 10.5. Note that we didn't use an intercept term.

Note that compared to OLS, in IV regressions, the Y_2 coefficient becomes insignificant. But among the three IV regressions, use of X_3 as the IV is making Y_2 coefficient grossly insignificant. Further analysis is needed for a confirmed conclusion. The 2SLS (which will be discussed later) may provide useful insights. The R^2 and adjusted R^2 remain almost same (around 0.85). However, for IV regression, R^2 is not a good measure of fit. Considering the difference in significance of Y_2 in OLS and IV method, the IV method is preferred (see reasons stated earlier).

As Y_1 equation is over-identified, the 2SLS is preferred over the IV method. The Stage 2 results are shown in Table 10.5 (column 7). The Stage 1 results show that both X_2 (coefficient = 3.62, t = 55.05) and X_3 (coefficient = –0.45, t = –96.60) are significantly related with Y_2. The R^2 value (0.99) is also very high (>0.90), indicating good model fit. So, both X_2 and X_3 are good predictors of Y_2. Further, the 2SLS results are similar to the IV method with two IVs (X_2 and X_3). The Y_2 coefficient is insignificant and the X_1 coefficient is significant. The R^2 and adjusted R^2 remain almost the same (around 0.85). As per theory, in the case of an over-identifying model, 2SLS is preferable over the IV method. So, the results of 2SLS should be used to explain the relationships and to draw insights.

Similarly, the Y_2 and Y_3 equations are solved using the IV method or 2SLS as deemed fit. The salient information from the results for Y_2 are given in Table 10.6.

Y_2 depends on X_2, X_3, and Y_1. The RHS of the equation contains one endogenous variable. It is just-identified. So, the IV method is preferred. From Table 10.6, it is seen that Y_2

TABLE 10.6
Parameter Estimates of The Regressors and Their t-values Based on OLS Regression, IV Method, and 2SLS Method for Y_2

Y Equation	RHS Variables In The Model	OLS Estimates (t-value)	IV Estimates (t-value) X_1 as IV	2 SLS Estimates (t-value)
Y_2	Y_1	2.555 (159.6)	2.545 (146.4)	2.546 (146.9)
	X_2	3.493 (339.3)	3.495 (337.5)	3.495 (337.6)
	X_3	–0.452 (–623.1)	–0.452 (–621.9)	–0.452 (–622)
	R^2	0.99	0.99	0.99
	(R_a^2)	(0.99)	(0.99)	(0.99)

TABLE 10.7
Parameter Estimates of The Regressors and Their t-values Based on OLS Regression, IV Method, and 2SLS Method for Y_3

Y Equation	RHS Variables In The Model	OLS Estimates (t-value)	IV Estimates (t-value) X_1, X_2, and X_3 as IV	X_1 and X_2 as IV	X_1 and X_3 as IV	X_2 and X_3 as IV	2 SLS Estimates (t-value)
Y_3	Y_1	1.005 (1.55)	1.211 (1.735)	–0.141 (–0.143)	1.398 (1.944)	10.774 (2.214)	1.213 (1.738)
	Y_2	0.512 (8.897)	0.507 (8.802)	0.934 (5.796)	0.375 (5.503)	0.449 (4.061)	0.507 (8.802)
	R^2	0.50	0.50	0.23	0.47	–0.72	0.50
	(R_a^2)	(0.49)	(0.49)	(0.21)	(0.46)	(–0.76)	(0.49)

is significantly explained by all the three explanatory variables, Y_1, X_2, and X_3. The R^2 value (0.99) is very high (>0.90), indicating good model fit. However, for curiosity the OLS- and 2SLS-based estimations are also provided in Table 10.6. Interestingly, the results of the IV method, OLS method, and 2SLS are similar. So, the endogeneity of Y_1 needs to be checked. Probably Y_1 does not suffer from endogeneity problem.

The parameter estimates of the regressors and their t-values based on OLS regression, the IV method, and 2SLS method for equation Y_3 are shown in Table 10.7.

Y_3 depends on Y_1 and Y_2. The RHS of the equation contains two endogenous variables and there are three X variables. Y_3 equation is over-identified. 2SLS is preferred. Y_3 is significantly influenced by Y_2 at 0.05 probability level. The R^2 value (0.50) is low, indicating poor model fit. The OLS- and IV-based estimations are also provided in the Table 10.7. Interestingly, the R^2 and R_a^2 values for the IV method (with X_2 and X_3 as IVs) are negative. This indicates that R^2 based goodness of fit is not preferable in the IV method.

Three Stage Least Squares (3SLS)

Three stage least squares (3SLS) is the extension of 2SLS to accommodate the cross-equation correlation effects. Instead of estimating one equation (for example, Equation 10.25) at a time, all the path equations together (Equation 10.24) are considered in estimation, and hence, the full information is used. As seen earlier, 2SLS can accommodate the endogeneity issue in estimation. To handle the cross-equation correlation issue, generalized least squares (GLS) is recommended. For GLS, readers may refer to Chapter 2. For details of 3SLS, reader may refer to Zellner and Theil (1962) and Green (2012).

If n observations on Y and X (hence, $Z = [Y\ X]$) are planned to be collected, we can rewrite Equation 10.4 in terms of the following matrix equations.

$$\begin{bmatrix} \mathbf{y}_1 \\ \mathbf{y}_2 \\ \vdots \\ \mathbf{y}_j \\ \vdots \\ \mathbf{y}_q \end{bmatrix} = \begin{bmatrix} \mathbf{Z}_1 & 0 & .. & 0 & .. & 0 \\ 0 & \mathbf{Z}_2 & .. & 0 & .. & 0 \\ \vdots & \vdots & \vdots & \vdots & \vdots & \vdots \\ 0 & 0 & .. & \mathbf{Z}_j & 0 & 0 \\ \vdots & \vdots & \vdots & \vdots & \vdots & \vdots \\ 0 & 0 & 0 & 0 & 0 & \mathbf{Z}_q \end{bmatrix} \begin{bmatrix} \Pi_1 \\ \Pi_2 \\ \vdots \\ \Pi_j \\ \vdots \\ \Pi_q \end{bmatrix} + \begin{bmatrix} \epsilon_1 \\ \epsilon_2 \\ \vdots \\ \epsilon_j \\ \vdots \\ \epsilon_q \end{bmatrix} \tag{10.34}$$

Note that $\mathbf{y}_j$ and ϵ_j, $j = 1, 2, \ldots, q$, all are $n \times 1$ vectors. Further, ϵ is a $n \times q$ matrix, and $E(\epsilon) = \mathbf{0}$. The covariance of ϵ is

$$E(\epsilon\,\epsilon^T) = \boldsymbol{\Sigma} \otimes \mathbf{I} = \begin{pmatrix} \sigma_{11}\mathbf{I} & .. & \sigma_{1j}\mathbf{I} & .. & \sigma_{1q}\mathbf{I} \\ .. & .. & .. & .. & .. \\ \sigma_{j1}\mathbf{I} & .. & \sigma_{jj}\mathbf{I} & .. & \sigma_{jq}\mathbf{I} \\ .. & .. & .. & .. & .. \\ \sigma_{q1}\mathbf{I} & .. & \sigma_{qj}\mathbf{I} & .. & \sigma_{qq}\mathbf{I} \end{pmatrix} \tag{10.35}$$

$\otimes$ is called the "Kronecker product." See the similarity of the equations with multivariate multiple linear regression (Chapter 9).

Let W be the required IV matrix. Then, following the procedure of seemingly unrelated regression model (Greene, 2012):

$$\hat{\Pi}_{IV.GLS} = [\mathbf{W}^T(\boldsymbol{\Sigma}^{-1} \otimes \mathbf{I})\mathbf{Z}]^{-1}\mathbf{W}^T(\boldsymbol{\Sigma}^{-1} \otimes \mathbf{I})\mathbf{Y} \tag{10.36}$$

The immediate requirement is to find out **W**. In 2SLS, it is seen that $\hat{\boldsymbol{Z}}_j$ is a good candidate for $\boldsymbol{W}_j$. So, **W** can be considered as

$$\mathbf{W} = \hat{\mathbf{Z}} = \begin{pmatrix} \hat{Z}_1 & 0 & .. & 0 & .. & 0 \\ 0 & \hat{Z}_2 & .. & 0 & .. & 0 \\ .. & .. & .. & .. & .. & .. \\ 0 & 0 & .. & \hat{Z}_j & 0 & 0 \\ .. & .. & .. & .. & .. & .. \\ 0 & 0 & .. & 0 & .. & \hat{Z}_q \end{pmatrix}, \ \hat{\mathbf{Z}}_j \text{ is the predicted value of } \mathbf{Z}_j,\ j = 1,2,...,q.$$

With the above information, the three stages of the 3SLS can be written as below (Zellner and Theil, 1962).

Stage 1: Develop a reduced form equation matrix, similar to the one shown in Equation 10.16. Estimate the reduced form parameter matrix **A** using OLS, and compute $\hat{\mathbf{Y}} = \hat{\mathbf{A}}\mathbf{X}$.
Stage 2: For each equation, compute $\hat{\Pi}_{j(2SLS)}$ and compute $\hat{\boldsymbol{\Sigma}}_j$ (see 2SLS).
Stage 3: Compute the 3SLS estimates using Equation 10.36 and by replacing **W** with $\hat{\mathbf{Z}}$ and **Y** with $\hat{\mathbf{Y}}$.

Accordingly,

$$\hat{\Pi}_{3SLS} = [\hat{\mathbf{Z}}^T(\hat{\boldsymbol{\Sigma}}^{-1} \otimes \mathbf{I})\hat{\mathbf{Z}}]^{-1}\hat{\mathbf{Z}}^T(\hat{\boldsymbol{\Sigma}}^{-1} \otimes \mathbf{I})\hat{\mathbf{Y}} \tag{10.37}$$

Note the differences between Equation 10.36 and Equation 10.37. Both **W** and **Z** are replaced by $\hat{\mathbf{Z}}$. This becomes possible as (**I–H**) is idempotent matrix. $\hat{\Pi}_{3SLS}$ is consistent and controlled for endogeneity and cross-equation correlations. So, $E(\hat{\boldsymbol{\Pi}}_{3SLS}) = \boldsymbol{\Pi}$ and the asymptotic covariance matrix of $\hat{\Pi}_{3SLS}$ is $[\hat{\mathbf{Z}}^T(\hat{\boldsymbol{\Sigma}}^{-1} \otimes \mathbf{I})\hat{\mathbf{Z}}]^{-1}$.

Example 10.8

Consider Example 10.7. Estimate the parameters of the model using 3SLS. Compare the results with that obtained in Example 10.7.

Solution:

The results of the 3SLS are shown in Table 10.8.

The 3SLS estimated coefficients, their t-values and level of significance are similar to the results of 2SLS. The R^2 value is also similar. As stated earlier, 3SLS is preferred over 2SLS to accommodate the cross-equation correlation effects of the system of equations. Based on the similarity of results between 2SLS and 3SLS, it may be concluded that the cross-equation correlation effects are insignificant.

Maximum Likelihood Estimation

Although 2SLS and 3SLS are widely used techniques for estimation of simultaneous equations, the use of maximum likelihood estimation (MLE) is also widely used, particularly after the emergence of structural equation modeling (SEM). SEM has two components: the measurement model and path model. We will discuss SEM in Chapter 14. However, it is necessary to mention here that

TABLE 10.8
Parameter Estimates of The Regressors Based On OLS Regression, 2SLS, and 3SLS Methods for Y_1, Y_2, and Y_3

Y equation	RHS Variables In The Model	OLS Estimates (t-value)	2 SLS Estimates (t-value)	3 SLS Estimates (t-value)
Y_1	X_1	–2.224 (–23.12)	–2.243 (–23.222)	–2.244 (–23.22)
	Y_2	0.007 (2.119)	0.0045 (1.281)	0.004 (1.23)
	R^2 (R_a^2)	0.86 (0.86)	0.86 (0.85)	0.85 (0.85)
Y_2	Y_1	2.555 (159.6)	2.546 (146.9)	2.55 (147.68)
	X_2	3.493 (339.3)	3.495 (337.6)	3.48 (372.03)
	X_3	–0.452 (–623.1)	–0.452 (-622)	–0.45 (–637.34)
	R^2 (R_a^2)	0.99 (0.99)	0.99 (0.99)	0.99 (0.99)
Y_3	Y_1	1.005 (1.55)	1.213 (1.738)	1.21 (1.74)
	Y_2	0.512 (8.897)	0.507 (8.802)	0.51 (8.80)
	R^2 (R_a^2)	0.50 (0.49)	0.50 (0.49)	0.50 (0.49)

the path model in SEM is similar to the path model of directly observed variables. Here, we will expose the readers to the basic requirements for MLE-based estimation. The additional assumption (for MLE) is that the error structure follows multivariate normality. The MLE can be used for a single equation at a time, similar to 2SLS, or for all equations simultaneously, similar to 3SLS. The former is known as the limited information maximum likelihood (LIML) estimator while the latter is the full information maximum likelihood (FIML) estimator. In this section, we discuss the FIML method, which is similar to the one used in confirmatory factor analysis (Timm, 2002).

Consider Equation 10.24 again. After algebraic manipulation, Equation 10.24 can be written as

$$Y = (\mathbf{I}-\boldsymbol{\beta})^{-1}\boldsymbol{\Gamma}X + (\mathbf{I}-\boldsymbol{\beta})^{-1}\epsilon \tag{10.38}$$

Let the covariance matrix of the observed X and Y variables be $\boldsymbol{\Sigma}$, where

$$\boldsymbol{\Sigma} = \begin{pmatrix} \boldsymbol{\Sigma}_{YY} & \boldsymbol{\Sigma}_{YX} \\ \boldsymbol{\Sigma}_{XY} & \boldsymbol{\Sigma}_{XX} \end{pmatrix} = \begin{pmatrix} E(\mathbf{YY}^T) & E(\mathbf{YX}^T) \\ E(\mathbf{XY}^T) & E(\mathbf{XX}^T) \end{pmatrix}$$

Following the procedure of normal equation approach, we get

$$\boldsymbol{\Sigma}(\boldsymbol{\theta}) = \begin{pmatrix} (\mathbf{I}-\boldsymbol{\beta})^{-1}(\boldsymbol{\Gamma\Phi\Gamma}^T+\boldsymbol{\Psi})(\mathbf{I}-\boldsymbol{\beta})^{-1^T} & (\mathbf{I}-\boldsymbol{\beta})^{-1}\boldsymbol{\Gamma\Phi} \\ \boldsymbol{\Phi\Gamma}^T(\mathbf{I}-\boldsymbol{\beta})^{-1^T} & \boldsymbol{\Phi} \end{pmatrix} \tag{10.39}$$

So, $\boldsymbol{\Sigma}(\boldsymbol{\theta})$ is a function of unknown path model parameters $\boldsymbol{\beta}$, $\boldsymbol{\Gamma}$, $\boldsymbol{\Phi}$, and $\boldsymbol{\Psi}$, where $\boldsymbol{\theta}$ represents a vector of these unknown parameters. If the path model is correctly specified, it can be hypothesized that $\boldsymbol{\Sigma} = \boldsymbol{\Sigma}(\boldsymbol{\theta})$. As there are q number of Y and p number of X variables in the path model, the population covariance matrix $\mathbf{\Sigma}$ has $v = (p + q)(p + q + 1)/2$ number of unique elements representing the variances and covariances among the X and Y variables. From, $\mathbf{\Sigma} = \mathbf{\Sigma}(\boldsymbol{\theta})$, there will be v number of equations available for estimation. If the number of parameters to be estimated for the path model considered is t, for identification, $v \geq t$. This is called order condition, described earlier. Note that the sufficient condition to be tested is the rank condition, the procedure described earlier is equally applicable here.

After data is collected, the sample covariance matrix ($\mathbf{S}$) can be computed. Note that the size of $\mathbf{S}$ will be equal to that of $\mathbf{\Sigma}(\boldsymbol{\theta})$. Now, using multivariate normality of the error structure, the MLE estimation finally minimizes the fit function given below.

$$F(\boldsymbol{\theta}) = \log|\mathbf{\Sigma}(\boldsymbol{\theta})| + trace(\mathbf{S}\mathbf{\Sigma}(\boldsymbol{\theta})^{-1}) - \log|\mathbf{S}| - (p+q) \tag{10.40}$$

Similar fit function is used in confirmatory factor analysis (CFA) and structural equation modeling (SEM). Readers may refer to Chapters 13 and 14 for derivation and further discussion. Equation 10.40 is a complex non-linear function. Numerical methods such as Newton-Rapson and Gauss-Newton algorithms are used to estimate $\boldsymbol{\theta}$.

10.5 OVERALL FIT AND SPECIFICATION TESTS

Goodness of fit tests, particularly for non-recursive equations, are not as straightforward as for multiple linear regression (MLR). If OLS is used to estimate the parameters of an equation, the t or F test as used in MLR can be used (Duncan, 1975). For example, for recursive equations, OLS is used to estimate the parameters. So, the regular t and F statistics-based hypothesis can be used for the recursive path model. Similarly, R^2 and its variants may also be used. This is because the exogeneity ($E(\in \mathbf{X}) = \mathbf{0}$) and sphericity ($V(\in /\mathbf{X}) = \sigma^2\mathbf{I}$) specifications are not violated. If any of the two specifications are not met, as observed in non-recursive equations, we need to modify the regular t of F tests, or find out alternative fit tests incorporating the specification errors. In addition, note that in one-equation-at-a-time estimation (such as 2SLS), it is assumed that the specification errors in other equations of the path model do not matter, provided we have the correct predetermined variables. In the following subsections, we discuss in brief the specification tests for path equations namely, (i) exogeneity or endogeneity tests and (ii) heteroscedasticity-robust tests. In addition, we also briefly discuss the tests for over-identifying restrictions and tests for weak instruments. For detailed discussion and derivation, there are good references available. Readers may refer to Davidson and MacKinnon (2004), Greene (2012), and Johnston and Dinardo (1996). When a path model is estimated using covariance structure analysis approach such as using Equation 10.40, different fit measures are used (see Chapter 14 for structural model fit measures).

10.5.1 Tests for Endogeneity

We will discuss the Hausman test here. The Hausman test is based on the following principles (Hausman, 1978):

1. Under the hypothesis of no misspecification (H_0), there exists a consistent and asymptotically efficient estimator, which under the hypothesis of misspecification (H_1) provides biased and inconsistent estimates. For example, if the assumptions of exogeneity and sphericity are met, OLS is a consistent and asymptotically efficient estimator; otherwise OLS is biased and inconsistent.
2. There exists another estimator, which is not asymptotically efficient under H_0 but not adversely affected by misspecification (H_1). So, this estimator is likely to produce consistent estimates under H_1.

3. If a path equation is estimated by both the estimators, the differences between the two estimates can be used to test the existence of misspecification such as endogeneity of the regressors.

The Hausman test can be formulated as follows. Let the parameter vector of interest is $\boldsymbol{\beta}$.

Step 1: Set null and alternate hypotheses (H_0 and H_1).

$$H_0 : \hat{\boldsymbol{\beta}} = \hat{\boldsymbol{\beta}}_0 \text{ (estimated under no misspecification errors)}$$
$$H_1 : \hat{\boldsymbol{\beta}} = \hat{\boldsymbol{\beta}}_1 \text{ (estimated under misspecification errors)}$$

If the difference between the two estimates (say, $\mathbf{d}$) is $\mathbf{d} = \hat{\boldsymbol{\beta}}_1 - \hat{\boldsymbol{\beta}}_0$. Then,

$$H_0 : \mathbf{d} = 0$$
$$H_1 : \mathbf{d} \neq 0$$

Under endogeneity, the OLS estimator is inconsistent but the instrumental variable (IV) method is consistent. So, we can write $\hat{\boldsymbol{\beta}}_1 = \hat{\boldsymbol{\beta}}_{IV}$, and $\hat{\boldsymbol{\beta}}_0 = \hat{\boldsymbol{\beta}}_{OLS}$.

Step 2: Define random variable of interest and obtain its expected value and variance under H_0.

The random variable of interest is $\mathbf{d}$ with $E(\mathbf{d}) = \mathbf{0}$. The variance of $\mathbf{d}$ is

$$V(\mathbf{d}) = V(\hat{\boldsymbol{\beta}}_1 - \hat{\boldsymbol{\beta}}_0) = V(\hat{\boldsymbol{\beta}}_1) - V(\hat{\boldsymbol{\beta}}_0) \tag{10.41}$$

For derivation of Equation 10.41, see Hausman (1978, Lemma 2.1 and its proof). Note that $\hat{\boldsymbol{\beta}}_1$ and $\boldsymbol{\beta}_0$ are correlated as these are estimated using the same data.

Step 3: Compute the test statistic and find out its probability distribution.

The test statistic under H_0 is m, given below, follows Chi-square distribution with K degrees of freedom where K is the number unknown parameters in $\boldsymbol{\beta}$, when no misspecification is present (Green, 2012).

$$m = (\hat{\boldsymbol{\beta}}_1 - \hat{\boldsymbol{\beta}}_0)^T [V(\hat{\boldsymbol{\beta}}_1) - V(\hat{\boldsymbol{\beta}}_0)]^{-1} (\hat{\boldsymbol{\beta}}_1 - \hat{\boldsymbol{\beta}}_0) \tag{10.42}$$

When the two estimators of interest are the OLS and IV, Equation 10.42 becomes

$$m = \frac{1}{\hat{\sigma}^2} (\hat{\boldsymbol{\beta}}_{IV} - \hat{\boldsymbol{\beta}}_{OLS})^T [(\hat{\mathbf{Z}}^T \hat{\mathbf{Z}})^{-1} - (\mathbf{Z}^T \mathbf{Z})^{-1}]^{-1} (\hat{\boldsymbol{\beta}}_{IV} - \hat{\boldsymbol{\beta}}_{OLS}) \tag{10.43}$$

Step 4: Compare the computed test statistic value with the threshold obtained from theoretical distribution and make a decision.

If computed m value is greater than the threshold (tabulated) m value for probability level α, i.e., $m \geq \chi^2_K(\alpha)$, reject H_0. The decision is the misspecification exists and hence, the preferred estimator is IV.

The Hausman test can be used in multiple ways such as (i) to test for model misspecification (i.e., endogeneity), (ii) to test whether the IV method is needed over OLS or not, (iii) to compare two different estimators such as 2SLS with 3SLS, and (iv) to test normality when comparing 3SLS to FIML.

Hausman (1978) has also given a generalized procedure for endogeneity of regressors which is given below. Consider Equation 10.25 again.

$$\mathbf{y}_j = \mathbf{Y}_j \boldsymbol{\beta}_j + \mathbf{X}_j \boldsymbol{\Gamma}_j + \boldsymbol{\epsilon}_j$$

Here, $\mathbf{Y}_j$ variables are probably correlated with $\boldsymbol{\epsilon}_j$. Let $\mathbf{W}_j$ be the appropriate IVs. Then,

Step 1: Using OLS, regress $\mathbf{Y}_j$ on $\mathbf{W}_j$ and obtain

$$\hat{\mathbf{Y}}_j = \mathbf{H}_j\mathbf{Y}_j, \text{ where } \mathbf{H}_j = \mathbf{W}_j(\mathbf{W}_j^T\mathbf{W}_j)^{-1}\mathbf{W}_j^T.$$

Step 2: Regress $\mathbf{y}_j$ as below (using OLS again).

$$\mathbf{y}_j = \mathbf{Y}_j\boldsymbol{\beta}_j + \hat{\mathbf{Y}}_j\boldsymbol{\tau}_j + \mathbf{X}_j\boldsymbol{\Gamma}_j + \boldsymbol{\epsilon}_j \tag{10.44}$$

Step 3: Test $H_0 : \boldsymbol{\tau}_j = \mathbf{0}$ (using F or t). If H_0 is rejected, at least one of the $\mathbf{Y}_j$ variables is endogenous.

Example 10.9

Consider Example 10.7. Conduct a test of endogeneity using the Hausman test and comment on the result.

Solution:

We will first show the computation of the Hausman test statistic for Equation Y_1. Here, the null and alternative hypotheses are

H_0: Both $\hat{\boldsymbol{\beta}}_{OLS}$ and $\hat{\boldsymbol{\beta}}_{2SLS}$ are consistent but $\hat{\boldsymbol{\beta}}_{2SLS}$ is more efficient.
H_1: Only $\hat{\boldsymbol{\beta}}_{2SLS}$ is consistent.

Using Equation 14.42, the test statistic m is computed as 47.49. Under H_0, the test statistic m follows Chi-square distribution with two degrees of freedom. The threshold $\chi_2^2(0.05) = 5.991 < 47.49$ (the computed test statistic). So, H_0 is rejected at 0.05 probability level of significance. Hence, 2SLS is a preferred method.

Similarly, the Hausman test can be performed for equations Y_2 and Y_3. The Hausman test results for equations Y_1, Y_2, and Y_3 are shown in Table 10.9.

From the Hausman test of endogeneity, it is observed that the variable Y_2 in the RHS of equation Y_1 has endogeneity effect. No endogeneity effect is observed for the Y_2 and Y_3

TABLE 10.9
Hausman Test Results for Equations Y_1, Y_2, and Y_3

Equation	Hypotheses (H_0 and H_1)	Degree of Freedom	Statistic	p-value	Decision
Y_1	H_0: Both $\hat{\boldsymbol{\beta}}_{OLS}$ and $\hat{\boldsymbol{\beta}}_{2SLS}$ are consistent but $\boldsymbol{\beta}_{2SLS}$ is more efficient H_1: Only $\hat{\boldsymbol{\beta}}_{2SLS}$ is consistent	2	47.49	4.86×10^{-11}	H_0 is rejected. The 2SLS results are preferred.
Y_2	H_0: Both $\hat{\boldsymbol{\beta}}_{OLS}$ and $\hat{\boldsymbol{\beta}}_{IV}$ are consistent but $\boldsymbol{\beta}_{IV}$ is more efficient H_1: Only $\hat{\boldsymbol{\beta}}_{IV}$ is consistent	3	1.977	0.577	Failed to reject H_0. Endogeneity effect is not significant at 0.05 probability level.
Y_3	H_0: Both $\hat{\boldsymbol{\beta}}_{OLS}$ and $\hat{\boldsymbol{\beta}}_{2SLS}$ are consistent but $\boldsymbol{\beta}_{2SLS}$ is more efficient H_1: Only $\hat{\boldsymbol{\beta}}_{2SLS}$ is consistent	2	4.233	0.120	Failed to reject H_0. Endogeneity effect is not significant at 0.05 probability level.

equations. Y_1 equation parameter estimates using OLS will differ from that of IV and 2SLS estimation, whereas for Y_2 and Y_3 equations, the difference will be negligible. The parameter estimates in Example 10.7 confirm this observation.

10.5.2 Tests for Over-Identifying Restrictions

Based on the conditions of identification, the j-th path equation is over-identified if the number of regressors (say, k_j) in the right-hand side (RHS) of the equation is less than the number of available instrumental variables (IVs, say K_j). The degree of over-identification is $K_j - k_j$. The key issue here is what could be the best set of valid or suitable instruments. As there are more instruments than required, several restrictions can be put while estimating the parameters of the path equation. Several statistics have been developed to test the over-identifying restrictions; for example, Anderson and Rubin (1949), Sargan (1958), Basmann (1960), and Byron (1974). We will discuss the Anderson-Rubin test and the Sargan test.

Anderson-Rubin Test

Anderson-Rubin (AR) test is a likelihood ratio (LR) test, where the ratio of the maximum likelihood value of a restricted (or reduced) model to the maximum likelihood value of an unrestricted (or full) model is used to test the fit of the model. For any path equation, we can create restricted and unrestricted models for IV estimators including 2SLS as follows:

- The restricted (or reduced) model may consider only those predetermined variables that are included in the path equation (provided the model is identified). In other words, if there are K_j number of IVs available for the path model and out of them, k_j number of IVs are needed to make the path equation just identified, the reduced model considers these k_j number of IVs as regressors. If $K_j > k_j$, there are $K_j - k_j$ over-identifying restrictions.
- The unrestricted (or full model) considers all the K_j number of IVs occurred in the path model

Consider Equation 10.25 again. Let $L(\hat{\beta}_j, \hat{\Gamma}_j)$ and $L(\hat{\beta}^*_j, \hat{\Gamma}^*_j)$ denote the maximum likelihood values of the unrestricted (or full) and restricted (or reduced) models, respectively. Then, the likelihood ratio for the j-th path equation is

$$\lambda_j = \frac{L(\hat{\beta}^*_j, \hat{\Gamma}^*_j)}{L(\hat{\beta}_j, \hat{\Gamma}_j)} \tag{10.45}$$

and the likelihood ratio (LR) test statistic for the j-th path equation is

$$\begin{aligned} LR_j &= -2\ln(\lambda_j) = 2[\ln L(\hat{\beta}_j, \hat{\Gamma}_j) - \ln L(\hat{\beta}^*_j, \hat{\Gamma}^*_j)] \\ LR_j &= n\left[\ln(\epsilon_j^{*T}\epsilon_j^*) - \ln(\epsilon_j^T\epsilon_j)\right] \end{aligned} \tag{10.46}$$

LR_j follows $\chi^2_{K_j - k_j}$, where k_j is the number of regressors (including both X and Y variables) in the RHS of the j-th path equation. Using this, the fit of the restricted model can be ascertained as a general hypothesis testing problem.

Alternatively, an artificial F-statistic can be used (Davidson and MacKinnon, 1999). The relevant test statistic under H_0 is

$$F_j = \frac{(\epsilon_j^{*T}\epsilon_j^* - \epsilon_j^T\epsilon_j)/K_j^*}{\epsilon_j^T\epsilon_j/(n - K_j)} \tag{10.47}$$

F_j follows F distribution with K_j^* numerator and $n–K_j$ denominator degrees of freedom, where K_j^* is the number of instrumental variables not included in the restricted model.

Example 10.10

Consider Example 10.7. Conduct AR test for the test of over-identifying restrictions and comment on the result.

Solution:

The AR test results based on F-statistics and Chi-square statistics are shown in Tables 10.10 and 10.11, respectively.

From the AR tests, it is found that for Y_1 equation, H_0 is accepted at 0.05 probability level, i.e., the restricted model (reduced model) is not inferior to the unrestricted model. So, for Y_1 equation, it can be concluded that either of the X_2 and X_3 can be used as the instrument variable for Y_2. For Y_3 equation, H_0 is rejected at 0.05 probability level, i.e., the restricted model (reduced model) is inferior to the unrestricted model. So, for Y_3 equation, all the three X variables should be considered as IV in 2SLS estimation. After a careful look into the computed test statistics, it is observed that for Y_3 equation, when X_1 or X_3 if omitted causes the restricted model to be inferior to the unrestricted model (all X_1, X_2, and X_3 are considered as IVs). When X_2 is omitted, the F-statistics shows that the restricted model is more or less equivalent to the unrestricted model. Hence, we can conclude that for the Y_3 equation, both X_1 and X_3 could be better IV as compared to X_2.

TABLE 10.10
AR Test Based On F-Statistics

Equation	IV for Restricted Model	Degree of Freedom		F-statistic	p-value
		Numerator	Denominator		
Y_1	Only X_2	1	97	3.323	0.0707
	Only X_3	1	97	2.598	0.110
Y_3	X_1 and X_2	1	97	53.10	8.52×10^{-11}
	X_1 and X_3	1	97	5.52	0.021
	X_2 and X_3	1	97	230.07	9.90×10^{-21}

As Y_2 is a just-identified equation, AR test is not needed.

TABLE 10.11
AR Test Based on Chi-square Statistics

Equation	IV for Restricted Model	Degree of Freedom	Chi-square Statistic	p-value
Y_1	Only X_2	1	3.4617	0.0628
	Only X_3	1	2.7322	0.098
Y_3	X_1 and X_2	1	43.6530	3.92×10^{-11}
	X_1 and X_3	1	5.5360	0.0186
	X_2 and X_3	1	123.3653	1.16×10^{-28}

As Y_2 is a just-identified equation, AR test is not needed.

Sargan Test

Sargan (1958) developed the test of over-identifying restriction using characteristic equations, popular in canonical correlation (Hotelling, 1936). We describe below the simplified steps for the 2SLS-based approach for the generic (i.e., j-th) path equation 10.25.

1. Estimate 2SLS estimators using Equation 10.32.
2. Compute the residual vector $\hat{\boldsymbol{\epsilon}}_j = \mathbf{y}_j - \mathbf{Z}_j \hat{\boldsymbol{\Pi}}_{j(2SLS)}$ and its variance as $\hat{\sigma}^2_{\epsilon_j} = \dfrac{\hat{\boldsymbol{\epsilon}}_j^T \hat{\boldsymbol{\epsilon}}_j}{n}$.
3. The Sargan test statistic is

$$S_j = \frac{\hat{\boldsymbol{\epsilon}}_j^T \mathbf{H}_j \boldsymbol{\epsilon}_j}{\hat{\sigma}^2_{\epsilon_j}} \tag{10.48}$$

Under H_0, S_j is asymptotically Chi-squared with k_j–q_j degrees of freedom, where q_j is the number of endogenous regressors in the j-th equation. In Equation 10.48, $\mathbf{H}_j$ represent hat matrix obtained in Stage 2 regression of the 2SLS estimation.

Alternatively, the Sargan test can be done as follows:

1. Estimate 2SLS estimators using Equation 10.32 and obtain $\hat{\boldsymbol{\epsilon}}_j = \mathbf{y}_j - \mathbf{Z}_j \hat{\boldsymbol{\Pi}}_{j(2SLS)}$.
2. Using OLS, regress $\hat{\boldsymbol{\epsilon}}_j$ on $\mathbf{Z}_j$ and compute its R_j^2.
3. The Sargan test statistic is

$$S_j = nR_j^2 \tag{10.49}$$

Under H_0, S_j is asymptotically Chi-squared with k_j–q_j degrees of freedom, where q_j is the number of endogenous regressors in the j-th equation.

Rejection of H_0 is not a preferable situation. It indicates that at least one of the IVs is not exogenous.

Example 10.11

Consider Example 10.7. Conduct a Sargan test and comment on the result.

Solution:

The results of the Sargan test using Equation 10.49 are shown in Table 10.12.

From the Sargan test, it is seen that H_0 is accepted for equation Y_1, whereas H_0 is rejected for Y_3 equation. The results are similar to that of AR test for both Y_1 and Y_3 equations. As stated earlier, rejection of H_0 is not a preferable situation. So, for Y_3, it may be concluded that one or more IVs are not exogenous in nature.

TABLE 10.12
Sargan Test Results

Equation	Degree of Freedom	Chi-square Statistic	p-value
Y_1	1	2.70	0.10
Y_3	1	16.74	4.99×10^{-5}

As Y_2 is a just-identified equation, Sargan test is not needed.

10.5.3 Heteroscedasticity-Robust Tests

As mentioned earlier, the OLS works well when both the erogeneity and sphericity specifications are met. If the sphericity ($V(\in/\mathbf{X}) = \sigma^2\mathbf{I}$) specification is satisfied, the error terms are homoscedastic. If $V(\in/\mathbf{X}) = \sigma^2\Omega$, where $\boldsymbol{\Omega}$ is a diagonal $n \times n$ matrix (n = sample size), the error terms are heteroscedastic. Under heteroscedastic conditions, the estimates (say $\hat{\boldsymbol{\Pi}}_{j(IV)}$), and its covariance matrix will take the following form:

$$\hat{\boldsymbol{\Pi}}_{j(IV)} = (\mathbf{W}_j^T \boldsymbol{\Omega}^{-1} \mathbf{Z}_j)^{-1} \mathbf{W}_j^T \boldsymbol{\Omega}^{-1} \mathbf{y}_j \tag{10.50}$$

$$Cov[\hat{\boldsymbol{\Pi}}_{j(IV)}] = \hat{\sigma}_j^2 (\mathbf{Z}_j^T \mathbf{W}_j (\mathbf{W}_j^T \mathbf{W}_j)^{-1} \mathbf{W}_j^T \mathbf{Z}_j)^{-1} \tag{10.51}$$

$\hat{\sigma}_j^2$ can be obtained from Equation 10.33. Using Equations 10.50–10.51, one can use Wald tests which are robust to heteroscedasticity (Davidson and MacKinnnon, 2004). For single linear restriction, i.e., $H_0 : \Pi_{jk} = \Pi_{0jk}$, the asymptotic t statistic is

$$t_{\Pi jk} = \frac{\hat{\Pi}_{jk} - \Pi_{0jk}}{\hat{\sigma}_{jk}} \tag{10.52}$$

Note that in Equation 10.52, the subscript j is used for j-th path equation and subscript k is used for the k-th explanatory variable in the j-th equation. Interestingly, $t_{\Pi jk}$ follows standard normal distribution.

For two or more linear restrictions, let $\boldsymbol{\Pi}_j = [\boldsymbol{\Pi}_{j1}, \boldsymbol{\Pi}_{j2}]$ and $H_0 : \boldsymbol{\Pi}_{j2} = \boldsymbol{\Pi}_{0j2}$, the suitable Wald statistic (WS) is

$$WS_{\Pi_{j2}} = (\boldsymbol{\Pi}_{j2} - \boldsymbol{\Pi}_{0j2})^T (Cov(\boldsymbol{\Pi}_{j2}))^{-1} (\boldsymbol{\Pi}_{j2} - \boldsymbol{\Pi}_{0j2}) \tag{10.53}$$

Note that $\boldsymbol{\Pi}_{j2}$ is a vector of parameters under test and $Cov(\boldsymbol{\Pi}_{j2})$ is the relevant sub-matrix obtained from $Cov(\boldsymbol{\Pi}_j)$.

10.5.4 Tests for Identifying Weak Instruments

In Section 10.4.2, when discussing the instrumental variable (IV) method, we have described the requirements for a variable to be considered as an IV. An IV must be uncorrelated with disturbance variable ($\in$), and correlated with the explanatory variables. The first requirement is justified based on the underlying theory pertaining to the context/application domain of the path model and the second requirement can be tested using different analyses. The IVs are called weak if the partial correlation between the IVs and the included endogenous regressors is low (Staiger and Stock, 1997). We will describe below an approach based on the F-test. For a more detailed explanation, the readers may refer Staiger and Stock (1997). Note that a weak IV causes several problems such as inconsistency and low precision in estimation.

Consider Equation 10.25 again. Suppose there are q_j number of IVs available for the endogenous variables and there are p_j number of X (exogenous) variables (i.e., q_j+p_j number of explanatory variables or regressors). Further, consider that Y_k is an endogenous variable included in the RHS of Equation 10.25. To test whether the q_j number of IVs are weak or not, weak instrument tests are used. Reader may refer Baltagi (2011) and Staiger and Stock (1997) for the tests of weak instruments.

If the Y_j equation has one endogenous regressor (say, Y_k) in the RHS, in 2SLS the first-stage regression coefficients of the appropriate instrumental variables (IVs) of Y_k (here Y_k is regressed over the IVs) can be tested to determine weather an IV is weak or not. The significance is checked using F-statistic. Staiger and Stock (1997) suggested that F_j value should be greater than 10 and the corresponding maximum bias is 10%. Further, if the j-th path equation needs only one IV, the t-test will be used instead of the F-test.

Example 10.12

Consider Example 10.7. Conduct the test for weak instruments for equations Y_1, Y_2, and Y_3 and comment on the result.

Solution:

From weak instrument test results, it is seen that the F-statistic value is very high (>>10) for all the endogenous regressors in all the three Y equations. Following Staiger and Stock (1997), it can be concluded that none of the IVs for all the three equations are weak.

There are more sophisticated methods to test weak IVs. Hahn and Hausman (2002) proposed a test based on forward and reverse estimation using 2SLS. Kleibergen (2002) provided several test statistics using Anderson and Rubin's work (1949). For brief discussion on these, see Greene (2012), and for a survey of weak IV tests, see Dufour (2003).

10.5.5 Coefficients of Determination (R^2)

For over-identifying path models with OLS estimates, coefficients of determination (R^2) and multiple correlation (r) for each path equation can be computed separately in the similar manner we did for multiple linear regression (MLR). For j-th path equation

$$R_j^2 = 1 - \frac{SSE_j}{SST_j} \tag{10.54}$$

and the residual path coefficient is

$$r_{jj} = \sqrt{1 - R_j^2} \tag{10.55}$$

Equation 10.54 can be used for instrumental variable (IV) methods. But it should be noted that R_{IV}^2 will be lower than R_{OLS}^2 and R_{IV}^2 could be negative (Maddala, 2007). The latter is an indication of selecting a wrong model or an under-identified model.

Example 10.13

Consider Example 10.7. Find out R_j^2 and r_{jj} and comment on the result.

Solution:

The R_j^2 and r_{jj} for the path equations Y_1, Y_2, and Y_3 are shown in Table 10.13.

In many social, management and administrative applications, we encounter low R^2 value. However, as Duncan (1975) pointed out, worrying too much about R^2 is not praiseworthy and the possible wrong attempts to improve R^2 are (i) increasing the number of causal links

TABLE 10.13
R_j^2 and r_{jj} For The Path Equations Y_1, Y_2, and Y_3

Equation	R_j^2	r_{jj}	Remarks
Y_1	0.86	0.37	86% variability of Y_1 is explained
Y_2	0.99	0.10	99% variability of Y_2 is explained
Y_3	0.50	0.71	50% variability of Y_3 is explained

and (ii) introducing new variables. The former may violate causal ordering and in the latter, a new variable may be an alternate measure of an existing variable. Instead, the entire model specifications are more important.

10.6 TEST OF INDIVIDUAL PATH COEFFICIENTS

If OLS is used to estimate the parameters, the *t* or *F* test as used in MLR can be used (Duncan, 1975). When IV estimators are used, care must be taken to use the appropriate statistics for the parameter test. As most of the time, the finite sample distributions of IV estimators are not known, large sample tests are recommended (Davidson and MacKinnon, 2004). The asymptotic *t*-test is used to test $H_0 : \Pi_{jk} = \Pi_{0jk}$ (see Equation 10.52). In Example 10.7, the estimates of the individual regression coefficient and its t-value for all the three *Y* equations are explained. It should be noted here that the interpretation for the test of regression parameters (coefficients) is similar to that of multiple linear regression (MLR).

10.7 CASE STUDY

Foundries are generally used for metal castings. Various processes, such as molding, core making, melting, pouring, solidification, cooling, and removal of mold, are involved during metal casting in foundries. Grey iron castings are the most commonly used ones because of their inherent properties. The melting process in a grey iron foundry is one of the most important foundry processes and has multiple inputs along with multiple process variables and quality variables which make it a multivariate process. Understanding the relationships (causal and correlation) between these multiple variables may be helpful in improving the stability of the melting process and the quality of the final product. This study demonstrates the use of the path model to model the relationship between several variables of a melting process in a grey iron foundry of an automobile ancillary unit in western India.

The details of the case study are available in the web resources of this book (Multivariate Statistical Modeling in Engineering and Management – 1st (routledge.com)).

10.8 LEARNING SUMMARY

A path model is used to explore the causal relationship between several exogenous and endogenous variables. There are some specific rules to drawing a path model. For example, arrows between the variables can be single-headed or bidirectional. Causal relationships are shown by straight arrows, while the correlations are represented by bidirectional curved arrows. In this chapter, we have discussed the conceptual path model followed by assumptions, types of model, model identification, parameter estimation, overall fit, and diagnostic tests for path modeling. A large number of solved examples are provided. Finally, a case study on the grey iron casting process in a manufacturing plant is presented.

In summary,

- The path model (i) assumes linear and additive relationships among the set of variables and (ii) statistically establishes the casual ordering (priorities) among the variables.
- In the recursive path model there is no reciprocal causation or feedback loop, whereas the same is relaxed in non-recursive path models.
- The error variables (residual variables) are uncorrelated with each other and with the predetermined variables.
- A path equation may be just-identified, under-identified, or over-identified. Model identification is essential before estimating the model parameters.
- A path model may suffer from endogeneity and cross-equation correlation problems. Under endogeneity situations, OLS-based regression is not applicable. For a just-identified path equation, the preferable method is the instrumentation variable (IV) method. Identification of IVs is therefore an important task.
- For an over-identified model, 2SLS regression is widely used to estimate the path model parameters. If cross-equation correlation occurs, 3SLS is used instead of 2SLS to account for the cross-equation correlations.
- Several diagnostic tests—namely, Hausman test, Anderson-Rubin test, Sargan test, and weak instrument test—are used to test the endogeneity, over-identifying restrictions, and for searching of strong instruments.
- The IV method and 2SLS are limited information methods as the estimation for one path equation is made at a time. The 3SLS method is known as full information method as it estimates all the path equations simultaneously.
- The above methods (IV, 2SLS, and 3SLS) of estimation are widely used in estimating simultaneous equations. Another popular approach is maximum likelihood estimation (MLE). When MLE is used to estimate one path equation at a time, it is known as limited information maximum likelihood (LIML). On the other hand, when MLE is used to estimate all the path equations simultaneously, the approach is known as full information maximum likelihood (FIML). The FIML approach of estimation will be discussed in Chapter 14.

EXERCISES

(A) Short conceptual questions

10.1 What is path modeling? In what ways it is different from multivariate multiple linear regression (MMLR)?

10.2 What are the different types of variables used in path model and what are the different types of relationships modeled?

10.3 What is a causal relationship? What are the requirements to be tested for a causal relation to be established?

10.4 Define recursive and non-recursive path models.

10.5 What are the generic steps of developing a path model?

10.6 State the assumptions of a path model.

10.7 Develop path equations for the following recursive path model.

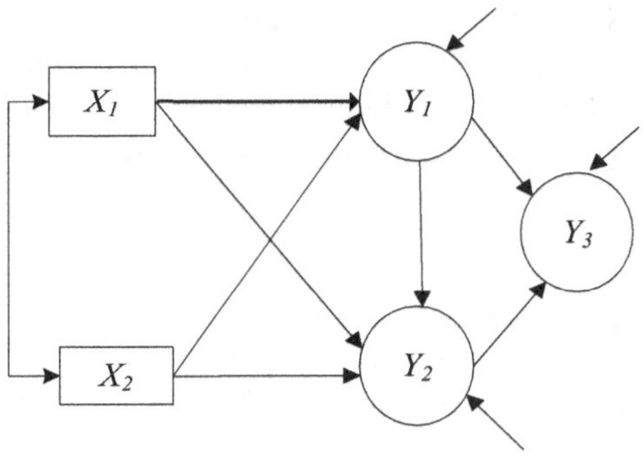

10.8 What is a predetermined variable (PV) in a path model? How do you identify a PV? For the path model given in question 10.7, identify the predetermined variables for Y_1, Y_2, and Y_3, separately.

10.9 Consider question 10.7. Develop the normal and reduced form equations for the path model.

10.10 Consider the following path model. In what ways it is different from the path model given in question 10.7? Develop path equations, its normal and reduced form equations for this path model.

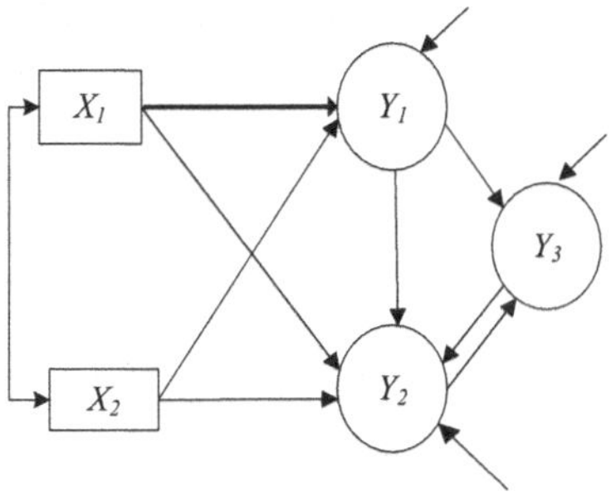

10.11 Consider the path model in question 10.10 and identify the parameter vectors and matrices to be estimated.

10.12 What is model identification? Explain the steps with examples. [Hint: consider the path model given in question 10.10.]

10.13 What are order condition and rank condition in a non-recursive path model? State an approach to test the rank condition.

10.14 What are the different types of methods available in estimating non-recursive path models? State the differences between the methods of estimation.

10.15 What is an instrumental variable (IV)? Explain with an example with reference to multiple linear regression (MLR).

10.16 Explain the steps involved in IV method of estimation of path model parameters.

10.17 What is the two stage least squares (2SLS) method? When it is used? In what ways it is different from IV method?

10.18 What is three stage least squares (3SLS)? When it is used? In what ways it is different from 2SLS?

10.19 What is test of endogeneity in path model? Explain Hausman test.

10.20 What is over-identifying restriction for path equation? Explain the Anderson-Rubin test.

10.21 What is the Sargan test? When it is used?
10.22 Explain with steps the test for identifying weak instruments.

(B) Long conceptual and/or numerical questions: see web resources (Multivariate Statistical Modeling in Engineering and Management – 1st (routledge.com))

NOTES

1 Both the independent variables and instrumental variables are denoted by IV. The independent (exogenous) variables are instrumental variables of their own.
2 All computations in this chapter were made using R and Excel, as and when required.

11 Principal Component Analysis

Principal component analysis (PCA) is a widely used statistical model for dimensionality reduction. PCA was developed by Hotelling in 1993 following the original work of Pearson (1901). When multivariate data (**X**) have inherent correlated structure, PCA transforms **X** to lower dimensions (**Z**). These transformed dimensions (**Z**) are known as principal components (PCs) and are independent in nature. The transformation can be linear or non-linear. In linear transformation, **Z** is weighted linear combination of **X**. PCA is extremely useful when the lower order transformed dimensions (**Z**) keep almost the full information as is contained in the original variables (**X**). The added feature with PCs is that the first PC contains the maximum information followed by the second PC and so on, in a decreasing order of magnitude. In this chapter, we provide an elaborate description of the linear PCA, starting from its purpose, conceptual model, extraction of PCs, number of PCs to be retained, test of variance, and orthogonality requirements. All the necessary mathematics are explained with examples. Finally, a case study on job stress of coke plant operation workers is presented.

11.1 CONCEPTUAL MODEL

Let us consider Figure 11.1, a bivariate scatter plot of variables X_1 and X_2 using n observations. If the population is characterized by the two variables X_1 and X_2, we can say that

- The population contains a wide range of values along both X_1 and X_2 dimensions.
- The variability (range of values) along X_2 (say σ_2^2) is more than along X_1 (say, σ_1^2) (i.e. $\sigma_2^2 > \sigma_1^2$).
- The two variables are correlated.

Now, let us assume that the same phenomenon can also be measured by another two dimensions, Z_1 and Z_2. Figure 11.2 depicts the situations. If we represent the same observations (as in Figure 11.1) along the Z_1–Z_2 coordinate system, we observe that

- The scatter contains a wide range of values along Z_1 but a low range of values along Z_2. So, the variability of Z_1 is much higher than the variability of Z_2.
- The variability of Z_1 is more than the variability of any of the original variables (X_1 and X_2).
- The new variables Z_1 and Z_2 are uncorrelated (independent).

Note that the Z_1–Z_2 plane is obtained by rotating the X_1–X_2 plane by angle θ anti-clockwise.

From the variance explanation point of view, Z_1 captures the maximum variance followed by Z_2 (which shares a small amount of variance). If the variance captured along Z_2 is small enough, then Z_1 alone is sufficient to explain the variance contained by the original variables X_1 and X_2 (in two dimensions). Under such conditions, the population can be characterized by only one dimension (Z_1). In other words, what we have observed about the population in the original two dimensions (X_1–X_2 plane) can be explained by the transformed dimension Z_1 alone. This is what is known as dimension reduction. PCA does it efficiently. The other transformed dimension Z_2 may not be ignored depending on the objectives of the study. If the objective is to get an uncorrelated variable

DOI: 10.1201/9781003303060-14

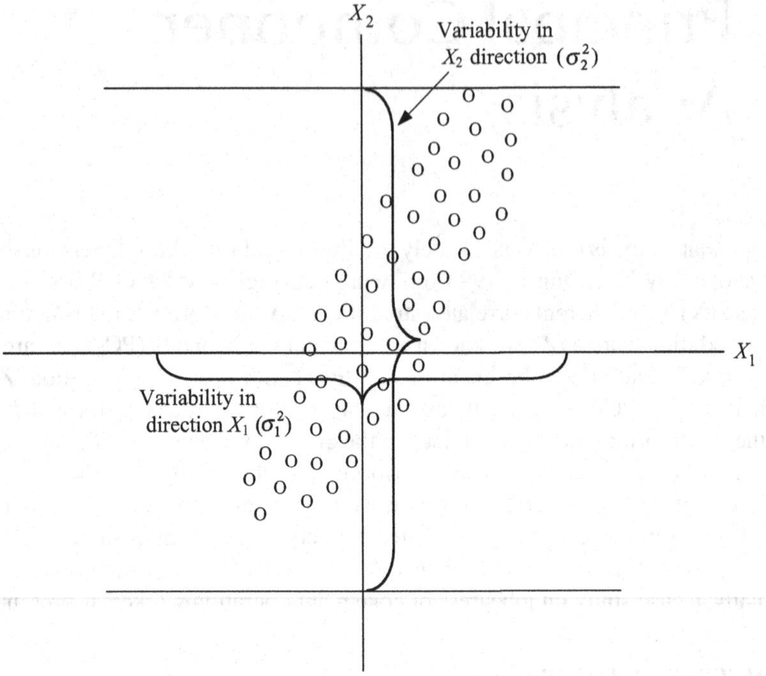

FIGURE 11.1 Scatter plot of two correlated variables along the X_1–X_2 plane. The plot shows that the variables X_1 and X_2 are positively correlated revealing a data structure (elliptical) amenable for data reduction.

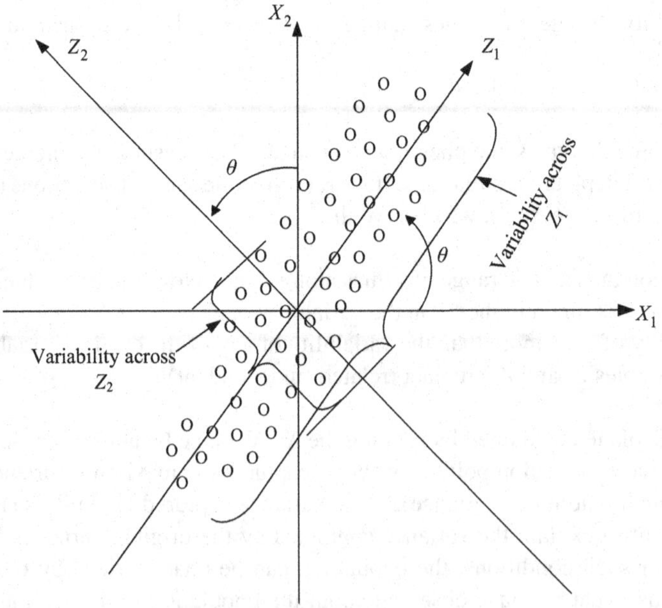

FIGURE 11.2 Scatter plot along Z_1–Z_2 plane. Z_1–Z_2 axes are obtained by rotating X_1–X_2 axes anticlockwise by an angle θ. The variability of Z_1 is comparatively higher than the variability of Z_2.

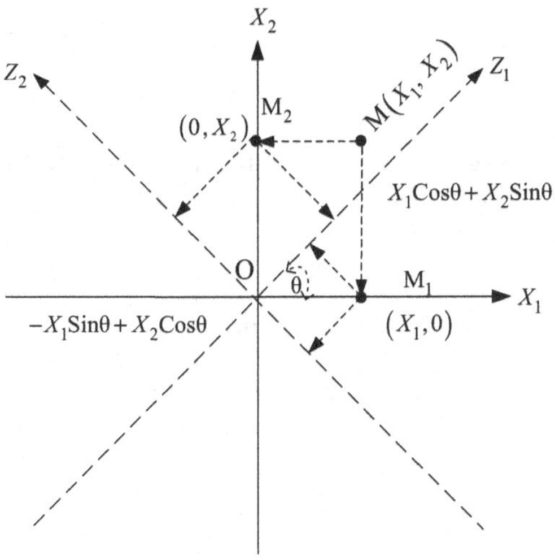

FIGURE 11.3 Geometry of two-dimensional transformation. $M(X_1, X_2)$ represents a point on the X_1–X_2 coordinate system. The projection of M on the Z_1–Z_2 coordinate system is shown in the figure.

vector for subsequent analysis such as multiple regression, both Z_1 and Z_2 could be treated as independent variables.

How can the data captured in the original X_1 and X_2 variables be transformed to the new variables Z_1 and Z_2? We will explain the simple geometry now. From Figure 11.2, it is seen that if we rotate X_1–X_2 axes by $\theta°$ anticlockwise, we get Z_1–Z_2 axes. Now, suppose there is an observation $M(X_1, X_2)$ in the original X_1 and X_2 plane (Figure 11.3). The corresponding points on the X_1 and X_2 axes are M_1 $(X_1,0)$ and M_2 $(0, X_2)$. Therefore, the length of OM_1(O is the origin) is X_1 and that of OM_2 is X_2. If we project these lengths into Z_1 and Z_2, we get the following information:

- The projected length of X_1 on Z_1 is $X_1 \cos\theta$.
- The projected length of X_2 on Z_1 is $X_2 \sin\theta$.
- The projected length of X_1 on Z_2 is $-X_1 \sin\theta$.
- The projected length of X_2 on Z_2 is $X_2 \cos\theta$.

Therefore, the coordinates of the point M on the Z_1 - Z_2 axes will be $M(Z_1, Z_2)$, where

$$\left.\begin{aligned} Z_1 &= X_1 \cos\theta + X_2 \sin\theta \\ Z_2 &= -X_1 \sin\theta + X_2 \cos\theta \end{aligned}\right\} \quad (11.1)$$

In matrix form, we can write

$$\begin{bmatrix} Z_1 \\ Z_2 \end{bmatrix} = \begin{bmatrix} \cos\theta & \sin\theta \\ -\sin\theta & \cos\theta \end{bmatrix} \begin{bmatrix} X_1 \\ X_2 \end{bmatrix} \quad (11.2)$$

$$\text{or } \boldsymbol{Z} = \mathbf{A}^T \boldsymbol{X}$$

where $\boldsymbol{Z} = \begin{bmatrix} Z_1 \\ Z_2 \end{bmatrix}, \boldsymbol{X} = \begin{bmatrix} X_1 \\ X_2 \end{bmatrix}$ and $\mathbf{A} = \begin{bmatrix} \cos\theta & -\sin\theta \\ \sin\theta & \cos\theta \end{bmatrix} = [\mathbf{a}_1 \ \ \mathbf{a}_2]$

such that, $\mathbf{a}_1 = \begin{bmatrix} \cos\theta \\ \sin\theta \end{bmatrix}$ and $\mathbf{a}_2 = \begin{bmatrix} -\sin\theta \\ \cos\theta \end{bmatrix}$

So,

$$Z_1 = \mathbf{a}_1^T x$$
$$Z_2 = \mathbf{a}_2^T x \quad (11.3)$$

Equation 11.3 defines the transformed variables (dimensions) from the original variables X_1 and X_2.

Let us see the properties of $\mathbf{a}_1$, $\mathbf{a}_2$ and $\mathbf{A}$, the transformation vectors along Z_1 and Z_2, and the transformation matrix, respectively.

$$\mathbf{a}_1^T\mathbf{a}_1 = \begin{bmatrix}\cos\theta & \sin\theta\end{bmatrix}\begin{bmatrix}\cos\theta \\ \sin\theta\end{bmatrix}$$
$$= \cos^2\theta + \sin^2\theta = 1$$

Similarly, $\mathbf{a}_2^T\mathbf{a}_2 = 1$

and

$$\mathbf{A}^{-1} = \frac{1}{|\mathbf{A}|}\left[\text{cofactor of } \mathbf{A}\right]^T$$
$$= \frac{1}{\cos^2\theta + \sin^2\theta}\begin{bmatrix}\cos\theta & -\sin\theta \\ \sin\theta & \cos\theta\end{bmatrix}$$
$$= \begin{bmatrix}\cos\theta & -\sin\theta \\ \sin\theta & \cos\theta\end{bmatrix} = \mathbf{A}^T$$

$$\therefore \mathbf{A}^T\mathbf{A} = \mathbf{A}^{-1}\mathbf{A} = \mathbf{I} = \mathbf{A}\mathbf{A}^T$$

From the above discussion, we have seen that a two-dimensional data matrix (**X**) can be transformed into two-dimensional orthogonal dimensions (**Z**). The same can be achieved for p-dimensional (p-variable) data matrix as given below.

$$\left.\begin{array}{lllll} Z_1 & = & \mathbf{a}_1^T x & = & a_{11}X_1 + a_{12}X_2 + \cdots + a_{1p}X_p \\ Z_2 & = & \mathbf{a}_2^T x & = & a_{21}X_1 + a_{22}X_2 + \cdots + a_{2p}X_p \\ \vdots & = & \cdots & = & \cdots\cdots\cdots\cdots\cdots\cdots\cdots\cdots \\ Z_j & = & \mathbf{a}_j^T x & = & a_{j1}X_1 + a_{j2}X_2 + \cdots + a_{jp}X_p \\ \vdots & = & \cdots & = & \cdots\cdots\cdots\cdots\cdots\cdots\cdots\cdots \\ Z_p & = & \mathbf{a}_p^T x & = & a_{p1}X_1 + a_{p2}X_2 + \cdots + a_{pp}X_p \end{array}\right\} \quad (11.4)$$

Where $\mathbf{a}_j^T\mathbf{a}_j = 1, j = 1,2,\ldots.., p$,

Equation 11.4 gives p-principal components $Z_1, Z_2, \ldots, Z_p$ obtained from p-original variables $X_1, X_2, \ldots, X_p$. The following need to be satisfied for Z_j, $j = 1, 2, \ldots, p$.

- $Z_1, Z_2, \ldots, Z_p$ are independent, i.e., $Cov(Z_j, Z_k) = 0, j \neq k$..
- $Var(Z_1) \geq Var(Z_2) \geq \cdots \geq Var(Z_p)$

Now, we will discuss about the mean (expected value) and variance of Z_j.

$$E(Z_j) = E(\mathbf{a}_j^T x) = \mathbf{a}_j^T E(x) = \mathbf{a}_j^T \mu \quad (11.5)$$

$$Var(Z_j) = Var(\mathbf{a}_j^T x) = \mathbf{a}_j^T Var(x)\mathbf{a}_j = \mathbf{a}_j^T \Sigma \mathbf{a}_j \quad (11.6)$$

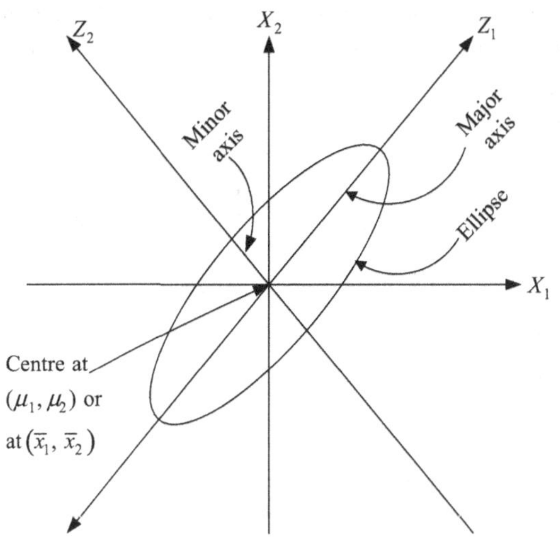

FIGURE 11.4 An ellipse covering the data points shown in Figure 11.1 (not to the scale). The new dimensions Z_1 and Z_2 are parallel to the major and minor axes of the ellipse, respectively. The center of the ellipse coincides with the mean vector of the two original variables (X_1 and X_2).

Where $\boldsymbol{\mu}$ and $\boldsymbol{\Sigma}$ are population mean vector of size $p\times1$ and population covariance matrix of size $p\times p$ for X, respectively. But it is seldom possible to get the values of the population mean vector and covariance matrix. Therefore, we consider the sample estimates of $\boldsymbol{\mu}$ and $\boldsymbol{\Sigma}$, namely $\bar{\mathbf{x}}$ and **S**, the sample mean vector and covariance matrix, respectively. When principal components (PCs) are extracted using the population covariance matrix ($\boldsymbol{\Sigma}$), they are known as population PCs and the analysis is known as population PCA. Similarly, when PCs are extracted using **S** (the sample covariance matrix), they are known as sample PCs and the analysis is known as sample PCA. PCA is an interdependence modeling technique.

11.2 EXTRACTING PRINCIPAL COMPONENTS

Looking at Figure 11.2 again, we can see that the data set resembles a pattern, and we can plot an ellipse covering the data points. Figure 11.4 depicts this.

The axes Z_1 and Z_2 represent the major and minor axes of the ellipse. Therefore, if we can define the ellipse and its axes, we can define the PCs Z_1 and Z_2. We have seen in Chapter 5 that an ellipse can be represented as:

$$(X-\boldsymbol{\mu})^T \boldsymbol{\Sigma}^{-1}(X-\boldsymbol{\mu}) \le \chi_p^2(\alpha) \tag{11.7}$$

$$(X-\bar{\mathbf{x}})^T \mathbf{S}^{-1}(X-\bar{\mathbf{x}}) \le \chi_p^2(\alpha) \tag{11.8}$$

For Equation 11.8 to hold, the sample size n should be large. It should be noted here that in Equations 11.7 and 11.8, the ellipse is centered at $\boldsymbol{\mu}$ and $\bar{\mathbf{x}}$, respectively.

The axes of the ellipse are defined as $\pm d\sqrt{\lambda_1}\,\mathbf{a}_1$ and $\pm d\sqrt{\lambda_2}\,\mathbf{a}_2$, where $(\lambda_1,\mathbf{a}_1)$ and $(\lambda_2,\mathbf{a}_2)$ are the eigenvalue–eigenvector pairs of the covariance matrix $\boldsymbol{\Sigma}$ (for Equation 11.7) and **S** (for Equation 11.8). d is $\sqrt{\chi_p^2(\alpha)}$. Therefore, we can say that if we know $(\lambda_j, \mathbf{a}_j)$ for $j = 1, 2, \ldots, p$, we know the PCs. If we know the population covariance matrix $\boldsymbol{\Sigma}$, we can obtain population PCs. But population $\boldsymbol{\Sigma}$ is seldom known.

When we use sample covariance matrix **S**, PCA gives sample PCs. No matter whatever matrix (**Σ** or **S**) we use, the estimation process of PCs is the same. We use the eigenvalue–eigenvector decomposition method for this purpose. We will describe only sample PCs.

Equation 11.4, in matrix form, can be written as

$$Z = \mathbf{A}^T X \tag{11.9}$$

where

$$\mathbf{A} = \begin{bmatrix} a_{11} & a_{21} & \cdots & a_{j1} & \cdots & a_{p1} \\ a_{12} & a_{22} & \cdots & a_{j2} & \cdots & a_{p2} \\ \vdots & \vdots & \cdots & \vdots & \cdots & \vdots \\ a_{1j} & a_{2j} & \cdots & a_{jj} & \cdots & a_{pj} \\ \vdots & \vdots & \cdots & \vdots & \cdots & \vdots \\ a_{1p} & a_{2p} & \cdots & a_{jp} & \cdots & a_{pp} \end{bmatrix} = \begin{bmatrix} \mathbf{a}_1 & \vdots & \mathbf{a}_2 & \vdots & \cdots & \vdots & \mathbf{a}_j & \vdots & \cdots & \vdots & \mathbf{a}_p \end{bmatrix}$$

Such that

$$\mathbf{a}_j = \begin{bmatrix} a_{j1} \\ a_{j2} \\ \vdots \\ a_{jj} \\ \vdots \\ a_{jp} \end{bmatrix}$$

As stated earlier, in PCA the first PC extracts the maximum variance of X followed by second PC and so on, subject to $\mathbf{a}_j^T\mathbf{a}_j = 1$, and $Cov\left(\mathbf{a}_j, \mathbf{a}_k\right) = 0,\ j \neq k,\ j,k = 1,2,\cdots p.$

As a generalization, we can write

$$Var(Z_j) = Var\left(\mathbf{a}_j{}^T X\right) = \mathbf{a}_j{}^T Var(X)\,\mathbf{a}_j = \mathbf{a}_j^T \mathbf{S} \mathbf{a}_j$$

where, $\mathbf{S} = \begin{bmatrix} s_{11} & s_{12} & \cdots & s_{1j} & \cdots & s_{1p} \\ s_{12} & s_{22} & \cdots & s_{2j} & \cdots & s_{2p} \\ \vdots & \vdots & \cdots & \vdots & \cdots & \vdots \\ s_{1j} & s_{2j} & \cdots & s_{jj} & \cdots & s_{jp} \\ \vdots & \vdots & \cdots & \vdots & \cdots & \vdots \\ s_{1p} & s_{2p} & \cdots & s_{jp} & \cdots & s_{pp} \end{bmatrix}$

So, we require to maximize $\mathbf{a}_j^T\mathbf{S}\mathbf{a}_j$ subject to $\mathbf{a}_j^T\mathbf{a}_j = 1$. This can be done by using Lagrange multipliers and maximizing L, as shown below.

$$L = \mathbf{a}_j^T\mathbf{S}\mathbf{a}_j - \lambda\left(\mathbf{a}_j^T\mathbf{a}_j - 1\right) \tag{11.10}$$

where, λ is the Lagrange multiplier.

Now, differentiating L with respect to $\mathbf{a}_j$, we get

$$\frac{\partial L}{\partial \mathbf{a}_j} = 2\mathbf{S}\mathbf{a}_j - 2\lambda \mathbf{a}_j.$$

$$\text{Putting } \frac{\partial L}{\partial \mathbf{a}_j} = 0, \text{ we get}$$

$$2\mathbf{S}\mathbf{a}_j - 2\lambda \mathbf{a}_j = 0$$

$$\text{or, } (\mathbf{S} - \lambda \mathbf{I})\mathbf{a}_j = 0 \tag{11.11}$$

Equation 11.11 is called the eigenvalue–eigenvector problem for the covariance matrix **S**. λ_j represents the eigenvalue and $\mathbf{a}_j$ represents the eigenvector corresponding to λ_j.

λ_j (in Equation 11.11) can be determined by solving the characteristic equation $|\mathbf{S} - \lambda \mathbf{I}| = 0$. For **S**, a $p\text{x}p$ matrix, the characteristic equation is a p-th order polynomial of λ and we have p-roots for λ, i.e., $\lambda_1, \lambda_2, \ldots \lambda_p$, in the order $\lambda_1 \geq \lambda_2 \geq \lambda_3 \geq \cdots \geq \lambda_p$. For **S**, a full rank and positive definite matrix, there will be p eigenvalues. Again solving Equation 11.11 for each of $\lambda_1, \lambda_2, \ldots \lambda_p$, we get the eigenvectors $\mathbf{a}_1$, $\mathbf{a}_2, \ldots, \mathbf{a}_p$, respectively.

Example 11.1

Obtain PCs for the following sample covariance matrix S.

$$\mathbf{S} = \begin{bmatrix} 100 & 50 & -30 \\ 50 & 64 & -20 \\ -30 & -20 & 49 \end{bmatrix}$$

Solution:

$$\boldsymbol{X} = \left[X_1, X_2, X_3\right]^T$$

Step 1: Find out $|\mathbf{S} - \lambda \mathbf{I}| = 0$, which yields (using R software)

$$\begin{array}{rcr|} \lambda_1 & = & 148.23 \\ \lambda_2 & = & 36.00 \\ \lambda_3 & = & 28.76 \end{array}$$

Step 2: Solving $(\mathbf{S} - \lambda_1 \mathbf{I})\mathbf{a}_1 = 0$ for $\mathbf{a}_1$ yields

$$\mathbf{a}_1 = \begin{bmatrix} 0.77 \\ 0.54 \\ -0.34 \end{bmatrix}$$

Step 3: Solving $(\mathbf{S} - \lambda_2 \mathbf{I})\mathbf{a}_2 = 0$ yields

$$\mathbf{a}_2 = \begin{bmatrix} 0.21 \\ 0.29 \\ 0.93 \end{bmatrix}$$

Step 4: Solving $(\mathbf{S}-\lambda_3\mathbf{I})\mathbf{a}_3 = 0$ yields

$$\mathbf{a}_3 = \begin{bmatrix} 0.60 \\ -0.79 \\ 0.11 \end{bmatrix}$$

So, the PCs are

$$Z_1 = \mathbf{a}_1^T X = 0.77X_1 + 0.54X_2 - 0.34X_3$$
$$Z_2 = \mathbf{a}_2^T X = -0.21X_1 + 0.29X_2 + 0.93X_3$$
$$Z_3 = \mathbf{a}_3^T X = 0.60X_1 - 0.79X_2 + 0.11X_3$$

Alternatively, singular value decomposition (SVD) can be used to obtain principal components (PCs). Here, the mean centered data matrix $\tilde{\mathbf{X}}(=\mathbf{X}-\mathbf{1}\bar{\mathbf{x}}^T)$ is decomposed as $\tilde{\mathbf{X}} = \underset{n\times n}{\mathbf{U}}\underset{n\times p}{\mathbf{D}}\underset{p\times p}{\mathbf{V}^T}$ (Krzanowski and Marriott, 2014a). The matrices $\underset{n\times n}{\mathbf{U}}$ and $\underset{p\times p}{\mathbf{V}}$ are orthogonal and $\mathbf{D}$ is diagonal. The diagonal elements of $\mathbf{D}$ are the singular values, which are the square roots of the eigenvalues.

Now, let $\mathbf{\Lambda}$ represents a diagonal matrix containing λ_j, $j = 1, 2, \cdots p$, at the diagonal line. Then,

$$\mathbf{\Lambda} = \begin{bmatrix} \lambda_1 & \cdots & 0 & \cdots & 0 \\ 0 & \cdots & \lambda_2 & \cdots & 0 \\ \vdots & \vdots & \vdots & \vdots & \vdots \\ 0 & \cdots & 0 & \cdots & \lambda_p \end{bmatrix}_{p\times p}$$

It can be proved that

$$\mathbf{S} = \mathbf{A\Lambda A}^T \tag{11.12}$$

where $\mathbf{A}$ is shown in Equation 11.9. A is idempotent matrix and hence, $\mathbf{A}^T\mathbf{A} = \mathbf{AA}^T = \mathbf{I}$.

$$\therefore \text{Trace of } \mathbf{S} = \textit{trace} \text{ of} (\mathbf{A\Lambda A}^T)$$

or,

$$\begin{aligned} \sum_{j=1}^{p} s_{jj} &= trace(\mathbf{A\Lambda A}^T) \\ &= trace(\mathbf{\Lambda AA}^T) \\ &= trace(\mathbf{\Lambda I}) \\ &= trace(\mathbf{\Lambda}) \\ &= \sum_{j=1}^{p} \lambda_j \end{aligned} \tag{11.13}$$

Therefore, the total variance contained in the original variables is equal to the sum total of λ_j values, $j = 1, 2, \ldots, p$.

Similarly, we can prove that the variance of the first PC, Z_1 is λ_1, second PC, Z_2 is λ_2 and so on. To verify this, consider Example 11.1.

$$Var(Z_1) = \mathbf{a}_1^T \mathbf{S}\mathbf{a}_1 = \begin{bmatrix} 0.77 & 0.54 & -0.34 \end{bmatrix} \begin{bmatrix} 100 & 50 & -30 \\ 50 & 64 & -20 \\ -30 & -20 & 49 \end{bmatrix} \begin{bmatrix} 0.77 \\ 0.54 \\ -0.34 \end{bmatrix} = 148.23 = \lambda_1,$$

$$Var(Z_2) = \mathbf{a}_2^T \mathbf{S}\mathbf{a}_2 = \begin{bmatrix} 0.21 & 0.29 & 0.93 \end{bmatrix} \begin{bmatrix} 100 & 50 & -30 \\ 50 & 64 & -20 \\ -30 & -20 & 49 \end{bmatrix} \begin{bmatrix} 0.21 \\ 0.29 \\ 0.93 \end{bmatrix} = 36.00 = \lambda_2, \text{and}$$

$$Var(Z_3) = \mathbf{a}_3^T \mathbf{S}\mathbf{a}_3 = \begin{bmatrix} 0.60 & -0.79 & 0.11 \end{bmatrix} \begin{bmatrix} 100 & 50 & -30 \\ 50 & 64 & -20 \\ -30 & -20 & 49 \end{bmatrix} \begin{bmatrix} 0.60 \\ -0.79 \\ 0.11 \end{bmatrix} = 28.76 = \lambda_3.$$

Now, we will explain the correlation coefficient of X_k and Z_j, $j = 1, 2, \ldots, p$. The correlation coefficient is r_{jk} (See Johnson and Wichern, 2002).

$$r_{jk} = \frac{a_{jk} \times \sqrt{\lambda_j}}{\sqrt{s_{kk}}}, j = 1,2,\ldots p, \text{ and } k = 1,2,\ldots p \tag{11.14}$$

For all (*j*, *k*) combinations, we will get a correlation matrix $\mathbf{R}_{zx}$ as shown below.

$$\mathbf{R}_{ZX} = \begin{bmatrix} \frac{a_{11}\sqrt{\lambda_1}}{\sqrt{s_{11}}} & \frac{a_{12}\sqrt{\lambda_1}}{\sqrt{s_{22}}} & \frac{a_{1p}\sqrt{\lambda_1}}{\sqrt{s_{pp}}} \\ \frac{a_{21}\sqrt{\lambda_2}}{\sqrt{s_{11}}} & \frac{a_{22}\sqrt{\lambda_2}}{\sqrt{s_{22}}} & \frac{a_{2p}\sqrt{\lambda_2}}{\sqrt{s_{pp}}} \\ \vdots & \vdots & \vdots \\ \frac{a_{p1}\sqrt{\lambda_1}}{\sqrt{s_{11}}} & \frac{a_{p2}\sqrt{\lambda_2}}{\sqrt{s_{22}}} & \frac{a_{pp}\sqrt{\lambda_p}}{\sqrt{s_{pp}}} \end{bmatrix} = \mathbf{\Lambda}^{\frac{1}{2}} \mathbf{A}^{\mathrm{T}} \mathbf{D}^{-\frac{1}{2}} \tag{11.15}$$

where

$$\mathbf{\Lambda} = \begin{bmatrix} \lambda_1 & 0 & \cdots & 0 \\ 0 & \lambda_2 & \cdots & 0 \\ \vdots & \vdots & \vdots & \vdots \\ 0 & 0 & \cdots & \lambda_p \end{bmatrix}, \mathbf{A} = \begin{bmatrix} a_{11} & a_{21} & \cdots & a_{p1} \\ a_{12} & a_{22} & \cdots & a_{p2} \\ \vdots & \vdots & \vdots & \vdots \\ a_{1p} & a_{2p} & \cdots & a_{pp} \end{bmatrix} \text{ and } \mathbf{D} = \begin{bmatrix} s_{11} & \cdots & 0 & \cdots & 0 \\ 0 & \cdots & s_{22} & \cdots & 0 \\ \vdots & \vdots & \vdots & \vdots & \vdots \\ 0 & \cdots & 0 & \cdots & s_{pp} \end{bmatrix}$$

Example 11.2

Consider Example 11.1. Compute the correlations coefficients between Z and X.

Solution:

From Equation 11.15, we know that $\mathbf{R}_{ZX} = \Lambda^{\frac{1}{2}} \mathbf{A}^{\mathrm{T}} \mathbf{D}^{-\frac{1}{2}}$. For Example 11.1, Λ, $\mathbf{A}$ and $\mathbf{D}$ are given below.

$$\Lambda = \begin{bmatrix} 148 & 0 & 0 \\ 0 & 36 & 0 \\ 0 & 0 & 28.76 \end{bmatrix}, \mathbf{A} = \begin{bmatrix} 0.77 & 0.21 & 0.6 \\ 0.54 & 0.29 & -0.79 \\ -0.34 & 0.93 & 0.11 \end{bmatrix} \text{ and } \mathbf{D} = \begin{bmatrix} 100 & 0 & 0 \\ 0 & 64 & 0 \\ 0 & 0 & 49 \end{bmatrix}.$$

$$\text{So, } \Lambda^{1/2} = \begin{bmatrix} 12.166 & 0 & 0 \\ 0 & 6 & 0 \\ 0 & 0 & 5.363 \end{bmatrix}, \mathbf{A}^{T} = \begin{bmatrix} 0.77 & 0.54 & -0.34 \\ 0.21 & 0.29 & 0.93 \\ 0.6 & -0.79 & 0.11 \end{bmatrix} \text{ and } \mathbf{D}^{-1/2} = \begin{bmatrix} 0.1 & 0 & 0 \\ 0 & 0.125 & 0 \\ 0 & 0 & 0.142857 \end{bmatrix}.$$

Therefore,

$$\mathbf{R}_{ZX} = \Lambda^{\frac{1}{2}} \mathbf{A}^{\mathrm{T}} \mathbf{D}^{-\frac{1}{2}} = \begin{bmatrix} 12.166 & 0 & 0 \\ 0 & 6 & 0 \\ 0 & 0 & 5.363 \end{bmatrix} \begin{bmatrix} 0.77 & 0.54 & -0.34 \\ 0.21 & 0.29 & 0.93 \\ 0.6 & -0.79 & 0.11 \end{bmatrix} \begin{bmatrix} 0.1 & 0 & 0 \\ 0 & 0.125 & 0 \\ 0 & 0 & 0.142857 \end{bmatrix}$$

$$= \begin{bmatrix} 0.937 & 0.821 & -0.591 \\ 0.126 & 0.218 & 0.797 \\ 0.322 & -0.530 & 0.084 \end{bmatrix}$$

$\tilde{X}_{p\times1}$ Let is a $p\times1$ vector representing the p number of standardized X variables. Note that $\mathbf{X}_{nxp}$ is a data matrix and $\tilde{X}_{p\times1}$ is a variable vector (standardized).

So far we have used covariance matrix **S** to compute the PCs. In many situations, it is required to extract PCs using correlation matrix, **R**. The situations are given below (Yang and Trewn, 2003).

1. When the variables $X_1, X_2, \ldots, X_p$ are measured in different units.
2. When the variables $X_1, X_2, \ldots, X_p$ have different meanings as well as varying magnitude.

Under such situations, if we use **S** for PCA, it is highly likely that some of the original variables X_1, $X_2, \ldots, X_p$ might exert more influence than others on the extracted PCs. It is preferable to use **R** in place of **S**.

The following steps are to be followed to get PCs using **R**. Let n multivariate observations on p number of X variables, say $\mathbf{X}_{n\times p}$ are collected.

Let $\tilde{X}_{p\times1}$ is a $p\times1$ vector representing the p number of standardized X variables.

Step 1: Convert the original data matrix $\mathbf{X}_{n\times p}$ into standardized data matrix $\tilde{\mathbf{X}}_{n\times p}$, where a general observation x_{ij} of $\mathbf{X}_{n\times p}$ is transformed into $\tilde{x}_{ij}$ for $\tilde{\mathbf{X}}_{n\times p}$ such that $\tilde{x}_{ij} = \dfrac{x_{ij} - \overline{x}_j}{\sqrt{s_{jj}}}$, $\overline{x}_j$ = mean of X_j, s_{jj} = variance of X_j and $\tilde{X}_j$ is the jth standardized variable, $i = 1,2,\ldots n$, and $j = 1,2,\ldots p$.

Step 2: Compute the correlation matrix **R**

$$\mathbf{R} = \frac{1}{n-1}\tilde{\mathbf{X}}^T\tilde{\mathbf{X}}$$

Alternatively, if covariance matrix **S** is given, **R** can be computed using Equation 4.19.

Step 3: Obtain eigenvalues λ_j and eigenvectors $\mathbf{a}_j$ for the correlation matrix **R** where $j = 1, 2, \ldots, p$ and $\lambda_1 \geq \lambda_2 \geq \cdots \geq \lambda_p$.

Step 4: The PCs are

$$\begin{aligned}
Z_1 &= \mathbf{a}_1^T\tilde{X} = a_{11}\tilde{X}_1 + a_{12}\tilde{X}_2 + \cdots a_{1p}\tilde{X}_p \\
Z_2 &= \mathbf{a}_2^T\tilde{X} = a_{21}\tilde{X}_1 + a_{22}\tilde{X}_2 + \cdots a_{2p}\tilde{X}_p \\
\vdots &= \vdots \\
Z_p &= \mathbf{a}_p^T\tilde{X} = a_{p1}\tilde{X}_1 + a_{p2}\tilde{X}_2 + \cdots a_{pp}\tilde{X}_p
\end{aligned}$$

The sample variance for the j-th PC, i.e., Z_j is λ_j, $j = 1,2,\ldots p$ and the covariance between Z_j and Z_k i.e., $Cov\left(Z_j, Z_k\right), j \neq k$, is zero. Further, the total variance explained by the p number of PCs is

$$\sum_{j=1}^{p}\lambda_j = trace(\mathbf{R}) = p. \tag{11.16}$$

and the correlation coefficient between Z_j and $\tilde{X}_k$ is

$$r_{jk} = a_{jk}\sqrt{\lambda_j},\ j = 1,2,\ldots., p \text{ and } k = 1,2,\ldots., p. \tag{11.17}$$

Example 11.3

Obtain correlation matrix **R** from **S** (given in Example 11.1) and compute PCs using **R**.

Solution:

Here,

$$\mathbf{S} = \begin{bmatrix} 100 & 50 & -30 \\ 50 & 64 & -20 \\ -30 & -20 & 49 \end{bmatrix} \text{ and } \mathbf{D} = \begin{bmatrix} 100 & 0 & 0 \\ 0 & 64 & 0 \\ 0 & 0 & 49 \end{bmatrix}.$$

$$\text{Then, } \mathbf{R}=\mathbf{D}^{-1/2}\mathbf{S}\mathbf{D}^{-1/2} = \begin{bmatrix} 0.1 & 0 & 0 \\ 0 & 0.125 & 0 \\ 0 & 0 & 0.142857 \end{bmatrix}\begin{bmatrix} 100 & 50 & -30 \\ 50 & 64 & -20 \\ -30 & -20 & 49 \end{bmatrix}\begin{bmatrix} 0.1 & 0 & 0 \\ 0 & 0.125 & 0 \\ 0 & 0 & 0.142857 \end{bmatrix}$$

$$= \begin{bmatrix} 1.0000 & 0.6250 & -0.4286 \\ 0.6250 & 1.000 & -0.3571 \\ -0.4286 & -0.3571 & 1.0000 \end{bmatrix}$$

Step 1: Find out $|\mathbf{R}-\lambda\mathbf{I}|=0$

which yields (using R software)

$$\left.\begin{array}{l}\lambda_1 = 1.9504\\ \lambda_2 = 0.6813\\ \lambda_3 = 0.3684\end{array}\right|$$

Step 2: Solving $(\mathbf{R}-\lambda_1\mathbf{I})\mathbf{a}_1 = 0$ for $\mathbf{a}_1$ yields

$$\mathbf{a}_1 = \begin{bmatrix} 0.6216 \\ 0.5986 \\ -0.5053 \end{bmatrix}$$

Step 3: Solving $(\mathbf{R}-\lambda_2\mathbf{I})\mathbf{a}_2 = 0$ yields

$$\mathbf{a}_2 = \begin{bmatrix} 0.2557 \\ 0.4547 \\ 0.8532 \end{bmatrix}$$

Step 3: Solving $(\mathbf{R}-\lambda_3\mathbf{I})\mathbf{a}_3 = 0$ yields

$$\mathbf{a}_3 = \begin{bmatrix} 0.7405 \\ -0.6595 \\ 0.1295 \end{bmatrix}$$

Therefore, the three PCs are extracted, and these are

$$\begin{aligned} Z_1 &= \mathbf{a}_1^T\tilde{X} = 0.6216\tilde{X}_1 + 0.5986\tilde{X}_2 - 5053\tilde{X}_3 \\ Z_2 &= \mathbf{a}_2^T\tilde{X} = 0.2557\tilde{X}_1 + 0.4547\tilde{X}_2 + 0.8532\tilde{X}_3 \\ Z_3 &= \mathbf{a}_3^T\tilde{X} = 0.7405\tilde{X}_1 - 0.6595\tilde{X}_2 + 0.1295\tilde{X}_3 \end{aligned}$$

There are some other issues that attracted researchers' attention such as (i) use of correlation matrix, where all correlation coefficients are equal and positive and (ii) use of the first PC as a measure of size and the remaining PCs as a measure of shape, under (i). Krzanowski and Marriott (2014a) give a wonderful review on these counts.

11.3 SAMPLING DISTRIBUTION OF λ_j AND $\mathbf{a}_j$

As λ_j and $\mathbf{a}_j$, $j = 1,2,\ldots p$ are obtained using sample covariance matrix $\mathbf{S}$ (or sample correlation matrix $\mathbf{R}$), λ_j and $\mathbf{a}_j$ are random variables and as such follow certain probability distributions. Jolliffe (2002) has discussed the distribution for λ_j and $\mathbf{a}_j$.

Let us assume that the population covariance matrix $\mathbf{\Sigma}_{p\times p}$ for the variables $\boldsymbol{X}_{p\times 1}$ is known. Let θ_j and $\boldsymbol{\alpha}_j$, $j = 1,2,\ldots p$ be the eigenvalue–eigenvector pairs of $\mathbf{\Sigma}$. So, the j-th population PC is $Z_j = \alpha_{j1}X_1 + \alpha_{j2}X_2 + \cdots \alpha_{jp}X_p, j = 1,2,\ldots p$. When $\lambda_j, j = 1, 2, \ldots, p$ are obtained using sample $\mathbf{S}$, for a representative sample of size n,

$$E(\lambda_j) = \theta_j \tag{11.18}$$

$$Cov(\lambda_j, \lambda_k) = \begin{cases} \dfrac{2\theta_j^2}{n-1} & for \quad j = k \\ 0 & for \quad j \neq k \end{cases} \tag{11.19}$$

The approximate distribution of λ_j is

$$\lambda_j \sim N\left(\theta_j, \frac{2\theta_j^2}{n-1}\right) \tag{11.20}$$

Similarly, the j-th eigenvector $\mathbf{a}_j$ follows approximately multivariate normal distribution as

$$\mathbf{a}_j \sim N_p(\mathbf{a}_j, T_j) \tag{11.21}$$

$$where,\ T_j = \frac{\theta_j}{n-1} \sum_{\substack{k=1 \\ j \neq k}}^{p} \frac{\theta_k}{(\theta_k - \theta_j)^2} \mathbf{a}_k \mathbf{a}_k^T \tag{11.22}$$

As λ_j is normally distributed with mean θ_j and variance $\frac{2\theta_j^2}{n-1}$, then using central limit theorem, the quantity (the standardized λ_j) for large n

$\dfrac{\lambda_j - \theta_j}{\sqrt{\dfrac{2\theta_j^2}{n-1}}}$ follows unit normal $[Z(0,1)]$ distribution.

Figure 11.5 shows the unit normal distribution for the standardized λ_j.

So,

$$-Z_{\alpha/2} < \frac{\lambda_j - \theta_j}{\sqrt{\dfrac{2\theta_j^2}{n-1}}} < Z_{\alpha/2}$$

Taking the left side of the inequality first,

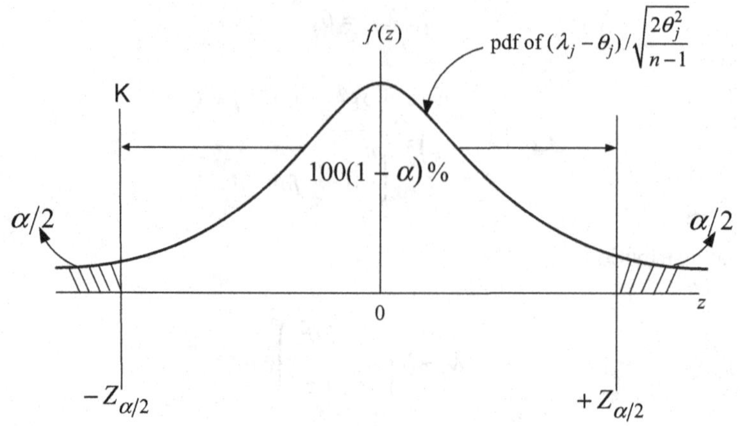

FIGURE 11.5 Distribution of standardized λ_j, which is Z distribution with mean 0 and standard deviation $\sqrt{\frac{2\theta_j^2}{n-1}}$. The hatched region represents the rejection region for H_0.

$$-Z_{\alpha/2} < \frac{\lambda_j - \theta_j}{\theta_j \sqrt{2/n-1}}$$

$$or, -\theta_j Z_{\alpha/2} \cdot \sqrt{\frac{2}{n-1}} < \lambda_j - \theta_j$$

$$or,\ \theta_j - \theta_j Z_{\alpha/2} \cdot \sqrt{\frac{2}{n-1}} < \lambda_j$$

$$or,\ \theta_j \left(1 - Z_{\alpha/2} \cdot \sqrt{\frac{2}{n-1}}\right) < \lambda_j$$

$$or,\ \theta_j < \frac{\lambda_j}{1 - Z_{\alpha/2} \cdot \sqrt{\frac{2}{n-1}}}$$

Similarly, taking the right side of the inequality, we get,

$$\theta_j > \frac{\lambda_j}{1 + Z_{\alpha/2} \cdot \sqrt{\frac{2}{n-1}}}$$

Therefore, the 100(1–α)% confidence interval for θ_j is

$$\frac{\lambda_j}{1 + Z_{\alpha/2} \cdot \sqrt{\frac{2}{n-1}}} < \theta_j < \frac{\lambda_j}{1 - Z_{\alpha/2} \cdot \sqrt{\frac{2}{n-1}}} \tag{11.23}$$

The derivation of 100(1–α)% confidence region for $\boldsymbol{\alpha}_j$ is mathematically complicated. The final result is (Jolliffe, 2002)

$$(n-1)\boldsymbol{\alpha}_j^T\left(\lambda_j\mathbf{S}^{-1}+\lambda_j^{-1}\mathbf{S}-2\mathbf{I}_p\right)\boldsymbol{\alpha}_j \le \chi_{p-1}^{2(\alpha)} \tag{11.24}$$

Where $\mathbf{I}_p$ is pxp identity matrix.

Example 11.4

Consider Example 11.1. Obtain the 100(1–α)% interval estimates of θ_j and 100(1–α)% CR for $\boldsymbol{\alpha}_j$, j = 1, 2, 3. Consider n = 30.

Solution:

(i) 100(1 – α)% confidence interval for θ_j

Here, $n = 30, Z_{0.025} = 1.96$ for $\alpha = 0.05$.

So, $1+Z_{0.025}\sqrt{\frac{2}{n-1}} = 1.5147$ and $1-Z_{0.025}\sqrt{\frac{2}{n-1}} = 0.4852.$

Now, using Equation 11.23, we get the following

$$95\%\ CI \text{ for } \theta_j, \text{j} = 1,2,3 \text{ is}$$

$$\frac{148.23}{1.5147} < \theta_1 < \frac{148.23}{0.4853}, or\ 97.86 < \theta_1 < 305.44,$$

$$\frac{36.00}{1.5147} < \theta_2 < \frac{36.00}{0.4853}, or\ 23.76 < \theta_2 < 74.18, \text{ and}$$

$$\frac{28.76}{1.5147} < \theta_3 < \frac{28.76}{0.4853}, or\ 18.99 < \theta_3 < 59.26,$$

(ii) 100(1–α)% confidence region for $\boldsymbol{\alpha}_j$

Now, let $\mathbf{m}_j = \lambda_j\mathbf{S}^{-1} + \lambda_j^{-1}\mathbf{S} - 2\mathbf{I}_p, j = 1,2,3.$

Then,

$$\mathbf{m}_1 = 148.23\begin{bmatrix}100 & 50 & -30\\ 50 & 64 & -20\\ -30 & -20 & 49\end{bmatrix}^{-1} + \frac{1}{148.23}\begin{bmatrix}100 & 50 & -30\\ 50 & 64 & -20\\ -30 & -20 & 49\end{bmatrix} - 2\begin{bmatrix}1 & 0 & 0\\ 0 & 1 & 0\\ 0 & 0 & 1\end{bmatrix}$$

$$= \begin{bmatrix}1.317 & -1.449 & 0.686\\ -1.449 & 2.295 & 0.348\\ 0.686 & 0.348 & 2.097\end{bmatrix}$$

Similarly,

$$\mathbf{m}_2 = \begin{bmatrix}1.419 & 0.955 & -0.617\\ 0.955 & 0.716 & -0.438\\ -0.617 & -0.438 & 0.276\end{bmatrix} \text{ and } \mathbf{m}_3 = \begin{bmatrix}1.990 & 1.392 & -0.871\\ 1.392 & 0.975 & -0.602\\ -0.871 & -0.602 & 0.434\end{bmatrix}$$

So, the 100(1–α)% confidence regions for $\boldsymbol{\alpha}_j$, j = 1, 2, 3 are $29(\boldsymbol{\alpha}_j^T \boldsymbol{m}_j \boldsymbol{\alpha}_j) \le \chi_{p-1}^{2(\alpha)}$, where $\boldsymbol{\alpha}_j = [\alpha_{j1}, \alpha_{j2}, \alpha_{j3}]^T$.

11.4 ADEQUACY TESTS FOR PCA

By adequacy test of PCA, we will explain

1. Bartlett's sphericity test to address the condition for PCA.
2. Different criteria for determining the number of PCs to be retained.
3. Hypothesis testing for variance explained by each of the PCs.

11.4.1 Bartlett's Sphericity Test

The basic condition for PCA is that there are substantial correlations among several of the variables (X) of interest so that their association can be captured in new but reduced dimensions which are orthogonal in nature. If the X variables are independent, PCA is not recommended. Under such condition PCA though create newer dimensions (Z), but no improvement in terms of reduction of dimensions will be achieved. For example, consider Figures 11.6a and 11.6b.

The observations on X_1 and X_2 in Figure 11.6a resemble a circle (as X_1 and X_2 are independent and have equal variance). Any two perpendicular diameters can represent two new dimensions (say Z_1 and Z_2) with equal explanation power of X_1 and X_2. Hence, no improvement is possible by extracting PCs. So, PCA is not recommended under similar conditions. On the contrary, in Figure 11.6b, the dimension Z is sufficient to represent the data observed on X_1 and X_2, as X_1 and X_2 are highly correlated (almost have perfect linear relations). PCA is highly recommended under such conditions. Bartlett (1950) has given a quantitative scheme to address the above-mentioned issue, known as the Bartlett sphericity test.

The Bartlett sphericity test is a statistical measure of the presence of significant correlations among some of the X variables (Heir et al., 1998). For a sample of size n involving p-variables, the Bartlett test statistic is

$$-\left[(n-1)-\left(\frac{2p+5}{6}\right)\right] \ln|\mathbf{R}|,$$

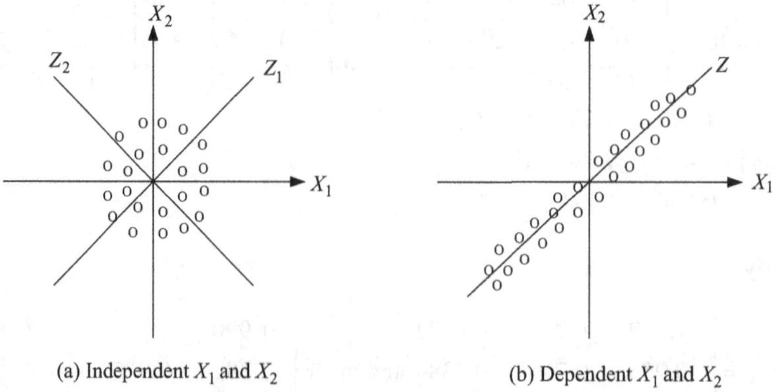

FIGURE 11.6 Condition for PCA. To get the advantage of PCA, the data should exhibit correlated structure. In Figure (a), the circular structure represents uncorrelated data, hence, PCA is not recommended. In Figure (b), the ellipse structure represents correlated data, hence, PCA is recommended.

which follows χ^2 distribution with p(p–1)/2 degree of freedom. So,

$$-\left[(n-1)-\left(\frac{2p+5}{6}\right)\right]\ln|\mathbf{R}| \sim \chi^2_{p(p-1)/2} \quad (11.25)$$

The principle applied in the Bartlett test is that if the p-variables $\left(\boldsymbol{X}_{p\times1}\right)$ are truly independent signifying no correlations among the p-variables, the correlation matrix $\mathbf{R}_{p\times p}$ becomes an identity matrix $\mathbf{I}_{p\times p}$, and the scatter plot of observations resembles a sphere (e.g., when $p = 2$, the scatter plot becomes a circle, see Figure 11.6a). This is known as sphericity. If the scatter plot defines sphericity, the correlation matrix **R** is **I** and then PCA is not useful from a dimensionality reduction point of view.

The null and alternative hypotheses in the Bartlett test are

$$H_0 : \mathbf{R} = \mathbf{I}$$
$$H_1 : \mathbf{R} \neq \mathbf{I} :$$

Now let us consider the two data sets (hypothetical) given in Figures 11.6a and 11.6b. Assume the correlation matrices as $\mathbf{R}_a$ and $\mathbf{R}_b$ as given below.

$$\mathbf{R}_a = \begin{bmatrix} 1 & 0 \\ 0 & 1 \end{bmatrix} \quad \text{and} \quad \mathbf{R}_b = \begin{bmatrix} 1 & 0.9 \\ 0.9 & 1 \end{bmatrix}$$

Then the determinants of $\mathbf{R}_a$ and $\mathbf{R}_b$ are 1.00 and 0.19, respectively. Accordingly, the Bartlett's sphericity test statistic for $\mathbf{R}_a$ and $\mathbf{R}_b$ are 0 and a large positive number, respectively. When **R** closes to **I** (H_0 is true), the Bartlett's sphericity test statistic value becomes low. However, the $\chi^2_{p(p-1)/2}$ distribution can be used to test the significance level (α) for the Bartlett's sphericity test statistic.

Example 11.5

Conduct Bartlett's sphericity test for the data set given by **S** in Example 11.1. Consider $n = 30$.

Solution:

From Example 11.3, we get

$$\mathbf{R} == \begin{bmatrix} 1.0000 & 0.6250 & -0.4286 \\ 0.6250 & 1.000 & -0.3571 \\ -0.4286 & -0.3571 & 1.0000 \end{bmatrix} \quad \text{and} \quad \det(\mathbf{R}) = 0.4895$$

Here, hypotheses are

$$H_0 : \mathbf{R} = \mathbf{I}$$
$$H_1 : \mathbf{R} \neq \mathbf{I}$$

Using Equation 11.25, we get

The computed test statistic = $-\left[(n-1)-\left(\frac{2p+5}{6}\right)\right] \ln|\mathbf{R}| = 19.4071$, and the tabulated $\chi^2_{p(p-1)/2}(\alpha) = \chi^2_3(0.05) = 7.81$.

So, H_0 is rejected. The covariance structure is not spherical. Hence, PCA is recommended.

11.4.2 Number of PCs to Be Extracted

There are several criteria that can be employed to determine the number of PCs to be extracted. The principle employed here is a tradeoff between the numbers of reduced dimensions versus the loss of information. In PCA, the total variance contained in the original data matrix $\mathbf{X}_{n\times p}$ is the sum of the variances of the p individual variables. If a few PCs can explain a substantial portion (say ≥ 90%) of the total variance, then those few PCs can be considered without significant loss of information. The criteria that are used for determining the number of PCs are explained below.

Cumulative Percentage of Total Variation

For p-variables $X_1, X_2, \ldots, X_p$ having the covariance matrix **S**, we can extract the maximum of p number of PCs, $Z_1, Z_2, \ldots, Z_p$. In Equation 11.13, it was proved that

$$\sum_{j=1}^{p} s_{jj} = \sum_{j=1}^{p} \lambda_j$$

where s_{jj} and λ_j are the variance contained in the variable X_j and $PC\ Z_j$, respectively for $j = 1, 2, \ldots, p$. If we retain the first m number of PCs out of p PCs, then the cumulative percentage of total variance explained by the first m number of PCs (V_m) is:

$$V_m = \sum_{j=1}^{m} \lambda_j \Big/ \sum_{j=1}^{p} \lambda_j, \quad j=1,2,\ldots m,\ldots,p, \ m \le p \tag{11.26}$$

The threshold value of V_m depends on the application of the PCs extracted. If reduction of dimensionality is the primary aim, then a value within 70–90% would be a cut-off (Jolliffe, 2002). If the aim is to get p-orthogonal variables for further analysis, then the cut-off will be 100%. But the maximum benefit is acquired when a small number of PCs (say, m = 2–5) explains a large amount of total variance (say, ≥ 90%) of a large number of original variables. However, if the first PC accounts for most of the variance, it leads to the so-called size dominance problem and the first PC becomes relatively unimportant as it fails to give meaningful information (Krzanowski and Marriott, 2014a). Researchers have suggested the use of other components as they hold interesting information.

Example 11.6

How many PCs can be extracted (for Example 11.1) to explain 60% of the total variance?

Solution:

From Table 11.1, it is seen that PC_1 alone can explain 70% of the total variance. So, the first PC is sufficient.

TABLE 11.1
PCs with Eigenvalues, % Variance Explained, Cumulative % of Total Variance for Example 11.1

PC	Eigenvalue	% variance	Cum %
PC_1	148.23	0.70	0.70
PC_2	36.01	0.17	0.87
PC_3	28.76	0.13	1

Kaiser's Rule (Eigenvalue Criteria)

Kaiser's rule (Kaiser, 1960) is primarily used to determine the number of PCs to be retained when the correlation matrix **R** is used. As stated earlier, using **R** implies using a covariance matrix of standardized variables each having mean zero and variance 1. For p variables, the total variance is $p \times 1 = p$. Kaiser's rule states that an extracted PC is important if it contains at least one variable's (standardized) variance, which is equal to 1. As the eigenvalues (λ_j, $j = 1, 2, \ldots p$) represent variance extracted, the first m number of PCs with eigenvalues greater than or equal to 1 ($\lambda_j \geq 1$, $j = 1, 2, \ldots m$) should be retained.

There are certain problems in using Kaiser's rule. The criterion $\lambda_j \geq 1$ retains only a few PCs, thus restricting the explained variance to be lower than the desired (in many cases). Krzanowski and Marriott (2014a) stated that this method lacks logical reasoning and pointed out that a truly independent variable may appear as a distinct PC with eigenvalue less than one. Jolliffe (1972) has shown that a cut-off for $\lambda = 0.70$ works well.

The Average Root

This method retains those PCs having eigenvalues (λ_j, $j = 1, 2, \ldots, p$) greater than the mean eigenvalue $\bar{\lambda}$ as

$$\bar{\lambda} = \frac{1}{p} \sum_{j=1}^{p} \lambda_j \tag{11.27}$$

Using Jolliffe's (1972) study, the cut-off for retaining a PC Z_j, $j = 1, 2, \ldots, p$ is $\lambda_j \geq 0.7\bar{\lambda}$.

The Broken Stick Method

If a stick of unit length is broken randomly into p-segments, then the expected length of the j-th longest segment is (Jolliffe, 2002)

$$l_j = \frac{1}{p} \sum_{k=j}^{p} \frac{1}{k} \tag{11.28}$$

If the proportion of variance explained by λ_j, $\dfrac{\lambda_j}{\sum_{j=1}^{p} \lambda_j} > l_j$, retain the j-th PC, Z_j. The procedure starts with λ_1 and continues until $\dfrac{\lambda_j}{\sum_{j=1}^{p} \lambda_j} \leq l_j$ for $j = m$ (say).

Example 11.7

Compute the number of PCs to be retained for the Example 11.6 problem using the criteria (i) Kaiser's rule, (ii) average root, and (iii) broken stick method.

Solution:

(i) The eigenvalues from the correlational matrix (**R**) are 2.85, 0.146 and 0.00 (zero). Based on Kaiser's rule ($\lambda_j \geq 1$), only one PC is to be retained. Similarly, following Jolliffe (1972), if we consider $\lambda_j \geq 0.7$, again only one PC is to be retained.

(ii) The average value of the eigenvalues obtained from **S** is 71. Only one eigenvalue (148.23) is greater than the average value (71). So, based on average root criteria, one PC is to be retained.

(iii) Using the broken stick method, the following results are obtained (see Table 11.2).

So, based on the broken stick criteria, one PC is to be retained.

Scree Plot

Scree plot is a graphical technique (Cattell, 1966) which plots eigenvalues (λ_j) along the ordinate (Y-axis) and the PC numbers along the abscissa (X-axis) as shown in Figure 11.7.

In order to determine the number of PCs to be retained, we visually observe the scree plot and identify the elbow point on the curve. The elbow point indicates that the variance explained by the PCs at this point and beyond is almost equal and corresponds to the random part of the model. The proportion of variance explained by each of these PCs is significantly smaller. So, the number of

TABLE 11.2
Results From Broken Stick Method

PC	Eigenvalue	l_j	% variance
PC_1	148.2345	0.61	$0.70 > l_j$
PC_2	36.00492	0.28	$0.17 < l_j$
PC_3	28.76054	0.11	

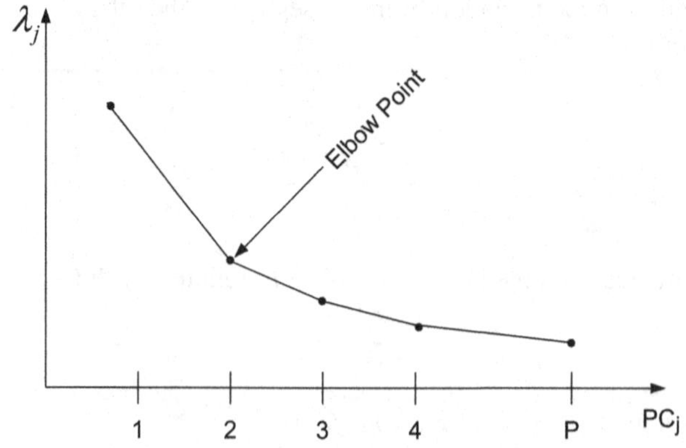

FIGURE 11.7 Scree plot. The elbow point indicates that the variance explained by the PCs at this point and beyond is almost equal.

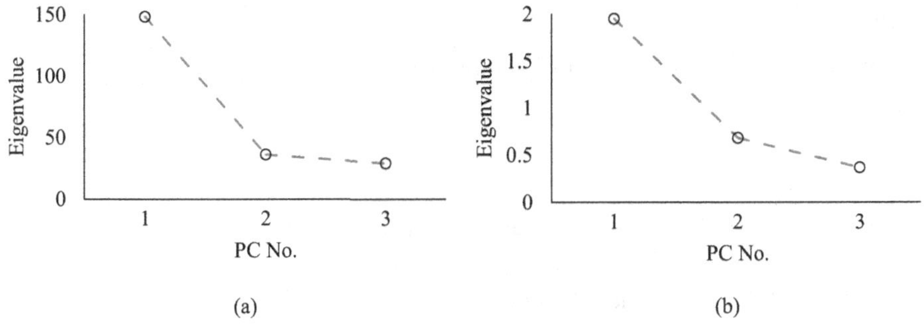

FIGURE 11.8 Scree plots for Example 11.1 using **S** (Figure (a)) and **R** (Figure (b)).

PCs to be retained is the first few PCs excluding the PCs at the elbow and beyond. From Figure 11.7, we see that the elbow point is at the second PC and hence one PC will be retained.

Example 11.8

Consider Example 11.1. Obtain the scree plot. How many PCs can be retained?

Solution:

The scree plots for the eigenvalues (obtained both from **S** and **R**) are shown in Figure 11.8. Both the plots show that the elbow starts from PC_2. Hence, only one PC is recommended.

11.4.3 Hypothesis Testing Concerning Insignificant Eigenvalues

In scree plots, we have seen that the PCs at elbow and beyond contain almost equal amount of variance. We can conduct hypothesis testing for the same. The null and alternative hypotheses are:

$H_0 : \lambda_{m+1} = \lambda_{m+2} = \cdots = \lambda_p$
$H_1 : \lambda_{m+j} \neq \lambda_{m+k}, j \neq k,$ for the at least one pair of eigenvalues from the last $(p\text{-}m)$ eigenvalues

The original hypothesis test is attributed to Bartlett (1950) which has been modified a number of times. Under H_0, the test statistic (Mardia et al., 1979 and Jolliffe, 2002)

$$n\left[(p-m)\ln\left(\bar{\lambda}_m\right) - \sum_{j=m+1}^{p} \ln\left(\lambda_j\right)\right] \sim \chi^2_{\frac{1}{2}(p-m-1)(p-m+2)} \tag{11.29}$$

Where,

$$\bar{\lambda}_m = \sum_{j=m+1}^{p} \frac{\lambda_j}{p-m}$$

We will reject H_0 if the computed statistic $n\left[(p-m)\ln\left(\bar{\lambda}_m\right) - \sum_{j=m+1}^{p} \ln\left(\lambda_j\right)\right] > \chi^{2(\alpha)}_{\frac{1}{2}(p-m-1)(p-m+2)}$.

If H_0 is accepted, the number of PCs to be retained is m.

Example 11.9

Test the hypotheses H_0: $\lambda_1 = \lambda_2 = \lambda_3$ for Example 11.1.

Here, the hypotheses are

$$H_0 : \lambda_1 = \lambda_2 = \lambda_3$$
$$H_1 : \lambda_j \neq \lambda_k, j \neq k, \text{for the at least one pair of eigenvalues}$$

Using Equation 11.29, we get the following results.

The computed test statistic $= n\left[(p-m)\ln\left(\bar{\lambda}_m\right) - \sum_{j=m+1}^{p} \ln\left(\lambda_j\right)\right] = 25.40$ (with $m = 0$), and the tabulated $\chi^2_{(p-m-1)(p-m+2)/2}(\alpha) = \chi^2_5(0.05) = 11.07$.

So, H_0 is rejected. All the three PCs do not explain variance equally.

Similarly considering $m = 1$, we get the following results.

$$H_0 : \lambda_2 = \lambda_3$$
$$H_1 : \lambda_2 \neq \lambda_3$$

The computed test statistic $= +n\left[(p-m)\ln\left(\bar{\lambda}_m\right) - \sum_{j=m+1}^{p} \ln\left(\lambda_j\right)\right] = 0.38$ (with $m = 1$), and the tabulated $\chi^2_{(p-m-1)(p-m+2)/2}(\alpha) = \chi^2_2(0.05) = 5.99$.

So, we fail to reject H_0. The PC_2 and PC_3 do explain variance equally.

11.5 PRINCIPAL COMPONENT SCORES

As seen earlier, the principal components are the linear combination of the original variables (i.e., $Z_j = \mathbf{a}_j^T x$) and uncorrelated. For subsequent analysis with PCs, we require to know the values that each PC holds for the n observations. Such n values are known as principal component scores. If PCA retains m principal components, the transformation of $\mathbf{X}_{n \times p}$ to $\mathbf{Z}_{n \times m}$ (the PC scores) can be represented as below.

$$\begin{array}{c} \\ 1 \\ 2 \\ \vdots \\ i \\ \vdots \\ n \end{array}\begin{array}{c} \begin{array}{cccc} X_1 & X_2 & \cdots & X_p \end{array} \\ \begin{bmatrix} x_{11} & x_{12} & \cdots & x_{1p} \\ x_{21} & x_{22} & \cdots & x_{2p} \\ \vdots & \vdots & \ddots & \vdots \\ x_{i1} & x_{i2} & \cdots & x_{ip} \\ \vdots & \vdots & \ddots & \vdots \\ x_{n1} & x_{n2} & \cdots & x_{np} \end{bmatrix} \end{array} \overset{PCA}{\Rightarrow} \begin{array}{c} \\ 1 \\ 2 \\ \vdots \\ i \\ \vdots \\ n \end{array}\begin{array}{c} \begin{array}{cccc} Z_1 & Z_2 & \cdots & Z_m \end{array} \\ \begin{bmatrix} z_{11} & z_{12} & \cdots & z_{1m} \\ z_{21} & z_{22} & \cdots & z_{2m} \\ \vdots & \vdots & \ddots & \vdots \\ z_{i1} & z_{i2} & \cdots & z_{im} \\ \vdots & \vdots & \ddots & \vdots \\ z_{n1} & z_{n2} & \cdots & z_{nm} \end{bmatrix} \end{array}$$

$$\text{or } \begin{bmatrix} \mathbf{x}_1 & \mathbf{x}_2 & \cdots & \mathbf{x}_p \end{bmatrix} \overset{PCA}{\Rightarrow} \begin{bmatrix} \mathbf{z}_1 & \mathbf{z}_2 & \cdots & \mathbf{z}_m \end{bmatrix}$$

The column vector $\mathbf{z}_j$, $j = 1, 2, \ldots, m$, represents the j-th principal component scores. Please note that $Z_j = \mathbf{a}_j^T X$ gives the j-th PC scores.

11.6 VALIDATION

By validation in PCA, we mean that the variance explained by each of the PCs, obtained from a sample covariance matrix **S**, will remain almost same for repeated samples (of equal size) taken from the same population. In order to confirm this, we may use one or more of the following techniques (Lattin et al., 2003).

1. Use of hold-out sample.
2. Jackknife validation.
3. Bootstrap validation.

In the hold-out sample method, a large sample is divided into two smaller samples. If the sample size is n, then usually $\frac{2}{3}n$ data points are used for initial PCA and remaining $\frac{1}{3}n$ data points, called the hold-out sample, are used to validate the PCs extracted initially. The hold-out sample (with size $\frac{1}{3}n$) is selected randomly from the original sample of size n. For a p-variable case, if $\lambda_j^{(m)}$ stands for the variance accounted for by the j-th PC of the original model (with $\frac{2}{3}n$ data points) and $\lambda_j^{(V)}$ stands for the same in the hold-out sample, then the extracted PCs can be said to be valid if $\lambda_j^{(m)} \simeq \lambda_j^{(v)}$ for $j = 1, 2, \ldots, p$. The symbol $\simeq$ represents nearly equal. Statistical testing can be conducted to test the difference between each pair of $\lambda_j^{(m)}$ and $\lambda_j^{(V)}$ for $j = 1, 2, \ldots, p$.

It is usually difficult, costly and time-consuming to collect a sufficiently large sample so that a hold-out sample can be kept for validation. In many cases, particularly dealing with large number of variables, it may not be possible to have a hold-out sample. The jackknife method (Kendall and Stuart, 1979) will become handy under such situations. The procedure is given below.

Step 1: Collect n multivariate observations on p-variables, i.e., $\mathbf{X}_{n\times p}$, where each multivariate observation can be written as $\mathbf{x}_i = \begin{bmatrix} x_{i1} & x_{i2} & \cdots & x_{ip} \end{bmatrix}^T, i = 1, 2, \ldots, n.$

Obtain PCs using the covariance matrix $\mathbf{S}_{p\times p}$ or correlation matrix $\mathbf{R}_{p\times p}$ obtained using the sample of n observations. There will be p number of PCs, $Z_1, Z_2, \ldots, Z_p$ with variances $\lambda_1, \lambda_2, \ldots, \lambda_p$, respectively (as discussed earlier).

Step 2: Remove the first multivariate observation $\mathbf{x}_1 = [x_{11} \quad x_{12} \quad \cdots \quad x_{1p}]^T$ from $\mathbf{X}_{n\times p}$, and treat the remaining n–1 observations $\mathbf{X}_{(n-1)\times p}$ as a separate sample, say Sample 1. Similarly, if we remove the second multivariate observations $\mathbf{x}_2 = [x_{21} \quad x_{22} \quad \cdots \quad x_{2p}]^T$ from $\mathbf{X}_{n\times p}$, we will get Sample 2 of size n–1. Thus by removing the third multivariate observations from $\mathbf{X}_{n\times p}$, we get Sample 3, and so on. When we complete this operation up to the n-th multivariate observation, we will get n jackknife samples, each having a size of n–1.

Step 3: Take the first jackknife sample (Sample 1) and extract p number of PCs. Compute the PC score for the first observation for $Z_1, Z_2, \ldots, Z_p$. The corresponding jackknife multivariate score vector is $\begin{bmatrix} z_{11} & z_{12} & \cdots & z_{1p} \end{bmatrix}^T$. Similarly, we repeat the process for remaining n–1 samples and compute the jackknife PC scores for the removed observations. This gives matrix $\mathbf{Z}_{n\times p}$, say jackknife matrix.

Step 4: Compute the variance of each of the PCs using the jackknife matrix obtained in Step 3. Let the extracted PCs be called jackknife PCs with jackknife variances $\lambda_1^{(J)}, \lambda_2^{(J)}, \ldots, \lambda_p^{(J)}$. Now, compare the jackknife variances $\lambda_1^{(J)}, \lambda_2^{(J)}, \ldots, \lambda_p^{(J)}$, with $\lambda_1, \lambda_2, \ldots, \lambda_p$, obtained in Step 1. If there are no substantial differences between for the first few PCs, we can expect that the PCA results can be generalized to larger applications in the broad area considered.

Bootstrap validation uses the resampling method to generate a large number of samples (say, 5,000 to 10,000 or more). The bootstrapped samples are of size n and are obtained by random sampling with replacement from the original sample of size n (say, $n = 100$). Each of the samples will be used to compute λ_j, $j = 1, 2, \ldots, p$. From such a large number of samples, a large number of λ_j, $j = 1, 2, \ldots, p$ values can be obtained, which may be used to plot histogram for each λ_j, $j = 1, 2, \ldots, p$. This gives the probability distribution for each λ_j and meaningful considerations can be met as post hoc analysis.

11.7 CASE STUDY

The operations in coke plants are hazardous in nature and the employees are subjected to several job-related stressors. The health and safety executive (HSE) provides seven job stress risk factors. They are demand (D), control (C), management support (MS), peer support (PS), relationship (RE), role (RO), and change (CH). Under many situations, it is difficult to address the seven risk stressors provided by HSE and it may be plausible to work with lower dimensions (less number of stressors without significant loss of information). This is a dimension reduction problem and in this study, we demonstrated the use of principal component analysis (PCA) in handling this problem.

The details of the case study is available in the web resources of this book (Multivariate Statistical Modeling in Engineering and Management – 1st (routledge.com)).

11.8 LEARNING SUMMARY

Principal component analysis (PCA) is a dimension reduction technique and has been widely used in almost all areas of applications from automobile manufacturing to banking while handling large amounts of data and variables. PCA was developed by Hotelling in 1933. In this chapter, we have described in detail PCA, which includes the conceptual model, assumptions, extraction of PCs and their interpretations, hypothesis testing, and number of PCs to be retained. Finally, a case study is discussed.

In summary,

- The principal components are the linear combinations of original variables ($\boldsymbol{X}$), represented as $\boldsymbol{Z} = \mathbf{A}^T \boldsymbol{X}$, where $\boldsymbol{X}$ is the vector of original variables, $\mathbf{A}$ is the transformation matrix, and $\boldsymbol{Z}$ represents the vector of PCs.
- The PCs are widely extracted using eigenvalue-eigenvector decomposition method.
- In PCA, the first PC extracts the maximum variance of X followed by second PC and so on, subject to $\mathbf{a}_j^T \mathbf{a}_j = 1$, and $Cov\left(\mathbf{a}_j, \mathbf{a}_k\right) = 0, j \neq k, j, k = 1, 2, \ldots, p$, where $\mathbf{a}_j$ represents the eigenvector corresponding to the j-th variable.
- The sampling distribution of eigenvalue is $\lambda_j \sim N\left(\theta_j, \frac{2\theta_j^2}{n-1}\right)$, where θ_j represents the population eigenvalue. Further, the eigenvector $\mathbf{a}_j \sim N_p\left(\mathbf{a}_j \cdot T_j\right)$. For details see Section 11.3.

- The 100(1–α)% confidence interval for θ_j is $\dfrac{\lambda_j}{1+Z_{\alpha/2}\cdot\sqrt{\dfrac{2}{n-1}}} < \theta_j < \dfrac{\lambda_j}{1-Z_{\alpha/2}\cdot\sqrt{\dfrac{2}{n-1}}}$ and the 100(1–α)% confidence region for $\boldsymbol{\alpha}_j$ is $(n-1)\boldsymbol{\alpha}_j^T\left(\lambda_j\mathbf{S}^{-1}+\lambda_j^{-1}\mathbf{S}-2\mathbf{I}_p\right)\boldsymbol{\alpha}_j \le \chi^{2(\alpha)}_{p-1}$. For details see Section 11.3.
- The precondition for PCA is that there should be substantial correlations among some of the original variables (*X*). For this purpose, the Bartlett sphericity test considers the following hypotheses:

$$H_0 : \mathbf{R} = \mathbf{I}$$
$$H_1 : \mathbf{R} \neq \mathbf{I}$$

The test statistic follows Chi-square distribution as given below.

$$-\left[(n-1)-\left(\frac{2p+5}{6}\right)\right]\ln|\mathbf{R}| \sim \chi^2_{p(p-1)/2}$$

- Several criteria such as cumulative percentage of variance explained, scree plot, Kaiser's rule, average root and the broken stick method are used to decide on the tradeoff between the numbers of PCs to be retained and the loss of information (in terms of cumulative percentage of variance unexplained).
- The hypothesis testing concerning insignificant eigenvalues is done using the following statistic:

$$n\left[(p-m)\ln\left(\bar{\lambda}_m\right)-\sum_{j=m+1}^{p}\ln\left(\lambda_j\right)\right] \sim \chi^2_{\frac{1}{2}(p-m-1)(p-m+2)}$$

- To conclude, this chapter gives reader a thorough overview of PCA with necessary skills for model-building and its verification. The PCA discussed in this chapter is a linear model. For non-linear PCA, the readers may consult principal curves.

EXERCISES

(A) Short conceptual questions

11.1 Why is PCA conducted? Explain the basic steps performed in PCA.

11.2 Consider a bivariate situation with original variables X_1 and X_2. Suppose the two principal components extracted are Z_1 and Z_2. Prove that $\boldsymbol{Z} = \mathbf{A}^T\boldsymbol{X}$, where, $\boldsymbol{X} = \begin{bmatrix} X_1 \\ X_2 \end{bmatrix}, \boldsymbol{Z} = \begin{bmatrix} Z_1 \\ Z_2 \end{bmatrix}, \mathbf{A} = \begin{bmatrix} \cos\theta & -\sin\theta \\ \sin\theta & \cos\theta \end{bmatrix}$.

11.3 If the *j*-th principal component of $\boldsymbol{X}_{(p\times 1)}$ is Z_j (where *j* = 1, 2, …, *p*), then show that $E(Z_j) = \mathbf{a}_j^T\boldsymbol{\mu}$ *and* $Var(Z_j) = \mathbf{a}_j^T\sum\mathbf{a}_j$, where $\boldsymbol{\mu}$ *and* $\boldsymbol{\Sigma}$ are the concerned mean vector and the covariance matrix.

11.4 Show that,

(a) $(\mathbf{S}-\lambda\mathbf{I})\mathbf{a}_j = \mathbf{0}$

(b) $\mathbf{S} = \mathbf{A}\boldsymbol{\Lambda}\mathbf{A}^T$,

(c) $\sum_{j=1}^{p} s_{jj} = \sum_{j=1}^{p}\lambda_j$

(d) $r_{jk} = \frac{a_{jk}\sqrt{\lambda_j}}{\sqrt{s_{kk}}}$; where $j = 1, 2, \ldots, p$; $k = 1, 2, \ldots, p$, r_{jk} = correlation soefficient between X_j and Z_k, a_{jk} = loading of X_j on Z_k,

(e) Sampling distribution of $\lambda_j \sim N\left(\theta_j, \frac{2\theta_j^2}{(n-1)}\right)$; where θ_j = j-th eigenvalue of population covariance matrix $\mathbf{\Sigma}$, and

(f) If λ_j is $N \sim \left(\theta_j, \frac{2\theta_j^2}{(n-1)}\right)$, then $\frac{\lambda_j}{1+Z_{\alpha/2}\sqrt{\frac{2}{(n-1)}}} \leq \theta_j \leq \frac{\lambda_j}{1-Z_{\alpha/2}\sqrt{\frac{2}{(n-1)}}}$.

11.5 What is Bartlett's sphericity test? Explain the steps. Why is Bartlett's sphericity test conducted in PCA?

11.6 How do you decide the number of principal components to be extracted? Explain cumulative % of total variation, Kaiser's rule, average root criterion, the broken stick method, and scree plot.

11.7 Conduct the following hypothesis test with reference to PCA.
$H_0 : \lambda_{m+1} = \lambda_{m+2} = \ldots = \lambda_p$, $H_1 : \lambda_{m+1} \neq \lambda_{m+k}$, $j \neq k$ for at least one pair (λ_j, λ_k) from the last $(p-m)$ number of λ values.

11.8 How do you validate the results of PCA? Explain with the help of holdout samples, jackknife validation, and bootstrap validation.

(B) Long conceptual and/or numerical questions: see web resources (Multivariate Statistical Modeling in Engineering and Management – 1st (routledge.com))

N.B.: R, MINITAB and MS Excel software are used for computation, as and when required.

12 Exploratory Factor Analysis

Exploratory factor analysis (EFA) is a data reduction technique similar to principal component analysis (PCA) with subtle differences in conceptualization, estimation to model adequacy testing including the purpose of the modeling. The development of factor analysis dates back to the classical works of Charles Spearman on association of two things (1904a) and general theory of intelligence (1904b). Although primarily developed in the domain of psychology, its applications spread across almost all disciplines where the issues involve the measurement of hidden things. The major purpose is to find out how many latent dimensions (or common factors) are involved in a group of measurements and what these dimensions are (Vincent, 1953). It exploits the covariance relationships of many observed variables (***X***) in search of the factors (*F*) and falls under covariance structure analysis. In this chapter, we discuss on the conceptual factor model with assumptions and certain useful results followed by three extraction methods. Then we discuss the model adequacy tests and different criteria to select the number of factors to be retained. Subsequently, factor rotation is described followed by the estimation of factor scores. Finally, a case study is presented which demonstrates the usefulness of EFA in solving real-world problems.

12.1 CONCEPTUAL MODEL

In real-life problem-solving, we come across many things. Some of them are directly observable and many cannot be measured directly. There are a few techniques available for measuring the unobservables and factor analysis is one of them. A few examples of directly unobserved variables include the mental ability of a person, the safety environment of any workplace, and supply chain coordination. These are known as hidden (or latent) variables, and measuring such hidden, unobserved variables is extremely important as these often represent the underlying causes of the phenomena we observe or experience. These unobserved variables are called factors or constructs. The factors are measured from observed symptoms, known as manifest variables (Figure 12.1).

Let us assume a situation, where workplace safety culture is monitored. It is known that if the workplace safety culture deteriorates, both unsafe conditions (UC) and unsafe acts (UA) increase. Here, "workplace safety culture" is an example of a factor, which is hidden, and the symptoms are UA and UC that are observed or manifested. In factor model terminology, we denote "workplace safety culture" in terms of F and the symptoms/manifest variables in terms of *X*.

Now, we mathematically formulate the factor model. Let ***X*** be a $p \times 1$ variable vector with mean $\boldsymbol{\mu}$, and covariance matrix, $\boldsymbol{\Sigma}$. Then

$$\boldsymbol{X} = \begin{bmatrix} X_1 \\ X_2 \\ \vdots \\ X_p \end{bmatrix}_{p\times 1},\ E(\boldsymbol{X}) = \boldsymbol{\mu} = \begin{bmatrix} \mu_1 \\ \mu_2 \\ \vdots \\ \mu_p \end{bmatrix}_{p\times 1},\ \text{and } Cov(\boldsymbol{X}) = \boldsymbol{\Sigma} = \begin{bmatrix} \sigma_{11} & \sigma_{12} & \cdots & \sigma_{1p} \\ \sigma_{12} & \sigma_{22} & \cdots & \sigma_{2p} \\ \vdots & \vdots & \ddots & \vdots \\ \sigma_{1p} & \sigma_{2p} & \cdots & \sigma_{Pp} \end{bmatrix}_{p\times p}$$

Let there are m hidden factors, $F_1, F_2, \ldots, F_m$ that cause[1] $X_1, X_2, \ldots, X_p$ to occur, then, pictorially we can represent the relationship of ***F*** with ***X*** as shown in Figure 12.2.

DOI: 10.1201/9781003303060-15

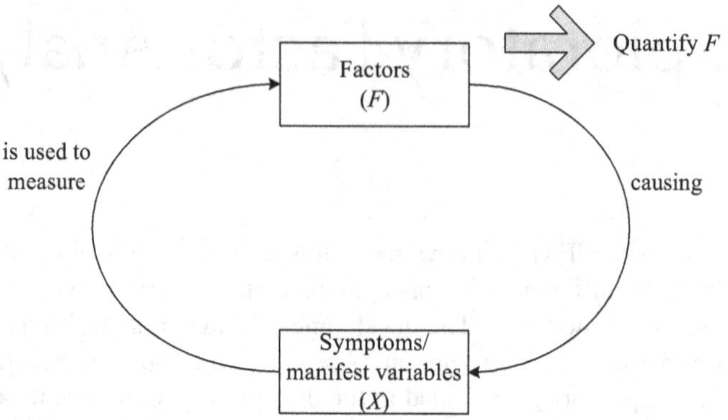

FIGURE 12.1 Concept of a factor. A factor (F) manifests several symptoms, termed as manifest variables (X). The symptoms/manifest variables (X) are used to measure the factor (F).

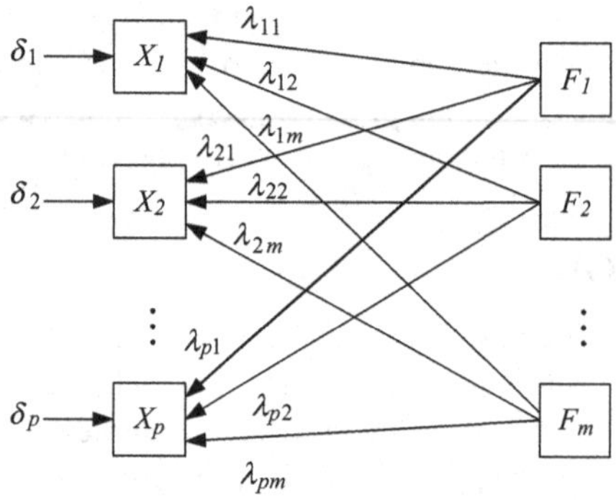

FIGURE 12.2 Orthogonal factor model. This figure visually demonstrates the relationship of a factor (F_m) with its manifest variables (X). Usually, a lower dimensional factors(say, *m* factors) are able to capture the variability of a higher dimensional manifest variables (say, *p*-variables; *p*>*m*). λ and δ represent factor loading and error, respectively.

Here, we assume that $F_1, F_2, \ldots, F_m$ are orthogonal, i.e., covariance between F_k and F_l for $k = 1, 2, \ldots, m$ and $l = 1, 2, \ldots, m$, $l \neq k$, is 0. Note that in Figure 12.2, there is no link between any two factors (***F***).

Further, note that each of the X variables is caused by the m orthogonal factors. In Figure 12.2, it is represented by the straight arrows emanating from the factors (***F***) and terminating at the manifest variables (***X***). See the arrowheads. Now, we can develop regression like equations for each observed variable X_j, $j = 1, 2, \ldots, p$. Following Figure 12.2, the p-linear equations are

$$\left.\begin{aligned} X_1 - \mu_1 &= \lambda_{11}F_1 + \lambda_{12}F_2 + \ldots + \lambda_{1m}F_m + \delta_1 \\ X_2 - \mu_2 &= \lambda_{21}F_1 + \lambda_{22}F_2 + \ldots + \lambda_{2m}F_m + \delta_2 \\ &\vdots \\ X_j - \mu_j &= \lambda_{j1}F_1 + \lambda_{j2}F_2 + \ldots + \lambda_{jm}F_m + \delta_j \\ &\vdots \\ X_p - \mu_p &= \lambda_{p1}F_1 + \lambda_{p2}F_2 + \ldots + \lambda_{pm}F_m + \delta_p \end{aligned}\right\} \tag{12.1}$$

So, in matrix form, Equation 12.1 can be written as

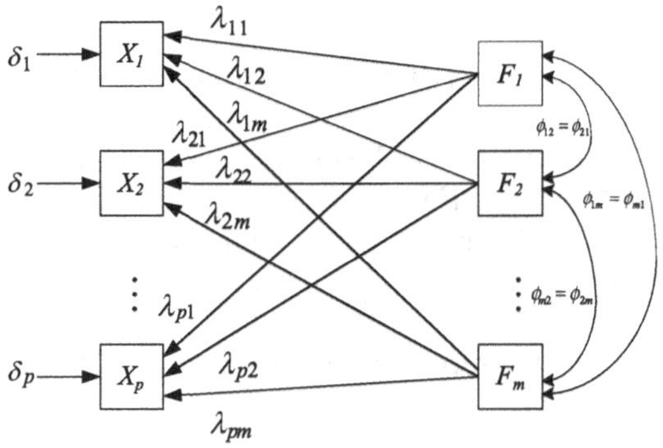

FIGURE 12.3 Oblique factor model. The oblique factors are correlated whereas orthogonal factors are uncorrelated (see Figure 12.2). The correlation between any two factors is represented by ϕ.

$$\begin{bmatrix} X_1-\mu_1 \\ X_2-\mu_2 \\ \vdots \\ X_j-\mu_j \\ \vdots \\ X_p-\mu_p \end{bmatrix}_{p\times1} = \begin{bmatrix} \lambda_{11} & \lambda_{12} \cdots \lambda_{1m} \\ \lambda_{21} & \lambda_{22} \cdots \lambda_{2m} \\ \vdots & \\ \lambda_{j1} & \lambda_{j2} \cdots \lambda_{jm} \\ \vdots & \\ \lambda_{p1} & \lambda_{p2} \cdots \lambda_{pm} \end{bmatrix}_{p\times m} \begin{bmatrix} F_1 \\ F_2 \\ \vdots \\ F_j \\ \vdots \\ F_p \end{bmatrix}_{m\times1} + \begin{bmatrix} \delta_1 \\ \delta_2 \\ \vdots \\ \delta_j \\ \vdots \\ \delta_p \end{bmatrix}_{p\times1} \tag{12.2}$$

$$\text{or } (X-\mu)_{(p\times1)} = \Lambda_{(p\times m)} F_{(m\times1)} + \delta_{(p\times1)}$$

$$\text{or } X-\mu = \Lambda F + \delta$$

Equation 12.2 is known as factor model of exploratory type. The word exploratory is used to describe the condition that the factors, $F_1, F_2, \ldots, F_m$ are not known in advance. The structure of hidden factors is explored using the observed or manifest variables (***X***). In Equation 12.2, apart from X, μ, and F, there are two other terms, Λ and δ, with subscripts relating to the concerned ***X*** and ***F***. These are called the factor loading matrix and error vector, respectively.

When the exploratory factor model (Equation 12.2) satisfies the orthogonality of the factors, it is called an orthogonal factor model. But, if the factors are not independent, then it becomes an oblique factor model. If the correlations among the factors are represented as $\phi_{21},\ldots,\phi_{m1}\ldots,\phi_{m2},\ldots etc.$, then the oblique factor model is as shown in Figure 12.3.

In exploratory factor analysis, we do not know which factors are causing which set of manifest variables. Therefore, the researchers consider a full set of relationships, i.e., each X is influenced by all the factors (***F***). Whereas, if it is known *a priori* which factors are influencing which X, and the researchers are interested to confirm the relationships, the relevant factor analysis is called confirmatory factor analysis (CFA). We will discuss CFA in Chapter 13 and its further applications in Chapter 14 under the measurement model of structural equation modeling.

In summary,

- The factors (***F***) are unobservable or hidden or latent and also known as constructs.
- The factors (***F***) manifest some symptoms that can be observed and measured, and are called manifest variables, (***X***).
- Factor analysis (FA) quantifies ***F*** using ***X***.
- Charles Spearman is called the father of factor analysis.
- Factor analysis is also a data reduction technique. It is also an interdependence modeling technique.
- The lesser the dimensions, the better the model, i.e., the lesser the number of factors, the better the model, provided the model is adequate.

12.2 ASSUMPTIONS

The assumptions pertaining to exploratory factor model (Equation 12.2) are given below.

- The factors ($\boldsymbol{F}$) are normally distributed with zero mean and unit variance. So, $\boldsymbol{F} \sim N_m(0,\mathbf{I})$, $E(\boldsymbol{F}) = \mathbf{0}_{m\times 1}$ and $Cov(\boldsymbol{F}) = \mathbf{I}_{m\times n}$. $\boldsymbol{F}$ is called common factors and $\boldsymbol{F}$ is orthogonal.
- The errors, $\boldsymbol{\delta}$, are a normal random vector with zero mean and constant variance. So, $\boldsymbol{\delta} \sim N_p(0,\psi)$, $E(\boldsymbol{\delta}) = \mathbf{0}_{p\times 1}$, and $Cov(\boldsymbol{\delta}) = \boldsymbol{\Psi}_{p\times p}$, where $\boldsymbol{\Psi}$ is a diagonal matrix. $\boldsymbol{\delta}$ is called specific factors.
- The common factors $\boldsymbol{F}$ and the specific factors $\boldsymbol{\delta}$ are uncorrelated. So, $Cov(\boldsymbol{F},\boldsymbol{\delta}) = 0$.

12.3 SOME USEFUL RESULTS

Based on the assumptions given above, we can obtain the all-important relationships of $\boldsymbol{\Sigma}$ with $\boldsymbol{\Lambda}$ and $\boldsymbol{\Psi}$, the heart of the factor analysis.

$$\begin{aligned}Cov(\boldsymbol{X}) = \boldsymbol{\Sigma} &= E\left[(\boldsymbol{X}-\boldsymbol{\mu})(\boldsymbol{X}-\boldsymbol{\mu})^T\right]\\ &= E\left[(\boldsymbol{\Lambda F}+\boldsymbol{\delta})(\boldsymbol{\Lambda F}+\boldsymbol{\delta})^T\right] \text{ (from equation 12.2)}\\ &= E\left[\boldsymbol{\Lambda F F}^T\boldsymbol{\Lambda}^T + \boldsymbol{\Lambda F}\boldsymbol{\delta}^T + \boldsymbol{\delta F}^T\boldsymbol{\Lambda}^T + \boldsymbol{\delta\delta}^T\right]\\ &= \boldsymbol{\Lambda} E(\boldsymbol{FF}^T)\boldsymbol{\Lambda}^T + \boldsymbol{\Lambda} E(\boldsymbol{F\delta}^T) + E(\boldsymbol{\delta F}^T)\boldsymbol{\Lambda}^T + E(\boldsymbol{\delta\delta}^T)\\ &= \boldsymbol{\Lambda I \Lambda}^T + \boldsymbol{\Lambda}\cdot\mathbf{0} + \mathbf{0}\cdot\boldsymbol{\Lambda}^T + \boldsymbol{\Psi} \quad \text{(from assumptions)}\\ &= \boldsymbol{\Lambda\Lambda}^T + \boldsymbol{\Psi}\end{aligned} \tag{12.3}$$

This is the fundamental relationship of orthogonal factor model.

Now, from Equation 12.3, we can write

$$\begin{bmatrix}\sigma_{11} & \sigma_{12} & \cdots & \sigma_{1p}\\ \sigma_{12} & \sigma_{22} & \cdots & \sigma_{2p}\\ \vdots & \vdots & \ddots & \vdots\\ \sigma_{1p} & \sigma_{2p} & \cdots & \sigma_{Pp}\end{bmatrix}_{p\times p} = \begin{bmatrix}\lambda_{11} & \lambda_{12} & \cdots & \lambda_{1m}\\ \lambda_{21} & \lambda_{22} & \cdots & \lambda_{2m}\\ \vdots & \vdots & \ddots & \vdots\\ \lambda_{p1} & \lambda_{p2} & \cdots & \lambda_{pm}\end{bmatrix}_{p\times m}\begin{bmatrix}\lambda_{11} & \lambda_{21} & \cdots & \lambda_{p1}\\ \lambda_{21} & \lambda_{22} & \cdots & \lambda_{p2}\\ \vdots & \vdots & \ddots & \vdots\\ \lambda_{1m} & \lambda_{2m} & \cdots & \lambda_{pm}\end{bmatrix}_{m\times p} + \begin{bmatrix}\Psi_{11} & 0 & \cdots & 0\\ 0 & \Psi_{22} & \cdots & 0\\ \vdots & \vdots & \ddots & \vdots\\ 0 & 0 & \cdots & \Psi_{pp}\end{bmatrix}$$

$$= \begin{bmatrix}\sum_{k=1}^{m}\lambda_{1k}^2 & \sum_{k=1}^{m}\lambda_{1k}\lambda_{2k} & \cdots & \sum_{k=1}^{m}\lambda_{1k}\lambda_{pk}\\ \sum_{k=1}^{m}\lambda_{1k}\lambda_{2k} & \sum_{k=1}^{m}\lambda_{2k}^2 & \cdots & \sum_{k=1}^{m}\lambda_{2k}\lambda_{pk}\\ \vdots & & & \\ \sum_{k=1}^{m}\lambda_{1k}\lambda_{pk} & \sum_{k=1}^{m}\lambda_{2k}\lambda_{pk} & \cdots & \sum_{k=1}^{m}\lambda_{pk}^2\end{bmatrix} + \begin{bmatrix}\Psi_{11} & 0 & \cdots & 0\\ 0 & \Psi_{22} & \cdots & 0\\ \vdots & & & \\ 0 & 0 & & \Psi_{pp}\end{bmatrix}$$

$$= \begin{bmatrix}\sum_{k=1}^{m}\lambda_{1k}^2 + \Psi_{11} & \sum_{k=1}^{m}\lambda_{1k}\lambda_{2k} & \cdots & \sum_{k=1}^{m}\lambda_{1k}\lambda_{pk}\\ \sum_{k=1}^{m}\lambda_{1k}\lambda_{2k} & \sum_{k=1}^{m}\lambda_{2k}^2 + \Psi_{22} & \cdots & \sum_{k=1}^{m}\lambda_{2k}\lambda_{pk}\\ \vdots & & & \\ \sum_{k=1}^{m}\lambda_{1k}\lambda_{pk} & \sum_{k=1}^{m}\lambda_{2k}\lambda_{pk} & \cdots & \sum_{k=1}^{m}\lambda_{pk}^2 + \Psi_{pp}\end{bmatrix} \tag{12.4}$$

The diagonal elements of the right-hand side (RHS) matrix of Equation 12.4 represent the variances of the respective manifest variables and the off-diagonal elements represent corresponding co-variances of the manifest variables.

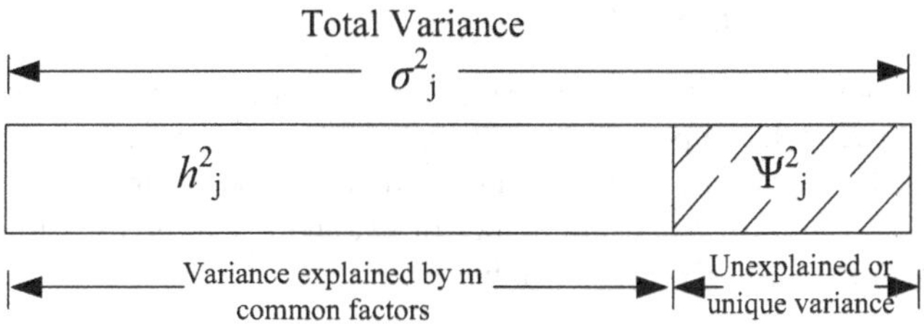

FIGURE 12.4 Portioning of σ_j^2. The variance of the j-th manifest variable (σ_j^2) comprises communality (h_j^2) and specific variance (Ψ_j^2). Communality is the variability explained by all the factors (common variance) and the specific variance is unique to a manifest variable.

So, from Equation 12.4, we can write

$$\sigma_{jj} = \sum_{k=1}^{m} \lambda_{jk}^2 + \Psi_{jj}, j = 1, 2, \ldots, p$$
$$\text{Or} \quad \sigma_j^2 = h_j^2 + \Psi_j^2 \tag{12.5}$$

where σ_j^2 = variance of X_j, $j = 1, 2, \ldots, p$, $h_j^2 = \sum_{k=1}^{m} \lambda_{jk}^2$ = communality, and Ψ_j^2 is called unique or specific variance of X_j. Further, h_j^2 is the portion of σ_j^2 explained by the m factors together, and Ψ_j^2 is the unexplained variance of X_j. So, in the factor model each observed variable's variance is decomposed into two parts (Figure 12.4). The objective of the factor model is to determine the number of common factors m and the elements of $\mathbf{\Lambda}$ and $\mathbf{\Psi}$ such that the communality, h_j^2 is maximized.

The covariance between $\boldsymbol{X}$ and $\boldsymbol{F}$ can be as follows:

$$\begin{aligned} Cov(\boldsymbol{XF}) &= E\left[(\boldsymbol{X} - \boldsymbol{\mu})(\boldsymbol{F} - E(\boldsymbol{F}))^T\right] \\ &= E\left[(\boldsymbol{X} - \boldsymbol{\mu})\boldsymbol{F}^T\right] \text{ as } E(\boldsymbol{F}) = \mathbf{0} \\ &= E\left[(\boldsymbol{\Lambda F} + \boldsymbol{\delta})\boldsymbol{F}^T\right] \text{ (from equation 12.2)} \\ &= E\left[\boldsymbol{\Lambda F F}^T + \boldsymbol{\delta F}^T\right] \\ &= E(\boldsymbol{\Lambda F F}^T) + E(\boldsymbol{\delta F}^T) \\ &= \boldsymbol{\Lambda} E(\boldsymbol{F F}^T) + \mathbf{0} \quad \text{(from assumptions)} \\ &= \boldsymbol{\Lambda} \end{aligned} \tag{12.6}$$

So, factor loading represents the covariance between a factor and its manifest variable.

12.4 FACTOR EXTRACTION METHODS

The factor model given in Equation 12.3 contains unknown parameter matrices $\mathbf{\Lambda}$ and $\mathbf{\Psi}$. To estimate these unknown matrices, we collect sample of size n and compute the sample covariance matrix **S**. If we consider that appropriate sampling strategy is adopted and representative sample is collected, then we expect **S** to be representative of $\mathbf{\Sigma}$. With this information, the common techniques employed for extraction of factors are

1. The principal component method
2. The principal factor method
3. The maximum likelihood method

All these methods finally yield some useful matrices, which are decomposed using eigenvalue and eigenvector decomposition or similar techniques.

12.4.1 The Principal Component Method

We described the extraction of principal components in Chapter 11. We have seen that the population covariance matrix $\mathbf{\Sigma}$ is decomposed into eigenvalues[2] (say Q_j, $j = 1, 2, \ldots, p$,) and eigenvectors ($\boldsymbol{e}_j$, $j = 1, 2, \ldots, p$,). If we assume that there are m hidden factors, and because of the property, $Q_1 \geq Q_2 \geq \ldots \geq Q_m \geq \ldots \geq Q_p$, the first m number of Q_j, and $\boldsymbol{e}_j$ can be used to estimate the factor loadings ($\mathbf{\Lambda}$) and specific factors ($\mathbf{\Psi}$) as shown below.

$$\mathbf{\Sigma} = \begin{bmatrix} \boldsymbol{\lambda}_1 : \boldsymbol{\lambda}_2 : \cdots : \boldsymbol{\lambda}_m \end{bmatrix} \begin{bmatrix} \boldsymbol{\lambda}_1^T \\ \boldsymbol{\lambda}_2^T \\ \vdots \\ \boldsymbol{\lambda}_m^T \end{bmatrix} + \mathbf{\Psi} = \mathbf{\Lambda}\mathbf{\Lambda}^T + \mathbf{\Psi} \tag{12.7}$$

Where $\sqrt{Q_j}\ \boldsymbol{e}_j = \boldsymbol{\lambda}_j$, $\sqrt{Q_j}\ \boldsymbol{e}_j^T = \boldsymbol{\lambda}_j^T$, and

$$\mathbf{\Psi} = \begin{bmatrix} \Psi_{11} & 0 & \cdots & 0 \\ 0 & \Psi_{22} & \cdots & 0 \\ \vdots & \vdots & \ddots & \vdots \\ 0 & 0 & \cdots & \Psi_{pp} \end{bmatrix}$$

The specific variance for X_j is Ψ_{jj}, and can be obtained using

$$\Psi_{jj} = \sigma_{jj} - \boldsymbol{\lambda}_j^T \boldsymbol{\lambda}_j = \sigma_{jj} - \sum_{k=1}^{m} \lambda_{jk}^2 \tag{12.8}$$

When $m = p$, the specific variance matrix $\mathbf{\Psi}$ is null (zero). However, the purpose of factor analysis is also to reduce dimensions. Hence, $m < p$ and the lower the value of m, the better the model is.

The above discussion is based on population parameters $\boldsymbol{\mu}$ and $\mathbf{\Sigma}$. They are not known and, hence, need to be estimated. If we collect a sample data, $\mathbf{X}_{n \times p}$, then, $\boldsymbol{\mu} = \bar{\mathbf{x}}$, and $\hat{\mathbf{\Sigma}} = \mathbf{S}$. The sample principal components are then obtained by decomposing $\mathbf{S}$ into its eigenvalues and eigenvectors and then using Equation 12.7 for factor loadings and Equation 12.8 for specific factors.

Example 12.1

Consider the following sample covariance matrix ($\mathbf{S}$). Obtain a suitable factor model using principal component method.

$$\mathbf{S} = \begin{bmatrix} 100 & 50 & -30 \\ 50 & 64 & -20 \\ -30 & -20 & 49 \end{bmatrix}$$

Solution:

From Example 11.1, $|\mathbf{S} - Q\mathbf{I}| = 0$ gives $Q_1 = 148.23$, $Q_2 = 36.00$, and $Q_3 = 28.76$, and their corresponding eigenvectors as

$$\boldsymbol{e}_1 = \begin{bmatrix} 0.77 \\ 0.54 \\ -0.39 \end{bmatrix};\ \boldsymbol{e}_2 = \begin{bmatrix} 0.21 \\ 0.29 \\ 0.93 \end{bmatrix};\ \text{and } \boldsymbol{e}_3 = \begin{bmatrix} 0.60 \\ -0.79 \\ 0.11 \end{bmatrix}$$

The largest eigenvalue, Q_1 (148.23) comprises around 70% of the total variances. So, $m = 1$ factor model may be explored. The parameters $\mathbf{\Lambda}$, and $\mathbf{\Psi}$ can be obtained using

Equations 12.7–12.8. Corresponding to the largest eigenvalue, Q_1 = 148.23, the eigenvector is $e_1 = [0.77 \;\; 0.54 \;\; -0.39]^T$.
Now,

$$\Lambda_{3\times1} = \begin{bmatrix} \lambda_{11} \\ \lambda_{21} \\ \lambda_{31} \end{bmatrix} = \begin{bmatrix} \sqrt{Q_1} e_{11} \\ \sqrt{Q_1} e_{21} \\ \sqrt{Q_1} e_{31} \end{bmatrix} = \sqrt{148.23} \begin{bmatrix} 0.77 \\ 0.54 \\ -0.39 \end{bmatrix} = \begin{bmatrix} 9.375 \\ 6.575 \\ -4.748 \end{bmatrix}$$

So,

$$\begin{aligned} X_1 - \mu_1 &= 9.375F_1 + \delta_1 \\ X_2 - \mu_2 &= 6.375F_1 + \delta_2 \\ X_3 - \mu_3 &= -4.748F_1 + \delta_3 \end{aligned}$$

The specific factor $\boldsymbol{\Psi}$ is a diagonal matrix and is

$$\boldsymbol{\Psi} = \begin{bmatrix} \Psi_{11} & 0 & 0 \\ 0 & \Psi_{22} & 0 \\ 0 & 0 & \Psi_{33} \end{bmatrix}$$

Where, with $m=1$

$$\Psi_{11} = s_{11} - \sum_{k=1}^{m} \lambda_{1k}^2 = 100 - \lambda_{11}^2 = 100 - (9.375)^2 = 12.11$$

$$\Psi_{22} = s_{22} - \sum_{k=1}^{m} \lambda_{2k}^2 = 64 - \lambda_{21}^2 = 64 - (6.575)^2 = 20.77$$

$$\Psi_{33} = s_{33} - \sum_{k=1}^{m} \lambda_{3k}^2 = 49 - \lambda_{31}^2 = 49 - (-4.748)^2 = 26.46$$

12.4.2 The Principal Factor Method

In the principal component method, we decompose the full sample covariance matrix to estimate $\boldsymbol{\Lambda}$, and then consider the first m eigenvalues and eigenvectors. Once $\boldsymbol{\Lambda}$ is known, then $\boldsymbol{\Psi}$ is estimated using Equation 12.7 or 12.8. In the principal factor method, we decompose $\mathbf{S} - \hat{\boldsymbol{\Psi}}$ instead of $\mathbf{S}$. But this requires $\hat{\boldsymbol{\Psi}}$ to be initialized. The fundamentals behind this approach are given below. From Equation 12.3, it can be seen that $\boldsymbol{\Sigma}$ can be reconstructed from $\boldsymbol{\Lambda}$, and $\boldsymbol{\Psi}$. Further, $\boldsymbol{\Lambda}$ and $\boldsymbol{\Psi}$ are estimated using $\mathbf{S}$ (sample covariance matrix). Therefore, if $\mathbf{S}$ is representative of $\boldsymbol{\Sigma}$, the discrepancy between $\mathbf{S}$ and $\boldsymbol{\Sigma}$ should be below a certain threshold value (convergence point). The discrepancy, say G, can be defined by $trace\left[(\mathbf{S}-\boldsymbol{\Sigma})^T(\mathbf{S}-\boldsymbol{\Sigma})\right]$.

So,

$$\begin{aligned} G &= trace\left[(\mathbf{S}-\boldsymbol{\Sigma})^T(\mathbf{S}-\boldsymbol{\Sigma})\right] \\ &= trace(\mathbf{S}-\boldsymbol{\Sigma})^2 \\ &= \left\|\mathbf{S}-\boldsymbol{\Sigma}\right\|^2 \\ &= \left\|\mathbf{S}-\boldsymbol{\Lambda}\boldsymbol{\Lambda}^T-\boldsymbol{\Psi}\right\|^2 \end{aligned} \tag{12.9}$$

In order to minimize G using least squares, we differentiate G in Equation 12.9 with respect to $\boldsymbol{\Lambda}$ and $\boldsymbol{\Psi}$. Then equating the resultant differential equations to zero, the estimating equations are

$$(\mathbf{S}-\hat{\boldsymbol{\Psi}})\hat{\boldsymbol{\Lambda}} = \hat{\boldsymbol{\Lambda}}(\hat{\boldsymbol{\Lambda}}^T\hat{\boldsymbol{\Lambda}}) \tag{12.10}$$

$$\hat{\boldsymbol{\Psi}} = diag(\mathbf{S}-\hat{\boldsymbol{\Lambda}}\hat{\boldsymbol{\Lambda}}^T) \tag{12.11}$$

The equations do not have closed form. Hence, an iterative process is adopted as follows. First, $\hat{\mathbf{\Psi}}$ is initialized and then $\hat{\mathbf{\Lambda}}$ is estimated using the m eigenvectors of $\mathbf{S}-\hat{\mathbf{\Psi}}$. Then, $\hat{\mathbf{\Psi}}$ is recomputed substituting the estimated $\hat{\mathbf{\Lambda}}$ in Equation 12.11. The process continued until convergence is achieved. In order to solve Equations 12.10 and 12.11, two things must be known: (i) a reasonable estimate of $\hat{\mathbf{\Psi}}$, and (ii) the values of *m*. Timm (2002) stated that the values of *m* could be the rank of $\mathbf{S}-\hat{\mathbf{\Psi}}$. Hardle and Simar (2003) pointed out that additional constraints can be put to make the factor model estimable. The common constraint is $\mathbf{\Lambda}^T\mathbf{\Psi}^{-1}\mathbf{\Lambda}$ is diagonal.

12.4.3 The Maximum Likelihood Method

The maximum likelihood estimation method assumes that the observations are independently sampled from a multivariate normal distribution with mean vector $\boldsymbol{\mu}$ and variance-covariance matrix of the form $\mathbf{\Sigma}=\mathbf{\Lambda}\mathbf{\Lambda}^T+\mathbf{\Psi}$.

If $\mathbf{x}_1, \mathbf{x}_2, \ldots, \mathbf{x}_n$ are the *n* multivariate random observations, $\mathbf{x}_i \sim N_p(\boldsymbol{\mu},\mathbf{\Sigma}) = N_p(\boldsymbol{\mu},\mathbf{\Lambda}\mathbf{\Lambda}^T+\mathbf{\Psi})$. The log likelihood function of the data matrix $\mathbf{X}_{n\times p}$ can be written as (Krzanowski and Marriott, 2014b).

$$L(\mathbf{X}|\boldsymbol{\mu},\mathbf{\Sigma}) = -\frac{np}{2}\log(2\pi) - \frac{n}{2}\log|\mathbf{\Sigma}| - \frac{n}{2}trace\left[\mathbf{\Sigma}^{-1}\mathbf{S}\right]$$

We can ignore the constant term $-\frac{np}{2}\log(2\pi)$ as it does not affect maximization of the log-likelihood.

So,

$$L(\mathbf{X}|\boldsymbol{\mu},\mathbf{\Sigma}) = -\frac{n}{2}\log|\mathbf{\Sigma}| - \frac{n}{2}trace\left(\mathbf{\Sigma}^{-1}\mathbf{S}\right)$$

Taking derivative of $L(\mathbf{X}|\boldsymbol{\mu},\mathbf{\Sigma})$ with respect to $\boldsymbol{\mu}$ and solving it, we get $\hat{\boldsymbol{\mu}}=\bar{\mathbf{x}}$.

Finally, putting $\mathbf{\Sigma}=\mathbf{\Lambda}\mathbf{\Lambda}^T+\mathbf{\Psi}$, we get

$$L(\mathbf{X}|\hat{\boldsymbol{\mu}},\mathbf{\Lambda}\mathbf{\Lambda}^T+\mathbf{\Psi}) = -\frac{n}{2}\left[\log|(\mathbf{\Lambda}\mathbf{\Lambda}^T+\mathbf{\Psi})| + trace\left\{(\mathbf{\Lambda}\mathbf{\Lambda}^T+\mathbf{\Psi})^{-1}\mathbf{S}\right\}\right] \quad (12.12)$$

Equation 12.12 is a complicated non-linear function of the model parameters $\mathbf{\Lambda}$ and $\mathbf{\Psi}$. Differentiating Equation 12.12 with respect to the parameters and equating them to zero yields, after some algebraic manipulations, the following two likelihood equations (Lawley, 1940):

$$(\hat{\mathbf{\Psi}}^{-\frac{1}{2}}\mathbf{S}\hat{\mathbf{\Psi}}^{-\frac{1}{2}})(\hat{\mathbf{\Psi}}^{-\frac{1}{2}}\hat{\mathbf{\Lambda}}) = (\hat{\mathbf{\Psi}}^{-\frac{1}{2}}\hat{\mathbf{\Lambda}})(\mathbf{I}+\hat{\boldsymbol{\theta}}) \quad (12.13)$$

$$\hat{\mathbf{\Psi}} = diag(\mathbf{S}-\hat{\mathbf{\Lambda}}\hat{\mathbf{\Lambda}}^T) \quad (12.14)$$

where, $\hat{\boldsymbol{\theta}}=\hat{\mathbf{\Lambda}}^T\hat{\mathbf{\Psi}}^{-1}\hat{\mathbf{\Lambda}}$.

Equations 12.13 and 12.14 can be solved iteratively with initial estimate of $\hat{\mathbf{\Psi}}$. The eigenvectors of $\hat{\mathbf{\Psi}}^{-\frac{1}{2}}\mathbf{S}\hat{\mathbf{\Psi}}^{-\frac{1}{2}}$ are the estimates of $\mathbf{\Lambda}$. If we keep *m* number of factors, then $\hat{\mathbf{\Lambda}}$ is the eigenvectors of the first *m* largest eigenvalues (see Chapter 3 for eigenvalue-eigenvector decomposition). Substituting the estimated $\mathbf{\Lambda}$, i.e., $\hat{\mathbf{\Lambda}}$, in Equation 12.14 gives improved values of $\hat{\mathbf{\Psi}}$. The process is repeated until convergence is achieved. When an iterative procedure is used for estimation, such as in principal factor method or MLE methods, selection of the initial values of $\hat{\mathbf{\Psi}}$, say $\hat{\mathbf{\Psi}}_0$ is very important. If covariance matrix ($\mathbf{S}$) is used for estimation, then $\hat{\mathbf{\Psi}}_0 = (diag(\mathbf{S}^{-1}))^{-1}$ or if correlation matrix ($\mathbf{R}$) is used for estimation, then $\hat{\mathbf{\Psi}}_0 = (diag(\mathbf{R}^{-1}))^{-1}$. Jöreskog (1963) suggested to use $\hat{\mathbf{\Psi}}_{jj0} = (1-m/2p)/s^{jj}$, where s^{jj} is the *j*-th diagonal element of $\mathbf{S}^{-1}$. Researchers also proposed to use $\hat{\mathbf{\Psi}}_{jj0} = s_j^2(1-R_j^2)$, where R_j^2 is the multiple coefficient of determination obtained after regressing X_j on all other $\boldsymbol{X}$. See Krzanowski and Marriott (2014b) for further details of MLE estimation and discussion.

Example 12.2

Consider the following sample correlation matrix (**R**) (see Table 12.1). Obtain $m = 2$ factor model using principal component analysis (PCA), principal factor method (PFM) and MLE. Compare and comment on the estimation.

Solution:

The results are shown in Table 12.2.

The following comments can be made based on the loadings, communalities, and variance explained about the three estimation methods.

- The pattern of loadings obtained from PCA and PFM is similar, with heavy loading to F_1. But MLE gives a different picture with higher loadings for X_1, X_2, and X_3 with F_1 and almost equal loadings for the rest with both F_1 and F_2.
- The total variance explained by PCA, PFM, and MLE (using the first two factors) are 4.59, 3.98, and 3.98, respectively with cumulative proportion of variance explained as 76.5%, 66.33% and 66.33%, respectively.

TABLE 12.1
Sample Correlation Matrix for Example 12.2

	X_1	X_2	X_3	X_4	X_5	X_6
X_1	1					
X_2	0.742	1				
X_3	0.714	0.656	1			
X_4	0.292	0.362	0.672	1		
X_5	0.336	0.324	0.594	0.602	1	
X_6	0.256	0.336	0.485	0.478	0.472	1

TABLE 12.2
Factors Obtained Using PCA, PFM, and MLE

Methods	PCA			PFM			MLE		
Manifest Variables	F_1	F_2	h_j^{2*}	F_1	F_2	h_j^{2*}	F_1	F_2	h_j^{2*}
X_1	0.75	–0.57	**0.89**	0.77	–0.56	**0.91**	0.94	–0.26	**0.94**
X_2	0.76	–0.5	**0.83**	0.72	–0.34	**0.63**	0.78	–0.05	**0.61**
X_3	0.92	–0.08	**0.85**	0.92	0.00	**0.85**	0.86	0.34	**0.85**
X_4	0.75	0.42	**0.74**	0.70	0.43	**0.67**	0.5	0.69	**0.73**
X_5	0.73	0.42	**0.71**	0.65	0.35	**0.55**	0.49	0.51	**0.51**
X_6	0.64	0.41	**0.58**	0.54	0.26	**0.36**	0.4	0.44	**0.35**
Explained $V(X_j)$	**3.48**	**1.11**		**3.18**	**0.8**		**2.87**	**1.11**	
Proportion explained (%)	**58**	**18.5**		**53**	**13.33**		**47.83**	**18.5**	
Cumulative proportion explained (%)	**58**	**76.5**		**53**	**66.33**		**47.83**	**66.33**	

- As the quantity and pattern of loadings differ from using different methods, which is attributed to the process of estimation, expert judgment may be sought for the selection of the solution from the theoretical and practical perspectives of the problem.

12.4.4 Choice of Methods of Estimation

If the manifest variables X follows multivariate normal distributions, the maximum likelihood method is the best choice as it provides several advantages such as scale invariance and use of likelihood ratio-based goodness of fit tests. But for non-normal X, the principle factor method can be preferred. Note that the principal components method is a special case of principal factor method when $\boldsymbol{\Psi} = \mathbf{0}$. Unfortunately, both the principal factor and principal component methods suffer from the lack of scale invariance. Lack of scale invariance relates to the condition when the estimated factor loadings and specific variances obtained from using the correlation matrix $\mathbf{R}$ differ from the standardized estimates obtained from using the covariance matrix $\mathbf{S}$. It is to be noted here that we often use $\mathbf{R}$ instead of $\mathbf{S}$, particularly when the X variables are measured in non-commensurate units.

12.4.5 Degrees of Freedom

The degrees of freedom (DOF) for a factor model could be computed as below.

$$\text{DOF} = (\text{No of parameters for } \boldsymbol{\Sigma} \text{ unconstrained}) - (\text{No of parameters for } \boldsymbol{\Sigma} \text{ constrained})$$

$$DOF = \left\{\frac{1}{2}p(p+1)\right\} - \left\{pm + p - \frac{1}{2}m(m-1)\right\} = \frac{1}{2}\left[(p-m)^2(p+m)\right] \tag{12.15}$$

The upper bound of the number of factors m can be obtained using Equation 12.15 for DOF≥0. When DOF=0, there will be unique solution. For DOF>0, the exact solution does not exist and the model is over-identified. It is preferable. For DOF<0, the factor model is under-identified. No matter, whatever estimation methods are used, the above important issues must be kept in mind. The DOF plays significant role in model adequacy test, and test of parameters.

12.5 MODEL ADEQUACY TESTS

The model adequacy test shows whether the factor model is able to explain the covariance relationships among the p-number of manifest variables in terms of the reduced number (m) of underlying factors, estimated using the sample data. Therefore, the first thing to check whether the data is factorable or not. The correlation matrix ($\mathbf{R}$) can be used for this purpose. If a good number of correlation coefficients are of high value and many are greater than or equal to 0.3, then we may go for factor analysis. However, for a large number of variables (usually so) the visual inspection of $\mathbf{R}$ with the focus on its individual elements is not trustworthy at all times. Several statistical tests were developed for this purpose. Some of them are explained below.

Bartlett's sphericity test

In Bartlett's sphericity test, the correlation matrix is tested. The procedure is as follows.

Step 1: Set the hypotheses

$$H_0 : \mathbf{R} = \mathbf{I}$$
$$H_a : \mathbf{R} \neq \mathbf{I}$$

Where, $\mathbf{I}$ is the identity matrix.

The logic behind this null hypothesis is that if there is no or insignificant correlations among the X variables, the geometry of $\mathbf{R}$ can be reasonably represented by a sphere.

Step 2: Compute the test statistic

The test statistic is

$$B_s = -[(n-1) - \frac{(2p+5)}{6}]\ln|\mathbf{R}| \tag{12.16}$$

where, **R** is the correlation matrix, p is the number of manifest variables, and n is the number of observations (sample size).

Step 3: Determine sampling distribution and make decision

The quantity B_s approximately follows χ^2 distribution with $p(p - 1)/2$ degrees of freedom. If the computed B_s is more than the tabulated $\chi^2_{p(p-1)/2}$ value for α probability level, then we reject H_0, and we recommend for factor analysis of the given matrix. See Chapter 11, for example on Bartlett's sphericity test.

Kaiser-Meyer-Olkin (KMO) test

The Kaiser-Meyer-Olkin (KMO) test is used for sampling adequacy and it considers the correlation matrix (**R**) of the manifest variables (***X***) as input. It indicates the proportion of variance of ***X*** that can be explained by the hidden factors (***F***).

The formula for the KMO test for the *j-th* manifest variable X_j is (Kaiser and Rice, 1974)

$$KMO_j = \frac{\sum_{k,(k\neq j)} r_{kj}^2}{\sum_{k,(k\neq j)} r_{kj}^2 + \sum_{k,(k\neq j)} u_{kj}^2} \tag{12.17}$$

Where,

$\mathbf{R} = \{r_{ij}\}$ is the correlation matrix, where $\{\ \}$ denotes all elements

$\mathbf{U} = \{u_{ij}\}$ is the anti-image correlation matrix

The anti-image matrix, **U** is computed as $\mathbf{U} = \mathbf{D}^{1/2}\mathbf{R}^{-1}\mathbf{D}^{1/2}$. **D** is a diagonal matrix and its diagonal elements are the inverse of diagonal elements of $\mathbf{R}^{-1}$.

The overall KMO is computed as below.

$$KMO = \frac{\sum\limits_{k\neq j}\sum r_{kj}^2}{\sum\limits_{k\neq j}\sum r_{kj}^2 + \sum\limits_{k\neq j}\sum u_{kj}^2} \tag{12.18}$$

The thumb rules for this test are as follows:

- KMO values between 0.8 and 1 indicate that the sampling is adequate.
- KMO values less than 0.6 indicate that the sampling is not adequate and that remedial action should be taken.
- KMO values close to zero means that there are anti image/large partial correlations compared to the sum of correlations.

Large sample likelihood ratio test

Bartlett's sphericity test shows whether the sample data can be used for factor analysis or not. It does not tell us anything about the nature of the factor model. For example, one may be interested to test whether Equation 12.3 is a right kind of parameterization or not. In this regard, an important measure is large sample likelihood ratio test, which is described below. However, it requires $\boldsymbol{X}$ to be multivariate normal.

The hypotheses to be tested are

$$H_0 : \boldsymbol{\Sigma} = \boldsymbol{\Lambda}\boldsymbol{\Lambda}^T + \boldsymbol{\Psi}$$
$$H_1 : \boldsymbol{\Sigma} \neq \boldsymbol{\Lambda}\boldsymbol{\Lambda}^T + \boldsymbol{\Psi} \quad (\boldsymbol{\Sigma} \text{ is unconstrained})$$

As stated earlier, the likelihood ratio test is based on the ratios of the likelihood maximized under H_1 to the likelihood maximized under H_0. See Section 12.4.3 for likelihood function for factor model. Lawley (1940) showed that the resultant test statistics is

$$LR = n\left[trace\left(\hat{\boldsymbol{\Sigma}}^{-1}\mathbf{S}\right) - \log\left|\hat{\boldsymbol{\Sigma}}^{-1}\mathbf{S}\right| - p\right] \tag{12.19}$$

where, $\hat{\boldsymbol{\Sigma}} = \hat{\boldsymbol{\Lambda}}\hat{\boldsymbol{\Lambda}}^T + \hat{\boldsymbol{\Psi}}$

LR follows asymptotic χ^2 distribution with $\frac{1}{2}\left[(p-m)^2 - (p+m)\right]$ degrees of freedom, if H_0 is true.

So, if the computed LR (as above) is equal or less that the tabulated $\chi^2_{\frac{1}{2}\left[(p-m)^2-(p+m)\right]}(\alpha)$ for a pre-defined α (usually 0.01 or 0.05), we conclude that the factor model is a good fit to the data.

Root mean square residual (RMSR) index

Timm (2002) has mentioned the use of root mean square residual (RMSR) index. The RMSR is

$$RMSR = \sqrt{\frac{trace(\mathbf{S} - \hat{\boldsymbol{\Lambda}}\hat{\boldsymbol{\Lambda}}^T - \hat{\boldsymbol{\Psi}})^2}{p(p-1)/2}} \tag{12.20}$$

The RMSR value should be as small as possible. A value of ≤ 0.05 is considered an acceptable value for many applications.

Example 12.3

Consider Example 12.2 and sample size, $n = 100$. Is the data adequate for exploratory factor analysis?

Solution:

We will present the results of Bartlett's sphericity test, KMO test, LR test, and RMSR-based test.

Bartlett's sphericity test

Using Equation 12.16, we get $B_s = 315.4435$. The DOF is 15. The tabulated $\chi^2_{15}(0.05) = 25$. As the computed $B_s = 315.4435 > 25$ (the tabulated value), H_0 is rejected. So, the correlation matrix is factorable. Alternatively, using the computed B_s, the p-value can be computed. If p-value is less than the recommended or threshold value, H_0 will be rejected. The p-value for $\chi^2_{15}(\alpha) = 315.44$ is $3.43 \times 10^{-58} (\approx 0.00) << 0.05$. So, H_0 is rejected.

KMO Test

Using Equations 12.17 and 12.18, we get overall KMO = 0.77, and individual variable's KMOs for X_1, X_2, X_3, X_4, X_5, and X_6 are 0.67, 0.81, 0.76, 0.74, 0.87, and 0.89, respectively. None of the values are less than 0.60 and most of them are close to 0.80 or more. So, KMO test suggests $\mathbf{R}$ can be factorable.

LR Test

Here we consider MLE solutions. The estimated $\hat{\Lambda}\hat{\Lambda}^T$, errors ($\hat{\Psi}$), and correlation matrix $\hat{\rho}$ are

$$\hat{\Lambda}\hat{\Lambda}^T = \begin{bmatrix} 0.95 & 0.75 & 0.72 & 0.29 & 0.33 & 0.26 \\ 0.75 & 0.61 & 0.65 & 0.36 & 0.36 & 0.29 \\ 0.72 & 0.65 & 0.86 & 0.66 & 0.59 & 0.49 \\ 0.29 & 0.36 & 0.66 & 0.73 & 0.60 & 0.50 \\ 0.33 & 0.36 & 0.59 & 0.60 & 0.50 & 0.42 \\ 0.26 & 0.29 & 0.49 & 0.50 & 0.42 & 0.35 \end{bmatrix}, \hat{\Psi} = \begin{bmatrix} 0.06 & 0.00 & 0.00 & 0.00 & 0.00 & 0.00 \\ 0.00 & 0.39 & 0.00 & 0.00 & 0.00 & 0.00 \\ 0.00 & 0.00 & 0.15 & 0.00 & 0.00 & 0.00 \\ 0.00 & 0.00 & 0.00 & 0.27 & 0.00 & 0.00 \\ 0.00 & 0.00 & 0.00 & 0.00 & 0.49 & 0.00 \\ 0.00 & 0.00 & 0.00 & 0.00 & 0.00 & 0.65 \end{bmatrix}$$

and $\hat{\rho} = \hat{\Lambda}\hat{\Lambda}^T + \hat{\Psi}$

$$\hat{\rho} = \begin{bmatrix} 1.01 & 0.75 & 0.72 & 0.29 & 0.33 & 0.26 \\ 0.75 & 1.00 & 0.65 & 0.36 & 0.36 & 0.29 \\ 0.72 & 0.65 & 1.01 & 0.66 & 0.59 & 0.49 \\ 0.29 & 0.36 & 0.66 & 1.00 & 0.60 & 0.50 \\ 0.33 & 0.36 & 0.59 & 0.60 & 0.99 & 0.42 \\ 0.26 & 0.29 & 0.49 & 0.50 & 0.42 & 1.00 \end{bmatrix}$$

Using Equation 12.19, and replacing **S** by **R** and $\hat{\Sigma}^{-1}$ by $\hat{\rho}^{-1}$ as **R** is used instead of **S**, we get the following:

$$\hat{\Sigma}^{-1}\mathbf{S} = \hat{\rho}^{-1}\mathbf{R} = \begin{bmatrix} 0.98 & -0.01 & -0.01 & -0.02 & 0.08 & -0.08 \\ 0.01 & 1.00 & 0.02 & 0.01 & -0.09 & 0.12 \\ -0.01 & 0.01 & 0.98 & 0.03 & -0.04 & -0.04 \\ 0.00 & 0.01 & 0.02 & 1.00 & 0.00 & -0.08 \\ 0.02 & -0.07 & 0.00 & 0.01 & 1.01 & 0.11 \\ -0.01 & 0.07 & -0.01 & -0.04 & 0.07 & 1.00 \end{bmatrix}$$

$\ln(|\hat{\rho}^{-1}\mathbf{R}|) = -0.053$, $trace(\hat{\rho}^{-1}\mathbf{R}) = 5.98$, and $LR = 3.23$. LR is distributed as χ^2_4. The tabulated value for α=0.05 is 9.49, which is more than the computed value. So, H_0 is accepted. The factor model is fit to the data.

RMSR-Based Test

Now, $(\mathbf{S} - \hat{\Lambda}\hat{\Lambda}^T - \hat{\Psi})^2 = (\mathbf{R} - \hat{\Lambda}\hat{\Lambda}^T - \hat{\Psi})^2 = (\mathbf{R} - \hat{\rho})^2 = (\mathbf{R} - \hat{\rho})^T(\mathbf{R} - \hat{\rho}) =$

$$\begin{bmatrix} 0.0003 & -0.0005 & 0.0001 & 0.0001 & -0.0002 & 0.0003 \\ -0.0005 & 0.0033 & -0.0003 & -0.0013 & 0.0021 & -0.0018 \\ 0.0001 & -0.0003 & 0.0002 & 0.0002 & -0.0005 & -0.0001 \\ 0.0001 & -0.0013 & 0.0002 & 0.0008 & -0.0015 & 0.0005 \\ -0.0002 & 0.0021 & -0.0005 & -0.0015 & 0.0039 & -0.0012 \\ 0.0003 & -0.0018 & -0.0001 & 0.0005 & -0.0012 & 0.0055 \end{bmatrix}$$

Using Equation 12.20, we get $RMSR = \sqrt{\dfrac{trace(\mathbf{S} - \hat{\Lambda}\hat{\Lambda}^T - \hat{\Psi})^2}{p(p-1)/2}} = 0.031$. The RMSR value is less than 0.05. So, the data is factorable.

12.6 NUMBER OF FACTORS

Once a factor model is accepted for a certain number of factors (say m), it is worthwhile to check the quality of the factors extracted. The quality of a factor in an explanatory factor model is determined by its ability in explaining manifest variables' variances. In addition, a factor must be interpretable in items of what it represents. The techniques employed in determining the number of factors are similar to those used for defining number of principle components described in Chapter 11. The commonly used criteria are given below (for explanation, see Chapter 11).

- Cumulative percentage of variance explained
- Average eigenvalue criteria
- Broken stick method
- Scree plot
- Kaiser's rule

Other methods include Horn's (1965) parallel analysis, minimum average partial (MAP) rule (Velicer, 1976), and the Hull method (Lorenzo-Seva et al., 2011). If the MLE method is used, then the likelihood ratio based approach can be adopted. Starting from $m = 1$, increase the m so long the goodness of fit measure becomes insignificant. However, a non-significant result does not guarantee the optimum value of m.

Another very important consideration while choosing m is the estimatibility condition of the model parameters. All the estimation methods finally consider as inputs the covariance (**S**) or the correlation (**R**) matrix. For a p-variable situation, both **S** and **R** contain $p(p + 1)/2$ unique elements of information. The number of parameters to be estimated are pm loadings and p specific variances. By imposing the constraint that $\mathbf{\Lambda}^T\mathbf{\Lambda}$ is a diagonal matrix, $m(m - 1)/2$ elements are put to zero. So, the resultant number of parameters to be estimated are $pm + p - m(m - 1)/2$. So, for estimatibility, $p(p + 1)/2 \geq pm + p - m(m - 1)$ or $(p - m)^2 \geq p + m$. As p is fixed, the m should be such that $(p - m)^2 \geq p + m$ is satisfied.

Example 12.4

Consider Example 12.2. Is the two factor model adequate?

Solution:

Both the scree plot (see Figure 12.5) and eigenvalue recommend a two-factor model. From proportion of total variance explained point of view, the two-factor model can explain 66.33% of the total variance (see Table 12.3). So, though the two-factor model is a good representation of the data, a three-factor model can be considered for better explanation of total variance.

12.7 FACTOR ROTATION

For all practical purposes, the factors must be interpretable. That means we can provide a name for each of the factors. This is possible only when the factors have a definite pattern of relationships with the observed or manifest variables ($\boldsymbol{X}$). The best way to find interpretable patterns is to find out, for every factor, high loadings for a specific subset of $\boldsymbol{X}$ and zero or negligible loadings for the rest. This can be obtained through factor rotation. Rotation can be orthogonal or oblique. In orthogonal rotation, the independence of the factors is maintained. But in oblique rotation, the factors will be correlated. Orthogonal rotation is also known as rigid rotation and oblique rotation as non-rigid rotation.

For factor rotation, two important concepts need explanation: how does rotation work and what could be the definite patterns that resemble simple structure for interpretation? Rotation is accomplished through transformation of the extracted factors, $\boldsymbol{F}$ (as a set of axes in the factor space)

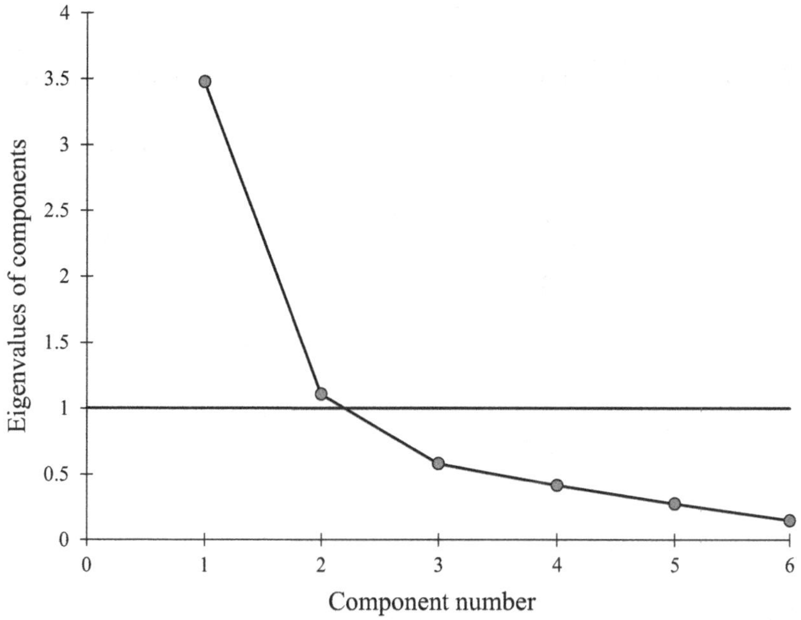

FIGURE 12.5 Scree plot for Example 12.2. The number of PCs having eigenvalues equal to one or greater than two. So, two factors may be retained.

TABLE 12.3
Model Adequacy for Example 12.2

Methods	PCA		PFM		MLE	
Explained $V(X_j)$	3.48	1.11	3.18	0.8	2.87	1.11
Proportion explained (%)	58	18.5	53	13.33	47.83	18.5
Cumulative proportion explained (%)	58	76.5	53	66.33	47.83	66.33

to new latent factors, say $\boldsymbol{F}^*$, where $\boldsymbol{F}^* = \mathbf{T}\boldsymbol{F}$ and $\mathbf{T}$ is called the transformation matrix. $\mathbf{T}$ is non-singular. In addition, if $\mathbf{T}$ is orthogonal, the transformation is known as orthogonal transformation. It can be shown that orthogonal rotation does not violate the fundamental relationship of factor analysis given in Equation 12.3. Rewriting Equation 12.2 using $\boldsymbol{F}^*$, we get.

$$X - \mu = \Lambda F^* + \delta = \Lambda \mathbf{T} F + \delta$$

Putting this in Equation 12.3, results in

$$\Sigma = \Lambda \mathbf{T}\mathbf{T}^T \Lambda^T + \Psi = \Lambda\Lambda^T + \Psi \text{ as } \mathbf{T}\mathbf{T}^T = \mathbf{I}$$

However, if we use oblique rotation (i.e., $E(\mathbf{T}\mathbf{T}^T) = \mathbf{\Phi}$) the $\Sigma = \Lambda\mathbf{\Phi}\Lambda^T + \Psi$, where $\mathbf{\Phi}$ represents the covariance structure of $\boldsymbol{F}$.

The simple structure concept dates back to Thurstone's five formal rules (Thurstone, 1947), where a simple structure hypothesis is "a large loading for one factor be opposite to a small loading for another factor" (Kaiser 1958). This indicates that we shall have a factor matrix with both large and small factor loadings in such a way that definite patterns of relationships among the manifest variables and factors can be obtained. In order to obtain the simple structure, an index of simplicity in terms of factor loadings is specified. Carroll (1953) first proposed an index using all factor

TABLE 12.4
Factor Loadings (λ_{jk}), Communality (h_j^2) and Total Variance Explained

Manifest Variables	Factors F_1	F_2	....	F_k	....	F_m	Communality (h_j^2)
X_1	λ_{11}	λ_{12}	...	λ_{1k}	...	λ_{1m}	$\sum_{k=1}^{m}\lambda_{1k}^2$
X_2	λ_{21}	λ_{22}	...	λ_{2k}	...	λ_{2m}	$\sum_{k=1}^{m}\lambda_{2k}^2$
.	.			.		.	.
.	.			.		.	.
.	.			.		.	.
X_j	λ_{j1}	λ_{j2}	...	λ_{jk}	...	λ_{jm}	$\sum_{k=1}^{m}\lambda_{jk}^2$
.	.			.		.	.
.	.			.		.	.
.	.			.		.	.
X_p	λ_{p1}	λ_{p2}	...	λ_{pk}	...	λ_{pm}	$\sum_{k=1}^{m}\lambda_{pk}^2$
Variance explained by k-th factor	$\sum_{j=1}^{p}\lambda_{j1}^2$	$\sum_{j=1}^{p}\lambda_{j2}^2$	...	$\sum_{j=1}^{p}\lambda_{jk}^2$	...	$\sum_{j=1}^{p}\lambda_{jm}^2$	

loadings and Neuhaus and Wrigley (1954) proposed the quartimax method for orthogonal structure. A little later Kaiser (1958) developed the varimax criterion for orthogonal structure. Both quartimax and varimax methods are described below.

Using Table 12.4, the index of simplicity can be defined by focusing on the columns or rows or both rows and columns. The normalized squared loading $\lambda_{jk}^2 / \sum_{k=1}^{m}\lambda_{jk}^2$ represents the proportion of variance of variable X_j explained by the factor F_k. A simple structure can be obtained by suitably rotating the factors such that $\lambda_{jk}^2 / \sum_{k=1}^{m}\lambda_{jk}^2$ becomes close to either 1 or 0. The varimax method uses this principle. Alternatively, we can focus on the rows instead of columns, where we simplify each row (i.e. manifest variables). The quartimax method uses this principle. The major limitation of quartimax is that it often gives a general factor. Equimax is a compromised method, where a trade-off between varimax and quartimax is made. It can be shown that in orthogonal rotation, the communality (h_j^2) of a variable (X_j) remains constant. Hence, the sum of all communalities also remains constant.

Quartimax rotation

In quartimax, the index of simplicity is defined as the variance of the squared loadings for the manifest variable (Neuhaus and Wrigley, 1954). So, using Table 12.4, the variance of squared loadings for the j-th row (Q_j) is

$$Q_j = \left[m\sum_{k=1}^{m}(\lambda_{jk}^2)^2 - \left(\sum_{k=1}^{m}\lambda_{jk}^2\right)^2 \right] / m^2$$

Summing over all rows, we get

$$Q = \sum_{j=1}^{p} \left\{ \left[m\sum_{k=1}^{m} (\lambda_{jk}^2)^2 - \left(\sum_{k=1}^{m} \lambda_{jk}^2 \right)^2 \right] / m^2 \right\} \tag{12.21}$$

The quartimax rotation is accomplished by maximizing Q in Equation 12.21.

Varimax rotation

In varimax (Kaiser, 1958), the index of simplicity is defined as the variance of squared factor loadings for a factor. Note the difference with quartimax. Then from Table 12.4, the variance of squared loadings for the k-th column is V_k, where

$$V_k = \left[p\sum_{j=1}^{p} (\lambda_{jk}^2)^2 - \left(\sum_{j=1}^{p} \lambda_{jk}^2 \right)^2 \right] / p^2$$

Summing over all the columns, we get

$$V = \sum_{k=1}^{m} \left\{ \left[p\sum_{j=1}^{p} (\lambda_{jk}^2)^2 - \left(\sum_{j=1}^{p} \lambda_{jk}^2 \right)^2 \right] / p^2 \right\} \tag{12.22}$$

The varimax rotation is accomplished by maximizing V in Equation 12.2.

Kaiser called Equation 12.22 raw varimax, as raw factor loadings are considered. Using normalized factor loadings, the normalized varimax, V_N, is obtained (Equation 12.23).

$$V_N = \sum_{k=1}^{m} \left\{ \left[p\sum_{j=1}^{p} (\lambda_{jk}^2 / \text{h}_j^2)^2 - \left(\sum_{j=1}^{p} \lambda_{jk}^2 / \text{h}_j^2 \right)^2 \right] / p^2 \right\} \tag{12.23}$$

As stated earlier, both quartimax and varimax are used for orthogonal rotation. If the factors are correlated, oblique rotation is recommended. The known methods of oblique rotation are *oblimin* and *promax* (Lawley and Maxwell, 1971).

Although the simple structure is an ultimate aim of factor rotation, the abovementioned methods do not always give a simple structure. Under many situations, expert opinion is required, particularly for verifying the factor loadings from its relevance and the rationality of being high or small (close to zero), and for selecting the number of relevant factors.

Example 12.5

Consider Example 12.2. Does factor rotation improve the interpretation?

Solution:

We consider the unrotated MLE estimates (see Table 12.2 in Example 12.2), as MLE gives a better picture of loadings for the two-factor model compared to the PCA and PFM methods. Although the pattern is clear for the manifest variables X_1, X_2, and X_3 with higher loading with F_1, it is not true for the other three variables (X_4, X_5, and X_6). See Table 12.5.

TABLE 12.5
Unrotated and Rotated Factors

Methods	Unrotated (MLE)			Varimax (MLE)			Quartimax (MLE)		
Manifest Variables	F_1	F_2	h_j^{2*}	F_1	F_2	h_j^{2*}	F_1	F_2	h_j^{2*}
X_1	**0.94**	−0.26	0.94	**0.96**	0.17	0.94	**0.96**	0.11	0.94
X_2	**0.78**	−0.05	0.61	**0.73**	0.29	0.61	**0.74**	0.25	0.61
X_3	**0.86**	0.34	0.85	**0.63**	0.68	0.85	**0.67**	0.64	0.85
X_4	0.5	**0.69**	0.73	0.16	**0.84**	0.73	0.21	**0.83**	0.73
X_5	0.49	**0.51**	0.51	0.23	**0.68**	0.51	0.27	**0.66**	0.51
X_6	0.4	**0.44**	0.35	0.18	**0.56**	0.35	0.21	**0.55**	0.35
Explained $V(X_j)$	2.87	1.11		1.95	2.04		2.08	1.91	
Proportion explained (%)	47.83	18.50		32.50	34		34.67	31.83	
Cumulative proportion explained (%)	**47.83**	**66.33**		**66.50**	**34**		**34.67**	**66.50**	

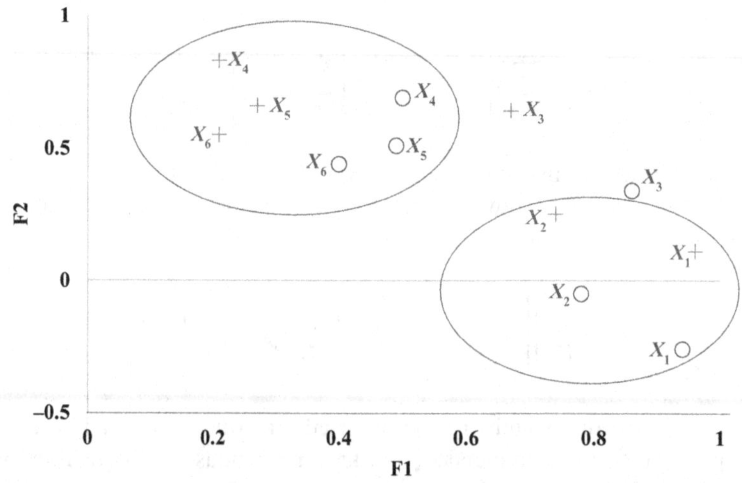

FIGURE 12.6 Patterns for unrotated (o) and quartimax rotated (+) factor loadings. After rotation, the loading patterns have improved. The variables X_4, X_5, and X_6 moved more towards F_2, X_1 and X_2 remained highly loaded with F_1 (but the loading sign changed from negative to positive). The loading pattern for X_3 did not improve and this may be an indication for inclusion of another factor (F_3).

The following findings can be generated from the factor rotation:

- Both varimax and quartimax rotations give better pattern than the unrotated loadings.
- The plot (Figure 12.6) suggests oblique rotation.
- The orthogonal rotation retains the original communality and cumulative proportion of variance explained by the selected number of factors (*m*). Note that if *m* changes, the cumulative proportion of variance explained also changes.

12.8 FACTOR SCORES

Once the number of factors is determined and the factors loadings are obtained, the factor model given in Equation 12.2 is known. Further given the $\boldsymbol{X}$ observations for the i-th individual/item in Equation 12.2, one may be interested to estimate the $\boldsymbol{f}$ values for the same i-th individual/item, which is known as factor score for the i-th individual/item. If the sample data comprises n individuals/items, we can obtain n factor-scores. The factor scores can be used in subsequent analysis. However, the calculation of factor scores is not easy, as $\boldsymbol{f}$ cannot be written as a linear function of $\boldsymbol{X}$ without knowing $\boldsymbol{\delta}$ (see Equation 12.2). For the i-th individual/item, Equation 12.2 can be written as

$$\mathbf{x}_i = \boldsymbol{\mu} + \Lambda \boldsymbol{f}_i + \boldsymbol{\delta}_i, \quad i = 1,2,\ldots,n$$

Here, we intend to estimate the factor scores, $\boldsymbol{f}_i$, for each of the observations.

Three different methods are commonly used for estimating the factor scores (Hershbeger, 2005). They are

- Principal component scores
- Estimation by least squares approach
- Estimation by regression

12.8.1 Estimation by Principal Component Scores

In this method, the linear model in the following form is assumed.

$$\tilde{\boldsymbol{X}}_{p\times 1} = \Lambda_{p\times m} \boldsymbol{f}_{m\times 1} \tag{12.24}$$

where, $\tilde{\boldsymbol{X}}$ is standardized, i.e., $Cov(\tilde{\boldsymbol{X}}) = \mathbf{S} = \mathbf{R} = Cor(\tilde{\boldsymbol{X}})$.

From Equation 12.24, we get

$$\boldsymbol{f}_{m\times 1} = \Lambda^{-1}_{p\times m} \tilde{\boldsymbol{X}}_{p\times 1} \tag{12.25}$$

The formula of obtaining k-th principal component scores for the i-th individual is as follows (see Hershberger, 2005):

$$f_{ik} = \sum_{j=1}^{p} \frac{\lambda_{jk}}{Q_k} x_{ij}, \text{ i=1,2,....,n} \tag{12.26}$$

Where λ_{jk} is the factor loading between the k-th factor and j-th manifest variable and Q_k is the eigenvalue of the k-th principal component. If correlation matrix is used, x_{ij} will be replaced by z_{ij}, where z_{ij} is the standardized observation.

12.8.2 Estimation By Least Squares Approach

Bartlett (1937b) proposed the weighted least squares based method to obtain factor scores. Here, the sum of the squared errors is weighted by the respective specific variance. The factor scores for the i-th individual, $\boldsymbol{f}_i$, $i = 1, 2, \ldots, n$, can be estimated by minimizing the following equation.

$$g(\boldsymbol{f}_i) = (\mathbf{x}_i - \bar{\mathbf{x}} - \hat{\Lambda}\boldsymbol{f}_i)^T \Psi^{-1} (\mathbf{x}_i - \bar{\mathbf{x}} - \hat{\Lambda}\boldsymbol{f}_i) \tag{12.27}$$

where, $\mathbf{x}_i$ is the observed vector of manifest variables for the i-th individual and $\bar{\mathbf{x}}$ is the sample mean vector.

The solution obtained after minimization is

$$\hat{f}_i = (\hat{\Lambda}^T \hat{\Psi}^{-1} \hat{\Lambda})^{-1} \hat{\Lambda}^T \hat{\Psi}^{-1} (\mathbf{x}_i - \bar{\mathbf{x}}) \tag{12.28}$$

When the correlation matrix is factored, Equation 12.28 becomes

$$\hat{f}_i = (\hat{\Lambda}^T \hat{\Psi}^{-1} \hat{\Lambda})^{-1} \hat{\Lambda}^T \hat{\Psi}^{-1} \mathbf{z}_i \tag{12.29}$$

where, $\mathbf{z}_i$ is the i-th standardized observation vector.

12.8.3 Estimation By Regression Method

This approach, proposed by Thomson (1951), is used to estimate the factor scores based on maximum likelihood estimation. A vector of the observed data is taken into account, and then supplemented by the factor loadings for the observation. Both the common factor $\boldsymbol{F}$ and specific factor $\boldsymbol{\delta}$ follow normality. In addition, joint distribution of the observation $\mathbf{x}_i$ and the factor $\boldsymbol{f}_i$, $i = 1, 2, \ldots, n$, is

$$\begin{pmatrix} \mathbf{x}_i \\ f_i \end{pmatrix} \sim N\left[\begin{pmatrix} \boldsymbol{\mu} \\ \mathbf{0} \end{pmatrix} \begin{pmatrix} \Lambda\Lambda^T + \Psi & \Lambda \\ \Lambda^T & \mathbf{I} \end{pmatrix}\right]$$

Based on the above formula, we can calculate the conditional expectation of the factor $\boldsymbol{f}_i$, given the data $\mathbf{x}_i$ as

$$E(\boldsymbol{f}_i | \mathbf{x}_i) = \Lambda^T (\Lambda\Lambda^T + \Psi)^{-1} (\mathbf{x}_i - \boldsymbol{\mu}) \tag{12.30}$$

By substituting the estimates of Λ and Ψ_i in Equation 12.29, we get

$$\hat{f}_i = \hat{\Lambda}^T (\hat{\Lambda}\hat{\Lambda}^T + \hat{\Psi})^{-1} (\mathbf{x}_i - \bar{\mathbf{x}}) \tag{12.31}$$

When correlation matrix is factored, Equation 12.31 becomes

$$\hat{f}_i = \hat{\Lambda}^T (\hat{\Lambda}\hat{\Lambda}^T + \hat{\Psi})^{-1} \mathbf{z}_i \tag{12.32}$$

where, $\mathbf{z}_i$ is the i-th standardized observation vector.

Anderson and Rubin (1956) revised Bartlett's method to get uncorrelated factor scores; however, it involves more tedious calculations. Which method is the best? Hershbeger (2005) pointed out that the factor scores obtained from the methods described above are highly correlated, but they often produce different values.

Example 12.6

Consider Example 12.2 and compute factor scores.

Solution:

We will use least squares and regression methods. Note, we have factored correlation matrix. Both the methods require standardized observation vectors for estimation of factor scores (see Equations 12.29 and 12.32). Let we are interested to estimate the factor score for the i-th and $(i + 1)$-th individuals with standardized observation vectors as given below.

$$\mathbf{z}_i = [-.6\ \ .5\ \ 0\ \ .3\ \ -.4\ -.2]^T$$
$$\mathbf{z}_{i+1} = [.8\ \ -.7\ \ .3\ \ .2\ \ \ .6\ \ \ .5]^T$$

Using the estimated parameters, we get

$$\hat{\Lambda}^T\hat{\Psi}^{-1}=\begin{bmatrix}0.94 & 0.78 & 0.86 & 0.5 & 0.49 & 0.4\\ -0.26 & -0.05 & 0.34 & 0.69 & 0.51 & 0.44\end{bmatrix}\begin{bmatrix}16.67 & 0 & 0 & 0 & 0 & 0\\ 0 & 2.57 & 0 & 0 & 0 & 0\\ 0 & 0 & 6.67 & 0 & 0 & 0\\ 0 & 0 & 0 & 3.65 & 0 & 0\\ 0 & 0 & 0 & 0 & 2.04 & 0\\ 0 & 0 & 0 & 0 & 0 & 1.55\end{bmatrix}$$

$$=\begin{bmatrix}15.67 & 2.00 & 5.73 & 1.83 & 1.00 & 0.62\\ -4.33 & -0.13 & 2.27 & 2.52 & 1.04 & 0.68\end{bmatrix}$$

$$(\hat{\Lambda}^T\hat{\Psi}^{-1}\hat{\Lambda})^{-1}=\begin{bmatrix}0.044 & 0.002\\ 0.002 & 0.222\end{bmatrix}$$

From Equation 12.29, we get

$$\boldsymbol{F}=\begin{bmatrix}f_i\\ f_{i+1}\end{bmatrix}=(\hat{\Lambda}^T\hat{\Psi}^{-1}\hat{\Lambda})^{-1}\hat{\Lambda}^T\hat{\Psi}^{-1}\begin{bmatrix}\mathbf{z}_i\\ \mathbf{z}_{i+1}\end{bmatrix}$$

$$=\begin{bmatrix}0.044 & 0.002\\ 0.002 & 0.222\end{bmatrix}\begin{bmatrix}15.67 & 2.00 & 5.73 & 1.83 & 1.00 & 0.62\\ -4.33 & -0.13 & 2.27 & 2.52 & 1.04 & 0.68\end{bmatrix}\begin{bmatrix}-.6 & .8\\ .5 & -.7\\ 0 & .3\\ .3 & .2\\ -.4 & .6\\ -.2 & .5\end{bmatrix}=\begin{bmatrix}-.36 & .62\\ .60 & -.25\end{bmatrix}$$

So, using *least squares method*, the factor scores on the i-th and $(i+1)$-th individuals for the first and second factors are –0.36, 0.62, and 0.60, –0.25, respectively.

Now, using *regression method* (Equation 12.32), the factor scores on the i-th and $(i+1)$-th individuals for the first and second factors are

$$\boldsymbol{F}=\begin{bmatrix}f_i\\ f_{i+1}\end{bmatrix}=\hat{\Lambda}^T(\hat{\Lambda}\hat{\Lambda}^T+\hat{\Psi})^{-1}\begin{bmatrix}\mathbf{z}_i\\ \mathbf{z}_{i+1}\end{bmatrix}$$

$$=\begin{bmatrix}0.94 & 0.78 & 0.86 & 0.5 & 0.49 & 0.4\\ -0.26 & -0.05 & 0.34 & 0.69 & 0.51 & 0.44\end{bmatrix}\begin{bmatrix}3.14 & -1.40 & -1.98 & 0.75 & 0.15 & 0.12\\ -1.40 & 2.40 & -0.43 & -0.10 & -0.06 & -0.04\\ -1.98 & -0.43 & 4.31 & -1.51 & -0.68 & -0.44\\ 0.75 & -0.10 & -1.51 & 2.34 & -0.56 & -0.36\\ 0.15 & -0.06 & -0.68 & -0.56 & 1.80 & -0.16\\ 0.12 & -0.04 & -0.44 & -0.36 & -0.16 & 1.45\end{bmatrix}\begin{bmatrix}-.6 & .8\\ .5 & -.7\\ 0 & .3\\ .3 & .2\\ -.4 & .6\\ -.2 & .5\end{bmatrix}$$

$$=\begin{bmatrix}-0.35 & 0.59\\ 0.49 & -0.20\end{bmatrix}$$

Note, the factors scores are similar.

12.9 DIFFERENCE WITH PRINCIPAL COMPONENT ANALYSIS

Both factor analysis and principal component analysis (PCA) are data reduction techniques and it has been seen earlier that PCA is also used as an estimation method in factor analysis. At the outset, therefore, readers may be confused in understanding the differences between the two techniques. Krzanowski and Marriott (2014b) lucidly demonstrated the differences as shown below.

- Factor analysis is model-based. The covariance/correlation structure (among the manifest variables) can be tested statistically. For example, if multivariate normality of the manifest variables (*X*) is satisfied, the likelihood ratio based test can be conducted to test the structure $\Sigma = \Lambda\Lambda^T + \Psi$. On the other hand, PCA tries to explain the variances of *X* without any underlying model and hypotheses testing.
- Factor analysis considers two types of variances, i.e., common and specific variances but PCA does not. If specific variances are small, then both PCA and common factor method yield similar results. But it is not true when we encounter large specific variances. While PCA tries to absorb the specific variances into the components, factor analysis separates them, thereby yielding a different interpretation.
- The relationship between the results of factor analysis obtained using the covariance matrix (**S**) and correlation matrix (**R**) can be easily established. For example, if the loadings and specific variance obtained using **S** are $\{\hat{\lambda}_{jk}\}$ and $\{\hat{\Psi}_{jj}\}$, respectively, then the corresponding estimates from **R** are $\{\hat{\lambda}_{jk} / \sqrt{s_{jj}}\}$ and $\{\hat{\Psi}_{jj} / s_{jj}\}$, respectively. The same is not true for PCA.

12.10 VALIDATION

The fundamental relationship of orthogonal factor model given in Equation 12.3 demonstrates that if we know the factor loading matrix $\mathbf{\Lambda}_{p\times m}$ and specific variance matrix $\mathbf{\Psi}_{p\times p}$, the population co variance structure $\mathbf{\Sigma}_{p\times p}$ can be reproduced. $\hat{\mathbf{\Lambda}}_{p\times m}$ and $\hat{\mathbf{\Psi}}_{p\times p}$ are usually estimated using sample covariance matrix $\mathbf{S}_{p\times p}$ from a sample of size *n*. The key question of validation here is whether the estimates $\hat{\Lambda}$ and $\hat{\mathbf{\Psi}}$ from **S** will be able to reproduce $\mathbf{\Sigma}$ and, if so, how to test this. Note that population $\mathbf{\Sigma}$ is seldom known. Lattin et al. (2003) have discussed some of the approaches such as (i) use of hold-out sample, (ii) assessment of sampling variation using bootstrapping, and (iii) fine-tuning of factor solution by dropping some of the key manifest variables that are less stable. The central issue here is the comparison of the factor loadings obtained using the original sample with that obtained by using other samples (e.g., hold-out or bootstrap samples). Let $\hat{\mathbf{\Lambda}}_0$ be the loading matrix from the original sample and $\hat{\mathbf{\Lambda}}_1$ the loading matrix from any other sample. Theoretically, for generalizability of factor model $\hat{\lambda}_0 \approx \hat{\lambda}_1$, the larger the departures, the worse the model. In search of simple structure, several approaches are provided where a binary coding scheme is adopted.

Let λ^c_{jk} denote the coded loading between X_j and F_k, $j = 1, 2, \ldots, p$ and $k = 1, 2, \ldots, m$. Further, put $\lambda^c_{jk} = 1$ if λ_{jk} is the maximum of the set $\{\lambda_{j1}, \lambda_{j2}, \ldots, \lambda_{jk}, \ldots, \lambda_{jm}\}$, else $\lambda^c_{jk} = 0$. This binary coding is done for all the X_j, $j = 1, 2, \ldots, p$. Now, if we consider $\mathbf{\Lambda}^c$ the coded loading matrix, then it contains only 1 and 0, with only one 1 in each row (note: the *j*-th λ represents the loading pattern for *j*-th manifest variable, X_j). Now this simple loading matrix for the original sample (used for model-building) and test sample (used for validation) should exhibit a similar pattern. Lattin et al. (2003) explained this lucidly with an example. There have been numerous application studies of EFA, where authors have reported various decision criteria before accepting EFA results as a valid explanation of the population covariance structure. These include (i) selection of manifest variables, (ii) type of input matrix (covariance or correlation), (iii) selection of extraction method, (iv) use of rotation, (v) identification of meaningful factors, (vi) sample size, (vii) quality of data, (viii) missing data, and (ix) cross-validation. During decision-making on the above, whenever required, expert opinion may be sought. Although many issues of EFA are still dealt with subjectively, the utility of EFA has been established for more than a century.

12.11 CASE STUDY

12.11.1 Background

Performance study of employees is often done to see whether there is a need to improve the competency of the employees and the condition of the working environment. One of the major performance-influencing factors is job demand. The job stress theory states that excessive job demand may lead to burnout and low job demand imposes less challenge, and both are responsible to lower performance. Further, employees vary in perceiving job demand. Hence, job demand is very difficult to measure. In this study, we demonstrate how exploratory factor analysis can be used to identify the dimensions of job demand. The study was conducted in an underground coal mine. It was semi-mechanized mine and operated six days in a week and three shifts per day. The average coal production was 1,000 metric tons and the number employed was around 900 at the time of the study.

12.11.2 Variables and Data

The study focuses on measuring job demand faced by mine workers. From literature of job stress, it was found that job demand (JD) is one important factor and being unobservable, it is measured using questionnaire survey. Based on literature survey, 14 questions (Q_1–Q_{14}) were prepared as shown in Table 12.6 below. A three-point Likert scale was used to measure the responses. Against every question in Table 12.6, the response could be "rarely," "sometimes," or "always." The numerical values assigned to these responses are 1, 2, and 3 respectively. However, some questions are in reverse order and accordingly the numerical values are also reversed. The sample size is 140 and the mean responses with their standard deviations are shown in Table 12.6. Note that each question is one manifest variable. So, we have 14 manifest variables.

The 140 responses (as sample size is 140) on 14 variables are used to compute the correlation matrix for the manifest variables (Q_1–Q_{14}). The correlation matrix is shown in Table 12.7. From visual inspection of the correlation matrix, it is not clear whether the correlation matrix is factorable or not. However, the table contains some high and many low correlation coefficients.

12.11.3 Data Analysis

First, we used Bartlett's test (Equation 12.16) and KMO test (Equations 12.17–12.18) to see whether the data is adequate for factor analysis or not. The results are given in Table 12.8. Both the tests recommend the use of factor analysis. But the overall MSA based on KMO is 0.60, which is just the threshold value and the item's MSA is also not on the higher side, except Q_{10} and Q_{14}. It not

TABLE 12.6
Manifest Variables (Questions Administered)

No.	Question	Mean (Std. Dev.)
Q1	Are you required to work at a rapid pace?	1.89 (0.70)
Q2	Do you have too much to do?	1.98 (0.75)
Q3	Does your work require physical endurance?	2.48 (0.59)
Q4	Does your work require quick decisions?	1.78 (0.67)
Q5	Does your work require maximum attention?	2.51 (0.62)
Q6	Does your work require great precision of movement?	1.27 (0.53)
Q7	Does your work require complex decisions?	1.42 (0.62)
Q8	Do you have to repeat the same work procedure at intervals of a few minutes?	2.46 (0.72)
Q9	Do you perform work tasks for which you need more training?	1.46 (0.54)
Q10	Are your skills and knowledge useful in your work?	2.16 (0.53)
Q11	Is your work challenging in a positive way?	2.21 (0.56)
Q12	Do you consider your work meaningful?	2.20 (0.58)
Q13	Is it possible to have social contacts with co-workers while you are working?	1.62 (0.52)
Q14	Are errors in your work associated with a risk of personal injury?	2.74 (0.49)

TABLE 12.7
Correlation Matrix

	Q1	Q2	Q3	Q4	Q5	Q6	Q7	Q8
Q1	1.00							
Q2	0.72	1.00						
Q3	0.20	0.20	1.00					
Q4	0.19	0.22	–0.11	1.00				
Q5	–0.08	–0.02	0.00	0.10	1.00			
Q6	0.16	0.21	0.02	0.19	0.19	1.00		
Q7	0.01	0.07	–0.08	0.43	–0.06	0.09	1.00	
Q8	0.00	–0.09	–0.11	–0.19	0.06	0.05	–0.37	1.00
Q9	–0.03	0.06	0.07	0.21	0.13	–0.02	0.12	–0.23
Q10	–0.15	–0.12	–0.13	0.06	–0.11	0.03	0.21	–0.13
Q11	–0.09	–0.01	–0.04	0.11	–0.08	0.08	0.20	–0.11
Q12	–0.01	–0.02	–0.09	0.25	0.04	0.15	0.16	–0.17
Q13	–0.12	–0.06	0.03	–0.06	0.07	–0.09	0.01	0.14
Q14	0.17	0.14	0.03	0.11	0.10	0.11	0.10	–0.03

TABLE 12.7
Correlation Matrix (Continued)

	Q9	Q10	Q11	Q12	Q13	Q14
Q9	1.00					
Q10	–0.08	1.00				
Q11	–0.01	0.40	1.00			
Q12	0.14	0.27	0.52	1.00		
Q13	–0.17	–0.10	–0.18	–0.23	1.00	
Q14	0.10	0.02	0.09	0.19	–0.13	1.00

TABLE 12.8
Bartlett and KMO Tests

Bartlett Test
Bartlett statistic (Bs) = 337.87
Tabulated Chi-square for α=0.05 with 91 degrees of freedom=70
H_0: **R=I** is rejected. Conclusion: The data is factorable.

Kaiser-Meyer-Olkin (KMO) factor adequacy
Overall MSA=0.6; Conclusion: The data is factorable.
MSA for each item

Q_1	Q_2	Q_3	Q_4	Q_5	Q_6	Q_7	Q_8	Q_9	Q_{10}	Q_{11}	Q_{12}	Q_{13}	Q_{14}
0.54	0.56	0.59	0.67	0.49	0.64	0.61	0.6	0.56	0.75	0.61	0.65	0.59	0.74

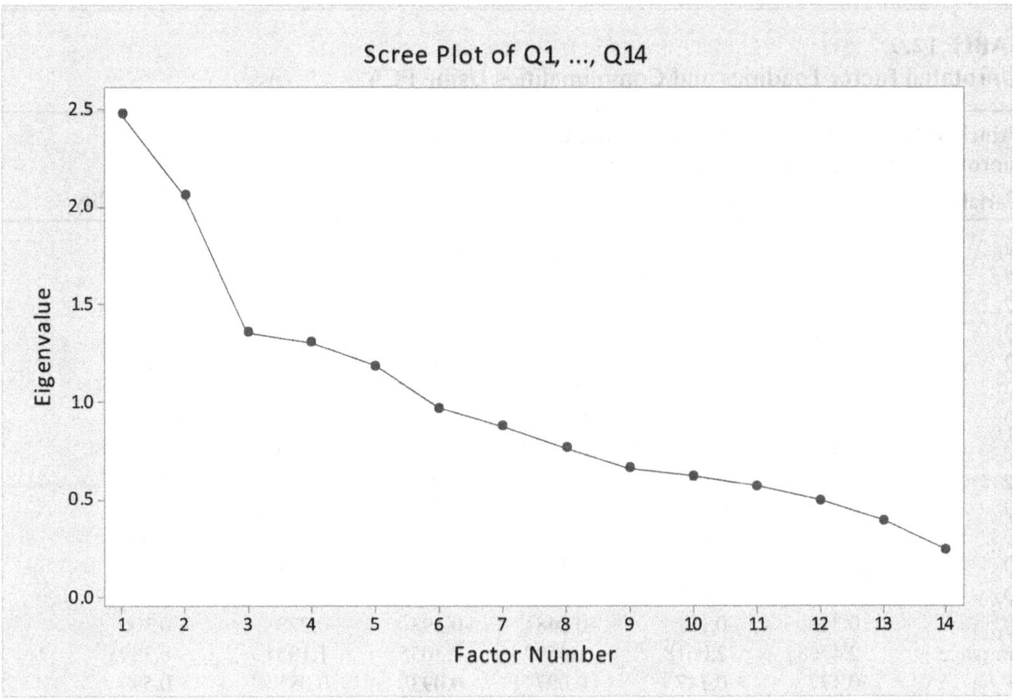

FIGURE 12.7 Scree plot for job demand factors. Following Kaiser's rule, five factors having eigenvalue equal to or greater than 1.0 may be retained.

surprising as we have considered a three-point Likert scale and the participants are mine workers with a lower education level. So, the factor analysis results to be used consciously.

Figure 12.7 represents the scree plot for the correlation matrix (Table 12.7). Following Kaiser's rule, we can consider five factors, as the first five factors have eigenvalue more than 1.0. Using PCA and MLE methods, the factor loadings are estimated. Tables 12.9 and 12.10 show the estimated factor loadings, communalities, and variance explained by the factors using PCA and MLE, respectively.

The salient features of the results obtained is summarized below.

- The unrotated factor loadings do not provide clear patterns. In both PCA and ML estimation, the first factor is loaded with most of the manifest variables (questions). Considering 0.30 as significant loading, it is seen that in PCA, factor 1 is significantly loaded with Q_4, and Q_7–Q_{14}, while in MLE, factor 1 is significantly loaded with Q_4, Q_7–Q_8, and Q_{10}–Q_{12}. For other factors, compared to PCA, MLE gives better structures.
- PCA with the first five factors explains 59.90% cumulative variance of the manifest variables, while the MLE method with five factors explains 40.05% cumulative variance of the manifest variables.
- From communality point of view, the PCA performs slightly better than the MLE method.
- However, the primary aim of EFA is to obtain a reduced number of factors with better (simplified) structure. So, the unrotated MLE has better edge than the unrotated PCA.

We further performed varimax and quartimax rotation for both the abovementioned cases (i.e., PCA and MLE-based parameter estimation). Tables 12.11 and 12.12 show the estimated factor loadings,

TABLE 12.9
Unrotated Factor Loadings and Communalities Using PCA

Principal Component Factor Analysis of The Correlation Matrix
Unrotated Factor Loadings and Communalities

Variable	Factor1	Factor2	Factor3	Factor4	Factor5	Communality
Q_1	0.257	–0.794	–0.31	0.119	–0.079	0.813
Q_2	0.327	–0.777	–0.214	0.124	–0.106	0.783
Q_3	–0.037	–0.401	–0.04	0.254	0.389	0.379
Q_4	0.608	–0.158	0.339	–0.057	–0.383	0.66
Q_5	0.015	–0.094	0.414	–0.676	0.035	0.638
Q_6	0.32	–0.253	–0.117	–0.52	–0.223	0.501
Q_7	0.585	0.112	0.323	0.283	–0.413	0.711
Q_8	–0.474	–0.055	–0.36	–0.493	–0.166	0.628
Q_9	0.313	–0.081	0.58	–0.002	0.468	0.661
Q_{10}	0.394	0.504	–0.327	0.112	–0.138	0.548
Q_{11}	0.561	0.408	–0.4	–0.036	0.076	0.649
Q_{12}	0.65	0.3	–0.183	–0.228	0.177	0.63
Q_{13}	–0.38	–0.002	0.206	0.051	–0.541	0.482
Q_{14}	0.356	–0.19	–0.048	–0.298	0.228	0.306
Variance	2.4768	2.0618	1.3599	1.3055	1.1854	8.3895
% Var	0.177	0.147	0.097	0.093	0.085	0.599

TABLE 12.10
Unrotated Factor Loadings and Communalities Using MLE

Maximum Likelihood Factor Analysis of The Correlation Matrix
Unrotated Factor Loadings and Communalities

Variable	Factor1	Factor2	Factor3	Factor4	Factor5	Communality
Q_1	–0.037	–0.938	0.042	0.036	0.003	0.884
Q_2	0.05	–0.771	–0.026	–0.037	0.028	0.599
Q_3	–0.114	–0.217	–0.01	–0.064	0.258	0.131
Q_4	0.497	–0.24	–0.248	–0.177	–0.246	0.458
Q_5	–0.02	0.062	–0.048	–0.38	–0.263	0.22
Q_6	0.161	–0.196	0.068	–0.14	–0.293	0.175
Q_7	0.665	–0.051	–0.449	0.187	–0.055	0.685
Q_8	–0.431	0.031	0.245	0.072	–0.41	0.42
Q_9	0.219	–0.007	–0.168	–0.496	0.215	0.369
Q_{10}	0.417	0.16	0.222	0.274	0.03	0.325
Q_{11}	0.573	0.084	0.46	0.138	0.033	0.568
Q_{12}	0.596	0.012	0.428	–0.184	–0.035	0.575
Q_{13}	–0.204	0.12	–0.219	0.163	–0.175	0.161
Q_{14}	0.186	–0.184	0.078	–0.166	–0.057	0.105
Variance	1.9015	1.7047	0.8596	0.6708	0.5376	5.6742
% Var	0.136	0.122	0.061	0.048	0.038	0.405

TABLE 12.11
Varimax Rotated Factor Loadings and Communalities Using PCA

Principal Component Factor Analysis of The Correlation Matrix
Rotated Factor Loadings and Communalities
Varimax Rotation

Variable	Factor1	Factor2	Factor3	Factor4	Factor5	Communality
Q_1	–0.043	–0.898	0.047	–0.05	–0.031	0.813
Q_2	–0.047	–0.867	0.153	–0.078	0.017	0.783
Q_3	–0.121	–0.386	–0.222	0.184	0.364	0.379
Q_4	0.075	–0.174	0.727	–0.302	0.063	0.66
Q_5	–0.183	0.194	0.006	–0.739	0.145	0.638
Q_6	0.166	–0.269	0.118	–0.588	–0.202	0.501
Q_7	0.149	–0.002	0.824	0.079	0.049	0.711
Q_8	–0.136	0.031	–0.486	–0.305	–0.528	0.628
Q_9	–0.037	0.076	0.182	–0.186	0.765	0.661
Q_{10}	0.625	0.17	0.208	0.199	–0.213	0.548
Q_{11}	0.796	0.041	0.102	0.032	–0.053	0.649
Q_{12}	0.73	0.042	0.145	–0.229	0.146	0.63
Q_{13}	–0.491	0.17	0.208	0.052	–0.408	0.482
Q_{14}	0.259	–0.228	–0.034	–0.368	0.224	0.306
Variance	1.9937	1.966	1.6858	1.3911	1.3529	8.3895
% Var	0.142	0.14	0.12	0.099	0.097	0.599

TABLE 12.12
Varimax Rotated Factor Loadings and Communalities Using MLE

Maximum Likelihood Factor Analysis of The Correlation Matrix
Rotated Factor Loadings and Communalities
Varimax Rotation

Variable	Factor1	Factor2	Factor3	Factor4	Factor5	Communality
Q_1	–0.935	–0.042	0.03	0.054	–0.066	0.884
Q_2	–0.761	–0.028	0.099	–0.048	–0.088	0.599
Q_3	–0.241	–0.081	–0.105	–0.185	0.145	0.131
Q_4	–0.17	0.106	0.506	–0.12	–0.384	0.458
Q_5	0.107	–0.106	–0.056	–0.102	–0.429	0.22
Q_6	–0.16	0.108	0.08	0.071	–0.355	0.175
Q_7	0.005	0.139	0.813	–0.066	–0.004	0.685
Q_8	0.036	–0.142	–0.385	0.459	–0.199	0.42
Q_9	0.012	–0.006	0.097	–0.569	–0.189	0.369
Q_{10}	0.138	0.482	0.197	0.119	0.144	0.325
Q_{11}	0.053	0.742	0.115	0.032	0.016	0.568
Q_{12}	0.005	0.692	0.078	–0.16	–0.253	0.575
Q_{13}	0.142	–0.288	0.064	0.231	0.029	0.161
Q_{14}	–0.171	0.151	0.053	–0.104	–0.2	0.105
Variance	1.649	1.4505	1.1706	0.7126	0.6915	5.6742
% Var	0.118	0.104	0.084	0.051	0.049	0.405

communalities, and variance explained by the factors using varimax rotation for PCA and MLE, respectively.

The salient features of the results obtained after varimax rotation can be summarized below.

- The varimax rotation provides clear patterns for both PCA and MLE. Again, considering 0.30 as significant loading, it is seen that in PCA, factor 1 is significantly loaded with Q_7–Q_{14}, while in MLE, factor 1 is significantly loaded with Q_1–Q_3. For other factors, similar clear patterns are observed. However, compared to PCA, MLE gives better structures. For example, in varimax rotated PCA, Q_4 is significantly loaded with factor 3 (loading = 0.727) and factor 4 (loading = –0.302), Q_8 is significantly loaded with factor 3 (loading = –0.486), factor 4 (loading = –0.305) and factor 5 (loading = –0.528), and Q_{13} is significantly loaded with factor 1 (loading = –0.491) and factor 5 (loading = –0.408). But in varimax rotated MLE, Q_4 is significantly loaded with factor 3 (loading = 0.506) and factor 5 (loading = –0.384), Q_8 is significantly loaded with factor 3 (loading = –0.385), and factor 4 (loading = 0.459). For further simplification, if we assign a variable to the factor with the maximum loading, the rotated factor loadings can uniquely define the factors with their manifest variables.
- The cumulative variance explained and the communality did not change after varimax rotation in both the PCA or MLE-based estimation. This is in line with the theory.

12.11.4 Results and Discussion

Table 12.13 shows the five factors with their manifest variables and probable name of each of the factors. Naming the factors is a very important task. It is possible if the factor analysis provides clear pattern of loadings and the manifest variables assigned to a factor are theoretically justified as well as practically relevant. The manifest variables assigned to the factors in both the PCA and MLE methods with varimax rotation enable us to name the factors. However, we prefer here the MLE-based assignment and accordingly, the factors with their manifest/indicator variables (questions) and name are shown in the first, fourth, and fifth columns of Table 12.13.

As question 13 was loaded lowly (–0.288, Table 12.12), we perform the varimax rotated MLE excluding Q_{13} from the analysis. Table 12.14 shows the results. However, the results do not provide any better patterns, hence, we preferred to go by the results of varimax rotated MLE without excluding any questions.

Factor loadings indicate how much a variable is explained by a factor. When estimated using correlation matrix, it ranges from –1 to +1, and a variable with loadings close to –1 or 1 indicates the strong influence of the factor on the variable. Further, loadings close to 0 represent weak influence on the variable. Loading plot is used to pictorially represent these influences. In Figure 12.8, obtained

TABLE 12.13
Number and Name of Factors

	Factors After Varimax Rotation			
	PCA		MLE	
Factor	Questions	Name	Questions	Name
F_1	Q_{10}–Q_{13}	Skill and development	Q_1–Q_3	Physical labor
F_2	Q_1–Q_3	Physical labor	Q_{10}–Q_{13}	Skill and development
F_3	Q_4, Q_7, Q_8	Mental load	Q_4, Q_7	Mental load
F_4	Q_5, Q_6, Q_{14}	Attention and concentration	Q_8, Q_9	Training and procedure
F_5	Q_9	Training needs	Q_5, Q_6	Attention and concentration

TABLE 12.14
Varimax Rotated Factor Loadings and Communalities Using MLE After Excluding Q_{13}

Maximum likelihood factor analysis of the correlation matrix
Rotated factor loadings and communalities
Varimax rotation
Note: Question 13 (Q_{13}) is excluded from the analysis

Variable	Factor1	Factor2	Factor3	Factor4	Factor5	Communality
Q_1	0.897	–0.088	–0.037	–0.034	0.033	0.815
Q_2	0.798	–0.044	–0.07	0.087	–0.015	0.651
Q_3	0.237	–0.079	0.232	0.255	0.042	0.183
Q_4	0.21	0.11	–0.625	0.084	–0.301	0.544
Q_5	–0.064	–0.073	0.003	–0.011	–0.452	0.214
Q_6	0.211	0.115	–0.136	–0.117	–0.252	0.153
Q_7	0.012	0.178	–0.664	0.254	0.09	0.545
Q_8	–0.019	–0.129	0.25	–0.644	–0.101	0.505
Q_9	–0.002	–0.006	–0.085	0.402	–0.339	0.284
Q_{10}	–0.131	0.485	–0.175	–0.015	0.219	0.331
Q_{11}	–0.008	0.782	–0.061	0.017	0.054	0.619
Q_{12}	0.037	0.671	–0.104	0.091	–0.257	0.537
Q_{14}	0.194	0.146	–0.051	0.049	–0.198	0.103
Variance	1.6465	1.4136	1.0269	0.7466	0.6509	5.4843
% Var	0.127	0.109	0.079	0.057	0.05	0.422

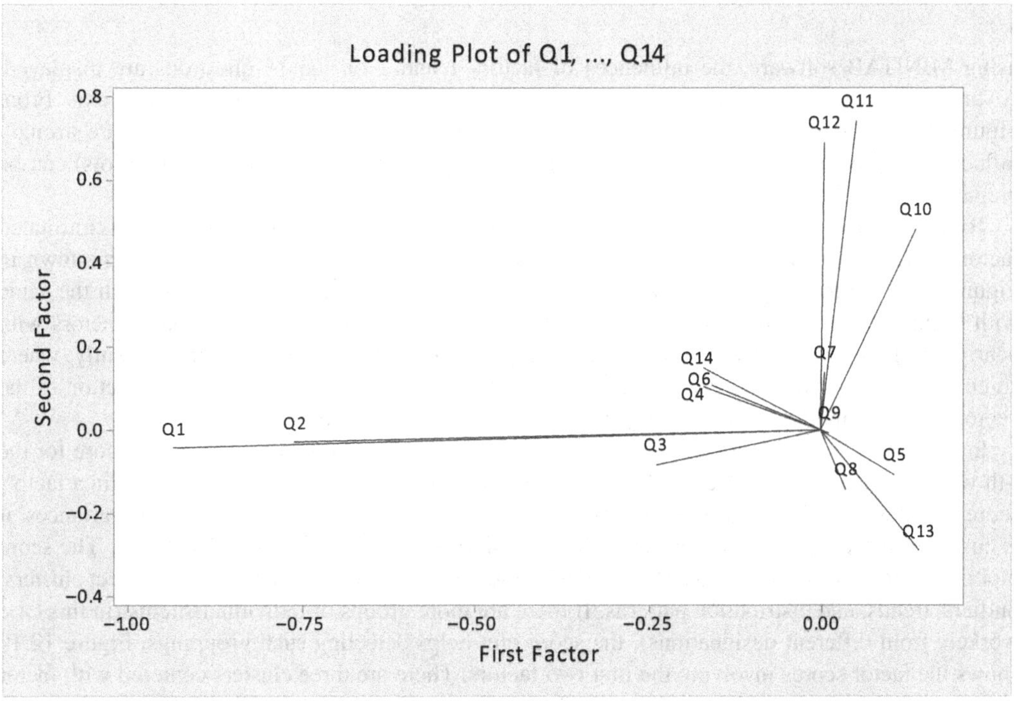

FIGURE 12.8 Rotated (varimax) factor loadings plot for the first two factors (F_1 and F_2). Q_1–Q_3 have higher loading with F_1 and Q_{10}–Q_{13} have higher loading with F_2 (see Table 12.12 for loading values).

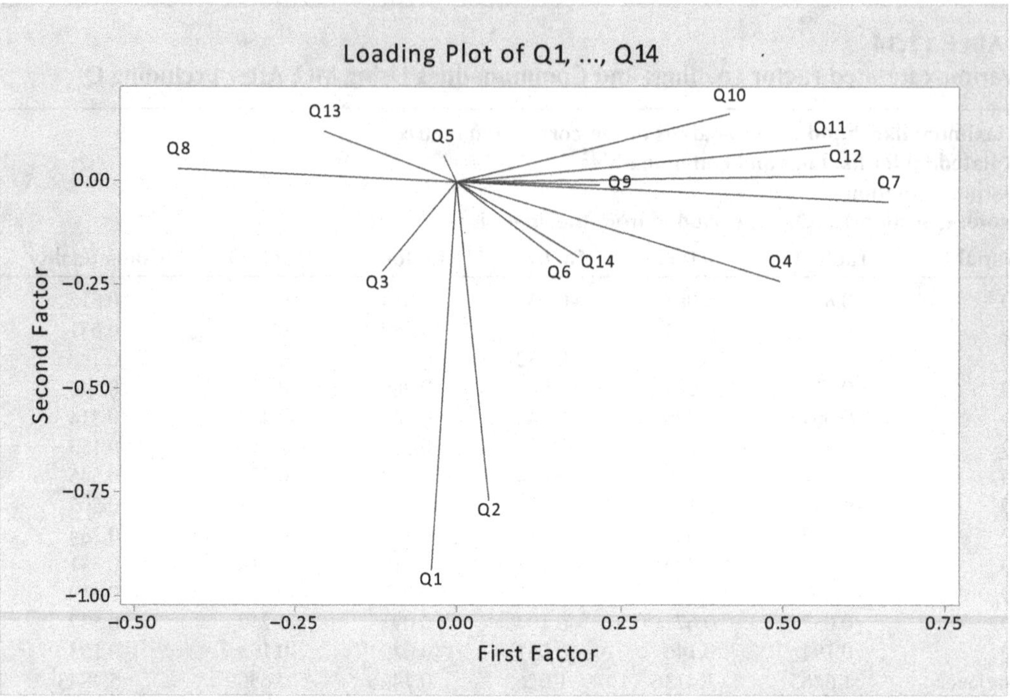

FIGURE 12.9 Unrotated factor loadings plot for the first two factors (F_1 and F_2). The unrotated factor loadings do not provide clear patterns.

using MINITAB software, the influences of factors 1 and 2 on the 14 questions are displayed. A variable with strong influence by a factor will lie far from the origin along that factor axis. From Figure 12.8, it can be said that Q_1 and Q_2 are strongly influenced by factor 1 and Q_{10}–Q_{12} are strongly influenced by factor 2. For a five-factor model ${}^5C_2$=10 such bivariate plots (loading plots) can be prepared and the influences can be visually compared.

To see the effect of rotation, one can compare the loading plots of rated factors vis-à-vis unrotated factors. The unrotated factor loading plot for factor 1 and factor 2 (based on MLE) is shown in Figure 12.9. See the length of the vectors representing the questions and compared with the same with Figure 12.8 (rotated factor loading plot). While in Figure 12.9, there are many vectors with nearly equal length (indicating poor patterns or structure), Figure 12.8 shows it differently, where some vectors are of high length and some have low vector length. In addition, the direction of the vectors is also important. Readers may explore this by observing Figures 12.8 and 12.9.

In Section 12.8, we described factor scores and their computation. Here, the factor score for the i-th worker is a vector of values of the five factors and with n=140 workers, we can obtain a factor-score matrix of size 140×5. As the first two factors usually explain the major share of variances, it is customary to plot the scores of the second factor versus the scores of the first factor. The score plot is used to understand the data structure for subsequent analysis. It can be used to detect clusters, outliers, trends, and distribution patterns. If there are more groups of individuals/items (in this case workers from different designations), the score plot helps detecting such groupings. Figure 12.10 shows the factor scores involving the first two factors. There are three clusters centered with factor scores −1.50, 0.00, and 1.50 for factor 1. Factor 1 is named as physical labor and it is a proxy measure of job load. So, it can be concluded that the 140 workers might belong to three different job

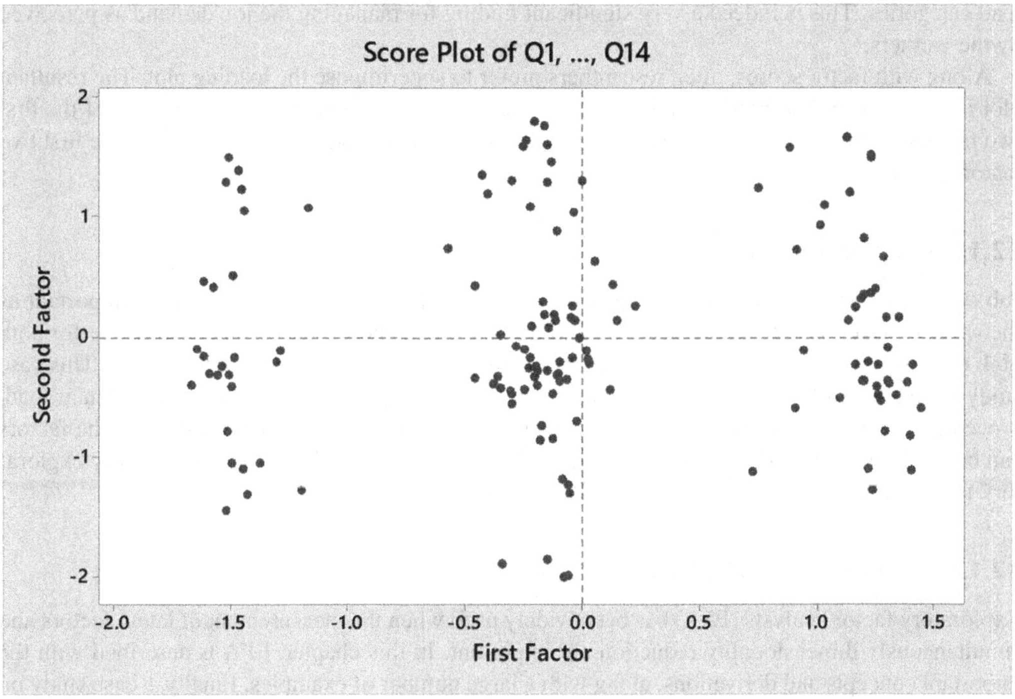

FIGURE 12.10 Factor scores of factors 1 (F_1) and 2 (F_2). The scatter shows that F_1 and F_2 are independent.

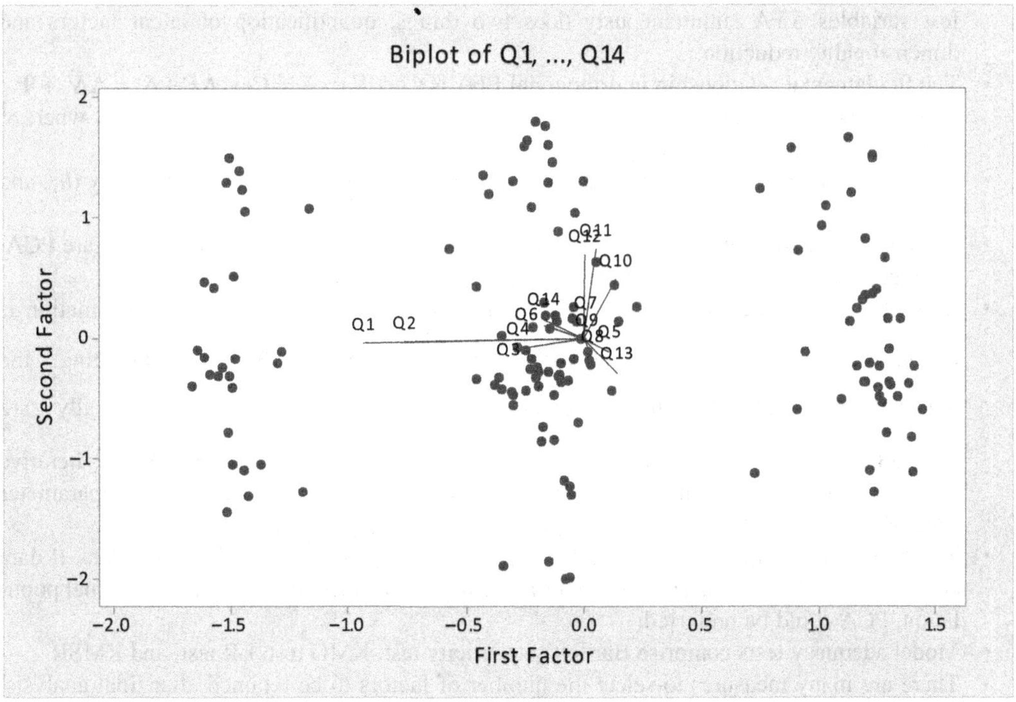

FIGURE 12.11 Biplot of Q_1-Q_{14} using factors 1 (F_1) and 2 (F_2). Biplot visually demonstrates both factor scores (dots) and factor loadings (vectors).

load categories. This is indeed a very significant finding for managing the job demand as perceived by the workers.

Along with factor scores, often researchers prefer to superimpose the loading plot. The resultant plot is known as biplot, which can be used to assess the data structure and the loadings of the first two factors. In Figure 12.11, the factor scores (dots) and factor loadings (vectors) using the first two factors are shown.

12.11.5 Conclusions

Job demand is a key issue in industrial work settings and for mining it is even more important as the workers have to work under difficult conditions. Understanding the factors responsible for both high and low job demand helps in managing job stress as well as worker's performance. This case study shows five job demand factors, namely physical labor, skill and development, mental load, procedure and training, and attention and concentration. The indicators against each of the factors can be used for intervention design. More importantly, this case study shows how to apply exploratory factor analysis in real-life problem-solving.

12.12 LEARNING SUMMARY

Exploratory factor analysis (EFA) has been widely used when the measurement of latent factors and simultaneously dimensionality reduction are important. In this chapter, EFA is described with the important concepts and derivations, along with a large number of examples. Finally, a case study on mine workers' job demand is presented. In summary,

- EFA is an exploratory data-driven model. The researcher, *a priori*, does not have the model structure and hence, tries to get one that will explain the covariance relationship of the manifest variables. EFA simultaneously does two things: quantification of latent factors and dimensionality reduction.
- The fundamental relationship in orthogonal EFA is $Cov(\mathbf{X}) = \mathbf{\Sigma} = Cov(\mathbf{\Lambda} F + \boldsymbol{\delta}) = \mathbf{\Lambda}\mathbf{\Lambda}^T + \mathbf{\Psi}$. EFA can also be oblique in nature with fundamental relationship $\mathbf{\Sigma} = \mathbf{\Lambda}\boldsymbol{\phi}\mathbf{\Lambda}^T + \mathbf{\Psi}$, where ϕ represents the covariance among the factors (F).
- EFA splits the total variance of a manifest variable (σ_j^2) into two parts: communality (h_j^2) and specific variance (ψ_j^2). Algebraically, $\sigma_j^2 = h_j^2 + \psi_j^2$.
- Parameter estimation comprises different techniques. The three popular techniques are PCA, principal factor method (PFM) and maximum likelihood estimation (MLE).
- In MLE method, the log-likelihood function to be maximized for parameter estimation is $L(X|\hat{\mu}, \mathbf{\Lambda}\mathbf{\Lambda}^T + \mathbf{\Psi}) = -\frac{n}{2}\left[\log|(\mathbf{\Lambda}\mathbf{\Lambda}^T + \mathbf{\Psi})| + tr\{(\mathbf{\Lambda}\mathbf{\Lambda}^T + \mathbf{\Psi})^{-1}\mathbf{S}\}\right]$. After differentiating the log-likelihood function, the resulting equations to be solved numerically are $(\hat{\mathbf{\Psi}}^{-\frac{1}{2}}\mathbf{S}\hat{\mathbf{\Psi}}^{-\frac{1}{2}})(\hat{\mathbf{\Psi}}^{-\frac{1}{2}}\hat{\mathbf{\Lambda}}) = (\hat{\mathbf{\Psi}}^{-\frac{1}{2}}\hat{\mathbf{\Lambda}})(\mathbf{I} + \hat{\boldsymbol{\theta}})$ and $\hat{\mathbf{\Psi}} = diag(\mathbf{S} - \hat{\mathbf{\Lambda}}\hat{\mathbf{\Lambda}}^T)$, where, $\hat{\boldsymbol{\theta}} = \hat{\mathbf{\Lambda}}^T\hat{\mathbf{\Psi}}^{-1}\hat{\mathbf{\Lambda}}$. Iterative numerical estimation methods such as Newton-Rapson method are used for parameter estimation.
- The selection of estimation method primarily governs by the distribution of the data. If data come from multivariate normal population, the MLE is a better choice. For non-normal population, PCA could be preferred.
- Model adequacy tests comprise Bartlett's sphericity test, KMO test, LR test, and RMSR.
- There are many measures to select the number of factors to be retained after final analysis. The most popular measures are cumulative percentage of variance explained, Kaiser's rule and scree plot.

- Quite often factor rotation is needed. The popular methods are varimax, and quartimax for orthogonal rotation, and oblimin for oblique rotation. Factor rotation helps in identifying simple structure so that naming of factors can be made.
- Factor scores are the predicted values of the factors for each of the observations. If there are n observations and the factor model comprises m factors, EFA provides a factor score matrix of size $n \times m$. Factor scores are used in subsequent analysis. Factor scores are usually obtained using least square or regression approaches.
- Although EFA looks to be similar to PCA, there are many differences. See Section 12.9.

Being a well-developed field in statistics, there are many good books and numerous case applications on EFA in the literature. Interested readers may use them for further information.

EXERCISES

(A) Short conceptual questions

12.1 What is exploratory factor analysis (EFA)? Is it different from PCA?

12.2 In what ways is EFA different from confirmatory factor analysis?

12.3 State the assumptions of EFA.

12.4 Prove that in the population factor model, $\mathbf{\Sigma} = \mathbf{\Lambda\Lambda}^T + \mathbf{\Psi}$ [the notations represent their usual meaning].

12.5 What is commonality? If σ_{jj} is the variance of the j-th original variable, then show that, $\sigma_{jj} = j^{\text{th}}\text{communality} + j^{\text{th}}$ specific variance.

12.6 Let $\boldsymbol{X} = [X_1, X_2, \ldots, X_p]^{\mathrm{T}}$ and $\mathbf{\Sigma} = Cov\ (\boldsymbol{X})$. Further, $m\ (< p)$ oblique factors are extracted. Show that $\mathbf{\Sigma} = \mathbf{\Lambda\phi\Lambda}^T + \mathbf{\Psi}$ [the notations represent their usual meaning].

12.7 What is factor rotation? Why it is used?

12.8 Show that the variance explained by a m-factor model remains unchanged even after rotation [Hint: $\mathbf{\Sigma}=\mathbf{\Lambda}^* \mathbf{\Lambda}^{*T}+\mathbf{\Psi}$].

12.9 Name three methods of orthogonal rotation of factors. Compare them.

12.10 What is factor score? Show that using weighted least square method, the estimated i-th factor score is $\hat{f}_i = (\hat{\mathbf{\Lambda}}^T\hat{\mathbf{\Psi}}^{-1}\hat{\mathbf{\Lambda}})^{-1}\hat{\mathbf{\Lambda}}^T\hat{\mathbf{\Psi}}^{-1}(\mathbf{x}_i - \bar{\mathbf{x}})$ for $i = 1, 2, \ldots, n$.

(B) Long conceptual and/or numerical questions: see web resources (Multivariate Statistical Modeling in Engineering and Management – 1st (routledge.com))

NOTES

1 Factor analysis is not a causal model. It is an interdependence model that is used to identify and quantify hidden factors, usually lesser in dimensions. The use of 'cause' here is to explain the inner meaning of a factor.

2 Note that eigenvalues are denoted by Q_j, $j = 1, 2, \ldots, p$.

N.B.: R, MINITAB and MS Excel software are used for computation, as and when required.

13 Confirmatory Factor Analysis

In exploratory factor analysis (Chapter 12), the focus is to identify a small number of unknown factors (***F***) which cause the manifest variables (***X***) to happen. As a result, a complete set of relationships among the manifest variables and all the factors are considered. On the contrary, if a researcher a *priori* has knowledge of the number of factors and the set of indicator/manifest variables against each of the factors, a structure different than EFA is obtained. The focus of this analysis is to confirm the hypothesized structure. Confirmatory factor analysis (CFA) is used for this purpose. However, as Mulaik (1988) pointed out, CFA is not a general purpose hypothesis-testing technique; rather it tests a specific kind of model. Like EFA, it also exploits the covariance relationships of many observed variables (***X***) and falls under covariance structure analysis. In this chapter, we will see the various aspects of CFA starting from model conceptualization, model identification, parameter estimation, and model adequacy tests. Finally, a case study is demonstrated for real-life problem-solving.

13.1 CONCEPTUAL MODEL

Suppose, *a priori*, we know m factors ($\xi_{m\times 1}$) and the subset of manifest variable $\boldsymbol{X}_{p\times 1}$ against each of the $\boldsymbol{\xi}$. Then, Figure 12.2 (in Chapter 12) will look like Figure 13.1. Note that we have used $\boldsymbol{\xi}$ to represent factors instead of ***F***. This is to keep parity with structural equation modeling (SEM), as CFA is a component of SEM (see Chapter 14). If we carefully observe both Figure 12.2 (in Chapter 12) and Figure 13.1, we get the following differences between EFA and CFA.

- In EFA, each X is related to all factors $\boldsymbol{\xi}$, whereas, in CFA a clear pattern of relationships exists (zero or non-zero loading) between each ξ and the ***X*** variables. For example, in Figure 13.1, $\xi_{\cdot 1}$ is related with $X_1, X_2, \ldots, X_k$ only ($k < p$), $\xi_{\cdot 2}$ is related with $X_{k+1}, X_{k+2}, \ldots, X_{k+l}$ only, and so on.
- The factors are generally orthogonal in EFA, but mostly oblique in CFA. The covariance matrix $\boldsymbol{\Phi}$ represents this obliqueness.
- The purpose of EFA is to determine the number of factors that can explain the covariance structure of ***X***. Whereas CFA is to test a pre-specified structure developed by the researcher.
- From a factor loadings point of view, CFA provides a much restricted model, where restriction means imposing zero or fixed loadings between $\boldsymbol{\xi}$ and ***X***.

Let a researcher identifies $m = 3$ factors ($\boldsymbol{\xi}$), with k (= 3) manifest variables (X_1, X_2, X_3) loaded with ξ_1, l (= 2) manifest variables (X_4, X_5) loaded with ξ_2, and l'(= 3) manifest variables (X_6, X_7, X_8) loaded with ξ_3. The hypothesized confirmatory factor model (Figure 13.1) representing the situation will like the form shown in Figure 13.2.

The regression type equations for all the eight ***X*** variables in Figure 13.2 can be written as

$$\begin{aligned}
X_1 &= \lambda_{11}\xi_1 && && + \delta_1 \\
X_2 &= \lambda_{21}\xi_1 && && + \delta_2 \\
X_3 &= \lambda_{31}\xi_1 && && + \delta_3 \\
X_4 &= && \lambda_{42}\xi_2 && + \delta_4 \\
X_5 &= && \lambda_{52}\xi_2 && + \delta_5 \\
X_6 &= && && \lambda_{63}\xi_3 + \delta_6 \\
X_7 &= && && \lambda_{73}\xi_3 + \delta_7 \\
X_8 &= && && \lambda_{83}\xi_3 + \delta_8
\end{aligned} \tag{13.1}$$

DOI: 10.1201/9781003303060-16

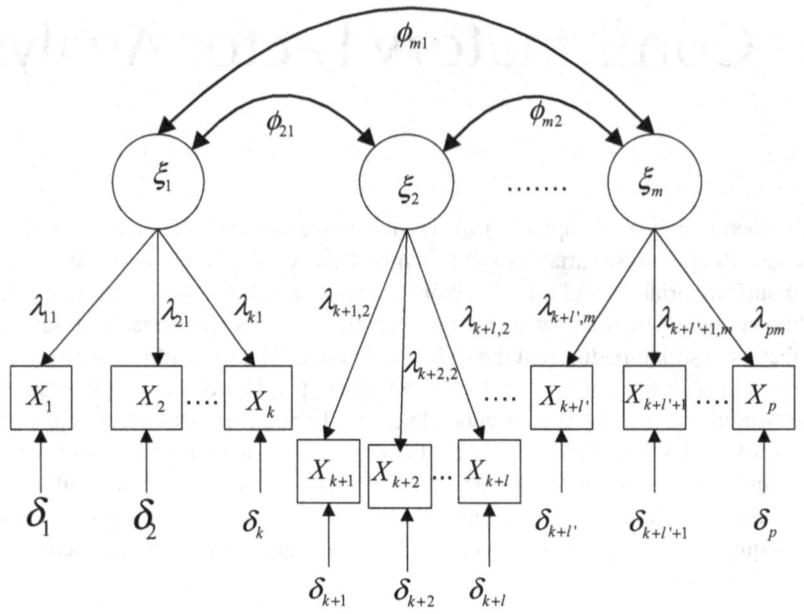

FIGURE 13.1 Confirmatory factor model with m factors (ξ) and p indicator/manifest variables (**X**). Each λ represents a factor loading, each δ is an error term, and each ϕ is the covariance between two factors.

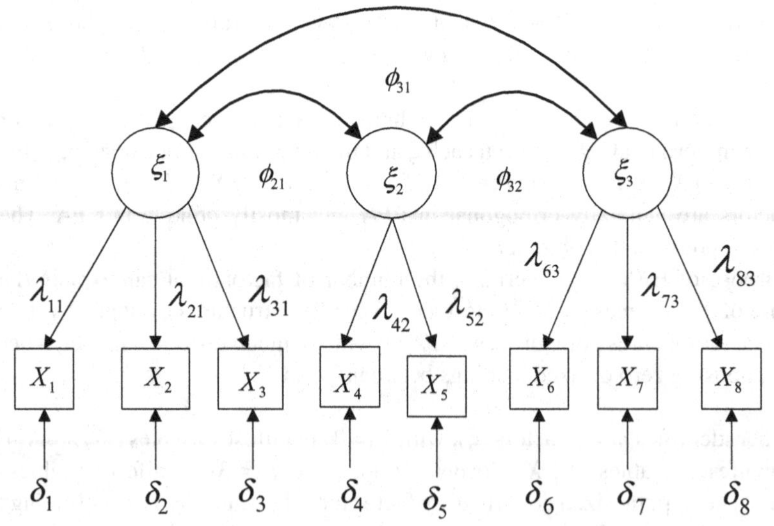

FIGURE 13.2 Confirmatory factor model with three factors and eight manifest variables. Each λ represents a factor loading, each δ is an error term, and each ϕ is the covariance between two factors.

In matrix form,

$$\begin{bmatrix} X_1 \\ X_2 \\ X_3 \\ X_4 \\ X_5 \\ X_6 \\ X_7 \\ X_8 \end{bmatrix} = \begin{bmatrix} \lambda_{11} & 0 & 0 \\ \lambda_{21} & 0 & 0 \\ \lambda_{31} & 0 & 0 \\ 0 & \lambda_{42} & 0 \\ 0 & \lambda_{52} & 0 \\ 0 & 0 & \lambda_{63} \\ 0 & 0 & \lambda_{73} \\ 0 & 0 & \lambda_{83} \end{bmatrix} \begin{bmatrix} \xi_1 \\ \xi_2 \\ \xi_3 \end{bmatrix} + \begin{bmatrix} \delta_1 \\ \delta_2 \\ \delta_3 \\ \delta_4 \\ \delta_5 \\ \delta_6 \\ \delta_7 \\ \delta_8 \end{bmatrix} \text{ or } \boldsymbol{X}_{8\times1} = \boldsymbol{\Lambda}_{8\times3}\boldsymbol{\xi}_{3\times1} + \boldsymbol{\delta}_{8\times1} \tag{13.2}$$

Considering Figure 13.1, the general equation for CFA is

$$\boldsymbol{X}_{p\times1} = \boldsymbol{\Lambda}_{p\times m}\boldsymbol{\xi}_{m\times1} + \boldsymbol{\delta}_{p\times1} \tag{13.3}$$

Here, $\boldsymbol{\Lambda}_{p\times m}$ contains many zeros based on the loadings pattern of CFA. In Figure 13.2, $\boldsymbol{\Lambda}_{8\times3}$ has only eight loadings to be estimated, as the loading patterns impose 16 zero loadings. In EFA, all the 24 loadings are required to be estimated.

The CFA falls under covariance structure analysis. There are three covariance matrices involved in a CFA model and they are

$$\begin{aligned} Cov(\boldsymbol{X}) &= \boldsymbol{\Sigma}_{p\times p} \\ Cov(\boldsymbol{\xi}) &= \boldsymbol{\Phi}_{m\times m} \\ Cov(\boldsymbol{\delta}) &= \boldsymbol{\theta}_{\boldsymbol{\delta}_{p\times p}} \end{aligned}$$

The assumptions of CFA are as follows:

$$\begin{aligned} &E(\boldsymbol{\delta}) = \mathbf{0} \\ &Cov(\boldsymbol{\delta}) = \boldsymbol{\theta}_{\boldsymbol{\delta}} \text{ (symetric)} \\ &\boldsymbol{\delta} \sim \text{N}_p(\mathbf{0}, \boldsymbol{\theta}_{\boldsymbol{\delta}}) \\ &Cov(\boldsymbol{\xi}, \boldsymbol{\delta}) = \mathbf{0}. \end{aligned} \tag{13.4}$$

Following the relevant assumptions, we can compute the covariance between a factor and its manifest variables. Multiplying Equation 13.3 by $\boldsymbol{\xi}$ and then taking expectation, we get

$$Cov(\boldsymbol{X\xi}) = E(\boldsymbol{X\xi}^T) = \Lambda E(\boldsymbol{\xi\xi}^T) + E(\boldsymbol{\delta\xi}^T) = \boldsymbol{\Lambda\Phi} + \mathbf{0} = \boldsymbol{\Lambda\Phi} \tag{13.5}$$

13.2 ESTIMATION OF PARAMETERS

CFA has two stages for estimation as follows:

1. Model identification: This tests whether all the parameters of the model can be estimated uniquely or not.
2. Model estimation: This computes the parameter values using sample data.

13.2.1 Model Identification

In CFA, the parameters to be estimated are $\boldsymbol{\Lambda}_{p\times m}$, $\boldsymbol{\Phi}_{m\times m}$, and $\boldsymbol{\theta}_{\boldsymbol{\delta}}$. The numbers of parameters to be estimated are:

- In $\mathbf{\Lambda}_{p\times m}$, the maximum number of parameters are $p \times m$ if all the loadings are considered (like EFA). However, it is to be noted that as CFA considers restricted loadings, the actual number of loading parameters to be estimated are much less than $p \times m$, say $l, (l < p \times m)$.
- For $\mathbf{\Phi}_{m\times m}$, the number of parameters to be estimated are $m(m + 1)/2$.
- For $\mathbf{\theta}_\delta$, the number of parameters to be estimated are p, as $\mathbf{\theta}_\delta$ is assumed to be diagonal. If covariances among $\boldsymbol{\delta}$ are allowed, the maximum number of $\mathbf{\theta}_\delta$ parameters to be estimated are $p(p + 1)/2$. However, $\mathbf{\theta}_\delta$ is generally assumed to be diagonal.

So, the total number of parameters t to be estimated (with diagonal $\mathbf{\theta}_\delta$) are

$$t = l + m(m+1)/2 + p \tag{13.6}$$

Now, these parameters are estimated using the fundamental relationships of CFA, given below.

$$\begin{aligned} Cov(\boldsymbol{X}) = \mathbf{\Sigma} &= Cov(\mathbf{\Lambda}\boldsymbol{\xi}+\boldsymbol{\delta}) \\ &= \mathbf{\Lambda} Cov(\boldsymbol{\xi})\mathbf{\Lambda}^T + Cov(\boldsymbol{\delta}) \\ &= \mathbf{\Lambda}\mathbf{\Phi}\mathbf{\Lambda}^T + \mathbf{\theta}_\delta \end{aligned} \tag{13.7}$$

Now, denoting the complete set of t parameters by a vector $\mathbf{\theta}$, $\mathbf{\Sigma}_{p\times p}$ can be written as $\mathbf{\Sigma}(\mathbf{\theta})$. In other words, the population covariance matrix of the manifest variables ($\boldsymbol{X}$) is a function $\mathbf{\Sigma}(\mathbf{\theta})$ of the t-parameters of the hypothesized model. We can further state that each element of $\mathbf{\theta}$ vector can be expressed as a function of one or more elements of $\mathbf{\Sigma}$, i.e., $\sigma_{jk}, j,k = 1,2,..,p$. Further, $\mathbf{\Sigma}$ has $p(p+1)/2$ unique elements (as $\sigma_{jk} = \sigma_{kj}$), which can be obtained from sample covariance matrix (**S**).

Essentially, we can see that we have a set of t parameters to be estimated from $p(p+1)/2$ equations, which leads to the following three situations:

1. $t > p(p+1)/2$, a situation when the parameter vector $\mathbf{\theta}$ cannot be estimated. This situation is called under-identification.
2. $t = p(p+1)/2$, a situation when $\mathbf{\theta}$ can be uniquely estimated, but not preferable as the solution lacks generalizability. This situation is known as exact identification.
3. $t < p(p+1)/2$, a situation when $\mathbf{\theta}$ have multiple solutions. This situation is known as over-identification. Researcher should aim to have an over-identified model as it ensures generalization.

For a CFA to be estimable, the necessary condition is that the model must fall under situations 2 or 3, but for all practical purpose, situation 3 is desirable. The abovementioned condition is known as *order condition* for model identification which is necessary but not sufficient. The sufficient condition is called *rank condition*, which algebraically determines whether a model is identified or not. But it is a complex process and requires the knowledge of matrix algebra. See Chapter 10, Section 10.4.1, for a discussion on rank condition.

Example 13.1

Identify whether the confirmatory factor model given in Figure 13.2 is estimable or not.

Solution:

From Equations 13.2 and the covariance structure of the factor model the number of parameters to be estimated are (i) for $\mathbf{\Lambda}$: 8 ($\lambda_{11}, \lambda_{21}, \lambda_{31}, \lambda_{42}, \lambda_{52}, \lambda_{63}, \lambda_{73}, \lambda_{83}$), (ii) for $\mathbf{\Phi} : 3\times(3+1)/2 = 6$, and (iii) for $\mathbf{\theta}_\delta : 8$ ($\theta_{\delta_{11}}, \theta_{\delta_{22}}, ..., \theta_{\delta_{88}}$). So, $t = 8 + 6 + 8 = 22$. Now, number of unique elements of $\mathbf{\Sigma} = p(p+1)/2 = 8(8+1)/2 = 36$. As $t = 22 < 36$ the model satisfies the order condition for estimation. It will be estimable if the rank condition is also satisfied. However, considering t = 22 is much less than 36, for most practical purposes, the model may satisfy the rank condition.

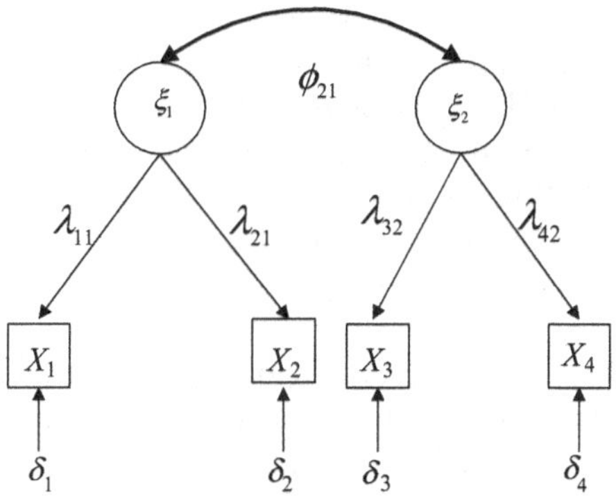

FIGURE 13.3 Factor model for Example 13.2. Both the latent factors (ξ_1 and ξ_2) have two indicators each. The notations represent their usual meanings (as defined in this chapter).

Example 13.2

Consider the following factor model (see Figure 13.3). It is identifiable?

Solution:

Here,

$$\Lambda = \begin{bmatrix} \lambda_{11} & 0 \\ \lambda_{21} & 0 \\ 0 & \lambda_{32} \\ 0 & \lambda_{42} \end{bmatrix}, \Phi = \begin{bmatrix} \phi_{11} & \phi_{12} \\ \phi_{12} & \phi_{22} \end{bmatrix} \text{ and } \boldsymbol{\theta}_\delta = \begin{bmatrix} \theta_{\delta_{11}} & 0 & 0 & 0 \\ 0 & \theta_{\delta_{22}} & 0 & 0 \\ 0 & 0 & \theta_{\delta_{33}} & 0 \\ 0 & 0 & 0 & \theta_{\delta_{44}} \end{bmatrix}$$

The number of parameters to be estimated, $\Lambda : 4\ (\lambda_{11}, \lambda_{21}, \lambda_{32}, \lambda_{42})$, $\Phi : 2 \times (2+1)/2 = 3$, and $\boldsymbol{\theta}_\delta : 4\ (\theta_{\delta_{11}}, \theta_{\delta_{22}}, \theta_{\delta_{33}}, \theta_{\delta_{44}})$. So, $t = 4 + 3 + 4 = 11$. Now, number of the unique elements of Σ is $p(p + 1)/2 = \frac{4 \times 5}{2} = 10$. As $t = 11 > 10$, the model is unidentifiable.

There are certain useful tricks to make an unidentified model identifiable. There are three possibilities for each of the parameters of CFA, fixed, constant, or free. The fixed parameters are assigned a particular value, usually 0 or 1. $\lambda_{jk} = 0$ indicates the non-inclusion of the j-th manifest variable as an indicator of the k-th factor (based on the formulation of the hypotheses). $\lambda_{jk} = 1$ provides a scale to the latent factor ξ_k. As $\boldsymbol{\xi}$ is latent, its origin and scale are unknown. Origin is taken care of by considering mean-subtracted variables and the scale can be fixed by putting at least one $\lambda_{jk} = 1$ in every column of Λ. The constant parameters are those whose values are unknown but can be equated with other parameters of the model. The parameters which are neither fixed nor constrained are called free parameters and their values are to be estimated. So, in order to make an unidentified model identifiable, the following guidelines can be used:

1. Put at least one $\lambda_{jk} = 1$ in each column of Λ. This provides a scale to the latent variable ξ_k with respect to δ_j.
2. Let $E(\xi_k) = 1, k = 1, 2, ... m$. This makes the latent variables standardized. This option also provides a scale to $\xi_k, k = 1, 2, ... m$.
3. Set some λ_{jk} to zero or to be equal to other values.

So, in Example 13.2, if we set $\lambda_{11} = 1$, the model becomes uniquely identified ($t = 10$ = number of knowns). In addition, if we put λ_{32} also $= 1$, the model becomes over-identified. This also satisfies the recommendation to put one λ_{jk} in each column of $\boldsymbol{\Lambda}$ to 1.

The necessary conditions described earlier using the number of parameters (t) to be estimated is also called t-rule (Bollen, 1989). Bollen (1989, Chapter 7) further described two-indicator and three-indicator rules for both single and multi-factor model identification. These rules are used for model sufficiency test and should be used once t-rule is satisfied.

- **Two-indicator rule**: This is used for multifactor models. The prescriptions for sufficiency are (i) at least two indicators per factor, (ii) one and only one non-zero value in each row of the loading matrix ($\boldsymbol{\Lambda}_{p\times m}$), (iii) at least one non-zero off-diagonal value in each row of the factor correlation/covariance matrix ($\boldsymbol{\Phi}_{m\times m}$), and (iv) diagonal $\boldsymbol{\Theta}_{\delta_{p\times p}}$ matrix.
- **Three-indicator rule**: This can be used for both one and multifactor models. For a one-factor model, the sufficient conditions are (i) at least three indicators per factor with non-zero loadings, and (ii) diagonal $\boldsymbol{\Theta}_{\delta_{p\times p}}$. For a multifactor model, the sufficient conditions are (i) at least three indicators per factor, (ii) one and only one non-zero value in each row of the loading matrix ($\boldsymbol{\Lambda}_{p\times m}$), and (iii) diagonal $\boldsymbol{\Theta}_{\delta_{p\times p}}$. Note that while the two-indicator rule puts restrictions on $\boldsymbol{\Phi}_{m\times m}$, the three-indicator rule does not.

13.2.2 Model Estimation

The parameter estimation process of CFA is similar to EFA. Let $\boldsymbol{X} \sim N_p(\mathbf{0}, \boldsymbol{\Sigma})$. Note $E(\boldsymbol{X}) = \mathbf{0}$, p = number of X variables. Then the probability density of drawing a particular multivariate observation $\mathbf{x}_i$ is $f(\mathbf{x}_i) = \dfrac{1}{(2\pi)^{p/2}|\boldsymbol{\Sigma}|^{1/2}} e^{-\frac{1}{2}[\mathbf{x}_i^T\boldsymbol{\Sigma}^{-1}\mathbf{x}_i]}$.

The joint density of $\mathbf{x}_1, \mathbf{x}_2, \ldots, \mathbf{x}_i, \ldots, \mathbf{x}_N$ or likelihood $L(\mathbf{0}, \boldsymbol{\Sigma})$, is (note: sample size = N)

$$L(\mathbf{0},\boldsymbol{\Sigma}) = \prod_{i=1}^{N}\left[\frac{1}{(2\pi)^{p/2}|\boldsymbol{\Sigma}|^{1/2}} e^{-\frac{1}{2}[\mathbf{x}_i^T\boldsymbol{\Sigma}^{-1}\mathbf{x}_i]}\right] = (2\pi)^{\frac{-Np}{2}} |\boldsymbol{\Sigma}|^{\frac{-N}{2}} e^{-\frac{1}{2}\sum_{i=1}^{N}[\mathbf{x}_i^T\boldsymbol{\Sigma}^{-1}\mathbf{x}_i]}$$

Taking logarithm of $L(\mathbf{0}, \boldsymbol{\Sigma})$, we get log likelihood, or $\ell n(L) = \dfrac{-Np}{2}\ell n(2\pi)\dfrac{-N}{2}\ell n|\boldsymbol{\Sigma}| - \dfrac{1}{2}\sum_{i=1}^{N}\mathbf{x}_i^T\boldsymbol{\Sigma}^{-1}\mathbf{x}_i$.

Now $-\dfrac{1}{2}\sum_{i=1}^{N}\mathbf{x}_i^T\boldsymbol{\Sigma}^{-1}\mathbf{x}_i$ can be written as $\dfrac{-N}{2}trace\left(\dfrac{1}{N}\mathbf{x}\boldsymbol{\Sigma}^{-1}\mathbf{x}^T\right)$. Again, $\dfrac{-N}{2}trace\left(\dfrac{1}{N}\mathbf{x}\boldsymbol{\Sigma}^{-1}\mathbf{x}^T\right) = \dfrac{-N}{2}trace\left(\dfrac{1}{N}\mathbf{x}^T\mathbf{x}\boldsymbol{\Sigma}^{-1}\right) = \dfrac{-N}{2}trace\left(\mathbf{S}\boldsymbol{\Sigma}^{-1}\right)$ as $\dfrac{1}{N}\mathbf{x}^T\mathbf{x}$ can be approximated to $\mathbf{S}$ for large N.

So,

$$\ell n(L) = \frac{-Np}{2}\ell n(2\pi) - \frac{N}{2}\ell n|\boldsymbol{\Sigma}| - \frac{N}{2}trace\left(\mathbf{S}\boldsymbol{\Sigma}^{-1}\right)$$

To estimate $\boldsymbol{\Sigma}$, we require maximizing $\ell n(L)$. $\dfrac{-Np}{2}\ell n(2\pi)$ is a constant term and it can be omitted while maximizing $\ell n(L)$ as the outcome of the maximization process is unaffected by the constant term. So, the resultant log likelihood equation is

$$\ell n(L)=\frac{-N}{2}\left[\ell n|\mathbf{\Sigma}|+trace\left(\mathbf{S\Sigma}^{-1}\right)\right] \tag{13.8}$$

As the elements of $\mathbf{\Sigma}$, i.e., $\mathbf{\Sigma}(\mathbf{\theta})$ is not known and we use the sample covariance matrix ($\mathbf{S}$) to estimate them, it is expected that for perfect fit $\mathbf{S}=\mathbf{\Sigma}(\mathbf{\theta})$. In real situations, $\mathbf{S}-\mathbf{\Sigma}(\mathbf{\theta})$ should be as small as possible. If we put $\mathbf{S}=\mathbf{\Sigma}$ in Equation 13.8, the resultant equation becomes

$$\ell n(L)=\frac{-N}{2}\left[\ell n|\mathbf{S}|+trace\left(\mathbf{SS}^{-1}\right)\right]=\frac{-N}{2}\left[\ell n|\mathbf{S}|+trace\left(\mathbf{I}\right)\right]=\frac{-N}{2}\left(\ell n|\mathbf{S}|+p\right) \tag{13.9}$$

Considering $\mathbf{S}-\mathbf{\Sigma}(\mathbf{\theta})\rightarrow\mathbf{0}$, our objective function becomes minimizing the difference $\mathbf{S}-\mathbf{\Sigma}(\mathbf{\theta})$, which is the fit function $F(\mathbf{\theta})$. So,

$$F(\mathbf{\theta})=\frac{-N}{2}\left(\ell n|\mathbf{S}|+p\right)+\frac{N}{2}\left[\ell n|\mathbf{\Sigma}|+trace\left(\mathbf{S\Sigma}^{-1}\right)\right]=\frac{N}{2}\left[\ell n|\mathbf{\Sigma}|+trace\left(\mathbf{S\Sigma}^{-1}\right)-\ell n|\mathbf{S}|-p\right] \tag{13.10}$$

Ignoring the constant term ($N/2$), we can define the fit function to be minimized as

$$F(\mathbf{\theta})=\ell n|\mathbf{\Sigma}(\mathbf{\theta})|+trace\left(\mathbf{S\Sigma}(\mathbf{\theta})^{-1}\right)-\ell n|\mathbf{S}|-p \tag{13.11}$$

For a perfect fit $F(\mathbf{\theta})$ must be zero. It is also to be noted here that $F(\mathbf{\theta})$ comprises constrained non-linear system of equations. Iterative numerical methods such as Newton-Rapson and Gauss-Newton algorithms are used to minimize $F(\mathbf{\theta})$.

As stated earlier, MLE requires the assumption of multivariate normality be satisfied. With moderate sample size, MLE is robust against moderate violation of normality (Tanaka, 1984). But, there could be problem with asymptotic standard errors and χ^2 test statistic. Anderson and Gerbing (1988) stated that for peaked normal distribution the null hypothesis could be overly rejected but for flat normal distribution, the result will be the opposite. Bentler (1985) and Jöreskog and Sorbom (1987) provided the following general fit function:

$$F(\mathbf{\theta})=[\mathbf{s}-\boldsymbol{\sigma}(\mathbf{\theta})]^T\mathbf{U}^{-1}[\mathbf{s}-\boldsymbol{\sigma}(\mathbf{\theta})] \tag{13.12}$$

In Equation 13.12, $\mathbf{s}$ is $p^*\times 1$ vector of unique elements of $\mathbf{S}$, $\boldsymbol{\sigma}(\mathbf{\theta})$ is the corresponding vector of $\mathbf{\Sigma}(\mathbf{\theta})$ and $\mathbf{U}$ is a $p^*\times p^*$ weight matrix, where $p^*=p(p+1)/2$. Minimizing $F(\mathbf{\theta})$, the parameters are estimated. This method is called generalized least square (GLS). The other methods include weighted least squares (WLS) and asymptotically distribution free (ADF) method (Browne, 1984). Equation 13.12 can be solved using some iterative algorithms.

Example 13.3

Obtain the fit function $F(\mathbf{\theta})$ for the confirmatory factor model given in Example 13.2.

Solution:

Let first compute the predicted covariance matrix $\Sigma(\mathbf{\theta})$. We know $\mathbf{\Sigma}(\mathbf{\theta})=\mathbf{\Lambda\Phi\Lambda}^T+\mathbf{\theta}_\delta$. From Example 13.2, we get,

$$\mathbf{\Lambda}=\begin{bmatrix}\lambda_{11} & 0\\ \lambda_{21} & 0\\ 0 & \lambda_{32}\\ 0 & \lambda_{42}\end{bmatrix}_{4\times 2}, \quad \mathbf{\Phi}=\begin{bmatrix}\phi_{11} & \phi_{12}\\ \phi_{12} & \phi_{22}\end{bmatrix}$$

So,

$$\Lambda_{4\times2}\Phi_{2\times2} = \begin{bmatrix} \lambda_{11}\phi_{11} & \lambda_{11}\phi_{12} \\ \lambda_{21}\phi_{11} & \lambda_{21}\phi_{12} \\ \lambda_{32}\phi_{12} & \lambda_{32}\phi_{22} \\ \lambda_{42}\phi_{12} & \lambda_{42}\phi_{22} \end{bmatrix}$$

$$\Lambda\Phi\Lambda^T = \begin{bmatrix} \lambda_{11}\phi_{11} & \lambda_{11}\phi_{12} \\ \lambda_{21}\phi_{11} & \lambda_{21}\phi_{12} \\ \lambda_{32}\phi_{12} & \lambda_{32}\phi_{22} \\ \lambda_{42}\phi_{12} & \lambda_{42}\phi_{22} \end{bmatrix}_{4\times2} \begin{bmatrix} \lambda_{11} & \lambda_{21} & 0 & 0 \\ 0 & 0 & \lambda_{32} & \lambda_{42} \end{bmatrix}_{2\times4}$$

$$= \begin{bmatrix} \lambda_{11}^2\phi_{11} & \lambda_{11}\lambda_{21}\phi_{11} & \lambda_{11}\phi_{12}\lambda_{32} & \lambda_{11}\phi_{12}\lambda_{42} \\ \lambda_{11}\lambda_{21}\phi_{11} & \lambda_{21}^2\phi_{11} & \lambda_{21}\phi_{12}\lambda_{32} & \lambda_{21}\phi_{12}\lambda_{42} \\ \lambda_{32}\phi_{12}\lambda_{11} & \lambda_{32}\phi_{12}\lambda_{21} & \lambda_{32}\phi_{22}\lambda_{32} & \lambda_{32}\phi_{22}\lambda_{42} \\ \lambda_{42}\phi_{12}\lambda_{11} & \lambda_{42}\phi_{12}\lambda_{22} & \lambda_{42}\phi_{22}\lambda_{32} & \lambda_{42}\phi_{22}\lambda_{42} \end{bmatrix}$$

Let $\boldsymbol{\theta}_\delta = \text{diag}\left(\theta_{\delta_{11}} \quad \theta_{\delta_{22}} \quad \theta_{\delta_{33}} \quad \theta_{\delta_{44}}\right)$. So, $\boldsymbol{\Sigma}(\boldsymbol{\theta})$ is

$$\boldsymbol{\Sigma}(\boldsymbol{\theta}) = \begin{bmatrix} \lambda_{11}^2\phi_{11}+\theta_{\delta_{11}} & \lambda_{11}\lambda_{21}\phi_{11} & \lambda_{11}\lambda_{32}\phi_{12} & \lambda_{11}\lambda_{12}\phi_{12} \\ \lambda_{11}\lambda_{21}\phi_{11} & \lambda_{21}^2\phi_{11}+\theta_{\delta_{22}} & \lambda_{21}\lambda_{32}\phi_{12} & \lambda_{21}\lambda_{42}\phi_{12} \\ \lambda_{11}\lambda_{32}\phi_{12} & \lambda_{21}\lambda_{32}\phi_{12} & \lambda_{32}^2\phi_{22}+\theta_{\delta_{33}} & \lambda_{32}\lambda_{42}\phi_{22} \\ \lambda_{11}\lambda_{42}\phi_{12} & \lambda_{21}\lambda_{42}\phi_{12} & \lambda_{32}\lambda_{42}\phi_{22} & \lambda_{42}^2\phi_{22}+\theta_{\delta_{44}} \end{bmatrix}$$

As the number of parameters to be estimated above is 11 and number of known unique elements are 10 (Example 13.2), we require fixing some parameters. In addition, we need to solve the scalability problem. So, we put $\lambda_{11} = 1$ and $\lambda_{32} = 1$. This makes the number of parameters to be estimated as 9(< 10). The resultant $\boldsymbol{\Sigma}(\boldsymbol{\theta})$ is

$$\boldsymbol{\Sigma}(\boldsymbol{\theta}) = \begin{bmatrix} \phi_{11}+\theta_{\delta_{11}} & \lambda_{21}\phi_{11} & \phi_{12} & \lambda_{42}\phi_{12} \\ \lambda_{21}\phi_{11} & \lambda_{21}^2\phi_{11}+\theta_{\delta_{22}} & \lambda_{21}\phi_{12} & \lambda_{21}\lambda_{42}\phi_{12} \\ \phi_{12} & \lambda_{21}\phi_{12} & \phi_{22}+\theta_{\delta_{33}} & \lambda_{42}\phi_{22} \\ \lambda_{42}\phi_{12} & \lambda_{21}\lambda_{42}\phi_{12} & \lambda_{42}\phi_{22} & \lambda_{42}^2\phi_{22}+\theta_{\delta_{44}} \end{bmatrix}$$

It is to be noted here that $\boldsymbol{\Sigma}(\boldsymbol{\theta})$ is a symmetric square matrix. From sample, **S** will be computed, where

$$\mathbf{S} = \begin{bmatrix} s_{11} & s_{12} & s_{13} & s_{14} \\ s_{12} & s_{22} & s_{23} & s_{24} \\ s_{13} & s_{23} & s_{33} & s_{34} \\ s_{14} & s_{24} & s_{34} & s_{44} \end{bmatrix}$$

So, the fit function is the Equation 13.11 with $\boldsymbol{\Sigma}(\boldsymbol{\theta})$ and S as developed above. The system of equations that is to be solved can be obtained by equating the elements of **S** with $\boldsymbol{\Sigma}(\boldsymbol{\theta})$, and then using some iterative procedure as mentioned earlier.

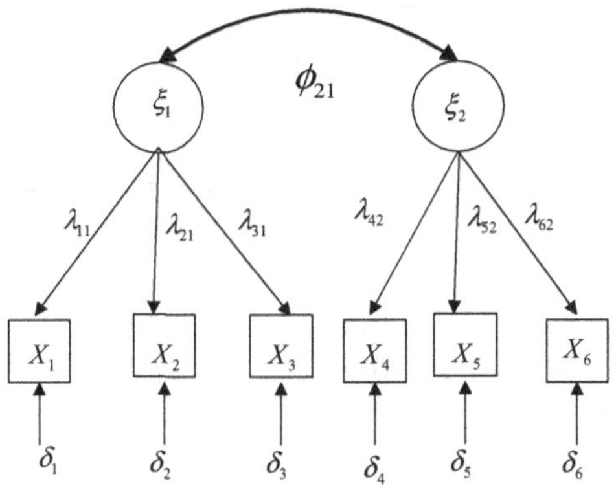

FIGURE 13.4 Factor model for Example 13.4. Both the latent factors (ξ_1 and ξ_2) have three indicators each. The notations represent their usual meanings (as defined in this chapter).

Example 13.4

Consider the model (see Figure 13.4) and the sample covariance matrix given below. Sample size, $N = 76$. Use MLE, estimate the parameters.

	X_1	X_2	X_3	X_4	X_5	X_6
X_1	139.56					
X_2	62.4	91.84				
X_3	52.37	76.75	92.54			
X_4	−25.02	−18.21	−18.77	86.39		
X_5	−20.93	−6.16	−13.25	68.46	78.51	
X_6	−22.09	−12.72	−15.53	57.93	62.61	70.65

Solution:

Here,

$$\Lambda = \begin{bmatrix} \lambda_{11} & 0 \\ \lambda_{21} & 0 \\ \lambda_{31} & 0 \\ 0 & \lambda_{42} \\ 0 & \lambda_{52} \\ 0 & \lambda_{62} \end{bmatrix}, \Phi = \begin{bmatrix} \phi_{11} & \phi_{21} \\ \phi_{21} & \phi_{22} \end{bmatrix} \text{ and } \theta_\delta = \begin{bmatrix} \theta_{\delta_{11}} & 0 & 0 & 0 & 0 & 0 \\ 0 & \theta_{\delta_{22}} & 0 & 0 & 0 & 0 \\ 0 & 0 & \theta_{\delta_{33}} & 0 & 0 & 0 \\ 0 & 0 & 0 & \theta_{\delta_{44}} & 0 & 0 \\ 0 & 0 & 0 & 0 & \theta_{\delta_{55}} & 0 \\ 0 & 0 & 0 & 0 & 0 & \theta_{\delta_{66}} \end{bmatrix}$$

The number of parameters to be estimated is $t = 6 + 3 + 6 = 15$ and the number unique elements in S is $6 \times 7/2 = 21$. So, using t-rule the model satisfies the order condition or necessary condition. Further, to maintain the scale of the factors, $\lambda_{11} = \lambda_{42} = 1$. This makes $t = 13$ and the model is over-identified. For a test of sufficient condition, the three-indicator rule can be applied, which is also satisfied. So, the model is identified.

The MLE results, using R package, are shown below.

TABLE 13.1
Sample Correlation Matrix

	X_1	X_2	X_3	X_4	X_5	X_6
X_1	1					
X_2	0.55	1				
X_3	0.46	0.83	1			
X_4	–0.23	–0.2	–0.21	1		
X_5	–0.2	–0.07	–0.16	0.83	1	
X_6	–0.22	–0.16	–0.19	0.74	0.84	1

$$\hat{\Lambda} = \begin{bmatrix} 1 & 0 \\ \hat{\lambda}_{21} & 0 \\ \hat{\lambda}_{31} & 0 \\ 0 & 1 \\ 0 & \hat{\lambda}_{52} \\ 0 & \hat{\lambda}_{62} \end{bmatrix} = \begin{bmatrix} 1.00 & 0.00 \\ 1.41 & 0.00 \\ 1.23 & 0.00 \\ 0.00 & 1.00 \\ 0.00 & 1.07 \\ 0.00 & 0.91 \end{bmatrix}, \hat{\Phi} = \begin{bmatrix} \hat{\phi}_{11} & \hat{\phi}_{21} \\ \hat{\phi}_{21} & \hat{\phi}_{22} \end{bmatrix} = \begin{bmatrix} 43.468 & -6.922 \\ -6.922 & 63.035 \end{bmatrix}$$

$$\hat{\theta}_\delta = \begin{bmatrix} 94.255 & & & & & \\ 0 & 3.856 & & & & \\ 0 & 0 & 25.25 & & & \\ 0 & 0 & 0 & 22.218 & & \\ 0 & 0 & 0 & 0 & 5.165 & \\ 0 & 0 & 0 & 0 & 0 & 17.034 \end{bmatrix}$$

Example 13.5

Consider Example 13.4 and the sample correlation matrix in Table 13.1. Use MLE, estimate the parameters and compare the results with Example 13.4.

Solution:

The MLE results, using correlation matrix, are shown below.

$$\hat{\Lambda} = \begin{bmatrix} 0.558 & 0.00 \\ 0.970 & 0.00 \\ 0.844 & 0.00 \\ 0.00 & 0.853 \\ 0.00 & 0.960 \\ 0.00 & 0.863 \end{bmatrix}, \hat{\Phi} = \begin{bmatrix} \hat{\phi}_{11} & \hat{\phi}_{21} \\ \hat{\phi}_{21} & \hat{\phi}_{22} \end{bmatrix} = \begin{bmatrix} 1.000 & -0.132 \\ -0.132 & 1.000 \end{bmatrix}$$

$$\hat{\boldsymbol{\theta}}_\delta = \begin{bmatrix} 0.675 & & & & & \\ 0 & 0.047 & & & & \\ 0 & 0 & 0.274 & & & \\ 0 & 0 & 0 & 0.259 & & \\ 0 & 0 & 0 & 0 & 0.066 & \\ 0 & 0 & 0 & 0 & 0 & 0.242 \end{bmatrix}$$

When the correlation matrix (**R**) is used in place of covariance matrix (**S**), as shown in Example 13.4, the loadings are standardized (i.e., lie between –1 to 1). Similarly, $\boldsymbol{\Phi}$ gives correlations among the factors and $\boldsymbol{\theta}_\delta$ represents the standardized variance (with maximum value of 1).

13.3 MODEL ADEQUACY TESTS

Once a confirmatory factor model is fitted, the next step is how to evaluate a model and how to select among competing models. Marsh and Balla (1994) proposed the use of theoretical basis, inspection of estimated parameters, goodness of fit, parsimony of fit measures, ease of interpretability, and comparison of competing model performances as the criteria for evaluating a covariance structure model. Hair et al. (1998, 2010) suggested investigating offending estimates, model re-specifications, overall model fit, and fit of individual factors (they refer to it as measurement model fit). The theoretical basis can be established from prior empirical researches, actual observations in the field, experimental results, or from the confirmatory theories. Based on all three, what Hair et al. (1998, 2010) term as "big picture" can be addressed with a series of relationships describing the phenomenon of interests.

Marsh and Balla (1994) provided a general approach using three sets of criteria: (i) solution quality, (ii) test of parameters, and (iii) goodness of fit indices. The quality of solution can be defined by model identifiability, convergence of estimation procedure (as iterative process is used) and offending estimates. Anderson and Gerbing (1988) considered these under model re-specification. They suggested the use of convergence and discriminant validity of each of the manifest variables. Offending estimates include negative variance estimates, standardized coefficients close to zero, and very large standard errors. Van Driel (1978) proposed to examine the confidence interval of the negative estimate to understand whether it results from sampling error. If the interval contains positive values and spreads over a feasible range of proper estimates, the negative variance is due to sampling error. Gerbing and Anderson (1987) proposed to fix such parameters close to zero, a small positive value such as 0.005. Dillon et al. (1987) experimented with the Haywood case (negative error variances) in different settings, such as putting offending estimates to zero and forcing error variances to be positive, and concluded that in the case of sampling fluctuating, setting the parameter value close to zero works reasonably well. Recently Kolenikov and Bollen (2012) provided tests to show how structural misspecification leads to negative variance estimates (or Haywood cases).

There are three types of fit indices used in CFA. They are absolute fit indices, relative fit indices and parsimonious fit indices. Marsh and Balla (1994) have documented 22 goodness of fit indices for confirmatory factor analysis under the abovementioned three categories.

13.3.1 Absolute Fit Indices

An absolute fit index answers the question: Is the residual or unexplained variance remaining after model fitting appreciable? The null hypothesis that is tested using absolute fit indices is $H_0 : \boldsymbol{\Sigma} = \boldsymbol{\Sigma}(\boldsymbol{\theta})$ against the alternate hypothesis $H_1 : \boldsymbol{\Sigma} \neq \boldsymbol{\Sigma}(\boldsymbol{\theta})$. If H_0 is true, the quantity $\chi^2 = (N-1)F_{\min}(\boldsymbol{\theta})$ follows χ^2_v, where $v = p(p+1)/2 - t$ degrees of freedom. Note that t is the number of parameters to be estimated (also known as free parameters). It is to be further noted that the above statement is true when data comes from a multivariate normal population. In addition, H_0 is testable only if $diag\boldsymbol{\Sigma}(\boldsymbol{\theta}) = diag(\mathbf{S})$ or ($diag(\mathbf{R})$, if $\mathbf{R}$ is used instead of $\mathbf{S}$).

What should be the value of χ_v^2? Theoretically, it should be zero as $F(\boldsymbol{\theta})$ is a function that tradeoffs between $\mathbf{S}$ and $\boldsymbol{\Sigma}(\boldsymbol{\theta})$. For a perfect fit, $\mathbf{S} = \boldsymbol{\Sigma}(\boldsymbol{\theta})$, $F(\boldsymbol{\theta}) = 0$. Statistically, the cut-off value of χ^2 is $\chi_v^2(\alpha)$ where α is the probability level or level of significance. However, χ^2, as described above is not recommended owing to (i) its tendency to over-fit in case of complex models, and (ii) provide wrong results when sample size increases (Kaplan, 1990). Further, χ^2 varies as sample size (*N*) varies. McDonald and Marsh (1990) demonstrated that indices that depend on *N* are biased. It leads to the following two situations: (i) almost all models will be accepted if sample size is small and (ii) any model can be rejected if sample size is sufficiently large. For these reasons, researchers use scaled χ^2 -statistic which is χ^2 divided by its degrees of freedom (*df*). Theoretically, $\chi^2 / \text{df} = 1$ as $E\left(\chi_{df}^2\right) = df$. The recommended value of χ^2 / df is 2 to 5. However, the *df* does not depend on *N* whereas χ^2 does. So, the effect of sample size is not adjusted. The other absolute fit measures are goodness of fit index (GFI), root mean square error of approximation (RMSEA), and root mean square residual (RMSR).

Example 13.6

Consider Example 13.4. Conduct χ^2 test and comment on the model fit.

Solution:

We use Equation 13.11 to compute *F(θ)*, using the estimated parameter values. The predicted covariance matrix is (see Equation 13.7) $\hat{\boldsymbol{\Sigma}} = \hat{\boldsymbol{\Lambda}}\hat{\boldsymbol{\Phi}}\hat{\boldsymbol{\Lambda}}^T + \hat{\boldsymbol{\theta}}_\delta$.

$$\hat{\boldsymbol{\Lambda}}\hat{\boldsymbol{\Phi}}\hat{\boldsymbol{\Lambda}}^T = \begin{bmatrix} 1.00 & 0.00 \\ 1.41 & 0.00 \\ 1.23 & 0.00 \\ 0.00 & 1.00 \\ 0.00 & 1.07 \\ 0.00 & 0.91 \end{bmatrix} \begin{bmatrix} 43.468 & -6.922 \\ -6.922 & 63.035 \end{bmatrix} \begin{bmatrix} 1.00 & 1.41 & 1.23 & 0.00 & 0.00 & 0.00 \\ 0.00 & 0.00 & 0.00 & 1.00 & 1.07 & 0.91 \end{bmatrix}$$

$$= \begin{bmatrix} 43.47 & 61.42 & 53.60 & -6.92 & -7.41 & -6.33 \\ 61.42 & 86.79 & 75.73 & -9.78 & -10.48 & -8.94 \\ 53.60 & 75.73 & 66.08 & -8.53 & -9.14 & -7.80 \\ -6.92 & -9.78 & -8.53 & 63.04 & 67.51 & 57.61 \\ -7.41 & -10.48 & -9.14 & 67.51 & 72.30 & 61.70 \\ -6.33 & -8.94 & -7.80 & 57.61 & 61.70 & 52.66 \end{bmatrix}$$

$$\text{So, } \hat{\boldsymbol{\Sigma}} = \hat{\boldsymbol{\Lambda}}\hat{\boldsymbol{\Phi}}\hat{\boldsymbol{\Lambda}}^T + \hat{\boldsymbol{\theta}}_\delta = \begin{bmatrix} 137.72 & 61.42 & 53.60 & -6.92 & -7.41 & -6.33 \\ 61.42 & 90.64 & 75.73 & -9.78 & -10.48 & -8.94 \\ 53.60 & 75.73 & 91.33 & -8.53 & -9.14 & -7.80 \\ -6.92 & -9.78 & -8.53 & 85.25 & 67.51 & 57.61 \\ -7.41 & -10.48 & -9.14 & 67.51 & 77.47 & 61.70 \\ -6.33 & -8.94 & -7.80 & 57.61 & 61.70 & 69.69 \end{bmatrix}$$

$$F(\boldsymbol{\theta}) = \ell n\left|\hat{\boldsymbol{\Sigma}}(\boldsymbol{\theta})\right| + tr\left(\mathbf{S}\hat{\boldsymbol{\Sigma}}(\boldsymbol{\theta})^{-1}\right) - \ell n|\mathbf{S}| - p = 23.008 + 6.080 - 22.913 - 6 = 0.175.$$

$$\chi_v^2 = (N-1)F(\boldsymbol{\theta}) = (76-1)\times 0.175 = 13.125.$$

χ^2_v follows χ^2 distribution with $v = p(p+1)/2 - t = 8$ degrees of freedom. The tabulated value for $\alpha = 0.05$ is 15.51. So, H_0 cannot be rejected. Further, the χ^2_v / v is 1.64, which is not within the desired range.

Goodness of Fit Index (GFI)

GFI is a measure of the amount of variances and covariances that are explained by the proposed model (Jöreskog and Sorbom, 1981). GFI works like R^2 does in multiple linear regression. Consider the formula, $R^2 = 1 - SSE / SST$. Here *SSE* is the sum of the squared errors and *SST* is the sum of squares total. R^2 ranges between 0 to 1 depending on the value of *SSE*. $R^2 = 0$ when *SSE* = *SST*, and 1 when *SSE* = 0. Analogously, GFI can be written as

$$GFI = 1 - \frac{trace\left[\left(\mathbf{\Sigma}^{-1}\mathbf{S} - \mathbf{I}\right)^2\right]}{trace\left[\left(\mathbf{\Sigma}^{-1}\mathbf{S}\right)^2\right]} \tag{13.13}$$

If a model is perfectly fit, i.e., each estimated value of $\mathbf{\Sigma}$ equals to the corresponding element of $\mathbf{S}$, then $\mathbf{\Sigma}^{-1}\mathbf{S}$ becomes $\mathbf{I}$, resulting in GFI = 1. On the other hand, if the model fails to explain any true covariance between the measured variables, the second term in the right-hand sides of equation 13.13 approaches 1, resulting in GFI = 0. Note that GFI does not depend on sample size (*N*).

A GFI value greater than or equal to 0.90 is desirable. However, it is very difficult to get GFI ≥ 0.90 for highly complex and dynamic systems, particularly in social and management studies. A conservative value less than 0.9 can also be considered.

Root Mean Square Residual (RMSR)

The following formula defines RMSR.

$$RMSR = \sqrt{2\sum_{j=1}^{p}\sum_{k=1}^{j}\left(s_{jk} - \hat{\sigma}_{jk}\right)^2 / p(p+1)} \tag{13.14}$$

In *RMSR*, each element (s_{jk}) of sample covariance matrix ($\mathbf{S}$) is compared with the corresponding estimate ($\hat{\sigma}_{jk}$) of the predicted population covariance matrix ($\mathbf{\Sigma}$). So, the error value is $s_{jk} - \hat{\sigma}_{jk}$. The smaller the RMSR, the better the model.

Standardized Root Mean Square Residual (SRMSR)

The following formula defines SRMSR.

$$SRMSR = \sqrt{2\sum_{j=1}^{p}\sum_{k=1}^{j}\left[\left(s_{jk} - \hat{\sigma}_{jk}\right) / (s_{jj}s_{kk})\right]^2 / p(p+1)} \tag{13.15}$$

SRMSR lies between 0 and 1 and SRMSR ≤ 0.08 is considered to be desirable (Hu and Bentler, 1999).

Root Mean Square Error of Approximation (RMSEA)

RMSEA uses the property of χ^2 distribution and can be considered as a safeguard to the use of χ^2 statistic (Hair et al., 1998, 2010). The formula is

$$RMSEA = \sqrt{\frac{\left|\chi^2 - df\right|}{(N-1)df}} \tag{13.16}$$

As $E(\chi^2) = df$, $|\chi^2 - \mathrm{df}|$ approaches zero for a perfect fit. This makes RMSEA ideally to be zero. RMSEA usually lies in between 0.03 and 0.08. $RMSEA \le 0.06$ is considered to be a very good fit (Hu and Bentler, 1999).

Other indices include AIC (Akaike, 1974), Browne and Cudeck's (1989) gross validation index (C_k), McDonald's (1989) non-centrality parameters index (D_k), Steiger's (1989) estimate of population GFI (GFI*). For a brief description of these indices, see Marsh and Balla (1994).

13.3.2 Relative Fit Indices (RFI)

Relative fit indices (RFI) answer the question: How well does a particular model perform in explaining a covariance matrix compared with (a range of) other possible models? Essentially, a series of nested models including the null and saturated models are tested. Two models are nested if the set of parameters of one model is the subset of the parameters of the other. χ^2 distribution-based significance test is generally conducted to identify the superior model. However, there could be a large number of nested models and the model specification may vary from researchers to researchers. Therefore, a baseline or reference model is required. Most of the relative fit indices establish as a baseline a "worst fitting" model. Hair et al. (1998, 2010) pointed out that the most common baseline model is the null model which assumes that the observed variables are uncorrelated. So, in null model, the covariance matrix (Σ) is diagonal. The relative fit indices commonly used are normed fit index (NFI), Tucker-Lewis index (TLI), comparative fit index (CFI), and relative non-centrality index (RNI).

To get a quantitative value of the relative fit measures, χ^2 value of the proposed model is compared with that of the null model. The weighted improvement in χ^2 represents the index value. Marsh et al. (1988) categorize RFI into two types based on their quantification, as described below.

Let χ_0^2, v_0, and χ_v^2, v represent the model χ^2 value and the df for the null and proposed models, respectively. The type-I model quantifies the model fit using the form $\frac{\chi_v^2 - \chi_0^2}{\chi_0^2}$ or $\frac{\chi_0^2 - \chi_v^2}{\chi_0^2}$. The type-II model uses the form $\frac{\chi_v^2 - \chi_0^2}{E_v - \chi_0^2}$ or $\frac{\chi_0^2 - \chi_v^2}{\chi_0^2 - E_v}$. The term E_v represents the expected value of the standalone index if the proposed model is true (Marsh and Balla, 1994).

Tucker Lewis Index (TLI)

The first relative fit index (RFI) was defined by Tucker and Lewis (1973) as

$$TLI = \frac{\chi_0^2 / v_0 - \chi_v^2 / v}{(\chi_0^2 / v_0) - 1} \tag{13.17}$$

TLI ranges from 0 to 1 and the desirable value is 0.90 or more.

Normed Fit Index (NFI)

Bentler and Bonett (1980) provided NFI. It is also known as the Bentler-Bonett index (BBI). NFI is defined as

$$NFI = \frac{\chi_0^2 - \chi_v^2}{\chi_0^2} \tag{13.18}$$

NFI varies from 0 to 1, and the desirable value is 0.90 or more.

Comparative Fit Index (CFI)

Bentler (1990) proposed CFI as

$$CFI = 1 - \frac{\chi_v^2 - v}{\chi_0^2 - v_0} \tag{13.19}$$

The term $\chi_v^2 - v$ accounts for the non-centrality parameters of the model's lack of fit. *CFI* varies from 0 to 1 and the desirable value is 0.90 or more.

General Normed Fit Index (GNFI)

Mulaik et al. (1989) proposed GNFI as

$$GNFI = \frac{F_0 - F_v}{F_0 - F_h} \tag{13.20}$$

Where $F_m = (N-1)F_{m,\min}(\boldsymbol{\theta})$. Putting $m = 0, v$ or h, we get values F_0, F_v, and F_h. Note that F_h is the lack of fit for an intermediate model in the sequence from F_0 to F_v. GNFI varies from 0 to 1 and the desirable value is 0.90 or more.

Relative Non-Centrality Index (RNI)

McDonald and Marsh (1990) stated RNI as

$$RNI = \frac{(\chi_0^2 - v_0) - (\chi_v^2 - v)}{(\chi_0^2 - v_0)} \tag{13.21}$$

See that RNI is equal to CFI (Equation 13.19). RNI varies from 0 to 1 and the desirable value is 0.90 or more.

13.3.3 Parsimony Fit Indices

The parsimony fit indices capture the goodness of fit of a proposed model while adjusting the number of parameter to be estimated. It is similar to adjusted R^2, $\left(\text{i.e., } R_a^2\right)$ in multiple linear regression. Parsimony is achieved by multiplying a fit index (as described above) to the ratio of the *df* of the proposed model to the *df* of the null model (Mulaik et al., 1989). This ratio is called parsimony ratio. The commonly used parsimony fit indices are adjusted goodness of fit index (AGFI) and parsimony normed fit index (PNFI).

Adjusted Goodness of Fit Index (AGFI)

Consider GFI given in Equation 13.13. The numerator in the second term of the equation is the trace value of the square of differences between $\boldsymbol{\Sigma}^{-1}\mathbf{S}$ and $\mathbf{I}$ which is a measure of the error component of the model. The *df* available for its computation is v. The denominator of the second term is the total variability (like SST in regression). The number of distinct elements available for its computation is $p(p+1)/2$. Now, if we divide the numerator and denominator of the second term in Equation 13.13, by their *df*, we define AGFI.

$$AGFI = 1 - \frac{trace\left[\left(\boldsymbol{\Sigma}^{-1}\mathbf{S} - \mathbf{I}\right)^2 / v\right]}{trace\left[\left(\boldsymbol{\Sigma}^{-1}\mathbf{S}\right)^2 / \frac{1}{2}p(p+1)\right]} = 1 - \frac{p(p+1)\,trace\left[\left(\boldsymbol{\Sigma}^{-1}\mathbf{S} - \mathbf{I}\right)^2\right]}{2v\,trace\left[\left(\boldsymbol{\Sigma}^{-1}\mathbf{S}\right)^2\right]} \tag{13.22}$$

TABLE 13.2
Goodness of Fit Tests Using The Absolute, Relative, and Parsimonious Fit Indices

Type of Measure	Fit Index	Computed Value	Cut-off Value	Goodness of Fit of The Model
Absolute fit indices	GFI	0.95	≥ 0.90	Very good
	RMSR	6.92	As low as possible	As covariance matrix is used, the value is low.
	SRMSR	0.07	≤ 0.08	Very good
	RMSEA	0.09	≤ 0.06	Reasonably good
Relative fit indices	TLI	0.97	≥ 0.90	Very good
	NFI	0.96	≥ 0.90	Very good
	CFI/RNI	0.98	≥ 0.90	Very good
Parsimonious fit indices	AGFI	0.87	≥ 0.90	Good
	PNFI	0.51	Higher the better	Inconclusive

AGFI varies from 0 to 1 and its desirable value is 0.90 or more.

Parsimony normed Fit Index (PNFI)

$$PNFI = \frac{v}{v_0} NFI \tag{13.23}$$

Higher values of PNFI are better. For alternative models, a change in PNFI of the order of 0.06 to 0.09 is significant (Williams and Anderson, 1994).

In similar manner, the parsimony fit indices for other measures can be computed.

Example 13.7

Consider Example 13.4. Comment on the model fit based on goodness of fit indices.

Solution:

Table 13.2 shows the results of the goodness of fit tests using the absolute, relative and parsimonious fit indices.

Based on the computed values of the fit measures, it can be concluded that the model is a very good fit to the data. It is to be mentioned here that all fit measures may not be equally supportive.

13.4 TEST OF PARAMETERS

In CFA, the parameters to be estimated are $\mathbf{\Lambda}_{p\times m}$, $\mathbf{\Phi}_{m\times m}$, and $\mathbf{\theta}_{\delta}$. These are estimated by numerically minimizing Equation 13.11 based on certain iterative process such as the Newton-Rapson algorithm. If $\mathbf{\theta}$ represents the parameter vector to be estimated, then $\hat{\mathbf{\theta}}$ is the estimated parameter values at which the iterative process converges. For unbiased estimation, $E(\hat{\mathbf{\theta}}) = \mathbf{\theta}$. However, $\hat{\mathbf{\theta}}$ is a point estimate. To get its interval values, we need to know its asymptotic covariance matrix. If MLE is used (Equation 13.11), the asymptotic covariance matrix of $\hat{\mathbf{\theta}}$ is (Bollen, 1989)

$$Cov(\hat{\mathbf{\theta}}) = \frac{2}{N-1}\left[E\left(\frac{\partial^2 F(\mathbf{\theta})}{\partial\mathbf{\theta}\partial\mathbf{\theta}^T}\right)\right]^{-1}$$

If t number of parameters are to be estimated, i.e., $\boldsymbol{\theta}$ is a vector of size $t \times 1$, then $Cov(\hat{\boldsymbol{\theta}})$ is a $t \times t$ matrix. Its diagonal elements represent the variances of $\hat{\boldsymbol{\theta}}$ and off-diagonal elements are the covariances.

Once the estimates $\hat{\boldsymbol{\theta}}$ and its variances are known, the asymptotic Z-test can be performed to test the significance of each of the parameters. Under $H_0 : \boldsymbol{\theta}_t = \mathbf{0}$, the ratio of the estimate by its standard error follows approximate unit normal (Z) distribution. But due to the inherent complexity of CFA (e.g., covariance matrix is analyzed) coupled with the use of iterative procedure for estimation, CFA often produces improper solutions that cannot be accepted from the conceptual or theoretical or practical perspectives. These are also called offending estimates that include negative variance estimates, standardized coefficients close to zero, and very large standard errors. Among many reasons responsible for offending estimates, the three most contributing reasons are (i) sampling fluctuations or sampling errors, (ii) presence of outliers, and (iii) selection of a mis-specified model or theoretically wrong model. The first step to check offending estimates is the use of standardized coefficients. For factor loadings, $\hat{\lambda}_{jk}$, the standardized coefficients, say $\hat{\lambda}^s_{jk}$, can be obtained by using the following formula.

$$\hat{\lambda}^s_{jk} = \hat{\lambda}_{jk} \left[\frac{\hat{\phi}_{kk}}{\hat{\sigma}_{jj}} \right]^{\frac{1}{2}} \tag{13.24}$$

where, $\hat{\phi}_{kk}$ is the variance of the k-th factor (ξ_k) and $\hat{\sigma}_{jj} = s_{jj}$ is the variance of the j-th manifest variable (Xj). Unusual standardized coefficients need to be examined.

Researchers also suggested to use multiple correlation coefficient ($R^2_{X_j}$) for estimating the contribution of each of the manifest variables as a measure of goodness of fit of the manifest variables to its factors.

$$R^2_{X_j} = 1 - \frac{\hat{\theta}_{\delta_{jj}}}{\hat{\sigma}_{jj}} \tag{13.25}$$

TABLE 13.3
The Parameters, Their Standard Errors and t-test Statistics for Example 13.8

Parameter	ML Estimate	Standard Error (SE)	t-value = Estimate/SE	Significance	Remarks
λ_{11}	1	—	—	—	Fixed to 1.00
λ_{21}	1.413	0.282	5.01	0	Significant
λ_{31}	1.233	0.231	5.342	0	Significant
λ_{42}	1	—	—	—	Fixed to 1.00
λ_{52}	1.071	0.092	11.651	0	Significant
λ_{62}	0.914	0.089	10.246	0	Significant
$\theta_{\delta_{11}}$	94.255	16.019	5.884	0	Significant
$\theta_{\delta_{22}}$	3.856	9.491	0.406	0.685	**Not significant**
$\theta_{\delta_{33}}$	25.25	8.294	3.044	0.002	Significant
$\theta_{\delta_{44}}$	22.218	4.628	4.8	0	Significant
$\theta_{\delta_{55}}$	5.165	3.431	1.505	0.132	**Not significant**
$\theta_{\delta_{66}}$	17.034	3.678	4.632	0	Significant
ϕ_{11}	43.468	16.975	2.561	0.01	Significant
ϕ_{22}	63.035	13.664	4.613	0	Significant
ϕ_{21}	−6.922	6.477	−1.069	0.285	**Not significant**

p-value = 0 stands for very low value (say <0.01).

TABLE 13.4
Standardized Loading and Coefficient of Determination for X_1–X_6

Variables	X_1	X_2	X_3	X_4	X_5	X_6
Standardised loading	0.558	0.972	0.845	0.854	0.960	0.863
Coefficient of determination	0.325	0.958	0.727	0.743	0.934	0.759

$R^2_{X_j}$ lies in between 0 and 1. The higher the $R^2_{X_j}$, the better the fit. It is similar to R^2 in multiple linear regression.

Example 13.8

Consider Example 13.4. Test the significance of the parameters of the model and goodness of fit of the manifest variables.

Solution:

The parameters, their standard errors and t-test statistics are given in Table 13.3.

Three parameters $\theta_{\delta_{22}}$, $\theta_{\delta_{55}}$, and ϕ_{21} are not significant, indicating that their estimated value could be zero. This is due to high standard error in comparison to their mean estimate. As the problem is hypothetical, no reason could be assigned. But for practical studies, if such high values are unacceptable, then model respecification may be looked into.

The standardized loading and coefficient of determination for all the six indicators are shown in Table 13.4.

There is no substantial departure in the standardized loadings from fundamentals and the coefficient of determination is very good for all the manifest variables except X_1.

13.5 FIT INDICES FOR INDIVIDUAL FACTORS

The model adequacy tests described in Section 13.3 evaluate the overall fit of the confirmatory factor model, i.e., the degree to which the manifest variables (***X***) represent the hypothesized factors. Section 13.4 describes the significance of the parameters, i.e., whether the influence as observed is by chance or not. However, it assumes that the fit of the individual factor with its indicators (i.e., manifest variables loaded to this factor) is assured. So, there is a need to test the fit of individual factors, particularly from measurement point of view. Hair et al. (1998) recommended the tests for unidimensionality, reliability, and variance extracted. In addition, one has to ensure factor (more popularly construct) validity.

13.5.1 Construct Validity

Validity defines whether a variable does measure what it is supposed to measure (Bollen, 1989). There are four types of validity namely content validity, criterion validity, construct validity, and convergent and discriminant validity. Each of them is very important as errors therein pass through all the other stages of CFA resulting into erroneous conclusions. For details of discussion, readers may refer to Bollen (1989, Chapter 4). *Content validity* is a theoretical and practical issue and is assured subjectively. It basically defines the concepts and their domain. *Criterion validity* is assured through correlation between the criterion variable and a measure supposed to be available at the same time (known as concurrent validity) or at a later time (known as predictive validity). Bollen (1989) defines *construct validity* as "whether a measure relates to other observed variables in a way that is consistent with theoretically and derived predictions." *Convergent validity* determines that the indicators (or the manifest variables) of a factor (or a construct) are highly correlated with the factor. So, the estimated loadings of the factor to its indicators should be significant to ensure convergent validity. *Discriminant validity*, on the other hand, assesses whether two different factors are distinctly different or not. Discriminant validity can be tested using χ^2 tests following the procedure given by Jöreskog (1971). Two competing models, one with free ϕ_{jk} (the correlation between factor j and factor k) and fixed $\phi_{jk} = 1$, are considered. The factors are not perfectly correlated if χ^2 value of the former model is significantly lower than that of the later model (with $\phi_{jk} = 1$). The discriminant validity can also be checked by observing the confidence interval for ϕ_{jk} (Anderson and Gerbing, 1988). If the interval includes 1.0, discriminant validity is violated.

13.5.2 Unidimensionality

As discussed earlier (see Figure 13.1), each of the factors (in CFA) usually has a pre-specified set of manifest variables (indicators) and it is assumed that each set of variables measure the underlying factor. Hence, it is expected that the correlation coefficients among these variables (within a set) will be high and are directed towards a single dimension (the factor) only. Unidimensionality tests are performed to confirm this. Anderson and Gerbing (1988) discussed two criteria, internal and external consistency for unidimensionality. Following Anderson and Gerbing (1988), if X_1, X_2, X_3, and X_4 are the indicators of a common factor ξ, this internal consistency is achieved if the following equality holds (ρ stands for correlation)

$$\frac{\rho_{X_1X_3}}{\rho_{X_1X_4}} = \frac{\rho_{X_2X_3}}{\rho_{X_2X_4}} \tag{13.26}$$

To apply Equation 13.26, there should be at least four indicators per factor. For a two-indicator factor ξ with X_1 and X_2 as indicators, internal consistency holds when

$$\rho_{X_1X_2} = \rho_{X_1\xi} \cdot \rho_{X_2\xi} \tag{13.27}$$

For external consistency, Anderson and Gerbing (1988) have mentioned the following product rule. Consider two factors ξ_j and ξ_k, and one indicator X_j for ξ_j and one indicator X_k for ξ_k then, the product rule implies

$$\rho_{X_jX_k} = \rho_{X_j\xi_j} \times \rho_{\xi_j\xi_k} \times \rho_{\xi_kX_k} \tag{13.28}$$

Ziegler and Hagemann (2015), in an editorial, discussed the tests for unidimensionality along with their pitfalls. They suggested the use of EFA, CFA, and item response technique (IRT).

Example 13.9

Consider Example 13.4. Do the indicators internally and externally consistent?

Solution:

We first examine the correlation matrix as given in Table 13.5.

The correlation coefficients among the three indicators X_1, X_2, and X_3 for the factor ξ_1 and X_4, X_5, and X_6 for ξ_2 are on the higher side. But this is not true when we look into the correlation coefficient between the indicators of the two groups, i.e., X_1, X_2, or X_3 versus X_4, X_5, or X_6 (the shaded portion in the correlation matrix). This is an indication that the indicators are reasonably consistent with their factor.

TABLE 13.5
Correlation Matrix for Example 13.9

	X_1	X_2	X_3	X_4	X_5	X_6
X_1	1					
X_2	0.55	1				
X_3	0.46	0.83	1			
X_4	–0.23	–0.2	–0.21	1		
X_5	–0.2	–0.07	–0.16	0.83	1	
X_6	–0.22	–0.16	–0.19	0.74	0.84	1

TABLE 13.6
Test of Internal Consistency

Test of Internal Consistency		Product Rule		Observed Correlation
Factor	Variable (X)	$\hat{\rho}_{X\xi}$	$\hat{\rho}_{X_j\xi} \times \hat{\rho}_{X_k\xi}$	$Cor(X_j, X_k)$
Factor-1 (ξ_1)	X_1	0.558	$0.558 \times 0.972 = 0.54$	$Cor(X_1, X_2) = 0.55$
	X_2	0.972	$0.972 \times 0.845 = 0.83$	$Cor(X_2, X_3) = 0.83$
	X_3	0.845	$0.845 \times 0.558 = 0.46$	$Cor(X_1, X_3) = 0.46$
Factor-2 (ξ_2)	X_4	0.854	$0.854 \times 0.960 = 0.83$	$Cor(X_4, X_5) = 0.83$
	X_5	0.960	$0.960 \times 0.863 = 0.84$	$Cor(X_5, X_6) = 0.84$
	X_6	0.863	$0.863 \times 0.854 = 0.74$	$Cor(X_4, X_6) = 0.74$

TABLE 13.7
Test of External Consistency

Test of External Consistency (Product Rule; Equation 13.28)			
$\hat{\rho}_{\xi_1\xi_2} = \phi_{21} = -0.132$	$\hat{\rho}_{X_1\xi_1} = 0.558$	$\hat{\rho}_{X_2\xi_1} = 0.972$	$\hat{\rho}_{X_3\xi_1} = 0.845$
$\hat{\rho}_{X_4\xi_2} = 0.854$	–0.063	–0.110	–0.095
$\hat{\rho}_{X_5\xi_2} = 0.960$	–0.071	–0.123	–0.107
$\hat{\rho}_{X_6\xi_2} = 0.863$	–0.064	–0.111	–0.096

In order to use internal consistency, Equation 13.26 requires at least four indicators per factor. So it cannot be applied, as both the factors ξ_1 and ξ_2 have three indicators. Applying the product rule (Equation 13.27), let us see the internal consistency. From Table 13.6, we can conclude that the product rule (Equation 13.27) is satisfied. Hence, the indicators are internally consistent.

The external consistency is not perfectly followed in terms of magnitude of the correlations (see Table 13.7). The sign of relationships satisfies.

13.5.3 Reliability

Reliability is a measure of internal consistency (Hair et al., 1998). For multi-indicator factor, Cronbach's alpha (α) is often used as a measure of reliability. If J number of indicators are used to measure a factor (ξ_k), then the α for ξ_k is α_k (Nunnally, 1978) and

$$\alpha_k = \frac{J\bar{C}_k}{[\bar{V}_k + (J-1)\bar{C}_k]} \tag{13.29}$$

where, $\bar{C}_k$ is the average inter-indicator covariances of the J indicators of ξ_k, and $\bar{V}_k$ is the average variance of the J indicators. $\bar{C}_k$ and $\bar{V}_k$ are computed as below.

$$\bar{C}_k = \frac{1}{J(J-1)} \sum_{j=1}^{J} \sum_{k \neq j}^{J} \sigma_{jk}$$
$$\bar{V}_k = \frac{1}{J} \sum_{j=1}^{J} \sigma_{jj} \tag{13.30}$$

Cronbach's α varies from 0 to 1. For most of the practical purposes $\alpha_k \geq 0.70$ is considered to be a good fit (Nunnally, 1978); however, $\alpha_k \geq 0.90$ is always desirable.

Using the standardized factor loadings $(\hat{\lambda}^s_{jk}, j = 1,2,\ldots,J)$ for the factor ξ_k obtained after performing CFA, the construct reliability (C_{ξ_k}) can be computed as

$$C_{\xi_k} = \frac{(\Sigma_{j=1}^{J} \hat{\lambda}^s_{jk})^2}{(\Sigma_{j=1}^{J} \hat{\lambda}^s_{jk})^2 + \Sigma_j^J \theta^s_{\delta_{jj}}} \tag{13.31}$$

where $\theta^s_{\delta_{jj}}$ is the variance of the measurement error for the j-th manifest variable (X_j).

C_{ξ_k} value should be more than 0.50 for an acceptable reliability.

Another measure of reliability is variance extracted, say V_{ξ_k}, and is computed as

$$V_{\xi_k} = \frac{\Sigma_{j=1}^{J} (\hat{\lambda}^s_{jk})^2}{\Sigma_{j=1}^{J} (\hat{\lambda}^s_{jk})^2 + \Sigma_j^J \theta^s_{\delta_{jj}}} \tag{13.32}$$

The value of V_{ξ_k} should also exceed 0.50. For further reading on C_{ξ_k} and V_{ξ_k}, readers may refer to Hair et al. (1998) and Fornell and Larcker (1981). Other important issues of CFA include model specification such as setting the metric of the factors, member of manifest variables (indicators) per factor, sample size, and model diagnostics.

Example 13.10

Consider Example 13.4. Comment on the reliability of the indicators.

TABLE 13.8
Cronbach's α, Construct Reliability (C_{ξ_k}) and Variance Extracted (V_{ξ_k})

	Computation of Cronbach α			Construct Reliability	Variance Extracted
Factors	$\bar{C}_k$	$\bar{V}_k$	α_k	C_{ξ_k}	V_{ξ_k}
Factor-1 (ξ_1)	63.84	107.980	0.813	0.850	0.664
Factor-2 (ξ_2)	63.000	78.517	0.924	0.927	0.810

Solution:

Using Equations 13.29 and 13.30, the Cronbach α values are shown in Table 13.8. The Cronbach α values show a very good reliability of the indicators.

Both construct reliability (C_{ξ_k}) and variance extracted (V_{ξ_k}) values are very high compared to the minimum recommended value (see the last two columns of Table 13.8). This indicates high reliability of the indicators in measuring the factors concerned.

13.6 MODEL RESPECIFICATION

Model respecification is a very important step in CFA. The model respecification is needed under the following (but not limited to) conditions:

- If a model is under-identified even after following standard guidelines for fixing and constraining parameters of the model.
- When the parameter estimation process fails to converge.
- If the model produces improper solutions such as offending estimates (negative variances or Haywood cases).
- If the model is not fit to the data. It includes both the overall model fit and individual factor fit.
- When the modification index (MI) suggests model respecification, but it should be backed with sound theoretical and practical perspectives. For MI readers may refer Jöreskog and Sörbom (1984).

For application of model respecification, see Section 13.7.4.

13.7 CASE STUDY

13.7.1 Background

Employees' safety practices are important to the running of a workplace. Safety management theory states that employees' safety practices can be grouped into three broad categories, namely workers' safety practices, safety officers' safety practices, and higher authorities' safety practices. These dimensions are called *factors* or *constructs*. In this study, we demonstrate how confirmatory factor analysis (CFA) can be used to confirm and quantify the dimensions of safety practices. The study was conducted in an underground coal mine. It was semi-mechanized mine and operated six days a week and three shifts per day. The average coal production was 1,000 metric tons and the number employed was around 900 at the time of the study.

13.7.2 Variables and Data

The study focuses on measuring safety practice dimensions followed in underground coal mines. Based on literature survey and discussion with mine safety personnel, 16 questions (Q_1–Q_{16}) were prepared under three factors as shown in Table 13.9 below (Khanzode, 2010). A three-point Likert scale is used to measure the responses. Against every question in Table 13.9, the response could be "rarely," "sometimes," or "always." The numerical values assigned to these responses are 1, 2, and 3 respectively. However, some questions are in reverse order and accordingly the numerical values are also reversed.

Structured personal interviews with the workers were made for data collection. The participants were randomly selected through a stratified sampling scheme, where the stratification was done based on job designation, age, and accident experience. The interviews were conducted confidentially with each worker in the mine office away from the workplace to ensure a conducive environment. The interviews were conducted during the beginning and the end of shift hours when the individuals could spare time for discussions. The sample size is

TABLE 13.9
Factors and Their Indicators (Manifest Variables)

Factors	Indicators	Indicator Description (Questions)	Mean (Std. Dev.)
Factor-1 (WSP)	Q_1	Do miners wear the special shoes given by the company before going to the underground?	2.83 (0.38)
	Q_2	Does everybody utilize the knowledge of first aid training?	2.54 (0.59)
	Q_3	Does overman / mine sirdar check that everybody reaches a safe place before blasting?	2.83 (0.38)
	Q_4	Does ventilation officer regularly check the ventilation system?	2.51 (0.52)
	Q_5	Does mine sirdar / overman regularly test the roof and traveling roadways?	2.72 (0.45)
	Q_6	Does shot firer give warning before blasting?	2.91 (0.29)
	Q_7	Are winding engine and winding ropes regularly examined?	2.64 (0.48)
	Q_8	Have you seen anybody doing risky work in the mine?	2.04 (0.42)
Factor-2 (SSP)	Q_9	Do safety personnel inspect regularly to ensure that safe practices are being followed?	2.39 (0.55)
	Q_{10}	Are most of the areas in your mine inspected regularly?	2.49 (0.62)
	Q_{11}	Do the safety inspectors make use of a written checklist?	2.09 (0.68)
	Q_{12}	Are safety meetings conducted by mine sirdar /overman / section-in-charge with the workers?	2.01 (0.74)
Factor-3 (ASP)	Q_{13}	Does higher authority in your mine reward good safety performance?	1.46 (0.54)
	Q_{14}	Has your mine initiated or aided family or community effects to instil workers safety consciousness?	1.26 (0.49)
	Q_{15}	Are the workers praised openly for their safety performance?	1.41 (0.56)
	Q_{16}	Are any disciplinary actions taken against the workers who do not use required protective devices?	2.15 (0.66)

140 and the mean responses with their standard deviations are shown in Table 13.9. Note that each question is a manifest variable. So, we have 16 manifest variables or indicators. Further, there are three factors namely workers' safety practices (WSP), safety officers' safety practices (SSP), and higher authorities' safety practices (ASP).

It is expected that some of the questions will be correlated as per the principles of the confirmatory factor analysis. The correlation matrix for the 16 questions (indicators or manifest variables) is shown in Table 13.10. The correlation coefficient values greater than 0.30 are in bold. By visual inspection, we could not make much of it, as there are many correlation coefficients with low values. But careful observations reveal a pattern exists. For example, Q_1 (question 1) has more than 0.30 correlation coefficient with Q_2 and Q_7; Q_2 with Q_1 and Q_4; and Q_4 with Q_2, Q_5, and Q_7. From theoretical and practical perspectives, we know that factor 1 (WSP) has eight indicators from Q_1 to Q_8. Similar pattern exists for factor 2 (SSP) and factor 3 (ASP) with their indicators. Hence, we do expect that there are three factors and we want to confirm it through CFA.

TABLE 13.10
Correlation Matrix of 16 Manifest Variables

	Q_1	Q_2	Q_3	Q_4	Q_5	Q_6	Q_7	Q_8	Q_9
Q_1	1								
Q_2	**0.316**	1							
Q_3	–0.006	0.188	1						
Q_4	0.191	**0.352**	0.154	1					
Q_5	0.225	0.159	0.182	**0.365**	1				
Q_6	0.116	0.04	0.246	0.076	0.185	1			
Q_7	**0.365**	0.285	0.089	**0.371**	0.258	0.065	1		
Q_8	–0.051	–0.077	0.084	–0.117	–0.061	0.202	–0.006	1	
Q_9	–0.02	–0.144	0.224	0.028	0.185	0.141	–0.054	0.032	1
Q_{10}	0.087	–0.098	0.18	–0.045	0.083	0.216	0.1	0.097	**0.446**
Q_{11}	0.169	0.152	0.141	0.182	0.148	0.113	0.27	–0.135	0.237
Q_{12}	0.137	0.015	0.112	0.188	0.142	0.106	**0.317**	0.137	0.271
Q_{13}	0.11	0.116	0.145	0.258	0.151	0.138	0.266	–0.042	0.133
Q_{14}	0.124	–0.057	0.124	0.194	0.133	0.068	0.249	0.06	–0.004
Q_{15}	0.026	0.01	0.026	0.027	–0.003	0.057	0.1	–0.001	0.038
Q_{16}	0.046	–0.171	0.133	–0.014	0.094	–0.002	0.106	0.006	0.216

	Q_{10}	Q_{11}	Q_{12}	Q_{13}	Q_{14}	Q_{15}	Q_{16}
Q_{10}	1						
Q_{11}	**0.445**	1					
Q_{12}	0.252	**0.396**	1				
Q_{13}	0.085	0.241	**0.36**	1			
Q_{14}	–0.042	–0.002	0.19	**0.418**	1		
Q_{15}	–0.002	–0.017	0.229	**0.32**	**0.326**	1	
Q_{16}	0.242	0.131	0.099	0.106	0.172	0.224	1

13.7.3 Analysis and Results

The CFA path diagram for the case study considered is given in Figure 13.5. The parameters of the model (not shown in the figure), i.e., the loading matrix $\mathbf{\Lambda}$, covariances among factors $\mathbf{\Phi}$, and covariance matrix of errors, $\mathbf{\Theta}_\delta$ are shown below.

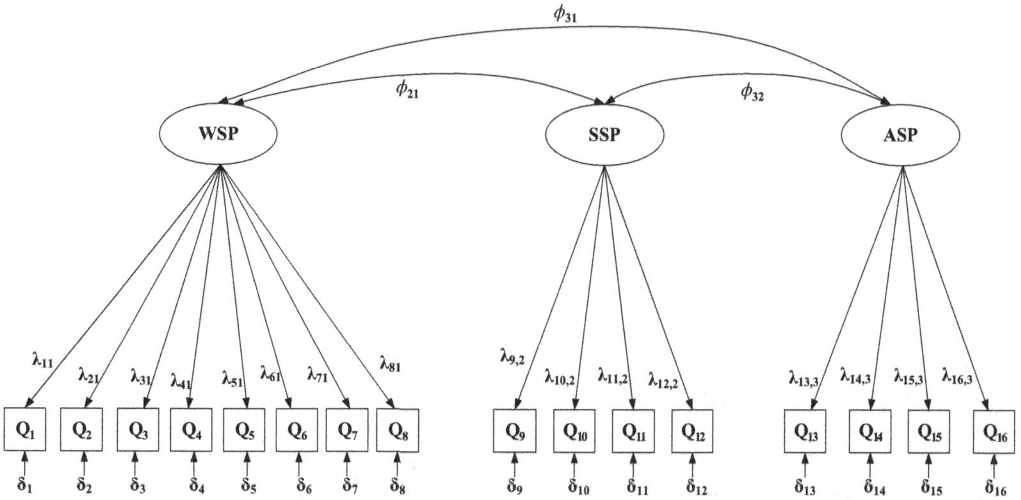

FIGURE 13.5 Confirmatory factor model (path diagram) for the safety practices in mines. There are three factors, WSP (measured using eight questions), SSP (measured using four questions), and ASP (measured using four questions). Notations represent their usual meanings.

$$\Lambda = \begin{bmatrix} \lambda_{11} & 0 & 0 \\ \lambda_{21} & 0 & 0 \\ \lambda_{31} & 0 & 0 \\ \lambda_{41} & 0 & 0 \\ \lambda_{51} & 0 & 0 \\ \lambda_{61} & 0 & 0 \\ \lambda_{71} & 0 & 0 \\ \lambda_{81} & 0 & 0 \\ 0 & \lambda_{92} & 0 \\ 0 & \lambda_{10,2} & 0 \\ 0 & \lambda_{11,2} & 0 \\ 0 & \lambda_{12,2} & 0 \\ 0 & 0 & \lambda_{13,3} \\ 0 & 0 & \lambda_{14,3} \\ 0 & 0 & \lambda_{15,3} \\ 0 & 0 & \lambda_{16,3} \end{bmatrix} \qquad \Phi = \begin{bmatrix} 1 & & \\ \phi_{21} & 1 & \\ \phi_{31} & \phi_{32} & 1 \end{bmatrix}$$

$$\boldsymbol{\theta}_\delta = \begin{bmatrix} \theta_{\delta_{11}} & & & & & & \\ 0 & \theta_{\delta_{22}} & & & & & \\ 0 & 0 & \theta_{\delta_{33}} & & & & \\ . & . & . & . & & & \\ . & . & . & & . & & \\ . & . & . & & & . & \\ 0 & 0 & 0 & . & . & . & \theta_{\delta_{16,16}} \end{bmatrix}$$

Note that $\boldsymbol{\theta}_\delta$ is diagonal.

The CFA model is estimated using the CFA module of the software R. Maximum likelihood estimation method is chosen. The goodness of fit measures of the proposed model (Figure 13.5) is given in Table 13.11. The parameters of the model $\boldsymbol{\Lambda}_{16\times3}, \boldsymbol{\theta}_{\delta_{16\times16}}, and\ \boldsymbol{\Phi}_{3\times3}$ are shown in Tables 13.12–13.14, respectively and in Figure 13.6.

Overall Fit of The Proposed Model

From Table 13.11, it is seen that the χ^2-statistic is significant and hence the null hypothesis is rejected. χ^2 / df is 1.73 which is not in the range of 2 to 5. So, the χ^2 statistic does not support the proposed model. As stated earlier, the χ^2-statistic is not an acceptable fit measure for CFA. The absolute fit indices GFI = 0.86, RMSR = 0.09, SRMSR = 0.09, and RMSEA = 0.07 show reasonable fit with marginal discrepancy from their acceptable good fit values. All the relative fit indices (see Table 13.11) are greater than 0.70 but less than 0.80. These are less than their acceptable good fit (0.90) but considering the subjectivity involved during data collection as well the fuzziness of responses, the fit may be substantiated with practical value proposition.

The AGFI value is 0.82 which is reasonably good. On the other hand, the $PNFI = 0.50$ is inconclusive. Overall, it can be concluded that the proposed model is not a good fit to accept and at the

TABLE 13.11
Fit Measures

Type of Measure	Fit Index	Computed Value (Original Model)	Computed Value (Revised Model)	Remarks
Chi-square based	χ^2_v	174.455 ($df = 101$, p-value = 0.00)	124.91 ($df = 85$, p-value = 0.00)	(i) Original model is reasonably fit. (ii) The revised model is good fit. (iii) PNFI is inconclusive
	χ^2_v / v	1.73	1.47	
Absolute fit indices	GFI	0.86	0.90	
	RMSR	0.09	0.08	
	SRMSR	0.09	0.08	
	RMSEA	0.07	0.06	
Relative fit indices	TLI	0.72	0.84	
	NFI	0.72	0.84	
	CFI	0.76	0.87	
	RNI	0.76	0.87	
Parsimonious fit indices	AGFI	0.82	0.85	
	PNFI	0.50	0.56	

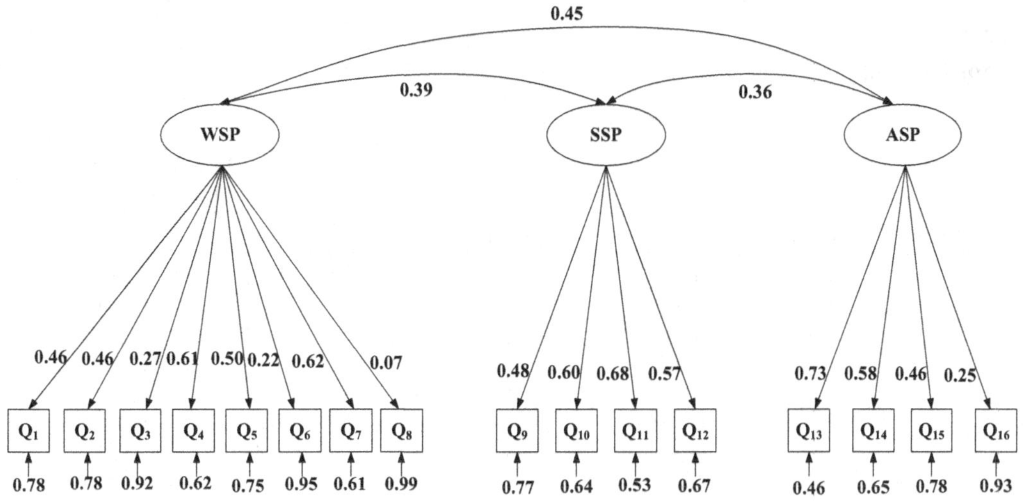

FIGURE 13.6 Confirmatory factor model with estimates for the safety practices in mines. There are three factors, WSP (measured using eight questions), SSP (measured using four questions), and ASP (measured using four questions). The numeric figures are the parameter estimates for standardized factor loadings, standardized error variance, and correlation coefficients between the factors.

same time, not such a bad model to reject completely. Model respecifications may result into finding a better model.

Test of Model Parameters

Table 13.12 shows the factor loadings, their standard errors, *p*-value, and 95% confidence intervals. All the loading parameters except λ_{81} (for question 8) are statistically significant. As a customary, a model without question 8 may be tested. The error variances with their standard errors, *p*-value, and 95% confidence interval (CI) are given in Table 13.13. There are no offending estimates (negative error variances) and all the estimates of error variances are statistically significant. One interesting observation is that the error variances are high in value indicating that high specific variance for each of the questions (manifest variables).

Table 13.14 shows the correlations among the factors WSP, SSP, and ASP. All the correlation coefficients are more than 0.30 and statistically significant. Finally, Table 13.15 shows the coefficient of determination for each of the 16 questions. None of them are of larger value indicating of lower common variances extracted by the factor allocated. For questions Q_3, Q_6, Q_8, and Q_{16}, the coefficient of determination is less than 0.10.

Test of Fit of Individual Factors

Table 13.16 shows Cronbach's α_k, construct reliability C_{ξ_k} and variance extracted V_{ξ_k}, where k stands for the k-th factor, $k = 1, 2$, and 3.

None of the α is 0.70 or more, but close to 0.60. Similarly, the constructs reliability values are also not high enough but more than threshold value (0.50). Further, the variance extracted by each of these factors is low. Collectively, we can conclude that the fit of the individual factors is not acceptable for the proposed (original) model.

TABLE 13.12
Factor Loadings

Parameters	Factor loadings: Estimate	Std. Err	*Z*-value	*p*-value	95% Confidence Intervals (CI): Lower	Upper
λ_{11}	0.457	0.095	4.786	0	0.27	0.644
λ_{21}	0.457	0.095	4.786	0	0.27	0.644
λ_{31}	0.268	0.098	2.728	0.006	0.075	0.461
λ_{41}	0.608	0.093	6.537	0	0.426	0.79
λ_{51}	0.497	0.095	5.242	0	0.311	0.682
λ_{61}	0.218	0.099	2.203	0.028	0.024	0.411
λ_{71}	0.62	0.093	6.669	0	0.437	0.802
λ_{81}	−0.068	0.1	−0.684	0.494	−0.264	0.127
λ_{92}	0.475	0.096	4.959	0	0.287	0.663
$\lambda_{10,2}$	0.595	0.095	6.291	0	0.41	0.78
$\lambda_{11,2}$	0.679	0.095	7.167	0	0.493	0.865
$\lambda_{12,2}$	0.572	0.095	6.044	0	0.387	0.758
$\lambda_{13,3}$	0.73	0.104	6.997	0	0.525	0.934
$\lambda_{14,3}$	0.582	0.1	5.835	0	0.386	0.777
$\lambda_{15,3}$	0.457	0.098	4.644	0	0.264	0.65
$\lambda_{16,3}$	0.245	0.1	2.44	0.015	0.048	0.441

p-value = 0 stands for very low value (say <0.01).

TABLE 13.13
Error Variances

θ_δ	Error variance: Estimate	Standard Error	Test statistic: *Z*	*p*-value	95% CI: Lower	Upper
$\theta_{\delta_{11}}$	0.784	0.105	7.448	0	0.578	0.991
$\theta_{\delta_{22}}$	0.784	0.105	7.449	0	0.578	0.991
$\theta_{\delta_{33}}$	0.921	0.114	8.1	0	0.698	1.144
$\theta_{\delta_{44}}$	0.623	0.099	6.291	0	0.429	0.817
$\theta_{\delta_{55}}$	0.746	0.103	7.222	0	0.544	0.949
$\theta_{\delta_{66}}$	0.945	0.115	8.196	0	0.719	1.172
$\theta_{\delta_{77}}$	0.609	0.099	6.164	0	0.415	0.803
$\theta_{\delta_{88}}$	0.988	0.118	8.351	0	0.756	1.22
$\theta_{\delta_{99}}$	0.767	0.105	7.303	0	0.561	0.973
$\theta_{\delta_{10,10}}$	0.639	0.101	6.311	0	0.44	0.837
$\theta_{\delta_{11,11}}$	0.532	0.103	5.173	0	0.33	0.733
$\theta_{\delta_{12,12}}$	0.665	0.102	6.553	0	0.466	0.864
$\theta_{\delta_{13,13}}$	0.46	0.123	3.735	0	0.219	0.701
$\theta_{\delta_{14,14}}$	0.654	0.108	6.075	0	0.443	0.866
$\theta_{\delta_{15,15}}$	0.784	0.108	7.286	0	0.573	0.995
$\theta_{\delta_{16,16}}$	0.933	0.115	8.122	0	0.708	1.158

p-value = 0 stands for very low value (say <0.01).

TABLE 13.14
Correlation Matrix of Factors (Standard Errors are Within Bracket)

Factors	WSP	SSP	ASP
WSP	1		
SSP	0.39 (0.11)	1	
ASP	0.449 (0.11)	0.364 (0.12)	1

TABLE 13.15
R-square and Theta-delta For Manifest Variables (Questions)

		Original Model		Revised Model	
Factors	Question	R^2	Theta-delta	R^2	Theta-delta
WSP	Q1	0.210	0.784	0.213	0.781
	Q2	0.210	0.784	0.212	0.782
	Q3	0.072	0.921	0.069	0.924
	Q4	0.372	0.623	0.362	0.633
	Q5	0.248	0.746	0.234	0.76
	Q6	0.048	0.945	0.047	0.947
	Q7	0.387	0.609	0.407	0.589
	Q8	0.005	0.988	—	—
SSP	Q9	0.228	0.767	0.084	0.909
	Q10	0.357	0.639	0.248	0.747
	Q11	0.465	0.532	0.767	0.231
	Q12	0.330	0.665	0.347	0.648
ASP	Q13	0.537	0.46	0.507	0.489
	Q14	0.341	0.654	0.353	0.642
	Q15	0.210	0.784	0.234	0.76
	Q16	0.060	0.933	0.052	0.941

TABLE 13.16
Test of Individual Factors

Factors (ξ_k)	Cronbach α_k	Construct Reliability C_{ξ_k}	Variance Extracted V_{ξ_k}
WSP (ξ_1)	0.57	0.59	0.19
SSP (ξ_2)	0.67	0.67	0.35
ASP (ξ_3)	0.59	0.59	0.29

13.7.4 Model Re-Specifications

As discussed earlier, model re-specification is needed if a model is unidentified, or there is failure to converge during estimation, or a solution with offending estimates, or lack of fit, or a combination of these. After careful investigation of the proposed model and results, the salient features of the model and its solutions are given below.

- Model is over-identified. It is a desirable phenomenon.
- The estimation process converges normally. We have used R package for estimation. The estimation converges after 76 iterations.
- No negative variances are observed.
- The model is not a good fit to this data. Model respecification might help.
- The loading parameter λ_{81} is insignificant. It might be excluded from the model.

In search of a better fit model, a modification index (MI) is often used. The MI works on the relationships that are excluded in the model. For example, some of the indicators might be loaded with two factors. The MI suggests which of the excluded parameters could be included in the model. It examines the reduction in the χ^2 value if the excluded parameter is included. If the reduction is significant, then the parameter (relationship) can be estimated. However, this suggestion should be verified from the theoretical and practical perspectives of the model. Usually, large MI for an excluded parameter supports its inclusion. But the literature is inconclusive on the value of the MI to be considered as large. One guideline is the "reduction in χ^2-value" that is significant at certain probability level of significance (usually 0.01). However, relying solely on χ^2-value is not recommended.

Hair et al. (1998) suggest using standardized MI value of 3.846 as the threshold for modification but this seems to be low. MacCullum (1986) stated that even with having adequate theoretical justification, many researchers are not willing to revise a model based on MI. We therefore consider $MI \geq 10$ as significantly large and this suggests inclusion of two parameters in the proposed (original) factor model. These parameters as $\theta_{\delta_{9,10}}$ (with MI=14.04), and $\lambda_{12,3}$ (with MI=12.93) i.e., the covariance between error terms of Q_9 and Q_{10}, and the loading of factor 3 (ASP) on question Q_{12}. We run two revised models: (i) the proposed model with inclusion of $\theta_{\delta_{9,10}}$ and $\lambda_{12,3}$ (based on MI), and (ii) the revised model in (i) but excluding $\lambda_{8,1}$, as this parameter is insignificant. The results favor the use of the revised model (ii). The revised model (ii) is shown in Figure 13.7. The goodness of fit and the estimated parameters are shown in Tables 13.11 and 13.17, respectively.

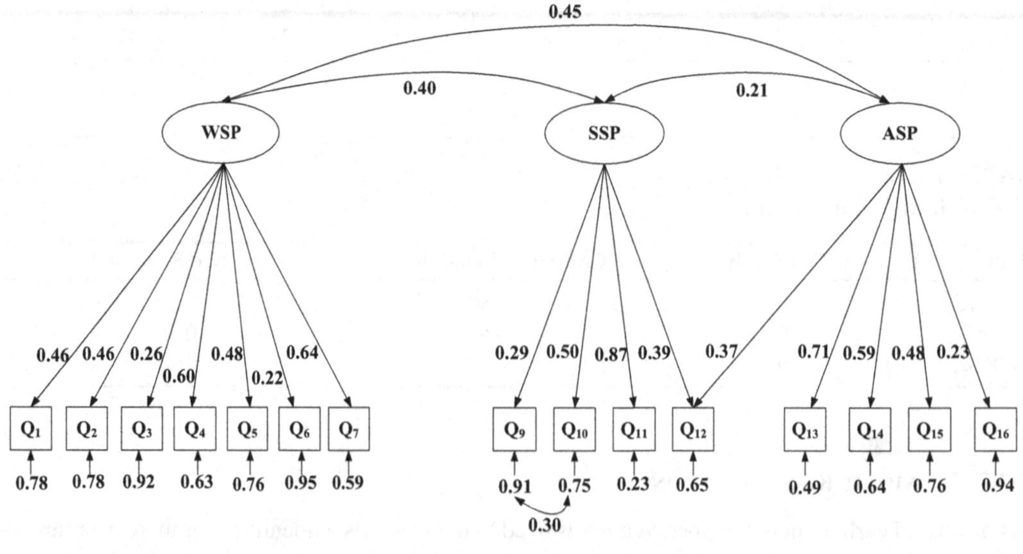

FIGURE 13.7 Confirmatory factor model with estimates (based on MI). The differences of this figure with Figure 13.5 are as follows: (i) Q_8 is excluded, (ii) Q_{12} becomes indicator (manifest) variable for both ASP and SSP, and (iii) correlated error terms between Q_9 and Q_{10}.

13.7.5 Discussions and Insights

First, we will compare the revised model with the proposed model in terms of model fit (see Table 13.11), parameter estimates (Tables 13.12–13.15 and Table 13.17), and test of individual factors (Table 13.16 and Table 13.19). From Table 13.11, it is seen that all the fit measures are improved in the revised model as compared to the proposed model. The revised model can be considered as a better fit, while the proposed (original) is reasonably or moderately fit.

TABLE 13.17
Parameter Estimates For The Revised Model (Figure 13.7)

	Parameter Estimates		Test Statistic		95% CI	
Parameter	**Estimate**	**Std. Error**	***Z***	***p*-value**	**Lower**	**Upper**
λ_{11}	0.46	0.095	4.85	0	0.274	0.646
λ_{21}	0.459	0.095	4.836	0	0.273	0.645
λ_{31}	0.262	0.098	2.673	0.008	0.07	0.454
λ_{41}	0.6	0.093	6.481	0	0.418	0.781
λ_{51}	0.482	0.094	5.102	0	0.297	0.667
λ_{61}	0.215	0.098	2.187	0.029	0.022	0.408
λ_{71}	0.635	0.092	6.901	0	0.455	0.816
λ_{92}	0.289	0.097	2.972	0.003	0.099	0.48
$\lambda_{10,2}$	0.496	0.102	4.872	0	0.296	0.695
$\lambda_{11,2}$	0.873	0.129	6.777	0	0.62	1.125
$\lambda_{12,2}$	0.387	0.097	3.974	0	0.196	0.578
$\lambda_{12,3}$	0.367	0.095	3.853	0	0.18	0.553
$\lambda_{13,3}$	0.71	0.098	7.242	0	0.517	0.902
$\lambda_{14,3}$	0.592	0.096	6.143	0	0.403	0.781
$\lambda_{15,3}$	0.482	0.097	4.987	0	0.293	0.672
$\lambda_{16,3}$	0.227	0.1	2.276	0.023	0.031	0.422
$\theta_{\delta_{11}}$	0.781	0.105	7.456	0	0.576	0.986
$\theta_{\delta_{22}}$	0.782	0.105	7.462	0	0.577	0.988
$\theta_{\delta_{33}}$	0.924	0.114	8.12	0	0.701	1.147
$\theta_{\delta_{44}}$	0.633	0.098	6.435	0	0.44	0.826
$\theta_{\delta_{55}}$	0.76	0.104	7.338	0	0.557	0.964
$\theta_{\delta_{66}}$	0.947	0.115	8.204	0	0.72	1.173
$\theta_{\delta_{77}}$	0.589	0.098	6.039	0	0.398	0.78
$\theta_{\delta_{99}}$	0.909	0.113	8.064	0	0.688	1.13
$\theta_{\delta_{10,10}}$	0.747	0.11	6.806	0	0.532	0.962
$\theta_{\delta_{11,11}}$	0.231	0.195	1.187	0.235	−0.151	0.613
$\theta_{\delta_{12,12}}$	0.648	0.094	6.913	0	0.465	0.832
$\theta_{\delta_{13,13}}$	0.489	0.11	4.451	0	0.274	0.705
$\theta_{\delta_{14,14}}$	0.642	0.104	6.196	0	0.439	0.845
$\theta_{\delta_{15,15}}$	0.76	0.105	7.208	0	0.554	0.967
$\theta_{\delta_{16,16}}$	0.941	0.115	8.168	0	0.715	1.167
$\theta_{\delta_{9,10}}$	0.303	0.084	3.59	0	0.138	0.469
ϕ_{11}	1	0	NA	NA	1	1
ϕ_{22}	1	0	NA	NA	1	1
ϕ_{33}	1	0	NA	NA	1	1
ϕ_{21}	0.396	0.109	3.642	0	0.183	0.61
ϕ_{31}	0.454	0.109	4.174	0	0.241	0.667
ϕ_{32}	0.212	0.116	1.82	0.069	−0.016	0.44

NA: Not applicable, p-value = 0 stands for very low value (say <0.01).

It is interesting to see that most of the parameters, that are unaffected by the change of the structure (relationships) in the revised model, have similar estimated parameter values. This indicates the robustness of the CFA. The parameters whose value changes in the revised model are shown in Table 13.18. In general, although the magnitude (strength) of the parameter is changed, its direction (pattern) remains unchanged.

The fit indices for the individual factors for the revised model are given in Table 13.19. The result is not encouraging, while comparing these values with those of the original model (Table 13.16), there is improvement in Cronbach's α and construct reliability for the factor WSP but in other measures no noticeable improvement is found.

This triggers discussion on the overall acceptability of the model. From overall goodness of fit, the revised model is better than the original model with fit measures closer to the acceptable level, indicating that the revised model can be used for all practical purposes. The individual fits, on the contrary, show a poorer fit with reasonably acceptable Cronbach's α and construct reliability but with poorer amount of variance extracted. Applied researches often face this problem. We prefer to accept this model as a starting point to explain the safety practices in mines but recommend searching for better model theoretically and practically, along with better questionnaire design and sampling strategy.

The insights that can be further drawn are as follows:

- The three-factor model is a moderate fit to the data. The revised model provides better fit and can be used as a baseline model to start with for analyzing safety practices in mines. A search for a better model and better data is recommended.

TABLE 13.18
Parameters with Change in Values in The Revised Model

Parameters	Estimated Value: Original Model	Estimated Value: Revised Model	Remarks (Revised Model with Respect to The Original Model)
λ_{92}	0.475	0.289	• No change in direction.
$\lambda_{11,2}$	0.679	0.873	• All changes are statistically significant.
$\lambda_{12,2}$	0.572	0.387	• λ_{92} decreases. Possible reason could be inclusion of $\theta_{\delta_{9,10}}$.
$\lambda_{12,3}$	—	0.367	• $\lambda_{12,2}$ decreases. Possible reason could be the shared loading with $\lambda_{11,3} = 0.367$.
$\theta_{\delta_{9,10}}$	0.767	0.909	• $\theta_{\delta_{99}}$ increases as λ_{92} decreases.
$\theta_{\delta_{11,11}}$	0.532	0.231	• $\theta_{\delta_{11,11}}$ decreases as $\lambda_{11,2}$ increases.
$\theta_{\delta_{9,10}}$	—	0.303	• The decrease of ϕ_{32} may be attributed to the model respecification.
ϕ_{32}	0.364	0.212	

TABLE 13.19
Test of Individual Factors in The Revised Model

Factors (ξ_k)	Cronbach α_k	Construct Reliability C_{ξ_k}	Variance Extracted V_{ξ_k}
WSP (ξ_1)	0.65	0.64	0.22
SSP (ξ_2)	0.67	0.62	0.35
ASP (ξ_3)	0.56	0.62	0.25

- An initial hypothesized confirmatory factor model is likely to undergo modification and model respecification. Testing several alternative models is recommended. The best model is to be chosen based on goodness of fit, quality of solution, and applicability of the model from theoretical and practical perspectives.
- CFA is a robust model and has flexibility in choosing alternative models and provides several avenues to modify and to test the quality of the models that are built in.

13.8 LEARNING SUMMARY

In this chapter, CFA is described with the important concepts and derivations, model identification, estimation, and model adequacy tests along with a case study. In addition, a good amount of solved examples is given to demonstrate the concepts and computation. In summary,

- CFA is model-driven and comprises several hypotheses. The researcher, *a priori*, knows the model structure. The model is confirmed based on sample data analysis.
- Although the factors are known *a priori*, they are unobserved and quantified after parameter estimation. The quantification is usually done by developing the covariance or correlation matrix of the factors.
- The fundamental relationship in CFA is $Cov(\boldsymbol{X}) = \boldsymbol{\Sigma} = Cov(\boldsymbol{\Lambda\xi}+\boldsymbol{\delta}) = \boldsymbol{\Lambda\Phi\Lambda}^T + \boldsymbol{\theta}_\delta$.
- Parameter estimation consists of model identification and model estimation. The necessary and sufficient conditions for model identification are known as order condition and rank condition, respectively. The necessary condition is also known as t-rule. The sufficient condition requires complex mathematics to understand, however, simple two-indicator and three-indicator rules are also proposed.
- Usually MLE method is used for estimation. The governing function to be minimized for model identification is $F(\boldsymbol{\theta}) = \ell n|\boldsymbol{\Sigma}(\boldsymbol{\theta})| + tr\left(\mathbf{S}\boldsymbol{\Sigma}(\boldsymbol{\theta})^{-1}\right) - \ell n|\mathbf{S}| - p$. Iterative numerical estimation methods such as the Newton-Rapson method are used for parameter estimation. For ill-constructed models and/or data, estimation process may fail to converge.
- Model adequacy tests comprise χ^2_ν statistic, and absolute, relative, and parsimonious fit measures. The χ^2_ν statistic is less reliable and multiple fit measures to be satisfied for overall model fit to the data.
- The statistical test of parameters is similar to other multivariate models like multiple linear regression; however, Z-test is used. The coefficient of determination is used to quantify the contribution of individual manifest variable to its factors.
- The fit of individual factors is measured based on factor (construct) validity, unidimensionality, and reliability measures. Cronbach's α, construct reliability, and variance extracted by a factor are mostly used.
- CFA often undergoes thorough model respecification. It is a recommended step but must be backed up by adequate theoretical and practical perspectives for the revised model as the original proposed model is based on sound theory, practice, and evidence.

Being a well-developed field in statistics, there are many good books and numerous case applications on CFA available.

EXERCISES

(A) Short conceptual questions

13.1 What is confirmatory factor analysis (CFA)? Is it different from EFA? If so, how?

13.2 State the assumptions of CFA.

13.3 Prove that in CFA, $\boldsymbol{\Sigma} = \boldsymbol{\Lambda\Phi\Lambda} + \boldsymbol{\Psi}$ [the notations represent their usual meaning].

13.4 What is model identification? Explain the order and rank conditions for model identification in CFA.

13.5 If a confirmatory factor model is unidentified, how do you make it identifiable? Explain t-rule, two-indicator and three-indicator rules for confirmatory factor model identification.

13.6 Derive $F(\boldsymbol{\theta}) = \ell n|\boldsymbol{\Sigma}(\boldsymbol{\theta})| + tr\left(\mathbf{S}\boldsymbol{\Sigma}(\boldsymbol{\theta})^{-1}\right) - \ell n|\mathbf{S}| - p$. How do you estimate θ vector?

13.7 What is degree of freedom (*df*)? How do you compute *df* in CFA? Explain the situations with *df* = *0*, *df* < *0*, and *df* > *0*.

13.8 What is model adequacy test? How do you assess it for confirmatory factor model?

13.9 What is Chi-square test of model adequacy of CFA? Why it is not recommended for CFA?

13.10 What is absolute fit index? Why it is called "absolute"? Explain the absolute fit indices used in CFA.

13.11 What is relative fit index? Why it is called "relative"? Explain the relative fit indices used in CFA.

13.12 What is parsimonious fit index? Why it is called "parsimonious"? Explain the parsimonious fit indices used in CFA.

13.13 There are plenty of model fit measures developed for CFA. Why so many? How do you choose the best set?

13.14 CFA suffers from improper solutions. What does it mean? How do you overcome the situation?

13.15 What is offending estimate? A model with very good overall fit might suffer from offending estimates—do you agree? What are the possible reasons for offending estimates?

13.16 What is component or individual fit in CFA? Why it is used? Explain some of the component fit measures.

13.17 Explain content validity, criterion validity, and construct validity.

13.18 What is unidimensionality? Why it is important in CFA? How do you test unidimensionality of a factor?

13.19 What is internal consistency of indicators of a factor? How do you assess it? Explain the product rules of the internal consistency test.

13.20 What is external consistency of indicators in CFA? How do you assess it? Explain the product rules of the external consistency test.

13.21 Suppose $\boldsymbol{X} = [X_1, X_2, \ldots, X_p]^T$ and $\boldsymbol{\Sigma} = Cov(\boldsymbol{X})$. Further, suppose there are m $(< p)$ oblique factors. Using the fundamental relationship $\boldsymbol{\Sigma} = \boldsymbol{\Lambda\Phi\Lambda} + \boldsymbol{\Psi}$, prove the product rules given in Equations 13.26–13.28.

13.22 What is meant by convergent and discriminant validity in CFA? How do you test this using factor loadings?

13.23 What is Cronbach's α? Why it is used? How do you compute it? Explain the guidelines for the use of Cronbach's α.

13.24 What is construct reliability? Why it is used? How do you compute it? Explain the guidelines for the use of construct reliability.

13.25 How do you use variance extracted to compute reliability in CFA? Explain the guidelines for its use.

13.26 What is model respecification in CFA? Respecification improves model fit and individual factor fit – explain how.

(B) Long conceptual and/or numerical questions: see web resources and Softwares used. (Multivariate Statistical Modeling in Engineering and Management – 1st (routledge.com))

N.B.: R and MS Excel software are used for computation, as and when required.

14 Structural Equation Modeling

Structural equation modeling (SEM) is probably the most versatile tool to capture multivariate relationships of several variables while maintaining the sequence of relationships in a structural framework, commensuration with the multivariate nature of the problem that mimics the reality. This makes SEM a complex statistical technique involving multiple equations. SEM comprises two multivariate models, namely measurement model and structural model. The measurement model is used to quantify the latent variables (factors) and the structural model is used to evaluate the relationships among the latent variables. The confirmatory factor model (Chapter13) and the path model (Chapter10) provide the necessary mathematical foundation for the measurement model and structural models, respectively. In this chapter, we describe the SEM in detail, starting from prerequisites, model conceptualization, model identification, parameter estimation, model adequacy tests, model respecification, and other important issues. Finally, a case study is elaborately discussed linking all the useful concepts and modeling strategies pertinent to the SEM.

14.1 PREREQUISITE AND MODELING STRATEGY

Structure equation modeling (SEM) is a complex multivariate modeling technique involving multiple equations that depict the structural relationships among variables of interest. SEM, in its complete form, consists of two kinds of multivariate models namely path model (Chapter 10) and confirmatory factor model (Chapter 13). The path model part and the confirmatory factor model part of the SEM are known as structural model, and measurement model, respectively. We use the terms structural model or path model, and confirmatory factor model or measurement model interchangeably in this chapter. The prerequisites[1] for SEM are shown in Figure 14.1. SEM is characterized by two groups of manifest variables, explanatory or independent group (***X***) and explained or dependent group (***Y***). Both the ***X*** and ***Y*** groups of manifest variables are caused by certain underlying factors (constructs), which are latent or unobservable. The latent factors that are quantified by the explanatory manifest variables (***X***) are termed as exogenous latent variables (ξ) and the latent factors, quantified by the explained manifest variables (***Y***) are termed as endogenous latent variables (η). The measurement of the exogenous latent variables (ξ) and the endogenous latent variables (η) follows the procedures of the confirmatory factor model. The dependence relationships of η and ξ are modeled following the procedures of the path model. The estimation of parameters may be done using a single stage (both measurement and structural model simultaneously) or using a two-stage process (first, estimation of measurement model parameters followed by the estimation of the structural model parameters).

Both the path model[2] and the confirmatory factor model explain the covariance structure of the variables of interest and SEM does the same. All these models are therefore grouped under covariance structure modeling. The other name of SEM is linear structural relation (LISREL), primarily due to the famous LISREL software (Jöreskog and Sorbom, 1988). From Figure 14.1, it can be said that researchers interested in SEM must have good knowledge of multiple linear regression (Chapter 8), path model (Chapter 10), and confirmatory factor model (Chapter 13).

Suppose there are p number of explanatory manifest variables (***X***) and q number of explained manifest variables (***Y***). Further, consider that there are m and n number of exogenous (ξ) and endogenous (η) latent variables that can be measured by the p number of ***X*** and q number of ***Y*** variables, respectively. Pictorially, we can represent this as in Figures 14.2(a) and 14.2(b). In addition, let the n number of η variables are dependent on the m number of ξ variables in a particular

DOI: 10.1201/9781003303060-17

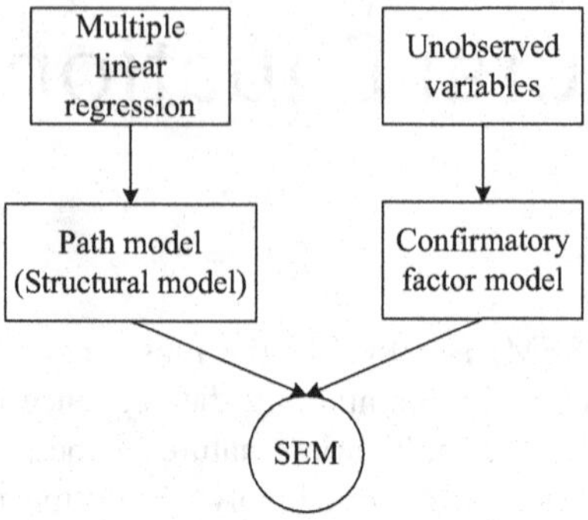

FIGURE 14.1 Prerequisite to structural equation modeling. Both confirmatory factor model and path model are integrated in SEM.

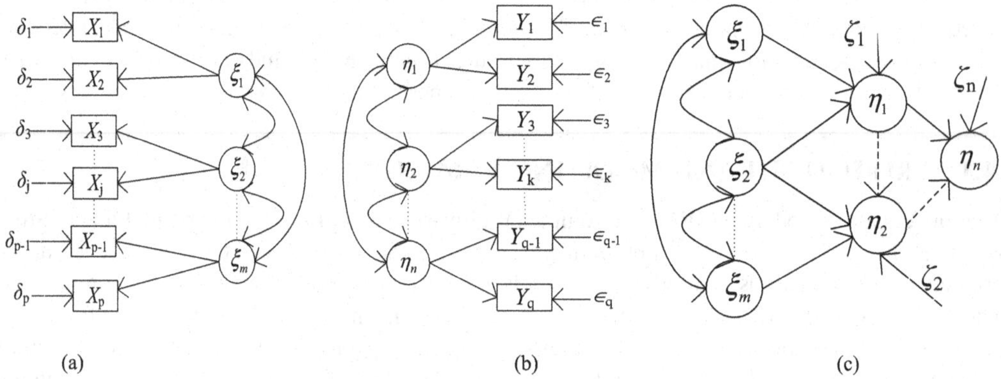

FIGURE 14.2 (a) A hypothetical confirmatory factor model for X, (b) A hypothetical confirmatory factor model for Y, and (c) A path model.

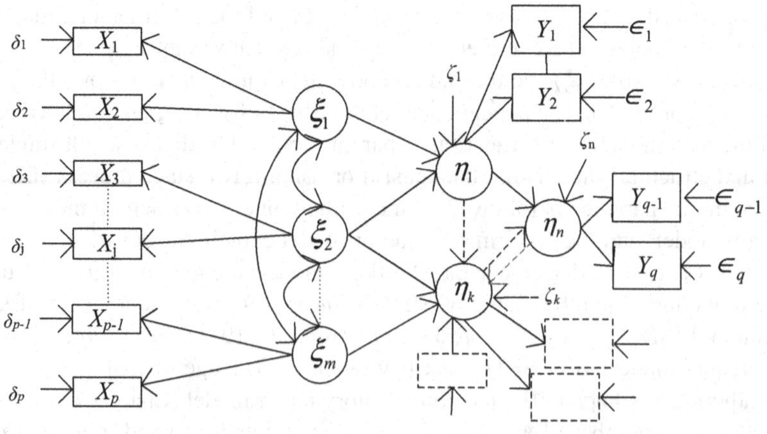

FIGURE. 14.3 A hypothetical SEM model with $p + q$ manifest variables and $m + n$ latent variables. For explanation of notations see Section 14.2. The SEM integrates the path model and the confirmatory factor model (see Figure 14.2) together.

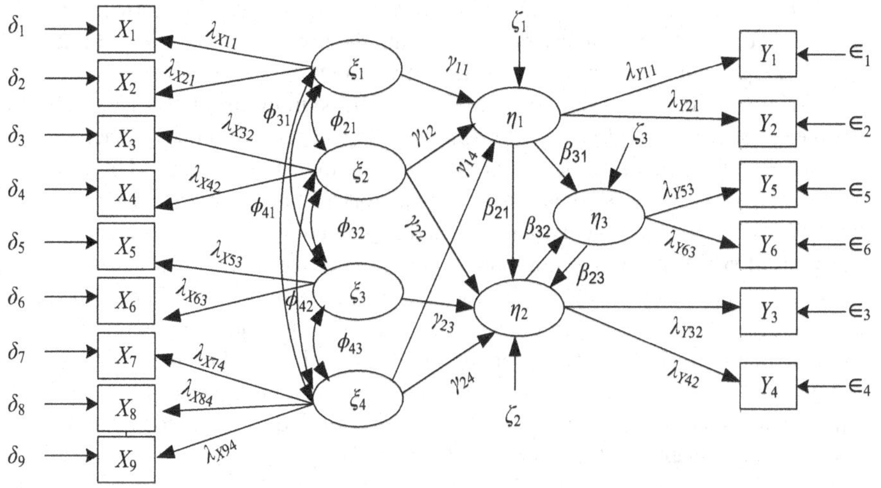

FIGURE 14.4 A hypothetical SEM model with $p = 9$, $q = 6$, $m = 4$, and $n = 3$.

Note: see Section 14.2.

structure (similar to the path model). Figure 14.2(c) shows one of many structural relations that can be thought of. The SEM combines three models into a single model as shown in Figure 14.3.

Let us further consider a SEM model representing a particular phenomenon, where a researcher identifies a finite set of $\boldsymbol{X}$ (say $p = 9$) and $\boldsymbol{Y}$ (say $q = 6$) variables and their respective underlying latent factors $\boldsymbol{\xi}$ (say $m = 4$) and $\boldsymbol{\eta}$ (say $n = 3$). Based on the theoretical and practical perspectives and due consideration with experts, she develops the SEM, as shown in Figure 14.4. The key questions she is interested to answer are given below.

1. Can the hypothesized factors (constructs) be adequately measured by the manifest variables (MVs)?
2. Is the given model able to confirm the hypothesized relationships among the latent factors?
3. Is the hypothesized model the best one or are there similar or better models? If so, how can she identify them?
4. Can she ignore the proposed model? How does she respecify the model for its improvement? What additional insights can she draw from of this improvement?

The questions under items 1 and 2 are answered to confirm the hypothesized SEM based on statistical significance. This strategy is known as confirmatory modeling strategy and much of the SEM applications are of this type. Questions under item 3 help the researcher to identify competing models and hence are known as competing models strategy. Hair et al. (1998, 2010) discussed several perspectives of competing models strategy such as equivalent models, TETRAD program (Glymour et al., 1987), and nested models. Finally, questions under item 4 help the researcher in respecifying the proposed model, both from the measurement of factors and their dependency modeling point of view. This strategy is called model development strategy. In this chapter, we seek answers to all these questions using the abovementioned three strategies.

14.2 VARIABLES, RELATIONSHIPS, TERMINOLOGIES, AND NOTATIONS

SEM is complex and has many variables with simultaneous relationships. It is essential to clearly define and categorize the variables and relationships. Several terminologies and notations are used for this purpose. These are listed below.

Latent variables: Latent means hidden or unobserved. In SEM, the latent variables (LVs) are the hidden factors or constructs that are measured using the manifest variables. In Figures 14.3 and 14.4, η and ξ represent the latent variables. Latent variables are often called constructs or factors.

Manifest variables: Manifest variables (MVs) are the symptoms of the LVs. The MVs are observables. The MVs are also known as indicators or items, particularly when questionnaire surveys are used for data collection. It is assumed that the MVs are caused by their respective LVs. In SEM, two types of MVs are used: explanatory (***X***) and explained (***Y***). In Figures 14.3 and 14.4, ***X*** and ***Y*** represent the explanatory and explained MVs, respectively.

Exogenous latent variables: Exogenous latent variables or exogenous constructs/factors are those which act as the predictors or the causes for other variables in the model. Their behavior is not modeled using other constructs/factors/variables; rather from a modeling point of view, they are independent in nature. The latent exogenous variables, such as ξ in Figure 14.4, are also known as exogenous constructs/factors. As the explanatory manifest variables (***X***) are used to measure/quantify the latent exogenous variables (ξ), ***X*** are also termed as exogenous manifest variables.

Endogenous latent variables: Endogenous latent variables or endogenous constructs/factors are those that act as outcomes, or dependent or explained variables (or constructs) in the system. Their behavior is modeled. In Figure 14.4, the η variables are endogenous. As the explained manifest variables (***Y***) quantify the η, ***Y*** are also termed as endogenous manifest variables.

Error or residual variables: In dependent relationships, the dependence cannot be completely explained by a set of predictors or causal variables. Even the behavior of a small system cannot be explained with exact values as there could be many uncontrollable variables affecting the behavior. Further, there could be errors in measurement and in sampling. In modeling dependence relationships, therefore, the effects of uncontrollable variables and noise or randomness are attributed to error variables, which are random in nature. The error variable is also called the residual variable as it quantifies the residual effect that is otherwise not specified/explained in the model. In SEM, there are three types of error variables and they are related to the endogenous latent variables (η), the explained manifest variables (***Y***), and the explanatory manifest variables (***X***). In Figures 14.3 and 14.4, these are denoted by ζ, ϵ, and δ, respectively. The error variables play an important role in model estimation, assessing overall fit of a model, and testing the significance of individual parameters of a model.

Several relationships are estimated and tested in SEM. One set of relationships arises from the structural (path) model part of SEM and another set arises from the CFA (measurement) part of SEM. The relationships captured in the path model and the CFA are explained in Chapter 10 and Chapter 13, respectively. For convenience, these are further discussed below in the context of SEM.

- **Casual relationships:** The relationship between one ξ variable to a η variable or one η variable to another η variable is assumed to be casual. It is anticipated that the dependent or endogenous latent variable η is caused by one or more exogenous latent variables ξ and/or other endogenous latent variables η. In the path model (Chapter 10), similar casual relationships are captured, with the only difference being that both endogenous and exogenous variables are directly observed. In Figures 14.3 and 14.4, the straight arrow line emanating from a ξ or a η variable and terminating at another η variable (with the arrowhead at the terminating η) represent a causal relationship. Another set of causation involves the latent constructs, ξ or η and their manifest variables, ***X*** or ***Y***, respectively. See straight arrow lines from ξ to ***X*** and η to ***Y***, respectively (in Figures 14.3 or 14.4). Although the fundamentals of CFA are based on the fact that the factors are responsible for the symptoms to occur, the "casual effect" term is not used in CFA (or in EFA). The factor loadings, which are basically the correlations between a factor and its indicators (or manifest variables), are used to denote this relationship.
- **Correlated relationships:** The covariance (or correlation) between any two exogenous latent variables (ξ) is allowed in the structural model of the SEM. The double-headed curved arrow

(see Figure 14.3 or 14.4) represents this covariance (or correlation). When evaluating the relationship of a particular ξ variable with a particular η variable, which are not casually related (not connected through straight arrows), a correlated relationship may exist between them because of the covariance (or correlation) of the concerned ξ variable with another ξ variable if the latter is casually connected with the concerned η variable. In Figure 14.4, for example, the relationship between ξ_1 and η_2 through ξ_2 is correlated in nature.

- **Direct and indirect relationships:** In the structural model of SEM, when an endogenous latent variable η is directly caused by an exogenous latent variable ξ or by another η (represented by a single arrow line between them), the relationship is direct causal relationship. In Figure 14.4, for example, the links between ξ_1 and η_1 or η_1 and η_2 or ξ_2 and η_2 represent the direct casual relationships. On the other hand, when the casual effect is achieved via another η, the relationship is indirect casual. For example, in Figure 14.4, one such indirect causal relationship is between ξ_1 and η_2 via η_1. If the intermediate variable is a ξ variable, then the relationship is indirect correlated relationship (e.g., the effect of ξ_3 and η_1 via ξ_2 in Figure 14.4). It can be stated that the casual effect of a latent variable (ξ or η) to another endogenous latent variable (η) may be decomposed into one or more direct as well as indirect casual and correlated relationships, known as direct and indirect effects. Total effect is the sum of direct and indirect effects. We will discuss this later. See Section 14.6.3.
- **Spurious relationships:** When the effect of one variable on the other (endogenous) is attributed to a third variable (common to both), which is not controlled for, the effect represents spurious relationship.

As SEM involves many relationships with different types of variables (latent exogenous, $\boldsymbol{\xi}$; latent endogenous, $\boldsymbol{\eta}$; explanatory manifest, $\boldsymbol{X}$; explained manifest, $\boldsymbol{Y}$; and error variables for $\boldsymbol{X}$, $\boldsymbol{Y}$, and $\boldsymbol{\eta}$), it uses a large number of notations to represent it mathematically. With the help of Figures 14.3 and 14.4, we describe them below. The notations are categorized into three groups: (i) variables, (ii) effect parameters (factor loadings and causal parameters), and (iii) variance-covariance (dispersion) matrices.

For a general SEM model, the notations are as follows:

- **Latent variables**

  $\boldsymbol{\eta}_{(n\times1)} = \left[\eta_1, \eta_2, \ldots, \eta_n\right]^T$ = $(n\times1)$ vector of latent endogenous variables (or constructs or factors). In Figure 14.4, $n = 3$. Note: n is the number of η variables in the model.

  $\boldsymbol{\xi}_{(m\times1)} = [\xi_1, \xi_2, \ldots, \xi_m]^T$ = $(m\times1)$ vector of latent exogenous variables (or constructs or factors). In Figure 14.4, $m = 4$. Note: m is the number of ξ variables in the model.
- **Manifest variables**

  $\boldsymbol{Y}_{(q\times1)} = [Y_1, Y_2, Y_3, \ldots, Y_q]^T$ = $(q\times1)$ vector of explained manifest variables. In Figure 14.4, $q = 6$. Note: q is the number of Y variables in the model.

  $\boldsymbol{X}_{(p\times1)} = [X_1, X_2, X_3, \ldots, X_p]^T$ = $(p\times1)$ vector of explanatory manifest variables. In Figure 14.4, $p = 9$. Note: p is the number of X variables in the model.
- **Error variables**

  $\boldsymbol{\epsilon}_{(q\times1)} = [\epsilon_1, \epsilon_2, \epsilon_3, \ldots, \epsilon_q]^T$ = $(q\times1)$ vector of error variables for $\boldsymbol{Y}$. In Figure 14.4, $q = 6$.

  $\boldsymbol{\delta}_{(p\times1)} = [\delta_1, \delta_2, \delta_3, \ldots, \delta_p]^T$ = $(p\times1)$ vector of error variables for $\boldsymbol{X}$. In Figure 14.4, $p = 9$.

  $\boldsymbol{\zeta}_{(n\times1)} = [\zeta_1, \zeta_2, \zeta_3, \ldots, \zeta_n]^T$ = $(n\times1)$ vector of error variables for $\boldsymbol{\eta}$. In Figure 14.4, $n = 3$.
- **Factor loadings**

  $\lambda_{X_{ik}}$ = The k-th exogenous factor (ξ_k) loading on the i-th explanatory manifest variable (X_i), where $i = 1, 2, \ldots, p$, and $k = 1, 2, \ldots, m$.

  $\lambda_{Y_{jl}}$ = The l-th endogenous factor (η_l) loading on the j-th explained manifest variable (Y_j), where $j = 1, 2, \ldots, q$, and $l = 1, 2, \ldots, n$.

  $\boldsymbol{\Lambda}_{X_{(p\times m)}}$ = $(p \times m)$ matrix of factor loadings involving $\boldsymbol{\xi}$ and $\boldsymbol{X}$. In Figure 14.4, $\boldsymbol{\Lambda}_X$ is (9×4) matrix.

$\Lambda_{Y_{(q \times n)}}$ = $(q \times n)$ matrix of factor loadings involving η and Y. In Figure 14.4, Λ_Y is (6×3) matrix.

- **Causal parameters**

  γ_{lk} = The path coefficient between the l-th η (i.e., η_l) and the k-th ξ (i.e., ξ_k), showing the influence of ξ_k on η_l; $k = 1, 2, \ldots, m$ and $l = 1, 2, 3, \ldots, n$.

  $\beta_{ll'}$ = The path coefficient between the l-th η (i.e., η_l) and l'-th η (i.e., $\eta_{l'}$), showing the influence of $\eta_{l'}$ on η_l; $l, l' = 1, 2, \ldots, n$ and $l \neq l'$.

  $\mathbf{\Gamma}_{(n \times m)}$ = $(n \times m)$ matrix of path coefficients for the effects of $\boldsymbol{\xi}$ on $\boldsymbol{\eta}$. In Figure 14.4, $\mathbf{\Gamma}$ is a (3×4) matrix.

  $\boldsymbol{\beta}_{(n \times n)}$ = $(n \times n)$ matrix of path coefficients showing the influences of endogenous latent variables $\boldsymbol{\eta}$ on each other. In Figure 14.4, $\boldsymbol{\beta}$ is a (3×3) matrix.

- **Dispersion matrices**

  In addition, SEM uses the variance-covariance matrices of the latent exogenous variables ($\boldsymbol{\xi}$) and the error variables, $\boldsymbol{\zeta}$, $\boldsymbol{\epsilon}$, and $\boldsymbol{\delta}$, as given below.

  $\mathbf{\Phi}_{(m \times m)}$ = $(m \times m)$ variance-covariance matrix of the m latent exogenous variables ($\boldsymbol{\xi}_{(m \times 1)}$). In Figure 14.4, $\mathbf{\Phi}$ is a (4×4) matrix.

  $\mathbf{\Psi}_{(n \times n)}$ = $(n \times n)$ variance-covariance matrix of $\boldsymbol{\zeta}_{(n \times 1)}$. For Figure 14.4, $\mathbf{\Psi}$ is a (3×3) matrix.

  $\boldsymbol{\theta}_{\epsilon_{(q \times q)}}$ = $(q \times q)$ variance-covariance matrix of $\boldsymbol{\epsilon}_{(q \times 1)}$. For Figure 14.4, $\boldsymbol{\theta}_\epsilon$ is a (6×6) matrix.

  $\boldsymbol{\theta}_{\delta_{(p \times p)}}$ = $(p \times p)$ variance-covariance matrix of $\boldsymbol{\delta}_{(p \times 1)}$. For Figure 14.4, $\boldsymbol{\theta}_\delta$ is a (9×9) matrix.

14.3 MODEL EQUATIONS AND DISPERSION MATRICES

As stated earlier, SEM encompasses three different sub-models, the central path model and two factor models. The central path model is called the structural model and the two factor models together are called the measurement model. So, SEM involves three sets of equations, also known as systems of equations, one set for each sub-model.

14.3.1 Structural Model Equations

Consider Figure 14.4. The structural model variables are η_1, η_2, and η_3, representing the endogenous factors, and ξ_1, ξ_2, ξ_3, and ξ_4, representing the exogenous factors. The regression equations for the structural model of the SEM in Figure 14.4 are as follows:

$$\begin{aligned} \eta_1 &= \gamma_{11}\xi_1 + \gamma_{12}\xi_2 + \gamma_{14}\xi_4 + \zeta_1 \\ \eta_2 &= \beta_{21}\eta_1 + \beta_{23}\eta_3 + \gamma_{22}\xi_2 + \gamma_{23}\xi_3 + \gamma_{24}\xi_4 + \zeta_2 \\ \eta_3 &= \beta_{31}\eta_1 + \beta_{32}\eta_2 + \zeta_3 \end{aligned} \tag{14.1}$$

In matrix form

$$\begin{bmatrix} \eta_1 \\ \eta_2 \\ \eta_3 \end{bmatrix} = \begin{bmatrix} 0 & 0 & 0 \\ \beta_{21} & 0 & \beta_{23} \\ \beta_{31} & \beta_{32} & 0 \end{bmatrix} \begin{bmatrix} \eta_1 \\ \eta_2 \\ \eta_3 \end{bmatrix} + \begin{bmatrix} \gamma_{11} & \gamma_{12} & 0 & \gamma_{14} \\ 0 & \gamma_{22} & \gamma_{23} & \gamma_{24} \\ 0 & 0 & 0 & 0 \end{bmatrix} \begin{bmatrix} \xi_1 \\ \xi_2 \\ \xi_3 \\ \xi_4 \end{bmatrix} + \begin{bmatrix} \zeta_1 \\ \zeta_2 \\ \zeta_3 \end{bmatrix}$$

or

$$\boldsymbol{\eta}_{3\times1} = \boldsymbol{\beta}_{3\times3}\boldsymbol{\eta}_{3\times1} + \mathbf{\Gamma}_{3\times4}\boldsymbol{\xi}_{4\times1} + \boldsymbol{\zeta}_{3\times1} \tag{14.2}$$

For general SEM, Equation 14.2 becomes

$$\begin{aligned}&\eta_{n\times1}=\boldsymbol{\beta}_{n\times n}\eta_{n\times1}+\boldsymbol{\Gamma}_{n\times m}\xi_{m\times1}+\zeta_{n\times1}\\ \text{or}\quad&(\mathbf{I}-\boldsymbol{\beta})_{n\times n}\eta_{n\times1}=\boldsymbol{\Gamma}_{n\times m}\xi_{m\times1}+\zeta_{n\times1}\end{aligned} \tag{14.3}$$

14.3.2 Exogenous Factor Model Equations

Consider Figure 14.4. There are two factor models: one in the left (involving $\boldsymbol{\xi}$ and $\boldsymbol{X}$) and another in the right (involving $\boldsymbol{\eta}$ and $\boldsymbol{Y}$). The factor model involving $\boldsymbol{\xi}$ and $\boldsymbol{X}$ is called exogenous factor model of the SEM. Following the procedure given in Chapter 13 (Confirmatory factor model), the system of equations for Figure 14.4 are

$$\begin{aligned}X_1&=\lambda_{X_{11}}\xi_1+\delta_1\\ X_2&=\lambda_{X_{21}}\xi_1+\delta_2\\ X_3&=\lambda_{X_{32}}\xi_2+\delta_3\\ X_4&=\lambda_{X_{42}}\xi_2+\delta_4\\ X_5&=\lambda_{X_{53}}\xi_3+\delta_5\\ X_6&=\lambda_{X_{63}}\xi_3+\delta_6\\ X_7&=\lambda_{X_{74}}\xi_4+\delta_7\\ X_8&=\lambda_{X_{84}}\xi_4+\delta_8\\ X_9&=\lambda_{X_{94}}\xi_4+\delta_9\end{aligned} \tag{14.4}$$

In matrix form

$$\boldsymbol{X}_{9\times1}=\boldsymbol{\Lambda}_{X\,(9\times4)}\,\boldsymbol{\xi}_{4\times1}+\boldsymbol{\delta}_{9\times1} \tag{14.5}$$

For general SEM, Equation 14.5 becomes

$$\boldsymbol{X}_{p\times1}=\boldsymbol{\Lambda}_{X\,(p\times m)}\,\boldsymbol{\xi}_{m\times1}+\boldsymbol{\delta}_{p\times1} \tag{14.6}$$

14.3.3 Endogenous Factor Model Equations

Consider Figure 14.4 again. The factor model in the right side of Figure 14.4 (involving $\boldsymbol{\eta}$ and $\boldsymbol{Y}$) involves the following equations:

$$\begin{aligned}Y_1&=\lambda_{Y_{11}}\eta_1+\epsilon_1\\ Y_2&=\lambda_{Y_{21}}\eta_1+\epsilon_2\\ Y_3&=\lambda_{Y_{32}}\eta_2+\epsilon_3\\ Y_4&=\lambda_{Y_{42}}\eta_2+\epsilon_4\\ Y_5&=\lambda_{Y_{53}}\eta_3+\epsilon_5\\ Y_6&=\lambda_{Y_{63}}\eta_3+\epsilon_6\end{aligned} \tag{14.7}$$

In matrix form

$$\boldsymbol{Y}_{6\times1}=\boldsymbol{\Lambda}_{Y(6\times3)}\boldsymbol{\eta}_{3\times1}+\boldsymbol{\epsilon}_{6\times1} \tag{14.8}$$

For general SEM, Equation 14.8 becomes

$$\boldsymbol{Y}_{q\times1} = \boldsymbol{\Lambda}_{Y(q\times n)}\eta_{n\times1} + \boldsymbol{\epsilon}_{q\times1} \tag{14.9}$$

14.3.4 DISPERSION MATRICES

Equations 14.3, 14.6, and 14.9 are the three fundamental equations of the SEM whose parameters are to be estimated (will be discussed later). Other important parameters to be estimated are the elements of the dispersion matrices $\boldsymbol{\Phi}_{m\times m}, \boldsymbol{\Psi}_{n\times n}, \boldsymbol{\theta}_{\epsilon_{q\times q}}$ and $\boldsymbol{\theta}_{\delta_{p\times p}}$ (see Section 14.2 for definitions). For Figure 14.4, these matrices are

$$\boldsymbol{\Phi}_{4\times4} = \begin{bmatrix} \phi_{11} & & & \\ \phi_{21} & \phi_{22} & & \\ \phi_{31} & \phi_{32} & \phi_{33} & \\ \phi_{41} & \phi_{42} & \phi_{43} & \phi_{44} \end{bmatrix} \qquad \boldsymbol{\Psi}_{3\times3} = \begin{bmatrix} \psi_{11} & 0 & 0 \\ 0 & \psi_{22} & 0 \\ 0 & 0 & \psi_{33} \end{bmatrix}$$

$$\boldsymbol{\theta}_{\epsilon_{6\times6}} = \begin{bmatrix} \theta_{\epsilon_{11}} & 0 & 0 & 0 & 0 & 0 \\ 0 & \theta_{\epsilon_{22}} & 0 & 0 & 0 & 0 \\ 0 & 0 & \theta_{\epsilon_{33}} & 0 & 0 & 0 \\ 0 & 0 & 0 & \theta_{\epsilon_{44}} & 0 & 0 \\ 0 & 0 & 0 & 0 & \theta_{\epsilon_{55}} & 0 \\ 0 & 0 & 0 & 0 & 0 & \theta_{\epsilon_{66}} \end{bmatrix}$$

and

$$\boldsymbol{\theta}_{\delta_{9\times9}} = \begin{bmatrix} \theta_{\delta_{11}} & 0 & 0 & 0 & 0 & 0 & 0 & 0 & 0 \\ 0 & \theta_{\delta_{22}} & 0 & 0 & 0 & 0 & 0 & 0 & 0 \\ 0 & 0 & \theta_{\delta_{33}} & 0 & 0 & 0 & 0 & 0 & 0 \\ 0 & 0 & 0 & \theta_{\delta_{44}} & 0 & 0 & 0 & 0 & 0 \\ 0 & 0 & 0 & 0 & \theta_{\delta_{55}} & 0 & 0 & 0 & 0 \\ 0 & 0 & 0 & 0 & 0 & \theta_{\delta_{66}} & 0 & 0 & 0 \\ 0 & 0 & 0 & 0 & 0 & 0 & \theta_{\delta_{77}} & 0 & 0 \\ 0 & 0 & 0 & 0 & 0 & 0 & 0 & \theta_{\delta_{88}} & 0 \\ 0 & 0 & 0 & 0 & 0 & 0 & 0 & 0 & \theta_{\delta_{99}} \end{bmatrix}$$

Note that for the SEM, given in Figure 14.4, $\boldsymbol{\Psi}_{3\times3}, \boldsymbol{\theta}_{\epsilon_{6\times6}}$ and $\boldsymbol{\theta}_{\delta_{9\times9}}$ are diagonal matrix. Although it is desirable, it may not always happen. If we allow correlation among the error variables, these matrices will have non-zero off-diagonal elements. For example, if we allow correlation between δ_6 and δ_9, the corresponding element in $\boldsymbol{\theta}_\delta$ will be $\theta_{\delta_{69}}$ or $\theta_{\delta_{96}}$.

14.4 ASSUMPTIONS

All the assumptions of the path model and confirmatory factor model are valid in the SEM. Important assumptions are (Krzanowski and Marriott, 2014b):

- All the error variables $\in$, δ, and ζ have zero mean vector, uncorrelated with each other and with other random variables in the model.
- Often the error variables are multivariate normal. This is important for maximum likelihood estimation (MLE).
- In the equation $(\mathbf{I}-\boldsymbol{\beta})\eta = \boldsymbol{\Gamma}\xi + \xi$ (Equation 14.3), the matrix $(\mathbf{I}-\boldsymbol{\beta})$ is non-singular.
- η, ξ, X, and Y are measured about their means.
- All casual relationships are considered only in the path model and they are linear. Note that advanced SEM can accommodate non-linear relationships.

14.5 FUNDAMENTAL RELATIONSHIPS

Like factor analysis, SEM does explain the covariance structure of the variables comprising the model. It not only is capable of modeling the covariance relationships among the latent variables (ξ and η) participating in the structural model but also of modeling the covariance relationships among the manifest variables (X and Y) participating in the measurement model, with the help of the model parameters. So, the covariances can be represented by the parameters of the SEM model. These relationships are called the fundamental relationships in SEM.

14.5.1 Fundamental Relationships for the Structural Model

Let $\boldsymbol{\Sigma}_{LV}$ represent the $(m+n)\times(m+n)$ covariance matrix of the $m+n$ latent variables (factors) represented by $\xi_{m\times 1}$ and $\eta_{n\times 1}$. By partitioning $\boldsymbol{\Sigma}_{LV}$, we can write $\boldsymbol{\Sigma}_{LV} = \begin{bmatrix} \boldsymbol{\Sigma}_{\eta\eta^T} & \boldsymbol{\Sigma}_{\eta\xi^T} \\ \boldsymbol{\Sigma}_{\xi\eta^T} & \boldsymbol{\Sigma}_{\xi\xi^T} \end{bmatrix} = \begin{bmatrix} E(\eta\eta^T) & E(\eta\xi^T) \\ E(\xi\eta^T) & E(\xi\xi^T) \end{bmatrix}$, E stands for the expected value.

With little algebraic manipulation, Equation 14.3 can be written as

$$\eta = (\mathbf{I}-\boldsymbol{\beta})^{-1}\boldsymbol{\Gamma}\xi + (\mathbf{I}-\boldsymbol{\beta})^{-1}\zeta \tag{14.10}$$

Following the assumptions of the SEM (see Section 14.4), $E(\xi\zeta^T) = E(\zeta\xi^T) = \mathbf{0}$ and $\mathbf{I}-\boldsymbol{\beta}$ is non-singular, $\eta\eta^T = \left[(\mathbf{I}-\boldsymbol{\beta})^{-1}\boldsymbol{\Gamma}\xi + (\mathbf{I}-\boldsymbol{\beta})^{-1}\zeta\right]\left[(\mathbf{I}-\boldsymbol{\beta})^{-1}\boldsymbol{\Gamma}\xi + (\mathbf{I}-\boldsymbol{\beta})^{-1}\zeta\right]^T$, here T stands for transpose.

Let $(\mathbf{I}-\boldsymbol{\beta})^{-1} = \mathbf{D}$, then

$$\begin{aligned}
\eta\eta^T &= (\mathbf{D}\,\boldsymbol{\Gamma}\xi + \mathbf{D}\zeta)(\mathbf{D}\,\boldsymbol{\Gamma}\xi + \mathbf{D}\zeta)^T \\
&= (\mathbf{D}\,\boldsymbol{\Gamma}\xi + \mathbf{D}\zeta)\xi^T\boldsymbol{\Gamma}^T\mathbf{D}^T + (\mathbf{D}\,\boldsymbol{\Gamma}\xi + \mathbf{D}\zeta)\zeta^T\mathbf{D}^T \\
&= \mathbf{D}\boldsymbol{\Gamma}\xi\xi^T\boldsymbol{\Gamma}^T\mathbf{D}^T + \mathbf{D}\zeta\xi^T\boldsymbol{\Gamma}^T\mathbf{D}^T + \mathbf{D}\boldsymbol{\Gamma}\xi\zeta^T\mathbf{D}^T + \mathbf{D}\zeta\zeta^T\mathbf{D}^T
\end{aligned}$$

Taking expectation

$$\begin{aligned}
E(\eta\eta^T) &= E(\mathbf{D}\boldsymbol{\Gamma}\xi\xi^T\boldsymbol{\Gamma}^T\mathbf{D}^T) + E(\mathbf{D}\zeta\xi^T\boldsymbol{\Gamma}^T\mathbf{D}^T) + E(\mathbf{D}\boldsymbol{\Gamma}\xi\zeta^T\mathbf{D}^T) + E(\mathbf{D}\zeta\zeta^T\mathbf{D}^T) \\
&= \mathbf{D}\boldsymbol{\Gamma}E(\xi\xi^T)\boldsymbol{\Gamma}^T\mathbf{D}^T + \mathbf{D}E(\zeta\xi^T)\boldsymbol{\Gamma}^T\mathbf{D}^T + \mathbf{D}\boldsymbol{\Gamma}E(\xi\zeta^T)\mathbf{D}^T + \mathbf{D}E(\zeta\zeta^T)\mathbf{D}^T \\
&= \mathbf{D}\boldsymbol{\Gamma}\boldsymbol{\Phi}\boldsymbol{\Gamma}^T\mathbf{D}^T + \mathbf{0} + \mathbf{0} + \mathbf{D}\boldsymbol{\Psi}\mathbf{D}^T \quad [\text{as } E(\xi\zeta^T) = E(\zeta\xi^T) = \mathbf{0}] \\
&= \mathbf{D}\boldsymbol{\Gamma}\boldsymbol{\Phi}\boldsymbol{\Gamma}^T\mathbf{D}^T + \mathbf{D}\boldsymbol{\Psi}\mathbf{D}^T \\
&= \mathbf{D}(\boldsymbol{\Gamma}\boldsymbol{\Phi}\boldsymbol{\Gamma}^T + \boldsymbol{\Psi})\mathbf{D}^T
\end{aligned}$$

Now, putting $\mathbf{D} = (\mathbf{I}-\boldsymbol{\beta})^{-1}$ in the above equation we get,

$$E(\eta\eta^T) = (\mathbf{I}-\boldsymbol{\beta})^{-1}(\boldsymbol{\Gamma}\boldsymbol{\Phi}\boldsymbol{\Gamma}^T + \boldsymbol{\Psi})(\mathbf{I}-\boldsymbol{\beta})^{-1^T} \tag{14.11}$$

Similarly,

$$\eta\xi^T = (\mathbf{D\Gamma}\xi + \mathbf{D}\zeta)\xi^T = \mathbf{D\Gamma}\xi\xi^T + \mathbf{D}\zeta\xi^T$$

$$\begin{aligned}\therefore E(\eta\xi^T) &= E(\mathbf{D\Gamma}\xi\xi^T) + E(\mathbf{D}\zeta\xi^T)\\ &= \mathbf{D\Gamma}E(\xi\xi^T) + \mathbf{D}E(\zeta\xi^T)\\ &= \mathbf{D\Gamma\Phi} + \mathbf{0} \qquad \text{as } E(\zeta\xi^T) = \mathbf{0}\\ &= (\mathbf{I}-\boldsymbol{\beta})^{-1}\mathbf{\Gamma\Phi}\end{aligned} \tag{14.12}$$

and,

$$E(\xi\eta^T) = E(\xi\xi^T)\mathbf{\Gamma}^T\mathbf{D}^T + E(\xi\zeta^T)\mathbf{D}^T = \mathbf{\Phi\Gamma}^T(\mathbf{I}-\boldsymbol{\beta})^{-1^T} \tag{14.13}$$

So,

$$\mathbf{\Sigma}_{LV} = \begin{bmatrix}(\mathbf{I}-\boldsymbol{\beta})^{-1}(\mathbf{\Gamma\Phi\Gamma}^T+\mathbf{\Psi})(\mathbf{I}-\boldsymbol{\beta})^{-1^T} & (\mathbf{I}-\boldsymbol{\beta})^{-1}\mathbf{\Gamma\Phi}\\ \mathbf{\Phi\Gamma}^T(\mathbf{I}-\boldsymbol{\beta})^{-1^T} & \mathbf{\Phi}\end{bmatrix} \tag{14.14}$$

Example 14.1

Consider the following structural model (see Figure 14.5) and obtain $\mathbf{\Sigma}_{LV}$.

Solution:

Here the structural equation is

$$\eta_1 = \gamma_{11}\xi_1 + \gamma_{12}\xi_2 + \zeta_1$$

It has one η variable and two $\boldsymbol{\xi}$ variables. So, $\mathbf{\Sigma}_{LV}$ is a 3×3 matrix.

Following the procedure given above,

$$\mathbf{\Sigma}_{LV} = \begin{bmatrix}E(\eta\eta^T) & E(\eta\xi^T)\\ E(\xi\eta^T) & E(\xi\xi^T)\end{bmatrix}$$

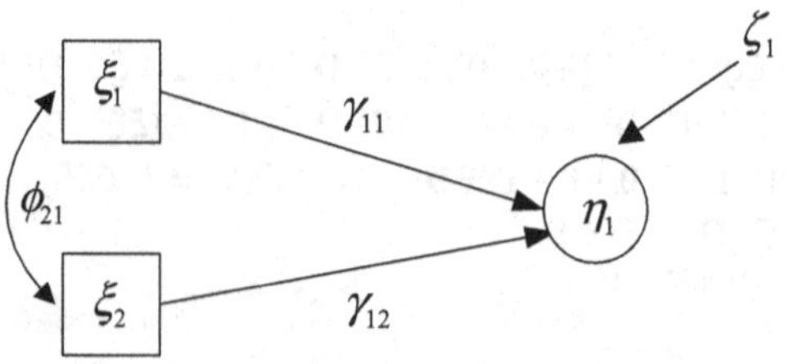

FIGURE 14.5 The structural model for Example 14.1. In the figure, η_1 represents the endogenous latent variable and $\xi = [\xi_1\ \xi_2]^T$ represents the exogenous latent variables. γ represents the path coefficient, ζ_1 represents the error variable, and ϕ represents covariance between exogenous latent variables.

(i) $\boldsymbol{\eta}\boldsymbol{\eta}^T = \eta_1^2 = (\gamma_{11}\xi_1 + \gamma_{12}\xi_2 + \zeta_1)^2$ [as there is only one n in the equation]

$$= \gamma_{11}^2\xi_1^2 + \gamma_{12}^2\xi_2^2 + \zeta_1^2 + 2\gamma_{11}\gamma_{12}\xi_1\xi_2 + 2\gamma_{11}\xi_1\zeta_1 + 2\gamma_{12}\xi_2\zeta_1$$

$$\therefore E(\boldsymbol{\eta}\boldsymbol{\eta}^T) = E(\eta_1^2) = \gamma_{11}^2E(\xi_1^2) + \gamma_{12}^2E(\xi_2^2) + \mathrm{E}(\zeta_1^2) + 2\gamma_{11}\gamma_{12}E(\xi_1\xi_2) + 2\gamma_{11}E(\xi_1\zeta_1) + 2\gamma_{12}E(\xi_2\zeta_1)$$

$$= \gamma_{11}^2\phi_{11} + \gamma_{12}^2\phi_{22} + \psi_{11} + 2\gamma_{11}\gamma_{12}\phi_{21} \text{ [as other terms vanish]}$$

(ii) $\boldsymbol{\eta}\boldsymbol{\xi}^T = \underset{1\times1}{\eta_1}\begin{bmatrix}\xi_1 & \xi_2\end{bmatrix}_{1\times2} = \begin{bmatrix}\eta_1\xi_1 & \eta_1\xi_2\end{bmatrix}$

So, $E(\boldsymbol{\eta}\boldsymbol{\xi}^T) = \begin{bmatrix}E(\eta_1\xi_1) & E(\eta_1\xi_2)\end{bmatrix}$

$$E(\eta_1\xi_1) = \mathrm{E}(\gamma_{11}\xi_1^2 + \gamma_{12}\xi_2\xi_1 + \zeta_1\xi_1) = \gamma_{11}E(\xi_1^2) + \gamma_{12}E(\xi_2\xi_1) + E(\zeta_1\xi_1)$$

$$= \gamma_{11}\phi_{11} + \gamma_{12}\phi_{21} + 0 = \gamma_{11}\phi_{11} + \gamma_{12}\phi_{21}$$

Similarly,

$$E(\eta_1\xi_2) = \mathrm{E}(\gamma_{11}\xi_1\xi_2 + \gamma_{12}\xi_2^2 + \zeta_1\xi_2) = \gamma_{11}\phi_{21} + \gamma_{12}\phi_{22} + 0 \quad \text{(assuming } \phi_{21} = \phi_{12}\text{)}$$

$$= \gamma_{11}\phi_{21} + \gamma_{12}\phi_{22}$$

So, $E(\eta_1\boldsymbol{\xi}^T) = \begin{bmatrix}\gamma_{11}\phi_{11} + \gamma_{12}\phi_{21} & \gamma_{11}\phi_{21} + \gamma_{12}\phi_{22}\end{bmatrix}$

(iii) $\boldsymbol{\xi}\boldsymbol{\eta}^T = \boldsymbol{\xi}\eta_1 = \begin{bmatrix}\xi_1 \\ \xi_2\end{bmatrix}\eta_1 = \begin{bmatrix}\xi_1\eta_1 \\ \xi_2\eta_1\end{bmatrix}$

So, $\mathrm{E}(\boldsymbol{\xi}\boldsymbol{\eta}^T) = \begin{bmatrix}\mathrm{E}(\xi_1\eta_1) \\ \mathrm{E}(\xi_2\eta_1)\end{bmatrix}$

Now,

$$E(\xi_1\eta_1) = \mathrm{E}(\gamma_{11}\xi_1^2 + \gamma_{12}\xi_1\xi_2 + \xi_1\zeta_1) = \gamma_{11}\phi_{11} + \gamma_{12}\phi_{21} + 0 = \gamma_{11}\phi_{11} + \gamma_{12}\phi_{21}$$

Similarly, $E(\xi_2\eta_1) = \gamma_{11}\phi_{21} + \gamma_{12}\phi_{22}$

So, $\mathrm{E}(\boldsymbol{\xi}\boldsymbol{\eta}^T) = \begin{bmatrix}\gamma_{11}\phi_{11} + \gamma_{12}\phi_{21} \\ \gamma_{11}\phi_{21} + \gamma_{12}\phi_{22}\end{bmatrix}$

(iv) $\boldsymbol{\xi}\boldsymbol{\xi}^T = \begin{bmatrix}\xi_1 \\ \xi_2\end{bmatrix}_{2\times1}\begin{bmatrix}\xi_1 & \xi_2\end{bmatrix}_{1\times2} = \begin{bmatrix}\xi_1^2 & \xi_1\xi_2 \\ \xi_2\xi_1 & \xi_2^2\end{bmatrix}$

$$\therefore E(\boldsymbol{\xi}\boldsymbol{\xi}^T) = \begin{bmatrix}E(\xi_1^2) & E(\xi_1\xi_2) \\ E(\xi_2\xi_1) & E(\xi_2^2)\end{bmatrix}$$

$$= \begin{bmatrix}\phi_{11} & \phi_{21} \\ \phi_{21} & \phi_{22}\end{bmatrix}$$

So, $\boldsymbol{\Sigma}_{LV}$ is

$$\boldsymbol{\Sigma}_{3\times3} = \begin{bmatrix}\gamma_{11}^2\phi_{11} + \gamma_{12}^2\phi_{22} + 2\gamma_{11}\gamma_{12}\phi_{21} + \psi_{11} & \gamma_{11}\phi_{11} + \gamma_{12}\phi_{21} & \gamma_{11}\phi_{21} + \gamma_{12}\phi_{22} \\ \gamma_{11}\phi_{11} + \gamma_{12}\phi_{21} & \phi_{11} & \phi_{21} \\ \gamma_{11}\phi_{21} + \gamma_{12}\phi_{22} & \phi_{21} & \phi_{22}\end{bmatrix}$$

Example 14.2

Consider the following structural model (see Figure 14.6) and obtain Σ_{LV}.

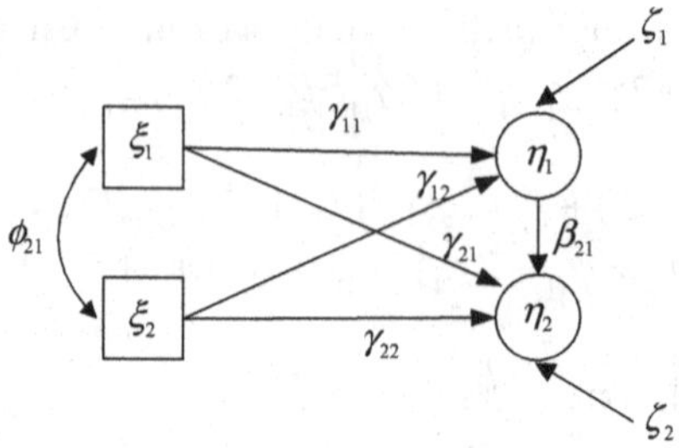

FIGURE 14.6 The structural model for Example 14.2. In the figure, $\boldsymbol{\eta} = [\eta_1\ \eta_2]^T$ represents the endogenous latent variables and $\boldsymbol{\xi} = [\xi_1\ \xi_2]^T$ represents the exogenous latent variables. β_{21} is the path coefficient for the relationship between η_1 and η_2. Each γ represents a path coefficient for the relationships between an exogenous latent variable (ξ) and an endogenous latent variable (η). $\zeta = [\zeta_1\ \zeta_2]^T$ represents the error vector. ϕ_{21} represents covariance between ξ_1 and ξ_2.

Solution:

The path equations are

$$\eta_1 = \gamma_{11}\xi_1 + \gamma_{12}\xi_2 + \zeta_1$$

$$\eta_2 = \beta_{21}\eta_1 + \gamma_{21}\xi_1 + \gamma_{22}\xi_2 + \zeta_2$$

Here, $\boldsymbol{\eta} = \begin{bmatrix} \eta_1 \\ \eta_2 \end{bmatrix}$, $\boldsymbol{\xi} = \begin{bmatrix} \xi_1 \\ \xi_2 \end{bmatrix}$ and $\boldsymbol{\zeta} = \begin{bmatrix} \zeta_1 \\ \zeta_2 \end{bmatrix}$

From Equation 14.3, we know,

$$\boldsymbol{\eta} = \boldsymbol{\beta\eta} + \boldsymbol{\Gamma\xi} + \boldsymbol{\zeta}$$

Here, $\boldsymbol{\beta}_{2\times 2} = \begin{bmatrix} 0 & 0 \\ 0 & \beta_{21} \end{bmatrix}$ and $\boldsymbol{\Gamma} = \begin{bmatrix} \gamma_{11} & \gamma_{12} \\ \gamma_{21} & \gamma_{22} \end{bmatrix}$

$$\therefore \mathbf{I} - \boldsymbol{\beta} = \begin{bmatrix} 1 & 0 \\ 0 & 1 \end{bmatrix} - \begin{bmatrix} 0 & 0 \\ 0 & \beta_{21} \end{bmatrix} = \begin{bmatrix} 1 & 0 \\ 0 & 1-\beta_{21} \end{bmatrix}$$

$$\therefore (\mathbf{I} - \boldsymbol{\beta})^{-1} = \frac{1}{|\mathbf{I} - \boldsymbol{\beta}|}\,\text{adjoint}(\mathbf{I} - \boldsymbol{\beta}) = \frac{1}{1-\beta_{21}}\begin{bmatrix} 1-\beta_{21} & 0 \\ 0 & 1 \end{bmatrix} = \begin{bmatrix} 1 & 0 \\ 0 & \dfrac{1}{1-\beta_{21}} \end{bmatrix}$$

Using Equation 14.14, $\boldsymbol{\Sigma}$ is

$$\Sigma = \begin{bmatrix} \boldsymbol{\Sigma}_{\eta\eta^T} & \boldsymbol{\Sigma}_{\eta\xi^T} \\ \boldsymbol{\Sigma}_{\xi\eta^T} & \boldsymbol{\Sigma}_{\xi\xi^T} \end{bmatrix} = \begin{bmatrix} (\mathbf{I}-\boldsymbol{\beta})^{-1}(\boldsymbol{\Gamma\Phi\Gamma}^T + \boldsymbol{\Psi})(\mathbf{I}-\boldsymbol{\beta})^{-1^T} & (\mathbf{I}-\boldsymbol{\beta})^{-1}\boldsymbol{\Gamma\Phi} \\ \boldsymbol{\Phi\Gamma}^T(\mathbf{I}-\boldsymbol{\beta})^{-1^T} & \boldsymbol{\Phi} \end{bmatrix}$$

Further, $\mathbf{\Phi} = \begin{bmatrix} \phi_{11} & \phi_{21} \\ \phi_{21} & \phi_{22} \end{bmatrix}$, and $\mathbf{\Psi} = \begin{bmatrix} \psi_{11} & 0 \\ 0 & \psi_{22} \end{bmatrix}$

(i) Find out $\mathbf{\Sigma}_{\eta\eta^T}$

$$\mathbf{\Gamma\Phi\Gamma}^T + \mathbf{\Psi} = \begin{bmatrix} \gamma_{11} & \gamma_{12} \\ \gamma_{21} & \gamma_{22} \end{bmatrix}\begin{bmatrix} \phi_{11} & \phi_{21} \\ \phi_{21} & \phi_{22} \end{bmatrix}\begin{bmatrix} \gamma_{11} & \gamma_{21} \\ \gamma_{12} & \gamma_{22} \end{bmatrix} + \begin{bmatrix} \psi_{11} & 0 \\ 0 & \psi_{22} \end{bmatrix}$$

$$= \begin{bmatrix} \gamma_{11} & \gamma_{12} \\ \gamma_{21} & \gamma_{22} \end{bmatrix}\begin{bmatrix} \phi_{11}\gamma_{11} + \phi_{21}\gamma_{12} & \phi_{11}\gamma_{21} + \phi_{21}\gamma_{22} \\ \phi_{21}\gamma_{11} + \phi_{22}\gamma_{12} & \phi_{21}\gamma_{21} + \phi_{22}\gamma_{22} \end{bmatrix} + \begin{bmatrix} \psi_{11} & 0 \\ 0 & \psi_{22} \end{bmatrix}$$

$$= \begin{bmatrix} \phi_{11}\gamma_{11}^2 + \phi_{21}\gamma_{11}\gamma_{12} + \phi_{21}\gamma_{11}\gamma_{12} + \phi_{22}\gamma_{12}^2 + \psi_{11} & \phi_{11}\gamma_{11}\gamma_{21} + \phi_{21}\gamma_{11}\gamma_{22} + \phi_{21}\gamma_{12}\gamma_{21} + \phi_{22}\gamma_{12}\gamma_{22} \\ \phi_{11}\gamma_{11}\gamma_{21} + \phi_{21}\gamma_{12}\gamma_{21} + \phi_{21}\gamma_{11}\gamma_{22} + \phi_{22}\gamma_{12}\gamma_{22} & \phi_{11}\gamma_{21}^2 + \phi_{21}\gamma_{21}\gamma_{22} + \phi_{21}\gamma_{21}\gamma_{22} + \phi_{22}\gamma_{22}^2 + \psi_{22} \end{bmatrix}$$

$$= \begin{bmatrix} g_{11} & g_{12} \\ g_{21} & g_{22} \end{bmatrix} \text{ (say)}$$

So, $\mathbf{\Sigma}_{\eta\eta^T} = (\mathbf{I}-\boldsymbol{\beta})^{-1}(\mathbf{\Gamma\Phi\Gamma}^T + \mathbf{\Psi})(\mathbf{I}-\boldsymbol{\beta})^{-1^T}$

$$= \begin{bmatrix} 1 & 0 \\ 0 & \dfrac{1}{1-\beta_{21}} \end{bmatrix}\begin{bmatrix} g_{11} & g_{12} \\ g_{21} & g_{22} \end{bmatrix}\begin{bmatrix} 1 & 0 \\ 0 & \dfrac{1}{1-\beta_{21}} \end{bmatrix} = \begin{bmatrix} 1 & 0 \\ 0 & \dfrac{1}{1-\beta_{21}} \end{bmatrix}\begin{bmatrix} g_{11} & g_{12} \\ \dfrac{g_{21}}{1-\beta_{21}} & \dfrac{g_{22}}{1-\beta_{21}} \end{bmatrix} = \begin{bmatrix} g_{11} & g_{12} \\ \dfrac{g_{21}}{(1-\beta_{21})^2} & \dfrac{g_{22}}{(1-\beta_{21})^2} \end{bmatrix}$$

(ii) Find out $\mathbf{\Sigma}_{\eta\xi^T}$

$$\mathbf{\Sigma}_{\eta\xi^T} = (\mathbf{I}-\boldsymbol{\beta})^{-1}\mathbf{\Gamma\Phi}$$

$$= \begin{bmatrix} 1 & 0 \\ 0 & \dfrac{1}{1-\beta_{21}} \end{bmatrix}\begin{bmatrix} \gamma_{11} & \gamma_{12} \\ \gamma_{21} & \gamma_{22} \end{bmatrix}\begin{bmatrix} \phi_{11} & \phi_{21} \\ \phi_{21} & \phi_{22} \end{bmatrix} = \begin{bmatrix} \gamma_{11} & \gamma_{12} \\ \dfrac{\gamma_{21}}{1-\beta_{21}} & \dfrac{\gamma_{22}}{1-\beta_{21}} \end{bmatrix}\begin{bmatrix} \phi_{11} & \phi_{21} \\ \phi_{21} & \phi_{22} \end{bmatrix}$$

$$= \begin{bmatrix} \phi_{11}\gamma_{11} + \phi_{21}\gamma_{12} & \phi_{21}\gamma_{11} + \phi_{22}\gamma_{12} \\ \dfrac{\phi_{11}\gamma_{21} + \phi_{21}\gamma_{22}}{1-\beta_{21}} & \dfrac{\phi_{21}\gamma_{21} + \phi_{22}\gamma_{22}}{1-\beta_{21}} \end{bmatrix}$$

(iii) Find out $\mathbf{\Sigma}_{\xi\eta^T}$

$$\mathbf{\Sigma}_{\xi\eta^T} = \mathbf{\Phi}\mathbf{\Gamma}^T(\mathbf{I}-\boldsymbol{\beta})^{-1^T}$$

$$= \begin{bmatrix} \phi_{11} & \phi_{21} \\ \phi_{21} & \phi_{22} \end{bmatrix} \begin{bmatrix} \gamma_{11} & \gamma_{21} \\ \gamma_{12} & \gamma_{22} \end{bmatrix} \begin{bmatrix} 1 & 0 \\ 0 & \dfrac{1}{1-\beta_{21}} \end{bmatrix} = \begin{bmatrix} \phi_{11} & \phi_{21} \\ \phi_{21} & \phi_{22} \end{bmatrix} \begin{bmatrix} \gamma_{11} & \dfrac{\gamma_{21}}{1-\beta_{21}} \\ \gamma_{12} & \dfrac{\gamma_{22}}{1-\beta_{21}} \end{bmatrix}$$

$$= \begin{bmatrix} \phi_{11}\gamma_{11} + \phi_{21}\gamma_{12} & \dfrac{\phi_{11}\gamma_{21}}{1-\beta_{21}} + \dfrac{\phi_{21}\gamma_{22}}{1-\beta_{21}} \\ \phi_{21}\gamma_{11} + \phi_{22}\gamma_{12} & \dfrac{\phi_{21}\gamma_{21}}{1-\beta_{21}} + \dfrac{\phi_{22}\gamma_{22}}{1-\beta_{21}} \end{bmatrix}$$

(iv) $\mathbf{\Phi} = \begin{bmatrix} \phi_{11} & \phi_{21} \\ \phi_{21} & \phi_{22} \end{bmatrix}$

Now using (i), (ii), (iii), and (iv) described above, we can get $\mathbf{\Sigma}_{4\times4}$ as

$$\mathbf{\Sigma}_{4\times4} = \begin{pmatrix} g_{11} & g_{12} & \phi_{11}\gamma_{11} + \phi_{21}\gamma_{12} & \phi_{21}\gamma_{11} + \phi_{22}\gamma_{12} \\ \dfrac{g_{21}}{(1-\beta_{21})^2} & \dfrac{g_{22}}{(1-\beta_{21})^2} & \dfrac{\phi_{11}\gamma_{21} + \phi_{21}\gamma_{22}}{1-\beta_{21}} & \dfrac{\phi_{21}\gamma_{21} + \phi_{22}\gamma_{22}}{1-\beta_{21}} \\ \phi_{11}\gamma_{11} + \phi_{21}\gamma_{12} & \dfrac{\phi_{11}\gamma_{21}}{1-\beta_{21}} + \dfrac{\phi_{21}\gamma_{22}}{1-\beta_{21}} & \phi_{11} & \phi_{21} \\ \phi_{21}\gamma_{11} + \phi_{22}\gamma_{12} & \dfrac{\phi_{21}\gamma_{21}}{1-\beta_{21}} + \dfrac{\phi_{22}\gamma_{22}}{1-\beta_{21}} & \phi_{21} & \phi_{22} \end{pmatrix}$$

14.5.2 Fundamental Relationships for the Complete SEM Model

Let $\mathbf{\Sigma}_{MV}$ represents the $(q+p)\times(q+p)$ covariance matrix of the $q+p$ manifest variables (MV), represented by Y and X, respectively. $\mathbf{\Sigma}_{MV}$ can be partitioned as below.

$$\mathbf{\Sigma}_{MV} = \begin{bmatrix} \Sigma_{YY} & \Sigma_{YX} \\ \Sigma_{XY} & \Sigma_{XX} \end{bmatrix} = \begin{bmatrix} E(\boldsymbol{YY}^T) & E(\boldsymbol{YX}^T) \\ E(\boldsymbol{XY}^T) & E(\boldsymbol{XX}^T) \end{bmatrix}$$

Using Equations 14.3, 14.6, and 14.9, and the fundamental relationships of the structural model (Equation 14.14), we can find out the relationships of $\mathbf{\Sigma}_{MV}$ with the SEM model parameters.

Using Equation 14.9 and $\Sigma_{YY} = E(YY^T)$, we find

$$\begin{aligned} Y &= \Lambda_Y \eta + \epsilon \\ So,\ YY^T &= (\Lambda_Y \eta + \epsilon)(\Lambda_Y \eta + \epsilon)^T = \Lambda_Y \eta\eta^T \Lambda_Y^T + \Lambda_Y \eta \epsilon^T + \epsilon \eta^T \Lambda_Y^T + \epsilon\epsilon^T \\ So,\ E(YY^T) &= \Lambda_Y E(\eta\eta^T)\Lambda_Y^T + \Lambda_Y E(\eta\epsilon^T) + E(\epsilon\eta^T)\Lambda_Y^T + E(\epsilon\epsilon^T) \\ &= \Lambda_Y E(\eta\eta^T)\Lambda_Y^T + E(\epsilon\epsilon^T) \qquad [as\ E(\eta\epsilon^T) = E(\epsilon\eta^T) = \mathbf{0}] \end{aligned}$$

Using Equation 14.11 and $E(\epsilon\epsilon^T) = \boldsymbol{\theta}_\epsilon$, we get

$$\boldsymbol{\Sigma}_{YY} = E(YY^T) = \Lambda_Y (\mathbf{I} - \boldsymbol{\beta})^{-1}(\boldsymbol{\Gamma\Phi\Gamma}^T + \boldsymbol{\Psi})(\mathbf{I} - \boldsymbol{\beta})^{-1^T} \Lambda_Y^T + \boldsymbol{\theta}_\epsilon \tag{14.15}$$

Using Equations 14.6, 14.9, and $\Sigma_{YX} = E(YX^T)$, we get

$$\begin{aligned} YX^T &= (\Lambda_Y \eta + \epsilon)(\Lambda_X \xi + \boldsymbol{\delta})^T = \Lambda_Y \eta\xi^T \Lambda_X^T + \Lambda_Y \eta \boldsymbol{\delta}^T + \epsilon\xi^T \Lambda_X^T + \epsilon \boldsymbol{\delta}^T \\ So,\ \ E(YX^T) &= \Lambda_Y E(\eta\xi^T)\Lambda_X^T + \Lambda_Y E(\eta\boldsymbol{\delta}^T) + E(\epsilon\xi^T)\Lambda_X^T + E(\epsilon\boldsymbol{\delta}^T) \\ &= \Lambda_Y E(\eta\xi^T)\Lambda_X^T \quad [\text{as other term vanishes, see assumptions}] \end{aligned}$$

Putting the value of $E(\eta\xi^T)$ from Equation 14.12

$$\boldsymbol{\Sigma}_{YX} = E(YX^T) = \Lambda_Y (\mathbf{I} - \boldsymbol{\beta})^{-1} \boldsymbol{\Gamma\Phi}\Lambda_X^T \tag{14.16}$$

In similar manner,

$$\begin{aligned} \Sigma_{XY} &= E(XY^T) = E[(\Lambda_X \xi + \boldsymbol{\delta})(\Lambda_Y \eta + \epsilon)^T] \\ &= \Lambda_X E(\xi\eta^T)\Lambda_Y^T \quad [\text{as other terms vanish}] \end{aligned}$$

Putting the value of $E(\xi\eta^T)$ from Equation 14.13

$$\Sigma_{XY} = E(XY^T) = \Lambda_X \boldsymbol{\Phi\Gamma}^T (\mathbf{I} - \boldsymbol{\beta})^{-1^T} \Lambda_Y^T \tag{14.17}$$

and,

$$\begin{aligned} \Sigma_{XX} &= E(XX^T) = E[(\Lambda_X \xi + \boldsymbol{\delta})(\Lambda_X \xi + \boldsymbol{\delta})^T] \\ &= \Lambda_X E(\xi\xi^T)\Lambda_X^T + E(\boldsymbol{\delta\delta}^T) = \Lambda_X \boldsymbol{\Phi}\Lambda_X^T + \boldsymbol{\theta}_\delta \end{aligned} \tag{14.18}$$

Finally,

$$\begin{aligned} \boldsymbol{\Sigma}_{MV} &= \begin{bmatrix} \Sigma_{YY} & \Sigma_{YX} \\ \Sigma_{XY} & \Sigma_{XX} \end{bmatrix} \\ &= \begin{bmatrix} \Lambda_Y (\mathbf{I} - \boldsymbol{\beta})^{-1}(\boldsymbol{\Gamma\Phi\Gamma}^T + \boldsymbol{\Psi})(\mathbf{I} - \boldsymbol{\beta})^{-1^T} \Lambda_Y^T + \boldsymbol{\theta}_\epsilon & \Lambda_Y (\mathbf{I} - \boldsymbol{\beta})^{-1} \boldsymbol{\Gamma\Phi}\Lambda_X^T \\ \Lambda_X \boldsymbol{\Phi\Gamma}^T (\mathbf{I} - \boldsymbol{\beta})^{-1^T} \Lambda_Y^T & \Lambda_X \boldsymbol{\Phi}\Lambda_X^T + \boldsymbol{\theta}_\delta \end{bmatrix} \end{aligned} \tag{14.19}$$

Equation 14.19 comprises eight matrices $\boldsymbol{\beta}$, $\boldsymbol{\Gamma}$, Λ_Y, Λ_X, $\boldsymbol{\Psi}$, $\boldsymbol{\Phi}$, $\boldsymbol{\theta}_\delta$, and $\boldsymbol{\theta}_\epsilon$. Now $\boldsymbol{\Sigma}_{MV}$ has $(p+q)(p+q+1)/2$ distinct elements, say σ_{ij}, $i = 1, 2, \ldots, p + q$ and $j = 1, 2, \ldots, p + q$. Each σ_{ij} is a function of some elements of the eight parameter matrices. If $\boldsymbol{\theta}$ represents the set of all the parameters of the eight parameter matrices, then we can write $\sigma_{ij} = \sigma_{ij}(\boldsymbol{\theta})$, which is a non-linear function. From the path model and the CFA, we have seen that knowledge on many of the parameters (e.g., factor loadings, path coefficients, covariances, etc.) are known to the researchers *a priori* from theoretical perspectives, practical experiences, empirical evidences, etc. Such information is extremely valuable and should also be used while specifying the SEM model. Accordingly, the parameters can be grouped into three categories: fixed, constrained and free (Krzanowski and Marriott, 2014b).

- Fixed parameters are generally assigned a value 0 or 1, or some other value, where substantial evidences (practical or empirical) support this value. The zero value is assigned when the corresponding variable is excluded in the hypothesized relationships. In single indicator case (for the measurement model), the factor loading is generally taken as 1.0. But considering measurement error, the value could be between 0.90 and 1.0.
- Constrained parameters are those whose exact values are not known but can be put equal to 1.0 or as the function of some other parameters. Constraint parameters generally arise out of scaling of variables, either by standardizing the variance of ξ and η or by fixing one element in each column of $\mathbf{\Lambda}_X$ and $\mathbf{\Lambda}_Y$ to 1.0. In addition, ξ and η being latent, their origin is set to zero, assuming $E(\xi) = E(\eta) = \mathbf{0}$.
- Free parameters are those whose values are estimated.

14.6 PARAMETER ESTIMATION

14.6.1 Model Identification

Like the path model and CFA, the parameter estimation process comprises two steps: identification and estimation. The process of estimation is also similar to the path model and CFA. However, SEM, being an integrated model of the path model and CFA, is more general and, hence, involves more parameters. From Equation 14.19, we can see that there are eight parameter matrices, $\boldsymbol{\beta}$, $\boldsymbol{\Gamma}$, $\boldsymbol{\Lambda}_{\Psi}$, $\boldsymbol{\Lambda}_{\Xi}$, $\boldsymbol{\Psi}$, $\boldsymbol{\Phi}$, $\boldsymbol{\theta}_{\delta}$, and $\boldsymbol{\theta}_{\epsilon}$. A parameter vector $\boldsymbol{\theta}$ is created which comprises the elements of the eight parameters matrices. So $\boldsymbol{\Sigma}_{MV}$ is $\boldsymbol{\Sigma}(\boldsymbol{\theta})$. Bollen (1989) mentioned the global and local identification issues. The model is globally identified, if $\boldsymbol{\Sigma}(\theta_1) = \boldsymbol{\Sigma}(\theta_2)$ implies $\theta_1 = \theta_2$. If $\boldsymbol{\Sigma}_{MV} = \{\sigma_{ij}\}$ and $\boldsymbol{\Sigma}(\boldsymbol{\theta}) = \{\sigma_{ij}(\boldsymbol{\theta})\}$, where $\{\}$ implies the set of all elements, then for identification, $\sigma_{ij} = \sigma_{ij}(\boldsymbol{\theta})$. As there are $\frac{1}{2}(p+q)(p+q+1)$ number of unique elements in $\boldsymbol{\Sigma}_{MV}$ and in $\boldsymbol{\Sigma}(\boldsymbol{\theta})$, the model has $\frac{1}{2}(p+q)(p+q+1)$ number of simultaneous equations. Now, if $\boldsymbol{\theta}$ comprises t number of parameters to be estimated, the t-rule (Bollen, 1989) for model identification is $t \le \frac{1}{2}(p+q)(p+q+1)$.

This is the necessary condition, but is not sufficient. The rules for sufficiency are the two-step rule and the MIMIC (multiple indicators and multiple causes) rule (Bollen, 1989). The two-step procedure is discussed here. For MIMIC, readers may see Bollen (1989) and Jöreskog and Goldberger (1975).

In the two-step rule, the SEM model is identified in two steps: first, the measurement model (CFA) is identified, followed by the structural model. Both the models are separately identified based on the applicable rules. Unless the measurement model is found identified, the structural model identification is not done. Bollen (1989) and Bollen and Davis (2009) provided the procedure, which is discussed below.

Measurement Model Identification

Both the $\boldsymbol{X}$ and $\boldsymbol{Y}$ variables are treated as $\boldsymbol{X}$ variables, and both the ξ and η variables are treated as ξ variables. So, $\boldsymbol{X}$ is $(p+q)\times 1$ vector of manifest variables and ξ is a $(m+n)\times 1$ vector of latent factors. It is same as CFA with the difference in the number of manifest variables (p + q, instead of p or q) and the number of factors ($m + n$, instead of m or n). Setting $p' = p + q$ and $m' = m + n$, the factor model is

$$\boldsymbol{X}_{p'\times 1} = \boldsymbol{\Lambda}_{X,p'\times m'}\boldsymbol{\xi}_{m'\times 1} + \boldsymbol{\delta}_{p'\times 1}$$

and the fundamental relationships are

$$\boldsymbol{\Sigma}_{p'\times p'} = \boldsymbol{\Lambda}_{X,p'\times m'}\boldsymbol{\Phi}_{m'\times m'}\boldsymbol{\Lambda}^{T}_{X,m'\times p'} + \boldsymbol{\theta}_{\delta,p'\times p'}$$

All the CFA rules such as *t*-rule, three-indicator or two-indicator rules of identification are applied here (see Chapter 13, Section 13.2.1). For more details, see Bollen (1989).

Structural Model Identification

Once the measurement model is identified, the next step of the two-step process is the identification of the structural model. Here, the latent factors $\boldsymbol{\xi}$ and $\boldsymbol{\eta}$ are considered as observed (like we see in the path model, Chapter 10). All the path model rules such as *t*-rule, null β rule, recursive rule, rank order, and rank conditions can be used as per requirements. For more details, see Bollen (1989).

If the two-step process succeeds, i.e., both the measurement and structural models are identified, then SEM (the complete model) is considered to be identified. In case it fails, the global identification can be done empirically using the rank of the score matrix and Fisher's information matrix (Timm, 2002).

Bollen and Davis (2009) demonstrated identification rules for structural models with casual indicators. They developed two new rules, namely 2 + emitted paths and exogenous *X* rule. They further stated that the two-step rule is a special case of piece-wise identification strategy, where the overall model is broken into smaller pieces (models) and identification is done sequentially.

Example 14.3

Consider the hypothetical SEM model with $p = 6$, $q = 6$, $m = 3$, and $n = 3$ shown in Figure 14.7. Is the model identifiable?

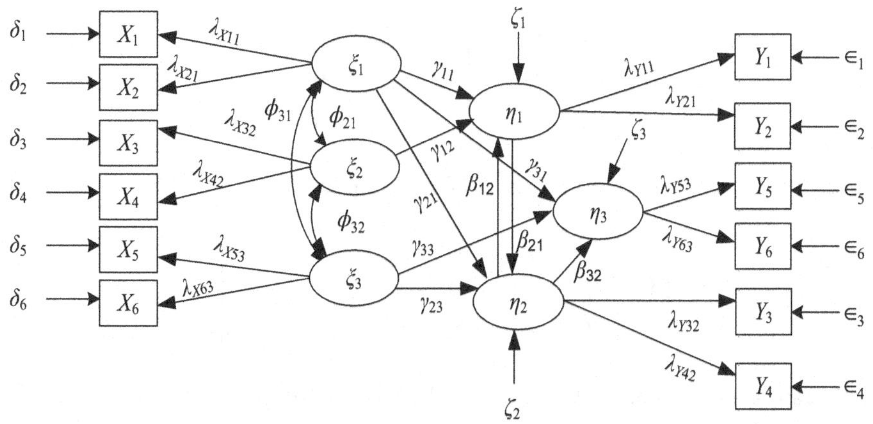

FIGURE 14.7 The structural equation model (SEM) for Example 14.3. In the figure, $\boldsymbol{\eta} = [\eta_1\ \eta_2\ \eta_3]^T$ represents the endogenous latent variables and $\boldsymbol{\xi} = [\xi_1\ \xi_2\ \xi_3]^T$ represents the exogenous latent variables. $\boldsymbol{Y} = [Y_1\ Y_2\ Y_3\ Y_4\ Y_5\ Y_6]^T$ and $\boldsymbol{X} = [X_1\ X_2\ X_3\ X_4\ X_5\ X_6]^T$ are the manifest variables for the endogenous latent variables ($\boldsymbol{\eta}$) and exogenous latent variables ($\boldsymbol{\xi}$), respectively. The other terms denote their usual meanings.

Solution:

For measurement model, number of explanatory manifest variables (p) is 6, and the number of explained manifest varaibles (q) is 6. So,

$$\frac{1}{2}(p+q)(p+q+1)=\frac{1}{2}(6+6)(6+6+1)=78$$

Now, based on the t-rule, the total number of parameters to be estimated (t) in the hypothesized model is 45.

As $t=45<\frac{1}{2}(p+q)(p+q+1)=78$, the order condition is satisfied. Further, all the latent factors in the model have two indicator variables. Hence, the model also satisfies the indicator rules. As the measurement model satisfies both the t-rules and indicator rules, the model in general is identifiable. However, the rank condition of the covariance matrix is required to be tested if estimation problem arises.

For structural model, the number of exogenous latent variables (m) = 3, and the number of endogenous latent varaibles (n) = 3.

$$\frac{1}{2}(m+n)(m+n+1)=\frac{1}{2}(3+3)(3+3+1)=21$$

Now, based on the t-rule, the total number of parameters to be estimated (t) in the hypothesized model is 18. As $t=18<\frac{1}{2}(m+n)(m+n+1)=21$, the order condition is satisfied.

Now we will test the rank condition for the structural model

$$\eta_1=\gamma_{11}\xi_1+\gamma_{12}\xi_2+\beta_{12}\eta_2+\zeta_1$$
$$\eta_2=\gamma_{21}\xi_1+\gamma_{23}\xi_3+\beta_{21}\eta_1+\zeta_2$$
$$\eta_3=\gamma_{31}\xi_1+\gamma_{33}\xi_3+\beta_{32}\eta_2+\zeta_3$$

So,

$$\boldsymbol{\beta}=\begin{bmatrix}0 & \beta_{12} & 0\\ \beta_{21} & 0 & 0\\ 0 & \beta_{32} & 0\end{bmatrix};\ \boldsymbol{\Gamma}=\begin{bmatrix}\gamma_{11} & \gamma_{12} & 0\\ \gamma_{21} & 0 & \gamma_{23}\\ \gamma_{31} & 0 & \gamma_{33}\end{bmatrix}$$

Hence, the matrix **M** will be (see Chapter 10 for description of **M**)

$$\mathbf{M}=\begin{bmatrix}1 & -\beta_{12} & 0 & -\gamma_{11} & -\gamma_{12} & 0\\ -\beta_{21} & 1 & 0 & -\gamma_{21} & 0 & -\gamma_{23}\\ 0 & -\beta_{32} & 1 & -\gamma_{31} & 0 & -\gamma_{33}\end{bmatrix}$$

To test the rank condition for η_1, we have the following $\mathbf{M}_1$ matrix.

$$\mathbf{M}_1=\begin{bmatrix}0 & 0\\ 0 & -\gamma_{23}\\ 1 & -\gamma_{33}\end{bmatrix}$$

Rank ($\mathbf{M}_1$) is 2, which is equal to $(n-1)=2$. So, the structural equation η_1 is identifiable.

Similarly, for η_2, we get

$$\mathbf{M}_2 = \begin{bmatrix} 0 & -\gamma_{12} \\ 0 & 0 \\ 1 & 0 \end{bmatrix}$$

Rank ($\mathbf{M}_2$) = 2, which is equal to $(n - 1) = 2$. So, the structural equation η_2 is identifiable.

Similarly, for η_3, we get

$$\mathbf{M}_3 = \begin{bmatrix} 1 & -\gamma_{12} \\ -\beta_{21} & 0 \\ 0 & 0 \end{bmatrix}$$

Rank ($\mathbf{M}_3$) = 2, which is again equal to $(n - 1) = 2$. So, the structural equation η_3 is identifiable.

As both the measurement and structural models are identified, the overall model is identifiable. It is to be noted here that the most widely used approach for parameter estimation is minimization of the fit function given in Section 14.6.2. The function F (see Equations 14.20–14.22) is minimized with respect to $\boldsymbol{\theta}$ (the parameter vector to be estimated). As an analytical solution (exact solution) is not available for F, numerical optimization methods are used. In the numerical optimization approach, the model-induced covariance (or correlation) matrix is compared with the sample covariance (or correlation) matrix. With an initial guess of $\boldsymbol{\theta}$, an iterative procedure is followed until convergence is achieved. Therefore, the rank of the covariance or correlation matrix, the number of unique elements in the matrix, the number of parameters to be estimated, and the nature of the fundamental relationships collectively determine whether the model can be estimated or not.

14.6.2 Model Estimation

Once the model is found to be identified, the free parameters of the model can be estimated using one of the many parameter estimation techniques; however, the three commonly used techniques are maximum likelihood (ML), generalized least squares (GLS), and unweighted least squares (ULS). The ML estimation is discussed in detail in Chapter 13 (CFA). Several other estimation techniques are discussed in Chapter 10 (path model). The fundamental functions to be minimized under ML, GLS, and ULS are given below.

$$F_{ML} = ln|\boldsymbol{\Sigma}(\boldsymbol{\theta})| - ln|\mathbf{S}| + trace\{\mathbf{S}\boldsymbol{\Sigma}^{-1}(\boldsymbol{\theta})\} - (p+q) \tag{14.20}$$

$$F_{GLS} = \frac{1}{2} trace\{[I - \boldsymbol{\Sigma}(\boldsymbol{\theta})\mathbf{S}^{-1}]^2\} \tag{14.21}$$

$$F_{ULS} = \frac{1}{2} trace\{[\mathbf{S} - \boldsymbol{\Sigma}(\boldsymbol{\theta})]^2\} \tag{14.22}$$

The function F is minimized with respect to $\boldsymbol{\theta}$. Each of the above functions is non-linear and devoid of an analytical solution (exact solution). So, these functions are minimized numerically. Specialized commercial software like AMOS SPSS, LISREL, EQS, and open-source R software may be used.

Example 14.4

Consider the hypothetical SEM model given in Example 14.3. The hypothetical correlation matrix for the 12 manifest variables are given in Table 14.1. Estimate the parameters using MLE method.

Solution:

Both the measurement and structural model parameters were estimated using the "lavaan" package in R (open source software). The results are given in Tables 14.2 and 14.3.

From Table 14.3, it is observed that the estimate of γ_{21} is greater than one, which is an offending estimate for the structural model as the correlation matrix is the input to the model for parameter estimation. Although it is desirable to have the absolute value of the parameter estimates to be less than equal to one, researchers have often pointed out the deviation of it. In this case, γ_{21} estimate is 1.012 which is marginally greater than one, and for all practical

TABLE 14.1
Correlation Matrix for Example 14.4

	X_1	X_2	X_3	X_4	X_5	X_6	Y_1	Y_2	Y_3	Y_4	Y_5	Y_6
X_1	1											
X_2	0.42	1										
X_3	0.03	0.11	1									
X_4	0.01	0.05	0.45	1								
X_5	0.01	0.01	0.01	0.01	1							
X_6	0.03	0.02	0.01	0.01	0.46	1						
Y_1	0.29	0.26	0.04	0.01	0.28	0.23	1					
Y_2	0.28	0.32	0.06	0.01	0.2	0.22	0.43	1				
Y_3	0.22	0.25	0.08	0.04	0.18	0.25	0.28	0.24	1			
Y_4	0.14	0.28	0.1	0.06	0.38	0.19	0.2	0.17	0.34	1		
Y_5	0.12	0.22	0.12	0.08	0.3	0.14	0.22	0.19	0.18	0.1	1	
Y_6	0.16	0.29	0.14	0.06	0.35	0.17	0.27	0.23	0.2	0.12	0.42	1

TABLE 14.2
Factor Loadings in the Measurement Model

Latent Variable	Manifest Variable	Parameters	Estimate
η_1	Y_1	λ_{Y11}	0.671
	Y_2	λ_{Y21}	0.639
η_2	Y_3	λ_{Y32}	0.5
	Y_4	λ_{Y42}	0.678
η_3	Y_5	λ_{Y53}	0.586
	Y_6	λ_{Y63}	0.715
ξ_1	X_1	λ_{X11}	0.554
	X_2	λ_{X21}	0.756
ξ_2	X_3	λ_{X32}	0.931
	X_4	λ_{X42}	0.482
ξ_3	X_5	λ_{X53}	0.857
	X_6	λ_{X63}	0.535

TABLE 14.3
Structural Path Coefficients of the Structural Model

Latent Endogenous Variable	Latent Variable	Parameters	Estimate	Estimate (After Fixing $\gamma_{21} = 1$)
η_1	ξ_1	γ_{11}	0.21	0.214
	ξ_2	γ_{12}	–0.075	–0.074
	η_2	β_{12}	0.774	0.767
η_2	ξ_1	γ_{21}	**1.012**	1
	ξ_3	γ_{23}	0.916	0.911
	η_1	β_{21}	–0.759	–0.746
η_3	ξ_1	γ_{31}	0.86	0.86
	ξ_3	γ_{33}	0.96	0.96
	η_2	β_{32}	–0.706	–0.706

purposes it could be considered as one. To verify, the model is again estimated after fixing γ_{21} to one (see column five in Table 14.3). The other parameter estimates remain almost the same.

14.6.3 Direct, Indirect, and Total Effects

In Chapter 10 (Section 10.3.2) we explained direct, indirect, and total effects of one explanatory variable (exogenous) on an explained variable (endogenous). In the SEM, similar decomposition can be done. The structural model gives the direct, indirect, and total effects for the latent variables, where total effect is equal to the sum of direct and indirect effects. From the path diagram (e.g., Figure 14.4), we see that the following effects occur:

- η vs η: one η variable is affected by another η variable in the model. The effect can be direct (e.g., β_{31}, the direct effect of η_1 on η_3) or indirect (e.g., $\beta_{21}.\beta_{32}$, the indirect effect of η_1 on η_3 through η_2) and total (e.g., $\beta_{31} + \beta_{21}.\beta_{32}$, the total effect of η_1 on η_3). See Figure 14.4 for these direct and indirect paths. So, the total effect is the sum of the direct effect and all the indirect effects.
- ξ vs η: one η variable is affected by a ξ variable in the model. For example, in Figure 14.4, the direct effect of ξ_1 on η_1 is γ_{11} and there are many indirect effects; $\phi_{21}\gamma_{12}$ is one of them. The total effect is the sum of the direct effect and all the indirect effects.
- η vs Y: one Y variable is affected by a η variable in the model. The direct effect comes from the factor loading. For example, the direct effect of η_3 on Y_5 in Figure 14.4 is $\lambda_{Y_{53}}$. The indirect effect comes via other η variables in the model. For example (in Figure 14.4), the indirect effect of η_1 on Y_5 is through the path $\eta_1 \rightarrow \eta_2 \rightarrow \eta_3 \rightarrow Y_5$ with indirect effect of $\beta_{21}.\beta_{32}.\lambda_{Y_{53}}$. Again, the total effect is the sum of the direct effect and all indirect effects.
- $\boldsymbol{\eta}$ vs $\boldsymbol{X}$: There is no effect (direct or indirect) of $\boldsymbol{\eta}$ on $\boldsymbol{X}$.
- $\boldsymbol{\xi}$ vs $\boldsymbol{Y}$: There is no direct effect of $\boldsymbol{\xi}$ on $\boldsymbol{Y}$ but indirect effects come through $\boldsymbol{\eta}$, or through $\boldsymbol{\eta}$ and $\boldsymbol{\xi}$, depending on the model structure. For example, in Figure 14.4 one of the indirect effect components of ξ_1 on Y_1 is $\gamma_{11}.\lambda_{Y_{11}}$ (through η_1). So, in this case, the total effect is equal to sum of all the indirect effects.
- $\boldsymbol{\xi}$ vs $\boldsymbol{X}$: $\boldsymbol{X}$ is directly affected by $\boldsymbol{\xi}$ and has no indirect effects. For example, the direct effect of ξ_2 on X_3 is $\lambda_{X_{32}}$ and indirect effects are null (see Figure 14.4). So, the total effect is only the direct effect.

TABLE 14.4
Decomposition of Effects in SEM

	Effect on:		
	η	Y	X
Effect of η:			
Direct effect	$\boldsymbol{\beta}$	$\boldsymbol{\Lambda}_Y$	0
Indirect effect	$(\mathbf{I}-\boldsymbol{\beta})^{-1}-\mathbf{I}-\boldsymbol{\beta}$	$\boldsymbol{\Lambda}_Y(\mathbf{I}-\boldsymbol{\beta})^{-1}-\boldsymbol{\Lambda}_Y$	0
Total effect	$(\mathbf{I}-\boldsymbol{\beta})^{-1}-\mathbf{I}$	$\boldsymbol{\Lambda}_Y(\mathbf{I}-\boldsymbol{\beta})^{-1}$	0
Effects of ξ:			
Direct effect	$\boldsymbol{\Gamma}$	0	$\boldsymbol{\Lambda}_X$
Indirect effect	$(\mathbf{I}-\boldsymbol{\beta})^{-1}\boldsymbol{\Gamma}-\boldsymbol{\Gamma}$	$\boldsymbol{\Lambda}_Y(\mathbf{I}-\boldsymbol{\beta})^{-1}\boldsymbol{\Gamma}$	0
Total effect	$(\mathbf{I}-\boldsymbol{\beta})^{-1}\boldsymbol{\Gamma}$	$\boldsymbol{\Lambda}_Y(\mathbf{I}-\boldsymbol{\beta})^{-1}\boldsymbol{\Gamma}$	$\boldsymbol{\Lambda}_X$

Source: Bollen, 1989

As the SEM involves many effect parameters even for a simpler model, the computation of indirect and total effects using matrix multiple is recommended. However, to get a feel of these effect components, the path diagram and reduced form equations (see Chapter 10) are to be studied carefully. There are two popular ways of computing total effects, namely infinite sum of power of coefficient matrices and reduced from coefficients. Table 14.4 shows the direct, indirect, and total effects of $\boldsymbol{\xi}$ and $\boldsymbol{\eta}$ on $\boldsymbol{\eta}$, $\boldsymbol{Y}$ and $\boldsymbol{X}$ variables.

Example 14.5

Consider Example 14.4. Estimate the direct, indirect and total effects.

Solution:

The direct, indirect, and total effect relationships are shown in Table 14.5.

The direct, indirect, and total effects, as obtained in Table 14.5, need to be explained further. As stated earlier, an endogenous latent variable could be affected directly and/or indirectly by another endogenous or exogenous variable in the model. The total effect is the sum total of the direct and indirect effects. Therefore, while explaining the causal relationship of one latent variable with another, a question may be raised as to which effects should be considered. Researchers have suggested considering the total effect as the variables considered in the model jointly influence the relationship as per the structure of the model. For example, ξ_1 has a direct effect coefficient of 0.214 on η_1 and an indirect effect coefficient of 0.767, resulting in the total effect coefficient of 0.981 (see Figure 14.8(a)). Further, ξ_1 has a direct effect coefficient of 0.860 on η_3 and an indirect effect coefficient of –0.593, resulting into the total effect coefficient of 0.267 (see Figure 14.8(b) below). Now, if we compare the effect of ξ_1 on η_1 and η_3 based on direct effect only, ξ_1 has more influence on η_3 as compared to that of η_1. But if we compare the same based on total effect, ξ_1 has less influence on η_3 as compared to that of η_1. Interestingly, this kind of decomposition of relationship is possible in structural (path) model because of its structure, which is not possible in multiple (MLR) or multivariate multiple linear regressions (MMLR) as the later models compute the direct relationships only. Therefore, if the structural model is the appropriate model for the problem to be solved and the analyst instead uses MLR or MMLR, the model will be grossly misspecified.

TABLE 14.5
Direct, Indirect, and Total Effect Relationships for Example 14.5

Endogenous Variables	Other Latent Variables	Direct	Indirect	Total
η_1	ξ_1	0.214	0.767	0.981
	ξ_2	–0.074	NA	–0.074
	ξ_3	NA	0.699	0.699
	η_2	0.767	NA	0.767
	η_3	NA	NA	NA
η_2	ξ_1	1	–0.160	0.840
	ξ_2	NA	0.055	0.055
	ξ_3	0.911	NA	0.911
	η_1	–0.746	NA	–0.746
	η_3	NA	NA	NA
η_3	ξ_1	0.860	–0.593	0.267
	ξ_2	NA	–0.039	–0.039
	ξ_3	0.960	–0.643	0.317
	η_1	NA	0.527	0.527
	η_2	–0.706	NA	–0.706

NA: Not applicable

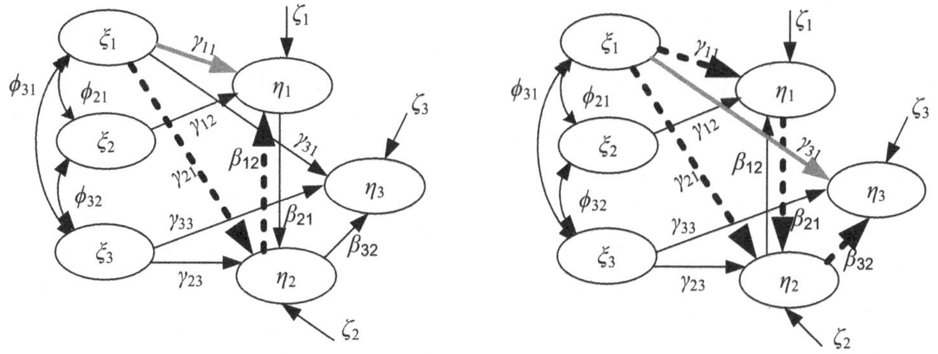

(a) Direct and indirect effects of ξ_1 on η_1 (b) Direct and indirect effects of ξ_1 on η_3

FIGURE 14.8 Visual representation of the direct and indirect effects for Example 14.5. Bold solid arrow indicates direct effect and each of the paths with bold dashed arrows represents the indirect effect.

14.7 EVALUATING MODEL FIT

The χ^2-goodness of fit statistics and different fit measures for covariance structure model are discussed in Chapter 13 (CFA). These are equally applicable in SEM. Table 14.6 summarizes these fit measures.

- The χ^2-goodness of fit statistics are sensitive to distributional assumptions and the ML-based estimation under multivariate normality are most widely used. Due to the inherent limitations of the χ^2-goodness of statistics, although many fit measures were proposed, it is not true that these fit measures will produce similar results under all conditions. The issues related to model misspecification, sample size, estimation method, and violation of assumptions are to be carefully handled.

TABLE 14.6
SEM Goodness of Fit Measures

Fit Measures	Formula for Computation	Guidelines of Use
Goodness of fit index (GFI)	$\text{GFI} = 1 - \dfrac{trace\left[(\mathbf{\Sigma}^{-1}\mathbf{S} - \mathbf{I})^2\right]}{trace\left[(\mathbf{\Sigma}^{-1}\mathbf{S})^2\right]}$	GFI varies from 0 to 1 with desirable value ≥ 0.90.
Root mean square residual (RMSR)	$\text{RMSR} = \sqrt{2\sum_{j=1}^{p}\sum_{k=1}^{j}\left(s_{jk} - \hat{\sigma}_{jk}\right)^2 / p(p+1)}$	The smaller the RMSR the better the model.
Standardized root mean square residual (SRMSR)	$\text{SRMSR} = \sqrt{2\sum_{j=1}^{p}\sum_{k=1}^{j}\left[\left(s_{jk} - \hat{\sigma}_{jk}\right)/(s_{jj}s_{kk})\right]^2 / p(p+1)}$	SRMSR lies between 0 and 1 and SRMSR ≤ 0.08 is desirable.
Root mean square error of approximation (RMSEA)	$\text{RMSEA} = \sqrt{\dfrac{\lvert\chi^2 - df\rvert}{(N-1)df}}$	RMSEA usually lies between 0.03 and 0.08 and RMSEA ≤ 0.06 is desirable.
Tucker-Lewis index (TLI)	$\text{TLI} = \dfrac{\chi_0^2 / v_0 - \chi_v^2 / v}{(\chi_0^2 / v_0) - 1}$	TLI ranges from 0 to 1 and the desirable value is 0.90 or more.
Normed fit index (NFI)	$\text{NFI} = \dfrac{\chi_0^2 - \chi_v^2}{\chi_0^2}$	NFI varies from 0 to 1, and the desirable value is 0.90 or more.
Comparative fit index (CFI)	$\text{CFI} = 1 - \dfrac{\chi_v^2 - v}{\chi_0^2 - v_0}$	CFI varies from 0 to 1 and the desirable value is 0.90 or more.
General normed fit index (GNFI)	$\text{GNFI} = \dfrac{F_0 - F_v}{F_0 - F_h}$	GNFI varies from 0 to 1 and the desirable value is 0.90 or more.
Relative non-centrality index (RNI)	$\text{RNI} = \dfrac{(\chi_0^2 - v_0) - (\chi_v^2 - v)}{(\chi_0^2 - v_0)}$	RNI varies from 0 to 1 and the desirable value is 0.90 or more.
Adjusted parsimony fit index (AGFI)	$\text{AGFI} = 1 - \dfrac{trace\left[(\mathbf{\Sigma}^{-1}\mathbf{S} - \mathbf{I})^2 / v\right]}{trace\left[(\mathbf{\Sigma}^{-1}\mathbf{S})^2 / \frac{1}{2}p(p+1)\right]} = 1 - \dfrac{p(p+1)trace\left[(\mathbf{\Sigma}^{-1}\mathbf{S} - \mathbf{I})^2\right]}{2v\, trace\left[(\mathbf{\Sigma}^{-1}\mathbf{S})^2\right]}$	AGFI varies from 0 to 1 and its desirable value is 0.90 or more.
Parsimony normed fit index (PNFI)	$\text{PNFI} = \dfrac{v}{v_0}\text{NFI}$	Higher values of PNFI is better. For alternative models, a change in PNFI of the order of 0.06 to 0.09 is significant.

- The examination of the offending estimates such as negative error variances, standardized coefficients approaching value of 1 or more, and large standard errors is equally important in the SEM as in the CFA or in the path model.

Another important measure is coefficient of determination R^2. It is similar to R^2 in multiple regression. In SEM, we have three sets of dependence equations, i.e., one in the structural (path) model and two in the measurement model.

- R_Y^2 and $R_{Y_j}^2$: R_Y^2 computes the total variance of the $\boldsymbol{Y}$ variables explained by the model and $R_{Y_j}^2$ does the same for the Y_j variable only.

$$R_Y^2 = 1 - \frac{\left|\hat{\boldsymbol{\theta}}_{\epsilon}\right|}{\left|\hat{\boldsymbol{\Sigma}}_{YY}\right|} \tag{14.23}$$

$$R_{Y_j}^2 = 1 - \frac{\hat{\theta}_{\epsilon_{jj}}}{\hat{\sigma}_{Y_j}^2} \tag{14.24}$$

- R_X^2 and $R_{X_j}^2$: Similar to R_Y^2 and $R_{Y_j}^2$ but computed for the $\boldsymbol{X}$ variables.

$$R_X^2 = 1 - \frac{\left|\hat{\boldsymbol{\theta}}_{\delta}\right|}{\left|\hat{\boldsymbol{\Sigma}}_{XX}\right|} \tag{14.25}$$

$$R_{X_j}^2 = 1 - \frac{\hat{\theta}_{\delta_{jj}}}{\hat{\sigma}_{X_j}^2} \tag{14.26}$$

- R_η^2 and $R_{\eta_j}^2$: These are estimated for the endogenous latent variables $\boldsymbol{\eta}$.

$$R_\eta^2 = 1 - \frac{\left|\hat{\boldsymbol{\Psi}}\right|}{\left|\hat{\boldsymbol{\Sigma}}_{\eta\eta}\right|} \tag{14.27}$$

$$R_{\eta_j}^2 = 1 - \frac{\hat{\Psi}_{jj}}{\hat{\sigma}_{\eta_j}^2} \tag{14.28}$$

All these R^2 values should lie in between 0 and 1.

Example 14.6

Is the SEM model in Example 14.4 fit to the data?

Solution:

The fit measures for both the measurement and structural models are shown in Table 14.7.

From Table 14.7, it is seen that for the measurement model, the χ^2-statistic is insignificant and the χ^2 / df is 1.190, which does not satisfy the range (2 to 5), as suggested by researchers. The absolute fit indices GFI = 0.975, RMSR = 0.035, SRMSR = 0.035, and RMSEA = 0.025 exhibit a good fit of the measurement model to the data. All the relative fit indices (see Table 14.7) are greater than 0.90, indicating very good fit of the measurement model to the data. The parsimonious fit index AGFI (0.951) is a good fit and PNFI (0.550) is moderately fit to the data. Overall, the measurement model is a good fit to the data.

For the structural model, it is seen (Table 14.7) that the χ^2-statistic is significant but the χ^2 / df is 18.450 which does not satisfy the range (2 to 5), as suggested by researchers. However, χ^2 statistic is not preferred. The absolute fit indices, GFI = 0.996, RMSR = 0.042, and SRMSR = 0.042, exhibit a reasonable fit of the structural model to the data. However, RMSEA = 0.241 is not within the suggested limit. The relative fit indices are greater than 0.79 indicating moderate fit of the model to the data. The parsimonious fit index, AGFI (0.980), also exhibits a good fit. The PNFI is 0.310 and a revised structural model with less number of parameters may improve it.

14.8 TEST OF MODEL PARAMETERS

If the overall SEM as well as the structural and the measurement models do exhibit acceptable fit to the data, then the significance of each of the model parameters is tested using Z-test or t-test as these practically serve the purpose in all situations. We are primarily interested to test the parameters of the $\boldsymbol{\beta}$, $\boldsymbol{\Gamma}$, Λ_Y, and Λ_X matrices.

Often the standardized coefficient is used, as this is scale invariance and is bounded in between –1 to 1. It provides an opportunity to compare the effects of two or more independent

TABLE 14.7
Fit Measures for the Measurement and Structural Models

Type of Measure	Fit Index	Measurement Model	Structural Model
Chi-square based	χ^2_v	46.413 (df = 39, p-value = 0.193)	73.801 (df = 4, p-value = 0.000)
	χ^2_v / v	1.190	18.450
Absolute fit indices	GFI	0.975	0.996
	RMSR	0.035	0.042
	SRMSR	0.035	0.042
	RMSEA	0.025	0.241
Relative fit indices	TLI	0.979	0.799
	NFI	0.931	0.930
	CFI	0.988	0.933
	RNI	0.988	0.933
Parsimonious fit indices	AGFI	0.951	0.980
	PNFI	0.550	0.310

variables (IVs) against a particular dependent variable, even if the IVs are measured in different units. Let $\hat{\lambda}^s_{jk}, \hat{\gamma}^s_{jk}, \hat{\beta}^s_{jk}$ and $\hat{\lambda}_{jk}, \hat{\gamma}_{jk}, \hat{\beta}_{jk}$ be the sets of factor loadings, effect of ξ_k (exogenous) variable on η_j, and effect of η_k (endogenous) variable on η_j, as standardized (denoted by superscript s) and unstandardized estimates, respectively. Then

$$\begin{aligned} \hat{\lambda}^s_{jk} &= \hat{\lambda}_{jk}\left(\frac{\hat{\sigma}_{kk}}{\hat{\sigma}_{jj}}\right)^{1/2} \\ \hat{\gamma}^s_{jk} &= \hat{\gamma}_{jk}\left(\frac{\hat{\sigma}_{kk}}{\hat{\sigma}_{jj}}\right)^{1/2} \\ \hat{\beta}^s_{jk} &= \hat{\beta}_{jk}\left(\frac{\hat{\sigma}_{kk}}{\hat{\sigma}_{jj}}\right)^{1/2} \end{aligned} \tag{14.29}$$

Example 14.7

Consider Example 14.4. Test the parameters of the model and comment on the results.

Solution:

As correlation matrix is used to estimate the parameters, all the estimates are standardized (see Tables 14.8 and 14.9). R software is used to estimate the model parameters. The parameters are statistically significant.

14.9 OTHER IMPORTANT ISSUES IN SEM

14.9.1 Sample Size

The χ^2-based goodness of fit as well as the test of parameters of SEM is dependent on sample size (N). Marsh and Balla (1994) stated that as the hypothesized models are the approximations of their realities, based on the χ^2-statistics ($\chi^2 = (N-1)F_m(\boldsymbol{\theta})$), any model can be accepted with small

TABLE 14.8
Test of Parameters of Measurement Model for Example 14.7

Latent Variable	Manifest Variable	Parameters	Estimate	Std. Error	Z-value	P(>|Z|)
η_1	Y_1	λ_{Y11}	0.671	0.067	9.965	0.000
	Y_2	λ_{Y21}	0.639	0.066	9.606	0.000
η_2	Y_3	λ_{Y32}	0.5	0.067	7.409	0.000
	Y_4	λ_{Y42}	0.678	0.075	9.081	0.000
η_3	Y_5	λ_{Y53}	0.586	0.066	8.911	0.000
	Y_6	λ_{Y63}	0.715	0.069	10.312	0.000
ξ_1	X_1	λ_{X11}	0.554	0.064	8.596	0.000
	X_2	λ_{X21}	0.756	0.07	10.846	0.000
ξ_2	X_3	λ_{X32}	0.931	0.237	3.93	0.000
	X_4	λ_{X42}	0.482	0.132	3.646	0.000
ξ_3	X_5	λ_{X53}	0.857	0.071	12.134	0.000
	X_6	λ_{X63}	0.535	0.063	8.488	0.000

TABLE 14.9
Test of Parameters of Structural Model for Example 14.7

Latent Endogenous Variable	Latent Variable	Parameters	Estimate	Std. Error	Z-value	P(>|Z|)
η_1	ξ_1	γ_{11}	0.214	0.065	3.306	0.001
	ξ_2	γ_{12}	–0.074	0.05	–1.475	0.14
	η_2	β_{12}	0.767	0.075	10.237	0.000
η_2	ξ_1	γ_{21}	1	NA	NA	NA
	ξ_3	γ_{23}	0.911	0.046	19.974	0.000
	η_1	β_{21}	–0.746	0.051	–14.502	0.000
η_3	ξ_1	γ_{31}	0.86	0.041	20.812	0.000
	ξ_3	γ_{33}	0.96	0.043	22.478	0.000
	η_2	β_{32}	–0.706	0.051	–13.718	0.000

sample size and any model can be rejected with large sample size. Sample size depends on many factors such as model size (i.e., number of parameters), distributional assumptions, reliability of measurements, missing data, and the strength of relationships among the variables (Muthén and Muthén, 2002). Although many fit indices were developed, most of them are χ^2-based and do not provide identical results under all situations. Small sample size may lead to non-convergence and improper solutions (Anderson and Gerbing, 1988). However, there is no conclusive guidelines for appropriate sample size determination. The following guidelines are cited in different literatures.

- If t is the number of free parameters to be estimated, Hair et al. (1998) recommends $t \le N \le 10t$. If data departs from normality, the recommended N is 15t.
- If ML estimation is used, $N \ge 150$ is recommended for reasonably small standard error of estimates (Gerbing and Anderson, 1985). Ding et al. (1995) recommended N to be within *100–150* and Hoelter (1983) suggested $N = 200$ as the critical sample size.
- For non-normal situations, researchers suggested N should be much more than 200, preferably of the order *400* or *500*. Under such case, the χ^2-distribution-based statistics are not recommended as fit measure.
- Using a Monte Carlo study, Muthén and Muthén (2002) stated that for the measurement model (CFA) with normal population the recommended $N = 150$ without missing values and $N = 175$ with missing values but for the non-normal population, the recommended N with and without missing data is *265* and *375*, respectively.

14.9.2 Input Matrix

As seen in Section 14.6.2, the SEM uses sample covariance matrix (**S**) and minimizes one of the three functions (Equations 14.20–14.22) to estimate the parameters. Alternatively, sample correlation matrix **R** (i.e., standardized **S**) can be used. Which one should be preferred? It depends on the purpose of the study. If the purpose is to compare different populations or groups, covariance matrix is preferred. Correlation matrix, on the other hand, is used when the pattern of relationships between the factors and/or direct comparisons of the coefficients within a model are of importance. Another advantage of using correlation matrix is that it automatically constrains more parameters (owing to its diagonal elements equal to 1) and hence, increases the degree of freedom for parameter estimation. However, there has been debate on the use of **S** or **R**. As SEM was developed based on the statistical theory of covariance matrix and as the standardized solutions can also be obtained post hoc, the use of **S** is advocated by many. In conclusion, both **S** and **R** are used as input to SEM and the selection should be guided by the purpose of the study and analyst's convenience from interpretation point of view. A preferred approach may be the use of both **S** and **R** and take the advantages of both in interpreting the results. Note that the standardized coefficients derived from the results of the SEM using **S** may not match with that obtained using **R**. Bentler and Lee (1983) stated that when **R** is used, the asymptotic standard errors and χ^2-goodness of fit tests are not correct unless adjusted during estimation.

A related issue, which is often overlooked by researchers, is the computation of covariance and correlation between two variables when one or both are non-metric in nature. Data can be metric (measured using ratio or interval scale) or non-metric (measured using ordinal or nominal scale). The Pearson product moment correlation is developed for metric data. The types of correlations between two variables, say *X* and *Y* are given below.

- **Pearson product moment correlation:** When both *X* and *Y* are measured using metric scale (ratio or interval).
- **Tetrachoric correlation:** When both *X* and *Y* are binary, i.e., each is measured with two categories (dichotomous).
- **Polychoric correlation:** When both *X* and *Y* are polytomous, i.e., each is measured with three or more categories.
- **Biserial correlation:** When one of *X* and *Y* is metric and another one is binary (dichotomous).
- **Polyserial correlation:** When one of *X* and *Y* is metric and another one is polytomous.

Many times researchers use indicator coding for nominal and ordinal variables and report only Pearson product moment correlations, ignoring the nature of the measurement scale. For large sample size (say $N \geq 200$), the difference is negligible and therefore, in almost all practical situations, if the sample size is large, the Pearson product moment correlation may serve the purpose.

14.9.3 Estimation Process

Hair et al. (1998) mentioned four estimation processes: direct estimation, bootstrapping, simulation, and jackknifing. Direct estimation is the traditional method, where one-time estimation using a sample of certain size (say *N*) is made. The model adequacy tests and parameter tests are made using the procedures described earlier in this chapter.

Bootstrapping is a resampling technique where the original sample (of size *N*) is considered as the pseudo-population and then randomly *K* samples (*K* is very large), each of size *N* are collected from the pseudo-population. The hypothesized SEM model is estimated using certain technique (i.e., ML, ULS, or GLS) for each of the *K* samples. So, for each of the free parameters of the model, we have *K* values of estimate. If γ is the parameter to be estimated, then we have *K* number of γ values. The final γ value, as estimate of γ, is the mean of *K*-values.

$$\gamma = \sum_{k=1}^{K} \gamma_k / K$$

The variance of γ can be estimated as

$$V(\gamma) = \frac{1}{K-1}\sum_{k=1}^{K}(\gamma_k - \gamma)^2$$

Using the mean and variance of the parameter estimated, appropriate sampling distribution can be obtained and then confidence interval of the parameters can be computed. It is to be noted here that the original sample must be representative of the population.

The K samples derived from the original sample preserve the original characteristics of the sample (pseudo-population). No attempt is made to change any of the characteristics of the original sample. The simulation model, on the other hand, not only generates multiple samples, but at the same time modifies certain characteristics of the original sample systematically based on the research needs.

The jackknife procedure randomly creates N new samples, each of size $N - 1$ by omitting one unique observation each time from the original sample. Then the proposed SEM model is run for each of the N samples; thus, resulting with N values of each of the free parameters of the model. The final estimate will be the average of all the N values. As every new sample has one unique omitted observation from the original sample, jackknife is also used for identifying influential observations.

Paxton et al. (2001) did an extensive study on the use of Monte Carlo simulation (MCA) in SEM and provided a nine-step procedure for design and implementation of MCA in SEM. However, use of MCA in SEM dates back to the 1980s. See Anderson and Gerbing (1984), Hu and Bentler (1999), and Muthen and Kaplan (1992).

14.9.4 Model Respecification

It might be experienced by researchers that a hypothesized SEM, well-grounded with theoretical and practical perspectives, may suffer from non-convergence or improper solutions. Among other things, a fundamentally misspecified model may be the reason for this. Hence, almost all hypothesized SEM models require respecifying the model. For example, after estimating the proposed model if we find offending estimates, its reasons must be identified and rectified. The rectified model is a respecified one and we need to test it again. As stated earlier, the SEM can be estimated as a single model or a two-step process, i.e., estimation of the measurement model followed by estimation of the structural model when the measurement model was accepted. If the two-step process is used, the SEM model respecification also comprises two steps, i.e., respecification of the measurement until it is accepted. Then the structural model will be estimated and respecified if needed. The two-step model specification is recommended by many researchers and most of the time it gives superior results, at least from the model respecification point of view. If the measurement model is correct, it is quite likely that the structural model has fewer problems, as the covariance or correlation matrix of the latent factors, as the input to the structural model, are obtained from the measurement model.

The following guidelines may help researchers in respecifying the measurement and structural models (Tarka, 2018):

- Omission of important parameters: This results when one or more parameters of the hypothesized (proposed) model are excluded from estimation. Researchers may use forward specification research to correct this type of specification error. The forward specification search starts with a more constrained model and then gradually relaxes it towards a more general model (Chou and Bentler, 1993). The Lagrange multiplier test or the expected parameter change test can be performed to detect the errors.
- Inclusion of more parameters than needed: This error is rectified using backward specification search, i.e., starting with a more general model and then gradually constraining it. The error is detected using the Wald test.

- Omission of important variables: This leads to biased parameter estimation and incorrect standard errors. This is more due to lack of domain knowledge or unavailability of appropriate measurement. Researchers use the Hausman test (Hausman, 1978) and Rosenbaum's (1986) sensitivity test to rectify the error.
- For measurement models with unacceptable fit, Anderson and Garbing (1988) specified four ways to respecify the model: (i) relate the indicator (manifest variable) to another factor, (ii) delete this indicator, (iii) relate the indicator to multiple factors, and (iv) use correlated measurement errors.
- The patterns of residuals can be used for model respecification. Normalized residuals greater than 2.58 are questionable. Alternatively, probability plot of the normalized residuals can be used. The residuals should not be significantly away from the straight line.
- Use of similarity coefficients between indicators are recommended by Anderson and Gerbing (1982). The similarity coefficient ranges from –1.0 to 1.0 and greater values (in absolute) represent better consistency (both internal and external). Alternative indicators of a factor should have similarity coefficient of 0.80 or more.
- Use modification index (MI) proposed by Jöreskog and Sorbom (1984) to improve the model fit. The MI is the expected decrease in the χ^2-statistic of the model when a fixed parameter is freed. If the reduction of χ^2 value is 3.84 or more, it is recommended to estimate the parameter instead of fixing it to a certain value (such as 0 or 1).
- Saris et al. (1987) proposed expected parameter change (EPC) statistic as

$$\text{EPC}(\theta_j) = \theta_j - \theta_{j0} = \frac{\text{MI}}{d\theta_j}$$

where θ_j = the parameter to be fixed,

θ_{j0} = value of θ_j under H_0, and

$d\theta_j = -(N-1)\dfrac{2F}{2\theta_j}$, evaluated at θ_j

The difference between the EPC and the MI is that the EPC measures the change in the parameter value, whereas the MI measures the change in the overall model χ^2 fit.

Saris et al. (1987) provided the following guidelines for model specification (see Kaplan, 1990, for further discussion and Bollen (1990) for further insights):

- Case I: Large MI and large EPC: Estimate the parameter provided the theoretical and practical perspectives support it.
- Case II: Large MI but small EPC: Do not free the parameter.
- Case III: Small MI but large EPC: No decision can be taken as large EPC can be attributed to other causes, such as sampling variability.
- Case IV: Small MI and small EPC: Do not free the parameter.

14.10 CASE STUDY

14.10.1 BACKGROUND

Employees' job risk perception is important as it psychologically affects the safety behavior of workers in the workplace. Several studies have demonstrated the usefulness of understanding the factors shape workers' perception regarding safety in the workplace. An employee's job risk perception might be influenced by co-workers' attitude to safety, supervisors' safety practices, management safety practices, and safety program effectiveness. These factors are often latent in nature and, hence, not directly observable. These are usually measured using soft instruments like questionnaire surveys. While modeling the relationships of latent variables that are measured using the soft instruments, two types of quantification become necessary: (i) quantification of the latent factors, and (ii) estimation of the relationships among the latent factors. Structural equation modeling is the most suited technique in this situation. In this study, we demonstrate how structural equation modeling (SEM) can be used to perform the quantification issues mentioned above. The study was conducted in a steel plant. Following Hayes et al. (1998), the required questionnaire were prepared and a survey was administered to 76 crane operators. Three observations were removed due to inconsistency in the responses.

14.10.2 Variables and Data

The study focuses on the factors influencing employees' job risk perception in a steel plant. The five latent factors were observed with a set of ten questions each, making the survey 50 questions long. The survey is based on the questionnaire of Hayes et al. (1998). As the study deals with job risk perception among employees, the question of "safety" was omitted from further analysis. A five-point Likert scale was used to measure the responses. Against every question in Table 14.10, the response could be "strongly disagree," "disagree," "undecided," "agree," or "strongly agree." The numerical values assigned to these responses are 1, 2, 3, 4, or 5, respectively. In order to make the model less complex, each set of ten questions under each factor were further classified under sub-factors as per the experience of the domain expert (Table 14.10). The scores for each of the sub-factors were computed based on the corresponding indicators using confirmatory factor analysis (CFA). The sub-factors were then considered as manifest variables for the five underlying factors.

The correlation matrix for the sub-factors is shown in Table 14.11.

14.10.3 Conceptual Model

The analysis can be done in a single stage or in two stages comprising measurement model and structural model. As the researchers suggested the two-stage modeling as a prefferred approach, in this study we adopted a two-stage modeling approach. The measurement model is used to compute the correlation matrix for the five latent factors namely, co-workers' attitude to safety (η_1), job risk perception (η_2), supervisors' safety practices (η_3), management safety practices (ξ_1), and safety program effectiveness (ξ_2). The correlation matrix is then used in the structural model to estimate the cause and effect relationship among the latent factors (see Figure 14.9). The measurement and structural models are described in detail below.

For the measurement model, following Section 14.6.1, the number of explanatory manifest variables (p) = 5, number of explained manifest varaibles (q) = 8.

$$\frac{1}{2}(p+q)(p+q+1) = \frac{1}{2}(5+8)(5+8+1) = 91$$

Now, based on the t-rule, the total number of parameters to be estimated (t) in the hypothesized model = 41.

As $t = 41 < \frac{1}{2}(p+q)(p+q+1) = 91$, the order condition is satisfied. Further, all the latent factors in the model have two or more indicator variables. Hence, the model satisfies the indicator rules. As the measurement model satisfies both the t-rules and indicator rules, the model will be identifiable

TABLE 14.10
Factors, Sub-factors, and Indicators for Questionnaire Survey

Factor	Sub-factors	Indicators	Sample Mean (std. dev.)
Job risk perception (JRP)	Hazard perception (Y_1)	Hazardous	3.93 (0.06)
		Risky	3.82 (0.96)
		Chance of death	2.29 (1.30)
		Scary	2.16 (1.26)
	Safety perception (Y_2)	Dangerous	3.92 (1.19)
		Could get hurt easily	2.27 (1.03)
		Unsafe	2.81 (1.15)
	Health perception (Y_3)	Unhealthy	2.90 (1.23)
		Fear of health	2.79 (1.32)
Co-workers' attitude to safety (CAS)	Safety rules (Y_4)	Ignore safety rules	1.58 (1.01)
		Pay attention to safety rules	4.51 (0.80)
		Follow safety rules	4.55 (0.69)
		Do not pay attention	1.66 (0.89)
	Care for others (Y_5)	Do not care about others safety	1.45 (0.75)
		Look out for others safety	4.56 (0.67)
		Encourage others to be safe	4.58 (0.62)
	Workplace safety (Y_6)	Take chances with safety	2.62 (1.51)
		Keep work area clean	4.34 (0.82)
		Safety oriented	4.44 (0.62)
Supervisor safety practices (SSP)	Supervisor attitude towards worker safety behavior (Y_7)	Praise safe work behaviors	4.16 (0.93)
		Encourage safe behaviors	4.11 (0.98)
		Reward safe behaviors	3.11 (1.35)
	Safety rules enforcement (Y_8)	Keep workers informed of safety rules	4.10 (0.97)
		Update safety rules	3.70 (1.20)
		Enforce safety rules	4.01 (1.07)
		Involves workers in setting safety goals	3.33 (1.40)
		Discuss safety issues with others	3.63 (1.25)
		Train workers to be safe	3.85 (1.04)
		Act on safety suggestion	3.45 (1.31)
Management safety practices (MSP)	Safety practice (X_1)	Provide enough safety programs	3.78 (1.06)
		Conducts frequent safety inspection	3.15 (1.29)
		Investigate safety problems quickly	3.15 (1.28)
		Rewards safe workers	3.10 (1.18)
	Providing safe environment (X_2)	Provide safe equipment	3.88 (1.01)
		Provide safe working condition	3.64 (1.07)
		Responds quickly to safety concerns	3.22 (1.23)
		Helps maintain clean work area	4.04 (0.79)
	Safety communication (X_3)	Provide safety information	3.88 (0.85)
		Keep workers informed of hazards	3.89 (1.06)

(continued)

TABLE 14.10 (Continued)
Factors, Sub-factors, and Indicators for Questionnaire Survey

Factor	Sub-factors	Indicators	Sample Mean (std. dev.)
Safety programs effectiveness (SPE)	Usefulness of the program (X_4)	Worthwhile	4.52 (0.60)
		Useful	4.53 (0.58)
		Important	4.19 (0.91)
		Does not apply to my workplace	2.77 (1.36)
		Does not work	2.01 (1.20)
		Helps prevents accidents	4.51 (0.58)
		Effective in reducing injuries	4.23 (0.81)
	Quality of conduct (X_5)	Good	4.42 (0.69)
		First rate	4.21 (1.07)
		Unclear	1.66 (1.08)

TABLE 14.11
Correlation Matrix for the Sub-factors

	Y_1	Y_2	Y_3	Y_4	Y_5	Y_6	Y_7	Y_8	X_1	X_2	X_3	X_4	X_5
Y_1	1												
Y_2	0.57	1											
Y_3	0.46	0.43	1										
Y_4	0.03	0.20	0.14	1									
Y_5	0.05	0.19	0.17	0.96	1								
Y_6	0.02	0.13	0.05	0.70	0.72								
Y_7	−0.02	−0.02	0.04	−0.31	−0.33	−0.36	1						
Y_8	−0.13	−0.09	−0.07	−0.26	−0.30	−0.36	0.65	1					
X_1	−0.29	−0.18	−0.32	−0.12	−0.15	−0.18	0.44	0.67	1				
X_2	−0.38	−0.15	−0.41	−0.24	−0.26	−0.29	0.45	0.67	0.65	1			
X_3	−0.14	−0.02	−0.16	−0.31	−0.32	−0.39	0.60	0.67	0.66	0.61	1		
X_4	0.01	0.08	0.01	−0.44	−0.43	−0.41	0.54	0.54	0.29	0.33	0.60	1	
X_5	−0.04	−0.03	−0.07	−0.61	−0.59	−0.54	0.59	0.61	0.35	0.40	0.68	0.92	1

unless the covariance matrix is non-invertible, which will emerge during parameter estimation, if it exists.

For the structural model, following Section 14.6.1, the number of exogenous latent variables (m) = 2, number of endogenous latent varaibles (n) = 3.

$$\frac{1}{2}(m+n)(m+n+1)=\frac{1}{2}(2+3)(2+3+1)=15$$

Now, based on the t-rule, the total number of parameters to be estimated (t) in the hypothesized model = 16.

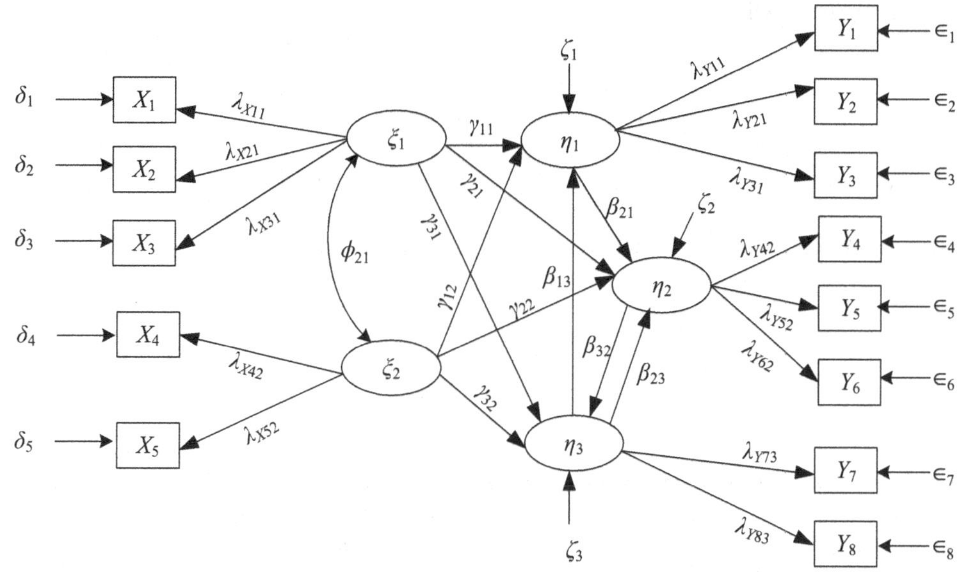

FIGURE 14.9 Conceptual structural equation model (combined measurement and structural). **Manifest variables:** X_1: Safety practices, X_2: Providing safe environment, X_3: Safety communication, X_4: Usefulness of the safety programs, X_5: Quality of conduct of safety programs, Y_1: Safety rules, Y_2: Care for others, Y_3: Workplace safety, Y_4: Hazard perception, Y_5: Safety perception, Y_6: Health perception, Y_7: Supervisors' attitude towards workers' safety behaviour, Y_8: Safety rules enforcement. **Latent variables:** ξ_1: Management safety practices, ξ_2: Safety programs effectiveness, η_1: Co-workerss' attitude to safety, η_2: Job risk perception, η_3: Supervisors' safety practices.

As $t = 16 > \frac{1}{2}(\mathrm{m}+n)(\mathrm{m}+n+1) = 15$, the order condition is not satisfied. The structural model is not identifiable. To make the structural model identifiable, one or more of the suggestions given by researchers may be adopted, such as (i) fixing of some of the parameters to constant value, (ii) omitting some of the paths, or (iii) equating some of the parameters to others. If we use the correlation matrix of the latent variables (obtained from the measurement model), the diagonal elements of the ϕ matrix (correlations among latent exogenous variables) become 1. In this case, the value of t will be reduced to 14, making the model identifiable based on order condition.

Following the procedure given in Section 10.4.1 of Chapter 10, the rank conditions for the structural model are tested

$$\eta_1 = \beta_{13}\eta_3 + \gamma_{11}\xi_1 + \gamma_{12}\xi_2 + \zeta_1$$
$$\eta_2 = \beta_{21}\eta_1 + \beta_{23}\eta_3 + \gamma_{21}\xi_1 + \gamma_{22}\xi_2 + \zeta_2$$
$$\eta_3 = \beta_{32}\eta_2 + \gamma_{31}\xi_1 + \gamma_{32}\xi_2 + \zeta_3$$

So,

$$\boldsymbol{\beta} = \begin{bmatrix} 0 & 0 & \beta_{13} \\ \beta_{21} & 0 & \beta_{23} \\ 0 & \beta_{32} & 0 \end{bmatrix}; \; \boldsymbol{\Gamma} = \begin{bmatrix} \gamma_{11} & \gamma_{12} \\ \gamma_{21} & \gamma_{22} \\ \gamma_{31} & \gamma_{32} \end{bmatrix}$$

Hence, the matrix **M** will be

$$\mathbf{M} = \begin{bmatrix} 1 & 0 & -\beta_{13} & -\gamma_{11} & -\gamma_{12} \\ -\beta_{21} & 1 & -\beta_{23} & -\gamma_{21} & -\gamma_{22} \\ 0 & -\beta_{32} & 1 & -\gamma_{31} & -\gamma_{32} \end{bmatrix}$$

To test the rank condition for η_1, we have the following **M** matrix.

$$\mathbf{M}_1 = \begin{bmatrix} 0 \\ 1 \\ -\beta_{32} \end{bmatrix}$$

Rank ($\mathbf{M}_1$) is 1, which is less than $(n-1) = 2$. So, the structural equation η_1 is under-identified.

Similarly, for η_2, we will not get $\mathbf{M}_2$ matrix as there are no missing endogenous or exogenous variables. Hence, the rank of $\mathbf{M}_2$ will be 0, which is less than $(n-1) = 2$. So, the structural equation η_2 is under-identifiable.

Similarly, for η_3, we get

$$\mathbf{M}_3 = \begin{bmatrix} 1 \\ -\beta_{21} \\ 0 \end{bmatrix}$$

Rank ($\mathbf{M}_3$) = 1, which is again less than $(n-1) = 2$. So, the structural equation η_3 is also under-identifiable.

None of the structural equations are identified and the overall structural model is under-identified. The identification options for the structural model will be discussed later in Section 14.10.5. As stated earlier, we adopt a two-stage process for estimation and as the measurement model is identifiable, we discuss first the parameter estimation, model fit, and findings of the measurement model below.

14.10.4 Measurement Model

Overall Fit of the Measurement Model

The measurement model is shown in Figure 14.10. The overall fit of the model and R^2 values of the manifest variables are shown in Tables 14.12 and 14.13, respectively.

From Table 14.12, it is seen that the χ^2-statistic is significant but the χ^2 / df is 1.679 which does not satisfy the range (2 to 5), as suggested by researchers. The absolute fit indices GFI = 0.850, RMSR = 0.077, SRMSR = 0.078, and RMSEA = 0.096 exhibit a reasonable fit of the model to the data. The relative fit indices (see Table 14.12) except NFI are greater than 0.90 indicating very good fit of the model to the data. The parsimonious fit indices AGFI (0.756) and PNFI (0.633) are moderately fit. From Table 14.13, it is evident that the R^2 values of all the manifest variables (except Y_2 and Y_3) are more than 0.50, indicating that the proportion of the variance of the manifest variables explained by the latent variables is more than 50%. The proportions of variance of Y_2 and Y_3 explained by the latent variable CAS are 0.445 and 0.443, respectively. Overall, the measurement model is a reasonably good fit and therefore could be accepted for further model-building.

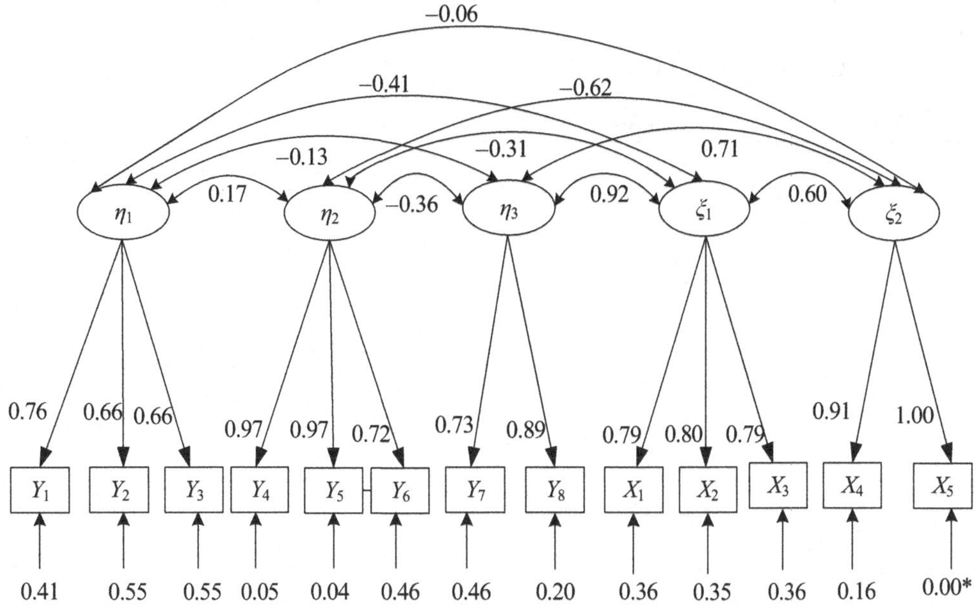

FIGURE 14.10 Measurement model. There are three endogenous latent variables ($\boldsymbol{\eta}$), two exogenous latent variables ($\boldsymbol{\xi}$), eight endogenous manifest variables (***Y***) and five exogenous manifest variables (***X***). The numeric figures represent the parameter estimates (for factor loadings, error variances, and covariance among the latent variables).

TABLE 14.12
Fit Measures

Type of Measure	Fit Index	Computed Value
Chi-square based	χ^2_v	93.999 ($df = 56$, p-value = 0.001)
	χ^2_v / v	1.679
Absolute fit indices	GFI	0.850
	RMSR	0.077
	SRMSR	0.078
	RMSEA	0.096
Relative fit indices	TLI	0.926
	NFI	0.882
	CFI	0.947
	RNI	0.947
Parsimonious fit indices	AGFI	0.756
	PNFI	0.633

TABLE 14.13
R-square for Manifest Variables

Latent Variables	Manifest Variables	R^2
CAS (η_1)	Y_1	0.586
	Y_2	0.445
	Y_3	0.443
JRP (η_2)	Y_4	0.954
	Y_5	0.96
	Y_6	0.531
SSP (η_3)	Y_7	0.536
	Y_8	0.798
MSP (ξ_1)	X_1	0.64
	X_2	0.642
	X_3	0.635
SPE (ξ_2)	X_4	0.839
	X_5*	1.000

*Error variance of X_5 is set to zero to avoid negative error variance.

TABLE 14.14
Parameter Estimates

Latent Variable	Manifest Variable	Parameters	Estimate	Std. Error	Z-value	P(>|Z|)
CAS (η_1)	Y_1	λ_{Y11}	0.761	0.119	6.392	0.000
	Y_2	λ_{Y21}	0.662	0.12	5.522	0.000
	Y_3	λ_{Y32}	0.661	0.12	5.51	0.000
JRP (η_2)	Y_4	λ_{Y42}	0.97	0.085	11.402	0.000
	Y_5	λ_{Y52}	0.973	0.085	11.475	0.000
	Y_6	λ_{Y62}	0.723	0.1	7.209	0.000
SSP (η_3)	Y_7	λ_{Y73}	0.727	0.104	7.016	0.000
	Y_8	λ_{Y83}	0.887	0.097	9.153	0.000
MSP (ξ_1)	X_1	λ_{X11}	0.794	0.1	7.937	0.000
	X_2	λ_{X21}	0.796	0.1	7.958	0.000
	X_3	λ_{X31}	0.791	0.1	7.895	0.000
SPE (ξ_2)	X_4	λ_{X42}	0.909	0.089	10.268	0.000
	X_5	λ_{X52}	0.993	0.082	12.083	0.000

Parameter Estimation

The parameters of the measurement model, error variances, and correlations among latent variables, are shown in Tables 14.14 to 14.16, respectively.

Test of Model Parameters

Table 14.14 shows the parameter estimates, their standard errors, and p-values of manifest variables. All the estimates are statistically significant. The error variances with their standard errors, Z-value, and p-values are given in Table 14.15. The results shown in Tables 14.14 and 14.15 are obtained after setting the only offending estimate (negative error variance for X_5) to zero in the model. All the

TABLE 14.15
Error Variances

	Error Variances		Test Statistic	
Error Variable	**Estimate**	**Standard Error**	**Z-value**	**P(>|Z|)**
ϵ_1	0.408	0.123	3.306	0.001
ϵ_2	0.548	0.123	4.47	0.000
ϵ_3	0.55	0.123	4.482	0.000
ϵ_4	0.046	0.026	1.753	0.080
ϵ_5	0.039	0.026	1.516	0.130
ϵ_6	0.463	0.079	5.892	0.000
ϵ_7	0.457	0.087	5.264	0.000
ϵ_8	0.199	0.071	2.784	0.005
δ_1	0.355	0.074	4.773	0.000
δ_2	0.353	0.074	4.759	0.000
δ_3	0.36	0.075	4.799	0.000
δ_4	0.159	0.026	6.042	0.000
δ_5	0.00*	* Fixed to zero to avoid negative error variance for δ_5		

TABLE 14.16
Correlation Matrix of Latent Variables

Latent Variables	CAS (η_1)	JRP (η_2)	SSP (η_3)	MSP (ξ_1)	SPE (ξ_2)
CAS (η_1)	1				
JRP (η_2)	0.166	1			
SSP (η_3)	–0.129	–0.355	1		
MSP (ξ_1)	–0.414	–0.305	0.917	1	
SPE (ξ_2)	–0.064	–0.618	0.709	0.597	1

estimates of error variances (except for Y_4 and Y_5) are statistically significant (p<0.05). But Y_4 and Y_5 error variances are significant at $p = 0.08$ and $p = 0.13$, respectively.

Table 14.16 shows the correlations among the five latent variables namely, CAS, JRP, SSP, MSP, and SPE. All the correlation coefficients (absolute values; except correlation of CAS with JRP, SSP, and SPE) are more than 0.30 and statistically significant.

14.10.5 Structural Model

Model Identification

As the structural model is not identified (see Section 14.10.3), the two widely used options to make the structural model identifiable are (i) inclusion of new latent variables in the model, and/or (ii) removal of some of the existing paths of the model. As option (i) was not available and the experts were in favor of exploring the relationships with the latent variables already included in the model, option (ii) seems plausible. Based on the practical experiences, a closer look into the η_2 equation (see Section 14.10.3) reveals that two paths, one from ξ_1 and another from ξ_2 may be removed, as the experts believed that the workers might not understand the safety program effectiveness in reducing their job risk directly. Therefore, these two paths are omitted. It is to be noted here that the model is still capable of estimating the effect of safety programs on job risk perception indirectly through η_1

TABLE 14.17
Results of Nine Structural Models

Path Removed from Equation		Model		Goodness	Relationship	
η_1	η_3	No.	Offending Estimates	of Fit (GFI)	Obtained*	*p*-value
$\xi_1 \rightarrow \eta_1$	$\xi_1 \rightarrow \eta_3$	M1	• Some of the error variances are more than one (for η_2 and η_1) • The absolute value of estimated coefficient for $\xi_2 \rightarrow \eta_3$ is > 1.00	GFI = 1.00	• SPE to CAS + • SSP to CAS – • SSP to JRP – • CAS to JRP + • SPE to SSP + • JRP to SSP +	• 0.273 • 0.103 • 0.000* • 0.162 • 0.000 • 0.000
	$\xi_2 \rightarrow \eta_3$	M2	• Nil	GFI = 0.931	• SPE to CAS + • SSP to CAS – • SSP to JRP – • CAS to JRP + • JRP to SSP – • MSP to SSP +	• 0.708 • 0.241 • 0.008* • 0.268 • 0.651 • 0.000*
	$\eta_2 \rightarrow \eta_3$	M3	• Nil	GFI = 0.970	• SPE to CAS + • SSP to CAS – • SSP to JRP – • CAS to JRP + • SPE to SSP + • MSP to SSP +	• 0.733 • 0.305 • 0.002* • 0.265 • 0.000* • 0.000*
$\xi_2 \rightarrow \eta_1$	$\xi_1 \rightarrow \eta_3$	M4	• Some of the error variances are more than one (for η_2 and η_3) • The absolute values of estimated coefficient for $\xi_1 \rightarrow \eta_1$, $\eta_3 \rightarrow \eta_1$, $\eta_1 \rightarrow \eta_2$, $\eta_3 \rightarrow \eta_2$, $\eta_2 \rightarrow \eta_3$, $\xi_2 \rightarrow \eta_3$ are > 1.00	GFI = 0.997	• MSP to CAS – • SSP to CAS + • SSP to JRP – • CAS to JRP – • JRP to SSP + • SPE to SSP +	• 0.000* • 0.000* • 0.596 • 0.609 • 0.095 • 0.058
	$\xi_2 \rightarrow \eta_3$	M5	• The absolute values of estimated coefficient for $\xi_1 \rightarrow \eta_1$ and $\eta_3 \rightarrow \eta_1$ are > 1.00	GFI = 1.00	• MSP to CAS – • SSP to CAS + • SSP to JRP – • CAS to JRP + • MSP to SSP + • JRP to SSP -	• 0.000* • 0.000* • 0.482 • 0.129 • 0.000* • 0.145
	$\eta_2 \rightarrow \eta_3$	M6	• The absolute values of estimated coefficient for $\xi_1 \rightarrow \eta_1$ and $\eta_3 \rightarrow \eta_1$ are > 1.00	GFI = 0.997	• MSP to CAS – • SSP to CAS + • SSP to JRP – • CAS to JRP + • SPE to SSP + • MSP to SSP +	• 0.000* • 0.000* • 0.002* • 0.265 • 0.000* • 0.000*
$\eta_3 \rightarrow \eta_1$	$\xi_1 \rightarrow \eta_3$	M7	• Some of the error variances are more than one (for η_2) • The absolute value of estimated coefficient for $\xi_2 \rightarrow \eta_3$ is > 1.00	GFI = 0.790	• SPE to CAS + • MSP to CAS – • SSP to JRP – • CAS to JRP + • JRP to SSP + • SPE to SSP +	• 0.027* • 0.000* • 0.000* • 0.654 • 0.000* • 0.000*

TABLE 14.17 (Continued)
Results of Nine Structural Models

Path Removed from Equation η_1	η_3	Model No.	Offending Estimates	Goodness of Fit (GFI)	Relationship Obtained*	*p*-value
	$\xi_2 \rightarrow \eta_3$	M8	• Nil	GFI = 0.952	• SPE to CAS + • MSP to CAS – • SSP to JRP – • CAS to JRP + • MSP to SSP + • JRP to SSP -	• 0.027* • 0.000* • 0.016* • 0.294 • 0.000* • 0.599
	$\eta_2 \rightarrow \eta_3$	M9	• Nil	GFI = 0.975	• SPE to CAS + • MSP to CAS – • SSP to JRP – • CAS to JRP + • SPE to SSP + • MSP to SSP +	• 0.027* • 0.000* • 0.003* • 0.289 • 0.000* • 0.000*

* + indicates positive and – indicates negative relation.

and η_3. A similar exercise was done for the η_1 and η_3 equations. However, no paths leading to η_1 and η_3 could be removed. In order to make η_1 and η_3 equations identifiable, at least one structural path leading to them needs to be omitted. A simulation approach is adopted by removing one of several possible paths (options) at a time. The results of nine structural models to be estimated are shown in Table 14.17.

Model Respecification

All the nine structural models were estimated separately. It was found that five models, namely M1, M4, M5, M6, and M7, have offending estimates, where either variances of some of the errors and/or the absolute values of some of the path coefficients become greater than one. Four models, namely M2, M3, M8, and M9, do not exhibit any offending estimates. Further, the path coefficients of each of the nine models are then compared to find out the patterns of relationships which are shown in Table 14.18. As expected, most of the path relationships in most of the models exhibit similar patterns. It is interesting to note that the models without offending estimates show similar patterns of relationship for all the paths. Hence, it might be better to select the most plausible model from the models without offending estimates (i.e., M2, M3, M8, and M9). The goodness of fit index (GFI) of the nine models is also shown in Table 14.17 (column 4). The GFI for M2, M3, M8, and M9 is more than 0.90 and M9 has the highest (0.975). Further, in M9, all the path coefficients (except CAS to JRP, $\beta_{21} = 0.122$, $p = 0.289$) are significant and cover all the significant paths that are obtained in the other three models without offending estimates. In addition, while comparing the path relationships with their practical implications, the pattern of relationships in M9 is also justifiable. Considering all these, M9 is chosen as the final accepted model for the case study. It is to be highlighted here that the other models also deserve attention and an aggregated approach could be adopted in decision-making, in line with the philosophy of ensemble methods. The final model (M9) with estimated coefficients is shown in Figure 14.11.

Overall Fit of the Structural Model

The overall fit of the structural model and R^2 values of the manifest variables are shown in Tables 14.19 and 14.20, respectively.

TABLE 14.18
Comparison of Path Coefficients of the Nine Structural Models

Causal Paths	Tested in Models	Common Relationships	Other Observations
MSP to CAS	M4 to M9	Negative (–)	
SPE to CAS	M1 to M3 and M7 to M9	Positive (+)	
MSP to SSP	M2, M3, M5, M6, M8, and M9	Positive (+)	
SPE to SSP	M1, M3, M4, M6, M7, and M9	Positive (+)	
SSP to CAS	M4 to M6	Positive (+)	Models without offending estimates (M2 and M3) show negative relationship
	M1 to M3	Negative (–)	
JRP to SSP	M1, M4, and M7	Positive (+)	Models without offending estimates (M2 and M8) show negative relationship
	M2, M5, and M8	Negative (–)	
CAS to JRP	M1 to M3 and M5 to M9	Positive (+)	All models except M4 show positive relationship
	M4	Negative (–)	
SSP to JRP	M1 to M9	Negative (–)	

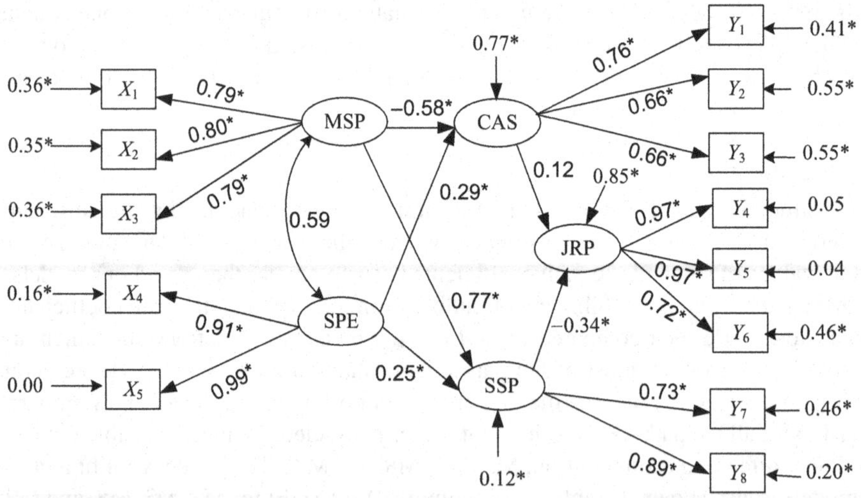

FIGURE 14.11 Final structural equation model (estimation in two steps). **Manifest variables:** X_1: Safety practices, X_2: Providing safe environment, X_3: Safety communication, X_4: Usefulness of the safety programs, X_5: Quality of conduct of safety programs, Y_1: Safety rules, Y_2: Care for others, Y_3: Workplace safety, Y_4: Hazard perception, Y_5: Safety perception, Y_6: Health perception, Y_7: Supervisors' attitude towards workers' safety behaviour, Y_8: Safety rules enforcement.

Note: The error variances are shown in the figure above.

From Table 14.19, it is seen that the χ^2-statistic is significant but the χ^2 / df is 24.514 which does not satisfy the range (2 to 5), as suggested by researchers. However, the χ^2-statistic is not preferred. The absolute fit indices are GFI = 0.975, RMSR = 0.110, and SRMSR = 0.112 and they exhibit a reasonable fit of the model to the data. Although RMSR and SRMSR values lower than 0.05 and 0.09, respectively, are preferable the fit indices depend on many factors (e.g., data distribution, input

TABLE 14.19
Fit Measures

Type of Measure	Fit Index	Computed Value
Chi-square based	χ_v^2	73.542 (*df* = 3, *p*-value = 0.000)
	χ_v^2 / v	24.514
Absolute fit indices	GFI	0.975
	RMSR	0.110
	SRMSR	0.112
Relative fit indices	NFI	0.716
	CFI	0.718
	RNI	0.718
Parsimonious fit indices	AGFI	0.877

TABLE 14.20
R^2 for Latent Endogenous Variables

Endogenous Variables	R^2
CAS	0.223
SSP	0.882
JRP	0.155

matrix, and sample size). The marginally higher values of RMSR and SRMSR may be attributed to low sample size (N = 73). The relative fit indices (see Table 14.19) are greater than 0.70, indicating moderate fit of the model to the data. The parsimonious fit index, AGFI (0.877), exhibits a good fit. From Table 14.20, it is seen that the R^2 value for SSP is 0.882, whereas the same for CAS and JRP is low. Overall, the structural model could be accepted considering the subjectivity involved in questionnaire survey-based data collection followed by derivation of the manifest variables using confirmatory factor analysis where some proportion of variability was lost (could not be explained by the factor). As the sample size for this study is very low, to improve the overall fit of the structural model, more respondents need to be included or, alternatively, resampling techniques such as bootstrapping could have been adopted.

Test of Model Parameters

The parameters of the final model (M9) and error variances are shown in Tables 14.21 and 14.22, respectively.

From Table 14.21, it can be observed that all the regression coefficients (except β_{21}) are statistically significant. The error variances of the latent endogenous variables are also found significant (see Table 14.22).

Direct, Indirect, and Total Effects

The direct, indirect, and total effects for job risk perception (JRP) could be obtained from the structural model. It is important as the hypothesized model assumes that the effects of MSP and SPE on JRP are realized through CAS and SSP. As expected, the indirect effect of both MSP and SPE on JRP is negative (see Table 14.23), which indicates that management efforts reduce the perceived job risk through the actions of supervisor (measured by SSP). As the effect coefficient of CAS on JRP is not statistically significant, we cannot convincingly accept the indirect effect of the management efforts through co-workers' activities (measured by CAS).

TABLE 14.21
Parameter Estimates

Latent Endogenous Variable	Latent Variable	Parameters	Estimate	Std. Error	Z-value	P(>|Z|)
CAS (η_1)	MSP (ξ_1)	γ_{11}	–0.584	0.129	–4.54	0.000
	SPE (ξ_2)	γ_{12}	0.285	0.129	2.217	0.027
JRP (η_2)	CAS (η_1)	β_{21}	0.122	0.115	1.06	0.289
	SSP (η_3)	β_{23}	–0.339	0.115	–2.946	0.003
SSP (η_3)	MSP (ξ_1)	γ_{31}	0.767	0.05	15.298	0.000
	SPE (ξ_2)	γ_{32}	0.251	0.05	5.003	0.000

TABLE 14.22
Error Variances

	Error Variances		Test Statistic	
Error Variable	Estimate	Standard Error	Z-value	P(>|Z|)
ζ_1	0.766	0.127	6.042	0.000
ζ_2	0.848	0.140	6.042	0.000
ζ_3	0.117	0.019	6.042	0.000

TABLE 14.23
Direct, Indirect, and Total Effects

Endogenous Variables	Other Latent Variables	Direct Effect	Indirect Effect	Total Effect
CAS (η_1)	MSP (ξ_1)	–0.584	NA	–0.584
	SPE (ξ_2)	0.285	NA	0.285
	JRP (η_2)	NA	NA	NA
	SSP (η_3)	NA	NA	NA
JRP (η_2)	MSP (ξ_1)	NA	–0.331	–0.331
	SPE (ξ_2)	NA	–0.050	–0.050
	CAS (η_1)	0.122	NA	0.122
	SSP (η_3)	–0.339	NA	–0.339
SSP (η_3)	MSP (ξ_1)	0.767	NA	0.767
	SPE (ξ_2)	0.251	NA	0.251
	CAS (η_1)	NA	NA	NA
	JRP (η_2)	NA	NA	NA

NA: Not applicable.

14.10.6 Discussions and Insights

The two-stage SEM model has been demonstrated with a case study. The case study establishes the relationships among the variables influencing employees' job risk perception. The measurement model comprises five latent and 13 manifest variables. The variance of the each of the manifest variables is satisfactorily explained by the measurement model. Further, the other fit measures also exhibit that the measurement model is fit to the data. In the estimation of the model parameters, only one offending estimate (for the error variance of X_5) was observed and fixed to zero as suggested by many researchers. Following the two-stage process of estimation, the correlation matrix for the five latent variables was used to estimate the structural model parameters.

On the other hand, the proposed structural (path) model was found to be under-identified. In order to make the structural model identified, a simulation-based approach was adopted. This involved the omission of paths (one at a time) from the model and estimation of the model parameters. In the process, nine different structural models were estimated. Finally, the best model among the nine estimated models was considered as the final model (M9), which gives a better explanation of the relationships among the five latent variables from their implementation points of view. The causal relationships between the latent variables from M9 can be explained as follows:

1. The exogenous latent variables considered in the structural model are (i) management safety programs (MSP) and (ii) safety programs effectiveness (SPE), as these reflect the management concern towards the safety of the employees. Further, researchers also suggested that all the management efforts are the root causes of positive or negative safety environment in an organization.
2. The other important fact that the researchers pointed out is that management efforts are usually implemented through supervisors' safety consciousness, involvement, and practices as the supervisors are the bridge between the management and the workers at the workplaces. Hence, supervisors' safety practices (SSP) was considered as an endogenous latent variable and was directly influenced by MSP and SSP. It is observed that both the MSP and SPE has a positive and significant relationship with SSP. So, both MSP and SPE provide encouragement to the supervisors to be more proactive in implementing safety efforts set by the management.
3. Co-workers' attitude towards safety (CAS) is another important variable for maintaining and improving safety in any organization. It is also suggested by the researchers that CAS is influenced by management efforts such as MSP and SPE. In this model, the relationship between CAS and SPE is found to be positive, while the relationship between CAS and MSP is negative. This apparently contradicts the relationships. It necessitates the study team to revisit the manifest variables of CAS and their correlation with other manifest variables including those of MSP and SPE. It is interesting to note that the manifest variables of concern have negative correlations. Therefore, it can be concluded that in the perception study, the respondents actually provided information towards quantification of co-workers' negligence on the safety of others. So, in this study, CAS might be considered as a negative variable and, hence, its relationship with MSP and SPE could be hypothesized as negative. The positive relationship between CAS and SPE, as obtained in the final model, could be attributed to the complacent nature of the workers as the SPE may induce a sense of satisfaction that they are safe and as a result they care even less for their co-workers.
4. The relationship between supervisors' safety practices (SSP) and job risk perception (JRP), as expected, is negative and significant. The workers are happy with SSP and observe that SSP reduces job risk.
5. The relationship between CAS and JRP is positive but insignificant. As stated earlier, CAS is measuring co-workers' negligence to safety, it increases job risk. As the relationship is not significant, we cannot conclude it convincingly.

The method of estimation applied in this analysis is the maximum likelihood estimation (MLE) as this is the most popular method of estimation among researchers. Other estimation methods such as weighted least squares and regularization-based methods could have been tested. As stated earlier, we have adopted a two-stage methodology for the estimation of the structural equation model. Apart from this, a combined modeling approach (the measurement model and the structural model estimated together) could have been adopted. However, researchers have found more merits in a two-stage procedure. The sample size for this study is low (73), which might have affected the overall fit of the model. Further, the respecifications in the structural model were made in accordance with the knowledge of the domain experts. Alternatively, the model may be respecified as per the modifications suggested from the modification indices, which may be explored further.

14.11 LEARNING SUMMARY

In this chapter, structural equation modeling is described from the points of view of model conceptualization, model identification, parameter estimation, goodness of fit, and test of model parameters. In addition, the modeling process (two-stage or combined), model respecification, exploration of direct and indirect relationships, and some specific discussions for improvement of model adequacy and its implication are explained. All the important fundamental relationships and their mathematical treatments are made and explained with a number of solved examples. Finally, a case study on crane operators' perception of job safety is presented.

In summary,

- A structural equation model comprises a measurement model and a structural model. The measurement model is based on the confirmatory factor analysis (see Chapter 13 for more details). The structural model is similar to path model described in Chapter 10. However, the treatment of the structural model in this chapter is different from that presented in Chapter 10. The main difference lies in the estimation of structural model parameters.
- The measurement model quantifies the latent variables using their manifest (indicator) variables. There are two types of manifest variables namely, endogenous (Y) and exogenous (X). The measurement model provides the covariance (and correlation) matrix among the latent variables. In addition, it estimates the loading matrices ($\mathbf{\Lambda}_X$ and $\mathbf{\Lambda}_Y$).
- The fundamental relationship, tested in measurement model is $\mathbf{\Sigma}_{MV}=\mathbf{\Lambda}\ \mathbf{\Phi}\mathbf{\Lambda}^T+\mathbf{\theta}$, where $\mathbf{\Sigma}_{\text{MV}}$ denotes the covariance matrix of the manifest variables (MV), both $\boldsymbol{X}$ and $\boldsymbol{Y}$. Here, both the $\boldsymbol{Y}$ and $\boldsymbol{X}$ variables are treated alike and the relationships are modeled in a single confirmatory factor model.
- The structural model captures the causal relationships between the latent variables and correlated relationships among the exogenous latent variables. It also captures the contribution of error variables. The structural model also helps in identifying the causal (direct and indirect) and spurious relationships among the latent variables.
- The fundamental relationship, tested in structural model part of SEM is $\mathbf{\Sigma}_{LV}=\begin{bmatrix}(\mathbf{I}-\boldsymbol{\beta})^{-1}(\mathbf{\Gamma\Phi\Gamma}^T+\mathbf{\Psi})(\mathbf{I}-\boldsymbol{\beta})^{-1^T} & (\mathbf{I}-\boldsymbol{\beta})^{-1}\mathbf{\Gamma\Phi} \\ \mathbf{\Phi\Gamma}^T(\mathbf{I}-\boldsymbol{\beta})^{-1^T} & \mathbf{\Phi}\end{bmatrix}$, where $\mathbf{\Sigma}_{LV}$ is the covariance matrix among the latent variables (LV).
- The fundamental relationship, tested in the combined (full) structural equation model (SEM) is $\mathbf{\Sigma}_{MV}=\begin{bmatrix}\mathbf{\Lambda}_Y(\mathbf{I}-\boldsymbol{\beta})^{-1}(\mathbf{\Gamma\Phi\Gamma}^T+\mathbf{\Psi})(\mathbf{I}-\boldsymbol{\beta})^{-1^T}\mathbf{\Lambda}_Y^T+\boldsymbol{\theta}_\epsilon & \mathbf{\Lambda}_Y(\mathbf{I}-\boldsymbol{\beta})^{-1}\mathbf{\Gamma\Phi\Lambda}_X^T \\ \mathbf{\Lambda}_X\mathbf{\Phi\Gamma}^T(\mathbf{I}-\boldsymbol{\beta})^{-1^T}\mathbf{\Lambda}_Y^T & \mathbf{\Lambda}_X\mathbf{\Phi\Lambda}_X^T+\boldsymbol{\theta}_\delta\end{bmatrix}$, where $\mathbf{\Sigma}_{MV}$ is the covariance matrix among the manifest variables (MV).
- In SEM, eight matrices $\boldsymbol{\beta}$, $\mathbf{\Gamma}$, $\mathbf{\Lambda}_\Psi$, $\mathbf{\Lambda}_\Xi$, $\mathbf{\Psi}$, $\mathbf{\Phi}$, $\boldsymbol{\theta}_\delta$, and $\boldsymbol{\theta}_\epsilon$ are required to be estimated.
- The methods of parameter estimation include maximum likelihood estimation (MLE), unweighted least squares (ULS), and generalized least squares (GLS). However, MLE is preferred by the researchers.

- As SEM contains several parameters to be estimated which are dependent on the model structure, model identification is a very important task in SEM modeling. Both the order condition and rank condition are to be satisfied for the model to become estimable. Using the order and rank conditions, the SEM can be over-identified, just-identified, or unidentified. There are several ways to make an unidentified SEM identifiable, as discussed in this chapter.
- In SEM, the estimation process may be a single-stage process where the complete SEM (both measurement and structural model parameters simultaneously) is estimated. On the other hand, a two-stage process (where the measurement and structural models are estimated separately) may be adopted. There are several advantages of the two-stage process mentioned by the researchers. In general, the two-stage process provides flexibility in the modeling.
- The overall fit of the SEM (for combined model or two-stage models) is measured using three types of fit indices namely, absolute fit indices (e.g., GFI, RMSR), relative fit indices (e.g., CFI, NFI), and parsimonious fit indices (e.g., AGFI, PNFI). There is no single fit measure that can be relied upon for the adequacy of SEM. As a result, several measures are considered. It is to be noted here that all the measures may not provide identical decisions regarding the overall model fit.
- The SEM model parameters are tested in a similar line as followed in testing regression coefficients in multiple or multivariate multiple linear regressions.
- The SEM is a complicated model and it considers the covariance or correlation matrix as inputs to its parameter estimation. Therefore, it is likely that the analyst encounters offending estimates (e.g., negative error variances). Under such situation, model respecification is suggested and there are several guidelines discussed in this chapter for model respecification.
- The researchers also identified other important issues in SEM such as sample size, model structure, and data distribution.

EXERCISES

(A) Short conceptual questions

14.1 Define structural equation modeling (SEM). In what ways it is different from the confirmatory factor model and path model? What prerequisite knowledge base is required for developing a structural equation model?

14.2 Often researchers use SEM to test several hypotheses. Consider a general structural equation model [Hint: see Figure 14.3] and enlist the hypotheses that the researchers may be interested to test.

14.3 Explain the different types of variables used in SEM. A structural equation model is explored to understand several relationships among latent variables and between latent and manifest variables. State these relationships with an example.

14.4 In SEM, several dispersion matrices are estimated. What are these? Explain with an example.

14.5 Consider the general SEM equation, $\eta_{n\times1}=\boldsymbol{\beta}_{n\times n}\boldsymbol{\eta}_{n\times1}+\boldsymbol{\Gamma}_{n\times m}\boldsymbol{\xi}_{m\times1}+\boldsymbol{\zeta}_{n\times1}$. Define the notations used in this equation. Develop the fundamental relationships for the SEM.

14.6 SEM is composed of two component models, namely the measurement model and structural model. Decompose the fundamental relationships (obtained in question 14.5) to its component models.

14.7 Show that in the SEM, the variance of η (endogenous constructs) explained by the model is $(\mathbf{I}-\boldsymbol{\beta})^{-1}(\boldsymbol{\Gamma\Phi\Gamma}^T+\boldsymbol{\Psi})(\mathbf{I}-\boldsymbol{\beta})^{-1^T}$ [the notations represent their usual meaning].

14.8 How do you define goodness of fit of a SEM?

14.9 What is model identification in SEM? Explain with an example. If a SEM model is under-identified, what are the different approaches to make it identifiable?

14.10 Consider the following structural model. State the structural equations. Is the model identifiable?

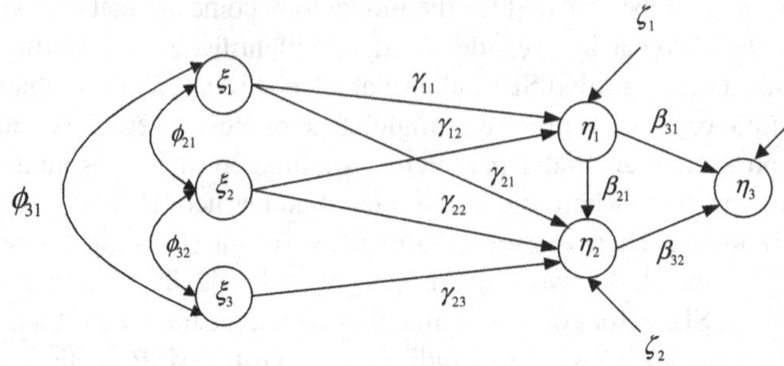

14.11 What fundamental function is minimized to estimate the SEM model parameters? State a few estimation methods and their corresponding fundamental function that are minimized while estimating the model parameters.

14.12 Consider the structural model of SEM. Explain direct, indirect, and total effects. How do you quantify them?

14.13 What are the different fit indices used to test the adequacy of SEM? What are the differences among the different fit indices?

14.14 What is coefficient of determination (R^2) in SEM?

14.15 How do you test model parameters in SEM? What ways it is different from that used in multiple linear regression (MLR)?

14.16 Sample size (N) is a critical issue in SEM. Explain, how Chi-square statistic is influenced by N? What are the recommendations for selection of acceptable sample size in SEM?

14.17 During parameter estimation, what input matrix is considered in SEM? What would be the differences in estimates if a correlation matrix is used as input matrix instead of a covariance matrix?

14.18 For SEM model estimation, two modeling approaches are proposed by the researchers. What are these approaches? Which one is preferred and when?

14.19 SEM often encounters model respecification. State the guidelines that could be adopted for model respecification under different situations.

14.20 What is modification index (MI) in SEM and how it is used? What ways it is different from expected parameter change (EPC)?

(B) Long conceptual and/or numerical questions: see web resources (Multivariate Statistical Modeling in Engineering and Management – 1st (routledge.com))

N.B.: R and MS Excel software are used for computation, as and when required.

NOTES:

1 Very important to understand and apply SEM. Readers may read the relevant chapters (8, 9 and 13) before reading this chapter.

2 See Maximum Likelihood Estimation of path model in Section 10.4 (Equation 10.38 and 10.39). In this Chapter, we estimate path model (or structural model part of SEM) using the approach of covariance structure modeling.

Bibliography

Aczel, A. D. & Sounderpandian, J. (2010). *Complete Business Statistics*, 6th edition. New York: Tata McGraw Hill.

Agresti, A. (2002). *An Introduction to Categorical Data Analysis*, 2nd edition. New York: John Wiley & Sons, Inc.

Akaike, H. (1974). A new look at the statistical model identification. *IEEE Transactions on Automatic Control*, 19: 716–723.

Anderson, J. C. & Gerbing, D. W. (1982). Some methods for respecifying measurement models to obtain unidimensional construct measurement. *Journal of Marketing Research*, 19(4): 453–460.

Anderson, J. C. & Gerbing, D. W. (1984). The effects of sampling error on convergence, improper solutions and goodness-of-fit indices for maximum likelihood confirmatory factor analysis. *Psychometrika*, 49: 155–173.

Anderson, J. C. & Gerbing, D. W. (1988). Structural equation modeling in practice: A review and recommended two-step approach. *Psychological Bulletin*, 103(3): 411–423.

Anderson, J. R. (1978). Arguments concerning representations for mental imagery. *Psychological Review*, 85(4): 249–277.

Anderson, T. W. (2003). *An Introduction to Multivariate Statistical Analysis*. New Delhi: Wiley.

Anderson, T. W. & Darling, D. A. (1952). Asymptotic theory of certain "goodness-of-fit" criteria based on stochastic processes. *Annals of Mathematical Statistics*, 23: 193–212.

Anderson, T. W. & Rubin, H. (1949). Estimation of the parameters of a single equation in a complete system of stochastic equations. *The Annals of Mathematical Statistics*, 20(1): 46–63.

Anderson, T. W. & Rubin, H. (1956). Statistical inference in factor analysis. In *Proceedings of the 3rd Berkeley Symposium on Mathematical Statistics and Probability*, Vol. 5, edited by J. Neyman. Berkeley: University of California Press, 111–150.

Banks, J., Carson, J. S., Nelson B. L., & Nicol, D. M. (2007). *Discrete-Event System Simulation*, 4th edition. New Delhi: Pearson Education.

Barnett, V. (1976). The ordering of multivariate data. *Journal of the Royal Statistical Society: Series A (General)*, 139(3): 318–344.

Barr, D. R. & Slezak, N. L. (1972). A comparison of multivariate normal generators. *Communication of the ACM*, 5(12): 1048–1049.

Bartlett, M. S. (1937a). Properties of sufficiency and statistical tests. *Proceedings of the Royal Society A*, 160: 268–282.

Bartlett, M. S. (1937b). The statistical conception of mental factors. *British Journal of Psychology*, 28: 97–104.

Bartlett, M. S. (1938). Further aspects of the theory of multiple regression. *Mathematical Proceedings of the Cambridge Philosophical Society*, 34: 33–40.

Bartlett, M. S. (1947). Multivariate analysis. *Supplement to the Journal of the Royal Statistical Society*, 9(2): 176–197.

Bartlett, M. S. (1950). Tests of significance in factor analysis. *British Journal of Psychology (Statistical Section)*, 3: 77–85.

Bartlett, M. S. (1954). A note on the multiplying factors for various chi square approximations. Journal of the Royal Statistical Society, 16: 296–298.

Bartlett, M. S. & Kendal, D. G. (1946). The statistical analysis of variance-heterogeneity and the logarithmic transformation. *Journal of the Royal Statistical Society*, 8: 128–138.

Basmann, R. L. (1960). On the asymptotic distribution of generalized linear estimators. *Econometrica: Journal of the Econometric Society*, 97–107.

Beaglehole, R., Bonita, R., & Kjellström, T. (1993). *Basic Epidemiology*. Geneva: World Health Organization.

Belsley, D. A., Kuh, E., & Welsch, R. E. (2004). *Regression Diagnostics: Identifying Influential Data and Sources of Collinearity*. New York: John Wiley & Sons.

Bentler, P. M. (1985). *Theory and Implementation of EQS: A Structural Equations Program*. Los Angeles: BMDP Statistical Software.

Bentler, P. M. (1990). Comparative fit indices in structural models. *Psychological Bulletin*, 107: 238–246.

Bentler, P. M. & Bonett, D. G. (1980). Significance tests and goodness of fit in the analysis of covariance structures. *Psychological Bulletin*, 88: 588–606.

Bentler, P. M. & Lee, S. Y. (1983). Covariance structures under polynomial constraints: Applications to correlation and alpha-type structural models. *Journal of Educational Statistics*, 8: 207–222.Beran, R. & Srivastava, M. S. (1985). Bootstrap tests and confidence regions for functions of a covariance matrix. *Annals of Statistics*, 13: 95–115.

Berk, K. N. (1977). Tolerance and condition in regression computations. *Journal of the American Statistical Association*, 72(360a): 863–866.

Berry, W. D. (1984). *Nonrecursive Causal Models*. London: Sage.

Bickel, P. J. & Doksum, K. A. (1981). An analysis of transformations revisited. *Journal of the American Statistical Association*, 76(374): 296–311.

Bjorck, A. (1996). Numerical Methods for Least Squares Problems. Philadelphia, PA: SIAM Publications.

Blom, G. (1958). *Statistical Estimates and Transformed Beta Variates*. New York: Wiley.

Bollen, K. A. (1989). *Structural Equations with Latent Variables*. New York: John Wiley & Sons.

Bollen, K. A. (1990). Overall fit in covariance structure models: Two types of sample size effects. *Psychological Bulletin*, 107(2): 256–259.

Bollen, K. A. & Davis, W. R. (2009). Two rules of identification for structural equation models. *Structural Equation Modeling: A Multidisciplinary Journal*, 16(3): 523–536.

Boudon, R. (1968). A new look at correlation analysis. In *Methodology in Social Research*, edited by H. M. Blalock & A. B. Blalock. New York: McGraw-Hill, 199–235.

Bowman, K. O. & Shenton, L. R. (1975). Omnibus test contours for departures from normality based on √ b1 and b2. *Biometrika*, 62(2): 243–250.

Box, G. E. P. (1949). A general distribution theory for a class of likelihood criteria. *Biometrika*, 36(3/4): 317–346.

Box, G. E. P. & Cox, D. R. (1964). An analysis of transformations. *Journal of the Royal Statistical Society: Series B (Methodological)*, 26(2): 211–243.

Box, G. E. P. & Muller, M. F. (1958). A note on the generation of random normal deviates. *Annals of Mathematical Statistics*, 29: 610–611.

Bray, J. H. & Maxwell, S. E. (1982). Analyzing and interpreting significant MANOVAs. *Review of Educational Research*, 52(3): 340–367.

Brown, M. B. & Forsythe, A. B. (1974). Robust tests for the equality of variances. *Journal of the American Statistical Association*, 69(346): 364–367.

Browne, M. M. (1984). Asymptotically distribution-free methods for the analysis of covariance structures. British Journal of Mathematical and Statistical Psychology, 37(1): 62–83.

Browne, M. W. & Cudeck, R. (1989). Single sample cross-validation indices for covariance structures. *Multivariate Behavioral Research*, 24: 445–455.

Byron, R. P. (1974). Testing structural specification using the unrestricted reduced form. *Econometrica: Journal of the Econometric Society*, 869–883.

Cai, T. T. (2017). Global testing and large-scale multiple testing for high-dimensional covariance structures. *Annual Review of Statistics and Its Application*, 4: 423–446.

Cai, T. T., Liang, T., & Zhou, H. H. (2015). Law of log determinant of sample covariance matrix and optimal estimation of differential entropy for high-dimensional Gaussian distributions. *Journal of Multivariate Analysis*, 137: 161–172.

Carmines, E. G. (1990). The statistical analysis of overidentified linear recursive models. *Quality and Quantity*, 24(1): 65–85.

Carrol, R. J. & Ruppert, D. (1988). *Transformation Weighting in Regression*. New York: Chapman & Hall.

Carroll, J. B. (1953). An analytical solution for approximating simple structure in factor analysis. *Psychometrika*, 18: 23–38.

Cattell, R. B. (1966). The scree test for the number of factors. *Multivariate Behavioral Research*, 1(2): 245–276.

Cheema J. R. (2014). A review of missing data handling methods in education research. *Review of Educational Research*, 84(4): 487–508.

Chou, C. P. & Bentler, P. M. (1993). Invariant standardized estimated parameter change for model modification in covariance structure analysis. *Multivariate Behavioral Research*, 28(1): 97–110.

Cochran, W. G. (1947). Some consequences when the assumptions for the analysis of variance are not satisfied. *Biometrics*, 3: 22–38.

Cohen, J., Cohen, P., West, S. G., & Aiken L. S. (2003). *Applied Multiple Regression/Correlation Analysis for the Behavioral Sciences*. London: L. E. Associates, Publishers.
Cook, R. D. & Weisberg, S. (1982). *Residuals and Influence in Regression*. New York: Chapman and Hall.
Cramer, H. (1946). *Mathematical Method of Statistics*. Princeton: Princeton University Press.
Cronbach, L. J. (1951). Coefficient alpha and the internal structure of tests. *Psychometrika*, 16: 297–334.
D'Agostino, R. B. (1986). *Goodness-of-Fit-Techniques*. Boca Raton, FL: CRC Press.
Daniel, C. & Wood, F. S. (1980). *Fitting Equations to Data: Computer Analysis of Multifactor Data*. New York: John Wiley & Sons.
Daniel, W. W. (2000). *Biostatistics: A Foundation for Analysis in the Health Sciences*, 7th edition. New York: John Wiley & Sons.
Davidson, R. & MacKinnon, J. G. (2004). *Econometric Theory and Methods*. New York: Oxford University Press.
Davison, A. C. & Hinkley, D. V. (1997). *Bootstrap Methods and their Application*. Cambridge: Cambridge University Press.
De Maesschalck, R., Jouan-Rimbaud, D., & Massart, D. L. (2000). The Mahalanobis distance. *Chemometrics and Intelligent Laboratory Systems*, 50(1): 1–18.
Dillon, W. R., Kumar, A., & Mulani, N. (1987). Offending estimates in covariance structure analysis: Comments on the causes of and solutions to Heywood cases. *Psychological Bulletin*, 101: 126–135.
Ding, D. J,, Martin, J. G., & Macklem, P. T. (1987). Effects of lung volume on maximal methacholine-induced bronchoconstriction in normal humans. *Journal of Applied Physiology*, 62(3): 1324–1330.
Ding, L., Velicer, W. F., & Harlow, L. L. (1995). Effects of estimation methods, number of indicators per factor, and improper solutions on structural equation modeling fit indices. *Structural Equation Modeling: A Multidisciplinary Journal*, 2(2): 119–143.
Dufour, J. M. (2003). Identification, weak instruments, and statistical inference in econometrics. *Canadian Journal of Economics/Revue canadienne d'économique*, 36(4): 767–808.
Duncan, O. D. (1975). *Introduction to Structural Equation Models*. New York: Academic Press.
Durbin, J. & Watson, G. S. (1951). Testing for serial correlation in least squares regression II. *Biometrica*, 38: 159–178.
Efron, B. (1992). Bootstrap methods: Another look at the jackknife. In *Breakthroughs in Statistics* (pp. 569–593). New York: Springer.
Efron, B. & Gong, G. (1983). A leisurely look at bootstrap, jackknife, and cross-validation. *The American Statistician*, 37(1): 36–48.
Efron, B. & Tibshirani, R. (1997). Improvements on cross-validation: The 632 + bootstrap method. *Journal of the American Statistical Association*, 92(438): 548–560.
Field, A. (2009). *Discovering Statistics Using SPSS*, 3rd edition. London: Sage.
Finn, J. D. (1974). *A General Model for Multivariate Analysis*. New York: Holt, Rinehart & Winston.
Finn, J. D. (2003). *A General Model for Multivariate Analysis*. New York: Holt, Rinehart & Winston.
Fornell, C. & Larcker, D. F. (1981). Evaluating structural equation models with unobservable variables and measurement error. *Journal of Marketing Research*, 18(1): 39–50.
Friendly, M. & Sigal, M. (2020). Visualizing tests for equality of covariance matrices. *The American Statistician*, 74(2): 144–155.
Gentle, J. E. (2009). *Computational Statistics*. New York: Springer.
Gerbing, D. W. & Anderson, J. C. (1985). The effects of sampling error and model characteristics on parameter estimation for maximum likelihood confirmatory factor analysis. *Multivariate Behavioral Research*, 20(3): 255–271.
Gerbing, D. W. & Anderson, J. C. (1987). Improper solutions in the analysis of covariance structures: Their interpretability and a comparison of alternate respecifications. *Psychometrika*, 52: 99–111.
Giraud, C. (2021). Introduction to High-Dimensional Statistics. New York: Chapman and Hall/CRC.
Glymour, C., Scheines, R., Spirtes, P., & Kelly, K. (1987). *Discovering Causal Structure*. Orlando, FL: Academic Press.
Gnanadesikan, R. & Kettenring, J. R. (1972). Robust estimates, residuals, and outlier detection with multiresponse data. *Biometrics*, 28(1): 81–124.
Goldberger, A. S. (1970). On Boudon's method of linear causal analysis. *American Sociological Review*, 35(1): 97–101.
Golub, G. H. & Loan, C. F. V. (2007). *Matrix Computations*, 3rd edition. New Delhi: Hindustan Book Agency.

Goodman, L. A. & Kruskal, W. H. (1954). Measures of association for cross classification. *Journal of the American Statistical Association*, 49: 732–754.

Greene, W. H. (2012). *Econometric Analysis*. Karnataka: Pearson Education India.

Gryna, F. M., Richard, C. H., Chua, J., & Defeo, A. (2006). *Juran's Quality Planning and Analysis for Enterprise Quality*, 5th edition. New York: McGraw-Hill Education.

Haase, R. F. & Ellis, M. V. (1987). Multivariate analysis of variance. *Journal of Counselling Psychology*, 34(4): 404–413.

Hahn, J. & Hausman, J. (2002). A new specification test for the validity of instrumental variables. *Econometrica*, 70(1): 163–189.

Hair, J. F., Anderson, R. E., Tatham, R. L., & Black, W. C. (1998). *Multivariate Data Analysis*, 5th edition. New York: Prentice Hall.

Hair, J. F., Black, W. C., Babin, B. J., & Anderson, E. (2010). *Multivariate Data Analysis: A Global Perspective*, 7th edition. Delhi: Pearson.

Hardle, W. & Simar, L. (2003). Applied Multivariate Statistical Analysis. New York: Springer.

Hausman, J. A. (1978). Specification tests in econometrics. *Econometrica: Journal of the Econometric Society*, 46(6): 1251–1271.

Hawkins, D. M. (1980). *Identification of Outliers*. London: Chapman and Hall.

Hayes, B. E., Perander, J., Smecko, T., & Trask, J. (1998). Measuring perceptions of workplace safety: Development and validation of the work safety scale. *Journal of Safety Research*, 29: 145–161.

Heise, D. R. (1969). Problems in path analysis and causal inference. *Sociological Methodology*, 1: 38–73.

Hershbeger, S. L. (2005). Factor score estimation. In *Encyclopedia of Statistics in Behavioral Science*, edited by B. S. Everitt & D. C. Howell. New York: John Wiley & Sons, 636–644.

Hines, W. W., Montgomery, D. C., Goldsman, D. M., & Borror, C. M. (2003). *Probability and Statistics in Engineering*. New Delhi: Pearson Education.

Hoelter, J. W. (1983). The analysis of covariance structures: Goodness-of-fit indices. *Sociological Methods & Research*, 11(3): 325–344.

Horn, J. L. (1965). A rationale and test for the number of factors in factor analysis. Psychometrika, 30(2): 179–185.

Hotelling, H. (1931). The generalization of student's ratio. *Annals of Mathematical Statistics*, 2: 360–378.

Hotelling, H. (1933). Analysis of a complex of statistical variables into principal components. *Journal of Educational Psychology*, 24(6): 417–441.

Hotelling, H. (1936). Relations between two sets of variables. *Biometrika*, 28: 321–335.

Hu, L. T. & Bentler, P. M. (1999). Cutoff criteria for fit indexes in covariance structure analysis: Conventional criteria versus new alternatives. *Structural Equation Modeling: A Multidisciplinary Journal*, 6(1): 1–55.

Hummel, T. J. & Sligo, J. R. (1971). Empirical comparison of univariate and multivariate analysis of variance procedures. *Psychological Bulletin*, 76: 49–57.

Jobson, J. D. (1991). *Applied Multivariate Data Analysis, Volume I: Regression and Experimental Design*. New York: Springer-Verlag.

John, J. A. & Draper, N. R. (1980). An alternative family of transformations. *Journal of the Royal Statistical Society: Series C (Applied Statistics)*, 29(2): 190–197.

Johnson, R. A. (2002). *Miller and Freund's Probability and Statistics for Engineers*, 8th edition. Chennai: Prentice Hall of India.

Johnson, R. A. & Wichern, D. W. (2002). *Applied Multivariate Statistical Analysis*, 5th edition. Delhi: PHI Learning Private Limited.

Johnson, R. A. & Wichern, D. W. (2013). *Applied Multivariate Statistical Analysis*, 6th edition. London: Pearson.

Johnston, J. & Dinardo, J. (1996). *Econometric Methods*. New York. McGraw-Hill Education.

Jolliffe, I. T. (1972). Discarding variables in a principal component analysis 1: Artificial data. *Journal of the Royal Statistical Society C*, 21(2): 160–173.

Jolliffe, I. T. (2002). *Principal Component Analysis*, 2nd edition. New York: Springer.

Jöreskog, K. G. (1963). Statistical Estimation in Factor Analysis. Stockholm: Almqvist and Wiksell.

Jöreskog, K. G. (1971). Statistical analysis of sets of congeneric tests. *Psychometrika*, 36: 109–133.

Jöreskog, K. G. & Goldberger, A. S. (1975). Estimation of a model with multiple indicators and multiple causes of a single latent variable. *Journal of the American Statistical Association*, 70: 631–639.

Jöreskog, K. G. & Sorbom, D. (1981). *LISREL V: Analysis of Linear Structural Relationships by Maximum Likelihood and Least Squares Methods*. Research Report 8 1–8, Department of Statistics, University of Uppsala, Sweden, 1981.

Jöreskog, K. G. & Sorbom, D. (1984). *LISREL-VI: Analysis of Linear Structural Relationships by Maximum Likelihood, Instrumental Variable, and Least Square Methods*. Mooresville, IN: Scientific Software.

Jöreskog, K. G. & Sorbom, D. (1987). *New Developments in LISREL*. Paper presented at the National Symposium on Methodological Issues in Causal Modeling, University of Alabama, Tuscaloosa.

Jöreskog, K. G. & Sorbom, D. (1988). *LISREL 7: A Guide to the Program and Applications*. Chicago: SPSS, Inc.

Kaiser, H. F. (1958). The varimax criterion for analytic rotation in factor analysis. *Psychometrika*, 23(3): 187–200.

Kaiser, H. F. (1960). The application of electronic computers to factor analysis. *Educational and Psychological Measurement*, 20(1): 141–151.

Kaiser, H. F. & Rice, J. (1974). Little jiffy, mark IV. *Educational and Psychological Measurement*, 34(1): 111–117.

Kaplan, D. (1990). Evaluating and modifying covariance structure models: A review and recommendation. *Multivariate Behavioral Research*, 25(2): 137–155.

Kendall, M. G. & Stuart, A. (1979). *The Advanced Theory of Statistics*, Vol. 2, 4th edition. London: Griffin.

Khanzode, V. V. (2010). *Modeling Risk of Occupational Injury*. Unpublished PhD thesis, Indian Institute of Technology Kharagpur.

Kleibergen, F. (2002). Pivotal statistics for testing structural parameters in instrumental variables regression. *Econometrica*, 70(5): 1781–1803.

Kolenikov, S. & Bollen, K. A. (2012). Testing negative error variances: Is a Heywood case a symptom of misspecification? *Sociological Methods & Research*, 41(1): 124–167.

Korn, E. L. & Graubard, B. I. (1990). Simultaneous testing of regression coefficients with complex survey data: Use of Bonferroni t statistics. *The American Statistician*, 44: 270–276.

Kraemer, H. C. (2006). Correlation coefficients in medical research: From product moment correlation to odds ratio. *Statistical Methods in Medical Research*, 15: 525–545.

Kraus, A. D. (2002). *Matrices for Engineers*. New York: Oxford University Press.

Krzanowski, W. & Marriott, F. (2014a, reprint). *Kendall's Library of Statistics 1: Multivariate Analysis, Part 1*. New Delhi: Wiley.

Krzanowski, W. & Marriott, F. (2014b, reprint). *Kendall's Library of Statistics 2: Multivariate Analysis, Part 2*. New Delhi: Wiley.

Lattin, J., Carroll, J. D., & Green, P. E. (2003). *Analysing Multivariate Data*. Brooks/Cole, Cengage Learning.

Lawley, D. N. (1940). The estimation of factor loadings by the method of maximum likelihood. *Proceedings of the Royal Society of Edinburgh Section A*, 40: 64–82.

Lawley, D. N. & Maxwell, A. E. (1971). *Factor Analysis as a Statistical Method*, 2nd edition. London: Butterworths.

Lehmer, D. H. (1951). *Proceedings of the Second Symposium on Large-Scale Digital Computing Machinery*. Cambridge, MA: Harvard University Press.

Lev, J. (1949). The point biserial coefficient of correlation. *Annals of Mathematical Statistics*, 20(1): 125–126.

Levene, H. (1960). Robust tests for equality of variances. In *Contributions to Probability and Statistics*, edited by I. Olkin. Palo Alto, CA: Stanford University Press, 278–292.

Lindsey, J. K. (2006). *Introduction to Applied Statistics: A Modelling Approach*. Oxford: Oxford University Press.

Lloyd, E. H. (1952). Least-squares estimation of location and scale parameters using order. *Biometrika*, 39(1/2): 88–95.

Lorenzo-Seva, U., Timmerman, M. E., & Kiers, H. A. (2011). The Hull method for selecting the number of common factors. *Multivariate Behavioral Research*, 46(2): 340–364.

MacCallum, R. (1986). Specification searches in covariance structure modeling. *Psychological Bulletin*, 100: 107–120.

Maddala, G. S. (2007). *Introduction to Econometrics*, 3rd edition. New York: Wiley.

Maiti, J. (2014). Applied Multivariate Statistical Modeling. Web-based course on National program on Technology Enhanced Learning, IIT Kharagpur. Available at https://nptel.ac.in/courses/110/105/110105060/.

Mandel, M. & Betensky, R. A. (2008). Simultaneous confidence intervals based on the percentile bootstrap approach. *Computational Statistics and Data Analysis*, 52(4): 2158–2165.

Manly, B. F. (1976). Exponential data transformations. Journal of the Royal Statistical Society: Series D (The Statistician), 25(1): 37–42.

Mardia, K. V. (1970). Measures of multivariate skewness and kurtosis with applications. *Biometrika*, 57(3): 519–530.

Mardia, K. V., Kent, J. T., & Bibby, J. M. (1979). *Multivariate Analysis*. London: Academic Press.

Marsh, H. W. & Balla, J. (1994). Goodness of fit in confirmatory factor analysis: The effects of sample size and model parsimony. *Quality and Quantity*, 28(2): 185–217.

Marsh, H. W., Balla, J. R., & McDonald, R. P. (1988). Goodness-of-fit indices in confirmatory factor analysis: The effect of sample size. *Psychological Bulletin*, 102: 391–410.

McDonald, R. P. (1989). An index of goodness-of-fit based on noncentrality. *Journal of Classification*, 6: 97–103.

McDonald, R. P. & Marsh, H. W. (1990). Choosing a multivariate model: Noncentrality and goodness-of-fit. *Psychological Bulletin*, 107: 247–255.

Menard, S. (1995). An introduction to logistic regression diagnostics. *Applied Logistic Regression Analysis*, 1: 58–79.

Mood, A. M. (1974). *Introduction to the Theory of Statistics*. New York: McGraw-Hill.

Mondal, S. C. (2011). *Modelling Robustness in Manufacturing Processes*. Unpublished PhD thesis, Indian Institute of Technology Kharagpur.

Mondal, S. C., Maiti, J., & Ray, P. K. (2013). Modelling robustness in serial multi-stage manufacturing processes. *International Journal of Production Research*, 51(21): 6359–6377.

Montgomery, D. C. (2012). *Design and Analysis of Experiments*. New York: John Wiley & Sons.

Montgomery, D. C., Peck, E. A., & Vining, G. G. (2003). *Introduction to Linear Regression Analysis*. New York: John Wiley & Sons.Montgomery, D. C., Peck, E. A., & Vining, G. G. (2006). *Introduction to Linear Regression Analysis*, 3rd edition. Delhi: Wiley-India.

Morris, M. D. (2011). *Design of Experiments: An Introduction Based on Linear Models*. New York: CRC Press.

Mulaik, S. A. (1988). Confirmatory factor analysis. In *Handbook of Multivariate Experimental Psychology*, 2nd edition, edited by J. R. Nesselroade & R. B. Cattell. New York: Plenum Press, 259–288.

Mulaik, S. A., James, L. R., Van Alstine, J., Bennett, N. Lind, S., & Stilwell, C. D. (1989). Evaluation of goodness-of-fit indices for structural equation models. Psychological Bulletin, 105(3): 430–445.

Muller, K. E. & Peterson, B. L. (1984). Practical methods for computing power in testing the multivariate general linear hypothesis. *Computational Statistics and Data Analysis*, 2: 143–158.

Muthen, B. & Kaplan, D. (1992). A comparison of some methodologies for the factor analysis of non-normal Likert variables: A note on the size of the model. British Journal of Mathematical and Statistical Psychology, 45(1): 19–30.

Muthén, L. K. & Muthén, B. O. (2002). How to use a Monte Carlo study to decide on sample size and determine power. *Structural Equation Modeling*, 9(4): 599–620.

Myers, R. H. (1990). *Classical and Modern Regression with Applications*. Belmont, CA: Duxbury Press.

Myung, J. (2003). Tutorial on maximum likelihood estimation. *Journal of Mathematical Psychology*, 47: 90–100.

Neuhaus, J. O. & Wrigley, C. (1954). The quartimax method: An analytical approach to orthogonal simple structure. *British Journal of Mathematical and Statistical Psychology*, 7: 81–91.

Nunnally, J. C. (1978). *Psychometric Theory*, 2nd edition. New York: McGraw-Hill.

Olkin, I. & Tate, R. F. (1961). Multivariate correlation models with mixed discrete and continuous variable. *The Annals of Mathematical Statistics*, 32(2): 448–465.

Olson, C. L. (1974). Comparative robustness of six tests in multivariate analysis of variance. *Journal of the American Statistical Association*, 69(348): 894–908.

Olson, C. L. (1976). On choosing a test statistic in multivariate analysis of variance. *Psychological Bulletin*, 83: 579–586.

Olsson, U., Drasgow, F., & Dorans, N. J. (1982). The polyserial correlation coefficient. *Psychometrika*, 47(3): 337–347.

Paxton, P., Curran, P. J., Bollen, K. A., Kirby, J., & Chen, F. (2001). Monte Carlo experiments: Design and implementation. *Structural Equation Modeling*, 8(2): 287–312.

Pearson, K. (1901). On lines and planes of closest fit to systems of points in space. Philosophical Magazine, 2(11): 559–572.

Pearson, K. (1909). On a new method of determining correlation between a measured character A, and a character B, of which only the percentage of cases wherein B exceeds (or falls short of) a given intensity is recorded for each grade of A. *Biometrika*, 7(1/2): 96–105.

Pillai, K. C. S. (1967). Upper percentage points of the largest root of a matrix in multivariate analysis. *Biometrika*, 54: 189–194.

Puri, M. L. & Sen, P. K. (1968). Nonparametric confidence regions for some multivariate location problems. *Journal of American Statistical Association*, 63(324): 1373–1378.

Rao, C. R. (1952). *Advanced Statistical Methods in Biometric Research*. New York: Wiley.

Rao, C. R. (1973). *Linear Statistical Inference and its Applications*, 2nd edition. New York: Wiley.

Ravindran, A., Phillips, D. T., & Solberg, J. J. (2006). *Operations Research: Principle and Practice*, 2nd edition. New York: Wiley.

Rencher, A. C. (2002). *Methods of Multivariate Analysis*. New York: John Wiley & Sons.

Rosenbaum, P. R. (1986). Dropping out of high school in the United States: An observational study. *Journal of Educational Statistics*, 11(3): 207–224.

Sakia, R. M. (1992). The Box-Cox transformation technique: A review. *Journal of the Royal Statistical Society: Series D (The Statistician)*, 41(2): 169–178.

Sargan, J. D. (1958). The estimation of economic relationships using instrumental variables. *Econometrica*, 26(3): 393–415.

Saris, W. E., Satorra, A., & Sorbom, D. (1987). The detection and correction of specification errors in structural equation models. In *Sociological Methodology*, edited by C. C. Clogg. Washington, DC: American Sociological Association, 105–130.

Scariano, S. M. & Davenport, J. M. (1987). The effects of violations of independence assumptions in the one-way ANOVA. *The American Statistician*, 41(2): 123–129.

Schatzoff, M. (1966). Sensitivity comparisons among tests of the general linear hypothesis. *Journal of the American Statistical Association*, 61(314): 415–435.

Scheffé, H. (1959). *The Analysis of Variance*. New York: Wiley.

Shapiro, S. S. & Wilk, M. B. (1965). An analysis of variance test for normality (complete samples). *Biometrika*, 52: 591–611.

Spearman, C. (1904a). Proof and measurement of the association of two things. *American Journal of Psychology*, 15(1): 72–101.

Spearman, C. (1904b). General intelligence objectively determined and measured. *American Journal of Psychology*, 15(2), 202–292.

Srivastava, M. S. & Du, M. (2008). A test for the mean vector with fewer observations than the dimension. *Journal of Multivariate Analysis*, 99: 386–402.

Staiger, D. & Stock, J. H. (1997). Instrumental variables regressions with weak instruments. *Econometrica*, 65: 557–586.

Steiger, J. H. (1989). *EzPATH: A Supplementary Module for SYSTAT and SYGRAPH*. Evanston, IL: SYSTAT.

Steiger, J. H. (1990). Structural model evaluation and modification: An interval estimation approach. *Multivariate Behavioral Research*, 25: 173–180.

Stokes, M. E., Davis, C. S., & Koch, G. G. (2012). *Categorical Data Analysis Using SAS*, 3rd edition. Cary, NC: SAS Publishing.

Strang, G. (2003). *Introduction to Linear Algebra*, 3rd edition. Wellesley, MA: Wellesley-Cambridge Press.

Strang, G. (2006). *Linear Algebra and its Applications*, 4th edition. Noida, UP: Thomson.

Tanaka, J. C. (1984). Some results on the estimation of covariance structure models. *Dissertation Abstracts International*, 45: 924B.

Tarka, P. (2018). An overview of structural equation modeling: Its beginnings, historical development, usefulness and controversies in the social sciences. *Quality & Quantity*, 52(1): 313–354.

Tate, R. F. (1955). Applications of correlation models for biserial data. *Journal of the American Statistical Association*, 50(272): 1078–1095.

Tatsuoka, M. M. (1969). Multivariate analysis. *Review of Educational Research*, 39: 739–743.

Thomson, G. H. (1951). *The Factorial Analysis of Human Ability*, 5th edition. New York: Houghton Mifflin.

Thurstone, L. L. (1947). *Multiple-Factor Analysis*. Chicago: University of Chicago Press.

Timm, N. H. (2002). *Applied Multivariate Analysis*. New York: Springer.

Tucker, L. & Lewis, C. (1973). A reliability coefficient for maximum likelihood factor analysis. *Psychometrika*, 38: 1–10.

Van Driel, O. P. (1978). On various causes of improper solutions of maximum likelihood factor analysis. *Psychometrika*, 43: 225–243.

Velicer, W. F. (1976). Determining the number of components from the matrix of partial correlations. *Psychometrika*, 41(3): 321–327.

Vincent, D. F. (1953). The origin and development of factor analysis. *Journal of the Royal Statistical Society. Series C (Applied Statistics)*, 2(2): 107–117.

Vinod, H. D. (1978). A survey of ridge regression and related techniques for improvements over ordinary least squares. *The Review of Economics and Statistics*, 60(1): 121–131.

Wacker, J. G. (1998). A definition of theory: research guidelines for different theory-building research methods in operations management. *Journal of Operations Management*, 16(4): 361–385.

Wall, F. J. (1967). *The Generalized Variance Ratio or U-Statistic*. Albuquerque, NM: The Dikewood Corporation.

Wang, L., Peng, B., & Li, R. (2015). A high-dimensional nonparametric multivariate test for mean vector. *Journal of the American Statistical Association*, 110(512): 1658–1669.

Wehrens, R., Putter, H., & Buydens, M. C. L. (2000). The bootstrap: A tutorial. *Chemometrics and Intelligent Laboratory Systems*, 54(1): 35–52.

Wilk, M. B. & Gnanadesikan, R. (1968). Probability plotting methods for the analysis for the analysis of data. *Biometrika*, 55(1): 1–17.

Wilks, S. S. (1932). Certain generalizations in the analysis of variance. *Biometrika*, 24: 471–194.

Williams, L. J. & Anderson, S. E. (1994). An alternative approach to method effects using latent variable models: Applications in organizational behavior research. *Journal of Applied Psychology*, 79(3): 323–331.

Wishart, J. (1928). The generalised product moment distribution in samples from a normal multivariate population. *Biometrika*, 32–52.

Wonnacott, R. J. & Wonnacott, T. H. (1970). *Econometrics*. New York: Wiley.

Wright, S. (1921). Systems of mating. I. The biometric relations between parent and offspring. *Genetics*, 6(2): 111–123.

Yang, K. & Trewn, J. (2003). *Multivariate Statistical Methods in Quality Management*. New York: McGraw-Hill Education.

Zellner, A. & Theil, H. (1962). Three-stage least squares: Simultaneous estimation of simultaneous equations. In H. Theil, *Contributions to Economics and Econometrics*. New York: Springer, 147–178.

Ziegler, M. & Hagemann, D. (2015). Testing the unidimensionality of items: pitfalls and loopholes. *European Journal of Psychological Assessment*, 31(4): 231–237.

Zoubir, A. M. & Iskandler, D. R. (2007). Bootstrap methods and applications. *IEEE Signal Processing Magazine*, July, 10–19.

Appendix A1: Cumulative Standard Normal Distribution

z	0	0.01	0.02	0.03	0.04	0.05	0.06	0.07	0.08	0.09
0	0.5000	0.5040	0.5080	0.5120	0.5160	0.5199	0.5239	0.5279	0.5319	0.5359
0.1	0.5398	0.5438	0.5478	0.5517	0.5557	0.5596	0.5636	0.5675	0.5714	0.5753
0.2	0.5793	0.5832	0.5871	0.5910	0.5948	0.5987	0.6026	0.6064	0.6103	0.6141
0.3	0.6179	0.6217	0.6255	0.6293	0.6331	0.6368	0.6406	0.6443	0.6480	0.6517
0.4	0.6554	0.6591	0.6628	0.6664	0.6700	0.6736	0.6772	0.6808	0.6844	0.6879
0.5	0.6915	0.6950	0.6985	0.7019	0.7054	0.7088	0.7123	0.7157	0.7190	0.7224
0.6	0.7257	0.7291	0.7324	0.7357	0.7389	0.7422	0.7454	0.7486	0.7517	0.7549
0.7	0.7580	0.7611	0.7642	0.7673	0.7704	0.7734	0.7764	0.7794	0.7823	0.7852
0.8	0.7881	0.7910	0.7939	0.7967	0.7995	0.8023	0.8051	0.8078	0.8106	0.8133
0.9	0.8159	0.8186	0.8212	0.8238	0.8264	0.8289	0.8315	0.8340	0.8365	0.8389
1	0.8413	0.8438	0.8461	0.8485	0.8508	0.8531	0.8554	0.8577	0.8599	0.8621
1.1	0.8643	0.8665	0.8686	0.8708	0.8729	0.8749	0.8770	0.8790	0.8810	0.8830
1.2	0.8849	0.8869	0.8888	0.8907	0.8925	0.8944	0.8962	0.8980	0.8997	0.9015
1.3	0.9032	0.9049	0.9066	0.9082	0.9099	0.9115	0.9131	0.9147	0.9162	0.9177
1.4	0.9192	0.9207	0.9222	0.9236	0.9251	0.9265	0.9279	0.9292	0.9306	0.9319
1.5	0.9332	0.9345	0.9357	0.9370	0.9382	0.9394	0.9406	0.9418	0.9429	0.9441
1.6	0.9452	0.9463	0.9474	0.9484	0.9495	0.9505	0.9515	0.9525	0.9535	0.9545
1.7	0.9554	0.9564	0.9573	0.9582	0.9591	0.9599	0.9608	0.9616	0.9625	0.9633
1.8	0.9641	0.9649	0.9656	0.9664	0.9671	0.9678	0.9686	0.9693	0.9699	0.9706
1.9	0.9713	0.9719	0.9726	0.9732	0.9738	0.9744	0.9750	0.9756	0.9761	0.9767
2	0.9772	0.9778	0.9783	0.9788	0.9793	0.9798	0.9803	0.9808	0.9812	0.9817
2.1	0.9821	0.9826	0.9830	0.9834	0.9838	0.9842	0.9846	0.9850	0.9854	0.9857
2.2	0.9861	0.9864	0.9868	0.9871	0.9875	0.9878	0.9881	0.9884	0.9887	0.9890
2.3	0.9893	0.9896	0.9898	0.9901	0.9904	0.9906	0.9909	0.9911	0.9913	0.9916
2.4	0.9918	0.9920	0.9922	0.9925	0.9927	0.9929	0.9931	0.9932	0.9934	0.9936
2.5	0.9938	0.9940	0.9941	0.9943	0.9945	0.9946	0.9948	0.9949	0.9951	0.9952
2.6	0.9953	0.9955	0.9956	0.9957	0.9959	0.9960	0.9961	0.9962	0.9963	0.9964
2.7	0.9965	0.9966	0.9967	0.9968	0.9969	0.9970	0.9971	0.9972	0.9973	0.9974
2.8	0.9974	0.9975	0.9976	0.9977	0.9977	0.9978	0.9979	0.9979	0.9980	0.9981
2.9	0.9981	0.9982	0.9982	0.9983	0.9984	0.9984	0.9985	0.9985	0.9986	0.9986
3	0.9987	0.9987	0.9987	0.9988	0.9988	0.9989	0.9989	0.9989	0.9990	0.9990
3.1	0.9990	0.9991	0.9991	0.9991	0.9992	0.9992	0.9992	0.9992	0.9993	0.9993
3.2	0.9993	0.9993	0.9994	0.9994	0.9994	0.9994	0.9994	0.9995	0.9995	0.9995

z	0	0.01	0.02	0.03	0.04	0.05	0.06	0.07	0.08	0.09
3.3	0.9995	0.9995	0.9995	0.9996	0.9996	0.9996	0.9996	0.9996	0.9996	0.9997
3.4	0.9997	0.9997	0.9997	0.9997	0.9997	0.9997	0.9997	0.9997	0.9997	0.9998
3.5	0.9998	0.9998	0.9998	0.9998	0.9998	0.9998	0.9998	0.9998	0.9998	0.9998
3.6	0.9998	0.9998	0.9999	0.9999	0.9999	0.9999	0.9999	0.9999	0.9999	0.9999
3.7	0.9999	0.9999	0.9999	0.9999	0.9999	0.9999	0.9999	0.9999	0.9999	0.9999
3.8	0.9999	0.9999	0.9999	0.9999	0.9999	0.9999	0.9999	0.9999	0.9999	0.9999

Appendix A2: Percentage Points of Student's t-distribution

	Alpha (α)									
DoF	**0.4**	**0.25**	**0.1**	**0.05**	**0.025**	**0.01**	**0.005**	**0.0025**	**0.001**	**0.0005**
1	0.325	1.000	3.078	6.314	12.706	31.821	63.657	127.321	318.309	636.619
2	0.289	0.816	1.886	2.920	4.303	6.965	9.925	14.089	22.327	31.599
3	0.277	0.765	1.638	2.353	3.182	4.541	5.841	7.453	10.215	12.924
4	0.271	0.741	1.533	2.132	2.776	3.747	4.604	5.598	7.173	8.610
5	0.267	0.727	1.476	2.015	2.571	3.365	4.032	4.773	5.893	6.869
6	0.265	0.718	1.440	1.943	2.447	3.143	3.707	4.317	5.208	5.959
7	0.263	0.711	1.415	1.895	2.365	2.998	3.499	4.029	4.785	5.408
8	0.262	0.706	1.397	1.860	2.306	2.896	3.355	3.833	4.501	5.041
9	0.261	0.703	1.383	1.833	2.262	2.821	3.250	3.690	4.297	4.781
10	0.260	0.700	1.372	1.812	2.228	2.764	3.169	3.581	4.144	4.587
11	0.260	0.697	1.363	1.796	2.201	2.718	3.106	3.497	4.025	4.437
12	0.259	0.695	1.356	1.782	2.179	2.681	3.055	3.428	3.930	4.318
13	0.259	0.694	1.350	1.771	2.160	2.650	3.012	3.372	3.852	4.221
14	0.258	0.692	1.345	1.761	2.145	2.624	2.977	3.326	3.787	4.140
15	0.258	0.691	1.341	1.753	2.131	2.602	2.947	3.286	3.733	4.073
16	0.258	0.690	1.337	1.746	2.120	2.583	2.921	3.252	3.686	4.015
17	0.257	0.689	1.333	1.740	2.110	2.567	2.898	3.222	3.646	3.965
18	0.257	0.688	1.330	1.734	2.101	2.552	2.878	3.197	3.610	3.922
19	0.257	0.688	1.328	1.729	2.093	2.539	2.861	3.174	3.579	3.883
20	0.257	0.687	1.325	1.725	2.086	2.528	2.845	3.153	3.552	3.850
21	0.257	0.686	1.323	1.721	2.080	2.518	2.831	3.135	3.527	3.819
22	0.256	0.686	1.321	1.717	2.074	2.508	2.819	3.119	3.505	3.792
23	0.256	0.685	1.319	1.714	2.069	2.500	2.807	3.104	3.485	3.768
24	0.256	0.685	1.318	1.711	2.064	2.492	2.797	3.091	3.467	3.745
25	0.256	0.684	1.316	1.708	2.060	2.485	2.787	3.078	3.450	3.725
26	0.256	0.684	1.315	1.706	2.056	2.479	2.779	3.067	3.435	3.707
27	0.256	0.684	1.314	1.703	2.052	2.473	2.771	3.057	3.421	3.690
28	0.256	0.683	1.313	1.701	2.048	2.467	2.763	3.047	3.408	3.674
29	0.256	0.683	1.311	1.699	2.045	2.462	2.756	3.038	3.396	3.659
30	0.256	0.683	1.310	1.697	2.042	2.457	2.750	3.030	3.385	3.646
40	0.255	0.681	1.303	1.684	2.021	2.423	2.704	2.971	3.307	3.551
60	0.254	0.679	1.296	1.671	2.000	2.390	2.660	2.915	3.232	3.460
120	0.254	0.677	1.289	1.658	1.980	2.358	2.617	2.860	3.160	3.373

Appendix A2: Percentage Points of Student's t-distribution

Appendix A3: Percentage Points of the Chi-square Distribution

	Alpha (α)								
DoF	**0.995**	**0.99**	**0.975**	**0.95**	**0.5**	**0.05**	**0.025**	**0.01**	**0.005**
1	0.000	0.000	0.001	0.004	0.455	3.841	5.024	6.635	7.879
2	0.010	0.020	0.051	0.103	1.386	5.991	7.378	9.210	10.597
3	0.072	0.115	0.216	0.352	2.366	7.815	9.348	11.345	12.838
4	0.207	0.297	0.484	0.711	3.357	9.488	11.143	13.277	14.860
5	0.412	0.554	0.831	1.145	4.351	11.070	12.833	15.086	16.750
6	0.676	0.872	1.237	1.635	5.348	12.592	14.449	16.812	18.548
7	0.989	1.239	1.690	2.167	6.346	14.067	16.013	18.475	20.278
8	1.344	1.646	2.180	2.733	7.344	15.507	17.535	20.090	21.955
9	1.735	2.088	2.700	3.325	8.343	16.919	19.023	21.666	23.589
10	2.156	2.558	3.247	3.940	9.342	18.307	20.483	23.209	25.188
11	2.603	3.053	3.816	4.575	10.341	19.675	21.920	24.725	26.757
12	3.074	3.571	4.404	5.226	11.340	21.026	23.337	26.217	28.300
13	3.565	4.107	5.009	5.892	12.340	22.362	24.736	27.688	29.819
14	4.075	4.660	5.629	6.571	13.339	23.685	26.119	29.141	31.319
15	4.601	5.229	6.262	7.261	14.339	24.996	27.488	30.578	32.801
16	5.142	5.812	6.908	7.962	15.338	26.296	28.845	32.000	34.267
17	5.697	6.408	7.564	8.672	16.338	27.587	30.191	33.409	35.718
18	6.265	7.015	8.231	9.390	17.338	28.869	31.526	34.805	37.156
19	6.844	7.633	8.907	10.117	18.338	30.144	32.852	36.191	38.582
20	7.434	8.260	9.591	10.851	19.337	31.410	34.170	37.566	39.997
25	10.520	11.524	13.120	14.611	24.337	37.652	40.646	44.314	46.928
30	13.787	14.953	16.791	18.493	29.336	43.773	46.979	50.892	53.672
40	20.707	22.164	24.433	26.509	39.335	55.758	59.342	63.691	66.766
50	27.991	29.707	32.357	34.764	49.335	67.505	71.420	76.154	79.490
60	35.534	37.485	40.482	43.188	59.335	79.082	83.298	88.379	91.952
70	43.275	45.442	48.758	51.739	69.334	90.531	95.023	100.425	104.215
80	51.172	53.540	57.153	60.391	79.334	101.879	106.629	112.329	116.321
90	59.196	61.754	65.647	69.126	89.334	113.145	118.136	124.116	128.299
100	67.328	70.065	74.222	77.929	99.334	124.342	129.561	135.807	140.169

Appendix A4: Percentage Points of the F-distribution ($\alpha = 0.10$)

F (α = 0.10)	Numerator Degrees of Freedom (DoF)																	
Denominator Degrees of Freedom (DoF)	1	2	3	4	5	6	7	8	9	10	12	15	20	24	30	40	60	120
1	39.86	49.50	53.59	55.83	57.24	58.20	58.90	59.44	59.86	60.20	60.71	61.22	61.74	62.00	62.27	62.53	62.79	63.06
2	8.526	9.000	9.162	9.243	9.293	9.326	9.349	9.367	9.381	9.392	9.408	9.425	9.441	9.450	9.458	9.466	9.475	9.483
3	5.538	5.462	5.391	5.343	5.309	5.285	5.266	5.252	5.240	5.230	5.216	5.200	5.184	5.176	5.168	5.160	5.151	5.143
4	4.545	4.325	4.191	4.107	4.051	4.010	3.979	3.955	3.936	3.920	3.896	3.870	3.844	3.831	3.817	3.804	3.790	3.775
5	4.060	3.780	3.619	3.520	3.453	3.405	3.368	3.339	3.316	3.297	3.268	3.238	3.207	3.191	3.174	3.157	3.140	3.123
6	3.776	3.463	3.289	3.181	3.108	3.055	3.014	2.983	2.958	2.937	2.905	2.871	2.836	2.818	2.800	2.781	2.762	2.742
7	3.589	3.257	3.074	2.961	2.883	2.827	2.785	2.752	2.725	2.703	2.668	2.632	2.595	2.575	2.555	2.535	2.514	2.493
8	3.458	3.113	2.924	2.806	2.726	2.668	2.624	2.589	2.561	2.538	2.502	2.464	2.425	2.404	2.383	2.361	2.339	2.316
9	3.360	3.006	2.813	2.693	2.611	2.551	2.505	2.469	2.440	2.416	2.379	2.340	2.298	2.277	2.255	2.232	2.208	2.184
10	3.285	2.924	2.728	2.605	2.522	2.461	2.414	2.377	2.347	2.323	2.284	2.244	2.201	2.178	2.155	2.132	2.107	2.082
11	3.225	2.860	2.660	2.536	2.451	2.389	2.342	2.304	2.274	2.248	2.209	2.167	2.123	2.100	2.076	2.052	2.026	2.000
12	3.177	2.807	2.606	2.480	2.394	2.331	2.283	2.245	2.214	2.188	2.147	2.105	2.060	2.036	2.011	1.986	1.960	1.932
13	3.136	2.763	2.560	2.434	2.347	2.283	2.234	2.195	2.164	2.138	2.097	2.053	2.007	1.983	1.958	1.931	1.904	1.876
14	3.102	2.726	2.522	2.395	2.307	2.243	2.193	2.154	2.122	2.095	2.054	2.010	1.962	1.938	1.912	1.885	1.857	1.828
15	3.073	2.695	2.490	2.361	2.273	2.208	2.158	2.119	2.086	2.059	2.017	1.972	1.924	1.899	1.873	1.845	1.817	1.787

F ($\alpha = 0.10$)	Numerator Degrees of Freedom (DoF)																	
	1	2	3	4	5	6	7	8	9	10	12	15	20	24	30	40	60	120
16	3.048	2.668	2.462	2.333	2.244	2.178	2.128	2.088	2.055	2.028	1.985	1.940	1.891	1.866	1.839	1.811	1.782	1.751
17	3.026	2.645	2.437	2.308	2.218	2.152	2.102	2.061	2.028	2.001	1.958	1.912	1.862	1.836	1.809	1.781	1.751	1.719
18	3.007	2.624	2.416	2.286	2.196	2.130	2.079	2.038	2.005	1.977	1.933	1.887	1.837	1.810	1.783	1.754	1.723	1.691
19	2.990	2.606	2.397	2.266	2.176	2.109	2.058	2.017	1.984	1.956	1.912	1.865	1.814	1.787	1.759	1.730	1.699	1.666
20	2.975	2.589	2.380	2.249	2.158	2.091	2.040	1.999	1.965	1.937	1.892	1.845	1.794	1.767	1.738	1.708	1.677	1.643
21	2.961	2.575	2.365	2.233	2.142	2.075	2.023	1.982	1.948	1.920	1.875	1.827	1.776	1.748	1.719	1.689	1.657	1.623
22	2.949	2.561	2.351	2.219	2.128	2.060	2.008	1.967	1.933	1.904	1.859	1.811	1.759	1.731	1.702	1.671	1.639	1.604
23	2.937	2.549	2.339	2.207	2.115	2.047	1.995	1.953	1.919	1.890	1.845	1.796	1.744	1.716	1.686	1.655	1.622	1.587
24	2.927	2.538	2.327	2.195	2.103	2.035	1.983	1.941	1.906	1.877	1.832	1.783	1.730	1.702	1.672	1.641	1.607	1.571
25	2.918	2.528	2.317	2.184	2.092	2.024	1.971	1.929	1.895	1.866	1.820	1.771	1.718	1.689	1.659	1.627	1.593	1.557
26	2.909	2.519	2.307	2.174	2.082	2.014	1.961	1.919	1.884	1.855	1.809	1.760	1.706	1.677	1.647	1.615	1.581	1.544
27	2.901	2.511	2.299	2.165	2.073	2.005	1.952	1.909	1.874	1.845	1.799	1.749	1.695	1.666	1.636	1.603	1.569	1.531
28	2.894	2.503	2.291	2.157	2.064	1.996	1.943	1.900	1.865	1.836	1.790	1.740	1.685	1.656	1.625	1.592	1.558	1.520
29	2.887	2.495	2.283	2.149	2.057	1.988	1.935	1.892	1.857	1.827	1.781	1.731	1.676	1.647	1.616	1.583	1.547	1.509
30	2.881	2.489	2.276	2.142	2.049	1.980	1.927	1.884	1.849	1.819	1.773	1.722	1.667	1.638	1.606	1.573	1.538	1.499
40	2.835	2.440	2.226	2.091	1.997	1.927	1.873	1.829	1.793	1.763	1.715	1.662	1.605	1.574	1.541	1.506	1.467	1.425
60	2.791	2.393	2.177	2.041	1.946	1.875	1.819	1.775	1.738	1.707	1.657	1.603	1.543	1.511	1.476	1.437	1.395	1.348
120	2.748	2.347	2.130	1.992	1.896	1.824	1.767	1.722	1.684	1.652	1.601	1.545	1.482	1.447	1.409	1.368	1.320	1.265

Appendix A5: Percentage Points of the F-distribution ($\alpha = 0.05$)

F ($\alpha = 0.05$)		Numerator Degrees of Freedom (DoF)																	
		1	2	3	4	5	6	7	8	9	10	12	15	20	24	30	40	60	120
Denominator Degrees of Freedom (DoF)	1	161.45	199.50	215.71	224.58	230.16	233.99	236.77	238.88	240.54	241.88	243.91	245.95	248.01	249.05	250.10	251.14	252.20	253.25
	2	18.51	19.00	19.16	19.25	19.30	19.33	19.35	19.37	19.39	19.40	19.41	19.43	19.45	19.45	19.46	19.47	19.48	19.49
	3	10.3	9.55	9.28	9.12	9.01	8.94	8.89	8.85	8.81	8.79	8.75	8.70	8.66	8.64	8.62	8.60	8.57	8.55
	4	7.709	6.944	6.591	6.388	6.256	6.163	6.094	6.041	5.999	5.964	5.912	5.858	5.803	5.774	5.746	5.717	5.688	5.658
	5	6.608	5.786	5.409	5.192	5.050	4.950	4.876	4.818	4.772	4.735	4.678	4.619	4.558	4.527	4.496	4.464	4.431	4.398
	6	5.987	5.143	4.757	4.534	4.387	4.284	4.207	4.147	4.099	4.060	4.000	3.938	3.874	3.841	3.808	3.774	3.740	3.705
	7	5.591	4.737	4.347	4.120	3.972	3.866	3.787	3.726	3.677	3.637	3.575	3.511	3.445	3.410	3.376	3.340	3.304	3.267
	8	5.318	4.459	4.066	3.838	3.687	3.581	3.500	3.438	3.388	3.347	3.284	3.218	3.150	3.115	3.079	3.043	3.005	2.967
	9	5.117	4.256	3.863	3.633	3.482	3.374	3.293	3.230	3.179	3.137	3.073	3.006	2.936	2.900	2.864	2.826	2.787	2.748
	10	4.965	4.103	3.708	3.478	3.326	3.217	3.135	3.072	3.020	2.978	2.913	2.845	2.774	2.737	2.700	2.661	2.621	2.580
	11	4.844	3.982	3.587	3.357	3.204	3.095	3.012	2.948	2.896	2.854	2.788	2.719	2.646	2.609	2.570	2.531	2.490	2.448
	12	4.747	3.885	3.490	3.259	3.106	2.996	2.913	2.849	2.796	2.753	2.687	2.617	2.544	2.505	2.466	2.426	2.384	2.341
	13	4.667	3.806	3.411	3.179	3.025	2.915	2.832	2.767	2.714	2.671	2.604	2.533	2.459	2.420	2.380	2.339	2.297	2.252
	14	4.600	3.739	3.344	3.112	2.958	2.848	2.764	2.699	2.646	2.602	2.534	2.463	2.388	2.349	2.308	2.266	2.223	2.178

F ($\alpha = 0.05$)	Numerator Degrees of Freedom (DoF)																
	1	2	3	4	5	6	7	8	9	10	12	15	20	24	30	40	60	120
15	4.543	3.682	3.287	3.056	2.901	2.790	2.707	2.641	2.588	2.544	2.475	2.403	2.328	2.288	2.247	2.204	2.160	2.114
16	4.494	3.634	3.239	3.007	2.852	2.741	2.657	2.591	2.538	2.494	2.425	2.352	2.276	2.235	2.194	2.151	2.106	2.059
17	4.451	3.592	3.197	2.965	2.810	2.699	2.614	2.548	2.494	2.450	2.381	2.308	2.230	2.190	2.148	2.104	2.058	2.011
18	4.414	3.555	3.160	2.928	2.773	2.661	2.577	2.510	2.456	2.412	2.342	2.269	2.191	2.150	2.107	2.063	2.017	1.968
19	4.381	3.522	3.127	2.895	2.740	2.628	2.544	2.477	2.423	2.378	2.308	2.234	2.155	2.114	2.071	2.026	1.980	1.930
20	4.351	3.493	3.098	2.866	2.711	2.599	2.514	2.447	2.393	2.348	2.278	2.203	2.124	2.082	2.039	1.994	1.946	1.896
21	4.325	3.467	3.072	2.840	2.685	2.573	2.488	2.420	2.366	2.321	2.250	2.176	2.096	2.054	2.010	1.965	1.916	1.866
22	4.301	3.443	3.049	2.817	2.661	2.549	2.464	2.397	2.342	2.297	2.226	2.151	2.071	2.028	1.984	1.938	1.889	1.838
23	4.279	3.422	3.028	2.796	2.640	2.528	2.442	2.375	2.320	2.275	2.204	2.128	2.048	2.005	1.961	1.914	1.865	1.813
24	4.260	3.403	3.009	2.776	2.621	2.508	2.423	2.355	2.300	2.255	2.183	2.108	2.027	1.984	1.939	1.892	1.842	1.790
25	4.242	3.385	2.991	2.759	2.603	2.490	2.405	2.337	2.282	2.236	2.165	2.089	2.007	1.964	1.919	1.872	1.822	1.768
26	4.225	3.369	2.975	2.743	2.587	2.474	2.388	2.321	2.265	2.220	2.148	2.072	1.990	1.946	1.901	1.853	1.803	1.749
27	4.210	3.354	2.960	2.728	2.572	2.459	2.373	2.305	2.250	2.204	2.132	2.056	1.974	1.930	1.884	1.836	1.785	1.731
28	4.196	3.340	2.947	2.714	2.558	2.445	2.359	2.291	2.236	2.190	2.118	2.041	1.959	1.915	1.869	1.820	1.769	1.714
29	4.183	3.328	2.934	2.701	2.545	2.432	2.346	2.278	2.223	2.177	2.104	2.027	1.945	1.901	1.854	1.806	1.754	1.698
30	4.171	3.316	2.922	2.690	2.534	2.421	2.334	2.266	2.211	2.165	2.092	2.015	1.932	1.887	1.841	1.792	1.740	1.683
40	4.085	3.232	2.839	2.606	2.449	2.336	2.249	2.180	2.124	2.077	2.003	1.924	1.839	1.793	1.744	1.693	1.637	1.577
60	4.001	3.150	2.758	2.525	2.368	2.254	2.167	2.097	2.040	1.993	1.917	1.836	1.748	1.700	1.649	1.594	1.534	1.467
120	3.920	3.072	2.680	2.447	2.290	2.175	2.087	2.016	1.959	1.910	1.834	1.750	1.659	1.608	1.554	1.495	1.429	1.352

Appendix A6: Percentage Points of the F-distribution ($\alpha = 0.01$)

F (α = 0.01)	Numerator Degrees of Freedom (DoF)																	
Denominator Degrees of Freedom (DoF)	1	2	3	4	5	6	7	8	9	10	12	15	20	24	30	40	60	120
1	4052	4999.5	5403	5624.6	5763.6	5859	5928.4	5981	6022.5	6056	6106	6157	6208.7	6234.6	6260.7	6286.8	6313	6339.4
2	98.50	99.00	99.17	99.25	99.30	99.33	99.36	99.37	99.39	99.40	99.42	99.43	99.45	99.46	99.47	99.47	99.48	99.49
3	34.12	30.82	29.46	28.71	28.24	27.91	27.67	27.49	27.35	27.23	27.05	26.87	26.69	26.60	26.50	26.41	26.32	26.22
4	21.20	18.00	16.69	15.98	15.52	15.21	14.98	14.80	14.66	14.55	14.37	14.20	14.02	13.93	13.84	13.75	13.65	13.56
5	16.26	13.27	12.06	11.39	10.97	10.67	10.46	10.29	10.16	10.05	9.89	9.72	9.55	9.47	9.38	9.29	9.20	9.11
6	13.75	10.92	9.78	9.15	8.75	8.47	8.26	8.10	7.98	7.87	7.72	7.56	7.40	7.31	7.23	7.14	7.06	6.97
7	12.25	9.55	8.45	7.85	7.46	7.19	6.99	6.84	6.72	6.62	6.47	6.31	6.16	6.07	5.99	5.91	5.82	5.74
8	11.26	8.65	7.59	7.01	6.63	6.37	6.18	6.03	5.91	5.81	5.67	5.52	5.36	5.28	5.20	5.12	5.03	4.95
9	10.56	8.02	6.99	6.42	6.06	5.80	5.61	5.47	5.35	5.26	5.11	4.96	4.81	4.73	4.65	4.57	4.48	4.40
10	10.04	7.56	6.55	5.99	5.64	5.39	5.20	5.06	4.94	4.85	4.71	4.56	4.41	4.33	4.25	4.17	4.08	4.00
11	9.65	7.21	6.22	5.67	5.32	5.07	4.89	4.74	4.63	4.54	4.40	4.25	4.10	4.02	3.94	3.86	3.78	3.69
12	9.33	6.93	5.95	5.41	5.06	4.82	4.64	4.50	4.39	4.30	4.16	4.01	3.86	3.78	3.70	3.62	3.54	3.45
13	9.07	6.70	5.74	5.21	4.86	4.62	4.44	4.30	4.19	4.10	3.96	3.82	3.66	3.59	3.51	3.43	3.34	3.25
14	8.86	6.51	5.56	5.04	4.69	4.46	4.28	4.14	4.03	3.94	3.80	3.66	3.51	3.43	3.35	3.27	3.18	3.09
15	8.68	6.36	5.42	4.89	4.56	4.32	4.14	4.00	3.89	3.80	3.67	3.52	3.37	3.29	3.21	3.13	3.05	2.96

Numerator Degrees of Freedom (DoF)

F (α = 0.01)	1	2	3	4	5	6	7	8	9	10	12	15	20	24	30	40	60	120
16	8.53	6.23	5.29	4.77	4.44	4.20	4.03	3.89	3.78	3.69	3.55	3.41	3.26	3.18	3.10	3.02	2.93	2.84
17	8.40	6.11	5.18	4.67	4.34	4.10	3.93	3.79	3.68	3.59	3.46	3.31	3.16	3.08	3.00	2.92	2.83	2.75
18	8.29	6.01	5.09	4.58	4.25	4.01	3.84	3.71	3.60	3.51	3.37	3.23	3.08	3.00	2.92	2.84	2.75	2.66
19	8.18	5.93	5.01	4.50	4.17	3.94	3.77	3.63	3.52	3.43	3.30	3.15	3.00	2.92	2.84	2.76	2.67	2.58
20	8.10	5.85	4.94	4.43	4.10	3.87	3.70	3.56	3.46	3.37	3.23	3.09	2.94	2.86	2.78	2.69	2.61	2.52
21	8.02	5.78	4.87	4.37	4.04	3.81	3.64	3.51	3.40	3.31	3.17	3.03	2.88	2.80	2.72	2.64	2.55	2.46
22	7.95	5.72	4.82	4.31	3.99	3.76	3.59	3.45	3.35	3.26	3.12	2.98	2.83	2.75	2.67	2.58	2.50	2.40
23	7.88	5.66	4.76	4.26	3.94	3.71	3.54	3.41	3.30	3.21	3.07	2.93	2.78	2.70	2.62	2.54	2.45	2.35
24	7.82	5.61	4.72	4.22	3.90	3.67	3.50	3.36	3.26	3.17	3.03	2.89	2.74	2.66	2.58	2.49	2.40	2.31
25	7.77	5.57	4.68	4.18	3.85	3.63	3.46	3.32	3.22	3.13	2.99	2.85	2.70	2.62	2.54	2.45	2.36	2.27
26	7.72	5.53	4.64	4.14	3.82	3.59	3.42	3.29	3.18	3.09	2.96	2.81	2.66	2.58	2.50	2.42	2.33	2.23
27	7.68	5.49	4.60	4.11	3.78	3.56	3.39	3.26	3.15	3.06	2.93	2.78	2.63	2.55	2.47	2.38	2.29	2.20
28	7.64	5.45	4.57	4.07	3.75	3.53	3.36	3.23	3.12	3.03	2.90	2.75	2.60	2.52	2.44	2.35	2.26	2.17
29	7.60	5.42	4.54	4.04	3.73	3.50	3.33	3.20	3.09	3.00	2.87	2.73	2.57	2.49	2.41	2.33	2.23	2.14
30	7.56	5.39	4.51	4.02	3.70	3.47	3.30	3.17	3.07	2.98	2.84	2.70	2.55	2.47	2.39	2.30	2.21	2.11
40	7.31	5.18	4.31	3.83	3.51	3.29	3.12	2.99	2.89	2.80	2.66	2.52	2.37	2.29	2.20	2.11	2.02	1.92
60	7.08	4.98	4.13	3.65	3.34	3.12	2.95	2.82	2.72	2.63	2.50	2.35	2.20	2.12	2.03	1.94	1.84	1.73
120	6.85	4.79	3.95	3.48	3.17	2.96	2.79	2.66	2.56	2.47	2.34	2.19	2.03	1.95	1.86	1.76	1.66	1.53

Index

D

E

F

N

O

P

Q

R

S

Printed in the United States
by Baker & Taylor Publisher Services